Analytical
Measurements
in Aquatic
Environments

ANALYTICAL CHEMISTRY SERIES

Series Editor

Charles H. Lochmüller
Duke University

Quality and Reliability in Analytical Chemistry, *George E. Baiulescu, Raluca-Ioana Stefan, Hassan Y. Aboul-Enein*

HPLC: Practical and Industrial Applications, Second Edition, *Joel K. Swadesh*

Ionic Liquids in Chemical Analysis, *edited by Mihkel Koel*

Environmental Chemometrics: Principles and Modern Applications, *Grady Hanrahan*

Quality Assurance and Quality Control in the Analytical Chemical Laboratory: A Practical Approach, *Piotr Konieczka and Jacek Namieśnik*

Analytical Measurements in Aquatic Environments, *edited by Jacek Namieśnik and Piotr Szefer*

ANALYTICAL CHEMISTRY SERIES

Analytical Measurements in Aquatic Environments

Edited by
Jacek Namieśnik
Piotr Szefer

CRC Press
Taylor & Francis Group
Boca Raton London New York

CRC Press is an imprint of the
Taylor & Francis Group, an **informa** business

Publishing

Co-published by IWA Publishing, Alliance House, 12 Caxton Street, London SW1H 0QS, UK
Tel. +44 (0) 20 7654 5500, Fax +44 (0) 20 7654 5555
publications@iwap.co.uk
www.iwapublishing.com
ISBN 1843393069
ISBN13 9781843393061

CRC Press
Taylor & Francis Group
6000 Broken Sound Parkway NW, Suite 300
Boca Raton, FL 33487-2742

First issued in paperback 2017

ISBN-13: 978-1-4200-8268-5 (hbk)
ISBN-13: 978-1-138-11652-8 (pbk)

Library of Congress Cataloging-in-Publication Data

Analytical measurements in aquatic environments / editors, Jacek Namiesnik, Piotr Szefer.
    p. cm. -- (Analytical chemistry series)
   Includes bibliographical references and index.
   ISBN 978-1-4200-8268-5 (hardcover : alk. paper)
   1. Aquatic ecology--Mathematical models. 2. Chimometrics. I. Namiesnik, Jacek, 1949- II. Szefer, Piotr. III. Title. IV. Series.

QH541.5.W3A565 2010
577.601'5118--dc22
                                                2009026755

Visit the Taylor & Francis Web site at
http://www.taylorandfrancis.com

and the CRC Press Web site at
http://www.crcpress.com

Contents

Preface vii

Editors ix

Contributors xi

Chapter 1 Strategy of Collecting Samples from an Aquatic Environment 1

Bogdan Zygmunt and Anna Banel

Chapter 2 Preservation and Storage of Water Samples 19

Marek Biziuk, Angelika Beyer, and Joanna Żukowska

Chapter 3 Application of Passive Sampling Techniques for Monitoring the Aquatic Environment 41

Graham A. Mills, Richard Greenwood, Ian J. Allan, Ewa Łopuchin, Janine Brümmer, Jesper Knutsson, and Branislav Vrana

Chapter 4 Modern Techniques of Analyte Extraction 69

Thaer Barri and Jan-Åke Jönsson

Chapter 5 Mineralization Techniques Used in the Sample Preparation Step 95

Henryk Matusiewicz

Chapter 6 Biota Analysis as a Source of Information on the State of Aquatic Environments 103

J.P. Coelho, A.I. Lillebø, M. Pacheco, M.E. Pereira, M.A. Pardal, and A.C. Duarte

Chapter 7 Speciation Analytics in Aquatic Ecosystems 121

A. de Brauwere, Y. Gao, S. De Galan, W. Baeyens, M. Elskens, and M. Leermakers

Chapter 8 Immunochemical Analytical Methods for Monitoring the Aquatic Environment 139

Javier Adrian, Fátima Fernández, Alejandro Muriano, Raquel Obregón, Javier Ramón, Nuria Tort, and M.-Pilar Marco

Chapter 9 Application of Biotests 189

Lidia Wolska, Agnieszka Kochanowska, and Jacek Namieśnik

Chapter 10 Total Parameters as a Tool for the Evaluation of the Load of Xenobiotics in the Environment 223

Tadeusz Górecki and Heba Shaaban El-Hussieny Mohamed

Chapter 11 Determination of Radionuclides in the Aquatic Environment 241

Bogdan Skwarzec

Chapter 12 Analytical Techniques for the Determination of Inorganic Constituents 259

Jorge Moreda-Piñeiro and Antonio Moreda-Piñeiro

Chapter 13 Analytical Techniques for the Determination of Organic and Organometallic Analytes 303

Erwin Rosenberg

Chapter 14 Introducing the Concept of Sustainable Development into Analytical Practice: Green Analytical Chemistry 353

Waldemar Wardencki and Jacek Namieśnik

Chapter 15 Chemometrics as a Tool for Treatment Processing of Multiparametric Analytical Data Sets 369

Stefan Tsakovski and Vasil Simeonov

Chapter 16 Quality Assurance and Quality Control of Analytical Results 389

Ewa Bulska

Chapter 17 Analytical Procedures for Measuring Precipitation Quality Used within the EMEP Monitoring Program 399

Wenche Aas

Chapter 18 Life Cycle Assessment of Analytical Protocols 413

Helena Janik and Justyna Kucińska-Lipka

Chapter 19 Preparation of Samples for Analysis: The Key to Analytical Success 431

Jacek Namieśnik and Piotr Szefer

Index 475

Preface

Even a cursory perusal of any analytical journal must lead one to the conclusion that trace and ultra-trace analyses is a domain of chemical analysis that is gaining in importance. This conclusion is corroborated not only by the feelings and opinions of analysts. According to the current IUPAC definition of the term "trace component," the limit from which we can talk about trace analysis is the concentration of 100 ppm (100 μg g^{-1}). Naturally, this limit is purely conventional and is not constant. As recently as 30 years ago, "trace analysis" was understood to denote activities aiming to determine components at a concentration level one order of magnitude higher, that is, below 1000 ppm, or 0.1%.

Even today, the determination of components at a concentration level of 100 ppm, including samples with complex matrices, poses no major problems and is routine in many laboratories. This is mainly due to the rapid development of instrumentation—the science of the construction and use of monitoring and measuring devices. Hence, we can expect the definition of the term "trace component" to change again soon.

There are three particular areas of science and technology that are spurring the development of analytical methods and techniques employed in the determination of low and very low analyte contents in samples of various kinds. They are

- Technologies of the production of high-purity materials; to date, the purity of the purest man-made material is denoted by 11 N, which means that the sum total of all the impurities it contains does not exceed 10^{-9}%, or 10 ppt.
- Genetic engineering and biotechnology.
- Environmental protection, including the chemistry of specific elements in the environment.

The determination of ever lower concentrations of analytes has brought into common use special ways of expressing such concentrations.

Ecotoxicological considerations and the efforts undertaken to achieve an increasingly accurate description of the state of the environment pose a great challenge to analytical chemists in terms of the necessity of determining still lower concentrations of various analytes in samples having complex and even nonhomogenous matrices. The task can be accomplished by following either of two approaches:

- The use of more sensitive and selective, or even specific, detectors: This approach is exemplified by the introduction of the photo-ionization detector (used in gas chromatography, [GC]), which is more sensitive and more selective than the flame-ionization detector hitherto commonly used in GC.
- The introduction to analytical procedures of an additional step: The isolation and/or enrichment of analytes prior to their final determination. This extra step enables the interference caused by the components of a primary matrix (due to matrix simplification) to be removed; more importantly, however, it allows the analyte concentration to be increased to a level above the detection limit of the method or the analytical instrument used. With this approach, routine determinations of analytes at the ppb level and even the determination of analytes at concentration levels down to a fraction of ppq become possible.

The term "analytics" is being used more and more frequently in the analytical chemistry literature. This newly coined expression emphasizes the interdisciplinary nature of methods of obtaining information about material systems, that is, methods that exceed the strict definition of analytical chemistry. Analytics, hitherto practiced mostly as analytical chemistry and, to a large extent identified with the work of chemists, has recently developed into a scientific discipline in its own right, whose role far exceeds chemistry alone and covers almost all branches of science and technology. Analytics has thus become an interdisciplinary science. This interdisciplinary nature is revealed through a variety of phenomena utilized at the measurement stage. Analytics is a scientific discipline embracing

- Various areas of chemistry (particularly physical chemistry and biochemistry)
- Physics
- Computer science
- Electronics, automation, and robotics
- Material science
- Biology
- Instrumentation
- Chemometrics.

This book consists of a set of chapters focused on the most important aspects of analytical procedures for the determination of both inorganic and organic constituents in samples taken from different parts of aquatic ecosystems. Special attention is paid to

- Handling of representative samples
- Samples of preservation techniques
- Extraction techniques
- Solvent-free sample preparation for analysis
- Application of biotests
- Green analytical chemistry—application of the concept of sustainability in analytical laboratories
- Application of the life cycle assessment approach
- Quality control and quality assurance of analytical results
- Enhanced techniques of sample preparation
- Hyphenated analytical techniques

We hope that this book will be a useful source of information for a wide spectrum of readers.

Jacek Namieśnik and Piotr Szefer

Editors

 **Jacek Namieśnik** received his MSc (1972), PhD (1978), and DSc (1985) degrees from the Gdańsk University of Technology (GUT). He has been employed at GUT since 1972. A full professor since 1998, he has also served as vice dean of the Chemical Faculty (1990–1996) and dean of the Chemical Faculty (1996–2000 and 2005–present). He has been the head of the Department of Analytical Chemistry since 1995, as well as chairman of the Committee of Analytical Chemistry of the Polish Academy of Sciences since 2007, and Fellow of the International Union of Pure and Applied Chemistry (IUPAC) since 1996. Dr. Namieśnik was director of the Center of Excellence in Environmental Analysis and Monitoring during 2003–2005. Among his scientific publications, there are seven books, over 300 papers, and more than 350 lectures and communications published in conference proceedings. He is the recipient of various awards, including Professor *honoris causa* from the University of Bucharest (Romania) (2000), the Jan Hevelius Scientific Award of Gdańsk City (2001), and the Prime Minister of Republic of Poland Award (2007). He has seven patents to his name and his research interests include environmental analytics and monitoring and trace analysis.

 **Piotr Szefer** received his MSc (1972), PhD (1978), and DSc (1990) degrees from the Medical University of Gdańsk (MUG). He was awarded Full Professorship in 2000. During 1990–2002, he was vice dean and dean of the Faculty of Pharmacy, MUG. Since 2000, he has been the head of the Department of Food Sciences, MUG. He has published approximately 200 papers, 17 book chapters, three books published by Elsevier and CRC Press\ Taylor & Francis, and approximately 300 symposial abstracts. He has been a member of approximately 30 national and international scientific associations and organizations (including nine editorial boards, e.g., The Science of the Total Environment), for example, the International Scientific Committee on Oceanic Research (SCOR) and WG Marine Board—European Science Foundation. He has visited 14 countries as a visiting professor or research scientist. Dr. Szefer has reviewed approximately 600 manuscripts for more than 60 journals. He received several scientific awards, for example, one from the Scientific Secretary of the Division VII of the Polish Academy of Sciences; nine awards from the Minister of Health; and a joint award from the Ministry of Environmental Protection, Natural Resources, and Forestry. His research is focused on food and marine chemistry, and bioanalytics.

Contributors

Wenche Aas
Chemical Coordinating Center of the European
    Monitoring and Evaluation Program
Norwegian Institute for Air Research
Kjeller, Norway

Javier Adrian
Networking Research Center on Bioengineering,
    Biomaterials and Nanomedicine
Spanish National Research Council
Barcelona, Spain

Ian J. Allan
Norwagian Institute for Water Research
Oslo, Norway

W. Baeyens
Department of Analytical and
    Environmental Chemistry
Vrije Universiteit Brussel
Brussels, Belgium

Anna Banel
Department of Analytical Chemistry
Gdańsk University of Technology
Gdańsk, Poland

Thaer Barri
Department of Analytical Chemistry
Lund University
Lund, Sweden

Angelika Beyer
Department of Analytical Chemistry
Gdańsk University of Technology
Gdańsk, Poland

Marek Biziuk
Department of Analytical Chemistry
Gdańsk University of Technology
Gdańsk, Poland

A. de Brauwere
Department of Analytical and
    Environmental Chemistry
Vrije Universiteit Brussel
Brussels, Belgium

Janine Brümmer
School of Biological Sciences
University of Portsmouth
Portsmouth, United Kingdom

Ewa Bulska
Faculty of Chemistry
Warsaw University
Warsaw, Poland

J.P. Coelho
Department of Chemistry
Centre for Environmental and Marine Studies
University of Aveiro
Aveiro, Portugal

A.C. Duarte
Department of Chemistry
Centre for Environmental and
    Marine Studies
University of Aveiro
Aveiro, Portugal

M. Elskens
Department of Analytical and
    Environmental Chemistry
Vrije Universiteit Brussel
Brussels, Belgium

Fátima Fernández
Networking Research Center on
    Bioengineering, Biomaterials
    and Nanomedicine
Spanish National Research Council
Barcelona, Spain

S. De Galan
Department of Analytical and
    Environmental Chemistry
Vrije Universiteit Brussel
Brussels, Belgium

Y. Gao
Department of Analytical and
    Environmental Chemistry
Vrije Universiteit Brussel
Brussels, Belgium

Tadeusz Górecki
Department of Chemistry
University of Waterloo
Waterloo, Ontario, Canada

Richard Greenwood
School of Biological Sciences
University of Portsmouth
Portsmouth, United Kingdom

Helena Janik
Polymer Technology Department
Gdańsk University of Technology
Gdańsk, Poland

Jan-Åke Jönsson
Department of Analytical Chemistry
Lund University
Lund, Sweden

Jesper Knutsson
Water Environment Transport
Chalmers University of Technology
Göteborg, Sweden

Agnieszka Kochanowska
Department of Analytical Chemistry
Gdańsk University of Technology
Gdańsk, Poland

Justyna Kucińska-Lipka
Polymer Technology Department
Gdańsk University of Technology
Gdańsk, Poland

M. Leermakers
Department of Analytical and
    Environmental Chemistry
Vrije Universiteit Brussel
Brussels, Belgium

A.I. Lillebø
Department of Chemistry
Centre for Environmental and Marine Studies
University of Aveiro
Aveiro, Portugal

Ewa Łopuchin
Department of Analytical Chemistry
Gdańsk University of Technology
Gdańsk, Poland

M.-Pilar Marco
Networking Research Center on
    Bioengineering, Biomaterials
    and Nanomedicine
Spanish National Research Council
Barcelona, Spain

Henryk Matusiewicz
Department of Analytical Chemistry
Poznań University of Technology
Poznań, Poland

Graham A. Mills
School of Pharmacy and Biomedical
    Sciences
University of Portsmouth
Portsmouth, United Kingdom

Heba Shaaban El-Hussieny Mohamed
Department of Chemistry
University of Waterloo
Waterloo, Ontario, Canada

Antonio Moreda-Piñeiro
Department of Analytical Chemistry,
    Nutrition, and Bromatology
University of Santiago de Compostela
Santiago de Compostela, Spain

Jorge Moreda-Piñeiro
Department of Analytical Chemistry
University of A Coruña
A Coruña, Spain

Alejandro Muriano
Networking Research Center on
    Bioengineering, Biomaterials
    and Nanomedicine
Spanish National Research Council
Barcelona, Spain

Jacek Namieśnik
Department of Analytical Chemistry
Gdańsk University of Technology
Gdańsk, Poland

Raquel Obregón
Networking Research Center on
    Bioengineering, Biomaterials
    and Nanomedicine
Spanish National Research Council
Barcelona, Spain

M. Pacheco
Department of Biology
Centre for Environmental and
    Marine Studies
University of Aveiro
Aveiro, Portugal

M.A. Pardal
Zoology Department
Institute of Marine Research
University of Coimbra
Coimbra, Portugal

M.E. Pereira
Department of Chemistry
Centre for Environmental and
    Marine Studies
University of Aveiro
Aveiro, Portugal

Javier Ramón
Networking Research Center on
    Bioengineering, Biomaterials
    and Nanomedicine
Spanish National Research Council
Barcelona, Spain

Erwin Rosenberg
Vienna University of Technology
Institute of Chemical Technologies and
    Analytics
Vienna, Austria

Vasil Simeonov
Faculty of Chemistry
University of Sofia
Sofia, Bulgaria

Bogdan Skwarzec
Department of Analytical Chemistry
University of Gdańsk
Gdańsk, Poland

Piotr Szefer
Department of Food Sciences
Medical University of Gdańsk
Gdańsk, Poland

Nuria Tort
Networking Research Center on
    Bioengineering, Biomaterials
    and Nanomedicine
Spanish National Research Council
Barcelona, Spain

Stefan Tsakovski
Faculty of Chemistry
University of Sofia
Sofia, Bulgaria

Branislav Vrana
Slovak National Water Reference
    Laboratory
Water Research Institute
Bratislava, Slovakia

Waldemar Wardencki
Department of Analytical Chemistry
Gdańsk University of Technology
Gdańsk, Poland

Lidia Wolska
Department of Analytical
    Chemistry
Gdańsk University of Technology
Gdańsk, Poland

Joanna Żukowska
Department of Analytical
    Chemistry
Gdańsk University of Technology
Gdańsk, Poland

Bogdan Zygmunt
Department of Analytical Chemistry
Gdańsk University of Technology
Gdańsk, Poland

1 Strategy of Collecting Samples from an Aquatic Environment

Bogdan Zygmunt and Anna Banel

CONTENTS

1.1 Introduction 1
1.2 General Considerations 2
    1.2.1 Types of Samples 3
        1.2.1.1 Discrete Sample 3
        1.2.1.2 Composite Sample 3
    1.2.2 Basic Sampling Patterns 5
1.3 Sampling-Related Uncertainty 5
1.4 Basic Aspects of Strategies for Sampling Water and Sediments from
    Aquatic Environments 8
    1.4.1 Water Samples 8
        1.4.1.1 Sample Size 8
        1.4.1.2 Sampling Location and Sampling Sites 9
        1.4.1.3 Sample Collection 9
        1.4.1.4 Sampling Frequency 11
    1.4.2 Sediments 11
1.5 Selection of Sampling Equipment 12
    1.5.1 Compatibility of Sampler Material with Water Samples 12
    1.5.2 Water Sampling Using Traditional Techniques 13
        1.5.2.1 Manual Surface Water Samplers 13
        1.5.2.2 Ground Water Samplers 14
    1.5.3 Automatic Water Sampling Systems 14
    1.5.4 Passive Samplers 14
    1.5.5 Sediment Samplers 15
1.6 Conclusions 16
Acknowledgment 16
References 16

1.1 INTRODUCTION

The aquatic environment has played a crucial role since the very beginning of civilization. Leonardo da Vinci compared water with blood when he said: "water is the blood of the soil." Indeed, "water" must have been one of the first words invented by human. Nowadays, the word water is commonly used in two meanings. First, in a purely chemical sense, it represents the simple chemical compound with the formula H_2O. Second, it stands for a wide range of different aqueous solutions, whose composition determines their usefulness or uselessness for a given purpose. The specific names of the solutions relate to the source or usage, for example, sea water, river water, ground water, rain water, drinking water, and irrigation water.

Water, always a vital commodity for humans, is used for drinking, cooking, agriculture, transport, and recreation, among other purposes. To be applicable to a given purpose or suitable for various forms of life, water must satisfy certain requirements. The characteristics of water include a number of physicochemical parameters that should be monitored. Generally, water in any compartment of the environment is not completely isolated from its surroundings, and components are exchanged between the liquid phase and its immediate neighborhood. In the majority of cases it is the bottom sediment that releases or takes in the substances discharged into a water body. Therefore, any determination of water quality in a given compartment should include sediment analysis.

Only a few physicochemical parameters can be measured by immersing the relevant instrument into a water body. In most cases a small fraction of a given water population or sample is collected and analyzed. The aim of taking samples or sampling is to extract a fraction of the water body that has chemical, physical, and biological properties identical to those of the bulk of the system to be studied. Ideally, all the characteristics of the sample, or at least, the parameters that are to be determined, should not change until the time of measurement. Only then can the results of the sample analysis be representative of the composition of the system under scrutiny.

If analytical data are to be the source of reliable and valuable information on an aquatic environment, the samples should not only be properly collected, preserved, transported, and stored, but should also be taken at the proper place, time, frequency, and so on. All these factors must be properly planned, taking into account both the characteristics of the aquatic environment to be sampled and the feasibility of the task.

This chapter looks at the sampling of different surface and ground waters. As sediments constitute an integral part of most aquatic environments, they too have been taken into consideration. It should be mentioned that precipitation and wet deposition also play an important role in the hydrological cycle, but the sampling problems are somewhat different and will not be dealt with here—they are comprehensively described elsewhere.[1,2]

1.2   GENERAL CONSIDERATIONS

Sampling is the critical step in the whole analytical process; indeed, it is often the weakest link in the procedure and needs special care if the analytical results obtained are to be a source of reliable analytical information on a system.[3]

While designing the sampling process it should be remembered that

- Samples that are not representative of the population studied are of little importance
- Poor sampling procedures yield unrepresentative samples, which may make a disproportionately large contribution to the uncertainty of the analytical results
- Sampling errors and analytical errors are independent of each other, so sampling-related errors cannot be corrected using laboratory blanks or control samples
- Sampling errors can rarely be corrected without resampling and analysis
- Contamination of samples and loss of analytes are common sources of errors in environmental measurements.

There are many possible means of sampling with regard to techniques, devices, methodologies, types of samples, and so on.

1.2.1 TYPES OF SAMPLES[4–6]

The two basic types of water sample are discrete samples and composite samples; in the majority of cases, each type supplies slightly different information on the water body in question. They are depicted graphically in Figure 1.1.

1.2.1.1 Discrete Sample

A discrete sample is a sample collected in a short period of time (generally <15 min) and deposited in a separate container. Discrete samples collected at a particular time and place represent only the composition of the source at that time and place. In fact, a single discrete sample gives only a snapshot of the situation. Nonetheless, discrete samples collected at many suitable intervals and locations can document compositional variations in time and space.

1.2.1.2 Composite Sample

A composite sample consists of a series of smaller samples collected at regular intervals over a period of time and deposited in the same container. To some extent, composite samples represent the average characteristics of the source during the sampling period—they are useful in determining average pollutant concentrations or pollutant loads. The laboratory analysis of such samples is more cost-effective but generally yields less information.

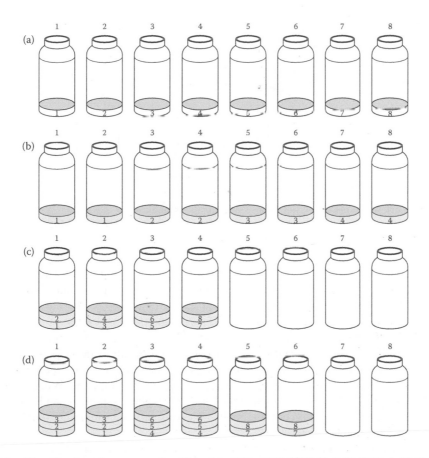

FIGURE 1.1 Classification of water samples. (Based on Dick, M.E. 1996. In: L.H. Keith (ed.), *Principles of Environmental Sampling*, 2nd edition, pp. 237–258. American Chemical Society, Washington.)

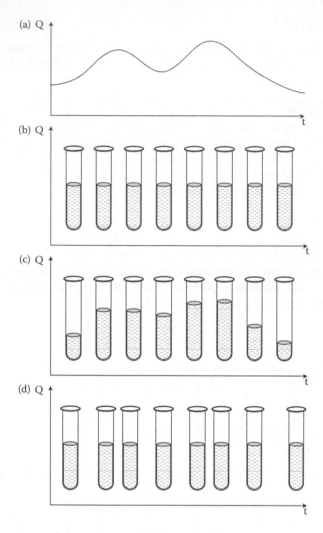

FIGURE 1.2   Different types of discontinuous sampling techniques.

Samples can be collected continuously or discontinuously. According to ISO 5667-10,[7] there are three main categories of discontinuous sampling (Figure 1.2). The samples can be analyzed separately or can be mixed to produce a composite sample. With the appropriate selection of sampling type, a composite sample can better characterize a given situation. In *time-proportional* sampling, equal volumes of water are collected at constant time intervals, but a composite sample produced in this way can yield a reliable average discharge or concentration only if these factors are relatively constant over the sampling period. In *discharge-proportional* sampling, the time intervals remain constant whereas the volume of each sample is proportional to the discharge. When the water flow rate changes, the so-called *quantity-proportional* or flow-weighted sampling can be useful. In this type, the volume of each sample is constant but the intervals between sampling events are inversely proportional to the flow rate.

Sometimes, harmful substances are released into a water body for short periods, but only at irregular intervals. In such situations *event-controlled* sampling is useful (Figure 1.3).[7] When some selected parameter (e.g., conductivity and temperature) reaches its threshold value, then an automatic pump sampler is switched on and a sample is collected.

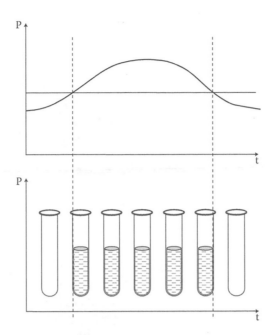

FIGURE 1.3 Event-controlled sampling.

1.2.2 BASIC SAMPLING PATTERNS

Once the sampling location has been established, the actual sampling sites must be selected to obtain reliable information on a given water body. The arrangement of sampling sites depends on the study's objectives and the location's complexity. The three basic patterns of sampling site selection are judgmental, systematic, and random (Figure 1.4).[6,8]

Combinations of any two are also used. The judgmental sampling pattern requires the smallest number of samples but the relative bias is the largest; the opposite holds for the random pattern, where the bias is the smallest but the number of samples is the largest. In scientific studies it is the judgmental approach that is most often applied, whereas for legal purposes absolutely random sampling is often needed.

1.3 SAMPLING-RELATED UNCERTAINTY

The total uncertainty of the analytical results, which can be expressed quantitatively as the variance, is the sum of the variances related to the successive steps of the analytical process. Depending on the analytical task, the contribution of the variances of the particular steps can differ from each other quite considerably. In the case of samples collected from an aquatic environment, the total variance of the analytical result (s_T^2) can be expressed as follows.

$$s_T^2 = s_a^2 + s_t^2 + s_s^2 + s_h^2 + s_p^2 + s_m^2,$$ (1.1)

where the subscripts denote the variances: a, spatial; t, temporal; s, sampling proper; h, transport and storage; p, preparation; and m, measurement.

If a sample is to describe a population averaged over space and time, spatial and temporal changes should be taken into account and included in the total variance of sampling. The later steps, that is, sample preservation, transport, storage, preparation for analysis, and the measurement proper, will be dealt with in the subsequent chapters of this book. Taylor[9] uses the terms "sample

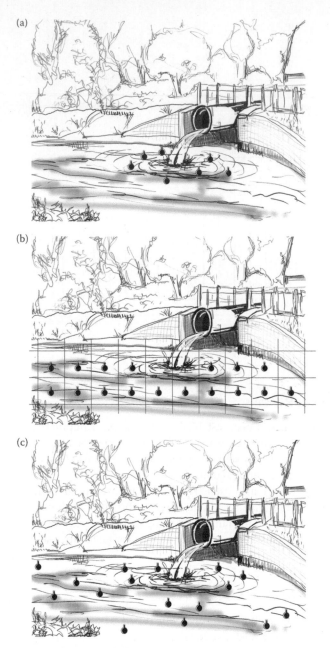

FIGURE 1.4 Three basic patterns of sampling site selection. (Based on USEPA. 1995. Superfund Program Representative Sampling Guidance, Vol. 5: Water and Sediment. Available at http://www.epa.gov/tio/ download/char/sf_rep_samp_guid_water.pdf and Keith, L.H. 1991. *Environmental Sampling and Analysis: A Practical Guide*. Lewis, Michigan.)

variance" (s^2_{sample}) for the total variability up to the time the sample is poured into a container, "sampling variance" (s^2_{sampling}) for the variability inherent in the sampling proper, and "population variance" ($s^2_{\text{population}}$) for spatial and temporal variability, which seems quite logical. However, in a large number of publications the term "sampling uncertainty" is used to express the variability of the spatial, temporal, and sample-collecting sources. This is the sense in which the term "sampling uncertainty" will be used in this chapter.

In environmental analysis, especially in the determination of organic pollutants, sampling is considered to be the most critical step, the one that most often makes the greatest contribution to the total uncertainty of analysis. Therefore, to reduce the uncertainty of the analytical result, the closest attention should be paid to the sampling process. The main sources of error in sampling can be found in Madrid and Zayas.[10]

The most important sources of uncertainty seem to be the heterogeneity of an analyte within the matrix and the nonhomogeneity of the matrix itself. The uncertainty in the data can be reduced to acceptable levels by increasing the sample volume, number of samples, density of sampling sites, and frequency of sampling. However, reducing the uncertainty by increasing the number of samples can be cost-ineffective, since uncertainty is inversely proportional to the square root of that number. But again, this only holds if the procedure has been optimized; if not, the required uncertainty is unattainable. This was demonstrated by Minkkinen,[11] who applied sampling theory in practice.

Liess and Schulz[12] have given a formula for predicting the number of samples required for an assumed uncertainty:

$$N = 4\left(\frac{s}{xd}\right)^2, \tag{1.2}$$

where s is the estimated standard deviation of the arithmetic mean of all single samples, x is the estimated arithmetic mean of all single samples, and d is the tolerable uncertainty of the result, for example, 20% ($d = 0.2$).

Samples are collected in order to obtain information on a given environmental compartment, and so the sampling sites should be located such that the collected samples represent the environmental compartment under study with respect to a selected parameter. The optimal selection of sampling sites depends on the aim of the program. It is often the case that the location of the sampling sites, among other factors, is predetermined, before the pollution status has been defined. The uncertainty in status is considerable even if it has been preceded by exploratory sampling, the aim of which is to establish species of concern, the approximate range of concentration, variability, and so on. Uncertainties will also result from inappropriate timing if temporal fluctuations occur.

Sample collection may be a source of uncertainty for quite a few reasons. Some target analytes may be deposited on the contact surfaces of the sampling device together with suspended matter. Volatiles may be released, be adsorbed on, or react with the sampler material or with external agents, or they may decompose in the presence of elevated temperatures, UV radiation, microbial activity, and so on. Hence, sampling should include a special quality control procedure for estimating and possibly reducing the uncertainty related to particular phenomena. The essential components of a sound quality control system, according to Keith,[8] are the consistent use of qualified personnel, reliable, and well-maintained equipment, appropriate calibration of standards, and the supervision of all operations by management and senior personnel. Analytical protocols should include all the steps, in order to check for possible contamination of the sample or loss of analytes that can lead to a change in the analyte concentration.

First, routine tests should be conducted to check the effectiveness of the cleaning of sampling devices and sample containers. This can be done in the laboratory by applying equipment or rinsate blanks. The blanks should be collected after each decontamination and before resampling, and where necessary, corrective action should be taken. The tests should be made to check whether the material, from which the device/container is made and which is in contact with the sample, does not adsorb or react with or release relatively significant amounts of target analytes.

Field blanks are used to provide routine contamination tests. They are samples that do not contain target analytes and have a matrix composition similar to that of the analyzed media. Examples of blanks are water collected from a nonpolluted water body, or deionized, or distilled water. Field blanks are delivered to the sampling site and treated in exactly the same way as real samples; they take account not only of the uncertainty of sampling but also of transport and storage.

The contribution made by the last two steps can be discovered by applying a field check sample, which is obtained by dividing the problem sample into two and spiking one subsample with a target analyte. Recovery is then determined under different conditions of light, temperature, pH, and so on, in order to select the best sample storage conditions.

Replicate samples are used to indicate sample uncertainty, which shows the contribution of sampling and sample handling to the overall uncertainty.

Thompson et al.[13] have shown quality control in sampling to be practicable in sampling procedures requiring a combination of sample increments to form a composite sample. It is very important, however, that the approach proposed does not require any extra time and resources beyond those used for the normal sampling procedure that precedes analysis.

1.4   BASIC ASPECTS OF STRATEGIES FOR SAMPLING WATER AND SEDIMENTS FROM AQUATIC ENVIRONMENTS

1.4.1   WATER SAMPLES

Independently of the monitoring aim, samples should be representative of the water studied. Sampling strategies, as well as the techniques, the requirements, and the recommendations for the routine collection of representative water samples, are described in the U.S. National Field Manual and elsewhere.[14–16] In the formulation of strategies for sampling water from an aquatic environment, several factors should be looked at. Which of these should be taken into account in a given situation depends on the objective of sampling and the type of water body to be analyzed. The most important factors are the selection of sampling location (general position in the water body) and sampling sites (exact position), the size of samples and the number to be taken at each site, the frequency and timing of sampling, and the manner in which the sample is taken. The nature of water also has an important influence on the sampling strategy, which depends on whether sampling is carried out for exploratory or monitoring purposes.

1.4.1.1   Sample Size

The sample volume depends on the target analytes and their expected concentrations, the quantification limit of the analytical procedure, the matrix composition, and the homogeneity of the medium. The sample volume should be such that after enrichment, the concentrations of target analytes are higher than the quantification limit (LOQ) of the final analytical procedure. If different procedures are used for different analytes, then the volume should be increased. It should be increased still further if replicate analyses are to be carried out. It is not uncommon that water contains some particulate matter (PM) of different compositions and surface areas. The distribution of particulate matter is by nature less homogenous than that of dissolved substances. This can introduce some inhomogeneity in the concentrations of soluble substances: some of these are adsorbed on the PM, and their content in the liquid phase decreases consequently. This is especially important when organic pollutants are characterized by high octanol–water partition coefficients. If only the dissolved fraction of an analyte is of interest, the results will depend strongly on the concentration of suspended matter, and larger samples will have to be taken to reach the required LOQ. If the analytical procedure for a given matrix is still undergoing development, then the sample volume must be larger still.

The accuracy and precision of the analytical procedure and the limiting uncertainty of the results determine the number of necessary replicate measurements and hence the minimum sample volume. On the other hand, the sample should only be as large as need be, since larger samples mean higher costs, not to mention the inconveniences of transport, storage, material use, and disposal. Taking into account present regulations and available analytical procedures, it can be estimated that approximately 100 mL samples should be sufficient for heavy metal determinations. However, if organic analytes are of interest, then samples of a volume of typically 1 L, but sometimes also 3 L, and in some cases even 20 L are required.[12] There are also situations where much larger (e.g., 100 L) samples must be collected in order to achieve the desired detection limit.

1.4.1.2 Sampling Location and Sampling Sites

The sampling location is selected in accordance with the measurement objectives. If it is the efficiency of a water treatment plant that is to be determined, the sampling sites should be located above and below the points of water entry to the plant. For studying the effect of effluent discharge on water quality in a river, water samples should be collected upstream and downstream of the outfall.

The selection of sampling site depends on the variability of the system and the type of information to be acquired. This has been described in guidelines and scientific journals. Dixon et al.[17] described a new approach to optimizing the selection of river sampling sites based on the geographical information system, graph theory, and a simulated annealing algorithm. For measuring the quantity of pollutants carried by a river, the sampling sites should represent the water body as a whole. Boundaries such as banks, surface, bottom, and the confluences of streams or other rivers should thus be avoided, since any samples collected there will generally be unrepresentative. For studying the effect of the discharge of wastewater, industrial effluents, and so on, on river water quality, samples should be taken downstream, at a location where mixing is complete. For monitoring the quality of water taken for a particular purpose, the sampling site should be situated close to the abstraction point. In the case of lakes and reservoirs, the heterogeneity related to thermal stratification, inflowing streams, morphology, and even wind needs to be taken into consideration during site selection. For obtaining a sample representative of the vertical cross-section of a stratified water body, for example, stratified estuaries, depth-integrating samplers can be used. Samples can be collected at different depths and mixed to obtain a composite sample. The surface water layer and the near-bottom layer can differ considerably, and so samples should be collected some distance from the bottom and surface. Some pollutants concentrate on the water surface, and so then the sea surface microlayer of approximately 50 μm thickness should be collected using special samplers.[18]

Ground water samples can be collected from monitoring or supply wells. Their location is not always straightforward—generally such water is sampled over hot spots and near locations following the subterranean stream in order to detect plume profile movements.[19] The depth of the well and the characteristics of the surrounding land surface and upstream activities can help in the interpretation of results.

1.4.1.3 Sample Collection

1.4.1.3.1 Surface Waters

How should samples be taken? Small streams are generally shallow, so samples can be taken manually from the bank or from some shallow spot in the stream simply by immersing a bottle in the water (Figure 1.5).

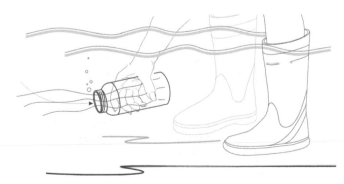

FIGURE 1.5 A hand-held open-mouth bottle sampler. (Based on Lane et al. 2003. U.S. Geological Survey Techniques of Water-Resources Investigations. Book 9, Chapter A2. Available at http://pubs.water.usgs.gov/twri9A2/ (accessed March 20, 2003).)

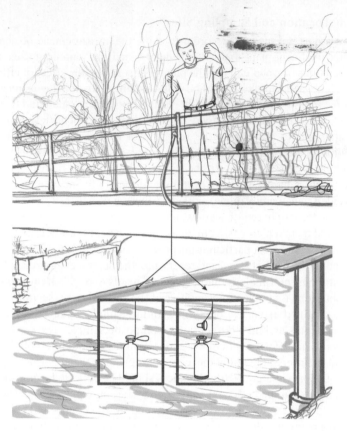

FIGURE 1.6 Sampling of water from given depths.

In the case of larger streams and rivers, samples can be taken from river banks, platforms, and bridges (Figure 1.6).

Water samples from large rivers, estuaries, and the sea can be collected in using flasks that are hand-held (polyethylene-gloved hand, mouth down) or attached to a 3–4 m telescopic tube, from platforms with the aid of a pump, and from oar-propelled rubber dinghies, and also with special bottles for discrete depth sampling (Figure 1.7).

FIGURE 1.7 Sampling of water at discrete depths with the Ruttner sampler.

The dinghy should be located upwind of the motor launch to prevent the sampled water from being polluted with exhaust gases from the latter. To sample water or sediments close to a dam, samplers can be immersed in water from helicopters;[21] their use for sampling was comprehensively described by Krinitz et al.[22]

1.4.1.3.2 Ground Waters

Ground water is a very complex matrix and can vary considerably from one aquifer to another; the physical characteristics of wells also differ. All these factors and also the target analytes must be considered when selecting sampling equipment that allows the representativeness and integrity of the samples to be maintained. Ground water samples must not contain particulate matter and must be protected from the air throughout the sampling process because aeration can drastically affect sample integrity.[23–25]

1.4.1.4 Sampling Frequency

The frequency of sampling is an important factor. In monitoring programs, the frequency and also locations are generally established by the regulatory agencies. The frequency depends on the purpose of sampling. Where changes in environmental parameters or quality are the object of study, the sampling frequency should be at least twice the frequency of the variation.[26] The multifarious aspects of sampling are discussed in a number of papers.[27,28]

1.4.2 SEDIMENTS

Pollutant adsorption on the solid material present in an aquatic environment depends strongly on the size of particles and their composition. Sediments containing particles 0.06–0.2 mm in size and larger have a rather low adsorption capability and pollutants are generally not associated with them. Fine-grained silts and clays have a much larger specific surface area and can therefore adsorb some organic and inorganic pollutants quite efficiently. These should be sampled and analyzed for their pollutant content in order to evaluate the threat to living organisms. In shallow waters sediments can be sampled manually with a spoon or a scoop (Figure 1.8) or with corers (Figure 1.9); in deep waters different kinds of dredges and special corers are operated from on board ships (Figure 1.10).

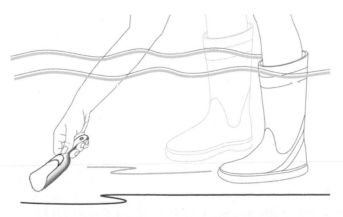

FIGURE 1.8 Collection of sediment samples from shallow water with a scoop.

FIGURE 1.9   Collection of sediment samples from shallow water with a manual corer.

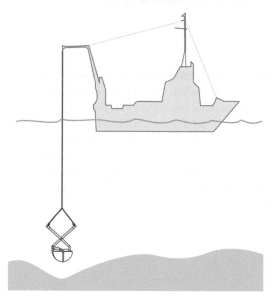

FIGURE 1.10   Collection of sediment samples from deep water with a dredge operated from on board a ship.

1.5   SELECTION OF SAMPLING EQUIPMENT

1.5.1   Compatibility of Sampler Material with Water Samples[29]

When selecting equipment for sampling water, the prime consideration is the material from which the parts in contact with the sample have been made. This can be various organic polymers, metals, and glass. The material selected should not release compounds interfering with the determination of target analytes and should neither adsorb nor react with these analytes. Hence, the nature of the analytes and the properties of the sampler material will be crucial factors in selecting the equipment for a given task. In general, organic polymers are incompatible with most organic analytes, which can dissolve in or be leached out of such materials, and this may lead to negative or positive errors, respectively.

Plastics such as fluorocarbon polymers (Teflon, Kynar, and Tefzel), polypropylene, polyethylene (linear), polyvinyl chloride, Silicone, and Nylon can all be used for inorganic sampling. Fluorocarbon polymers are completely inert to most inorganic analytes, Silicone is highly porous and relatively inert to the majority of such analytes, and the others are also relatively inert to them. Possible limitations are due to the fact that fluorocarbons can be a source of fluoride and Silicone a source of silicon-containing compounds. In general, organic materials should not be used for organic analytes, with the exception of fluorocarbon polymers, which sorb only certain organics, and Nylon, which can only be used for chlorofluorocarbons.

316-grade stainless steel, which has the highest resistance to corrosion, can be used for all organics, provided that it is not corroded. It can also be used for submersible pump casings, as long as its negative effects on inorganic compounds are minimized, for example, by using fluorocarbon polymers for sample-wetted components. If corroded, the material is a potential source of Cr, Ni, Fe, and possibly Mn and Mo. For surface water sampling, the equipment must have a plastic coating. The less corrosion-resistant 304-grade stainless steel can be used only for organics and then only if it is not corroded. Other metallic materials like brass, iron, copper, aluminum, and galvanized and carbon steels are unsuitable for inorganic compounds; however, these materials are compatible with organics as long as they are not corroded. They are routinely used for CFC monitoring.

Borosilicate glass and ceramics are inert materials and have great chemical stability; the weak point about glass, however, is its fragility. Glass can be used for both organics and inorganics, but one should be mindful of the fact that it is a potential source of boron and silicon and cannot be used if these elements are to be determined.

Obviously, the equipment must be thoroughly cleaned before sampling. The cleaning procedures depend mainly on the target analytes (analytes to be determined) and, to some extent, also on the matrix; there are a few protocols for the removal of possible interfering substances.[24] If the sample to be analyzed has a similar or a higher concentration of analytes than the sample analyzed before, then the same equipment can be used. A brief discussion of this problem regarding organic analytes is given in Hildebrandt et al.[19]

If gloves are used to handle equipment they should be disposable and powderless; before use, they should be examined visually for defects.

Namieśnik et al.[30] discuss at length the various problems encountered in environmental sampling and describe a large number of samplers for collecting water and sediments.

1.5.2 Water Sampling Using Traditional Techniques

A large number of samplers for sampling water from different aquatic environments have been designed and put on the market by various manufacturers. The use of automatic samplers is increasing since they have quite a few significant advantages over manual samplers, which are still widely used. Some can be used for grab or discrete samples, and others for composite samples. The sampling devices should permit rapid immersion in water, drift minimally from the vertical position, have a suitable closing/sealing mechanism to retain the sample, have the appropriate sample capacity, and be easy to use. Sampler selection depends, among other things, on the location of the sampling site, the depth at which samples should be taken, how far from the bottom it is situated, the size and type of sample, site accessibility, and the type of matrix.

1.5.2.1 Manual Surface Water Samplers

1.5.2.1.1 Samplers for Small Depths

For taking small samples close to the bank of a water body, a held-hand open-mouth bottle sampler can be used, provided that the depth and the water velocity are smaller than the minimum for depth-integrating samplers. Should large-volume samples be needed, for example, in the trace analysis of surface water for organics, the sample can be pumped through tubes into the proper container from platforms, from buoys far away from the vessel or from an oar-propelled rubber dinghy (upwind of the motor boat).

1.5.2.1.2   Deep-Water Samplers

For taking water from a selected depth, samplers with special systems enabling the container to be filled with water at the required depth and transporting the sample in undisturbed form to the analyst are used. The simplest sampler of this type is a weighted bottle stoppered with a cork connected to the bottle neck by a line used to open the bottle at the required depth.[12] During sampling, the water sample comes into contact with the air present in the bottle. This can be deleterious to many analytical tasks. Fortunately, there are a number of water samplers with which nonaerated samples can be collected at a selected depth: the Ruttner sampler, the biochemical oxygen demand (BOD) sampler, and the volatile organic compounds (VOC) sampler are among the most common.[20] To collect instantaneous discrete samples so-called thief samplers are used, which are available in different sizes, mechanical configurations, and construction material. The most commonly used are the Kremmer sampler, Van Dorn sampler, and the double check-valve bailer with a bottom-emptying device.[20]

1.5.2.2   Ground Water Samplers

The most important factors in selecting samplers for ground water are the type and location of the well, depth of water with respect to the land surface, the physical characteristics of the well, and target analytes. For ground water monitoring, pumps designed specifically for monitoring wells or pumps installed in supply wells, bailers, over point, and thief-type samplers are most commonly used. The above-mentioned equipment is described in the U.S. Geological Survey Manual.[20] Parker[31] compared the ability of different ground water sampling devices to deliver representative samples. He found that grab samplers, positive displacement devices, and suction-lift devices can, under certain conditions, alter the chemistry of ground water samples, and that gas-lift pumps, older types of submersible centrifugal pumps, and suction-lift devices should not be recommended for sampling sensitive analytes such as volatile organics and inorganics, or inorganics that undergo oxidation and precipitation reactions. The best recoveries of sensitive analytes were obtained with bladder pumps.

1.5.3   Automatic Water Sampling Systems

Though widely applied, manual sampling has certain drawbacks. The most important are the difficulty in performing event-triggered sampling, the inconvenience of collecting samples at certain hours or unpredictable times, and the dangers to which the person carrying out the sampling is exposed in some situations. For these and other less important reasons, automatic sampling is becoming increasingly popular and advances in their construction are taking place. The basic components of an automatic sampling system are presented in Figure 1.11.

Automatic sampling is less labor-intensive and often produces more consistent results; it is also to be recommended in the situations mentioned above.[5,32,33]

1.5.4   Passive Samplers

The limitations of conventional sampling in the monitoring of aquatic environments have led to the introduction of passive samplers, which are now being used more often. The samples taken with these devices allow very low concentrations of pollutants to be determined and episodic pollution events to be detected: both could be missed with the conventional approach. Passive samplers are designed to accumulate target analytes over a long time. The basic difference between passive and conventional sampling is that the former collects only selected groups of components, including

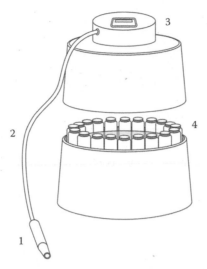

FIGURE 1.11   Basic components of an automatic sampler: 1—sample intake; 2—sample transport tube; 3—pump controller; and 4—sample bottles. (Based on Dick, M.E. 1996. In: L.H. Keith (ed.), *Principles of Environmental Sampling*, 2nd edition, pp. 237–258. American Chemical Society; and Dick, E.M. 1994. In: B. Markert (ed.), *Environmental Sampling for Trace Analysis*, pp. 255–278. VCH.)

target analytes (the best passive samplers collect only these), leaving the water behind, whereas the latter collect the whole sample, which often means a large sample volume. Martin et al.[34] compared the monitoring of ground water for its content of polycyclic aromatic hydrocarbons and benzene, toluene, and xylenes using time-integrating ceramic dosimeters with conventional water sampling. Their results show that ceramic dosimeters are suitable for monitoring aqueous pollutant concentrations over long periods of time without any artifacts arising from pumping, handling, and storing the water samples. The first passive samplers were developed for monitoring inorganic compounds in surface water in the mid-1970s.[35] Since the first application of a passive sampler for collecting organic pollutants in soil in 1985,[36] devices for sampling organics from aquatic environments have developed apace. The techniques and corresponding sampling devices have been comprehensively discussed in a number of reviews and original papers.[37–43] In all these devices, a barrier is applied, which allows the more or less selective transfer of target analytes from the aqueous matrix to the accumulating medium. Ongoing research into sampling devices aims to increase their selectivity, widen their range of application, and improve their robustness. However, passive sampling will no doubt gain broader acceptance in regulatory programs once quality assurance, quality control, and validation methods have been developed.

1.5.5   SEDIMENT SAMPLERS

As the data indicating the interdependence between pollution of the sediment and aqueous phase in the aquatic environment are increasing, so is information on sediment pollution clearly taking on a greater significance. A good selection of sediment samplers is commercially available; they are described in a number of papers and sampling guidelines.[6,44–53] The main factors to be taken into consideration when choosing a sampler are the depth of water and the sample type (grab or core).

For taking grab samples of the upper sediment layer in relatively calm and shallow waters, spoons or scoops can be used (Figure 1.8). These are simple and inexpensive, but the samples are poorly defined with respect to area and depth, and the finest particles may be washed out as the sample is retrieved through the water column. In deeper waters, dredges of various sizes and designs are used for collecting the upper layer of sediment; the integrity of the sample collected may be disturbed to a greater or a lesser extent. They generally require a boat or a barge fitted with a winch (Figure 1.10), although there are some that can be operated manually.

Vertical pollutant profiles in sediments can be determined in samples collected with multilevel sampling devices or corers. Manual corers are simple, easy to use, clean, and decontaminate (Figure 1.9); they can be deployed by hand or with the aid of a hammer. Though recommended for shallow water, corers can be extended for use in deeper waters. In general, they are applicable to clayey and sandy sediments, but the use of inserts is recommended for the latter. The main short-coming of this technique is that the samples are relatively small in size. For larger depths and up to moderately strong water currents, specially designed gravity corers, equipped with stabilizing fins and adjustable weights, are in use. Vibro-corers are another type, which enter the sediment with the aid of the vibration from an electric motor.

1.6   CONCLUSIONS

The sampling of the aquatic environment is the critical step in the determination of organic and inorganic pollutants, and hence the suitability of water for a given purpose and for living organisms. Very often, however, the sampling process itself contributes to the total uncertainty of the analytical results in the highest degree. To obtain reliable information on the quality of a given water compart-ment, the sampling program must be meticulously planned. The sampling strategy includes the selection of the sampling location and sampling sites, frequency of sampling, sample size, the num-ber of samples, and selection of sampling technique and equipment, although in some situations certain parameters are stipulated by the regulatory agencies. The above factors emerge from the purpose of sampling, which defines the permissible uncertainty of sampling and limit of quan-titation. The correct decisions on water use and management can only be made if the data produced are reliable. Therefore, quality control procedures must be applied to test and correct this crucial step of chemical analysis. In the optimization of the sampling process, the costs must also be taken into account.

ACKNOWLEDGMENT

The authors thank Professor Jacek Namiesńik for the fruitful discussions they were able to have with him.

REFERENCES

1. Skarzynska, K., Z. Polkowska, and J. Namiesnik. 2006. Samples handling and determination of physico-chemical parameters in rime, hoarfrost, dew, fog and cloud water samples—a review. *Pol. J. Environ. Stud.* 15: 185–209.
2. Skarzynska, K., Z. Polkowska, A. Przyjazny, and J. Namiesnik. 2007. Application of different sampling procedures in studies of composition of various types of runoff waters. *Crit. Rev. Anal. Chem.* 37: 91–105.
3. Cochran, W.G. 1977. *Sampling Techniques*, 3rd edition. Wiley, USA. Available at http://www.amazon.com/Sampling-Techniques-3rd-William-Cochran/dp/047116240X/ref=si3_rdr_bb_product.
4. ISO 6107-2. 2006. Water quality, Vocabulary. Part 2.
5. Dick, M.E. 1996. Automatic water and wastewater—sampling, Chapter 13. In: L.H. Keith (ed.), *Principles of Environmental Sampling*, 2nd edition, pp. 237–258. American Chemical Society, Washington.
6. USEPA. 1995. Superfund Program Representative Sampling Guidance, Vol. 5: Water and Sediment. Available at http://www.epa.gov/tio/download/char/sf_rep_samp_guid_water.pdf.
7. ISO 5667-10. 2004. Water quality—sampling. Part 10: Guidance on sampling of waste waters.
8. Keith, L.H. 1991. *Environmental Sampling and Analysis: A Practical Guide*. Lewis, Michigan.
9. Taylor, J.K. 1996. Defining the accuracy, precision, and confidence limits of sample data, Chapter 4. In: L.H. Keith (ed.), *Principles of Environmental Sampling*, 2nd edition, pp. 77–83. American Chemical Society, Washington.

10. Madrid, Y. and Z.P. Zayas. 2007. Water sampling: Traditional methods and new approaches in water sampling strategy. *Trends Anal. Chem.* 26: 293–299.
11. Minkkinen, P. 2004. Practical applications of sampling theory. *Chemom. Intell. Lab. Syst.* 74: 85–94.
12. Liess, M. and R. Schulz. 2000. Sampling methods in surface waters, Chapter 1. In: L.M.L. Nollet (ed.), *Handbook of Water Analysis*, pp. 1–24. Marcel Dekker, New York.
13. Thompson, M., B.J. Coles, and J.K. Douglas. 2002. Quality control of sampling: Proof of concept. *Analyst* 127: 174–177.
14. U.S. Geological Survey. 2006. Collection of water samples (version 2.0): U.S. Geological Survey Techniques of Water-Resources Investigations, Book 9, Chapter A4. Available at http://pubs.water.usgs.gov/twri9A4/ (accessed September 2006).
15. Fresenius, W., K.E. Quentin, and W. Schneider. 1988. *Water Analysis: A Practical Guide to Physicochemical and Microbiological Water Examination and Quality Assurance.* Springer, Germany.
16. Hermanowicz, W., J. Dojlido, W. Dozanska, B. Koziorowski, and J. Zerze. 1999. *Fizyko–chemiczne Badanie Wody i Ścieków.* Arkady, Warszawa.
17. Dixon, W., G.K. Smyth, and B. Chiswell. 1999. Optimized selection of river sampling sites. *Water Res.* 33: 971–978.
18. Wardencki, W. and J. Namiesnik, 2002. Sampling water and aqueous solutions, Chapter 2. In: J. Pawliszyn (ed.), *Sampling and Sample Preparation for Field and Laboratory. Fundamentals and New Directions in Sample Preparation*, pp. 33–60. Elsevier, Amsterdam.
19. Hildebrandt, A., S. Lacorte, and D. Barcelo. 2006. Sampling of water, soil and sediment to trace organic pollutants at a river-basin scale. *Anal. Bioanal. Chem.* 386: 1075–1088.
20. Lane, S.L., S. Flanagan, and F.D. Wilde. 2003. Selection of equipment for water sampling (version 2.0): U.S. Geological Survey Techniques of Water-Resources Investigations. Book 9, Chapter A2. Available at http://pubs.water.usgs.gov/twri9A2/ (accessed March 20, 2003).
21. Wiegel, S., A. Aulinger, R. Brockmeyer, H. Harms, J. Loffler, H. Reincke, R. Schmidt, B. Stachel, W. von Tumpling, and A. Wanke. 2004. Pharmaceuticals in the river Elbe and its tributaries. *Chemosphere* 57: 107–126.
22. Krinitz, J., B. Stachel, and H. Reincke. 2000. Stoffkonzentrationen in mittels Hubschrauber entnommenen Elbewasserproben (1979 bis 1998). Rapport Arbeitsgemeinschaft für die Reinhaltung der Elbe, Hamburg.
23. Kent, T.R. and K.E. Payne. 1996. Sampling groundwater monitoring wells special quality assurance and quality control considerations, Chapter 21. In: L.H. Keith (ed.), *Principles of Environmental Sampling*, 2nd edition, pp. 337–392. American Chemical Society, Washington.
24. USEPA REGION I. 1996. Low stress (low flow) purging and sampling procedure for the collection of ground water samples from monitoring wells (Groundwater Sampling, January 9, 2003).
25. Smith, J.S, D.P. Steele, J.M. Malley, and M.A. Bryant. 1996. Groundwater sampling, Chapter 22. In: L.H. Keith (ed.), *Principles of Environmental Sampling*, 2nd edition, pp. 393–398. American Chemical Society, Washington.
26. Barcelona, M.J. 1996. Overview of the sampling process, Chapter 2. In: L.H. Keith (ed.), *Principles of Environmental Sampling*, 2nd edition, pp. 41–61. American Chemical Society, Washington.
27. Shaw, R.W., M.V. Smith, and R.J. Pour. 1984. The effect of sample frequency on aerosol mean values. *J. Air Pollut. Control Assoc.* 34: 839–841.
28. Nelson, J.D. and R.C. Ward. 1981. Statistical considerations and sampling techniques for ground-water quality monitoring. *Ground Water* 19: 617–625.
29. Parker L.V. and T. Ranney. 2000. Decontaminating materials used in ground water sampling devices: Organic contaminants. *Ground Water Monit. Rev.* 20: 56–68.
30. Namiesnik, J., J. Łukasiak, and Z. Jamrogiewicz. 1995. *Pobieranie Próbek Środowiskowych Do Analizy.* PWN, Warszawa.
31. Parker, L.V. 1994. The effects of ground water sampling devices on water quality: A literature review, pp. 130–141. GWMR. Spring.
32. Dick, E.M. 1994. Water and wastewater sampling for environmental analysis, Chapter 12. In: B. Markert (ed.), *Environmental Sampling for Trace Analysis*, pp. 255–278. VCH, Weinheim.
33. USEPA. 1992. *NPDES Storm Water Sampling Guidance Document.* Available at http://www.epa.gov/npdes/pubs/owm0093.pdf.
34. Martin, H., B.M. Patterson, and G.B. Davis. 2003. Field trial of contaminant groundwater monitoring: Comparing time-integrating ceramic dosimeters and conventional water sampling. *Environ. Sci. Technol.* 37: 1360–1364.
35. Benes, P. and E. Steinnes. 1974. *In situ* dialysis for the determination of the state of trace elements in natural water. *Water Res.* 8: 947–953.

36. Coutant, R.W., R.G. Lewis, and J. Mulik. 1985. Passive sampling devices with reversible adsorption. *Anal. Chem.* 57: 219–223.
37. Stuer-Lauridsen, F. 2005. Review of passive accumulation devices for monitoring organic micro-pollutants in the aquatic environment. *Environ. Pollut.* 136: 503–524.
38. Gorecki, T. and J. Namiesnik. 2002. Passive sampling. *Trends Anal. Chem.* 21: 276–291.
39. Namiesnik, J., B. Zabiegala, A. Kot-Wasik, M. Partyka, and A. Wasik. 2005. Passive sampling and/or extraction techniques in environmental analysis: A review. *Anal. Bioanal. Chem.* 381: 279–301.
40. Vrana, B., G.A. Mills, I.J. Allan, E. Dominiak, K. Svensson, J. Knutsson, G. Morrison, and R. Greenwood 2005. Passive sampling techniques for monitoring pollutants in water. *Trends Anal. Chem.* 24: 845–868.
41. Bopp, S., W. Hansjorg, and K. Schirmer. 2005. Time-integrated monitoring of polycyclic aromatic hydro-carbons (PAHs) in groundwater using the ceramic dosimeter passive sampling device. *J. Chromatogr. A* 1072: 137–147.
42. Kot-Wasik, A., B. Zabiegala, M. Urbanowicz, E. Dominiak, A. Wasik, and J. Namiesńik. 2007. Advances in passive sampling in environmental studies. *Anal. Chim. Acta* 602: 141–163.
43. Vermeirssen, E.L.M., O. Korner, R. Schonenberger, M.J.F. Suter, and P. Burkhardt-Holm. 2005. Characterization of environmental estrogens in river water using a three pronged approach: Active and passive water sampling and the analysis of accumulated estrogens in the bile of caged fish. *Environ. Sci. Technol.* 39: 8191–8198.
44. Heim, S., M. Ricking, J. Schwarzbauer, and R. Littke. 2005. Halogenated compounds in a dated sediment core of the Teltow canal, Berlin: Time related sediment contamination. *Chemosphere* 61: 1427–1438.
45. Chang, B.V., C.S. Liao, and S.Y. Yuan. 2005. Anaerobic degradation of diethyl phthalate, di-*n*-butyl phthalate, and di-(2-ethylhexyl) phthalate from river sediment in Taiwan. *Chemosphere* 58: 1601–1607.
46. Borghini, F., J.O. Grimalt, J.C. Sanchez-Hernandez, R. Barra, C.J.T. Garcia, and S. Focardi. 2005. Organochlorine compounds in soils and sediments of the mountain Andean Lakes. *Environ. Pollut.* 136: 253–266.
47. Zhang, Q. and G. Jiang. 2005. Polychlorinated dibenzo-*p*-dioxins/furans and polychlorinated biphenyls in sediments and aquatic organisms from the Taihu Lake, China. *Chemosphere* 61: 314–322.
48. Zygmunt, B. and J. Namiesnik. 2002. Sampling selective solid materials, Chapter 3. In: J. Pawliszyn (ed.), *Sampling and Sample Preparation for Field and Laboratory. Fundamentals and New Directions in Sample Preparation*, pp. 61–86. Elsevier, Amsterdam.
49. USEPA. Sampling for contaminants in sediments and sediment pore water, measurement and monitoring technologies for the 21st Century [21M²]. Available at http://www.clu-in.org/programs/21m2/sediment/default.cfm.
50. Ohio, E.P.A. 2001. *Sediment Sampling Guide and Methodologies*, 2nd edition. Available at http://www.epa.state.oh.us/dsw/guidance/sedman2001.pdf.
51. Punning, M.J., T. Alliksaar, J. Terasmaa, and S. Jevrejeva. 2004. Recent patterns of sediment accumula-tion in a small closed eutrophic lake revealed by the sediment records. *Hydrobiologia* 529: 71–81.
52. Ricking, M., J. Schwarzbauer, and S. Franke. 2003. Molecular markers of anthropogenic activity in sedi-ments of the Havel and Spree Rivers (Germany). *Water Res.* 37: 2607–2617.
53. Schwarzbauer, J., M. Ricking, S. Franke, and W. Francke. 2001. Halogenated organic contaminants in sediments of the Havel and Spree Rivers (Germany). Part 5 of organic compounds as contaminants of the Elbe River and its tributaries. *Environ. Sci. Technol.* 35: 4015–4025.

2 Preservation and Storage of Water Samples

Marek Biziuk, Angelika Beyer, and Joanna Żukowska

CONTENTS

2.1 Introduction 19
2.2 Preservation of Water Samples for the Determination of Inorganic Compounds 19
2.3 Preservation of Water Samples for the Speciation Analysis of Metals 22
    2.3.1 Arsenic 25
    2.3.2 Chromium 26
    2.3.3 Mercury 26
    2.3.4 Selenium 26
    2.3.5 Tin 27
2.4 Preservation and Storage of Water Samples for the Determination of Organic Compounds 27
2.5 Summary 33
Acronyms and Abbreviations 33
References 33

2.1 INTRODUCTION

Accurate determinations of the numerous parameters and components in water samples are essential in many research programs. Nevertheless, no matter how accurate and sensitive the analytical devices and techniques applied during the analysis, the data will be useless unless special attention is given to counteracting the potential changes proceeding within a sample. A large number of substances contained in water can be expected to undergo various chemical, physical, and biological transformations. Such processes can alter the sample composition and consequently lead to unrepresentative results. Sample preservation is therefore necessary in order to inhibit reactions in the sample during the period between sampling and analysis. Unfortunately, analysis of water samples at the point of sample collection is often not possible, so sample preservation is imperative. A wide variety of techniques are therefore applied to minimize loss of target compounds from water samples during the sample holding time (Figure 2.1).

2.2 PRESERVATION OF WATER SAMPLES FOR THE DETERMINATION OF INORGANIC COMPOUNDS

The determination of inorganic compounds in water samples provides significant data for risk assessment in aquatic environments and for characterizing the chemical quality of water. The presence of inorganic chemicals in water is due to both natural processes and human activities. Most of these

Chemical methods	Physical methods
Addition of: • Acid, • Sulfite, • Solvent, • Toxic metal ions, • Azide, • Formaldehyde, • Other substances.	• Cold storage, • Frozen storage, • Dark storage, • Use of amber bottels (or other special types of bottles), • Filtration, • Collection of the sample without headspace, • Collection of the sample without changing oxidation-reduction (redox) conditions, • Real-time or nearly real-time isolation of the chemicals using LLE, SPE, or SPME, • Using coacervates.

FIGURE 2.1 Methods used to minimize loss of inorganic and organic chemicals from water samples.

compounds are water soluble, and they may be responsible for many disorders occurring in living organisms. The monitoring of water quality is thus of paramount importance. To control the levels of target compounds, accurate measurement is indispensable, and hence, sample contamination or degradation prior to analysis must be minimized. To this end, the establishment of a control program for assessing contamination risks at each stage in the analytical process is of crucial significance.

The containers used for transporting and storing water (intended for the determination of inorganic components) have to be carefully selected and cleaned to minimize the risk of possible contamination. The usual materials from which these containers are made include high-density polyethene (HDPE), polypropene (PP), fluorinated ethene propene copolymer (FEP), perfluoroalkoxy polymer (PFA), poly(tetrafluoroethene) (PTFE), or glass (e.g., Pyrex borosilicate glass).[1-6] When sample bottles are being selected, the type of analyte being measured, the conditions of preservation, and the physical characteristics of the container should be taken into consideration.

In general, the storage of water samples for the determination of inorganic components should be avoided if at all possible.[7] Nevertheless, the nature of the research and the lack of adequate equipment required for further analysis may preclude the immediate determination of the compounds in question. It is therefore of great interest to explore appropriate preservation and storage methods that enable unfavorable sample changes to be minimized. Maintaining the original sample composition until the analysis can be performed is the absolute requirement for successful preservation. Beyond this, there are many other factors that should be considered, such as compatibility with the analytical technique to be used for the final determination, simplicity, rapidity of implementation, and so on.[8,9]

Numerous approaches, both physical and chemical, are recommended in the literature for the preservation of water samples for inorganic chemicals analysis. In particular, the significance of freezing, pasteurization, as well as the use of chemical preservatives such as acids, sodium hydroxide, chloroform, formaldehyde, or mercuric chloride as alternative techniques has been emphasized.[8,10] Table 2.1 describes some recommended stabilization methods.

The large majority of known stabilization procedures are based on relatively few parameters, the most important of which are temperature, pH, and redox potential. Despite the different treatments, the final aim of these efforts is the same: All preservation processes are intended to (1) inhibit the growth and biological activity of microorganisms, a unique source of concentration changes of dissolved inorganic components; (2) diminish the volatility of sample components; (3) retard hydrolysis reactions, precipitation, and coprecipitation reactions; and (4) reduce absorption effects.[11,12] Unfortunately, there is no single, universal procedure applicable to the stabilization of all compounds (Table 2.2). The choice of procedure is therefore dependent primarily on factors relevant to the sample in question (e.g., salinity and dissolved chemical compounds), type of analyte, environmental impact of the method, and the particular preservation situation (e.g., costs and periods of storage).

TABLE 2.1

Advantages and Disadvantages of Selected Preservation Methods Used for Water Samples[10,13–17]

Water Preservation Methods	Advantage	Disadvantage
Physical		
Freezing	• No change of the sample matrix, and in consequence elimination of potential risk of contamination due to added chemicals • No risk connected with exposure to toxic substances • An alternative method for long-term sample storage • Causes the reduction of the activity of microorganisms present in the sample	• Necessity of maintenance of low temperature storage • Risk of losing the sample in the case of equipment failure • Possibility of occurrence of considerable variability in amount of compounds arising from the matrix, adsorption on suspended matter presented in the sample, formation of suspended matter during freezing, and adsorption on the vessel walls • Possibility of precipitation of colloids
Pasteurization	• Possibility of storage samples at room temperature over long periods	• Necessity of using a special sample bottles selected for their cap tightness and pressure resistance
Chemical		
Addition of poisons (biocides)	• Enables the long-term stabilization of inorganic compounds	• Risk of sample contamination and additional interference during analyses • Anions of acids may be a source of interference for the analytical method • Some of the poisons can hydrolyze organic compounds during storage or precipitate bacteria and proteins

TABLE 2.2

Examples of Preservation Methods of Water and Wastewater Samples for the Determination of Inorganic Compounds

Type of Compound	Type of Preservation	Maximum Holding Time	Reference
Ammonium nitrogen, free and ionized ammonia	Cooling to 2–6°C	6 h	International Organization for Standardization (ISO)
	Acidification to pH < 2 with H_2SO_4	28 days	EPA
Ammoniacal nitrogen, oxidized nitrogen	Storage on water ice	6 h	18
	Acidification to pH < 2 with H_2SO_4	6 h	
	Immediate freezing with dry ice and then stored in laboratory freezer	24 h	
	Prolonged refrigeration	54 h	
	Pasteurization and stored at room temperature	18 months	19

continued

TABLE 2.2   (continued)

Type of Compound	Type of Preservation	Maximum Holding Time	Reference
Kjeldahl nitrogen	Cooling to 2–6°C	6 h	ISO
	Acidification to pH < 2 with H_2SO_4	28 days	EPA
		6 h	18
	Storage on water ice	6 h	
	Immediate freezing with dry ice and then stored in laboratory freezer	24 h	
	Prolonged refrigeration	54 h	
Nitrate	Cooling to 2–6°C	48 h	EPA
	Acidification to pH < 2 with H_2SO_4	28 days	EPA
	Frozen at −40°C initially and then stored at −20°C	24 months	20
	Addition of mercuric chloride	26 months	10
Total phosphorus (TP) and filterable reactive phosphorus (FRP)	Storage on water ice	6 h	18
	Immediate freezing with dry ice and then stored in laboratory freezer	24 h	
	Prolonged refrigeration	54 h	
TP	Acidification to pH < 2 with H_2SO_4 and stored at room temperature	7 days	21
	Acidification to pH < 2 with H_2SO_4 and stored at 4°C	28 days	
FRP	Pasteurization and stored at room temperature	18 months	11
	Frozen at −16°C	4–8 years	22
Metals	Acidification to pH < 2 with HNO_3	6 months	American Public Health Association (APHA), ISO, EPA,
Nonionic surfactants	Addition of formaldehyde	48 h	ISO
Silicate	Addition of mercuric chloride	26 months	10
Sulfide	Addition of cadmium acetate or zinc acetate	7 days	EPA
	Addition of sodium hydroxide to pH > 12	28 days	APHA
Bromide, chlorine, chloride, fluoride	Cooling to 2–6°C	28 days	EPA
Iodine	Addition of sodium hydroxide to pH > 12	14 days	EPA

2.3   PRESERVATION OF WATER SAMPLES FOR THE SPECIATION ANALYSIS OF METALS

The preservation of chemical compounds, especially the individual chemical species in a sample, is an important consideration in analysis. At present, much effort is going into the search for effective procedures to prevent quantities and species of elements in their original state from undergoing chemical, biochemical, and photochemical changes.

Arsenic, chromium, mercury, selenium, and tin have been the object of numerous investigations. Because some of them are classified as probable human carcinogens[23–25] (strictly speaking, some of their species), the accurate assessment of concentration and speciation in environmental matrices is enormously important. Unfortunately, such factors as chemical reactions between species, low concentration, microbial activity, redox conditions, as well as the presence of other dissolved metal ions, may cause the amounts and distributions of chemical species in a sample to vary. In response to these problems, analytical research efforts have focused on developing techniques enabling the original valence state of the metals to be preserved. Table 2.3 lists some of these stabilization methods.

TABLE 2.3
Some Characteristics of Storage Methods for the Speciation Analysis in Water Samples

Type of Compound	Type of Sample	Type of Preservation and Storage of Samples, and Main Parameters		Maximum Holding Time	Reference
As(III), As(V)	Seepage water (rich in Fe and Mn)	The acidification to pH ≤ 2 with 0.01 mol L^{-1} H$_3$PO$_4$ and cooling to a temperature of 6°C, storing in darkness		3 months	26
	Well water (rich in Fe and Mn)	The addition of ethylenediaminetetraacetic acid (EDTA)		14–27 days	27
	Acid mine waters	The addition of EDTA		3 months	28
	Synthetic groundwater without Fe (pH 8.4)	The addition of preservative reagent and stored in brown PP bottles at room temperature (22–24°C)	H$_2$SO$_4$, H$_3$PO$_4$, and EDTA-HAc	7 days	29
	Synthetic groundwater with Fe(II) (pH 8.4)		H$_2$SO$_4$ and EDTA-HAc	28 days	
			H$_3$PO$_4$	7 days	
	Drinking waters		EDTA-HAc	>30 days	
	Groundwater samples (with neutral pH and poor in Fe)	Without any treatment		≤3 days	30
		The acidification with phosphoric acid and storage at 4°C		>3 days	
	Seepage water (rich in Fe and Mn)	The acidification with 0.01 mol L^{-1} H$_3$PO$_4$ and storage at 6°C		6 days	31
Total dissolved As and As(III)	Synthetic, double-distilled water (pH 3.7)	Without preservation	Dark experiment	16 days	32
			Light experiment		
		The addition of HCl	Dark experiment	45 days	
			Light experiment		
	Synthetic water with Fe(III)	The addition of HCl, Fe(II), and SO$_4$	Dark experiment	71 days	
As(III)	Natural waters	Filtration, acidification to pH < 2 with HCl, chilled at 4°C, and stored in the dark		—	
	Geothermal and acid mine waters	Filtration, acidification with HCl and stored in opaque bottles		19 months	
As(V)	Deionized water spiking with 40 μg As(V)	The storage in dark at 4°C in polyethylene (PE) containers		67 days	33
Cr(III), Cr(VI)	Aqueous reference material	The setting pH to 6.4 by HCO$_3^-$/H$_2$CO$_3$ buffer and storage at 5°C in quartz ampoules		228 days	34
	Aqueous sample	Freeze-drying, stored at −20°C and later reconstituted in HCO$_3$/H$_2$CO$_3$ buffer (pH 6.4) under a CO$_2$ blanket		88 days	35
Cr(VI)	Coastal seawater	The storage at room temperature and at natural pH		1 month	36
	Oceanic water	Deep freezing		8 months	
	Water sample	The storage of samples at a temperature of 4°C		24 h	37, 38

continued

TABLE 2.3 (continued)

Type of Compound	Type of Sample	Type of Preservation and Storage of Samples, and Main Parameters		Maximum Holding Time	Reference
Cr(III)	Water sample	The collection of samples into 500 mL or 1 L fluoropolymer, conventional or linear PE, polycarbonate (PC), or PP containers with lid	The addition of 1 mL chromium (III) extraction solution to 100 mL aliquot, vacuum filtration through 0.4 μm membrane, and the addition of 1 mL 10% HNO_3	—	39
Cr(IV)			The addition of 50% NaOH		
Total Hg	Freshwater and seawater	The collection of samples into a glass (with PTFE-lined lid) or PTFE bottles prerinsed with acid, and the addition of BrCl or 0.5% HCl		300 days	40
Methyl-Hg	Freshwater	The collection of filtered samples into a glass (with PTFE-lined lid) or PTFE bottles prerinsed with acid, and storage in the dark at 1–4°C		1 week	
		The acidification with 0.5% HCl or 0.2% H_2SO_4 and storage in the dark at 1–4°C		250 days	
Dimethyl-Hg and Hg(0)		The collection of unfiltered samples into a glass (with PTFE-lined lid) bottles and storage in the dark at 1–4°C		1 day	
Hg(II)		The collection of filtered samples into a glass (with PTFE-lined lid) bottles and storage in the dark at 1–4°C		2–5 days	
Dissolved/ particulate speciation		The collection of samples into a glass (with PTFE-lined lid) or PTFE bottles and storage in the dark at 1–4°C		2–5 days	
Dissolved inorganic and labile Hg	Estuarine water	The collection of unfiltered samples into a glass container and addition of 1% HNO_3		30 days	41
Inorganic and methyl-Hg	River water	Trapping in minicolumn packed with C_{18} modified with diethyldithiocarbamate and elution with 500 μL of 5% thiourea in 0.5% HCl		1 week	42
Total Hg	Water sample	The collection of samples into fluoropolymer or borosilicate glass bottles with fluoropolymer or fluoropolymer-lined caps	The acidification to pH > 2 with 0.5% high-purity HCl or 0.5% BrCl	—	39
Total and methyl-Hg			The addition of 0.5% high-purity HCl	—	
Se(IV), Se(VI)	Natural and distilled water	The acidification to pH 1.5 with H_2SO_4 and storage in PE or Pyrex containers at room temperature		4 months	43

continued

TABLE 2.3 (continued)

Type of Compound	Type of Sample	Type of Preservation and Storage of Samples, and Main Parameters	Maximum Holding Time	Reference
Total Se and Se(IV)	Seawater	The acidification to pH 2 with HCl and storage in PE or glass containers	4.5 months	44
SeCys, SeMet, TMSe	Aqueous matrix	The storage in the dark at pH 4.5 in Pyrex containers at both 4°C and 20°C	1 year	45
		The storage in the dark at 4°C in PP containers	6 weeks	46
Se(IV), Se(VI), SeMet	Aqueous mixtures	The storage at 3°C	2 weeks	47
Tributyltin chloride (TBT),	River water	The acidification to pH 4 with HNO_3 and storage in PE containers at 4°C in the dark	1 month	48
triphenyltin chloride (TPhT)	Aquatic solution	The acidification with HCl and storage in PE containers at 4°C in the dark	3 months	49
TPhT	Seawater	Storage on C_{18} cartridges at room temperature	60 days	50
TBT		The storage in the dark at 4°C in PC bottles or storage on C_{18} cartridges at room temperature.	7 months	
Organotin compounds	Water samples	The storage in 1-L brown glass bottles at 25°C	20 days	51

Filtration, acidification, chilling, and freezing are some of the numerous techniques for preserving inorganic species in water samples. Depending on the species of the target chemical element, however, the stabilization conditions may vary from one type of sample to another (see Table 2.3).

2.3.1 ARSENIC

The presence of high levels of arsenic in aquatic ecosystems is a consequence, firstly, of the weathering of arsenic-containing rocks and soils, and secondly, of rapid industrial growth, the intensive use of agricultural chemicals, and the urban activities of human beings, such as irresponsible sewage and waste disposal. In natural waters, arsenic is primarily present as trivalent arsenite As(III) and pentavalent arsenate As(V),[52] the former being more toxic and more mobile than the latter.[53] Because of the strong tendency for arsenic compounds to undergo changes, every effort should be made to preserve As(III)/As(V) speciation in water samples. The most frequently used pretreatment procedures for stabilizing arsenic species include filtration, acidification with inter alia sulfuric, phosphoric, or hydrochloric acids,[28,31,54] and the addition of a complexing agent such as EDTA;[55] all these procedures are carried out under controlled temperature and light conditions. Recent experiments on the effect of UV radiation on the stability of arsenic species in the presence of Fe(II) have shown that oxidation of As(III) to As(V) depends strongly on UV exposure. Samples should therefore be stored in the dark or in opaque propylene bottles until analysis.[29] Moreover, many researchers have reported problems with the preservation of arsenic species in iron-rich waters. Gallagher et al.[27] recommend preserving iron-rich water samples at a lower pH. In contrast, Bednar et al.[28] state that using EDTA without lowering the pH is sufficient to preserve As species. Several studies have shown that nitric acid can preserve As species from oxidation of As(III) to As(V).[28,56] Fanning,[57] however, reports that nitric acid is an oxidizing agent capable of undergoing photochemical reduction and should not be used to preserve redox species.

2.3.2 CHROMIUM

Chromium enters surface waters from natural sources, such as the weathering of rock and the wet precipitation and dry fallout from the atmosphere, as well as from the extensive use of this metal in, for example, chemical, metallurgical, and refractory industries.[58] The unsatisfactory disposal of industrial wastes also contributes to the presence of this element in the environment. Chromium can exist in several chemical forms, but only two of them—trivalent Cr(III) and hexavalent Cr(VI)—are stable enough to occur in the environment. Cr(VI) is approximately 100 times more toxic than Cr(III);[59] unfortunately, its compounds are usually highly soluble, mobile, and bioavailable compared to those of Cr(III). The accurate determination of each of these species is thus of the utmost importance, especially as regards the proper evaluation of physiological and toxicological effects. Water samples should be subjected to an appropriate preservation treatment that prevents species degradation and interconversion. For chromium species, refrigeration, minimal sample handling, and immediate analysis of water samples are generally preferable.[60,61] However, should long-term storage be unavoidable, it is enough to freeze the sample for chromium compounds to remain stable.[58] The Cr(III)/Cr(VI) ratio can also be preserved if the sample is stored at pH 9; under such conditions, the oxidizing potential of Cr(VI) is too low to oxidize reducers present in natural water.[62] The ratio can also be upheld at a nearly neutral pH, particularly under a CO_2 blanket.[36,63] Acid media should not, however, be used to preserve chromium species, since under such conditions Cr(VI) will undergo rapid reduction to Cr(III).

2.3.3 MERCURY

Mercury is one of the most dangerous water pollutants because of its accumulative and persistent character in the environment.[64] The impacts of volcanic activity, as well as the mining and smelting of mercury and other metal sulfide ores, fossil fuel combustion, and industrial processes involving the use of mercury are considered to be of the greatest magnitude.[65] It is well known that the reactivity and toxicity of Hg depends to a large extent on the species. All forms of mercury are poisonous,[66] but its organic forms (methyl and dimethyl mercury) exhibit the greatest toxicity. The efficiency of techniques for minimizing potential transformation and degradation processes need to be improved continually. The material from which a sample container is made may give rise to major changes in mercury speciation.[4,67,68] Yu and Yan[69] reviewed the application of various bottle samples, focusing particularly on improving the stability of mercury species during sample storage. Generally, PTFE and glass bottles are recommended; containers made from PE should not be used to store water samples with low levels of mercury as this material is permeable to mercury vapor from the atmosphere.[41]

The most common means of preserving Hg species include acidification with strong mineral acids (HCl, H_2SO_4, and HNO_3) and the addition of oxidizing ($KMnO_4$, $K_2Cr_2O_7$, H_2O_2, or Au^{3+}) or complexing agents (Cl^-, I^-, Br^-, CN^-, L-cysteine, and humic acid).[67,70–72] Nevertheless, although acidification with HCl works well for preserving both total and methyl mercury, this type of stabilization is not suitable for labile Hg(II) because of the possible dramatic loss to the container walls and oxidation of Hg(0) to Hg(II).[40] Furthermore, it seems inadvisable to use HNO_3 as a preservative for methyl mercury as this form of the metal may decompose, especially in the presence of halides.[40] Thus, no ideal method has been found for stabilizing all the original valence states of mercury in water samples.

2.3.4 SELENIUM

A contaminant of concern, selenium enters aquatic ecosystems from natural sources, and from anthropogenic ones such as coal mining and combustion, gold, silver, and nickel mining, metal smelting, oil transport, the utilization and refining of crude oil, municipal landfills, and agricultural irrigation.[73] In recent decades, considerable research effort has focused on selenium, the reason being the narrow margin between its toxicity and its role as an essential element in the human organism. Both organic (mainly as amino acids or volatile methylated compounds) and inorganic species (in the oxidation states—II, 0, IV, and VI)[74] of Se are present in the environment, each species with different toxicological properties. In general, inorganic forms are considerably more toxic than organic ones, and selenium (IV) is considerably more toxic than selenium (VI).[75] In view of the above, analysts should strive to achieve greater accuracy in the determination of selenium species. There are numerous reports on the preservation of selenium species; the significance as regards selenium stabilization in an aqueous medium of parameters such as pH, storage medium, temperature, and container material is discussed in a number of articles.[76,77] In aqueous solution selenium (VI) is generally more stable than selenium (IV), and is not so sensitive to the pH of the sample.[78] According to Héninger et al.,[76] aqueous samples containing selenium species are best stored in HCl in PTFE containers at 4°C.

2.3.5 TIN

As in the case of the elements mentioned above, the toxicity of tin compounds to living organisms depends largely on the oxidation state of the metal. Generally, the organic species are significantly more toxic than the inorganic ones, and trisubstituted organotins (TBTs) (especially butyl and phenyl species) are considered more toxic than the corresponding mono-, di-, or tetrasubstituted compounds.[79–81] The European Parliamentary Commission has included TBTs in the European list of most hazardous substances.[82] Because of their numerous applications in various sectors of industry,[83] and consequently, the real risks to the environment and human health from organotin exposure, the European Community has imposed the obligation to monitor the TBT content in aquatic environments.[84–86] Hence, their persistence and concentration are determined in many environmental matrices. Nevertheless, with regard to the quantitative speciation of tin compounds, a special effort should be made to gaining an understanding of the reactions occurring during the storage of water samples containing different tin species, for example, the activity of microorganisms, and changes in oxidation states and sorption. In order to prevent changes to the initial sample composition, a variety of preservation techniques have been put forward (Table 2.3). Burns et al.[87] recommend that for reliable results of speciation analysis, samples should be frozen and analyzed within the shortest possible time.

2.4 PRESERVATION AND STORAGE OF WATER SAMPLES FOR THE DETERMINATION OF ORGANIC COMPOUNDS

Water pollution by organic compounds, many of which are known to be toxic and carcinogenic, has given rise to considerable concern worldwide.[88] That is why water samples should be treated appropriately in order to prevent changes to particular sample components.

It is essential that both the identity and the concentration of the target organic compounds in water samples remain the same from sample collection to analyte determination. Minimizing change in and loss of target organic chemicals from water samples is important to the integrity of an investigation.[89]

Organic chemicals can be lost from water samples through volatilization, sorption, and conversion reactions. Volatilization can remove chemicals from the water phase to the air space in an unfilled bottle (headspace), sorption can remove them from the water to the walls and cap of the sample bottle, and conversion reactions can eliminate them from the water altogether. Conversions of possible concern are anaerobic or aerobic biodegradation, photolysis, abiotic hydrolysis, and abiotic redox reactions.[89]

The analytical procedures for determining organic compounds in water samples usually involve a number of steps, such as solvent extraction, chemical fractionation, sample cleanup, and solvent reduction, before the final analysis is undertaken. Regardless of the complexity of the analytical procedure, however, almost all water samples are stored in a container for some time between sample collection (or isolation of the organic compounds from the water) and final analysis, unless they are isolated in real time during collection.[89]

Organic compounds are the most common anthropogenic water pollutants,[8] and industrial, domestic, and agricultural wastewaters are major sources of water contamination with these substances. A variety of techniques for preserving and storing water samples containing organic compounds can be applied; which one is chosen depends mainly on the target chemical. The organic compounds most frequently found in water and wastewater samples are

- Pesticides
- Phenols
- Aliphatic and aromatic hydrocarbons
- Polynuclear aromatic hydrocarbons (PAHs)
- Surfactants
- Halogen compounds.

Pesticides come under the headings of both organic and inorganic compounds. Because of their different chemical and physical properties, they have been divided into groups—the inorganic, botanical, and synthetic organic pesticides.[90,91] The four basic types of synthetic organic pesticides are the chlorinated hydrocarbons, organophosphates, carbamates, and pyrethroids. The stability of pesticides from different groups has already been studied in order to specify the conditions of transport or temporary storage of samples before their analysis. To stabilize samples containing pesticides, different agents, such as pH modifiers, chelating agents, or microbial inhibitors can be used; a reduced temperature may also be applied.[92] In such a diverse group as pesticides, however, the agent required to stabilize certain compounds in the sample may affect the stability of others. By way of example, Ferrer and Barcelo[93] demonstrated that acidification may damage certain pesticides such as fenamiphos, whereas other compounds such as fenitrothion can only be kept stable in an acidic medium. Also, water samples containing acidic herbicides are usually acidified for preservation, and then stored at 4°C prior to extraction and analysis.[94] Table 2.4 lists some of the most commonly used recommendations for determining pesticides in water.

Like pesticides, phenols turn up in aquatic environments quite frequently because of their wide application in the chemical industry. The largest single use of phenol is in the manufacture of plastics, but it is also used in the synthesis of caprolactam (for making nylon 6 and other man-made fibers) and bisphenol A (for producing epoxy and other resins). Further uses are as a slimicide (a chemical that kills the bacteria and fungi found in aquatic slimes), as a disinfectant, and in medical products.[95] Because of their toxicity and environmental persistence, a number of phenols have been targeted by different monitoring programs, such as those of the US Environmental Protection Agency (EPA) and of the European Union (EU).[96] According to EPA instructions, water samples containing phenols must be extracted within seven days of collection and completely analyzed within 40 days of extraction.[97] Acidification is frequently recommended to preserve water samples for phenols determination (see Table 2.4).

Polychlorinated biphenyls (PCBs) have entered the natural environment via human agencies. The risks posed by their presence in the environment are a direct consequence of their physico-chemical properties—good solubility in nonpolar solvents, oils and fats, low vapor pressure, low electrical conductivity, high thermal conductivity, high ignition temperature, and very high resistance to chemical factors. These favorable properties are the reason for their wide application in industry.[98–100] EPA instructions state that if samples containing PCBs and organochlorine pesticides are not extracted within 72 h of collection, the pH of the sample should be adjusted to 5.0–9.0 with sodium hydroxide solution or sulfuric acid.[101] Water samples containing PCBs should be stored in amber glass containers, restricting the access of light, and preserved by freezing (see Table 2.4).

PAHs, a large group of organic compounds, have received considerable attention, since several of them have been shown to elicit carcinogenic and teratogenic reactions in experimental animals. Environmentally significant PAHs, from naphthalene to coronene,[88] are ubiquitous

TABLE 2.4
Methods for the Preservation of Water and Wastewater Samples for the Determination of Organic Compounds

Type of Compound	Type of Preservation and Storage of Samples, and Main Parameters	Maximum Holding Time or Time of Experiment	Reference
Pesticides			
Pesticides	The addition of extracting solvent and cooling to a temperature 2–5°C	Depends on the solvent used	102
Chloroorganic and phosphoroorganic pesticides, and derivatives of phenoxyacetic acid	The addition of mercury chloride (1 mL solution of 10 mg (mL)$^{-1}$ concentration per 1 L sample)	40 days	103
Organophosphorus insecticides in surface water	Trapping in large-particle-size graphitized carbon black cartridges and keeping at −20°C	2 months (recoveries approaching 70–134%)	104
Different group of polar pesticides in surface and tap water	Trapping in SPE cartridges packed with polymeric material and stored at 4–5°C or at laboratory temperature (no substantial differences)	7 weeks	105
Organophosphorus pesticides	Trapping in SPE cartridges filled with C_{18} and stored at −20°C	8 months	106
	Trapping in SPE cartridges filled with C_{18} and stored at 4°C	2 months	
	Trapping in SPE cartridges filled with C_{18} and stored at laboratory temperature	1 month	
Carbamate pesticides	The addition of sodium thiosulfate and monochloroacetic acid, cooling to a temperature of 4°C in amber vials	28 days	107
Herbicides, organochlorine, and organophosphoric insecticides	The collection of samples in glass bottles prerinsed with acetone and hexane, the addition of 1 mL of $HgCl_2$ 1% for preservation, storage at <10°C	7 days (than extraction), after extraction 14–28 days	108
Diquat and paraquat	The acidification with sulfuric acid to pH<2, addition of sodium thiosulfate and cooling to 4°C		109
Phenols	The acidification with sulfuric acid to pH<2, frozen, and stored in glass container	3–4 weeks	110,111
	The acidification with phosphoric acid to pH < 4, frozen, and stored in glass container		102
	Trapping in SPE cartridges packed with polymeric material (Isolute ENV+) and stored at −20°C	2 months	96
	Trapping in SPE cartridges packed with polymeric material (Isolute ENV+) and stored at 4°C	0.5 month	
PCBs	The preservation mostly through freezing, storing in amber glass containers, restricting the access of light	7 days	112
	The collection of samples in Pyrex borosilicate amber glass containers with caps lined with aluminum foil and storage in the dark at 4°C	48 h	EPA

continued

TABLE 2.4   (continued)

Type of Compound	Type of Preservation and Storage of Samples, and Main Parameters	Maximum Holding Time or Time of Experiment	Reference
PAHs	Storage of samples at a temperature of 4°C in amber or foil-wrapped bottles	7 days	113
	The addition of 1.0% SDSA and storing in glass containers	4 days—recovery values near to 100%	114
	Storage of solvent extracts or sorption tubes with analytes trapped, in freezer in glass vials closed with PTFE stoppers	1 month	115,116
Aliphatic hydrocarbons	Trapping in SPE phases, XAD-2 macroreticular resin and C_{18} and stored at room temperature	100 days	117
VOCs	Cooling to 4°C and the acidification with HCl to pH < 2	14 days if preserved with HCl, 7 days without HCl	118
	The acidification with sodium bisulfate or ascorbic acid, storing at 4°C	112 days	119
	Storage of samples at a temperature of 4°C	96 h (time of experiment)	120
Dioxines and furanes	Addition of sodium thiosulfate (if residual chlorine is present), cooling to 0–4°C, storage in the dark	There is no demonstrated maximum holding time, up to 1 year	121
Nonionic surfactants	Addition of formaline in concentration 1%	20 days	122
	Addition of formaldehyde in concentration 0.1%	6 days	
	Addition of mercury (II) (25 mg L^{-1}) or copper (II) (50 mg L^{-1})	6 days	
	Addition of chloroform together with refrigeration of a sample (4°C)	6 days	
	Dried ethyl acetate extracts of sewage samples mixed with chloroform (1:2)	28 months	123
Low-molecular-weight organic acids	Freezing	4 weeks	124
	Sterile filtration	1 day	
	Sterile filtration in combination with storage of the samples (natural waters) in the dark at 4°C	<30 days	
Dissolved organic carbon	Storage of samples (rainwater) at a temperature of 4°C	1 week	125
	The acidification with H_3PO_4 and storage at 4°C	15 months	126

pollutants formed, inter alia, from the combustion of fossil fuels; they are always present as a mixture of individual compounds. Because many PAHs and their derivatives are so danger-ous, the source, occurrence, transport, and fate of PAHs in waters have been extensively studied.[88,127,128] According to the literature, the main factor affecting the stability of PAHs in waters is their adsorption to containers.[129,130] This problem can be overcome by using contain-ers made from appropriate materials, by acidifying the water sample, or by adding acetonitrile at concentrations between 25% and 40% (v/v);[131] it should be noted, however, that these strategies are only successful in the very short term. Generally, samples must be frozen or refrigerated at 4°C from the time of collection until extraction. Because PAHs are light sensitive, samples and extracts should be stored in amber glass or foil-wrapped bottles to minimize photolytic decomposition.[123] Table 2.4 summarizes these and other appropriate preservation methods.

Water (especially drinking water) containing high levels of volatile organic compounds (VOCs) can also be harmful to human health because of their physical and chemical properties and biological effects. VOCs consist of aromatic and chlorinated compounds with boiling points below 200°C and are one of the classes of compounds most frequently found at hazardous waste sites. They are carcinogenic and/or mutagenic, even at low concentrations, and are environmentally persistent. Generally, for the determination of VOCs it is recommended to fill the sample storage container completely and then to freeze it (UKSCA).[120] In the case of dihaloacetonitriles, which are the most common class of volatile chlorination by-products after trihalomethanes, practically no decomposition took place during the storage period (96 days at 4°C).[132] For further recommendations, see Table 2.4.

Determination of noninorganic surfactants (NS) in environmental water samples or in samples relevant to environmental water quality (sewage, processing liquors in sewage treatment plants, and treated sewage) is difficult because of the complexity of the matrix, the multicomponent nature of the NS mixture in the aquatic environment, and the limited stability of samples.[133]

Natural waters also contain a wide range of low-molecular-weight organic molecules (LMWOM), which are regarded as products of the microbial decomposition of, primarily, plant matter; some have even been used as identifiers for certain genera of decomposers.[123,124]

Water samples can contain different organic contaminants that are known to be toxic and/or carcinogenic even at low concentrations (see Table 2.4). That is why it is so important to maintain vigilance and control of pollution in the aquatic environment to ensure that water quality standards are met.

Organic pollutants occur in low concentrations, so large volumes of water usually have to be sampled if suitable detection limits are to be achieved after an appropriate preconcentration step. As a result, the reduction in effort and costs of transporting and preserving such high-volume samples has become an important topic, especially in environmental monitoring programs.[134] For this reason, it is the concentrates of organic analytes following separation and enrichment rather than the original high-volume water samples that are stored.

Among the numerous techniques for separating and enriching organic compounds from water samples, the following are worthy of mention: solid-phase extraction (SPE), solid-phase microextraction (SPME), liquid–liquid extraction (LLE), and lyophilization.

SPE with a variety of sorbents is an effective sample handling method for the analysis of organic compounds in water.[135] It is used to extract compounds directly in the field for the short-term storage of analytes, thereby enabling the transfer of samples to the laboratory for analysis.[136] Other studies have investigated the storage of different groups of pesticides on a C_{18} silica precolumn,[137,138] polymeric sorbents,[93,139] or graphitized carbon black sorbents.[104] Fenitrothion (an organophosphorus pesticide), for example, was preconcentrated on a XAD-2 column, after which the samples remained stable for five weeks at room temperature.[140] Green and Le Pape[117] examined two solid phases—XAD-2 macroreticular resin and octadecane bonded to silica gel—and demonstrated the stability of hydrocarbons sorbed from water onto these types of solid phases. Hydrocarbons stored on these solid phases for periods of up to 100 days in the presence of an oleophilic bacterial population showed no evidence of biological degradation. In contrast, hydrocarbons stored in water samples containing the same bacteria showed pronounced degradation over much shorter storage periods. In the last few years a variety of sorbents have been proposed for the preservation of organic compounds (see Table 2.4). Figure 2.2 presents the advantages of SPE as a handling technique.

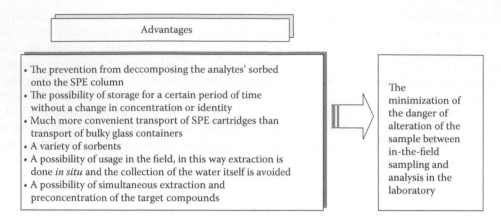

FIGURE 2.2   SPE as an effective sample handling method.[105,134]

SPME, a variant of SPE, involves the adsorption of organic pollutants from the matrix onto the solid-phase coating. The adsorbed analytes are then directly transferred to a gas chromatography (GC) injector using a modified syringe assembly or can be stored for a certain period of time.[141]

LLE has traditionally been used to separate and concentrate organic compounds from water samples. Because this technique has many drawbacks, there has been a general trend to replace LLE with SPE protocols.[141]

A quite novel approach to the preservation of organic compounds in water samples is the use of coacervates. These are water-immiscible liquid phases produced in colloidal solutions by the action of a dehydrating agent (e.g., changes in the temperature or pH of the solution, addition of an electrolyte, or addition of a water-miscible solvent in which the macromolecule is poorly soluble).[142] Luque et al.[117] studied the ability of coacervates to preserve organic compounds in order to determine their applicability for the extraction/preconcentration/preservation of pollutants in environmental monitoring programs. For this purpose, anionic micelle-based coacervates were used. The target pollutants were benzalkonium surfactants and polycyclic aromatic hydrocarbons, because of the instability problems they present in environmental water samples. Their stability in coacervates was investigated for a period of three months under different experimental conditions (Figure 2.3).

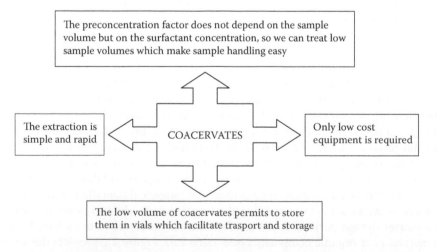

FIGURE 2.3   Advantages of using coacervates for the preservation of organic compounds.[134]

2.5 SUMMARY

Samples should be collected, transported, and stored in such a way that they remain in unchanged form until they are subjected to final analysis. A thorough knowledge of the conditions of sample preservation and storage for the determination of the compounds in question is therefore essential.

Numerous approaches, both physical and chemical, are recommended in the literature for the preservation of water samples prior to the analysis of inorganic and organic chemicals. Unfortunately, none of these methods is able to prevent analyte loss from different kinds of water matrices.

ACRONYMS AND ABBREVIATIONS

APHA	American Public Health Association
ASTM	American Society for Testing and Materials
EDTA	ethylenediaminetetraacetic acid
EPA	Environmental Protection Agency
EU	European Union
FEP	fluorinated ethene propene copolymer
FRP	filterable reactive phosphorus
GC	gas chromatography
HDPE	high-density polyethene
ISO	International Organization for Standardization
LLE	liquid–liquid extraction
LMWOM	low-molecular-weight organic molecules
NS	noninorganic surfactants
PAHs	polynuclear aromatic hydrocarbons
PC	polycarbonate
PCBs	polychlorinated biphenyls
PE	polyethylene
PFA	perfluoroalkoxy polymer
PP	polypropene
PTFE	poly(tetrafluoroethene)
SPE	solid-phase extraction
SPME	solid-phase microextraction
TBT	tributyltin chloride
TP	total phosphorus
TPhT	triphenyltin chloride
VOCs	volatile organic compounds

REFERENCES

1. Reimann, C., U. Siewers, H. Skarphagen, and D. Banks. 1999. Does bottle type and acid-washing influence trace element analyses by ICP-MS on water samples? A test covering 62 elements and four bottle types: High density polyethene (HDPE), polypropene (PP), fluorinated ethene propene copolymer (FEP) and perfluoroalkoxy polymer (PFA). *Sci. Total Environ.* 239: 111–130.
2. Reimann, C., A. Grimstvedta, B. Frengstada, and T.E. Finne. 2007. White HDPE bottles as source of serious contamination of water samples with Ba and Zn. *Sci. Total Environ.* 374: 292–296.
3. Hall, G.E., J.C. Pelchat, P. Pelchat, and J.E. Vaive. 2002. Sample collection, filtration and preservation protocols for the determination of 'total dissolved' mercury in waters. *Analyst* 127: 674–680.
4. Parker, J.L. and N.S. Bloom. 2005. Preservation and storage techniques for low-level aqueous mercury speciation. *Sci. Total Environ.* 337: 253–263.
5. Fadini, P.S. and W.F. Jardim. 2000. Storage of natural water samples for total and reactive mercury analysis in PET bottles. *Analyst* 125: 549–551.

6. Sekaly, A.L.R., C.L. Chakrabarti, M.H. Back, D.C. Gregoire, J.Y. Lu, and W.H. Schroeder. 1999. Stability of dissolved metals in environmental aqueous samples: Rideau River surface water, rain and snow. *Anal. Chim. Acta* 402: 223–231.

7. Grasshoff, K., M. Ehrhardt, and K. Kremling. 1983. *Methods of Seawater Analysis*, 2nd edition. Weinheim: Verlag Chemie.

8. Śliwka-Kaszyńska, M., A. Kot-Wasik, and J. Namieśnik. 2003. Preservation and storage of water samples. *Crit. Rev. Environ. Sci. Technol.* 33: 31–44.

9. ISO. 1995. Quality of water. Sampling: Principles rules at preservation storage of samples. EN ISO 5667-3.

10. Kattner, G. 1999. Storage of dissolved inorganic nutrients in seawater: Poisoning with mercuric chloride. *Mar. Chem.* 67: 61–66.

11. Aminot, A. and R. Kérouel. 1998. Pasteurization as an alternative method for preservation of nitrate and nitrite in seawater samples. *Mar. Chem.* 61: 203–208.

12. Greenberg, A.E., L.S. Clescerl, and A.D. Eaton. 1992. *Standard Methods for the Examination of Water and Wastewater*. Washington, DC: American Public Health Association.

13. Dore, J.E., T. Houlihan, D.V. Hebel, G. Tien, L. Tupas, and D.M. Karl. 1996. Freezing as a method of sample preservation for the analysis of dissolved inorganic nutrients in seawater. *Mar. Chem.* 53: 173–185.

14. Rinne, D. and P. Schmitt. 2003. On the problem of sample preservation by freezing. *Acta Hydrochim. Hydrobiol.* 31: 501–512.

15. Gardolinski, P.C.F.C., G. Hanrahan, E.P. Achterberg, M. Gledhill, A.D. Tappin, W.A. House, and P.J. Worsfold. 2001. Comparison of sample storage protocols for the determination of nutrients in natural waters. *Water Res.* 35: 3670–3678.

16. Henriksen, A. 1969. Preservation of water samples for phosphorus and nitrogen determination. *Vatten* 25: 247–254.

17. Maher, W. and L. Woo. 1998. Procedures for the storage and digestion of natural waters for the determination of filterable reactive phosphorus, total filterable phosphorus and total phosphorus. *Anal. Chim. Acta* 375: 5–47.

18. Kotlash, A.R. and B.C. Chessman. 1998. Effects of water sample preservation and storage on nitrogen and phosphorus determinations: Implications for the use of automated sampling equipment. *Water Res.* 32: 3731–3737.

19. Aminot, A. and R. Kérouel. 1997. Assessment of heat treatment for nutrient preservation in seawater samples. *Anal. Chim. Acta* 351: 299–309.

20. Clementson, L.A. and S.E. Wayte. 1992. The effect of frozen storage of open-ocean seawater samples on the concentration of dissolved phosphate and nitrate. *Water Res.* 26: 1171–1176.

21. Burke, P.M., S. Hill, N. Iricanin, C. Douglas, P. Essex, and D. Tharin. 2002. Evaluation of preservation methods for nutrient species collected by automatic samplers. *Environ. Monit. Assess.* 80: 149–173.

22. Avanzino, R.J. and V.C. Kennedy. 1993. Long-term frozen storage of stream water samples for dissolved orthophosphate, nitrate plus nitrite and ammonia analysis. *Water Resour. Res.* 29: 3357–3362.

23. Smith, A., E. Lingas, and M. Rahman. 2000. Contamination of drinking water by arsenic in Bangladesh: A public health emergency. *Bull. World Health Org.* 78: 1093–1103.

24. Environmental Protection Agency. 2006. Lead in drinking water. Available at http://www.epa.gov/safewater/lead/basicinformation.html.

25. Fairbrother, A., R. Wenstel, K. Sappington, and W. Wood. 2007. Framework for metals risk assessment. *Ecotoxicol. Environ. Saf.* 68: 145–227.

26. Daus, B., H. Weiss, J. Mattusch, and R. Wennrich. 2006. Preservation of arsenic species in water samples using phosphoric acid—limitations and long-term stability. *Talanta* 69: 430–434.

27. Gallagher, P.A., C.A. Schwegel, X. Wei, and J.T. Creed. 2001. Speciation and preservation of inorganic arsenic in drinking water sources using EDTA with IC separation and ICP-MS detection. *J. Environ. Monit.* 3: 371–376.

28. Bednar, A.J., J.R. Garbarino, J.F. Ranville, and T.R. Wildeman. 2002. Preserving the distribution of inorganic arsenic species in groundwater and acid mine drainage samples. *Environ. Sci. Technol.* 36: 2213–2218.

29. Samanta, G. and D.A. Clifford. 2005. Preservation of inorganic arsenic species in groundwater. *Environ. Sci. Technol.* 39: 8877–8882.

30. Kim, Y.T., H. Yoon, C. Yoon, and N.C. Woo. 2007. An assessment of sampling, preservation, and analytical procedures for arsenic speciation in potentially contaminated waters. *Environ. Geochem. Health* 29: 337–346.

31. Daus, B., J. Mattusch, R. Wennrich, and H. Weiss. 2002. Investigation on stability and preservation of arsenic species in iron rich water samples. *Talanta* 58: 57–65.
32. McCleskey, R.B., D.K. Nordstrom, and A.S. Meast. 2004. Preservation of water samples for arsenic (III/V) determinations: An evaluation of the literature and new analytical results. *Appl. Geochem.* 19: 995–1009.
33. Palacios, M.A., M. Gomez, C. Camara, and M.A. Lopez. 1997. Stability studies of arsenate, monomethyl-arsonate, dimethylarsinate, arsenobetaine and arsenocholine in deionized water, urine and clean-up dry residue from urine samples and determination by liquid chromatography with microwave-assisted oxidation-hydride generation atomic absorption spectrometric detection. *Anal. Chim. Acta* 340: 209–220.
34. Dyg, S., R. Cornelis, P. Quevauviller, B. Griepink, and J.M. Christensen. 1994. Development and inter-laboratory testing of aqueous and lyophilized Cr(III) and Cr(VI) reference materials. *Anal. Chim. Acta* 286: 297–308.
35. Mena, M.L., A. Morales-Rubio, A.G. Cox, C.W. McLeod, and P. Quevauviller. 1995. Stability of chromium species immobilized on microcolumns of activated alumina. *Quim. Anal.* 14: 164–168.
36. Sirinawin, W. and S. Westerlund. 1997. Analysis and storage of samples for chromium determination in seawater. *Anal. Chim. Acta* 356: 35–40.
37. U.S. Environmental Protection Agency, Environmental Monitoring and Support Laboratory. 1983. Methods for chemical analysis of water and wastes, EPA-600/4-79-020.
38. Norma EN-ISO 5667-3:2004. Jakość wody. Pobieranie próbek. Część 3: Wytyczne dotyczące utrwalania i postępowania z próbkami wody.
39. U.S. Environmental Protection Agency. 1996. *Method 1669: Sampling Ambient Water for Trace Metals at EPA Water Quality Criteria Levels*. Washington, DC:, Office of Water Engineering and Analysis Division (4303).
40. Parker, J.L. and N.S. Bloom. 2005. Preservation and storage techniques for low-level aqueous mercury speciation. *Sci. Total Environ.* 337: 253–263.
41. Plaschke, R., G. Dal Pont, and E.C.V. Butler. 1997. Mercury in waters of the Derwent estuary—sample treatment and analysis. *Mar. Pollut. Bull.* 34: 177–185.
42. Blanco, R.M., M.T. Villanueva, J.E.S. Uria, and A. Sanz-Medel. 2000. Field sampling, preconcentration and determination of mercury species in river waters. *Anal. Chim. Acta* 419: 137–144.
43. Cheam, V. and H. Agemian. 1980. Preservation and stability of inorganic selenium compounds at ppb levels in water samples. *Anal. Chim. Acta* 113: 237–245.
44. Measures, C.I. and J.D. Burton. 1980. Gas chromatographic method for the determination of selenite and total selenium in sea water. *Anal. Chim. Acta* 120: 177–186.
45. Olivas, R.M., P. Quevauviller, and O.F.X. Donard. 1998. Long term stability of organic selenium species in aqueous solutions. *Fresenius J. Anal. Chem.* 360: 512–519.
46. Pyrzyńska, K. 1998. Speciation of selenium compounds. *Anal. Sci.* 14: 479–483.
47. Lindemann, T., A. Prange, W. Dannecker, and B. Neidhart. 2000. Stability studies of arsenic, selenium, antimony and tellurium species in water, urine, fish and soil extracts using HPLC/ICP-MS. *Fresenius J. Anal. Chem.* 368: 214–220.
48. Bancon-Montigny, C., G. Lespes, and M. Potin-Gautier. 2001. Optimisation of the storage of natural freshwaters before organotin speciation. *Water Res.* 35: 224–232.
49. Quevauviller, P., M. Astruc, L. Ebdon, H. Muntau, W. Cofino, R. Morabito, and B. Griepink. 1996. A programme to improve the quality of butyltin determinations in environmental matrices. *Microchim. Acta* 123: 163–173.
50. Gómez-Ariza, J.L., I. Giráldez, E. Morales, F. Ariese, W. Cofino, and P. Quevauviller. 1999. Stability and storage problems in organotin speciation in environmental samples. *J. Environ. Monit.* 1: 197–202.
51. Bergmann, K., U. Röhr, and B. Neidhart. 1994. Examination of the different procedural steps in the determination of organotin compounds in water samples. *Fresenius J. Anal. Chem.* 349: 815–819.
52. Smedley, P., H.B. Nicolli, D.M.J. Macdonald, A.J. Barros, and J.O. Tullio. 2002. Hydrogeochemistry of arsenic and other inorganic constituents in groundwaters from La Pampa, Argentina. *Appl. Geochem.* 17: 259–284.
53. Francesconi, K.A. and D. Kuehnelt. 2002. Arsenic compounds in the environment. In: W.T. Jr. Frankenberger (ed.), *Environmental Chemistry of Arsenic*, pp. 51–94. New York: Marcel Dekker.
54. U.S. Environmental Protection Agency. 2001. Method 1632. Chemical speciation of arsenic in water and tissue by hydride generation quartz furnace atomic absorption spectrometry.
55. Gallagher, P.A., C.A. Schwegel, X. Wei, and J.T. Creed. 2001. Speciation and preservation of inorganic arsenic in drinking water sources using EDTA with IC separation and ICP-MS detection. *J. Environ. Monit.* 3: 371–376.

56. Garbarnio, J.R., A.J. Bednar, and M.R. Burchardt. 2002. Methods of analysis by the US geological survey national water quality laboratory—arsenic speciation in natural water samples using laboratory and field methods. U.S. Geological Survey, Water-Resources Investigations Reports 02-4144.

57. Fanning, J.C. 2000. The chemical reduction of nitrate in aqueous solution. *Coord. Chem. Rev.* 199: 159–179.

58. Kotaś, J. and Z. Stasicka. 2000. Chromium occurrence in the environment and methods of its speciation. *Environ. Pollut.* 107: 263–283.

59. Bag, H., A.R. Turker, M. Lale, and A. Tunceli. 2000. Separation and speciation of Cr(III) and Cr(VI) with *Saccharomyces cerevisiae* immobilized on sepiolite and determination of both species in water by FAAS. *Talanta* 51: 895–902.

60. Pantsar-Kallio, M. and P.K.G. Manninen. 1996. Speciation of chromium in waste waters by coupled column ion chromatography-inductively coupled plasma mass spectrometry. *J. Chromatogr. A* 750: 89–95.

61. DIO EX Technical Note 26. 1998. Determination of Cr (VI) in water, waste water, and solid waste extracts. Available at http://www1.dionex.com/en-us/webdocs/4428_TN26_16May07_LPN034398-02.pdfN.

62. Sperling, M., S. Xu, and B. Welz. 1992. Determination of chromium(III) and chromium(VI) in water using flow injection on-line preconcentration with selective adsorption on activated alumina and flame atomic absorption spectrometric detection. *Anal. Chem.* 64: 3101–3108.

63. Dyg, S., R. Cornelis, B. Griepink, and P. Quevauviller. 1994. Development of interlaboratory testing of aqueous and lyophilized Cr(III) and Cr(VI) reference materials. *Anal. Chim. Acta* 286: 297–308.

64. U.S. Department of Health and Human Services. 1993. Agency for Toxic Substances and Disease Registry, Atlanta, Georgia. Toxicological Profile for Mercury. Available at http://www.atsdr.cdc.gov/toxprofiles/tp46.pdf.

65. Jones, A.B. and D.G. Slotton. 1996. *Mercury Effects, Sources and Control Measures*. Richmond: San Francisco Estuary Regional Monitoring Program. Available at http://www.sfei.org/rmp/reports/mercury/mercury.pdf.

66. Gochfeld, M. 2003. Cases of mercury exposure, bioavailability, and absorption. *Ecotoxicol. Environ. Saf.* 56: 174–179.

67. Leermakers, M., P. Lansens, and W. Baeyens. 1990. Storage and stability of inorganic and methylmercury solutions. *Fresenius J. Anal. Chem.* 336: 655–662.

68. Bloom, N.S. 1995. Mercury as a case study of ultra-clean sample handling and storage in aquatic trace metal research. *Environ. Lab.* 3–4: 20–25.

69. Yu, L.-P. and X.-P. Yan. 2003. Factors affecting the stability of inorganic and methylmercury during sample storage. *Trends Anal. Chem.* 22: 245–253.

70. Heiden, R.W. and D.A. Aikens. 1983. Humic acid as a preservative for trace mercury (II) solutions stored in polyolefin containers. *Anal. Chem.* 55: 2327–2332.

71. Ahmed, R. and M. Stoeppler. 1986. Decomposition and stability studies of methylmercury in water using cold vapor atomic absorption spectrometry. *Analyst* 111: 1371–1374.

72. Hamlin, S.N. 1989. Preservation of samples for dissolved mercury. *J. Am. Water Res. Assoc.* 25: 255–262.

73. Lemly, A.D. 2004. Aquatic selenium pollution is a global environmental safety issue. *Ecotoxicol. Environ. Saf.* 59: 44–56.

74. Frankenberger, W.T. and S. Benson. 1994. *Selenium in the Environment*. New York: Marcel Dekker.

75. Perez-Corona, T., Y. Madrid, and C. Camara. 1997. Evaluation of selective uptake of selenium (Se(IV) and Se(VI)) and antimony (Sb(III) and Sb(V)) species by Baker's yeast cells (*Saccharomyces cerevisiae*). *Anal. Chim. Acta* 345: 249–255.

76. Héninger, I., M. Potin-Gautier, I. de Gregori, and H. Pinochet. 1997. Storage of aqueous solutions of selenium for speciation at trace level. *Fresenius J. Anal. Chem.* 357: 600–610.

77. Gomez Ariza, J.L., E. Morales, D. Sanchez-Rodas, and I. Giraldez. 2000. Stability of chemical species in environmental matrices. *Trends Anal. Chem.* 19: 200–209.

78. Pyrzyńska, K. 2002. Determination of selenium species in environmental samples. *Microchim. Acta* 140: 55–62.

79. Cooney, J.J. 1995. Organotin compounds and aquatic bacteria: A review. *Helgol. Mar. Res.* 49: 663–677.

80. Cooney, J.J. and S. Wuertz. 1989. Toxic effects of tin compounds on microorganisms. *J. Ind. Microbiol. Biotechnol.* 4: 375–402.

81. Gadd, G.M. 2000. Microbial interactions with tributyltin compounds: Detoxification, accumulation, and environmental fate. *Sci. Total Environ.* 258: 119–127.

82. Decision no. 2455/2001/EC of the European Parliament and of the Council of November 20, 2001, establishing the list of priority substances in the field of water policy and amending Directive 2000/60/EC (Text with EEA relevance). *Off. J. Eur. Commun.* L 331, 0001–0005.Available at http://eur-lex.europa.eu/ LexUriServ/site/en/oj/2001/l_331/l_33120011215en00010005.pdf.

83. Hoch, M. 2001. Organotin compounds in the environment: An overview. *Appl. Geochem.* 16: 719–743.

84. Final report prepared for European Commission Directorate—General Enterprise and Industry, impact assessment of potential restrictions on the marketing and use of certain organotin compounds, October 2007. Available at http://ec.europa.eu/enterprise/chemicals/docs/studies/organotins.pdf.

85. Lekkas, T.D., M. Kostopoulou, G. Kolokythas, N. Thomaidis, S. Golfinopoulos, A. Kotrikla, G. Pavlogeorgatos, et al. 2003. Optimization of analytical methods for the determination of trace concentrations of toxic pollutants in drinking and surface waters. *Global Nest J.* 5: 165–175.

86. Naddeo, V., T. Zarra, and V. Belgiorno. 2005. European procedures to river quality assessment. *Global Nest J.* 7: 306–312.

87. Burns, D.T., M. Harriott, and F. Glockling. 1987. The extraction, determination and speciation of tributyltin in seawater. *Fresenius J. Anal. Chem.* 327: 701–703.

88. Manoli, E. and C. Samara. 1999. Polycyclic aromatic hydrocarbons in natural waters: Sources, occurrence and analysis. *Trends Anal. Chem.* 18: 417–428.

89. Capel, P.D. and S.J. Larson. 1995. A chemodynamic approach for estimating losses of target organic chemicals from water during sample holding time. *Chemosphere* 30: 1097–1107.

90. Biziuk, M. 2001. *Pesticides—Occurrence, Determination and Decontamination.* Warszawa: WN-T.

91. White-Stevens, R. 1977. *Pesticides in the Environment.* Warszawa: PWRiL.

92. Deplagne, J., J. Vial, V. Pichon, B. Lalere, G. Hervouet, and M.C. Hennion. 2006. Feasibility study of a reference material for water chemistry: Long term stability of triazine and phenylurea residues stored in vials or on polymeric sorbents. *J. Chromatogr. A* 1123: 31–37.

93. Ferrer, I. and D. Barcelo. 1997. Stability of pesticides stored on polymeric solid-phase extraction cartridges. *J. Chromatogr. A* 778: 161–170.

94. Woudneh, M.B., M. Sekela, T. Tuominen, and M. Gledhill. 2007. Acidic herbicides in surface waters of Lower Fraser Valley, British Columbia, Canada. *J. Chromatogr. A* 1139: 121–129.

95. Agency for Toxic Substances and Disease Registry. 1989. Toxicological Profile for Phenol. Available at http://www.eco-usa.net/toxics/phenol.shtml (accessed July 28, 2008).

96. Castillo, M., D. Puig, and D. Barcelo. 1997. Determination of priority phenolic compounds in water and industrial effluents by polymeric liquid–solid extraction cartridges using automated sample preparation with extraction columns and liquid chromatography: Use of liquid–solid extraction cartridges for stabilization of phenols. *J. Chromatogr. A* 778: 301–311.

97. U.S. Environmental Protection Agency. 1996. Methods for organic chemical analysis of municipal and industrial wastewater. Method 604—Phenols. Washington, DC. Available at http://www.accustandard.com/asi/pdfs/epa_methods/604.pdf (accessed July 28, 2008).

98. WHO. 1993. *EHC—Environmental Health Criteria 140. Polychlorinated Biphenyls and Terphenyls*, 2nd edition. World Health Organization, Geneva. Available at http://www.inchem.org/documents/ehc/ ehc/ehc140.htm (accessed July 28, 2008).

99. Erickson, M.D. 1992. *Analytical Chemistry of PCBs.* Chelsea: Lewis.

100. Falandysz, J. 1999. *Polichlorowane bifenyle (PCBs) w środowisku: Chemia, analiza, toksyczność, stężenia i ocena ryzyka.* Gdańsk: Fundacja Rozwoju Uniwersytetu Gdańskiego.

101. U.S. Environmental Protection Agency. 1996. Methods for organic chemical analysis of municipal and industrial wastewater. Method 608—Organochlorine pesticides and PCBs. Washington, DC. Available at http://www.accustandard.com/asi/pdfs/epa_methods/608.pdf (accessed July 28, 2008).

102. International Organization for Standardization. 2003. ISO 5667-3. Water quality—sampling.Part 3: Guidance on the preservation and handling of water samples. Geneva.

103. U.S. Department of Commerce. 1989. EPA-600/4-88-039. *Test Methods for Evaluating Solid Waste Physical/Chemical Methods*, 3rd edition, U.S. Environmental Protection Agency, Washington, DC.

104. Sabik, H. and R. Jeannot. 2000. Stability of organophosphorus insecticides on graphitized carbon black extraction cartridges used for large volumes of surface water. *J. Chromatogr. A* 879: 73–82.

105. Liska, I. and K. Bilikova. 1998. Stability of polar pesticides on disposable solid-phase extraction precolumns. *J. Chromatogr. A* 795: 61–69.

106. Lacorte, S., N. Ehresmann, and D. Barcelo. 1995. Stability of organophosphorus pesticides on disposable solid-phase extraction precolumns. *Environ. Sci. Technol.* 29: 2834–2841.

107. Munch, J.W. 1995. EPA, Method 531.1—Measurement of N-methylcarbamoyloximes and N-methyl-carbamates in water by direct aqueous injection HPLC with post column derivatization, Revision 3.1. Washington, DC.

108. Fatta, D., S. Canna-Michaelidou, C. Michael, E. Demetriou Georgiou, M. Christodoulidou, A. Achilleos and M. Vasquez. 2007. Organochlorine and organophosphoric insecticides, herbicides and heavy metals residue in industrial wastewaters in Cyprus. *J. Hazard. Mater.* 145: 169–179.

109. U.S. Environmental Protection Agency. 1992. Environmental Monitoring Systems Laboratory. Methods for the determination of organic compounds in drinking water, Supplement II, 500 Series. EPA-600/R-92/129. Washington, DC.

110. U.S. Environmental Protection Agency. 1983. Environmental Monitoring and Support Laboratory. Methods for chemical analysis of water and wastes. EPA-600/4-79-020 (method 420.2). Washington, DC.

111. American Society for Testing and Materials. 1980. ASTM Annual Book of Standards. Part 31. D3370. Standard Practice for Sampling Water. Philadelphia.

112. U.S. Environmental Protection Agency. 1988. Environmental Monitoring Systems Laboratory. Methods for the determination of organic compounds in drinking water, 500 Series. EPA-600/4-88/039. Washington, DC.

113. U.S. Environmental Protection Agency. 1996. Methods for organic chemical analysis of municipal and industrial wastewater. Method 610—Polynuclear aromatic hydrocarbons. Washington, DC. Available at http://www.accustandard.com/asi/pdfs/epa_methods/610.pdf (accessed July 28, 2008).

114. Sicilia, D., S. Rubio, D. Perez-Bendito, N. Maniasso, and E.A.G. Zagatto. 1999. Anionic surfactants in acid media: A new cloud point extraction approach for the determination of polycyclic aromatic hydrocarbons in environmental samples. *Anal. Chim. Acta* 392: 29–38.

115. Law, R.J. and J.L. Biscaya. 1994. Polycyclic aromatic hydrocarbons (PAH)—problems and progress in sampling, analysis and interpretation. *Mar. Pollut. Bull.* 29: 235–241.

116. Rawa-Adkonis, M., L. Wolska, and J. Namiesnik. 2006. Analytical procedures for PAH and PCB determination in water samples—error sources. *Crit. Rev. Anal. Chem.* 36: 63–73.

117. Green, D.R. and D. Le Pape. 1987. Stability of hydrocarbon samples on solid-phase extraction columns. *Anal. Chem.* 59: 699–703.

118. U.S. Environmental Protection Agency. 2008. Test methods for evaluating solid waste. Physical/chemical methods. EPA SW-846 5030/8240. Available at http://www.epa.gov/epaoswer/hazwaste/test/main.htm (accessed July 28, 2008).

119. Maskarinec, M.P., L.H. Johnson, S.K. Holladay, R.L. Moody, C.K. Bayne, and R.A. Jenkins. 1990. Stability of volatile organic compounds in environmental water samples during transport and storage. *Environ. Sci. Technol.* 24: 1665–1670.

120. American Public Health Association, American Water Works Association, Water Pollution Control Federation. 1989. *Standard Methods for the Examination of Water and Wastewater*, 17th edition. Washington, DC: APHA.

121. U.S. Environmental Protection Agency. 1994. Methods for organic chemical analysis of municipal and industrial wastewater. Method 1613—Tetra- through octa-chlorinated dioxins and furans by isotope dilution HRGC/HRMS. Washington, DC. Available at http://www.accustandard.com/asi/pdfs/epa_methods/1613.pdf (accessed July 28, 2008).

122. Szymanski, A., Z. Swit, and Z. Lukaszewski. 1995. Studies of preservation of water samples for the determination of non-ionic surfactants. *Anal. Chim. Acta* 311: 31–36.

123. Karlsson, S., H. Wolrath, and J. Dahlen. 1999. Influence of filtration, preservation and storing on the analysis of low molecular weight organic acids in natural waters. *Water Res.* 33: 2569–2578.

124. Berdie, L., J.O. Grimalt, and E.T. Gjessing. 1995. Combined fatty acids and amino acids in the dissolved + colloidal fractions of the waters from a dystrophic lake. *Org. Geochem.* 23: 343–353.

125. Campos, M.L.A.M., R.F.P. Nogueira, P.R. Dametto, J.G. Francisco, and C.H. Coelho. 2007. Dissolved organic carbon in rainwater: Glassware decontamination and sample preservation and volatile organic carbon. *Atmos. Environ.* 41: 8924–8931.

126. Lara, L.B.L.S., P. Artaxo, L.A. Martinelli, R.L. Victoria, P.B. Camargo, and A. Krusche. 2001. Chemical composition of rainwater and anthropogenic influences in the Piracicaba River Basin, Southeast Brazil. *Atmos. Environ.* 35: 4937–4945.

127. WHO. 1998. *Polynuclear Aromatic Hydrocarbons in Guidelines for Drinking-water Quality*, Vol. 2, 2nd edition. Addendum to Health Criteria and Other Supporting Information, Geneva. Available at http://www.emro.who.int/ceha/pdf/Guidelines_DrinkingWater_Edition2_Volume2_Addendum.pdf (accessed July 28, 2008).

128. Chen, B., X. Xuan, L. Zhu, J. Wang, Y. Gao, K. Yang, X. Shen, and B. Lou. 2004. Distributions of poly-cyclic aromatic hydrocarbons in surface waters, sediments and soils of Hangzhou City, China. *Water Res.* 38: 3558–3568.

129. Kummerer, K., A. Eitel, U. Braun, P. Hubner, F. Daschner, G. Mascart, M. Milandri, F. Reinthaler, and J. Verhoef. 1997. Analysis of benzalkonium chloride in the effluent from European hospitals by solid-phase extraction and high-performance liquid chromatography with post-column ion-pairing and fluores-cence detection. *J. Chromatogr. A* 774: 281–286.

130. Wolska, L., M. Rawa-Adkonis, and J. Namiesnik. 2005. Determining PAHs and PCBs in aqueous sam-ples: Finding and evaluating sources of error. *Anal. Bioanal. Chem.* 382: 1389–1397.

131. Lopez Garcia, A., E. Blanco Gonzalez, J.I. Garcia Alonso, and A. Sanz-Medel. 1992. Potential of micelle-mediated procedures in the sample preparation steps for the determination of polynuclear aromatic hydrocarbons in waters. *Anal. Chim. Acta* 264: 241–248.

132. Nikolaou, A.D., S.K. Golfinopoulos, M.N. Kostopoulou, and T.D. Lekkas. 2000. Decomposition of diha-loacetonitriles in water solutions and fortified drinking water samples. *Chemosphere* 41: 1149–1154.

133. Szymanski, A. and Z. Lukaszewski. 2000. Initial separation and preservation for long-term storage of non-ionic surfactants from raw and treated sewage. *Water Res.* 34: 3635–3639.

134. Luque, N., S. Rubio, and D. Perez-Bendito. 2007. Use of coacervates for the on-site extraction/preserva-tion of polycyclic aromatic hydrocarbons and benzalkonium surfactants. *Anal. Chim. Acta* 584: 181–188.

135. Pichon, V. 2000. Solid-phase extraction for multiresidue analysis of organic contaminants in water. *J. Chromatogr. A* 885: 195–215.

136. Primus, T.M., D.J. Kohler, M. Avery, P. Bolich, M.O. Way, and J.J. Johnston. 2001. Novel field sampling procedure for the determination of methiocarb residues in surface waters from rice fields. *J. Argic. Food Chem.* 49: 5706–5709.

137. Penuela, G.A. and D. Barcelo. 1998. Application of C_{18} disks followed by gas chromatography tech-niques to degradation kinetics, stability and monitoring of endosulfan in water. *J. Chromatogr. A* 795: 93–104.

138. Penuela, G.A. and D. Barcelo. 1998. Photodegradation and stability of chlorothalonil in water studied by solid-phase disk extraction, followed by gas chromatographic techniques. *J. Chromatogr. A* 823: 81–90.

139. Aguilar, C., I. Ferrer, F. Borrull, R.M. Marce, and D. Barcelo. 1999. Monitoring of pesticides in river water based on samples previously stored in polymeric cartridges followed by on-line solid-phase extrac-tion-liquid chromatography–diode array detection and confirmation by atmospheric pressure chemical ionization mass spectrometry. *Anal. Chim. Acta* 386: 237–248.

140. Berkane, K., G.E. Caissie, and V.N. Mallet. 1977. The use of Amberlite XAD-2 resin for the quantitative recovery of fenitrothion from water—a preservation technique. *J. Chromatogr. A* 139: 386–390.

141. Puig, D. and D. Barcelo. 1996. Determination of phenolic compounds in water and waste water, *Trends Anal. Chem.* 15: 362–375.

142. McNaught, A.D. and A. Wilkinson. 1997. *IUPAC Compendium of Chemical Terminology.* Oxford: Blackwell Scientific Publications.

3 Application of Passive Sampling Techniques for Monitoring the Aquatic Environment

*Graham A. Mills, Richard Greenwood, Ian J. Allan,
Ewa Łopuchin, Janine Brümmer, Jesper Knutsson,
and Branislav Vrana*

CONTENTS

3.1   Introduction 41
3.2   Concept of Passive Sampling 42
     3.2.1   Equilibrium Sampling 43
     3.2.2   Kinetic Sampling 44
     3.2.3   Sampler Construction 45
         3.2.3.1   Sorption Phase 45
         3.2.3.2   Diffusion Barrier 45
     3.2.4   Modeling and Calibration of Passive Sampling Devices 46
     3.2.5   Factors Affecting Passive Sampling 47
     3.2.6   Biofouling 48
3.3   Recent Applications 50
     3.3.1   Hydrophobic Organic Compounds 50
     3.3.2   Polar Organic Compounds 50
     3.3.3   Volatile Organic Compounds 54
     3.3.4   Organometallic Compounds 54
     3.3.5   Metals 55
     3.3.6   Algal Toxins 55
     3.3.7   Applications in Ecotoxicity Assessment 56
     3.3.8   Applications in Biomonitoring 57
3.4   Quality Assurance, Quality Control and Validation 57
3.5   Use of Passive Sampling in Regulatory Monitoring 59
3.6   Future Trends 60
Acknowledgments 61
References 61

3.1 INTRODUCTION

A challenge for the environmental chemist is the development of reliable sampling and analytical procedures for the representative assessment of water quality.[1–4] Currently, most water monitoring methods rely on the collection at defined periods of discrete spot, grab, or bottle samples followed by instrumental analysis in the laboratory. As many pollutants can be present in the water phase at very low concentrations, the collection of large volumes of water is often needed for analysis. In addition, the analysis of spot water samples provides only a snapshot of the levels of pollutants at the time of sampling. This may not give an accurate assessment of water quality in situations where pollutant concentrations fluctuate widely and where episodic pollution events occur. Data from round robin exercises of the analysis of pollutants found in spot water samples showed that there are often recurring problems of interlaboratory reproducibility especially when the pollutants are present in trace concentrations.[5] Issues of poor reproducibility are often linked to the procedures used in collecting the water for analysis.

There are several options available to improve the sampling procedures and hence the quality of the water monitoring data obtained. One possibility is to significantly increase the frequency of sampling or to install automatic sampling systems that can collect numerous water samples over a given time period. The latter systems are often installed at sites of strategic importance (e.g., monitoring stations near transnational boundaries and locations for drinking water capitation), but they cannot be used in widespread monitoring networks. The cost of the equipment is high and a secure site is needed in which to locate the apparatus. Biomonitoring can be used as an alternative to spot sampling. A number of test species (e.g., mussels and fish) can be employed depending on the water body being investigated. These sentinel organisms are deployed for extended periods of time during which they accumulate pollutants from the surrounding water. Analysis of tissue extracts can give an indication of the level of waterborne contamination. A number of factors can increase the uncertainty of the results obtained using these procedures including metabolism, depuration rates, excretion, stress, viability, and the condition of test organisms. Furthermore, the extraction procedures required for the analysis of tissue samples are complex and time-consuming.

Another option is the use of passive sampling. These devices have been used since the early 1970s, when they were first used for the measurement of gaseous pollutants in air;[6] recently, they have shown much promise as useful tools for the measurement of a wide range of priority pollutants in water. Passive samplers avoid many of the above problems since they sample compounds *in situ* without affecting the concentration in bulk solution. The mass of analyte accumulated by a sampler reflects either the aqueous concentration with which the device is at equilibrium or the time-weighted average (TWA) concentration with which the sampler was exposed during deployment. Such devices can be exposed in the field for time periods up to several months. The technology is relatively inexpensive, simple to use and, as the devices have no mechanical parts, they do not require any external source of energy for their operation. They have been used in complex monitoring networks and can be deployed in remote areas without any infrastructure. The devices continuously sequester pollutants from the bulk water matrix and trap them in a suitable sorbent medium. A number of different devices are available commercially or as laboratory prototypes. Several reviews have been published describing the design, calibration procedures, figures of merit, and applications of the different samplers for monitoring the aquatic environment.[7–11] In addition, a book describing the semipermeable membrane device (SPMD)[12] and a general text on all passive sampling techniques for environmental monitoring[13] is available. Due to the extensive number of scientific publications in the field, this chapter is not meant to be exhaustive and concentrates only on developments and potential applications of the technology from the last 10 years.

3.2   CONCEPT OF PASSIVE SAMPLING

Passive sampling can be defined as any sampling technique based on the movement (by diffusion) of analyte molecules from the sampled medium to a receiving phase contained in a sampling device. This mass transfer process is driven by a difference in chemical potentials of the analyte in the two media. This process continues until equilibrium is reached in the system, or until the sampling process is stopped.[14] Analytes are retained in a suitable medium within the device, known as a receiving or sorption phase. This can be a solvent, chemical reagent, absorbent, or

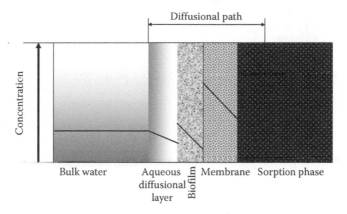

FIGURE 3.1 Concentration profiles in a passive sampling device. The driving force of accumulation is the difference in chemical potentials of the analyte between the bulk water and the sorption phase. The mass transfer of an analyte is governed by the overall resistance along the whole diffusional path, including contributions from the individual barriers (e.g., aqueous boundary layer, biofilm layer, and membrane).

porous adsorbent material. Unlike dynamic extraction methods (e.g., liquid–liquid extraction or solid-phase extraction), the aim is not to exhaustively extract the dissolved analyte molecules from the water phase.

The sorption of a pollutant molecule from the bulk water phase follows the pathway shown in Figure 3.1.

The uptake of an analyte by a passive sampling device is a multistage mass transfer process. First, water containing analyte molecules enter the space that protects the sampler from mechanical damage (usually a cage or well or cavity in the sampler housing). Here, transport is by convective processes. Molecules then diffuse through the aqueous boundary layer and biofilm layer (if present). Finally, analytes diffuse through the membrane and accumulate in the sorption phase, which has a high affinity for the compounds of interest. This general scheme can vary according to the specific construction of the sampling device.

The exchange kinetics between a passive sampler and the water phase can be described by a first-order, one-compartment mathematical model:

$$C_S(t) = C_W \frac{k_1}{k_2}(1 - e^{-k_2 t}) \tag{3.1}$$

where $C_S(t)$ is the concentration of the analyte in the sampler at exposure time t, C_W is the concentration of the analyte in the bulk water phase, and k_1 and k_2 are the uptake and offload rate constants, respectively. Two accumulation modes, either equilibrium or kinetic sampling, can be distinguished in the operation of a sampler (Figure 3.2).

3.2.1 EQUILIBRIUM SAMPLING

In equilibrium sampling, the exposure time is sufficiently long to permit the establishment of thermodynamic equilibrium between the water and sorption phases. In this case the dissolved analyte concentration can be estimated using the sorption phase–water partition coefficient (K_{SW}):

$$K_{SW} = \frac{C_S}{C_W} \tag{3.2}$$

The theory of equilibrium passive sampling devices has been published by Mayer et al.[15] The basic requirements of the equilibrium sampling approach are that equilibrium concentrations are

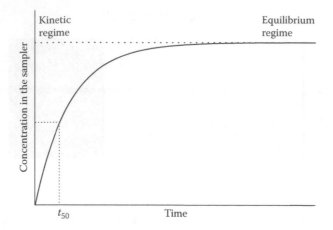

FIGURE 3.2   Passive sampling devices operate in two main regimes: kinetic and equilibrium.

reached after a known response time, and the device response time needs to be shorter than any fluctuations in concentration in the environmental medium. Analyte depletion from the sampled medium must not occur during the sampling process, as otherwise this would disturb the equilibrium of the system. The amount of analyte accumulated by a passive sampling device is independent of the sample volume. The analyte concentration measured by equilibrium samplers does not necessarily reflect all of the contamination events during the whole sampling period but provides a snapshot of the concentration representative for the equilibration period.

The sorption phase–water partition coefficient (K_{SW}) is the driving force for the uptake of compounds by passive sampling devices. Measurement of K_{SW} in the laboratory can be difficult, particularly for hydrophobic compounds, as results can be biased through errors in the measurement of their concentration in the water phase. To overcome this problem, a cosolvent approach has been developed. Here, compounds are equilibrated with samplers using a range of water/cosolvent mixtures (e.g., water/methanol) and the partition coefficients measured. The cosolvent increases the solubility of the analytes in the aqueous phase and also reduces the sorption to any solid phases present. Extrapolation of the curve of log partition coefficient versus percentage cosolvent to zero percent cosolvent yields the true partition coefficient of the compound between the sorption phase and the water. Smedes[16] and Yates et al.[17] successfully used this approach to determine the partition coefficients of a range of hydrophobic organic contaminants between silicone rubber materials and water. In some cases, the partition coefficient can be estimated using linear relationships between K_{SW} and the corresponding n-octanol–water partition coefficients.[12,17]

3.2.2   KINETIC SAMPLING

With kinetic or TWA sampling, it is assumed that the rate of mass transfer to the sorption phase is linearly proportional to the difference between the chemical activity of the contaminant in the water phase and that in the sorption phase. During the initial phase of sampler exposure, the rate of desorption of analyte from the sorption phase to water is negligible and the sampler works in the linear uptake mode. The amount of analyte accumulated is therefore linearly proportional to its TWA concentration in water, even for situations where aqueous concentrations fluctuate over time (Figure 3.2). In this case Equation 3.1 reduces to

$$C_S(t) = C_W k_1 t \tag{3.3}$$

Equation 3.3 can be rearranged to

$$m_S(t) = C_W R_S t \tag{3.4}$$

where $m_S(t)$ is the mass of analyte accumulated in the sorption phase after an exposure time t and R_S is the sampling rate. R_S may be interpreted as the volume of water cleared of analyte per unit of exposure time (e.g., mL h^{-1} or L day^{-1}) by the device. If R_S is known, C_W (the TWA concentration of a pollutant in the water phase) may be estimated from the exposure time (t) and the amount (m_S) of the analyte collected by the sorption phase. For devices operating in the TWA mode, R_S does not vary with aqueous concentration but is often affected by water flow (turbulence), water temperature, and the extent of biofouling of the diffusive surface.

TWA sampling can be used in situations where analyte concentrations are variable and can be used to measure episodic pollution events. As integrative samplers permit the measurement of concentrations over extended time periods, they can provide a more realistic picture of contaminant levels than can be achieved by the collection of discrete spot samples of water.

3.2.3 SAMPLER CONSTRUCTION

The selection of materials used for the construction of passive samplers is based on a number of criteria, including price, mechanical strength, the maximum sampling rate, and accumulation capacity.

3.2.3.1 Sorption Phase

Analytes may accumulate in the sorption phase either by *adsorption* onto the surface of solid sorbent materials or by *absorption* in absorbent liquids or polymers that behave like subcooled liquids.The advantage of solid adsorbents is the potential to select materials with a high affinity and selectivity for target analytes. However, the sorption capacity of adsorbents is usually limited, and the description of adsorption/desorption kinetics of analytes to adsorbents is complex. Typically, the adsorbent materials used in passive samplers are similar to those used in solid-phase extraction techniques.

Absorbents usually have a higher sampling capacity but a lower selectivity than adsorbents. Many materials with a noncrystalline polymeric structure have been employed as absorbents in sampling devices. This type of sampling device consists only of a sorption phase that is not separated from the bulk water matrix by an extra diffusion-limiting membrane. The sampling rate of analytes is determined by diffusion into the sorption phase polymeric material itself or by diffusion into the aqueous boundary layer at the sampler surface. Recently, Rusina et al.[18] evaluated the performance properties (e.g., release of oligomers, swelling effects in solvents, analyte diffusion coefficients, and partition coefficients) of a range of polymers used in single-phase samplers for the measurement of hydrophobic compounds. She proposed that silicone rubber is an attractive sorption phase owing to its high partition coefficient and low mass transfer resistance for hydrophobic analytes such as polynuclear aromatic hydrocarbons (PAHs).

3.2.3.2 Diffusion Barrier

An important performance characteristic of passive samplers that operate in the TWA regime is the diffusion barrier that is inserted between the sampled medium and the sorption phase. This barrier is intended to control the rate of mass transfer of analyte molecules to the sorption phase. It is also used to define the selectivity of the sampler and prevent certain classes (e.g., polar or nonpolar compounds) of analytes, molecular sizes, or species from being sequestered. The resistance to mass transfer in a passive sampler is, however, seldom caused by a single barrier (e.g., a polymeric membrane), but equals the sum of the resistances posed by the individual media (e.g., aqueous boundary layer, biofilm, and membrane) through which analyte diffuses from the bulk water phase to the sorption phase.[19] The individual resistances are equal to the reciprocal value of their respective mass transfer coefficients and are additive. They are directly proportional to the thickness of the barrier and are inversely proportional to the permeability (a product of diffusion coefficient and solubility in the medium) of the medium for a given analyte.[19] The dominant barrier to mass transfer for a particular compound is the one that contributes more than 50% to the total resistance. High sampling rates are obtained when the transport resistance of the membrane is much smaller than that of the aqueous boundary layer. Consequently, the product of membrane diffusion coefficient and membrane/water partition coefficient ($D_M K_{MW}$) is called the membrane permeability, and this can be used for ranking materials for use in the construction of passive sampling devices.[18]

Often, the dominant barrier to mass transfer is used to classify samplers into two categories, either diffusion-based or permeation-based devices.[14,20] In diffusion-based samplers, analyte molecules reach the sorption phase by diffusion through a static boundary layer of water contained as a well-defined gap between the bulk water phase and the sorption phase. In permeation-based samplers, compounds must also diffuse through a porous or nonporous membrane. However, this classification can be artificial, as many sampling devices, although designed as "permeation samplers," have later been shown to accumulate some analytes under water boundary layer control, and thus work as diffusion samplers. The SPMD and the Chemcatcher® device used for sampling hydrophobic compounds are examples of such behavior.[21,22]

Passive samplers are usually designed to maximize the amount of analyte sampled in order to detect low levels of pollutants present in water. Diffusion samplers mostly use a "tube" design, where the receiving phase is located inside a long, narrow inert tube or a capillary. The space between the edge of the sampler and the surface of the sorption phase is filled with a stagnant layer of the sampled water. Diffusion through this immobilized layer of water defines the sampling rate. To avoid fluctuations in the sampling rate, caused by disturbance of the diffusion distance by movement in the bulk water phase, such diffusion samplers are characterized by a low ratio of sampler surface area to diffusion distance. An example is the solid-phase microextraction (SPME) sampling device in which the fiber is retracted a known distance into its needle housing during operation.[23–25] Since mass transfer is directly proportional to surface area, tube-type samplers (characterized in most cases by a small surface area) are less sensitive than badge-type samplers with a large surface area. In water monitoring badge-type samplers predominate. In cases where the aqueous boundary layer controls uptake, sampling kinetics in flat samplers with a large surface area are more affected by fluctuations in water velocity. To minimize the impact of such fluctuations on mass transfer, a permeable polymeric membrane (e.g., used in SPMD,[21] membrane enclosed sorptive coating (MESCO),[26] and Chemcatcher[27] samplers) or a gel layer used in the diffusive gradient in thin films (DGT[28] sampler) are used to separate the sorption phase from the bulk water environment. However, the use of such membranes (or any other additional diffusion barriers) that reduce flow sensitivity also automatically reduce the maximum sampling rate and thereby result in decreased sensitivity.[19]

Several types of polymeric membrane have been used for the construction of passive samplers. Nonporous membranes include low-density polyethylene (LDPE),[21,22,29] polypropylene, polyvinyl-chloride,[27,30] polydimethylsiloxane (PDMS),[16,30–32] polyacrylate,[33] and other nonpolar polymers.[34] Microporous membranes include glass fiber,[27] regenerated cellulose[26,27,35] nylon, polysulfone,[27] polyethersulfone,[36] and polyacrylamide hydrogel.[28] The membrane acts in effect as a barrier; the dissolved analytes can pass through whereas particulates, microorganisms, and macromolecules with a size greater than the exclusion limit cannot. Without the protection of the membrane, there is an increased risk of deterioration of the sorption phase due to biofouling.

3.2.4 MODELING AND CALIBRATION OF PASSIVE SAMPLING DEVICES

A number of models has been developed to improve the understanding of the kinetics of analyte transfer to passive samplers.[9,12,19,37] These models are essential for understanding how the amount of analyte accumulated in a device relates to its concentration in the sampled aquatic environment as well as for the design and evaluation of laboratory calibration experiments. Models differ in the number of phases and simplifying assumptions that are taken into account, for example, the existence of (pseudo-) steady-state conditions, the presence or absence of linear concentration gradients within the phases, the way in which transport within the boundary layers is described, and whether or not the aqueous concentration is constant during the sampler exposure.

The substance-specific kinetic constants, k_1 and k_2, and partition coefficient K_{SW} (see Equations 3.1 and 3.2) can be determined in two ways. In theory, kinetic parameters characterizing the uptake of analytes can be estimated using semiempirical correlations employing mass transfer coefficients, physicochemical properties (mainly diffusivities and permeabilities in various media), and hydrodynamic parameters.[38,39] However, because of the complexity of the flow of water around passive sampling devices (usually nonstreamlined objects) during field exposures, it is difficult to estimate uptake parameters from first principles. In most cases, laboratory experiments are needed for the calibration of both equilibrium and kinetic samplers.

Calibration of equilibrium samplers depends on estimating K_{SW}. Tank exposure studies are used and the concentration of the analyte of interest in the two phases at equilibrium is measured. For highly nonpolar compounds bias can be introduced into the estimation of the partition coefficients since these compounds can bind to the walls of the calibration tank, to dissolved organic carbon (DOC), and to any suspended organic matter. This can lead to an overestimation of the freely dissolved fraction in the water phase. To overcome this problem, a cosolvent method using a range of concentrations of methanol in the external water phase has been used.[16] For kinetic samplers, the key parameter to be determined is the apparent water sampling rate (R_S) that has units of volume per unit time. This is directly related to the effective sampling area of the sampler and is affected by environmental variables such as temperature and turbulence at the face of the sampler. A number of methods have been used to estimate this parameter including static exposure, static exposure with renewal, and continuous flow; these have recently been reviewed by Stephens and Müller.[40] Ideally, these should provide estimates that cover the range of water temperatures and turbulence conditions (usually achieved by varying the stirring rate in the calibration tank), which would be found in typical field exposures. The problems of bias for very hydrophobic compounds are similar to those described for the equilibrium samplers, where the presence of dissolved or suspended organic carbon can reduce the effective (freely dissolved) concentration of analytes in the calibration tank.[40]

3.2.5 FACTORS AFFECTING PASSIVE SAMPLING

A number of methods has been developed to compensate for the effect of environmental variables on sampler performance. Booij et al.[41] and Huckins et al.[42] described a method to estimate the uptake kinetics in both laboratory and field situations by spiking sampling devices prior to exposure with a number of performance reference compounds (PRCs) that do not occur in the environment (usually deuterated analogs of the compounds being measured). Where factors influencing uptake kinetics affect the offloading kinetics of PRCs in an identical manner, the release rate of these compounds is a measure of the exchange kinetics between the sampler and water, and can be used to compensate for variations in environmental conditions during field exposures.[43] The PRC approach is applicable in situations only where the exchange kinetics are isotropic. This is the case when the overall uptake of target pollutants and release of PRCs are governed by first-order kinetics and the sum of the resistances to mass transfer across the sampler is equal in both directions. These characteristics are observed in samplers where the sorption phase consists of an immiscible liquid or a nonpolar polymeric film (a subcooled liquid).[42,44] The PRC approach may not be applicable, however, for samplers fitted with solid-phase sorbent receiving phases (e.g., the Polar Organic Chemical Integrative Sampler (POCIS) or the polar pollutant variant of the Chemcatcher sampler) because of the fundamental differences between solute partitioning and adsorption phenomena.[45] Also, as the solid-phase sorbents often act as an infinite sink with a very high sorption capacity, the selection of a PRC with a sufficient fugacity to enable it to be released from such samplers is problematic. Similar problems exist with devices for measuring metals as chelating agents are often used as receiving phases, and the element, once sequestered, can only be released by the addition of a strong acid.

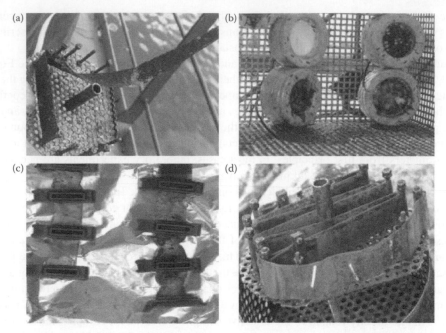

FIGURE 3.3 Examples of passive samplers affected by deposition of suspended particulate matter and bio-fouling: (a) silicone strip sampler, (b) nonpolar version of the Chemcatcher, (c) MESCO sampler fitted with a cellulose membrane, and (d) MESCO sampler fitted with polyethylene membrane (front) and SPMD (back).

3.2.6 BIOFOULING

Unprotected surfaces submersed in water eventually become colonized by bacteria and various flora and fauna that may ultimately form a biofilm (Figure 3.3).

The thickness and density of this biofilm vary not only from exposure to exposure but also from zone to zone on the same surface. The composition of biofilms depends on the properties of the aquatic system being measured as well as the properties of the material colonized by microorganisms. During passive sampling, buildup of a biofilm layer can increase the resistance to mass transfer of sampled analytes, thus reducing their sampling uptake rates. Moreover, if the microbial communities that develop on the surface of the sampler possess a potential for biodegradation, they can decompose the analytes in the water that is in contact with the biofilm. This would result in an increase in the concentration gradient between the sorption phase and the biofilm layer. As a result, compounds with low K_{SW} values may be released from the sorption phase back to the biofouling layer and subsequently degraded there. Such effects may result in a serious underestimation of analyte concentrations.

Colonizing organisms may also physically damage the surface of membranes that are made of a biodegradable material (e.g., cellulose). Paschke et al.[46] observed up to three times faster analyte exchange kinetics with the MESCO sampler in the field in comparison with a laboratory calibration study. Besides the effects of water turbulence, the increased exchange kinetics could be explained by degradation of the cellulose diffusion-limiting membranes during the field exposure, resulting in a significant loss of resistance to analyte exchange between the sampler and water. The effect of biofouling on sampling kinetics is shown in Figure 3.4.

LDPE membranes were prefouled for one month in water collected from a park fountain. Membranes became heavily fouled with a thick algal and bacterial film. The fouled membranes were used for the construction of Chemcatcher samplers fitted with sorption phases previously spiked with several PRCs. Samplers fitted with either fouled or unfouled membranes were simultaneously exposed (rotation speed: 40 rpm and water temperature: 11°C) in a laboratory flow through a calibration system

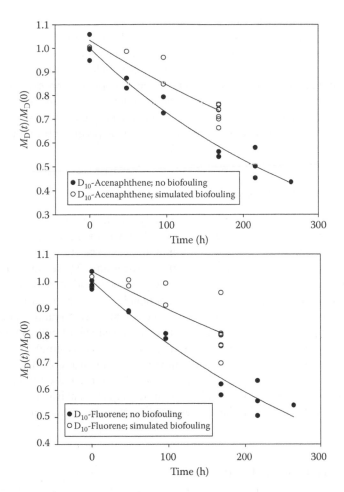

FIGURE 3.4 Biofouling reduces the exchange kinetics of PRCs (deuterated acenaphthene and fluorene) between the nonpolar Chemcatcher sampler (fitted with either fouled or unfouled LDPE membranes) and water. The experiment was performed in a laboratory flow-through calibration system at a water temperature of 11°C with simulated water turbulence of 40 rpm. $M_D(t)/M_D(0)$ is the fraction of the PRC remaining in the sampler during exposure.

as described by Vrana et al.[43] Offload kinetics of PRCs from fouled samplers were compared with those from unfouled samplers. The heavy degree of biofouling caused a reduction of up to 50% in the elimination rate of PRCs: although a significant effect, it is less dramatic than might be expected.

Huckins et al.[29] reported a 20–70% impedance in the uptake of PAHs in cases of severe biofouling on the surface of SPMDs. Their model describing the mass transfer in a biofilm indicated that it behaved like an immobilized water layer with a resistance that is independent of the biofilm/water partition coefficient. This would result in a similar mobility of compounds in the biofilm since this is independent of their hydrophobicity.[19] Similarly, Richardson et al.[47] observed that biofouling caused a reduction of up to 50% in the uptake of PAHs and organochlorine pesticides by SPMDs. It has been suggested by several authors that PRCs can be used to correct biofouling during deployment,[42,47] but more experimental evidence is needed.

To minimize the effects of excessive fouling, prolonged exposures of samplers should be avoided.[48] Biofouling may be reduced by the selection of suitable construction materials. For example, the polyethersulfone diffusion membranes used in POCIS and the polar version of the Chemcatcher are less prone to fouling than the LDPE membranes used in SPMDs. This may be due to the low surface energy properties of polyethersulfone, discouraging the initial onset of the biofouling process by creating unfavorable conditions for the settlement of colonizing microorganisms.[49] Alternatively, coating the membrane with a low surface energy material, for example, Nafion®, can be used to inhibit fouling.[50] Some solvent-filled membrane devices are protected from fouling by the slow diffusion of the solvent (e.g., *n*-hexane) from the sampler during field exposure. Such chemicals inhibit the growth of microorganisms. Protective screens made of copper or bronze mesh have also been shown to inhibit biofouling; however, they cannot be used when monitoring heavy metals. Recently, a novel approach using antibiotics added to the DGT device has been attempted in order to prevent the development of biofilms.[51] Attempts to inhibit fouling by applying antifouling agents to SPMDs prior to field deployments have proved unsuccessful.[52]

3.3   RECENT APPLICATIONS

Passive samplers can combine sampling, selective analyte isolation, preconcentration, and in some cases preservation of speciation in a single step. They eliminate the need for an external energy supply at the sampling site and allow the entire sampling process to be simplified. Following exposure, samplers are transported to the laboratory for further processing. These steps are often similar to those used in conventional sample extraction, sample preconcentration, and instrumental chemical analysis. Passive sampling technology is widely applicable. It can be used in monitoring studies aimed at screening for the presence or absence of pollutants, investigating temporal and spatial trends in levels of pollutants, speciation of contaminants, assessment of pollutant fate, measurement of TWA concentrations, and in biomimetic sampling (an approach to simulate contaminant uptake in biota). Several recently published reviews and monographs are available that describe the design, construction, and use of passive samplers. The reader should refer to these for a fuller description of the devices and their use for monitoring pollutants in the aquatic environment.[7–11,13,30,53] Table 3.1 summarizes the different devices that have been used to measure organic and inorganic contaminants in water.

3.3.1   HYDROPHOBIC ORGANIC COMPOUNDS

Hydrophobic organic pollutants include several groups of compounds, such as organochlorine pesticides, polychlorinated biphenyls (PCBs), polychlorinated dibenzo-*p*-dioxins, polychlorinated dibenzofurans, and PAHs. The concentrations of nonpolar pollutants dissolved in water are frequently very low, usually less than 1 ppb. This is due to their low aqueous solubilities. These compounds adsorb strongly onto particulate matter and are deposited in the sediment. Nevertheless, because of their persistence and potential for bioaccumulation, it is necessary to monitor the concentrations of these chemicals in water. A range of passive sampling devices has been developed and applied for monitoring hydrophobic organic pollutants in water.[14,22,34,53,78] Among these, the SPMD is the most mature technique for measuring hydrophobic organic contaminants.[21] The design of the SPMD was first published in 1990, since when over 200 studies have been reported.[79] Several reviews and one monograph on this technology are available.[11,30,80,81]

3.3.2   POLAR ORGANIC COMPOUNDS

Environmental research interests have recently extended from persistent hydrophobic organic chemicals to more hydrophilic organic compounds. The latter include some polar pesticides, many pharmaceuticals and personal care products, microbial toxins, and endocrine disrupting compounds.[4] Polar organic compounds are often present at low concentrations in the aquatic environment, which poses a problem for most conventional sampling and analytical procedures. Recently, considerable effort has been directed toward the development of extraction methods suitable for the preconcentration of polar organic compounds commonly found in water bodies. Many of these methods use

TABLE 3.1
Overview of Different Passive Sampling Devices Used for Monitoring Pollutants in Water

Sampler	Sorption Phase	Barrier to Diffusion[a]	Analytes	Sampling Purpose	Typical Deployment Period	Sample Preparation for Chemical Analysis	Reference
Ceramic dosimeter	Various solid sorbent materials	Microporous ceramic material	PAHs, BTEX, chlorinated hydrocarbons	Integrative sampling in groundwater	Up to 1 year	Solvent extraction or thermal desorption	54
Chemcatcher (organics version)	Various Empore™ sorbent disks	Nonporous LDPE membrane or porous polyethersulfone membrane	Nonpolar and polar organics	Integrative	14 days to 1 month	Solvent extraction	26
Chemcatcher (metals version)	Chelating and C_{18} Empore™ sorbent disks	Cellulose acetate membrane	Range of heavy metals included in priority pollutant lists, including organotin and mercury compounds	Integrative, speciation	2–4 weeks	Extraction with acid	55,56
Diffusion gradient in thin films (DGT)	Binding resin impregnated in a hydrogel	Hydrogel layer	Most heavy metal pollutants, phosphorus, sulfide, and radioactive metal isotopes	Integrative, speciation, screening, mimicking biological uptake	1 day to 4 weeks	Extraction with acid	57
Dosimeter	Granular activated carbon	Stagnant layer of water	BTEX and atrazine	Integrative	Up to 2 months	Solvent extraction	58
Ecoscope	Hexane (organics) chelating resin (metals)	Polyethylene membrane (organics), porous membrane (metals)	Hydrophobic organic compounds, heavy metals	Qualitative screening	2–4 weeks	Direct injection of solvent or after concentration or extraction with acid	59
Gaiasafe sampler	Paper strips impregnated with binding agents		Metals, anions, organic compounds	Screening	2 days to 2 months	Solvent extraction	60
Gore-Sorber	Hydrophobic adsorbent	Microporous membrane of polytetrafluoroethylene—GORE-TEX®	BTEX, MTBE, PAHs. VOCs, semi-VOCs	Equilibrium	14 days	Thermal desorption	61

continued

TABLE 3.1 (continued)

Sampler	Sorption Phase	Barrier to Diffusion[a]	Analytes	Sampling Purpose	Typical Deployment Period	Sample Preparation for Chemical Analysis	Reference
LDPE and silicone strips	Nonporous polyethylene or silicone materials	The sorption phase itself acts as a diffusion barrier	Hydrophobic organic compounds	Integrative	1 month	Solvent extraction	62,63
MESCO	Silicone rods or bars coated with silicone	Regenerated cellulose or LDPE	PAHs, PCBs, organochlorine pesticides	Integrative	2 weeks	Thermal desorption	25
Negligible depletion-SPME	Glass fibers coated with a thin layer of various sorbent materials	Water boundary layer	Hydrophobic chemicals, including PAHs, PCBs, petroleum hydrocarbons, organochlorine pesticides, aniline, phenols	Equilibrium	Hours	Thermal desorption in GC inlet	64
Passive sampler	Various granular adsorbent materials	Silicone-polycarbonate membrane	Chlorobenzenes, nitrobenzenes, and nitrotoluenes	Integrative	Up to 1 day	Solvent extraction	65
Polyethylene diffusion bags (PDB)	Water or gas	Nonporous LDPE membrane	Polar organic compounds, VOCs, metals, trace elements	Equilibrium sampling in groundwater	2 weeks	Conventional analysis of the receiving water phase	66
Passive integrative mercury sampler (PIMS)	A liquid reagent consisting of nitric acid and gold (Au^{3+}) solution	Nonporous LDPE membrane	Neutral mercury species	Preconcentration, screening	Weeks to months	Acid extraction	67
Permeation liquid membrane (PLM)	Aqueous solution	An immiscible liquid membrane containing specific metal ion carrier	Heavy metals	Bioavailable metal species	Hours	Analysis of a subsample from the acceptor solution	68
Passive *in situ* continuous extraction sampler (PISCES)	Hexane	Semipermeable membrane filter	PCBs	Integrative	2 weeks	Volume reduction of the receiving phase	69
POCIS	Various granular adsorbent materials	Polyethersulfone membrane	Herbicides and pharmaceuticals with log $K_{OW} < 3$	Integrative	Up to 2 months	Solvent extraction	35
Flux-proportional sampler	Various adsorbent materials	A porous cartridge permeable to water	Wide range of contaminants	Flux-proportional sampling in soil and groundwater	1 month	Solvent extraction	70

Name	Receiving phase	Membrane	Target compounds	Type	Sampling time	Analysis	Ref.
Sampler according to Kot-Wasik	Organic solvent	PDMS membrane	Phenols, acid herbicides, triazines	Integrative	1 month	Analysis of a subsample of solvent is taken and analyzed without further cleanup steps	71
Solvent-filled dialysis membranes	Hexane	Cellulose acetate membrane	Hydrophobic organic compounds	Integrative	1 month	Volume reduction of the receiving phase	
SPATT	Adsorbent polymeric resin	Membranes made of porous materials	Polar phytotoxins	Integrative	1 week	Solvent extraction	72
SPMD	Triolein	Nonporous LDPE membrane	Hydrophobic organic compounds	Integrative	1 month	Dialysis in organic solvents, size exclusion chromatography	29
Stabilized liquid membrane device (SLMD)	A water-insoluble organic complexing mixture diffuses to the exterior surface of the sampler through a polymeric membrane		Divalent metal ions	Preconcentration, in situ sampling, determination of labile metal ions in grab samples	Days to several weeks	Extraction with acid	73
Supported liquid membrane (SLM)	Aqueous solution	An immiscible immobilized liquid membrane containing specific metal ion carriers	Divalent metal ions	Integrative field sampling, preconcentration of trace elements, mimicking biological membranes	Days		74
Thin layer chromatographic plate	Various adsorbent materials	Water boundary layer	Organophosphates	Screening	1 month	Solvent extraction	75
LDPE bags containing trimethylpentane solvent (TRIMPS)	Trimethylpentane	Nonporous LDPE membrane	Pesticides	Integrative	1 month	Direct analysis of the receiving phase solvent	76
TWA-SPME	Glass fibers coated with a thin layer of various sorbent materials	Water boundary layer	BTEX, PAHs, organometallic compounds	Integrative	Up to several days	Thermal desorption in GC inlet	22,38,77

a In the aquatic environment, the barrier to diffusion will also include a water boundary layer and in some cases a biofilm layer.

specially formulated solid-phase extraction media and some of these can also serve as receiving phases for passive samplers for measurement of this class of pollutants.

The first sampler (POCIS) reported for this range of chemicals was developed at US Geological Survey. This sampler uses a triphasic sorbent mixture (Isolute ENV⁺ polystyrene divinylbenzene and Ambersorb 1500 carbon dispersed on S-X3 Biobeads) or the Oasis HLB sorbent as the receiving phase and a polyethersulfone diffusion-limiting membrane.[36] It has been used to monitor polar pesticides, prescription and over-the-counter drugs, steroids, hormones, antibiotics, and personal care products.[36] The POCIS samples hydrophilic chemicals from the dissolved water phase and permits the determination of TWA concentrations over extended periods (several weeks). To date, more than 100 hydrophilic organic contaminants have been identified in extracts obtained from the receiving phase of POCIS.[45] Another sampler suitable for the monitoring of a broad range of polar organic contaminants is one variant of the Chemcatcher. It consists of an Empore™ disk receiving phase overlain with a polyethersulfone diffusion-limiting membrane.[27,34] This device is suitable for monitoring chemicals with log $K_{OW} < 3$, and its functionality can be modulated by choosing the appropriate receiving phase disk chemistry within the Empore disk range, for example, either modified polystyrenedivinylbenzene copolymers (SDB-XC or SDB-RPS) or ion-exchange (Cation-SR or Anion-SR) phases. Mills et al.[82] recently reviewed the use of passive sampling devices for the measurement of pharmaceuticals and personal care products in different aquatic environments. A number of potentially valuable forensic applications of the technology were identified, such as estimating illicit drug usage from the population within a given wastewater catchment area. It was highlighted that more calibration data (in terms of uptake rates) for frequently detected pharmaceutical compounds and related products are needed for both the Chemcatcher and the POCIS.[83–85]

3.3.3 Volatile Organic Compounds

Passive samplers are widely used in monitoring volatile organic chemicals (VOCs) in groundwater. Such samplers have the potential to reduce costs of monitoring from the high levels associated with the use of pumps to sample the test wells. Moreover, the risk of loss of volatile analytes during sample transport and storage is substantially reduced once the compounds are accumulated in the sampler sorption phase.

Equilibrium-based passive samplers used in groundwater monitoring typically consist of a closed container composed of a semipermeable membrane containing a gas or water which is free of the target analytes. When these sampling devices are deployed in VOC-contaminated groundwater, the analytes diffuse inside the sampler until equilibrium is reached between VOC concentrations in the ambient water and the gas or water inside the device.[86] The equilibrated device is collected and the sample can either be immediately sealed in the sampler or be transferred to a vial, depending on requirements. An equilibration time of 1–7 days is typical for this type of sampler.[87–90] A database of diffusion-based equilibrium passive samplers and their applications in monitoring of VOCs is maintained at a website.[91]

Among the kinetic sampling devices, ceramic dosimeters have been used successfully for the long-term surveillance of VOCs.[92] They use a ceramic tube as the diffusion-limiting barrier that encloses a receiving phase consisting of solid sorbent beads. Over a three-month deployment in a contaminated aquifer, the ceramic dosimeter provided TWA concentrations of benzene, toluene, ethylbenzenes, xylenes, and naphthalenes. The levels obtained matched closely those found in spot water samples that were taken frequently over the trial period.[54]

3.3.4 Organometallic Compounds

Passive sampling devices have been used to measure a number of organometallic species, including those of lead, mercury, and tin. Følsvik et al.[93,94] and Harman et al.[95] reported the use of SPMDs for monitoring tributyltins. A variant of the Chemcatcher sampler has been developed and calibrated for the measurement of TWA concentration of organotin compounds.[96] Ouyang et al.[23] demonstrated the feasibility of using SPME as an *in situ* passive sampler for the determination of TWA concentrations of a range of organic chemicals in water. The SPME technique has been used for measuring organometallic species of lead, mercury, and tin in spot samples of water.[77] It may be possible to extend this technique for use as a passive sampler for organometallic compounds.

3.3.5 Metals

Besides the use of passive sampling for measurement of metal ions, there is also a significant interest in using these devices to characterize the speciation of metals in water, that is, differentiating between free, inorganic, and organic bound metal species and organometallic compounds (see Table 3.1). Among the passive sampling techniques for metals, the most widely used is the diffusive gradients in thin films (DGT) device.[28,97] Other devices based on the permeation of analytes through a variety of membrane materials, such as the metals version of the Chemcatcher, are finding an increasing range of applications (Table 3.1).[55] In passive samplers, metals are usually accumulated on a Chelex-based sorption phase. After deployment, analytes are extracted from the receiving phase of samplers using a strong acid (e.g., nitric acid) and concentrations are measured using a range of spectroscopic techniques.[57]

Information on speciation is important for the understanding of the fate and toxicity of metals in waters.[98] Recently developed passive sampling techniques allow a better understanding of the speciation of metals in the environment. Passive samplers are suitable for determining free hydrated metal ions and metal complexes with sufficiently high dissociation rates.[99–104] Advances in the understanding of the speciation of metals in the aquatic environment have been achieved using the DGT. Hydrogels of varying pore size are used, which can exclude or restrict (on a size basis) the diffusion of large organic complexes into the device.[99,100] Sampling selectivity can also be modified by varying the thickness of the hydrogel diffusion layer. Increasing the thickness allows the sampling of stable complexes since labile species will dissociate during the time required to diffuse across the layer.[101,103] Different binding phases can also be used to modify selectivity. Li et al.[105] investigated the effect of the binding strength (stability constant) of the sorption phase on the accumulation of metals. They investigated several binding phases, including Chelex 100 polyacrylamide hydrogel, poly(acrylamide-co-acrylic acid) gel, poly(acrylamidoglycolic acid-co-acrylamide) gel, Whatman P-81 cellulose phosphate ion-exchange membrane, and poly(4-styrenesulfonate) solution. Tests conducted using solutions of metals spiked with ethylenediaminetetraacetic acid (EDTA) or humic acid suggested that the DGT measures only free metal ions and inorganic metal complexes. Field trials at both freshwater and seawater sites showed that the DGT-labile metal concentrations measured by devices fitted with different binding phases can be significantly different. This suggested that the DGT-labile metal fractions sequestered were dependent on the binding strength of the receiving phase. DGT devices fitted with different binding phases that can compete to varying extents with various natural complexing ligands can be utilized to measure metal speciation in natural waters.

3.3.6 Algal Toxins

Mackenzie et al.[72] developed a simple and sensitive sampler for monitoring toxic algal blooms and shellfish contamination. This involved the adsorption of biotoxins onto porous synthetic resin-filled sachets (solid-phase adsorption toxin tracking: SPATT bags) followed by their extraction and analysis. The technique measured pectenotoxins and okadaic acid complex toxins, the levels of which are increased in seawater during algal blooms. This approach is simpler than the analysis of shellfish tissue extracts. Time-integrated sampling provided an estimate of biotoxin accumulation in filter feeders, and the high sensitivity of the method provided an early warning of the potential of contamination. Recently, Fux et al.[106] showed that divinylbenzene polymeric resins can be used as sorption phases in SPATT samplers for the detection of the lipophilic marine toxins okadaic acid, and dinophysistoxin-1 from *Prorocentrum lima*. Kohoutek et al.[107] used a modified POCIS fitted with a polycarbonate membrane and Oasis HLB sorbent for monitoring common and highly hazardous microcystins (cyanotoxins) in water. A seven-day exposure in the field was sufficient to detect the microcystins.

3.3.7 Applications in Ecotoxicity Assessment

A novel application is the use of extracts obtained from the elution of the sorption phase in bioassays to assess the ecotoxicity of accumulated pollutants. This has been attempted mainly with samplers used to sequester either nonpolar or polar organic chemicals.[108] This approach overcomes some of the problems in obtaining samples suitable for testing the biological effects of low levels of pollutants present in water. When testing extracts are obtained from a sorption phase, solvent compatibility with the chosen bioassay is important. A number of solvents can be toxic, inhibitory, or immiscible, and sometimes solvent exchange is needed before an extract can be assayed. Several different bioassays have been used with extracts from samplers. These include Microtox®, Mutatox®, mixed-function oxygenase induction as 7-ethoxyresorufin-*o*-deethylase (EROD) activity, sister chromatid exchange, vitellogenin induction, enzyme-linked immunosorbent assay, Ames mutagenicity test, the yeast estrogen assay (YES-screen), and yeast androgen assay (YAS-screen).[108–111] The use of these screening procedures can be a quick, cost-effective means of identifying problems that may otherwise require rigorous chemical analysis, resulting in greater expenditure of time and money.[112] They may also have a contribution in improving risk assessment and in optimizing remediation measures. The following examples highlight how extracts from the sorbent phase or the sorbent phase itself can be used to help in the assessment of the ecotoxicity of surface water and groundwater.

Rastall et al.[113] used extracts from SPMDs in a bioassay-directed chemical analysis scheme for the detection and identification of bioconcentratable hydrophobic estrogen receptor agonists (ERA) at riverine sites in Germany and the United Kingdom. An aliquot of the extract was fractionated using a reverse-phase high-performance liquid chromatography (HPLC) method that was calibrated to provide an estimation of target analyte hydrophobicity. Each fraction was then tested in the YES screen to determine any estrogenic activity. This activity was plotted against HPLC retention time (i.e., a measure of compound hydrophobicity). In fractions where significant activity was found, an aliquot was subsequently analyzed by gas chromatography (GC)/mass spectrometry (MS) in an attempt to identify the compounds responsible for the observed estrogenic activity. The study indicated that improvements to the analytical methodology would be required in order to identify all of the ERAs or other target analytes present in the active fractions. This would increase the applicability of the method to risk assessment and water quality monitoring programs.

Vermeirssen et al.[111] used POCIS for sampling polar estrogens. Masses of estrogens accumulated in POCIS and concentrations of these compounds in spot samples were determined using the YES screen (being expressed as 17-β-estradiol equivalents). Chemical analysis of selected compounds was also performed using LC/MS. They found that data from spot sampling, passive sampling, and bioaccumulation in caged fish were correlated and provided comparable values. POCIS provided an integrated and biologically meaningful measure of estrogenicity in that it accumulated estrogens in a pattern similar to that of brown trout.

A novel approach using the sorbent phase directly (with no extraction) in contact with cell lines from rainbow trout liver was developed by Schirmer et al.[114] and Bopp et al.[115] The cell lines were used to detect EROD activity from specific PAHs. A modified ceramic dosimeter (Toximeter sampler containing Biosilon 160–300 μm diameter polystyrene beads as the sorbent phase) was deployed for extended periods in groundwater known to be contaminated by PAHs. The response obtained using the cell lines incubated with Biosilon beads correlated with the concentrations of PAHs known to have EROD-inducing activity. The system is versatile as other cell lines can be used depending on the toxicological response of interest.

3.3.8 APPLICATIONS IN BIOMONITORING

Historically, biota have been used to obtain information on biologically relevant concentrations of pollutants in water. Various test organisms (e.g., mussels and fish) can be used depending on the water body being investigated. Test biota can be deployed for extended periods of time during which they bioaccumulate pollutants from the surrounding water. Analysis of tissues or lipid extracts can give an indication of the equilibrium concentration of waterborne contamination. Several factors can influence the levels of pollutants accumulated: they include metabolism, depuration rates, excretion, stress, viability, and the condition of test organisms. These factors can lead to uncertainty associated with the data. Furthermore, the extraction of analytes from the tissue of animals prior to instrumental analysis is complex and time-consuming. In spite of these difficulties, mussels have been extensively used (e.g., the Dutch National Institute of Coastal and Marine Management "mussel watch" program) in the monitoring of hydrophobic contaminants in both freshwater and marine environments[16] and a large historical dataset of pollutant levels is available.

Recently, Smedes[16] investigated whether passive sampling devices could be used to give similar information on biologically available concentrations of hydrophobic compounds (e.g., PAHs and PCB). The samplers comprised silicone rubber sheets (9.5 cm × 5.5 cm × 0.5 cm) spiked with a range of PRCs and were deployed in tandem with mussels (*Mytilus edulis*) for six weeks at several coastal sampling stations in the Netherlands. Results showed there was a close relationship (even over different seasons) between the concentrations found in mussels and those derived from the passive samplers. However, for the very hydrophobic compounds, there was some uncertainty as to whether equilibrium had been achieved by the end of the deployment period. The use of passive samplers shows considerable promise as an alternative to the use of biota in routine monitoring programs.

3.4 QUALITY ASSURANCE, QUALITY CONTROL AND VALIDATION

The application of appropriate quality control (QC) measures is essential in environmental monitoring projects using passive samplers. This is important in both sampler exposure and the subsequent sampler processing and instrumental analysis. In general, quality assurance (QA) measures should be implemented throughout all procedures including preparation, handling (transportation, deployment, and retrieval), and storage processes in accordance with ISO 5667-14.[116] The level of QC applied to passive sampling will vary depending on the project objectives and procedures involved. The recently published British Standards Institute Publicly Available Specification (BSI PAS-61) on the use of passive samplers for monitoring pollutants in water gives helpful advice on this issue.[117]

QC samples should address such issues as the purity of materials used to construct the device, and potential contamination during preparation, transport, deployment, retrieval, and subsequent storage. Furthermore, QC protocols are required for analyte extraction, further processing (preconcentration and isolation operations), and instrumental analysis. The samples related to the control requirements for passive sampler studies include fabrication controls, reagent blanks, field controls, and sampler recovery spikes. Fabrication controls are QC samples used to record contamination during manufacture, laboratory storage, processing, and analytical procedures. Field controls are QC samples used to record any contamination during transportation, deployment, and retrieval. In some cases, samplers spiked with PRCs serve as a special type of QC sample. These provide information about *in situ* uptake kinetics and reduce bias in the estimation of TWA concentrations when laboratory-derived calibration data are applied to the situation in the field.[42]

An example of the investigation of QC issues associated with the use of passive samplers has been published by DeVita and Crunkilton.[118] The results of their study showed that QC measures applied to SPMDs met or surpassed conventional guidelines (e.g., EPA Method 610 for PAHs in water) in terms of both precision and accuracy.

Only a small number of interlaboratory proficiency tests, required for full method validation, have been performed. In 2006, the International Council for the Exploration of the Sea (ICES) initiated a trial survey and intercalibration exercise for passive sampling involving 13 participating laboratories.[119] The study sampled freely dissolved hydrophobic contaminants in water and sediment. For water, only PDMS sheet passive samplers were used. In parallel, sediment samples were collected at the same locations and equilibrated with passive samplers in the laboratory. Analyses of the samplers used in the water and sediment studies by both the participating laboratories and a central reference laboratory allowed the trial to act as an analytical intercalibration study. Currently, only preliminary results on freely dissolved concentrations of phenanthrene and PCB-153 in seawater and sediment pore water samples at the 13 test sites across Europe have been reported.[119] This technique enabled the detection of nonpolar analytes in water at low pg L^{-1} levels.

Field studies in which the results obtained with passive samplers are compared with those obtained with conventional sampling techniques also increase the body of evidence that is available to underpin acceptance of the validity of passive sampling within a regulatory framework. However, assessing the accuracy of measurements made by passive samplers against other techniques may prove difficult, as the results may not be directly comparable. One possible reason for differences sometimes observed between spot sampling and passive sampling data is that the discontinuous spot samples may be taken at time points when concentrations of a pollutant were higher or lower than average and large fluctuations may occur in the intervals between samples. When the conventional sampling approach is used, any change in ambient concentration during the intervals between samples is undetected, but passive samplers can reflect such events. Spot (bottle or grab) sampling measures either total concentrations of pollutants or concentrations remaining in the sample after filtration (e.g., through a 0.45-μm filter). In comparison, passive samplers sequester only that fraction of contaminants that can diffuse into the sorption phase. This fraction is sometimes referred to as the freely available or bioavailable fraction. This can be misleading as this fraction will vary according to the design of the device. For example, samplers can be fitted with either porous or nonporous membranes for different applications. In the case of the DGT, different pore size gels (open pore or restricted pore) can be used to measure different molecular size fractions of metals and this can help in speciation studies.[120]

One approach to help overcome this disparity between spot and passive sampling data is to use additional water quality data. If average values of DOC, suspended particulate matter, and total organic carbon content are known, it may be possible to estimate the total concentration using empirical relationships that describe the distribution of a chemical between the different phases that may be present in an environmental water sample.[121] There is, however, uncertainty associated with this approach, as a number of assumptions are made in the calculations and a better understanding of the partitioning behavior of priority pollutants between the different phases is needed.

Several studies have compared the accuracy of pollutant concentrations measured using passive samplers with those obtained by active sampling methods. Ellis et al.[122] found close agreement between the concentrations of a number of organochlorine pesticides measured using SPMDs and those measured in spot water samples processed through a tangential-flow ultrafilter. However, many compounds detected in SPMD extracts were not found in the filtered spot samples. Rantalainen et al.[123] compared the levels and congener profiles of polychlorinated dibenzo-*p*-dioxins, dibenzo-furans, and non-*o*-chlorinated biphenyls in the water column measured using SPMDs with those collected by an Infiltrex® resin column sampler. Both the SPMD and Infiltrex sampler data were similar. Zeng et al.[124] found very good agreement between the results obtained using an equilibrium passive sampling device based on SPME and those obtained using an Infiltrex 100 *in situ* large-volume extractor for a range of PCBs and organochlorine pesticides. Axelman et al.[125] compared concentrations of dissolved PAHs at a site contaminated by discharges from an aluminum smelter. Measurements using SPMDs were similar to those obtained using an on-line filtration system. The PAH concentrations calculated from SPMD data showed a systematic deviation from the active sampling method that increased with compound hydrophobicity. The method used for calculating water concentrations from SPMD data was implicated as a potential source of systematic error. Several other field validation trials for measuring organic pollutants using different designs of passive sampler (Chemcatcher,[27,56,126,127] ceramic dosimeters,[54,114,128] POCIS,[36,45] SPMDs,[129,130] SPME,[131] TRIMPS,[132] and MESCO[44]) alongside frequent spot sampling have been reported. For inorganic pollutants, several field studies comparing the Chemcatcher and DGT with other sampling methods and theoretical speciation models have been undertaken.[120,133–139]

3.5 USE OF PASSIVE SAMPLING IN REGULATORY MONITORING

Until recently, checking water quality compliance with the regulatory provisions has been based on the chemical analysis of spot samples of water taken at a defined frequency. In Europe, the implementation of the European Union's Water Framework Directive (WFD, 2000/60/EC)[140] has resulted in an urgent need for cost-efficient monitoring tools that can provide the required information for the assessment of water quality. This legislation applies to the management of all water bodies (surface, ground, coastal, and lakes) with a target aim of achieving good water quality status by 2015. Three different monitoring approaches (surveillance, operational, and investigative) are defined within the Framework.[140]

The European Union (EU) assessment of water quality according to the provisions of the WFD includes the measurement of a number of quality elements and an evaluation of the presence of priority pollutants in these various water bodies. Based on the levels of priority substances measured, the water body is classified as having either "good" or "bad" chemical status. For the classification of the chemical status, environmental quality standards (EQS) have been set for priority organic pollutants in surface water. These include annual average and maximum acceptable concentration quality standards (AA-QS and MAC-QS, respectively). For compliance checking, total concentrations of priority organic pollutants in water are compared with EQS values. For heavy metals, the concentrations determined in filtered (through a 0.45-μm filter) samples of water are compared with EQS values. For a meaningful comparison with EQS, monitoring data should be representative of the pollutant levels in the water body.

Under the WFD, current monitoring practice is based on the regular (in most cases once a month) collection of spot water samples and their subsequent analysis for defined priority substances. This approach suffers from several drawbacks. Spot samples provide concentrations of pollutants only at the moment of sampling. Thus in water bodies characterized by marked temporal and spatial variability there is an increased risk of a false classification of the chemical status.[141] Further, EQS values for many priority compounds are very low (e.g., tributyltin cation, PAHs, and polybrominated diphenyl ethers) and the laboratory methods commonly used for the analysis of spot samples of water are often not sensitive enough to fulfill the required minimum performance criteria. Moreover, several studies have demonstrated a decrease in method reproducibility with decreasing concentrations of organic pollutants.[142] Therefore, conventional bottle sampling followed by instrumental analysis for the measurement of trace levels of organic and metal contaminants has severe limitations in terms of achievable limits of detection, reproducibility, confidence in the results obtained, and the ability to use the data on their own for decision-making within the context of the WFD.[141]

The WFD does not specify the techniques to be used for the monitoring of quality elements, including priority chemical substances. Among the alternative technologies available, passive samplers have the potential to be used in various regulatory monitoring programs aimed at assessing the levels of chemical pollutants. This has been recognized in the recently published technical guidance document for monitoring chemical substances within the context of the WFD.[143,144] A number of possible applications of passive sampling in the regulatory monitoring of surface waters were identified, and these scenarios were based on field demonstrations undertaken as part of the EU-funded SWIFT-WFD project.[141,145] However, passive samplers can only be used in surveillance and operational monitoring if they meet the Commission's requirements concerning defined minimum performance criteria for chemical monitoring methods and the quality of the analytical results.[146]

Passive samplers can be used alongside spot sampling in order to corroborate or contradict the data obtained. This approach can provide additional "weight-of-evidence" in water bodies where concentrations of contaminants are expected to fluctuate widely with time. Since one of the primary objectives of the WFD is the assessment of representative concentrations of pollutants in water bodies, the measurement of TWA concentrations over a period of several weeks using passive samplers seems to be a promising approach. Passive samplers can also effectively sample large volumes of water (the equivalent of tens of liters) over the deployment period. Therefore, it is possible at reasonable cost to measure much lower aqueous concentrations of pollutants than is possible using low-volume spot sampling procedures. This is particularly important for the very hydrophobic priority pollutants, for example, polybrominated diphenyl ethers, high-molecular-weight PAHs, some organochlorine pesticides, and organotin compounds. An additional use of passive samplers is in investigative monitoring. Samplers can be deployed at a number of sites to detect and hence locate unknown sources of pollution.

Besides the advantages that passive sampling may offer, it is important to recognize that in many cases these devices measure a different fraction of contaminants than that defined for the checking of EQS compliance within the WFD. This becomes especially important when monitoring very hydrophobic chemicals (log $K_{OW} > 4$), where a large fraction of the total amount present in a spot water sample is bound to colloids and particles. In contrast, most passive samplers used for monitoring hydrophobic compounds (e.g., SPMD, Chemcatcher, and silicone materials) measure only the truly dissolved fraction of these chemicals.

Nevertheless, for compliance checking, reliable estimation of the relevant (total or filtered) concentration from passive sampling data should be possible using additional information (e.g., DOC level and concentration of suspended particulate matter) on water quality at the monitoring site. These data can be used in models that describe the distribution of chemicals between water, colloidal, and particulate matter phases. With the exception of extreme situations (water bodies with a very high colloidal or particulate matter content), such calculations could provide a "worst case estimate" (the maximum concentration of a priority pollutant present in the water during the sampler deployment). If such estimates fall below the EQS limit, the water body would be classified by a "good" chemical status. When sampling trace concentrations of pollutants, the level of uncertainty in this approach will be lower than that associated with spot sampling techniques. However, the validity of this approach in compliance monitoring still needs to be demonstrated before it can be accepted by regulators, water quality managers, and other end users of the data.

3.6 FUTURE TRENDS

There are several future trends for the development of passive sampling techniques. The first is the development of devices that can be used to monitor "emerging" environmental pollutants. Recently, attention has shifted from hydrophobic persistent organic pollutants to compounds with a medium-to-high polarity, for example, polar pesticides, pharmaceuticals, and personal care products.[82,147,148] Novel materials will need to be tested as selective receiving phases (e.g., ionic liquids, molecularly imprinted polymers, and immunoadsorbents), together with membrane materials that permit the selective diffusion of these chemicals. The sample extraction and preconcentration methods used for these devices will need to be compatible with LC-MS analytical techniques.

The second trend is toward miniaturization of devices. Small devices are usually less expensive to use because of the lower costs of materials needed for their preparation and the reduced equipment requirements for their deployment. Lower volumes of solvents and reagents are consumed during their subsequent processing. Small samplers also offer the advantage of easy transportation to and from the sampling site. As miniaturized devices should not deplete the bulk matrix, they can be used in situations where space, volume, and the flow of water are limited, for example, in groundwater boreholes.[149] This trend can benefit from recent achievements in the area of solvent-free sample preparation techniques. Passive samplers based on *in situ* analyte preconcentration (e.g., SPME or stir-bar sorptive extraction) allow sample processing using thermal desorption-GC[22] or solvent microextraction followed by HPLC.[150] However, the practical application of SPME-based techniques in the *in situ* sampling of trace levels of aqueous contaminants still requires further enhancement in terms of robustness and sensitivity.

A major challenge in the future development of the technology is the calibration of devices to enable the quantitative assessment of the levels of pollutants present in water. A full evaluation of the various factors (e.g., water temperature, water turbulence, salinity, DOC level, pH, biofouling, and the presence of complex mixtures of contaminants) that may affect the performance of samplers is essential. Further steps are necessary for the reduction or control of the known impacts of environmental variables on sampler performance. For samplers where analytes are accumulated in the receiving phase by absorption mechanisms, PRCs have been successfully employed for improving the accuracy of the measurement of TWA concentrations of contaminants in the field. However, a better understanding of accumulation kinetics in samplers fitted with adsorbent-type receiving phases remains a challenge and requires further research.

Robust calibration data for passive sampling devices used in monitoring trace metals are essential for quantification of the various metal species and complexes found in different aquatic environments. This requires knowledge of the uptake kinetics of different metal moieties. Configuration of specific devices for monitoring well-defined fractions and species of metals will enhance their potential as regulatory tools and help to overcome some of the current limitations of spot sampling. The ability to model uptake parameters for passive samplers based on the physicochemical properties of the sampled compounds and their interactions with materials used in the construction of devices is also important. This may help to reduce the need for extensive laboratory-based calibration experiments. Further work is needed to model the performance of passive samplers when concentrations of pollutants fluctuate widely. This is important where a lag phase in the response of the sampler can affect the efficiency of the detection of a short pollution event. A better understanding of the factors determining the responsiveness of devices to changing water concentrations will help to improve sampler design in the future.

Development of biomimetic devices capable of simulating the accumulation of toxic chemicals in tissues of aquatic organisms will enable a reduction in the use of biomonitoring procedures in routine monitoring programs. It will also decrease the uncertainty associated with the data obtained, as this is based on highly variable samples of biological material. The combination of the deployment of passive samplers followed by the biological testing of sampler extracts with the aim of detecting and subsequently identifying toxicologically relevant compounds offers much potential. This approach can provide information concerning the relative toxicological significance of water-borne contaminants and hence help to improve risk assessments for different water bodies.

Finally, further development of QA/QC, method validation schemes, and standards for the use of passive sampling devices is urgently needed. Successful demonstration of the performance of passive samplers alongside conventional sampling schemes will help to facilitate the acceptance of passive sampling in routine regulatory monitoring programs in the future.

ACKNOWLEDGMENTS

We acknowledge the financial support of the European Commission (Contracts EVK1-CT-2002-00119; http://www.port.ac.uk/stamps/ and SSPI-CT-2003-502492; and http://www.swift-wfd.com/) and the Slovak Research and Development Agency (Contract SK-ZA-0006-07).

REFERENCES

1. Koester, C.J., S.L. Simonich, and B.K. Esser. 2003. Environmental analysis. *Anal. Chem.* 75: 2813–2829.
2. Koester, C.J. and A. Moulik. 2005. Trends in environmental analysis. *Anal. Chem.* 77: 3737–3754.
3. Pawliszyn, J. 2003. Sample preparation: Quo Vadis? *Anal. Chem.* 75: 2543–2558.
4. Richardson, S.D. and T.A. Ternes. 2005. Water analysis: Emerging contaminants and current issues. *Anal. Chem.* 77: 3807–3838.
5. Coquery, M., A. Morin, A. Becue, and B. Lepot. 2005. Priority substances of the European Water Framework Directive: Analytical challenges in monitoring water quality. *Trends Anal. Chem.* 24: 117–127.

6. Palmes, E.D. and A.F. Gunnison. 1973. Personal monitoring device for gaseous contaminants. *Am. Ind. Hyg. Assoc. J.* 34: 78–81.
7. Stuer-Lauridsen, F. 2005. Review of passive accumulation devices for monitoring organic micropollutants in the aquatic environment. *Environ. Pollut.* 136: 503–524.
8. Vrana, B., I.J. Allan, R. Greenwood, G.A. Mills, E. Dominiak, K. Svensson, J. Knutsson, and G. Morrison. 2005. Passive sampling techniques for monitoring pollutants in water. *Trends Anal. Chem.* 24: 845–868.
9. Ouyang, G. and J. Pawliszyn. 2007. Configurations and calibration methods for passive sampling techniques. *J. Chromatogr. A* 1168: 226–235.
10. Kot-Wasik, A., B. Zabiegała, M. Urbanowicz, E. Dominiak, A. Wasik, and J. Namieśnik. 2007. Advances in passive sampling in environmental studies. *Anal. Chim. Acta* 602: 141–163.
11. Esteve-Turrillas, F.A., A. Pastor, V. Yusa, and M. De La Guardia. 2007. Using semi-permeable membrane devices as passive samplers. *Trends Anal. Chem.* 26: 703–712.
12. Huckins, J.N., J.D. Petty, and K. Booij (eds). 2006. *Monitors of Organic Chemicals in the Environment: Semipermeable Membrane Devices.* New York: Springer.
13. Greenwood, R., G.A. Mills, and B. Vrana (eds). 2007. *Passive Sampling Techniques in Environmental Monitoring.* Amsterdam: Elsevier.
14. Gorecki, T. and J. Namiesnik. 2002. Passive sampling. *Trends Anal. Chem.* 21: 276–291.
15. Mayer, P., J. Tolls, L. Hermens, and D. Mackay. 2003. Equilibrium sampling devices. *Environ. Sci. Technol.* 37: 184A–191A.
16. Smedes, F. 2007. Monitoring of chlorinated biphenyls and polycyclic aromatic hydrocarbons by passive sampling in concert with deployed mussels. In: R. Greenwood, G.A. Mills, and B. Vrana (eds), *Passive Sampling Techniques in Environmental Monitoring*, pp. 407–448. Amsterdam: Elsevier.
17. Yates, K., I. Davies, L. Webster, P. Pollard, L. Lawton, and C. Moffat. 2007. Passive sampling: Partition coefficients for a silicone rubber reference phase. *J. Environ. Monit.* 9: 1116–1121.
18. Rusina, T.P., F. Smedes, J. Klanova, K. Booij, and I. Holoubek. 2007. Polymer selection for passive sampling: A comparison of critical properties. *Chemosphere* 68: 1344–1351.
19. Booij K., B. Vrana, and J. Huckins. 2007. Theory, modeling and calibration of passive samplers used in water monitoring. In: R. Greenwood, G.A. Mills, and B. Vrana (eds), *Comprehensive Analytical Chemistry*, pp. 141–169. Amsterdam: Elsevier.
20. Huckins, J.N., G.K. Manuweera, J.D. Petty, D. Mackay, and J.A. Lebo. 1993. Lipid-containing semipermeable membrane devices for monitoring organic contaminants in water. *Environ. Sci. Technol.* 27: 2489–2496.
21. Vrana, B., R. Greenwood, G. Mills, J. Knutsson, K. Svensson, and G. Morrison. 2005. Performance optimisation of a passive sampler for monitoring hydrophobic organic pollutants in water. *J. Environ. Monit.* 7: 612–620.
22. Ouyang, G., Y. Chen, and J. Pawliszyn. 2005. Time-weighted average water sampling with a solid-phase microextraction device. *Anal. Chem.* 77: 7319–7325.
23. Ouyang, G.F., W.N. Zhao, L. Bragg, Z.P. Qin, M. Alaee, and J. Pawliszyn. 2007. Time-weighted average water sampling in Lake Ontario with solid-phase microextraction passive samplers. *Environ. Sci. Technol.* 41: 4026–4031.
24. Ouyang, G., Z.N. Wennan, A. Mehran, and J. Pawliszyn. 2007. Time-weighted average water sampling with a diffusion-based solid-phase microextraction device. *J. Chromatogr. A* 1138: 42–46.
25. Vrana, B., P. Popp, A. Paschke, and G. Schüürmann. 2001. Membrane-enclosed sorptive coating. An integrative passive sampler for monitoring organic contaminants in water. *Anal. Chem.* 73: 5191–5200.
26. Kingston, J.K., R. Greenwood, G.A. Mills, G.M. Morrison, and B.L. Persson. 2000. Development of novel passive sampling system for the time-averaged measurement of a range of organic pollutants in aquatic environments. *J. Environ. Monit.* 2: 487–495.
27. Zhang, H. and W. Davison. 1995. Performance-characteristics of diffusion gradients in thin-films for the in-situ measurement of trace-metals in aqueous-solution. *Anal. Chem.* 67: 3391–3400.
28. Wennrich, L., B. Vrana, P. Popp, and W. Lorenz. 2003. Development of an integrative passive sampler for the monitoring of organic water pollutants. *J. Environ. Monit.* 5: 813–822.
29. Huckins, J.N., J.D. Petty, and K. Booij. 2006. Fundamentals of SPMD. In: J.N. Huckins, J.D. Petty, and K. Booij (eds), *Monitors of Organic Chemicals in the Environment: Semipermeable Membrane Devices*, pp. 29–41. New York: Springer.
30. Baltussen, E., P. Sandra, F. David, and C. Cramers. 1999. Stir bar sorptive extraction (SBSE), a novel extraction technique for aqueous samples: Theory and principles. *J. Microcolumn Sep.* 11: 737–747.
31. Pawliszyn, J. 2002. Sampling and sample preparation for field and laboratory. In: J. Pawliszyn (ed.), *Sampling and Sample Preparation for Field and Laboratory*, pp. 389–478. Amsterdam: Elsevier.

32. Vaes, H.J., C. Hamwijk, E.U. Ramos, H.J.M. Verhaar, and J.L.M. Hermens. 1996. Partitioning of organic chemicals to polyacrylate-coated solid phase. *Anal. Chem.* 68: 4458–4462.

33. Greenwood, R., G.A. Mills, B. Vrana, I. Allan, R. Aqilar-Martinez, and G. Morrison. 2007. Monitoring of priority pollutants in water using Chemcatcher passive sampling devices. In: R. Greenwood, G.A. Mills, and B. Vrana (eds), *Passive Sampling Techniques in Environmental Monitoring*, pp. 199–229. Amsterdam: Elsevier.

34. Södergren, A. 1987. Solvent-filled dialysis membranes simulate uptake of pollutants by aquatic organisms. *Environ. Sci. Technol.* 21: 855–859.

35. Alvarez, D.A., J.D., Petty, J.N. Huckins, et al. 2004. Development of a passive, in situ, integrative sampler for hydrophilic organic contaminants in aquatic environments. *Environ. Toxicol. Chem.* 23: 1640–1648.

36. Gale, R.W. 1998. Three-compartment model for contaminant accumulation by semipermeable membrane devices. *Environ. Sci. Technol.* 32: 2292–2300.

37. Chen Y. and J. Pawliszyn. 2007. Theory of solid phase microextraction and its application in passive sampling. In: R. Greenwood, G.A. Mills, and B. Vrana (eds), *Comprehensive Analytical Chemistry*, pp. 3–32. Amsterdam: Elsevier.

38. Chen, Y. and J. Pawliszyn. 2003. Time-weighted average passive sampling with a solid-phase micro-extraction device. *Anal. Chem.* 75: 2004–2010.

39. Sukola, K., J. Koziel, F. Augusto, and J. Pawliszyn. 2001. Diffusion-based calibration for SPME analysis of aqueous samples. *Anal. Chem.* 73: 13–18.

40. Stephens, S. and J.F. Müller. 2007. Techniques for quantitatively evaluating aquatic passive sampling devices. In: R. Greenwood, G.A. Mills, and B. Vrana (eds), *Passive Sampling Techniques in Environmental Monitoring*, pp. 329–349. Amsterdam: Elsevier.

41. Booij, K., H.M. Sleiderink, and F. Smedes. 1998. Calibrating the uptake kinetics of semipermeable membrane devices using exposure standards. *Environ. Toxicol. Chem.* 17: 1236–1245.

42. Huckins, J.N., J.D. Petty, J.A. Lebo, et al. 2002. Development of the permeability/performance reference compound (PRC) approach for in situ calibration of semipermeable membrane devices (SPMDs). *Environ. Sci. Technol.* 36: 85–91.

43. Vrana, B., G.A. Mills, E. Dominiak, and R. Greenwood. 2006. Calibration of the Chemcatcher passive sampler for the monitoring of priority organic pollutants in water. *Environ. Pollut.* 142: 333–343.

44. Vrana, B., A. Paschke, and P. Popp. 2006. Calibration and field performance of membrane-enclosed sorptive coating for integrative passive sampling of persistent organic pollutants in water. *Environ. Pollut.* 144: 296–307.

45. Alvarez, D.A., J.N. Huckins, J.D. Petty, T.L. Jones-Lepp, F. Stuer-Lauridsen, D.T. Getting, J.P. Goddard, and A. Gravell. 2007. Tool for monitoring hydrophilic contaminants in water: Polar Organic Chemical Integrative Sampler (POCIS). In: R. Greenwood, G.A. Mills, and B. Vrana (eds), *Passive Sampling Techniques in Environmental Monitoring*, pp. 171–197. Amsterdam: Elsevier.

46. Paschke, A., B. Vrana, P. Popp, L. Wennrich, H. Paschke, and G. Schüürmann. 2007. Membrane enclosed sorptive coating for the monitoring of organic compounds in water. In: R. Greenwood, G.A. Mills, and B. Vrana (eds), *Passive Sampling Techniques in Environmental Monitoring*, pp. 231–249. Amsterdam: Elsevier.

47. Richardson, B.J., P.K.S. Lam, G.J. Zheng, K.E. McCellan, and S.B. De Luca-Abbott. 2002. Biofouling confounds the uptake of trace organic contaminants by semi-permeable membrane devices. *Mar. Pollut. Bull.* 44: 1372–1379.

48. Tan, B.L.L., D.W. Hawker, J.F. Muller, F.D.L. Leusch, L.A. Tremblay, and H.F. Chapman. 2007. Comprehensive study of endocrine disrupting compounds using grab and passive sampling at selected wastewater treatment plants in South East Queensland, Australia. *Environ. Int.* 33: 654–669.

49. Alvarez, D.A., P.E. Stackelberg, J.D. Petty, et al. 2005. Comparison of a novel passive sampler to standard water-column sampling for organic contaminants associated with wastewater effluents entering a New Jersey stream. *Chemosphere* 61: 610–622.

50. Björklund Blom, L., G.M. Morrison, M.S. Rauch, G. Mills, and R. Greenwood. 2003. Metal diffusion properties of a Nafion-coated porous membrane in an aquatic passive sampler system. *J. Environ. Monit.* 5: 404–409.

51. Pichette, C., H. Zhang, W. Davison, and S. Sauve. 2007. Preventing biofilm development on devices using metals and antibiotics. *Talanta* 72: 716–722.

52. Ellis, G.S., J.N. Huckins, C.E. Rostad, C.J. Schmitt, J.D. Petty, and P. MacCarthy. 1995. Evaluation of lipid-containing semipermeable membrane devices (SPMDs) for monitoring organochlorine contaminants in the upper Mississippi River. *Environ. Toxicol. Chem.* 14: 1875–1884.

53. Seethapathy, S., T. Górecki, and X. Li. 2008. Passive sampling in environmental analysis. *J. Chromatogr. A* 1184: 234–253.

54. Martin, H., B.M. Patterson, and G.B. Davis. 2003. Field trial of contaminant groundwater monitoring: Comparing time-integrating ceramic dosimeters and conventional water sampling. *Environ. Sci. Technol.* 37: 1360–1364.

55. Persson, L.B., G.M. Morrison, J.U. Friemann, J. Kingston, G. Mills, and R. Greenwood. 2001. Diffusional behaviour of metals in a passive sampling system for monitoring aquatic pollution. *J. Environ. Monit.* 3: 639–645.

56. Blom, L.B., G.M. Morrison, J. Kingston, G.A. Mills, R. Greenwood, T.J.R. Pettersson, and S. Rauch. 2002. Performance of an in situ passive sampling system for metals in stormwater. *J. Environ. Monit.* 4: 258–262.

57. Davison, W. and H. Zhang. 2000. In situ measurement of labile species in water and sediments using DGT. *Ocean Sci. Technol.* (Chemical Sensors in Oceanography) 1: 283–302.

58. DiGiano, F.A., D. Elliot, and D. Leith. 1988. Application of passive dosimetry to the detection of trace organics. *Environ. Sci. Technol.* 22: 1365–1367.

59. http://www.alcontrol.com.

60. www.gaiasafe.de.

61. Blind, T., M. Gerschwitz, H. Gessler, H. Slama, H. Sorge, P. Göttzelmann, and M. Nallinger. 1994. Vergleichsuntersuchungen bei der Grundwasserproben-ahme: Passives Adsorptionsverfahren zur Erkundung organischer Kontaminationen. *Terra Tech.* 4: 26–28.

62. Booij, K., H.E. Hofmans, C.V. Fischer, and E.M. van Weerlee. 2003. Temperature-dependent uptake rates of nonpolar organic compounds by semipermeable membrane devices and low-density polyethylene membranes. *Environ. Sci. Technol.* 37: 361–366.

63. Carls, M.G., L.G. Holland, J.W. Short, R.A. Heintz, and S.D. Rice. 2004. Monitoring polynuclear aromatic hydrocarbons in aqueous environments with passive low-density polyethylene membrane devices. *Environ. Toxicol. Chem.* 23: 1416–1424.

64. Heringa, M.B. and J.L.M. Hermens. 2003. Measurement of free concentrations using negligible depletion-solid phase microextraction (nd-SPME). *Trends Anal. Chem.* 22: 575–587.

65. Lee, H.L. and J.K. Hardy. 1998. Passive sampling of monocyclic aromatic priority pollutants in water. *Int. J. Environ. Anal. Chem.* 72: 83–97.

66. Vroblesky, D.A. and W.T. Hyde. 1997. Diffusion samplers as an inexpensive approach to monitoring VOCs in ground water. *Ground Water Monit. Remediat.* 17: 177–184.

67. Brumbaugh, W.G., J.D. Petty, T.W. May, and J.N. Huckins. 2000. A passive integrative sampler for mercury vapour in air and neutral mercury species in water. *Chemosphere* 2: 1–9.

68. Slaveykova, V.I., N. Parthasarathy, J. Buffle, and K.J. Wilkinson. 2004. Permeation liquid membrane as a tool for monitoring bioavailable Pb in natural waters. *Sci. Total Environ.* 328: 55–68.

69. Litten, S., B. Mead, and J. Hassett. 1993. Application of passive samplers (PISCES) to locating a source of PCBs on the Black River, New York. *Environ. Toxicol. Chem.* 12: 639–647.

70. De Jonge, H. and G. Rothenberg. 2005. New device and method for flux-proportional sampling of mobile solutes in soil and groundwater. *Environ. Sci. Technol.* 39: 274–282.

71. Kot-Wasik, A. 2004. A new passive sampler as an alternative tool for monitoring of water pollutants. *Chem. Anal.* 49: 691–705.

72. Mackenzie, L., V. Beuzenberg, P. Holland, P. McNabb, and A. Selwood. 2004. Solid phase adsorption toxin tracking (Spatt): A new monitoring tool that simulates the biotoxin contamination of filter feeding bivalves. *Toxicon* 44: 901–918.

73. Brumbaugh, W.G., J.D. Petty, J.N. Huckins, and S.E. Manahan. 2002. Stabilized liquid membrane device (SLMD) for the passive, integrative sampling of labile metals in water. *Water Air Soil Pollut.* 133: 109–119.

74. Parthasarathy, N., J. Buffle, N. Gassama, and F. Cuenod. 1999. Speciation of trace metals in waters: Direct selective separation and preconcentration of free metal ion by supported liquid membrane. *Chem. Anal.* 44: 455–470.

75. Leblanc, C.J., W.M. Stallard, P.G. Green, and E.D. Schroeder. 2003. Passive sampling screening method using thin-layer chromatography plates. *Environ. Sci. Technol.* 37: 3966–3971.

76. Leonard, A.W., R.V. Hyne, and F. Pablo. 2002. Trimethylpentane-containing passive samplers for predicting time-integrated concentrations of pesticides in water. *Environ. Toxicol. Chem.* 21: 2591–2599.

77. De Smaele, T., L. Moens, P. Sandra, and R. Dams. 1999. Determination of organometallic compounds in surface water and sediment samples with SPME-CGC-ICPMS. *Mikrochim. Acta* 130: 241–251.

78. Namieśnik, J., B. Zabiegala, A. Kot-Wasik, M. Partyka, and A. Wasik. 2005. Passive sampling and/or extraction techniques in environmental analysis: A review. *Anal. Bioanal. Chem.* 381: 279–301.

79. http://wwwaux.cerc.cr.usgs.gov/spmd/SPMD_references.htm.
80. Lu, Y.B., Z.J. Wang, and J.N. Huckins. 2002. Review of the background and application of triolein-containing semipermeable membrane devices in aquatic environmental study. *Aquat. Toxicol.* 60: 139–153.
81. Petty, J.D., C.E. Orazio, J.N. Huckins, R.W. Gale, J.A. Lebo, J.C. Meadows, K.R. Echols, and W.L. Cranor. 2000. Considerations involved with the use of semipermeable membrane devices for monitoring environmental contaminants. *J. Chromatogr. A* 879: 83–95.
82. Mills, G.A., B. Vrana, I. Allan, D.A. Alvarez, J.N. Huckins, and R. Greenwood. 2007. Trends in monitoring pharmaceuticals and personal-care products in the aquatic environment by use of passive sampling devices. *Anal. Bioanal. Chem.* 387: 1153–1157.
83. Togola, A. and H. Budzinski. 2007. Development of polar organic integrative samplers for analysis of pharmaceuticals in aquatic systems. *Anal. Chem.* 79: 6734–6741.
84. MacLeod, S.L., E.L. McClure, and C.S. Wong. 2007. Laboratory calibration and field deployment of the polar organic chemical integrative sampler for pharmaceuticals and personal care products in wastewater and surface water. *Environ. Toxicol. Chem.* 26: 2517–2529.
85. Mazzella, N., J.-F. Dubernet, and F. Delmas. 2007. Determination of kinetic and equilibrium regimes in the operation of polar organic chemical integrative samplers application to the passive sampling of the polar herbicides in aquatic environments. *J. Chromatogr. A* 1154: 42–51.
86. Vroblesky, D.A. 2007. Passive diffusion samplers to monitor volatile organic compounds in groundwater. In: R. Greenwood, G.A. Mills, and B. Vrana (eds), *Passive Sampling Techniques in Environmental Monitoring*, pp. 295–309. Amsterdam: Elsevier.
87. Vroblesky, D.A. and T.R. Campbell. 2001. Equilibration times, compound selectivity, and stability of diffusion samplers for collection of ground-water VOC concentrations. *Adv. Environ. Res.* 5: 1–12.
88. Divine, C.E. and J.E. McCray. 2004. Estimation of membrane diffusion coefficients and equilibration times for low-density polyethylene passive diffusion samplers. *Environ. Sci. Technol.* 38: 1849–1857.
89. Ehlke, T.A., T.E. Imbrigiotta, and J.M. Dale. 2004. Laboratory comparison of polyethylene and dialysis membrane diffusion samplers. *Ground Water Monit. Remediat.* 24: 53–59.
90. Harrington, G.A., P.G. Cook, and N.I. Robinson. 2000. Equilibration times of gas-filled diffusion samplers in slow-moving ground water systems. *Ground Water Monit. Remediat.* 20: 60–65.
91. http://diffusionsampler.itrcweb.org/.
92. Martin, H., M. Piepenbrink, and P. Grathwohl. 2001. Ceramic dosimeters for time-integrated contaminant monitoring. *Process Anal. Chem.* 6: 68–73.
93. Følsvik, N., E.M. Brevik, and J.A. Berge. 2000. Monitoring of organotin compounds in seawater using semipermeable membrane devices. *J. Environ. Monit.* 2: 281–284.
94. Følsvik, N., E.M. Brevik, and J.A. Berge. 2002. Organotin compounds in a Norwegian fjord. A comparison of concentration levels in semipermeable membrane devices (SPMDs), blue mussels (*Mytilus edulis*) and water samples. *J. Environ. Monit.* 4: 280–283.
95. Harman, C., O. Bøyum, K.E. Tollefsen, K. Thomas, and M. Grung. 2008. Uptake of some selected aquatic pollutants in semipermeable membrane devices (SPMDs) and the polar organic chemical integrative sampler (POCIS). *J. Environ. Monit.* 10: 239–247.
96. Aguilar-Martínez, R., R. Greenwood, G.A. Mills, B. Vrana, M.A. Palacios-Corvillo, and M.M. Gómez-Gómez. 2008. Assessment of Chemcatcher passive sampler for the monitoring of inorganic mercury and organotin compounds in water. *Int. J. Environ. Anal. Chem.* 88: 75–90.
97. Warnken, K.W., H. Zhang, and W. Davison. 2007. In situ monitoring and dynamic speciation measurement in solution using DGT. In: R. Greenwood, G.A. Mills, and B. Vrana (eds), *Passive Sampling Techniques in Environmental Monitoring*, pp. 251–278. Amsterdam: Elsevier.
98. Buffle, J. and M.-L. Tercier-Waeber. 2005. Voltammetric environmental trace-metal analysis and speciation: From laboratory to in situ measurements. *Trends Anal. Chem.* 24: 172–191.
99. Zhang, H. and W. Davison. 2000. Direct in situ measurements of labile inorganic and organically bound metal species in synthetic solutions and natural waters using diffusive gradients in thin films. *Anal. Chem.* 72: 4447–4457.
100. Zhang, H. and W. Davison. 2001. In situ speciation measurements. Using diffusive gradients in thin films (DGT) to determine inorganically and organically complexed metals. *Pure Appl. Chem.* 73: 9–15.
101. Scally, S., W. Davison, and H. Zhang. 2003. In situ measurements of dissociation kinetics and labilities of metal complexes in solution using DGT. *Environ. Sci. Technol.* 37: 1379–1384.
102. Scally, S., H. Zhang, and W. Davison. 2004. Measurements of lead complexation with organic ligands using DGT. *Aust. J. Chem.* 57: 925–930.
103. Li, W., H. Zhao, P.R. Teasdale, and F. Wang. 2005. Trace metal speciation measurements in waters by the liquid binding phase DGT device. *Talanta* 67: 571–578.

104. Garmo, O.A., O., Royset, E., Steinnes, and T.P. Flaten. 2003. Performance study of diffusive gradients in thin films for 55 elements. *Anal. Chem.* 75: 3573–3580.
105. Li, W., H. Zhao, P.R. Teasdale, R. John, and F. Wang. 2005. Metal speciation measurement by diffusive gradients in thin films technique with different binding phases. *Anal. Chim. Acta* 533: 193–202.
106. Fux, E., C. Marcaillou, F. Mondeguer, R. Bire, and P. Hess. 2008. Field and mesocosm trials on passive sampling for the study of adsorption and desorption behaviour of lipophilic toxins with a focus on OA and DTX1. *Harmful Algae* 7: 574–583.
107. Kohoutek, J., P. Babica, L. Blaha, and B. Marsalek. 2008. A novel approach for monitoring of cyanobacterial toxins: Development and evaluation of the passive sampler for microcystins. *Anal. Bioanal. Chem.* 390: 1167–1172.
108. Sabaliunas, D., J.R. Lazutka, and I. Sabaliuniene. 2000. Acute toxicity and genotoxicity of aquatic hydrophobic pollutants sampled with semipermeable membrane devices. *Environ. Pollut.* 109: 251–265.
109. Johnson, B.T., J.N. Huckins, J.D. Petty, and R.C. Clark. 2000. Collection and detection of lipophilic chemical contaminants in water, sediment, soil, and air—SPMD-TOX. *Environ. Toxicol.* 15: 248–252.
110. Johnson, B.T., J.D. Petty, J.N. Huckins, K. Lee, and J. Gauthier. 2004. Hazard assessment of a simulated oil spill on intertidal areas of the St. Lawrence River with SPMD-TOX. *Environ. Toxicol.* 19: 329–335.
111. Vermeirssen, E.L.M., O. Korner, R. Schonenberger, M.J.F. Sutter, and P. Burkhardt-Holm. 2005. Characterization of environmental estrogens in river water using a three pronged approach: Active and passive water sampling and the analysis of accumulated estrogens in the bile of caged fish. *Environ. Sci. Technol.* 39: 8191–8198.
112. Petty, J.D., J.N. Huckins, and D.A. Alvarez, et al. 2004. A holistic passive integrative sampling approach for assessing the presence and potential impacts of waterborne environmental contaminants. *Chemosphere* 54: 695–705.
113. Rastall, A.C., D. Getting, J. Goddard, D.R. Roberts, and L. Erdinger. 2006. A biomimetic approach to the detection and identification of estrogen receptor agonists in surface waters using semipermeable membrane devices (SPMDs) and bioassay-directed chemical analysis. *Environ. Sci. Pollut. Res.* 13: 256–267.
114. Schirmer, K., S. Bopp, and J. Gerhardt. 2007. Use of passive sampling devices in toxicity assessment of groundwater. In: R. Greenwood, G.A. Mills, and B. Vrana (eds), *Passive Sampling Techniques in Environmental Monitoring*, pp. 393–405. Amsterdam: Elsevier.
115. Bopp, S.K., M.S. Mclachlan, and K. Schirmer. 2007. Passive sampler for combined chemical and toxicological long-term monitoring of groundwater: The ceramic taximeter. *Environ. Sci. Technol.* 41: 6868–6876.
116. ISO 5667-14. 1998. Water quality—sampling. Part 14: Guidance on quality assurance of environmental water-sampling and handling.
117. British Standards Institute (BSI). Publicly available specification: Determination of priority pollutants in surface water using passive sampling (PAS-61), May 2006.
118. DeVita, W.M. and R.L. Crunkilton. 1998. Quality control associated with use of semipermeable membrane devices. *Environ. Toxicol. Risk Assess.* 1333: 237–245.
119. www.passivesampling.net/.
120. Sigg, L., F. Black, J. Buffle, et al. 2006. Comparison of analytical techniques for dynamic trace metal speciation in natural freshwaters. *Environ. Sci. Technol.* 40: 1934–1941.
121. Burkhard, L.P. 2000. Estimating dissolved organic carbon partition coefficients for nonionic organic chemicals. *Environ. Sci. Technol.* 34: 4663–4667.
122. Ellis, G.S., J.N. Huckins, C.E. Rostad, C.J. Schmitt, J.D. Petty, and P. MacCarthy. 1995. Evaluation of lipid-containing semipermeable membrane devices (SPMDs) for monitoring organochlorine contaminants in the upper Mississippi River. *Environ. Toxicol. Chem.* 14: 1875–1884.
123. Rantalainen, A.L., M.G. Ikonomou, and I.H. Rogers. 1998. Lipid-containing semipermeable-membrane devices (SPMDs) as concentrators of toxic chemicals in the lower Fraser River, British Columbia. *Chemosphere* 37: 1119–1138.
124. Zeng, E.Y., D. Tsukada, and D.W. Diehl. 2004. Development of solid-phase microextraction-based method for sampling of persistent chlorinated hydrocarbons in an urbanized coastal environment. *Environ. Sci. Technol.* 38: 5737–5743.
125. Axelman, J., K. Naes, C. Näf, and D. Broman. 1999. Accumulation of polycyclic aromatic hydrocarbons in semipermeable membrane devices and caged mussels (*Mytilus edulis* L.) in relation to water column phase distribution. *Environ. Toxicol. Chem.* 18: 2454–2461.

126. Vrana, B., G.A. Mills, M. Kotterman, P. Leonards, K. Booij, and R. Greenwood. 2007. Modelling and field application of the Chemcatcher passive sampler calibration data for the monitoring of hydrophobic organic pollutants in water. *Environ. Pollut.* 145: 895–904.

127. Schäfer, R.B., A. Paschke, B. Vrana, R. Mueller, and M. Liess. 2008. Performance of the Chemcatcher® passive sampler when used to monitor 10 polar and 3 semi-polar pesticides in 16 Central European streams, and comparison with two other sampling methods. *Water Res.* 42: 2707–2717.

128. Bopp, S., H. Weiss, and K. Schirmer. 2005. Time-integrated monitoring of polycyclic aromatic hydro-carbons (PAHs) in groundwater using the ceramic dosimeter passive sampling device. *J. Chromatogr. A* 1072: 137–147.

129. Luellen, D.R. and D. Shea. 2002. Calibration and field verification of semipermeable membrane devices for measuring polycyclic aromatic hydrocarbons in water. *Environ. Sci. Technol.* 36: 1791–1797.

130. Vrana, B., A. Paschke, P. Popp, and G. Schüürmann. 2001. Use of semipermeable membrane devices (SPMDs): Determination of bioavailable organic contaminants in the industrial region of Bitterfeld, Saxony-Anhalt, Germany. *Environ. Sci. Pollut. Res.* 8: 27–34. 131. Ouyang, G., W. Zhao, L. Bragg, Z. Qin, M. Alaee, and J. Pawliszyn. 2007. Time-weighted average water sampling in Lake Ontario with solid-phase microextraction passive samplers. *Environ. Sci. Technol.* 41: 4026–4031.

132. Hyne, R.V., F. Pablo, M. Aistrope, A.W. Leonard, and N. Ahmad. 2004. Comparison of time-integrated pesticide concentrations determined from field-deployed passive samplers with daily river-water extrac-tions. *Environ. Toxicol. Chem.* 23: 2090–2098.

133. Alfaro-De la Torre, M.C., P.Y. Beaulieu, and A. Tessier. 2000. In situ measurement of trace metals in lake water using the dialysis and DGT techniques. *Anal. Chim. Acta* 418: 53–68.

134. Gimpel, J., H. Zhang, W. Davison, and A.C. Edwards. 2003. In situ trace metal speciation in lake surface waters using DGT, dialysis and filtration. *Environ. Sci. Technol.* 37: 138–146.

135. Zhang, H. 2004. In-situ speciation of Ni and Zn in freshwaters: Comparison between DGT measurements and speciation models. *Environ. Sci. Technol.* 38: 1421–1427.

136. Meylan, S., N. Odzak, R. Behra, and L. Sigg. 2004. Speciation of copper and zinc in natural freshwater: Comparison of voltammetric measurements, diffusive gradients in thin films (DGT) and chemical equi-librium models. *Anal. Chim. Acta* 510: 91–100.

137. Allan, I.J., J. Knutsson, N. Guigues, G.A. Mills, A.-M. Fouillac, and R. Greenwood. 2007. Evaluation of the Chemcatcher and DGT passive samplers for monitoring metals with highly fluctuating water concen-trations. *J. Environ. Monit.* 9: 672–681.

138. Dunn, R.J.K., P.R. Teasdale, J. Warnken, and J.M. Arthur. 2007. Evaluation of the in situ, time-integrated DGT technique by monitoring changes in heavy metal concentrations in estuarine waters. *Environ. Pollut.* 148: 213–220.

139. Unsworth, E.R., K.W. Warnken, and H. Zhang, W. Davison, F. Black, J. Buffle, J. Cao, et al. 2006. Model predictions of metal speciation in freshwaters compared to measurements by in situ techniques. *Environ. Sci. Technol.* 40: 1942–1949.

140. Directive 2000/60/EC of the European Parliament and of the Council of October 23, 2000, establishing a framework for Community action in the field of water policy. *Off. J. Eur. Comm.* L327: 1.

141. Allan, I.J., B. Vrana, R. Greenwood, G.A. Mills, J. Knutsson, A. Holmberg, N. Guigues, A.M. Fouillac, and S. Laschi. 2006. Strategic monitoring for the European Water Framework Directive. *Trends Anal. Chem.* 25: 704–715.

142. Coquery, M., A. Morin, A. Becue, and B. Lepot. 2005. Priority substances of the European Water Framework Directive: Analytical challenges in monitoring water quality. *Trends Anal. Chem.* 24: 117–127.

143. Quevauviller, P., U. Borchers, and B.M. Gawlik. 2007. Coordinating links among research, standardisa-tion and policy in support of water framework directive chemical monitoring requirements. *J. Environ. Monit.* 9: 915–923.

144. WFD chemical monitoring guidance for surface water. Available at http://circa.europa.eu/Public/irc/env/wfd/library?l=/framework_directive/chemical_monitoring/technical_2007pdf/_EN_1.0_&a=d.

145. Allan, I.J., B. Vrana, R. Greenwood, G.A. Mills, B. Roig, and C.A. Gonzalez. 2006. "Toolbox" for bio-logical and chemical monitoring requirements for the European Union's Water Framework Directive. *Talanta* 69: 302–322.

146. Commission of the European Communities. Commission directive laying down, pursuant to Directive 2000/60/EC of the European Parliament and of the Council, technical specifications for chemical analy-sis and monitoring of water status. Draft version of March 7, 2008.

147. Alvarez, D.A., J.D. Petty, J.N. Huckins, T.L. Jones-Leep, D.T. Getting, J.P. Goddard, and S.E. Manahan. 2004. Development of a passive, in situ, integrative sampler for hydrophilic organic contaminants in aquatic environments. *Environ. Toxicol. Chem.* 23: 1640–1648.

148. Alvarez, D.A., P.E. Stackelberg, J.D. Petty, et al. 2005. Comparison of a novel passive sampler to standard water-column sampling for organic contaminants associated with wastewater effluents entering a New Jersey stream. *Chemosphere* 61: 610–622.

149. Richardson, S.D. and T.A. Ternes. 2005. Water analysis: Emerging contaminants and current issues. *Anal. Chem.* 77: 3807–3838.

150. Popp, P., C. Bauer, M. Moder, and A. Paschke. 2000. Determination of polycyclic aromatic hydrocarbons in waste water by off-line coupling of solid-phase microextraction with column liquid chromatography. *J. Chromatogr. A* 897: 153–159.

4 Modern Techniques of Analyte Extraction

Thaer Barri and Jan-Åke Jönsson

CONTENTS

4.1 Introduction 70
4.2 Modern Miniaturized Nonmembrane-Based Extraction Techniques 70
    4.2.1 Miniaturized Liquid-Phase Extraction Techniques 70
        4.2.1.1 Background 71
        4.2.1.2 Dynamic Liquid-Phase Microextraction 71
        4.2.1.3 Continuous-Flow Microextraction 71
        4.2.1.4 Dispersive Liquid–Liquid Microextraction 72
    4.2.2 Miniaturized SPE Techniques 72
        4.2.2.1 Fiber-in-Tube SPE 72
        4.2.2.2 Microextraction in a Packed Syringe 73
        4.2.2.3 Inside-Needle SPE 73
    4.2.3 Solid-Phase Microextraction 73
        4.2.3.1 Calibration in SPME 74
        4.2.3.2 Recent Trends in SPME Applications 74
    4.2.4 Stir Bar Sorptive Extraction 75
        4.2.4.1 SBSE Development 75
        4.2.4.2 SBSE Modes of Operation 75
4.3 Analytical Techniques Based on Nonporous Polymeric Membranes 75
    4.3.1 Membrane Inlet (Introduction) Mass Spectrometry 76
    4.3.2 Membrane Extraction with Sorbent Interface 76
    4.3.3 Membrane-Assisted Solvent Extraction 78
4.4 Analytical Techniques Based on the Use of Liquid Membranes 79
    4.4.1 Supported Liquid Membrane (SLM) Extraction 79
        4.4.1.1 SLM Extraction Principle 79
        4.4.1.2 SLM Unit and System Configurations 79
        4.4.1.3 Transport Mechanisms in SLM 81
        4.4.1.4 Selectivity in SLM 83
        4.4.1.5 Equilibrium Sampling Through SLM 83
    4.4.2 Microporous Membrane Liquid–Liquid Extraction 84
        4.4.2.1 MMLLE Principle 84
        4.4.2.2 Miniaturization, Automation, and Hyphenation of MMLLE 84
    4.4.3 Two-Phase HF-LPME 87
        4.4.3.1 Development of Two-Phase HF-LPME 87
        4.4.3.2 Automation of Two-Phase HF-LPME 88
4.5 Summary and Future Outlook 88
Acronyms and Abbreviations 89
References 90

4.1 INTRODUCTION

The Rio Declaration on Environment and Development in 1992 emphasized the sustainable development of human life. The output of the Rio summit clearly outlined and recommended several principles that all countries are advised to recognize. A couple of these principles stressed the need for the reduction and elimination of unsustainable patterns of production and consumption of substances that cause severe environmental degradation or are found to be harmful to human health.[1] The role of chemistry in human sustainability, bearing in mind the global demand for environmentally benign chemical processes and products, stems from providing unique and cost-effective chemical procedures for pollution prevention. From this role, the concept of green chemistry has emerged and has been touching upon all aspects of chemical processes, starting from simple environmentally friendly chemical tests and ending with biofuel production scale-up processes. The green aspects of a chemical protocol stand upon 12 principles that aim at reducing or eliminating the use or generation of hazardous substances in the design, manufacture, analysis, and applications of chemical products.[2] The discipline "green analytical chemistry" is concerned with the elimination of solvents in chemical processes or the replacement of hazardous solvents with environmentally friendly ones. The green analytical procedures used nowadays in modern analyte extraction techniques are normally designed by system miniaturization,[3] which in turn promotes procedure automation and hyphenation to the separation techniques.[4] These modern sample preparation procedures utilize different formats of mainly a miniaturized liquid phase, solid phase, sorptive phase, a synthetic polymeric nonporous membrane phase, or a liquid membrane phase, that is, a liquid supported in a porous membrane material.

4.2 MODERN MINIATURIZED NONMEMBRANE-BASED EXTRACTION TECHNIQUES

The huge leap in the development of many dedicated analytical instruments has put the ball in the analytical chemists' court, as sample preparation is still considered the bottleneck of sample extraction and analysis.

The primary aim of sample preparation is the cleanup and concentration of the analytes of interest, rendering them in a form that is compatible with the analytical instrument. In this regard, the shortcomings of liquid–liquid extraction (LLE) are well known (such as the consumption of huge amounts of potentially toxic organic solvents), and have been the motivation toward developing new, alternative procedures. Solid-phase extraction (SPE) employing an adsorptive solid extractive phase in the form of a cartridge or a disk has, on the other hand, significantly reduced the amount of solvents utilized compared with LLE. Therefore, SPE has been shown to be a better choice in many extraction situations and has gained popularity in spite of its high-price products. However, SPE still generally consumes significant volumes of solvent and an extra concentration step of the extract is usually needed to bring the volume down. SPE can be automated, but the additional complexity in design and the high cost make the technique difficult to accept; indeed, in many cases, the technique is even unaffordable.

As a corollary to this, more direct sample preparation procedures have been the pursuit of many scientists, who believe that miniaturization of analytical techniques can be a key solution to many of the unwanted drawbacks of LLE and SPE. Currently, several miniaturized extraction systems have been investigated, which are based primarily on utilizing downsized liquid, solid, or membrane extraction phases.

4.2.1 MINIATURIZED LIQUID-PHASE EXTRACTION TECHNIQUES

Trends in miniaturization of liquid-phase extraction procedures have been spawned and given many different names; they refer to the same principle but with different, or sometimes the same, configurations. For example, the terms "solvent microextraction," "liquid-phase microextraction," "single-drop microextraction," "liquid–liquid microextraction," and even "liquid–liquid–liquid microextraction" are not uncommon in the liquid-based sample preparation literature. The principle of these techniques first stemmed from the pioneering paper "Analytical Chemistry in a Drop. Solvent Extraction in a Microdrop" published by Liu and Dasgupta in 1996 as an accelerated article in *Analytical Chemistry*. The authors skillfully described a drop-in-drop microextraction system in which an organic microdrop (1.3 µL) was suspended inside a flowing aqueous drop containing the analyte to be extracted.[5]

4.2.1.1   Background

Dasgupta's paper inspired researchers to design new analytical techniques. For instance, Jeannot and Cantwell employed a Teflon rod as a support for hanging a small drop of *n*-octane (8 µL). The probe (drop-attached rod) was immersed in a stirred aqueous sample for extraction. The probe was then withdrawn and the organic phase sucked out by a microsyringe needle that was afterwards used for gas chromatography (GC) injection. The technique was called solvent microextraction (SME), which, in that work, was thoroughly elaborated with regard to its kinetics, giving estimates of mass transfer and diffusion coefficients.[6]

The work described by Jeannot and Cantwell implied utilizing two different apparatuses for two different steps; a Teflon rod supporting the drop (for extraction) and a microsyringe (for injection of the extract into the analytical instrument). The work was further developed by incorporating only a microsyringe for extraction (as solvent drop holder) and extract injection (as sample injector) into the analytical instrument.[7]

To enhance extraction efficiency in the SME system (also called single-drop microextraction [SDME]), He and Lee developed a procedure termed liquid-phase microextraction (LPME),[8,9] which, in its static mode, resembles the SME system.

4.2.1.2   Dynamic Liquid-Phase Microextraction

LPME can also be run in a dynamic mode in which a specified volume of an aqueous sample containing the analytes is repeatedly withdrawn into and expelled from a microsyringe barrel pre-loaded with a microliter-volume organic solvent. The process is performed several times within a short period.[8–10] The microsyringe here functions as a microseparatory apparatus for extraction and as an injector into a GC. Therefore, after a preset extraction time, the sample is expelled from the syringe barrel and the organic solvent is injected into the GC. As compared to the static mode, dynamic liquid-phase microextraction (DLPME) demonstrated that after each sample withdrawal by the microsyringe plunger, a thin solvent film is formed on the interior surface of the syringe barrel. In addition to the considerable agitation of the two phases, facilitated mass transfer of analyte from the sample is enhanced. This system, for example, thus provided better enrichment (27-fold) than the static mode (12-fold) for two chlorobenzenes.[8] When a programmable syringe pump was employed for the repetitive movement of the microsyringe plunger, the DLPME gave higher enrichment factors (60–280) than static LPME (60–180) for six polycyclic aromatic hydrocarbons (PAHs) within nearly the same extraction time ($\approx$20 min).[9]

4.2.1.3   Continuous-Flow Microextraction

Another promising aspect of SME development has been directed toward achieving the continuous flow of an aqueous sample phase to a seemingly immobilized organic solvent drop, permitting the direct and continuous interaction of the drop with a fresh sample. The technique was called continuous-flow microextraction (CFME).[11,12] In CFME, a polyetheretherketone (PEEK) tube is used as a holder for both the extraction solvent drop and the sample delivery supply. The drop (3 µL) is introduced into the PEEK tube via a valve and is pushed by the sample aliquot inside the tube until it reaches the tube outlet, where it remains (a little off-center of the tube) as a solvent drop. Then, the sample volume (typically 3 mL) is pumped continuously (from the PEEK tube) around the drop at a slow flow rate (50 μL min⁻¹). This allows a fresh sample to interact continuously with the solvent drop; this is how the extraction proceeds. This extraction format has resulted in a highly sensitive performance with enrichment factors, for instance, between 260 and 1600 for nitroaromatic compounds and chlorobenzenes, which were detected at subfemtogram-per-milliliter levels in environmental waters.[11]

Few review articles have been published on microextraction procedures based on the use of a liquid-phase extractant.[13,14] One drawback of drop-based microextraction procedures is drop vulnerability; this relates to its instability and potential dislodgement, which could be caused by sample complexity, a long extraction time, and a fast stirring speed. As a result, precision will often suffer significantly.

4.2.1.4 Dispersive Liquid–Liquid Microextraction

Quite recently, a new concept of analyte extraction has been introduced. The principle of this procedure relies on the formation of a ternary-component liquid-phase system consisting of a dispersed extraction solvent, a dispersive solvent, and the aqueous sample. The dispersive solvent (acetone, acetonitrile, or methanol) has to be miscible with both the extraction solvent (carbon disulfide, chlorobenzene, or carbon tetrachloride) and the aqueous sample. Another important feature of this procedure is that the density of the extraction solvent must be higher than that of the aqueous sample. In addition, analyte partitioning to the extraction solvent should be high enough to achieve high analyte enrichment and recovery. In practice, a few microliters of an extraction solvent are mixed with an appropriate amount of a dispersive solvent, and the mixture is introduced rapidly to a small sample volume (5 mL) in a test tube with a conical bottom. This causes finely dispersed extraction solvent droplets to be formed, which typically results in an immensely large contact surface area between the analyte molecules and the microdroplets of the extraction solvent.[15] After a predetermined extraction time, the mixture is centrifuged, which causes the extraction solvent to fall to the bottom of the test tube. This results in a rapid extraction, on a time scale of seconds to a couple of minutes; indeed, such extraction is considered time-independent, which is one of its most important characteristics. Apart from its rapidity, the technique is easy to apply, with minimal consumption of organic solvents and very high enrichment and recovery of analyte. Dispersive liquid–liquid microextraction (DLLME) has been employed for the extraction of different classes of organic pollutants[16–18] as well as metal ions[19,20] in relatively clean surface water samples. A potential limitation of DLLME, however, is that the procedure cannot be used for the extraction of complex samples, such as leachate or wastewater samples.

4.2.2 Miniaturized SPE Techniques

Further miniaturization of the SPE technique permits a reduction in the amount of organic solvent used, on-line coupling to analytical instruments, fast analysis times and excellent sensitivity. Downsizing of SPE has been focused mainly on the use of fibers, beads, and adsorbents as extraction phases that are reproducibly packed in tubes, capillaries, syringes, needles, and even micropipette tips.

4.2.2.1 Fiber-in-Tube SPE

In 2002, Jinno and coworkers developed a fiber-in-tube SPE (FIT-SPE) configuration in which polymeric fibers packed in a piece of PEEK tubing a few centimeters long were utilized. The tubing was then inserted between a sample valve, connected to the sample and desorption solvent syringe pumps, and a liquid chromatography (LC) injection valve for on-line SPE and micro-LC analysis. A much smaller, grain-of-rice size PEEK extraction tube (0.5 cm) was also incorporated in the sample rotor of a microinjector valve for on-line SPE and LC analysis.[21] In FIT-SPE, the sample is pumped by the sample pump through the extraction tube at a slow flow rate (8–32 μL min⁻¹). The desorption solvent (for instance, methanol) is also delivered by the solvent pump after the positions of the switching and injection valves have been changed.

The technique was further improved by employing a polymer coating on the polymeric fibers packed in a fused silica capillary. The coating material was based on GC stationary phases. The polymer-coated fiber-packed capillary was used as the sample loop of the LC injection valve for the extraction of phthalate esters from river water and wastewater.[22] The coated-fiber extraction capillaries demonstrated a better extraction efficiency and lower limit of quantification (LOQ) than the uncoated-fiber capillaries. Also, the coated fibers were similarly packed in a PEEK tube, which was used as the injection loop or integrated in the rotor of an LC injection valve employed for the extraction of phthalates. The results clearly showed that an extraction with high selectivity could be established with an appropriate type of polymer coating.[23]

Compared with conventional particle-packed SPE cartridges, FIT-SPE provides an increased surface area for the extraction medium and a reduced pressure drop during extraction and desorption. Also, the undesirable plugging effect from insoluble materials in real samples can be very much diminished. Utilization of FIT-SPE has been discussed in a few review articles concerning the on-line coupling of miniaturized SPE to microcolumn liquid-phase separation techniques.[24,25]

4.2.2.2 Microextraction in a Packed Syringe

Microextraction in a packed syringe (MEPS) is a new technique of miniaturized SPE that is fully automated and can be connected on-line to high-performance liquid chromatography (HPLC) or GC without any modifications. This technique involves a minute quantity (1 mg) of solid packing as adsorption material that is inserted into a gas-tight syringe barrel (200–250 μL) as a plug. Usually, a small volume (10–1000 μL) of sample is withdrawn into and expelled from the syringe several times by an autosampler. After the sample has passed through the solid sorbent, the solid phase is washed with water and the analytes desorbed by an organic solvent, such as methanol or HPLC mobile phase (20–50 μL), directly into the instrument injector.

MEPS has so far been applied mainly to the analysis of drugs in biological samples; only one application for the extraction of PAHs in water has been published.[26] One of the major advantages of the MEPS design is that the packed syringe can be used many times over, for example, more than 400 times for water samples. Moreover, the technique permits a fast handling time in the analysis of PAHs in water, the speed enhancement being 15 and 100 times compared to the literature procedures of solid-phase microextraction (SPME) and stir bar sorptive extraction (SBSE), respectively; see Sections 4.2.3 and 4.2.4.

4.2.2.3 Inside-Needle SPE

The research on extraction in packed needles promptly fostered the development of a technique called in-tube extraction (ITEX). The automated ITEX device, utilized for GC analysis and developed by CTC Analytics AG (Zwingen, Switzerland), is based on an adsorbent-packed needle attached to a syringe-like robotic autosampler. Another automated device based on inside-needle SPE was designed and has been marketed commercially by Chromtech (Idstein, Germany) as "solid-phase dynamic extraction (SPDE)."[27] In SPDE, 7 or 50 μL of PDMS is coated on the inner surface of a few-centimeters-long needle. The coated needle is attached to a 2.5-mL gas-tight syringe and the whole setup utilized for direct immersion or headspace sampling. The sample is aspirated into and dispensed from the needle as many times as needed, depending on whether single or multiple SPDE is required. Afterwards, the solutes are thermally desorbed into GC. It is important to bear in mind that, as liquid PDMS is used in SPDE, the extraction mechanism here differs from the SPE mechanism and relates more to SPME and SBSE, as discussed in Sections 4.2.3 and 4.2.4.

4.2.3 SOLID-PHASE MICROEXTRACTION

SPME was developed by Arthur and Pawliszyn in 1990 as a viable alternative to LLE and SPE techniques that are labor-intensive and solvent demanding, although SPE requires significantly smaller quantities of solvents than LLE. The SPME device, commercially marketed by Supelco, consists of a fused silica fiber that is coated with a thin layer of liquid (usually PDMS) or solid polymeric material constituting the extraction phase. The coated fiber is incorporated into a syringe-like design, facilitating and integrating direct sampling, extraction, enrichment, cleanup, and analyte introduction into the analytical instrument in one step and one device. A discussion of SPME theory is beyond the scope of this chapter; the principles have been amply covered elsewhere.[28]

During the last 20 years or so, the SPME technique has probably reached the culmination of its development in terms of mode of operation, automation, miniaturization and interfacing to other instruments, innovation of new coating materials, calibration procedures, and fields of application. As a result of these developments, SPME has become the currently most commonly used microextraction technique in field and laboratory experiments of a multidisciplinary nature.[29] Accordingly, the following sections will merely record progress in SPME from different perspectives.

4.2.3.1 Calibration in SPME

One of the important aspects of any analytical method is its calibration and, therefore, much effort has been put into SPME calibration. As it is not always practicable to employ traditional calibration methods (external standards, internal standards, and standard addition) owing to the sometimes significant matrix effects in complex samples, equilibrium calibration has been suggested as an alternative. In SPME, however, it would normally take rather a long time to achieve equilibrium calibration. If sensitivity were not a concern in an analysis, reduction of extraction time would be desirable, that is, the extraction could be stopped before equilibrium; but this would thus demand a new approach to calibration. In this regard, as a way of circumventing matrix effects in environmental analysis, several diffusion-based calibration methods have been recently developed for quantification in SPME.[30]

One of these methods is called kinetic calibration, in which analyte absorption from the sample to the liquid coating (PDMS) on the fiber is related to analyte desorption from the coating to the sample. The isotropy of absorption and desorption in the kinetic calibration has been described by Chen et al.[31] In kinetic calibration, also called in-fiber standardization, desorption of a radio-labeled standard (preloaded on the fiber coating) into the sample is used to calibrate the extraction (absorption/adsorption in the case of a liquid/solid coating) of analyte from the sample into the fiber. This calibration approach considerably facilitates the use of SPME for the on-site field sampling of water, where the control of flow velocity or addition of a standard to the matrix is very difficult.

The new in-fiber standardization method has been applied for time-weighted average sampling of PAHs in Hamilton Harbor, Canada,[32] and the sampling of carbofuran and carbaryl in river water.[33] Calibration in SPME was the topic of two recent review papers dedicated to environmental analysis and passive sampling.[30,34] It was explicitly shown in these publications that complex matrices did not have any effect on SPME performance when the new in-fiber calibration approach was implemented. Although this type of calibration circumvents matrix effects in many applications, it is anticipated that matrices having different concentrations of interfering substances could still have an influence on SPME performance, because competitive equilibria of labile concentrations of interferences with the fiber coating would not be negligible.

4.2.3.2 Recent Trends in SPME Applications

As the technique inherently works under equilibrium conditions and the amount of analyte extracted is negligible (<10%), SPME as a passive sampling method has found applications in different disciplines in general and in environmental analysis in particular, such as estimating free bioavailable chemical concentrations and distribution constants.[35] SPME has, for instance, been used to measure free, rather than total, pollutant concentrations in environmental compartments. These free concentrations in aquatic environments are of the utmost importance, as it is believed that free pollutant fractions are bioavailable, causing toxicity[36,37] and bioaccumulation[38] in aquatic organisms. SPME fibers have also been employed as biomimetic sampling devices to predict the toxicity of a chemical from its chemical uptake and bioavailability.[39,40] Moreover, SPME can be implemented as a tool for understanding the sorption and distribution of a pollutant in environmental multicompartments such as humic substances and other binding phases), which can be vitally important for environmental science and ecotoxicological studies.[41,42]

4.2.4 STIR BAR SORPTIVE EXTRACTION

The conventional polymeric coating, PDMS, employed in SPME has a film thickness of 100 μm, which corresponds to a volume of about 0.5 μL for the whole fiber. In SPME, the thin PDMS film provides the highest enrichment when equilibrium between the film and sample is realized, the attainment of which depends largely on analyte hydrophobicity and distribution to the coating. With thin PDMS films, SPME does not generally afford quantitative exhaustive extraction, which renders SPME a less sensitive technique even for nonpolar compounds and an unsatisfactory sampling device that fails to extract polar analytes. Implementing SPME for total exhaustive extraction is conceivably plausible if the PDMS film thickness is increased dramatically.

4.2.4.1 SBSE Development

Accordingly, Baltussen et al.[43] fabricated a magnetic rod housed in a short piece of glass tubing, which can be envisioned as a small stir bar. The stir bar was coated with PDMS (55 or 219 μL), and then used for stirring (sampling) water samples (10–250 mL) containing volatile and semivolatile compounds. The technique, named "SBSE," in principle works like SPME and depends mainly on the analyte partition coefficient between n-octanol and water ($K_{o/w}$), but also on the sample/PDMS volume ratio. After SBSE sampling is complete, the stir bar is placed in a thermal desorption unit (TDU) for analyte desorption.

The TDU equipment (commercially available from Gerstel GmbH, Mülheim an der Ruhr, Germany) is fully automated and connected on-line to a GC equipped with a programmable temperature vaporizer (PTV) injector for simultaneous cryotrapping of the analytes before injection. Another approach for analyte desorption is to place the stir bar in a small volume of a conventional HPLC liquid (or mobile phase) for HPLC analysis. The SBSE stir bars are trademarked as Twisters™; they can also be purchased from Gerstel. For more detailed information on SBSE technology, the reader is referred to two recent review articles.[44,45]

Comparing SPME with SBSE, the latter embodies a magnetic stir bar inserted into a glass jacket that is coated with a much thicker PDMS coating (100–400 times more PDMS). Accordingly, a much better sensitivity for nonpolar and polar compounds is achievable and exhaustive quantitative recovery is more readily attainable in SBSE than in SPME. The SBSE technique has been developed extensively since its introduction in 1999, the focus in this respect being mainly on its mode of operation and applications; several attempts have been made at improving material coatings.

4.2.4.2 SBSE Modes of Operation

The mode of operation in SBSE resembles that in SPME but with a few distinctive differences. For instance, after carrying out an SBSE procedure, two Twisters (dual SBSE extraction) can be simultaneously desorbed in one thermal desorption tube in order to further enhance the sensitivity. In the so-called multishot mode, up to five Twisters were desorbed in one desorption tube with *in situ* derivatization for quantification of estrogens in river water.[46]

4.3 ANALYTICAL TECHNIQUES BASED ON NONPOROUS POLYMERIC MEMBRANES

The types of polymeric membranes that have attracted much interest for analytical applications and are nowadays in common use are characterized as nonporous membranes such as low-density polyethylene (LDPE), dense PP and PDMS silicone rubbers, and asymmetric composite membranes such as silicone/polycarbonate and silicone/porous polypropylene (PP) membranes. The nonporous polymeric membranes (flat sheet (FS) or hollow fiber (HF)), especially dense silicone rubbers, are employed for analyte extraction and preconcentration in several different configurations. The nonporous polymeric membrane can be used to separate two different or similar phases, which can be

1. An aqueous phase and a gaseous or vacuum phase
2. Two gaseous phases
3. An aqueous phase and an organic phase
4. Two aqueous phases.

The most important characteristic of nonporous membranes is that they are hydrophobic and contain no pores in the polymeric structure. This means that these membranes not only selectively act as a barrier to particles and polar species, but they also provide unique selectivity and specificity for the permeation and transport of a specific group of compounds that can readily solubilize and diffuse in the membrane material. The analyte extraction rate (permeability) in a nonporous membrane separation process is governed by the "solution–diffusion" mechanism, as commented on earlier.

The analytical extraction systems related to points 1 and 2 are pervaporation-based techniques (such as those mentioned in Sections 4.3.1 and 4.3.2). Extraction based on the membrane separation of an aqueous phase and an organic phase (point 3 above) will be dealt with in Section 4.3.3. As the system concerning point 4 is very rarely used, it will not be considered here.

4.3.1 MEMBRANE INLET (INTRODUCTION) MASS SPECTROMETRY

Nonporous polymeric membranes have been incorporated as sample inlets in mass spectrometry (MS) for the direct sampling of volatile and semivolatile organic compounds (VOCs and SVOCs). The technique of membrane inlet (introduction) mass spectrometry (MIMS) has achieved tremendous success in the last two decades in terms of instrumentation and applications. Figure 4.1 depicts the experimental setup of (i) in-sample membrane MIMS and (ii) direct insertion (near the ion source in MS) membrane MIMS.

The core development of MIMS has centered upon MS ionization techniques for the analysis of aromatic contaminants in water,[47] improving membrane extraction selectivity[48] and enhancing MIMS sensitivity by cooling and heating the membrane (trap and release MIMS).[49,50] A dedicated review has been written on the applications of MIMS in environmental analysis.[51]

4.3.2 MEMBRANE EXTRACTION WITH SORBENT INTERFACE

The use of silicone membranes as an interface in MIMS for direct extraction and analysis by MS has fostered their implementation for extraction purposes that can be combined off-line or on-line with other analytical instrumentation, such as GC. The technique of membrane extraction with sorbent interface (MESI) (Figure 4.2) employs the pervaporation principle in a nonporous polymeric membrane unit, where the membrane is used as a selective barrier for the extraction of VOCs and SVOCs in gaseous or liquid samples.

In MESI, the sample flows on one side of an FS or HF membrane and the analytes diffuse through the membrane to the other side, where a continuous flow of a carrier gas is applied. The analytes carried by the stripping gas are focused prior to GC injection via a sorbent interface consisting of a trap and a heating coil.

MESI with a cap sampling device has been applied for the continuous monitoring of VOCs in surface waters.[52] MESI has subsequently been improved in terms of technical developments, instrument configurations, and applications for on-site sampling of water.[53,54]

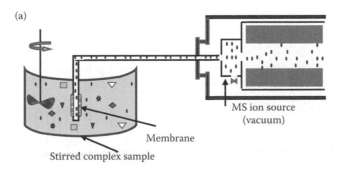

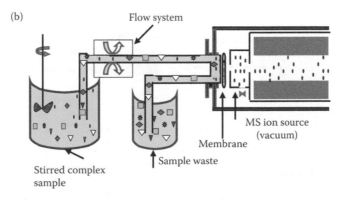

FIGURE 4.1 (a) MIMS systems with direct membrane sampling and (b) direct insertion probe in MS ion source.

MESI operation requires processing of the whole sample to be extracted and has to reach steady-state permeation, which usually takes a long time. Thus, a new technical modification of MESI, called pulse introduction (flow injection-type) membrane extraction (PIME), has been developed, in which the sample is introduced to the membrane as a pulse pushed by a stream of eluent (usually water).[55] This means that attaining a steady state is no longer crucial. PIME therefore provides not only a faster response and higher sensitivity, but also allows extraction of individual samples via discrete injections in addition to continuous on-line monitoring by sequential injection of a series of samples. Guo et al.[56] described a mathematical model for the PIME permeation process, which showed that (a) there was a trade-off between the sensitivity and the time lag (the time taken to complete the permeation process) and (b) a large sample volume and a low flow rate enhance the sensitivity but also increase the time lag.

It is important to remember that, as in all types of MESI designs, the time necessary to complete analyte permeation through the membrane can be fairly long, because the positive pressure of the stripping (carrier) gas on the acceptor side of the membrane slows down the permeation. The aqueous boundary layer formed on the membrane is believed to be the major contributor to mass transfer resistance. To increase extraction efficiency and speed up analysis (a shorter time lag), a stream of nitrogen gas can be introduced into the membrane prior to extraction and after sample elution. This will reduce, if not eliminate, the static aqueous boundary layer on either side of the membrane.[57] It was also found that using a gas as sample carrier (gas injection membrane extraction) instead of a water stream (as described previously) caused minimal axial mixing of the sample, which eliminated tailing in the permeation profiles, and significantly reduced the boundary layer (i.e., faster extraction—a shorter time lag) with no loss in sensitivity. Moreover, this approach required simpler instrumentation and operational procedures.[58]

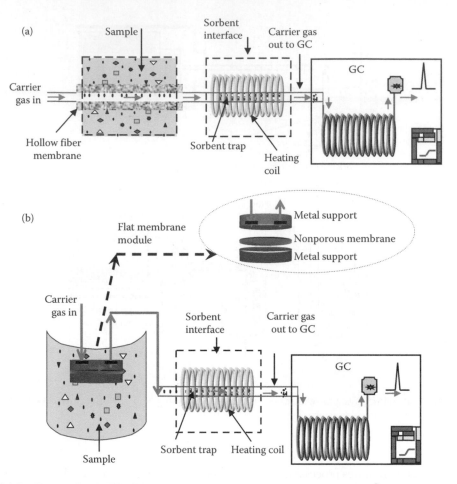

FIGURE 4.2   Two on-line MESI-GC setups with (a) HF and (b) FS membrane modules.

4.3.3   MEMBRANE-ASSISTED SOLVENT EXTRACTION

Another use of nonporous membranes in extraction sheds light on the exploitation of a membrane as a phase separator between an aqueous phase and an organic phase, thus forming a three-phase system with an aqueous–polymeric–organic configuration. Hauser et al. designed an experimental setup for the aqueous–polymeric–organic format carried out in a glass extraction cell. The new technique was called membrane-assisted solvent extraction (MASE). The extraction cell consisted of two compartments separated by a nonporous FS membrane of LDPE. One of the compartments was small, accommodating about 800 µL organic extractant, and the other was relatively large, housing a 10-mL sample of heavily chlorobenzene-contaminated groundwater.[59] After extraction, the large volume injection (LVI) of the extract into the GC was deemed necessary to enhance sensitivity.

A modified extraction cell containing a bag-shaped membrane made of LDPE, instead of an FS membrane, was designed to contain the extraction solvent for the extraction of polycyclic musk compounds and pharmaceuticals in wastewater.[60] The extraction cell was further developed in terms of membrane design and material. A dense nonporous PP membrane was preferably chosen as a membrane bag in the extraction cell, which was incorporated into a fully automated MASI device that is now commercially available from Gerstel (Mülheim an der Ruhr, Germany).

This device has been applied for the extraction of different classes of organic species in wastewater.[61]

4.4 ANALYTICAL TECHNIQUES BASED ON THE USE OF LIQUID MEMBRANES

4.4.1 SUPPORTED LIQUID MEMBRANE (SLM) EXTRACTION

4.4.1.1 SLM Extraction Principle

The first realization and application of SLM in analytical sample preparation was initiated by Audunsson[62] in 1986. The SLM extraction procedure involves partitioning uncharged (but ionizable) analytes from an aqueous sample phase (donor) to an immiscible organic liquid phase immobilized in a thin porous FS or HF polymeric membrane made of PP or PTFE (thus making the membrane nonporous). The analyte is then backextracted from the organic phase into another aqueous phase (acceptor) on the other side of the membrane. The ionizable analytes have to be in an extractable form in the aqueous sample before extraction so that they can dissolve in the organic phase. It has been discussed in Sections 4.2.3 and 4.2.4 that SPME and SBSE systems do not provide satisfactory extraction of polar species because the PDMS coating material used is hydrophobic and thus best suited for extracting nonpolar analytes (log $K_{o/w} > 4$). SLM extraction is considered superior for the extraction of polar species ($2 < \log K_{o/w} < 4$).

The analyte in the sample can be neutralized by adjusting the sample pH, or by adding an ion pairing or complexing agent in the sample or in the membrane liquid if the analytes happen to be permanently charged. Hence, the analytes in the sample diffuse through the membrane liquid, and, on the membrane/acceptor interface, are instantaneously trapped on the acceptor side by chemical means.

Analytes can be trapped in the acceptor solution if the sample pH is changed or an agent added that can selectively capture the analytes and make them nonextractable, preventing redissolution in the membrane liquid.

The performance of SLM extraction is characterized by two measures: the enrichment factor (E_e) and the extraction efficiency (E). E measures how much of the analyte in the sample is recovered on the acceptor side after extraction: $E = (C_A/C_S) \times (V_A/V_S)$, and E_e reflects how many times the concentration in the acceptor is increased compared to the initial sample concentration: $E_e = (C_A/C_S)$. C_A and C_S are the final acceptor and initial sample concentrations, respectively. Similarly, V_A and V_S are the acceptor and sample volumes, respectively. A more comprehensive elaboration of SLM principles[63] and mass transfer kinetics,[62,64] as well as the role of the octanol–water partition coefficient in SLM extraction of ionizable compounds,[65] is given elsewhere.

4.4.1.2 SLM Unit and System Configurations

The heart of the SLM extraction procedure is the SLM unit, in which analyte extraction, preconcentration, and sample cleanup take place in one single step. The design of the SLM unit can be engineered according to the intended use of the unit. For instance, if an SLM unit is needed for an automated and flowing sample extraction, it can be manufactured from two blocks of polytetrafluoroethylene (PTFE), polyvinylidine difluoride (PVDF), or titanium. However, for nonautomated, nonflowing SLM extraction, a short piece of porous HF membrane is used.

Three physical realizations of SLM modules have been reported: they are based on spiral, flat, and HF extraction units, as presented in Figure 4.3. Where a porous FS membrane is the liquid support in an automated flowing SLM configuration, flat and spiral modules are usually used, in which, respectively, a straight and a spiral machined groove is made in the inner two surfaces of the unit blocks.[66,67] When the membrane support is sandwiched between the two unit blocks, donor and acceptor channels are formed on either side of the membrane.

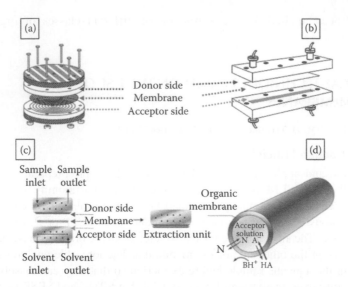

FIGURE 4.3 SLM extraction units with (a) spiral groove of 1 mL channel volume, (b) straight groove of 10 μL channel, (c) miniaturized straight groove of 1.65 μL channel volume, and (d) an HF membrane exemplifying an SLM extraction of an acidic analyte in an acidic sample containing the analyte (A), basic compounds (B), and neutral species (N).

Miniaturized and on-line flowing FS-SLM systems with 1.65 and 10 μL channel volumes have been applied, for example, for the extraction of simazine in mineral water[68] and haloacetic acids in water.[69] But not only do these flowing SLM systems yield low E values, they are usually prone to tubing blockage and analyte carryover. To circumvent these problems in FS-SLM, the HF-SLM configuration in the form of a nonautomated, nonflowing, off-line setup has been shown to be a better alternative, which has been investigated with regard to the extraction of dinitrophenols and chlorophenols in river and leachate water.[70,71]

One way of accomplishing this operational mode of HF-SLM was to seal one end of a short piece of HF (≈4 cm long) and fill the acceptor phase in the HF lumen through the other, open end. The HF was then immersed in an organic liquid to impregnate its pores, after which the HF device was added to the stirred sample for sampling. An HPLC syringe was used for filling the acceptor phase, supporting the HF during extraction, and emptying the HF lumen after extraction, as shown in Figure 4.4.[70] Owing to the high surface area-to-volume ratio of the HF (immersed in a stirred sample) and its one-off use, the HF-SLM procedure resulted in high E values with no carryover problems.

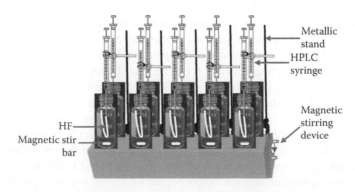

FIGURE 4.4 A nonautomated, nonflowing, off-line HF-SLM extraction setup.

Another format of the HF-SLM procedure was to employ a 15–20 cm long piece of HF as the extraction device. The HF lumen was then filled with acceptor solution using a microsyringe, and a loop made out of the HF. This loop was then soaked in an organic liquid before being added to the sample agitated on a shaker.[71]

There are several review papers on SLM extraction that touch upon off-line or on-line SLM hyphenation to analytical instruments as well as the fields of SLM applications.[67,72] The following discussion will focus mainly on other aspects of SLM, such as transport mechanisms, selectivity, and equilibrium sampling with SLM.

4.4.1.3 Transport Mechanisms in SLM

In SLM extraction, the transport mechanism is influenced primarily by the chemical characteristics of the analytes to be extracted and the organic liquid in the membrane into which the analytes will interact and diffuse. Analyte solubility in the membrane and its partition coefficient will have the main impact on separation and enrichment. Analyte transport in SLM extraction can be substantially categorized into two major types: one is diffusive transport (or simple permeation) and the other covers facilitated transport (or carrier-mediated transport).[73]

4.4.1.3.1 Diffusive Transport

Diffusive transport across an SLM membrane can be conceived of as five steps of analyte diffusion in a single phase and partitioning in two different phases: the analyte (a) diffuses in the bulk aqueous sample through the aqueous boundary layer to the sample/membrane interface; (b) partitions between the aqueous sample and the organic phase in the membrane in accordance with the partition coefficient, often approximated by the $K_{o/w}$ value of the analyte; (c) diffuses through the organic membrane phase along a concentration gradient until it approaches the membrane/acceptor interface; (d) partitions between the organic phase and the acceptor phase, where it is trapped; and (e) diffuses through the aqueous boundary layer to the bulk acceptor phase.

Figure 4.5 shows an example of the diffusive transport of an ionizable basic analyte being exhaustively extracted from an alkaline sample solution through an organic phase impregnated in the pores of an HF (dipped in the sample) into an acidic acceptor solution filling the HF lumen. The uncharged basic analyte (B) is selectively extracted into the organic liquid and, on the acceptor side, the acidic solution (pH = 3.3 units less than the analyte pKa value) irreversibly and chemically traps the analyte from the organic phase in a charged form (BH⁺).

On the other hand, acidic compounds (A⁻) in the sample will already be charged and will thus not be enriched. Moreover, the neutral species (N) could be extracted, but not enriched, and the size and charge of macromolecules (if they are ionizable) will prevent them from entering the membrane. But even if some macromolecules do enter the liquid membrane, their low diffusion coefficients will cause a slow rate of transport. Accordingly, enrichment of small basic compounds and

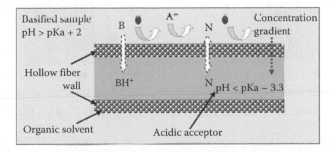

FIGURE 4.5 Schematic representation of selective SLM extraction of a small basic compound (B).

sample cleanup will be efficient. In a very similar manner, small ionizable acidic analytes can be extracted by reversing the pH in the sample and acceptor phases.[66]

4.4.1.3.2 Facilitated Transport

The other type of transport, facilitated transport, involves an analyte-specific carrier mixed in the organic membrane phase at a certain concentration. The solubility of the carrier in the surrounding aqueous phases has to be very low to prevent leakage, which would prevent specific analyte transport across the membrane.

As an example of carrier-mediated transport, Figure 4.6 (left) illustrates the coupled countertransport of an anionic analyte—aminomethyl phosphonic acid (AMPA) (a metabolite of the herbicide N-phosphonomethyl glycine (glyphosate))—in the sample and a cationic carrier—methyltrioctylammonium chloride (known as Aliquat 336)—in the membrane liquid.[74] This kind of transport basically involves the formation of an ion pair between the cationic carrier in the membrane and the anionic analyte in the sample. The ion pair diffuses through the organic membrane until it approaches the boundary with the acceptor side. There, the ion pair breaks up so that the anionic analyte is left on the acceptor side, whereas the cationic carrier takes a chloride ion in place of the analyte anion and returns in the form of a chloride complex. On the way back, the chloride anion is released on the sample side and transport continues. The driving force for such transport is the chloride anion gradient from the acceptor to the sample, which means that transport will ultimately cease when the chloride gradient no longer exists.

Anionic carriers in the membrane, such as di-2-ethylhexyl phosphoric acid (D2EHPA), have also been investigated for the coupled countertransport of cationic analytes such as metal ions from the sample side to the acceptor side. The driving force for this transport is, however, a proton gradient from the acceptor to the sample (Figure 4.6, right). This figure depicts schematically the transport of cationic metal ions from the sample to the acceptor side and the countertransport of a proton (pH gradient) in the opposite direction. Such a transport mode has been used, for example, to enrich permanently positively charged species, such as metal ions,[75] and bipyridilium herbicides (diquat and paraquat)[76] in river water.

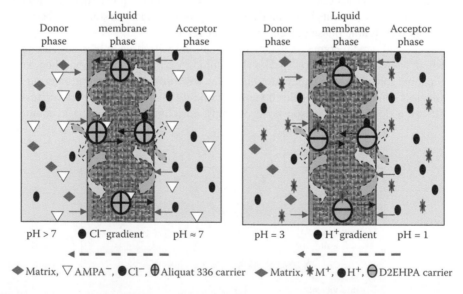

FIGURE 4.6 Illustration of the SLM-coupled countertransport of an anionic herbicide (glyphosate) metabolite, AMPA⁻ (left) and a cationic metal ion, M⁺ (right).

4.4.1.4 Selectivity in SLM

The discussion in Section 4.4.1.3 on transport mechanisms in SLM has manifestly demonstrated another facet of tuning analyte-selective extraction. For example, Figure 4.5 clearly demonstrates the selective extraction of a basic compound—all that is required here is a simple adjustment of the pH on either side of the membrane. Also, Figure 4.6 neatly illustrates the possibility of performing such selective extraction of anionic and cationic species in another transport mechanism that employs selective carriers. Thus, by fine-tuning the chemistry/composition of the sample, membrane liquid, and acceptor phases, analyte-selective extraction can be tailor-made.

The intention in this section is to further demonstrate SLM selectivity by highlighting a few worthwhile examples of tuning selectivity in the acceptor phase. SLM selectivity can be tuned in the acceptor phase in several ways, such as using a selective complexing agent that readily undergoes a chemical reaction with the analyte on the acceptor side (e.g., Cu^{2+} and a complexing ligand).[77,78]

Another means of achieving selective trapping in the acceptor is to suitably incorporate an analyte-specific antibody that will interact selectively with the analyte (antigen), forming an antigen–antibody complex that does not affect the analyte concentration gradient over the membrane.[79] This immuno-SLM (ISLM) system can be run in an on-line flowing configuration that integrates sampling, extraction, and detection by a suitable format of immunoassay methods. An FS-SLM unit with a channel volume of 10 μL has been typically used in ISLM for accommodating a free acceptor-dissolved antibody.[79–82] In order to achieve markedly high sensitivity and to minimize consumption of an expensive antibody, a miniaturized μSLM unit with a channel volume of 1.65 μL was implemented.[68] The acceptor channel in the μSLM unit was gold-coated, and the antibody was covalently bound to the gold surface via a self-assembled monolayer of sulfur-containing material. With this design, an LOD of 0.1 ng L^{-1} of simazine was obtained in mineral water.

A novel design of an ultrasensitive ISLM system with minimal consumption of antibody has been constructed on the basis of antibody immobilization on magnetic beads in a 10-μL acceptor channel. The position of the antibody molecules in the acceptor phase was meticulously controlled by two alternating and opposing magnetic fields generated by a current applied to either of two electromagnets placed above and below the acceptor channel of the SLM unit.[82] In this configuration, Tudorache et al. succinctly described the superior sensitivity obtained by the mobilized (forced movement up and down in the acceptor) over the immobilized antibody molecules, permitting a 2000 times concentration enrichment for simazine with an LOD of 13 pg L^{-1}.

Immunological trapping in ISLM has been used to quantify 4-nitrophenol in wastewater,[79] atrazine[80] and 2,4,6-trichlorophenol[81] in river water, and simazine in mineral and river water samples.[68,82]

4.4.1.5 Equilibrium Sampling Through SLM

As mentioned in Section 4.2.3.2, nonexhaustive SPME at equilibrium can be used for speciation studies and the sampling of freely dissolved nonpolar compounds. However, speciation and sampling of polar compounds in aqueous samples is extremely difficult with SPME. As an alternative, the appropriateness of SLM extraction of polar species makes the SLM technique uniquely suitable for performing speciation studies of such compounds, when SLM operation is carried out at equilibrium and under nonexhaustive conditions.

This SLM extraction mode, based on establishing an equilibrium between the undisturbed sample phase (to maintain natural equilibria) and the acceptor buffer phase, and working at nondepletive conditions ($E < 5\%$), has been termed "equilibrium sampling through membranes" (ESTM). Equilibrium sampling devices (ESDs) operating at negligible depletion of the sample, such as ESTM, will require a large volume ratio of sample to extractant (the acceptor solution in ESTM) if analyte depletion is excessive at small ratios.[35] ESDs are essentially beneficial for sensing free analyte fractions involved in secondary equilibria. These fractions are believed to be bioavailable for uptake by organisms in environmental compartments, thus causing toxicity.

The operational performance of an HF-based ESTM technique was first described by Liu et al.[71] A 15-cm piece of HF was filled with an acceptor buffer solution, after which the HF was made into a loop. This loop-like HF device was soaked in *n*-undecane, and then immersed in 1 L of a river or leachate water sample for extraction of freely dissolved chlorophenols. This HF-loop device was also employed for selective ESTM sampling of freely available Cu^{+2} in leachate water.[78] The selectivity stemmed from a selective liquid membrane (di-*n*-dihexyl ether) containing a carrier (crown ether/oleic acid) and a selective stripping agent in the acceptor solution.

4.4.2 MICROPOROUS MEMBRANE LIQUID–LIQUID EXTRACTION

4.4.2.1 MMLLE Principle

Microporous membrane liquid–liquid extraction (MMLLE) is a two-phase extraction setup. In MMLLE procedures, the membrane material and format (FS and HF), extraction units, and system configurations are identical to those described in SLM (Section 4.4.1.2).[63] The two-phase HF-MMLLE system is identical to that used in Section 4.4.3, although sometimes with minor differences. In contrast to three-phase SLM extraction, MMLLE employs a microporous membrane as a miniaturized barrier between two different phases (aqueous and organic). One of the phases is organic, filling both the membrane pores (thus making the membrane nonporous) and the compartment on one side of the membrane (acceptor side). The other phase is the aqueous sample on the other side of the membrane (donor side). In this way, the two-phase MMLLE system is highly suited to the extraction of hydrophobic compounds (log $K_{o/w} > 4$) and can thus be considered a technique complimentary to SLM in which polar analytes ($2 < \log K_{o/w} < 4$) can be extracted.

4.4.2.2 Miniaturization, Automation, and Hyphenation of MMLLE

As in the SLM systems, FS- and HF-MMLLE configurations can be run automatically in flowing modes and operated off-line or connected on-line to analytical instruments. Recently, a microfluidic chip-based FS-MMLLE system was reported.[83] In addition, miniaturized, nonautomated, nonflowing, off-line MMLLE systems are usually used with HF membranes. The emphasis in this section will be placed on these latter modes of MMLLE operation.

4.4.2.2.1 Automated, Flowing MMLLE

4.4.2.2.1.1 Off-Line Systems
Flowing FS- and HF-MMLLE systems are usually operated by pumping the sample solution on the sample side of the membrane, which can be done with a peristaltic or syringe pump. The stagnant or flowing organic solvent is supplied by another similar pump to the acceptor channel and membrane pores.

The impact of several factors on the MMLLE extraction yield of PAHs in water has been comprehensively studied using a flowing FS-MMLLE system and off-line analysis with GC-flame ionization detection (GC-FID).[84] The flowing FS-MMLLE procedure combined with off-line GC-mass spectrometry (GC-MS) analysis has been utilized for the extraction of nonionic and derivatized ionic organotin compounds in river water.[85]

4.4.2.2.1.2 On-Line Systems
Flowing MMLLE systems have been established in different layouts with automation and on-line hyphenation to GC and HPLC analysis. An automated on-line FS-MMLLE-GC system with a loop-type interface compatible with LVI was used for the extraction of pesticides and PAHs in surface waters.[86] In another study, pressurized hot water extraction (PHWE) was coupled on-line to a FS-MMLLE-GC-FID system and applied to the analysis of PAHs in soil, where MMLLE was used as a cleanup and concentration step of the PHWE extract prior to final GC analysis.[87] In addition, an HF-MMLLE setup was incorporated in PHWE and GC, resulting in an online PHWE-HF-MMLLE-GC system, where the HF membrane module contained 10–100 HFs. The system served for the extraction and analysis of PAHs in soil and sediments;

a significant improvement in extraction efficiency and low LOQs was obtained in comparison to the FS-MMLLE-based system.[88]

Flowing FS-MMLLE with on-line hyphenation to HPLC has also been investigated. Sandahl et al. were the first to interface FS-MMLLE with reversed-phase HPLC for the on-line extraction of methyl-thiophanate in natural water, obtaining an LOD of 0.5 µg L^{-1}.[89] Also, a parallel FS-SLM and FS-MMLLE design was coupled on-line to reverse-phase HPLC for the extraction of methyl-thiophanate (by MMLLE) and its metabolites (by SLM) in natural water.[90] In addition, on-line coupling of FS-MMLLE and normal-phase HPLC has been successfully applied in the determination of vinclozolin ($E_e = 118$ and LOD = 1 µg L^{-1}) in surface water[91] and of in-sample ion-paired cationic surfactants ($E_e > 250$ and LOD = 0.7–5 µg L^{-1}) in river water and wastewater samples.[92]

4.4.2.2.2 Extracting Syringe Device

As shown above, several attempts were made to establish automated flowing MMLLE systems that are interfaced on-line to an analytical instrument. The concept of the Extracting Syringe (ESy) stemmed from these earlier attempts. The ESy device automatically combines on-line micro-MMLLE and GC analysis in one step, as depicted in Figure 4.7. The ESy integrates automated sample pretreatment with simultaneous analyte enrichment, extract cleanup and injection, and analyte separation by GC in a closed system with minimal handling steps. The name "ESy" was given because, after extraction in the ESy has been completed, the ESy setup moves down, allowing the GC needle in the ESy to penetrate the GC septum and injector for whole extract injection.

In the ESy, a miniature FS membrane is supported by two small, identical pieces of PP plastic, constituting a miniaturized membrane unit called an "ESy extraction card" (see the inset in Figure 4.7), which is housed under mechanical pressure in a card holder. The two PP pieces have dimensions of 2 mm × 20 mm × 40 mm. The inner surface of each piece contains a machined groove defining a microchannel of 1.65 µL volume (0.125 mm depth × 0.6 mm width × 22 mm length). The very small piece of FS membrane (2 mm width × 22 mm length × 25 µm thickness) is fastened in

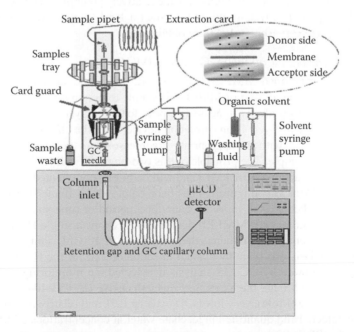

FIGURE 4.7 The on-line ESy-GC instrument with extraction card shown in the figure inset.

between the two PP pieces and is thus sandwiched when the two pieces are pressed together. In this ESy system, a solvent syringe pump is connected to the acceptor side of the ESy card so as to impregnate the membrane and fill the acceptor channel with organic solvent. Afterwards, the sample (1–3 mL) is continuously delivered by a syringe pump to the sample channel on the other side of the extraction card (donor side). When extraction is complete, the card holder moves down and its needle penetrates the GC septum for direct injection of the whole extract.

The FS-based ESy has been utilized in a small number of applications. For example, an ESy was coupled on-line with full automation to an on-column GC-electron capture detection (GC-ECD) system for the extraction and analysis of polychlorinated biphenyl (PCB) congeners in river water.[93] 1 mL samples with 40% acetonitrile containing 10 PCB congeners were automatically run and extracted at 100 µL min⁻¹. One of the major findings in this work pertained to the importance of acetonitrile as organic modifier in the sample and washing fluid. Acetonitrile as sample modifier played a central role in enhancing PCB enrichment; this enabled repeatable extractions and prevented adsorption onto glass container walls and ESy tubing, thereby minimizing carryover problems. When the optimal acetonitrile content was used in the sample and washing procedure, the on-line ESy-GC-ECD system yielded E_e values between 33 and 41 and LOD values between 2 and 3 ng L⁻¹ for all the PCBs studied.

In another application, Esy-GC-ECD was applied to the extraction of 14 organochlorine pesticides (OCPs) in spiked and contaminated complex samples, such as raw leachate water and soil–water slurry samples.[94] A downsized filtration vessel was deemed crucial for sample filtration after acidification and the addition of activated copper granules (to remove elemental sulfur) and 20% acetonitrile (to prevent adsorption and enhance enrichment). Under optimal conditions, extraction of a 3-mL leachate water sample dispensed at a flow rate of 100 µL min⁻¹ gave E_e values between 32 and 242 and LODs between 1 and 20 ng L⁻¹. It was also demonstrated that, since ESy extraction is dynamic and its extraction efficiency low, calculation of "relative recovery" was more relevant than "extraction efficiency" in all ESy applications.

Phthalate esters in river and leachate water were also extracted with an ESy-GC-FID system. Here, owing to the large variation in the polarity of the phthalate esters, 50% methanol had to be added to the samples as organic modifier in order to extract the most nonpolar phthalate esters (di-2-ethylhexylphthalate and di-n-octylphthalate); the relatively polar phthalate esters were extracted from unmodified samples. The time required to extract a 1-mL sample was 20 min; E_e values of 54–110-fold and LODs of 0.2–10 ng mL⁻¹ were obtained.[95]

4.4.2.2.3 Nonautomated, Nonflowing, Off-Line MMLLE

To simplify the above-mentioned MMLLE systems and, unlike the automated flowing MMLLE, the nonautomated, nonflowing design of MMLLE is simple to prepare manually and is an easy-to-use extraction procedure that is always done off-line prior to GC analysis. In this context, only a short piece of HF membrane is employed as an extraction device; after the HF lumen and pores[96] or only the pores[97] have been filled with an appropriate organic solvent, the membrane is immediately immersed in the aqueous sample. The principle of this two-phase HF-MMLLE system is also called HF liquid-phase microextraction (HF-LPME) and will be briefly commented on in the next section.

A very simple HF-MMLLE configuration has been employed by flame-sealing the two ends of the HFs. The HFs were then soaked in n-undecane for a period of time so as to allow them to fill with solvent; this makes simple HF-MMLLE devices. In this way, a single HF was utilized for the MMLLE of eight polybrominated diphenylethers (PBDEs) in 100 mL samples of tap, river, and leachate water. The analysis was done by manual injection of 2 µL of the HF lumen content into a splitless GC injector followed by GC-MS analysis in selected ion monitoring (SIM) mode. Under optimal HF-MMLLE conditions, the extraction was exhaustive (E = 57–104%), giving very good enrichment (E_e = 2800–5200-fold), very low LOD (<1.1 ng L⁻¹), and relative recoveries of 85–110%. Two PBDEs were detected and quantified in leachate water at concentrations of 3.5 ng L⁻¹ for BDE 153 and 23 ng L⁻¹ for BDE 183.[96]

In contrast to the exhaustive HF-MMLLE system mentioned above, a second HF-MMLLE configuration explored the possibility of using nonautomated, nonflowing, off-line HF-MMLLE for equilibrium sampling and nonexhaustive depletion of an analyte in environmental aqueous samples. To do this, the organic solvent was immobilized in the pores of a 1-cm HF membrane. This was made possible by using an HPLC syringe plunger inserted completely into the lumen of the 1-cm-long HF; after this, only the HF pores were impregnated in 2-heptanone ($\approx$3.3 μL). This arrangement allowed the HF lumen volume to be fully occupied by the plunger, so no solvent was filled into it. It could be argued that this format is not a membrane extraction technique, but it is useful nonetheless. The HF-in-plunger device was dipped in 500 mL river or aqueous sewage samples. With this large volume ratio between sample volume and organic solvent volume, nondepletive extraction conditions were established. Under equilibrium and no-depletion ($E = 2.5\%$) conditions, this experimental setup was employed for quantifying the free concentration of an ibuprofen degradation product in sewage water samples. Neither filtration nor sample pretreatment was needed, although there was a significant sewage water matrix effect. This type of HF-MMLLE design gave E_e values of over 2000 in the fiber pores and of over 300 after dilution, and LOD values of 7 ng L^{-1} in river water (downstream of the sewage treatment plant) and 14 ng L^{-1} in sewage water, where analyte concentrations of 26 and 40 ng L^{-1} were quantified.[97]

4.4.3 Two-Phase HF-LPME

The chemical principle of two-phase HF-LPME is identical to that in the FS- and HF-MMLLE systems discussed in the previous section. Thus, it is merely a case of different names for similar technical approaches. The intention of this section is to give a brief synopsis describing the most recent and important developments in this technique, and its potential automation.

4.4.3.1 Development of Two-Phase HF-LPME

The first setup of a two-phase HF-LPME procedure was established by Lee's group at the University of Singapore. A conventional GC-microsyringe needle fixed into one end of a 1.3-cm HF was used and the setup was employed for the extraction of triazine herbicides in water and soil–water slurry samples.[98] The GC syringe was utilized for supporting the HF, filling and collecting the organic solvent (3 μL) from inside the HF, and finally, for direct manual injection of the solvent into a GC system. The new procedure gave $E_e > 140$ (3–5 times higher than static-single drop LPME) for several triazines with LOD between 7 and 63 ng L^{-1} in water samples and 40 and 180 ng L^{-1} in soil–water slurry samples.

A few technical modifications have been introduced to the HF-LPME technique. Jiang et al.,[99] for example, extended the microsyringe-based HF-LPME to solvent bar microextraction (SBME), which exemplifies a more miniaturized version of the system described in the "nonautomated, nonflowing HF-MMLLE" section. In the SBME procedure, the organic solvent was confined within a short piece of an HF membrane (1.5 cm) sealed at both ends, thus no syringe was used. This extraction with organic solvent (1.5–4 μL) in a porous microbag of PP HF involved tumbling the solvent bag in a soil-slurry sample by using a stir bar, which resulted in faster mass transfer and higher enrichment factors for chlorobenzenes compared with syringe-based static HF-LPME. In another technical development of HF-LPME, Wang et al.[100] introduced a fiber-in-tube LPME device, where several PTFE HF membranes (14–28 HFs) packed into a short piece of PTFE tube (1–2 cm) were impregnated with an organic extraction solvent. After HF impregnation, the tip of a GC microsyringe was fitted into the packed tube and the whole set was immersed in a river water or wastewater sample for extraction of substituted benzenes. In another study, done by Lee et al.,[101] simultaneous in-fiber derivatization and HF-LPME of degradation products of chemical warfare agents in water was reported. This work adapted a 1:1 mixture of chloroform as extraction solvent and N-(tert-butyldimethylsilyl)-N-methyltrifluoroacetamide as derivatizing agent in the HF lumen. After extraction, 1 μL of this mixture was injected into GC-MS for analysis.

4.4.3.2 Automation of Two-Phase HF-LPME

The HF-LPME technique is seemingly difficult to automate. The first attempt at automating this technique was reported by Zhao et al., who developed a semiautomated dynamic two-phase HF-LPME procedure for the extraction of fluoranthene and pyrene in aqueous samples.[102] In this procedure, the organic solvent (3 μL) inside the HF (1.5 cm) was repeatedly withdrawn from and discharged into the HF (5 s waiting time after each withdrawing–discharging cycle) by a conventional GC syringe that was mounted onto a programmable syringe pump to control the movement of the syringe plunger. Compared with static HF-LPME, dynamic HF-LPME permitted much faster mass transfer of analytes through a thin organic film (as the solvent plug is withdrawn inside the syringe) to the solvent plug inside the HF (as the solvent plug is released into the HF), which resulted in higher enrichment factors. In another semiautomated two-phase HF-LPME system used for the extraction of pharmaceutical and endocrine disrupting compounds in sewage water samples, a 6-cm HF membrane filled with 40 μL of solvent was employed as a loop attached at one end to a stainless steel microfunnel guide, whereas the other end was kept unsealed and fixed in a small dent in the guide. This setup was then placed into a sample vial for the extraction of small sample volumes. On completion of the extraction, the sample vial was placed in a GC autosampler tray, and 1 μL of the extract was automatically taken from the HF and injected into the GC system by the autosampler.[103]

Very recently, Ouyang et al.[104] explored a fully automated two-phase HF-LPME device that was applied to the kinetic calibration of carbaryl extraction in water and red wine samples. The device encompassed a 1.8 cm-long microporous HF inserted into one end of a thin micropipette tip (0.5–10 μL) and firmly fixed to the needle-guiding tip (fixed in the sample vial) by heat. The other end of the micropipette was pressure-sealed, giving a final effective HF length of 1.5 cm. All extraction steps, including filling the extraction solvent (20 μL), transferring and agitating the sample, withdrawing and introducing the extraction phase into the GC injector, were performed automatically by a CTC CombiPal autosampler. With the aid of this device, the kinetics of HF-LPME absorption and desorption processes was extensively investigated. The kinetic calibration approach was successfully used to correct for matrix effects and showed that neither sample volume nor sampling time affected the feasibility of the calibration method. The automated HF-LPME technique has proved useful for determining the analyte distribution coefficient between a sample matrix and an extraction phase.

4.5 SUMMARY AND FUTURE OUTLOOK

The conclusions that can be drawn from the work presented in this chapter emphasize the miniaturization trends currently being pursued in modern analyte extraction techniques. The trends are focused mainly on miniaturization of the core idea of the extraction techniques utilized for analyte extraction from water samples. This has in turn promoted automation and hyphenation of extraction principles with on-line or off-line connection to analytical instruments.

For instance, miniaturized liquid-phase extraction techniques (Section 4.2.1) have become very popular because of their simplicity, ease of use, and low solvent demand. However, these procedures suffer significantly from the vulnerability of the solvent microdrop to dislodgement. To circumvent the shortcomings of these techniques, downsized adsorptive solid-phase and solvent-free sorptive extraction systems (Sections 4.2.2 through 4.2.4) and also membrane-based setups (Sections 4.3 and 4.4) have been pursued as alternatives. Among the SPE designs is the fully automated MEPS device, the applications of which can be extended to the extraction of environmental aqueous samples. The solvent-free SPME technique has demonstrated a high degree of robustness coupled with outstanding performance, as a result of which it has become very popular and is used in a broad diversity of applications. Indeed, the SPME technique has become prevalent and is implemented in many different scientific disciplines.

The polymeric membrane extraction systems, that is, MIMS, MESI, and PIME, have been extensively applied for sampling VOCs and SVOCs. In contrast, liquid membrane extraction configurations (SLM, MMLLE (ESy) and two-phase or three-phase HF-LPME have been widely and successfully employed in the extraction of a wide spectrum of analytes in different sample matrices. SLM extraction (also three-phase HF-LPME) selectively enriches ionizable polar to slightly polar analytes, whereas MMLLE is used primarily for extracting hydrophobic neutral analytes. The two techniques are thus regarded as complementary sampling procedures. The design of the flowing and automated setups of SLM and MMLLE are complicated, and there have been some teething troubles: analyte carryover, adsorption problems, and blockage of the narrow-diameter tubing used in flow systems have been reported. Hence, the simpler the SLM or MMLLE setups, the more robust and less problematic they will tend to be; for certain applications, however, setups based on nonautomated, nonflowing HF-based procedures might be preferable.

It is very probable that the SLM technique will become widely utilized for selective and sensitive extractions (immunoaffinity-based ISLM design). Moreover, the ESTM technique based on equilibrium SLM is likely to be extensively investigated in the context of speciation studies in environmental samples, with the major focus being on characterizing the interaction of analytes and macromolecules (such as HAs). Furthermore, the SLM format is well suited for studying different types of transport mechanisms.

The two-phase HF-MMLLE/HF-LPME procedure with potential automation is expected to achieve wide acceptance among researchers and experimentalists as it is direct and easy-to-use.

This chapter has shown that many extraction techniques available today are based on the same principle but have been given different names. An IUPAC initiative is therefore needed for the unification of the subject nomenclature in order to avoid misunderstandings and enable unambiguous knowledge to be disseminated to young scholars.

ACRONYMS AND ABBREVIATIONS

Aliquat 336	methyltrioctylammonium chloride
AMPA	aminomethyl phosphonic acid
CE	capillary electrophoresis
CFME	continuous-flow microextraction
D2EHPA	di-2-ethylhexyl phosphonic acid
DVB	divinylbenzene
ECD	electron capture detector
ESDs	equilibrium sampling devices
E_e	enrichment factor
E	extraction efficiency
ESTM	equilibrium sampling through membrane
ESy	extracting syringe
FID	flame ionization detector
FIT-SPE	fiber-in-tube solid-phase extraction
FS	flat sheet
GC	gas chromatography
HAs	humic acids
HF	hollow fiber
HF-LPME	hollow-fiber liquid-phase microextraction
HPLC	high-performance liquid chromatograph
ISLM	immuno-supported liquid membrane
LC	liquid chromatography
LDPE	low-density polyethylene
LLE	liquid–liquid extraction
LOD	limit of detection
$\log K_{o/w}$	logarithmic partition coefficient of analyte in octanol/water

LOQ	limit of quantification
LPME	liquid-phase microextraction
LVI	large volume injection
MASE	membrane-assisted solvent extraction
MEPS	microextraction in packed syringe
MESI	membrane extraction with sorbent interface
MIMS	membrane inlet (introduction) mass spectrometry
MMLLE	microporous membrane liquid–liquid extraction
MS	mass spectrometry
OCPs	organochlorine pesticides
PAHs	polycyclic aromatic hydrocarbons
PBDEs	polybrominated diphenyl ethers
PCBs	polychlorinated biphenyls
PDMS	polydimethyl siloxane
PEEK	polyetheretherketone
PHWE	pressurized hot water extraction
PIME	pulse introduction membrane extraction
PP	polypropylene
PTFE	polytetrafluoroethylene
PTV	programmable temperature vaporizer
PVDF	polyvinylidene fluoride
SBSE	stir bar sorptive extraction
SLM	supported liquid membrane
SME	solvent microextraction
SPDE	solid-phase dynamic extraction
SPE	solid-phase extraction
SPME	solid-phase microextraction
SVOCs	semivolatile organic compounds
TDU	thermal desorption unit
VOCs	volatile organic compounds

REFERENCES

1. UNESCO 2007. Rio declaration on environment and development, Rio de Janeiro, Brazil, June 3–14, 1992. Available at URL: http://www.unesco.org/education/information/nfsunesco/pdf/RIO_E.PDF (accessed April 5, 2009).
2. Anastas, P.T. and J.C. Warner. 1998. *Green Chemistry: Theory and Practice.* Oxford: Oxford University Press.
3. Ramos, L., J.J. Ramos, and U.A.T. Brinkman. 2005. Miniaturization in sample treatment for environmental analysis. *Anal. Bioanal. Chem.* 381: 119–140.
4. Hyötyläinen, T. 2007. Principles, developments and applications of on-line coupling of extraction with chromatography. *J. Chromatogr. A* 1153: 14–28.
5. Liu, H. and P.K. Dasgupta. 1996. Analytical chemistry in a drop. Solvent extraction in a microdrop. *Anal. Chem.* 68: 1817–1821.
6. Jeannot, M.A. and F.F. Cantwell. 1996. Solvent microextraction into a single drop. *Anal. Chem.* 68: 2236–2240.
7. Jeannot, M.A. and F.F. Cantwell. 1997. Mass transfer characteristics of solvent extraction into a single drop at the tip of a syringe needle. *Anal. Chem.* 69: 235–239.
8. He, Y. and H.K. Lee. 1997. Liquid-phase microextraction in a single drop of organic solvent by using a conventional microsyringe. *Anal. Chem.* 69: 4634–4640.
9. Hou, L. and H.K. Lee. 2002. Application of static and dynamic liquid-phase microextraction in the determination of polycyclic aromatic hydrocarbons. *J. Chromatogr. A* 976: 377–385.
10. Wang, Y., Y.C. Kwok, Y. He, and H.K. Lee. 1998. Application of dynamic liquid-phase microextraction to the analysis of chlorobenzenes in water by using a conventional microsyringe. *Anal. Chem.* 70: 4610–4614.

11. Liu, W. and H.K. Lee. 2000. Continuous-flow microextraction exceeding 1000-fold concentration of dilute analytes. *Anal. Chem.* 72: 4462–4467.
12. He, Y. and H.K. Lee. 2006. Continuous flow microextraction combined with high-performance liquid chromatography for the analysis of pesticides in natural waters. *J. Chromatogr. A* 1122: 7–12.
13. Wood, D.C., J.M. Miller, and C. Ingo. 2004. Headspace liquid microextraction. *LCGC North Am.* 22: 516–522.
14. Xu, L., C. Basheer, and H.K. Lee. 2007. Developments in single-drop microextraction. *J. Chromatogr. A* 1152: 184–192.
15. Rezaee, M., Y. Assadi, M.-R. Milani Hosseini, E. Aghaee, F. Ahmadi, and S. Berijani. 2006. Determination of organic compounds in water using dispersive liquid–liquid microextraction. *J. Chromatogr. A* 1116: 1–9.
16. Berijani, S., Y. Assadi, M. Anbia, M.-R. Milani Hosseini, and E. Aghaee. 2006. Dispersive liquid–liquid microextraction combined with gas chromatography–flame photometric detection: Very simple, rapid and sensitive method for the determination of organophosphorus pesticides in water. *J. Chromatogr. A* 1123: 1–9.
17. Garcia-Lopez, M., I. Rodriguez, and R. Cela. 2007. Development of a dispersive liquid–liquid microextraction method for organophosphorus flame retardants and plasticizers determination in water samples. *J. Chromatogr. A*, 1166: 9–15.
18. Fattahi, N., S. Samadi, Y. Assadi, and M. R. M. Hosseini. 2007. Solid-phase extraction combined with dispersive liquid–liquid microextraction-ultra preconcentration of chlorophenols in aqueous samples. *J. Chromatogr. A* 1169: 63–69.
19. Jiang, H., Y. Qin, and B. Hu. 2008. Dispersive liquid phase microextraction (DLPME) combined with graphite furnace atomic absorption spectrometry (GFAAS) for determination of trace Co and Ni in environmental water and rice samples. *Talanta* 74: 1160–1165.
20. Farajzadeh, M.A., M. Bahram, B.G. Mehr, and J.A. Jönsson. 2008. Optimization of dispersive liquid–liquid microextraction of copper (II) by atomic absorption spectrometry as its oxinate chelate: Application to determination of copper in different water samples. *Talanta* 75: 832–840.
21. Saito, Y., M. Imaizumi, T. Takeichi, and K. Jinno. 2002. Miniaturized fiber-in-tube solid-phase extraction as the sample preconcentration method for microcolumn liquid-phase separations. *Anal. Bioanal. Chem.* 372: 164–168.
22. Saito, Y., M. Nojiri, M. Imaizumi, Y. Nakao, Y. Morishima, H. Kanehara, H. Matsuura, K. Kotera, H. Wada, and K. Jinno. 2002. Polymer-coated synthetic fibers designed for miniaturized sample preparation process. *J. Chromatogr. A* 975: 105–112.
23. Saito, Y., M. Imaizumi, K. Ban, A. Tahara, H. Wada, and K. Jinno. 2004. Development of miniaturized sample preparation with fibrous extraction media. *J. Chromatogr. A* 1025: 27–32.
24. Saito, Y. and K. Jinno. 2003. Miniaturized sample preparation combined with liquid phase separations. *J. Chromatogr. A* 1000: 53–67.
25. Jinno, K., M. Ogawa, I. Ueta, and Y. Saito. 2007. Miniaturized sample preparation using a fiber-packed capillary as the medium. *Trends Anal. Chem.* 26: 27–35.
26. El-Beqqali, A., A. Kussak, and M. Abdel-Rehim. 2006. Fast and sensitive environmental analysis utilizing microextraction in packed syringe online with gas chromatography-mass spectrometry: Determination of polycyclic aromatic hydrocarbons in water. *J. Chromatogr. A* 1114: 234–238.
27. Lipinski, J. 2001. Automated solid phase dynamic extraction—extraction of organics using a wall coated syringe needle. *Fresenius J. Anal. Chem.* 369: 57–62.
28. Pawliszyn, J. 1997. *Solid-phase Microextraction, Theory and Practice.* New York: Wiley-VCH.
29. Pawliszyn, J. 2002. Sampling and sample preparation for field and laboratory. In: J. Pawliszyn (ed.), *Comprehensive Analytical Chemistry*, Vol. XXXVII, pp. 389–477. Amsterdam: Elsevier Science B.V.0. Ouyang, G. and J. Pawliszyn 2006. SPME in environmental analysis. *Anal. Bioanal. Chem.* 386: 1059–1073.
31. Chen, Y. and J. Pawliszyn. 2004. Kinetics and the on-site application of standards in a solid-phase microextraction fiber. *Anal. Chem.* 76: 5807–5815.
32. Zhao, W., G. Ouyang, M. Alaee, and J. Pawliszyn. 2006. On-rod standardization technique for time-weighted average water sampling with a polydimethylsiloxane rod. *J. Chromatogr. A* 1124: 112–120.
33. Zhou S.N., X. Zhang, G. Ouyang, A. Eshaghi, and J. Pawliszyn. 2007. On-fiber standardization technique for solid-coated solid-phase microextraction. *Anal. Chem.* 79: 1221–1230.
34. Ouyang, G. and J. Pawliszyn. 2007. Configurations and calibration methods for passive sampling techniques. *J. Chromatogr. A* 1168: 226–235.
35. Mayer, P., J. Tolls, J.L.M. Hermens, and D. Mackay. 2003. Equilibrium sampling devices. *Environ. Sci. Technol.* 37: 184A–191A.

36. Heringa, M.B., R.H.M.M. Schreurs, F. Busser, P.T. van der Saag, B. van der Burg, and J.L.M. Hermens. 2004. Toward more useful *in vitro* toxicity data with measured free concentrations. *Environ. Sci. Technol.* 38: 6263–6270.

37. Leslie, H.A., J.L.M. Hermens, and M.H.S. Kraak. 2004. Baseline toxicity of a chlorobenzene mixture and total body residues measured and estimated with solid-phase microextraction. *Environ. Toxicol. Chem.* 23: 2017–2021.

38. Leslie, H.A., T.L. Ter Laak, F.J.M. Busser, M.H.S. Kraak, and J.L.M. Hermens. 2002. Bioconcentration of organic chemicals: Is a solid-phase microextraction fiber a good surrogate for biota? *Environ. Sci. Technol.* 36: 5399–5404.

39. Lanno, R., T.W. La Point, J.M. Conder, and J.B. Wells. 2005. Application of solid-phase microextraction fibers as biomimetic sampling devices in ecotoxicology, Chapter 28. In: G.K. Ostrander (ed.), *Techniques in Aquatic Toxicology*, Vol. 2, pp. 511–522. Florida: Taylor & Francis-CRC Press.

40. Parkerton, T.F., M.A. Stone, and D.J. Letinski. 2000. Assessing the aquatic toxicity of complex hydrocarbon mixtures using solid phase microextraction. *Toxicol. Lett.* 112–113: 273–282.

41. Poerschmann, J., Z. Zhang, F.-D. Kopinke, and J. Pawliszyn. 1997. Solid phase microextraction for determining the distribution of chemicals in aqueous matrices. *Anal. Chem.* 69: 597–600.

42. Droge, S.T.J., T.L. Sinnige, and J.L.M. Hermens. 2007. Analysis of freely dissolved alcohol ethoxylate homologues in various seawater matrixes using solid-phase microextraction. *Anal. Chem.* 79: 2885–2891.

43. Baltussen, E., P. Sandra, F. David, and C. Cramers. 1999. Stir bar sorptive extraction (SBSE), a novel extraction technique for aqueous samples: Theory and principles. *J. Microcolumn Sep.* 11: 737–747.

44. Baltussen, E., C.A. Cramers, and P.J.F. Sandra. 2002. Sorptive sample preparation—a review. *Anal. Bioanal. Chem.* 373: 3–22.

45. David, F. and P. Sandra. 2007. Stir bar sorptive extraction for trace analysis. *J. Chromatogr. A* 1152: 54–69.

46. Kawaguchi, M., Y. Ishii, N. Sakui, N. Okanouchi, R. Ito, K. Inoue, K. Saito, and H. Nakazawa. 2004. Stir bar sorptive extraction with *in situ* derivatization and thermal desorption-gas chromatography-mass spectrometry in the multi-shot mode for determination of estrogens in river water samples. *J. Chromatogr. A* 1049: 1–8.

47. Oser, H., M.J. Coggiola, S.E. Young, D.R. Crosley, V. Hafer, and G. Grist. 2007. Membrane introduction/laser photoionization time-of-flight mass spectrometry. *Chemosphere* 67: 1701–1708.

48. Creaser, C.S., D.J. Weston, and B. Smith. 2000. In-membrane preconcentration/membrane inlet mass spectrometry of volatile and semivolatile organic compounds. *Anal. Chem.* 72: 2730–2736.

49. Mendes, M.A. and M.N. Eberlin. 2000. Trace level analysis of VOCs and semi-VOCs in aqueous solution using a direct insertion membrane probe and trap and release membrane introduction mass spectrometry. *Analyst* 125: 21–24.

50. Thompson, A.J., A.S. Creba, R.M. Ferguson, E.T. Krogh, and C.G. Gill. 2006. A coaxially heated membrane introduction mass spectrometry interface for the rapid and sensitive on-line measurement of volatile and semi-volatile organic contaminants in air and water at parts-per-trillion levels. *Rapid Commun. Mass Spectrom.* 20: 2000–2008.

51. Ketola, R.A., T. Kotiaho, M.E. Cisper, and T.M. Allen. 2002. Environmental applications of membrane introduction mass spectrometry. *J. Mass Spectrom.* 37: 457–476.

52. Luo, Y.Z. and J. Pawliszyn. 2000. Membrane extraction with a sorbent interface for headspace monitoring of aqueous samples using a cap sampling device. *Anal. Chem.* 72: 1058–1063.

53. Segal, A., T. Gorecki, P. Mussche, J. Lips, and J. Pawliszyn. 2000. Development of membrane extraction with a sorbent interface-micro gas chromatography system for field analysis. *J. Chromatogr. A* 873: 13–27.

54. Liu, X. and J. Pawliszyn. 2005. On-site environmental analysis by membrane extraction with a sorbent interface combined with a portable gas chromatograph system. *Int. J. Environ. Anal. Chem.* 85: 1189–1200.

55. Juan, A.S., X. Guo, and S. Mitra. 2001. On-site and on-line analysis of chlorinated solvents in ground water using pulse introduction membrane extraction gas chromatography (PIME-GC). *J. Sep. Sci.* 24: 599–605.

56. Guo, X. and S. Mitra. 1999. Theoretical analysis of non-steady-state, pulse introduction membrane extraction with a sorbent trap interface for gas chromatographic detection. *Anal. Chem.* 71: 4587–4593.

57. Guo, X. and S. Mitra. 1999. Enhancement of extraction efficiency and reduction of boundary layer effects in pulse introduction membrane extraction. *Anal. Chem.* 71: 4407–4412.

58. Kou, D., A. San Juan, and S. Mitra. 2001. Gas injection membrane extraction for fast on-line analysis using GC detection. *Anal. Chem.* 73: 5462–5467.

59. Hauser B. and P. Popp. 2001. Membrane-assisted solvent extraction of organochlorine compounds in combination with large-volume injection/gas chromatography-electron capture detection. *J. Sep. Sci.* 24: 551–560.

60. Einsle, T., H. Paschke, K. Bruns, S. Schrader, P. Popp, and M. Moeder. 2006. Membrane-assisted liquid–liquid extraction coupled with gas chromatography-mass spectrometry for determination of selected polycyclic musk compounds and drugs in water samples. *J. Chromatogr. A* 1124: 196–204.

61. Hauser, B., M. Schellin, and P. Popp. 2004. Membrane-assisted solvent extraction of triazines, organo-chlorine, and organophosphorus compounds in complex samples combined with large-volume injection-gas chromatography/mass spectrometric detection. *Anal. Chem.* 76: 6029–6038.

62. Audunsson, G. 1986. Aqueous/aqueous extraction by means of a liquid membrane for sample cleanup and preconcentration of amines in a flow system. *Anal. Chem.* 58: 2714–2723.

63. Jönsson, J.Å. and L. Mathiasson. 1999. Liquid membrane extraction in analytical sample preparation: I. Principles. *Trends Anal. Chem.* 18: 318–325.

64. Jönsson, J.Å., P. Lövkvist, G. Audunsson, and G. Nilvé. 1993. Mass transfer kinetics for analytical enrichment and sample preparation using supported liquid membranes in a flow system with stagnant acceptor liquid. *Anal. Chim. Acta* 277: 9–24.

65. Chimuka, L., L. Mathiasson, and J.Å. Jönsson. 2000. Role of octanol-water partition coefficients in extraction of ionisable organic compounds in a supported liquid membrane with a stagnant acceptor. *Anal. Chim. Acta* 416: 77–86.

66. Jönsson, J.Å. and L. Mathiasson. 2001. Membrane extraction techniques for sample preparation. In: E. Grushka (ed.), *Advances in Chromatography*, pp. 53–91. New York: Marcel Dekker.

67. Jönsson, J.Å. and L. Mathiasson. 2001. Membrane extraction in analytical chemistry. *J. Sep. Sci.* 24: 495–507.

68. Tudorache, M. and J. Emnéus. 2006. A micro-immuno supported liquid membrane assay (μ-ISLMA). *Biosens. Bioelectron.* 21: 1513–1520.

69. Wang, X., C. Saridara, and S. Mitra. 2005. Microfluidic supported liquid membrane extraction. *Anal. Chim. Acta* 543: 92–98.

70. Lezamiz, J. and J.Å. Jönsson. 2007. Development of a simple hollow fibre supported liquid membrane extraction method to extract and preconcentrate dinitrophenols in environmental samples at ng L^{-1} level by liquid chromatography. *J. Chromatogr. A* 1152: 226–233.

71. Liu, J.-F., J.Å. Jönsson, and P. Mayer. 2005. Equilibrium sampling through membranes of freely dis-solved chlorophenols in water samples with hollow fiber supported liquid membrane. *Anal. Chem.* 77: 4800–4809.

72. Jönsson, J.Å. and L. Mathiasson. 2000. Membrane-based techniques for sample enrichment. *J. Chromatogr. A* 902: 205–225.

73. Nilvé, G. 1992. Sample pretreatment methods for determination of acidic herbicides in water, with spe-cial emphasis on supported liquid membranes. PhD dissertation, Lund University, Sweden.

74. Dzygiel, P. and P. Wieczorek. 2001. Supported liquid membrane extraction of glyphosate metabolites. *J. Sep. Sci.* 24: 561–566.

75. Djane, N.-K., K. Ndung'u, F. Malcus, G. Johansson, and L. Mathiasson. 1997. Supported liquid mem-brane enrichment using an organophosphorus extractant for analytical trace metal determinations in river waters. *Fresenius J. Anal. Chem.* 358: 822–827.

76. Mulugeta, M. and N. Megersa. 2004. Carrier-mediated extraction of bipyridilium herbicides across the hydrophobic liquid membrane. *Talanta* 64: 101–108.

77. Romero, R. and J.Å. Jönsson. 2005. Determination of free copper concentrations in natural waters by using supported liquid membrane extraction under equilibrium conditions. *Anal. Bioanal. Chem.* 381: 1452–1459.

78. Romero, R., J.-F. Liu, P. Mayer, and J.Å. Jönsson. 2005. Equilibrium sampling through membranes of freely dissolved copper concentrations with selective hollow fiber membranes and the spectrophotomet-ric detection of a metal stripping agent. *Anal. Chem.* 77: 7605–7611.

79. Thordarson, E., J.Å. Jönsson, and J. Emnéus. 2000. Immunologic trapping in supported liquid membrane extraction. *Anal. Chem.* 72: 5280–5284.

80. Tudorache, M., M. Rak, P.P. Wieczorek, J.Å. Jönsson, and J. Emnéus. 2004. Immuno-SLM—a combined sample handling and analytical technique. *J. Immunol. Methods* 284: 107–118.

81. Tudorache, M. and J. Emnéus. 2005. Selective immuno-supported liquid membrane (ISLM) extraction, enrichment and analysis of 2,4,6-trichlorophenol. *J. Membr. Sci.* 256: 143–149.

82. Tudorache, M., M. Co, H. Lifgren, and J. Emnéus. 2005. Ultrasensitive magnetic particle-based immu-nosupported liquid membrane assay. *Anal. Chem.* 77: 7156–7162.

83. Cai, Z.-X., Q. Fang, H.-W. Chen, and Z.-L. Fang. 2006. A microfluidic chip based liquid–liquid extraction system with microporous membrane. *Anal. Chim. Acta* 556: 151–156.
84. Kuosmanen, K., M. Lehmusjärvi, T. Hyötyläinen, M. Jussila, and M.-L. Riekkola. 2003. Factors affecting microporous membrane liquid–liquid extraction. *J. Sep. Sci.* 26: 893–902.
85. Ndungu, K. and L. Mathiasson. 2000. Microporous membrane liquid–liquid extraction technique combined with gas chromatography mass spectrometry for the determination of organotin compounds. *Anal. Chim. Acta* 404: 319–328.
86. Lüthje, K.N.K., T. Hyötyläinen, and M.-L. Riekkola. 2004. On-line coupling of microporous membrane liquid–liquid extraction and gas chromatography in the analysis of organic pollutants in water. *Anal. Bioanal. Chem.* 378: 1991–1998.
87. Kuosmanen, K., T. Hyötyläinen, K. Hartonen, J.Å. Jönsson, and M.-L. Riekkola. 2003. Analysis of PAH compounds in soil with on-line coupled pressurised hot water extraction-microporous membrane liquid–liquid extraction-gas chromatography. *Anal. Bioanal. Chem.* 375: 389–399.
88. Kuosmanen, K., T. Hyötyläinen, K. Hartonen, and M.-L. Riekkola. 2003. Analysis of polycyclic aromatic hydrocarbons in soil and sediment with on-line coupled pressurised hot water extraction, hollow fibre microporous membrane liquid–liquid extraction and gas chromatography. *Analyst* 128: 434–439.
89. Sandahl, M., L. Mathiasson, and J.Å. Jönsson. 2000. Determination of thiophanate-methyl and its metabolites at trace level in spiked natural water using the supported liquid membrane extraction and the microporous membrane liquid–liquid extraction techniques combined on-line with high-performance liquid chromatography. *J. Chromatogr. A* 893: 123–131.
90. Sandahl, M., L. Mathiasson, and J.Å. Jönsson. 2002. On-line automated sample preparation for liquid chromatography using parallel supported liquid membrane extraction and microporous membrane liquid–liquid extraction. *J. Chromatogr. A* 975: 211–217.
91. Sandahl, M., E. Úlfsson, and L. Mathiasson. 2000. Automated determination of vinclozolin at the ppb level in aqueous samples by a combination of microporous membrane liquid–liquid extraction and adsorption chromatography. *Anal. Chim. Acta* 424: 1–5.
92. Norberg, J., E. Thordarson, L. Mathiasson, and J.Å. Jönsson. 2000. Microporous membrane liquid–liquid extraction coupled on-line with normal-phase liquid chromatography for the determination of cationic surfactants in river and waste water. *J. Chromatogr. A* 869: 523–529.
93. Barri, T., S. Bergström, J. Norberg, and J.Å. Jonsson. 2004. Miniaturized and automated sample pretreatment for determination of PCBs in environmental aqueous samples using an on-line microporous membrane liquid–liquid extraction-gas chromatography system. *Anal. Chem.* 76: 1928–1934.
94. Barri, T., S. Bergström, A. Hussen, J. Norberg, and J.-Å. Jönsson. 2006. Extracting syringe for determination of organochlorine pesticides in leachate water and soil-water slurry: A novel technology for environmental analysis. *J. Chromatogr. A* 1111: 11–20.
95. Bergstrom, S., T. Barri, J. Norberg, J.A. Jonsson, and L. Mathiasson. 2007. Extracting syringe for extraction of phthalate esters in aqueous environmental samples. *Anal. Chim. Acta* 594: 240–247.
96. Fontanals, N., T. Barri, S. Bergström, and J.-Å. Jönsson. 2006. Determination of polybrominated diphenyl ethers at trace levels in environmental waters using hollow-fiber microporous membrane liquid–liquid extraction and gas chromatography-mass spectrometry. *J. Chromatogr. A* 1133: 41–48.
97. Zorita, S., T. Barri, and L. Mathiasson. 2007. A novel hollow-fibre microporous membrane liquid–liquid extraction for determination of free 4-isobutylacetophenone concentration at ultra trace level in environmental aqueous samples. *J. Chromatogr. A* 1157: 30–37.
98. Shen, G. and H.K. Lee. 2002. Hollow fiber-protected liquid-phase microextraction of triazine herbicides. *Anal. Chem.* 74: 648–654.
99. Jiang, X. and H.K. Lee. 2004. Solvent bar microextraction. *Anal. Chem.* 76: 5591–5596.
100. Wang, J.-X., D.-Q. Jiang, and X.-P. Yan. 2006. Determination of substituted benzenes in water samples by fiber-in-tube liquid phase microextraction coupled with gas chromatography. *Talanta* 68: 945–950.
101. Lee, H.S.N., M.T. Sng, C. Basheer, and H.K. Lee. 2007. Determination of degradation products of chemical warfare agents in water using hollow fibre-protected liquid-phase microextraction with in situ derivatisation followed by gas chromatography-mass spectrometry. *J. Chromatogr. A* 1148: 8–15.
102. Zhao, L. and H.K. Lee. 2002. Liquid-phase microextraction combined with hollow fiber as a sample preparation technique prior to gas chromatography/mass spectrometry. *Anal. Chem.* 74: 2486–2492.
103. Müller, S., M. Möder, S. Schrader, and P. Popp. 2003. Semi-automated hollow-fibre membrane extraction, a novel enrichment technique for the determination of biologically active compounds in water samples. *J. Chromatogr. A* 985: 99–106.
104. Ouyang, G. and J. Pawliszyn. 2006. Kinetic calibration for automated hollow fiber-protected liquid-phase microextraction. *Anal. Chem.* 78: 5783–5788.

5 Mineralization Techniques Used in the Sample Preparation Step

Henryk Matusiewicz

CONTENTS

5.1 Introduction 95
5.2 Advanced Oxidation Processes for Water Sample Preparation 96
    5.2.1 UV Photo-Oxidation 97
    5.2.2 Ozone Oxidation 98
5.3 Microwave-Assisted Mineralization of Natural Waters 99
5.4 Conclusions 100
References 101

5.1 INTRODUCTION

Water is an essential resource for all living species, including human. Surface and subsurface water supplies accumulate many chemical constituents from both natural and anthropogenic sources. While knowledge about the effects of these accumulated impurities on biological, agricultural, and industrial systems is expanding, it is still limited in scope. It is therefore necessary to analyze a range of natural waters (surface waters, e.g., rivers, streams, lakes, reservoirs, oceans, and seas; precipitation, e.g., rain, dew, hail, and snow; groundwater), polluted waters (e.g., industrial effluents and sewage sludge), and purified waters (e.g., drinking water and distilled water). Many laboratories deal with the determination of heavy metals, carbon, nitrogen, and phosphorus in natural water samples. But such samples also contain numerous dissolved organic substances (mainly alcohols, aldehydes, carboxylic acids, and macromolecular compounds with several functional groups such as humic and fulvic acids) that could affect the complexation of the analyte or displace the retained metal complex from the stationary phase, so these competitive ligands need to be decomposed before instrumental analytical techniques can be applied. Therefore, the first step in the chemical analysis of a water sample is its proper preparation.

Metal concentrations are determined using molecular spectrophotometric, atomic spectrometric, and electrochemical techniques. All of these require samples to be homogenous, or at least to contain the smallest possible amounts of organic matter that could interfere with the metal determination by interacting with the metal ions and the analytical reagents. Traditionally, decomposition of the sample in elemental analysis requires it to be mineralized in order to remove the organic content.[1] Sample decomposition for total element determination therefore appears to be the recommended procedure on every occasion.

This chapter presents an overview of sample treatment procedures [e.g., decomposition, digestion, mineralization, oxidation, ultraviolet (UV) decomposition, and ozonation] and discusses a range of analytical techniques for the mineralization of natural aquatic systems. Other samples, such as effluents and sewage sludge, are beyond the scope of this contribution and will not be discussed here.

5.2 ADVANCED OXIDATION PROCESSES FOR WATER SAMPLE PREPARATION

Traditionally, pretreatment of a water sample in elemental analysis requires mineralization and/or dissolution of the matrix in order to remove the organic content. The presence of dissolved organic matter (DOM) in samples makes the application of electrochemical and atomic spectrometric techniques (AAS, OES, AFS) difficult or even impossible. Hence, the proper preparation of a sample, in particular the elimination of organic matter, is of great importance. Some analytical procedures include a wet digestion step, consisting of sample evaporation followed by sample heating with concentrated acids. However, there are several drawbacks to this classical approach: (i) the acids can themselves be a source of contamination, and they constitute a hazard; (ii) the sample preparation methods are time-consuming; and (iii) there is a chance of analyte loss through volatilization or retention by an insoluble residue.

In routine analytical laboratories, the use of advanced oxidation processes (AOPs) is an emerging alternative to conventional sample treatments[2] for analytical and environmental chemists. AOPs involve the *in situ* generation of highly potent chemical oxidants, such as the hydroxyl radical (OH*). Several processes have been applied in analytical sample pretreatment: homogenous UV irradiation, either by direct irradiation of the sample or photolysis mediated by an appropriate chemical reagent; ozone; and ultrasonic irradiation. A variety of AOPs ensures compliance of specific treatment requirements with optimum treatment technologies (Table 5.1).

Most AOPs take place at room temperature with the organic matter present in the sample. The results are various: decrease of DOM; inactivation of the sequestering agents present; destruction of the metal-organic compounds; diminished oxidant properties of the solution. The equipment is usually inexpensive, readily available, and does not introduce impurities in the sample. In addition, only minimal skills are required of the operator. As a result of the mild conditions used with AOPs (e.g., smaller quantities and lower concentrations of reagents and low temperature), the risk of analyte loss and the hazard to the chemist are less than with traditional procedures. Moreover, sample preparation is not so time-consuming. The acids and other reagents used in AOPs should be of sufficient purity to result in a minimal blank contribution to the final result. Most chemical reagent manufacturers supply a range of reagents suitable for low-level electrochemical and atomic spectroscopic analysis, and it is these that should be used. Blank values are dependent upon the type of AOP but

---

TABLE 5.1
Representative Advanced Oxidation Processes

UV/O_2	Photolysis
UV/H_2O_2	UV peroxide process
UV/O_3	Ozonolysis
$O_3/H_2O_2/UV$	Peroxon process
$UV/Fe^{3+}/O_2$	Fe^{3+}-catalyzed photolysis
$UV/H_2O_2/Fe^{3+}$	Photoassisted Fenton process
$UV/TiO_2/O_2$	Photocatalysis

are typically less than the criterion of detection, except for ubiquitous elements such as calcium, magnesium, sodium, and zinc.

5.2.1 UV Photo-Oxidation

UV digestion is a clean sample preparation method, not requiring large amounts of oxidants. UV photo-oxidation, that is, UV digestion, photolysis, UV irradiation, is utilized mainly in conjunction with uncontaminated or slightly contaminated natural water matrices (aqueous solutions). Liquids are decomposed by UV radiation in the presence of small amounts of hydrogen peroxide, acids (mainly HNO_3), or peroxydisulfate.[3] UV photo-oxidation minimizes the use of hot, hazardous reagents, for example, perchloric and sulfuric acids. DOM and the complexes of the chemical elements are decomposed to yield free metal ions. The digestion vessel should be placed as close as possible to the UV lamp (low, medium, and high pressure) to ensure a high photon flux; the lamp has an emission spectrum with a maximum at 254 nm. In photolysis, the digestion mechanism involves the formation, initiated by the UV radiation, of OH* radicals from both water and hydrogen peroxide molecules.[3] These reactive radicals are able to oxidize to carbon dioxide and water, the organic matter present in simple matrices containing up to about 100 mg/L of carbon. Complete elimination of the matrix effect is, of course, possible only with simple matrices or by combining photolysis with other digestion techniques such as ozonation. The method does not oxidize all the organic components that could be present in water; chlorinated phenols, nitrophenols, hexachlorobenzene, and similar compounds are only partly oxidized. Effective cooling of the sample is essential, because losses may otherwise be incurred with highly volatile elements. The addition of hydrogen peroxide may need to be repeated several times to produce a clear sample solution. Modern UV digestion systems are commercially available (Table 5.1).[3]

Photochemical operations offer several routes of hydroxyl radical formation by UV irradiation. The formation of hydroxyl radicals by irradiation of samples doped with hydrogen peroxide or ozone is the state-of-the-art in water treatment. Two comprehensive reviews cover the historical development of the UV photo-oxidation technique as a pretreatment step in the inorganic analysis of natural waters, its principles and the equipment available, and its principal applications in the analytical field.[3,4] They include tables summarizing the elements determined, the analytical techniques used, and the sample matrices studied.

Only a few of the numerous UV irradiation-based digestion methods to have been published meet the requirement of being both efficient and error-free. UV photo-oxidation appears to be irreplaceable for water analysis, because of the low concentration of the elements investigated; the application of other means of digestion could cause contamination. In these UV methods, a small amount of hydrogen peroxide, the source of oxygen-free radicals, has to be added to the acidified sample. In general, the sample needs to be boiled from 1 to 4 h, depending on the organic content in the water, in order to destroy this matter without producing perceptible errors.

UV digestion is a clean sample preparation method, as it does not require the use of large amounts of oxidants. Furthermore, the method is effective and can be readily incorporated into flow injection manifolds. On-line photo-oxidation (UV irradiation) has also been used for the digestion of organic compounds in waters. The sample flows through a tube (PTFE, quartz) coiled around a fixed UV lamp(s) in the presence of O_3, H_2O_2, $K_2S_2O_8$, or HNO_3, in which the UV irradiation acts as a catalyst. A short review of such flow systems has appeared recently.[3] Flow systems are becoming more popular in analysis because of their ease of automation, speed, small volume of sample, and elegance; the future in this respect looks promising.

Apart from acidification, UV irradiation is the only preliminary preparative step in voltammetric determinations; it may be required to degrade organic substances binding trace metals in the form of inert complex species. The analysis of samples from natural waters may be significantly affected by the binding capacity of such dissolved organic substances.

Table 5.2 lists the most relevant analytical applications of UV photo-oxidation.

TABLE 5.2
Published Selected UV Photo-Oxidation Techniques

Sample Type	Condition of Photo-Oxidation	Elements Determined	Analytical Technique	Reference
Natural water	UV lamp 150 W, $t = 12$ h, pH 2, addition of H_2O_2	Cu, Hg	DPASV-RDE-Au	5
Fresh water	UV lamp 900 W, a few drops of 30% H_2O_2 and 0.004 M H_2SO_4, $t = 1.5$–2 h	P	Colorimetric determination by molybdenum blue method	6
Water	UV lamp 1200 W, flow system, $t = 15$ min, addition of $K_2S_2O_8$ or H_2O_2, variable pH	As	HG-AAS	7
Fresh water	UV lamp 6 W, flow system, $t = 40$ min	Cd, Cu, Pb	FAAS, ICP-MS	8
River water	UV sample pretreatment, flow system, 0.14 M $K_2S_2O_8$	Hg	FI-CV-AAS	9
Sea water	UV lamp 700 W, flow system, $t = 100$ s, 0.1 M H_2SO_4/0.1 M NaOH	As	FI-HG-AAS	10
Sea water	UV lamp 8 W, flow system, 0.5% $K_2Cr_2O_7$/8% HCl	Hg	FI-CV-AAS	11
Sea water	UV lamp 15 W ($t > 24$ h) and 400 W ($t = 30$ min)	Cd, Cu, Hg, Pb	FAAS, CV-AAS	12
Natural water	UV lamp 150 W, $t = 2$–3 h, pH 2, addition of H_2O_2	Cd, Cu, Pb, Zn	DPASV-HMDE	13
River water, snow	UV lamp 150 W, pH 1	Bi, Sb	DPASV-HMDE	14
River water	UV lamp 1000 W, $t = 2$–10 min, pH 2, addition of H_2O_2	Cd, Cu, Pb, Zn	DPASV-HMDE	15
Natural waters	UV lamp 500 W, $t = 30$ min, pH 2, addition of H_2O_2	Cd, Co, Cu, Ni, Pb, Zn	DPASV	16
Sea water	In-line UV-digestion, lamp 100 (1000) W, addition of H_2O_2, pH 8.1; 2.5	Cr, Cu, Ni	ASV-HMDE	17
River and pond waters	UV lamp 30 W, $t = 20$ min, addition of $K_2S_2O_8$	Hg	CV-AAS	18

Notes: DP-ASV, *d*ifferential-*p*ulse *a*nodic *s*tripping *v*oltammetry; RDE-Au, *r*otating *g*old *e*lectrode; HG-AAS, *h*ydride *g*eneration *a*tomic *a*bsorption *s*pectrometry; FAAS, *f*lame *a*tomic *a*bsorption spectrometry; FI-CV-AAS, *f*low-*i*njection *c*old-*v*apor *a*tomic *a*bsorption *s*pectrometry; and HMDE, *h*anging *m*ercury *d*rop *e*lectrode.

5.2.2 Ozone Oxidation

Ozone oxidation (ozonation) is very effective in sample treatment and remarkably active in destroying natural organic compounds prior to elemental determination.[19] Ozone readily decomposes to produce radicals, which are believed to be responsible for the majority of the observed reactivity. It reacts at room temperature with the organic matter present in the sample and also renders inactive any sequestering agents present. It can be generated from the oxygen present in the air, an inexpensive and readily available procedure, and does not harbor impurities. Ozone is produced by either silent electrical discharge or the UV radiation method, the former being the more usual method.

TABLE 5.3
Trace Metal Determination after Ozone Sample Treatment

Sample Type	Condition of Ozonation	Elements Determined	Analytical Technique	Reference
Natural water	Samples in acetate buffer, pH 4.7, 1 h ozonation	Cd, Pb	ASV	20
River water	Samples acidified to pH 2 with HCl, ozonation time ranges from 30 to 60 min	Cu	ASV	21
Geothermal water		Hg	CV-AAS	22
Water	Ozonation time ranges from 15 to 60 min, sample in 0.2 M HCl	Hg	CV-AAS	23
Water	A small ozonizer was developed, 0.5 mL, ozonation time 30 s, pH > 7	Hg	CV-AAS	24

Notes: ASV, *anodic stripping voltammetry* and CV-AAS, *cold-vapor atomic absorption spectrometry.*

When aqueous solutions are treated with ozone for several hours, there is significant sample evaporation: the rate of this process depends on the initial sample volume and the average temperature of the sample during the period of ozonolysis. The time required for ozone application is considerably longer than that needed with other AOPs, such as UV irradiation.

Table 5.3 sets out the applications of ozone for the determination of elements in water; there are only a few such applications, despite the efficiency of ozone in decomposing organic matter.

5.3 MICROWAVE-ASSISTED MINERALIZATION OF NATURAL WATERS

Wet digestion with chemical oxidants such as nitric acid, sulfuric acid, perchloric acid, or hydrogen peroxide has been used to decompose DOM in aqueous solutions. A serious drawback of this approach, however, is the need to add large concentrations of oxidants to the sample, which must then be decreased after digestion (by evaporation or sample dilution) to meet the conditions required for analysis. It is preferable to use a less complex digestion method that gives complete and consistent recovery but is still compatible with the analytical methodology and with the metal to be analyzed. Nitric acid decomposes most water samples in an appropriate form, and nitrate is also an acceptable matrix for most chemical analysis techniques. As a general rule, HNO_3 is only used for easily oxidized natural water samples.

Relatively few studies have investigated the application of microwave-assisted digestion for the determination of elements in water samples (Table 5.4). This is because natural water samples (rivers, raw, and drinking waters), containing minimal amounts of solid material, require a relatively mild wet digestion pretreatment procedure; in contrast, samples that contain significant amounts of solid material such as effluents and sewage sludge require more vigorous procedures. However, optimal conditions cannot be achieved when microwave digestion is applied to the analysis of water, since comparatively large volumes of sample are used and, as a consequence, the digestion reagents are much diluted. Their diminished decomposition power may be insufficient to mineralize completely all the components of the sample. This limitation is not always evident if the sample solutions are eventually analyzed by atomic absorption or plasma excitation, as these techniques possess a considerable inherent decomposing power. They are, therefore, capable of evaluating incompletely digested sample solutions.

TABLE 5.4
Microwave-Assisted Digestion Procedures for Water Samples

Sample Type	Elements Determined	Reagents Used	Microwave System	Digestion Method	Analysis Technique	Reference
Water	Se	HCl	Prolabo Microdigest 301	On-line digestion	HG-AAS	25
Water	Se	HCl	Prolabo Microdigest 301	On-line digestion	HG-AFS	26
Environmental waters	As, Bi, Hg, Pb, Sn	$KBrO_3$-KBr-HCl	Prolabo Maxidigest MX-350	On-line digestion	FI-CV-AAS, HG-AAS	27
Water	Se	KBr-HCl	Milestone MLS-1200 MEGA	On-line pretreatment	HG-AAS	28
Mineral and sea waters	As	$K_2S_2O_8$-NaOH	Domestic oven	On-line oxidation	HG-AAS	29
Surface waters	Al, As, Be, Ca, Cu, Cd, Fe, K, Li, Mg, Mn, Na, Pb, Sn, Sr, Zn	HNO_3	CEM MDS-81D	Pressure digestion	FAAS, GF-AAS	30
Geothermal fluids	Fe	H_2O_2	Domestic oven	Flow system	ET-AAS	31
Rain water	Pd	HCl-HNO_3-HF, $HClO_4$, aqua regia	Multiwave 3000	Microwave-UV digestion	FI-ET-AAS	32
Lake, river, and rain waters	Hg	HCl	Prolabo Maxidigest MX-350	On-line pretreatment	CV-AAS	33

Notes: HG-AAS, *h*ydride *g*eneration *a*tomic *a*bsorption *s*pectrometry; HG-AFS, *h*ydride *g*eneration *a*tomic *f*luorescence *s*pectrometry; FI-CV-AAS, *f*low-*i*njection *c*old-*v*apor *a*tomic *a*bsorption *s*pectrometry; FAAS, *f*lame *a*tomic *a*bsorption *s*pectrometry; GF-AAS, *g*raphite *f*urnace *a*tomic *a*bsorption *s*pectrometry; and ET-AAS, *e*lectrothermal *a*tomic *a*bsorption *s*pectrometry.

5.4   CONCLUSIONS

Owing to the very low concentration of trace elements in natural waters, decomposition of interfering substances is recommended in order to eliminate matrix effects. Under these conditions, the total concentration of trace elements can be measured with an acceptable accuracy by skilled analysts in suitable laboratories.

In the analytical laboratory, AOPs and microwave-assisted digestion techniques are now important and powerful tools in the pretreatment of natural water samples for elemental spectrometry. AOPs are effective at extracting a large number of inorganic analytes; their ability to decompose the organic matter in a sample as well as simplicity, economy, and safety are further advantages. Ozonolysis as a water sample treatment procedure eliminates mainly the surface-active properties of organic matter rather than degrading the matter itself. Compared with the state-of-the-art H_2O_2/UV process, which must be operated with a surplus of oxidizer, microwave-assisted digestion can be operated at moderate microwave power. A novel microwave-assisted high-temperature UV digestion procedure has been developed for the accelerated decomposition of interfering dissolved organic carbon prior to trace element determination in water samples. It is a technique that has significantly improved the performance of UV digestion (oxidation) and is especially useful in ultratrace analysis because of its extremely low risk of contamination.[34,35]

REFERENCES

1. Matusiewicz, H. 2003. Wet digestion methods. In: Z. Mester and R. Sturgeon (eds), *Sample Preparation for Trace Element Analysis*, pp. 193–233. Amsterdam: Elsevier.

2. Capelo-Martínez, J.L., P. Ximénez-Embún, Y. Madrid, and C. Cámara. 2004. Advanced oxidation processes for sample treatment in atomic spectrometry. *Trends Anal. Chem.* 23: 331–340.

3. Golimowski, J. and K. Golimowska. 1996. UV-photooxidation as pretreatment step in inorganic analysis of environmental samples. *Anal. Chim. Acta* 325: 111–133.

4. Maher, W. and L. Woo. 1998. Procedures for the storage and digestion of natural waters for the determination of filterable reactive phosphorus, total filterable phosphorus and total phosphorus. *Anal. Chim. Acta* 375: 5–47.

5. Sipos, L., J. Golimowski, P. Valenta, and H.W. Nürnberg. 1979. New voltammetric procedure for the simultaneous determination of copper and mercury in environmental samples. *Fresenius Z. Anal. Chem.* 298: 1–8.

6. Henriksen, A. 1970. Determination of total nitrogen, phosphorus and iron in fresh water by photo-oxidation with ultraviolet radiation. *Analyst* 95: 601–608.

7. Atallah, R.H. and D.A. Kalman. 1991.On-line photo-oxidation for the determination of organoarsenic compounds by atomic-absorption spectrometry with continuous arsine generation. *Talanta* 38: 167–173.

8. Guéguen, C., C. Belin, B.A. Thomas, F. Monna, P.Y. Favarger, and J. Dominik. 1999. The effect of freshwater UV-irradiation prior to resin preconcentration of trace metals. *Anal. Chim. Acta* 386: 155–159.

9. Wurl, O., O. Elsholz, and J. Baasner. 2000. Monitoring of total Hg in the river Elbe: FIA-device for on-line digestion. *Fresenius J. Anal. Chem.* 366: 191–195.

10. Cabon, J.Y. and N. Cabon. 2000. Determination of arsenic species in seawater by flow injection hydride generation *in situ* collection followed by graphite furnace atomic absorption spectrometry. Stability of As(III). *Anal. Chim. Acta* 418: 19–31.

11. Wurl, O., O. Elsholz, and R. Ebinghaus. 2001. On-line determination of total mercury in the Baltic Sea. *Anal. Chim. Acta* 438: 245–249.

12. Vasconcelos, M.T.S.D. and M.F.C. Leal. 1997. Speciation of Cu, Pb, Cd and Hg in waters of the Oporto coast in Portugal, using pre-concentration in a chelamine resin column. *Anal. Chim. Acta* 353: 189–198.

13. Golimowski, J., A.G.A. Merks, and P. Valenta. 1990. Trends in heavy metal levels in the dissolved and particulate phase in the Dutch Rhine-Meuse (MAAS) delta. *Sci. Total Environ.* 92: 113–127.

14. Postupolski, A. and J. Golimowski. 1991. Trace determination of antimony and bismuth in snow and water samples by stripping voltammetry. *Electroanal.* 3: 793–797.

15. Labuda, J., D. Saur, and R. Neeb. 1994. Anodic stripping voltammetric determination of heavy metals in solutions containing humic acids. *Fresenius J. Anal. Chem.* 348: 312–316.

16. Kolb, M., P. Rach, J. Schäfer, and A. Wild. 1992. Investigation of oxidative UV photolysis. I. Sample preparation for the voltammetric determination of Zn, Cd, Pb, Cu, Ni and Co in waters. *Fresenius J. Anal. Chem.* 342: 341–349.

17. Achterberg, E.P. and C.M.G. van den Berg. 1994. In-line ultraviolet-digestion of natural water samples for trace metal determination using an automated voltammetric system. *Anal. Chim. Acta* 291: 213–232.

18. Nagashima, K., T. Murata, and K. Kurihara. 2002. Pretreatment of water samples using UV irradiation-peroxodisulfate for the determination of total mercury. *Anal. Chim. Acta* 454: 271–275.

19. Clem, R.G. and A.F. Sciamanna. 1975. Styrene impregnated, cobalt-60 irradiated, graphite electrode for anodic stripping analysis. *Anal. Chem.* 47: 276–280.

20. Clem, R.G. and A.T. Hodgson. 1978. Ozone oxidation of organic sequestering agents in water prior to the determination of trace metals by anodic stripping voltammetry. *Anal. Chem.* 50: 102–110.

21. Filipović-Kovačević, Ž. and L. Sipos. 1998. Voltammetric determination of copper in water samples digested by ozone. *Talanta* 45: 843–850.

22. Sakamoto, H., J. Taniyama, and N. Yonehara. 1997. Determination of ultra-trace amounts of total mercury by gold amalgamation-cold vapor AAS in geothermal water samples by using ozone as pretreatment agent. *Anal. Sci.* 13: 771–775.

23. Anthemidis, A.N., G.A. Zachariadis, and J.A. Stratis. 2004. Development of a sequential injection system for trace mercury determination by cold vapour atomic absorption spectrometry. *Talanta* 64: 1053–1057.

24. Sasaki, K. and G.E. Pacey. 1990. The use of ozone as the primary digestion reagent for the cold vapor mercury procedure. *Talanta* 50: 175–181.

25. Pitts, L., P.J. Worsfold, and J. Hill. 1994. Selenium speciation—a flow injection approach employing on-line microwave reduction followed by hydride generation-quartz furnace atomic absorption spectrometry. *Analyst* 119: 2785–2788.

26. Pitts, L., A. Fisher, P. Worsfold, and S.J. Hill. 1995. Selenium speciation using high-performance liquid chromatography-hydride generation atomic fluorescence with on-line microwave reduction. *J. Anal. At. Spectrom.* 10: 519–520.

27. Tsalev, D.L., M. Sperling, and B. Welz. 1992.On-line microwave sample pre-treatment for hydride generation and cold vapour atomic absorption spectrometry. Part 2. Chemistry and applications. *Analyst* 117: 1735–1741.

28. González LaFuente, J.M., M.L. Fernández Sánchez, J.M. Marchante-Gayón, J.E. Sánchez Uria, and A. Sanz-Medel. 1996. On-line focused microwave digestion-hydride generation of inorganic and organic selenium. *Spectrochim. Acta Part B* 51: 1849–1857.

29. López-Gonzales, M.A., M.M. Gómez, C. Cámara, and M.A. Palacios. 1994. On-line microwave oxidation for the determination of organoarsenic compounds by high-performance liquid chromatography-hydride generation atomic absorption spectrometry. *J. Anal. At. Spectrom.* 9: 291–295.

30. Paukert, T. and Z. Sirotek. 1993. A study of the microwave treatment of water samples from the Elbe River, Bohemia, Czech Republic. *Chem. Geol.* 107: 133–144.

31. Burguera, J.L., M. Burguera, and C.E. Rondon. 1998. Automatic determination of iron in geothermal fluids containing high dissolved sulfur-compounds using flow injection electrothermal atomic absorption spectrometry with an on-line microwave radiation precipitation–dissolution system. *Anal. Chim. Acta* 366: 295–303.

32. Limbeck, A. 2006. Microwave-assisted UV-digestion procedure for the accurate determination of Pd in natural waters. *Anal. Chim. Acta* 575: 114–119.

33. Welz, B., D.L. Tsalev, and M. Sperling. 1992. On-line microwave sample pretreatment for the determination of mercury in water and urine by flow-injection cold-vapour atomic absorption spectrometry. *Anal. Chim. Acta* 261: 91–103.

34. Florian, D. and G. Knapp. 2001. High-temperature, microwave-assisted UV digestion: A promising sample preparation technique for trace element analysis. *Anal. Chem.* 73: 1515–1520.

35. Matusiewicz, H. and E. Stanisz. 2007. Characteristics of a novel UV-TiO_2-microwave integrated irradiation device in decomposition processes. *Microchem. J.* 86: 9–16.

6 Biota Analysis as a Source of Information on the State of Aquatic Environments

J.P. Coelho, A.I. Lillebø, M. Pacheco, M.E. Pereira, M.A. Pardal, and A.C. Duarte

CONTENTS

6.1 Introduction 103
6.2 Choice of Species 104
    6.2.1 Primary Producers 105
    6.2.2 Suspension Feeders 106
    6.2.3 Sediment Dwellers 107
    6.2.4 Pelagic Species 108
6.3 Assessment Strategies 109
    6.3.1 Accumulation and Partitioning of Contaminants in Plants 110
    6.3.2 Differential Tissue Analysis in Animals 110
6.4 Supplementary Methodologies 112
    6.4.1 Oxidative Stress 112
    6.4.2 Metallothioneins 113
6.5 Conclusions 115
References 115

6.1 INTRODUCTION

An increased international awareness toward an assessment of the current status of aquatic environments and their protection has emerged as a result of multiple anthropogenic pressures, such as human population growth, progressive industrialization, and intensive agricultural practices. Initially, such an assessment was focused on the abiotic fractions of the environment, the dissolved fraction of the water column, and sediments. But the monitoring of dissolved contaminants is methodologically challenging, primarily because of limitations in analytical methods (capability of quantification), and also as a consequence of the typically low concentrations, not to mention the tendency for samples to become contaminated during collection or analysis.[1] Furthermore, levels of dissolved contaminants are extremely variable and depend not only on season,[2,3] but on factors such as tidal cycles[4] and river flow[5] as well. So a really accurate environmental assessment would require intensive, costly, and time-consuming monitoring programs.[1,2] Finally, we need to mention that total dissolved contaminant concentrations are in fact not an accurate measure either of the fraction available to organisms or of that responsible for toxicity and bioaccumulation problems.

The same is true for sediments. Total contaminant concentrations in sediments do not reflect their bioavailability to organisms, which can be affected by numerous variables such as sediment geochemistry, particle size, or organic matter content.[1,6–8] Nevertheless, the fact that contaminants accumulate in sediments is an advantage for monitoring, given that the relevant methodologies are more reliable and the matrix less prone to contamination.[1] Moreover, sediment contaminant levels are more stable and may provide a historical record of contaminant load.[9–11] The extent to which sediment contaminants are available to aquatic organisms is a current research topic in which different methodologies are applied—from sequential extraction methods[8,12] to studies on assimilation efficiencies and gut juice extractions.[7,13]

In view of the limitations associated with the monitoring of dissolved and sediment-bound contaminants, the use of biota as a source of information on the quality of aquatic environments was suggested, and it has now become a widely accepted methodology for assessing contaminant bioavailability. Such an approach provides a time-integrated measure of contaminant availability to organisms, is easily measurable owing to the generally higher levels found in tissues, and is not prone to accidental contamination; it can therefore remove some of the limitations associated with water and sediment monitoring strategies.[1]

The recent approval and implementation of the European Water Framework Directive further emphasizes the role of biota as a tool for assessing aquatic environmental quality, in that it strives not only for the improvement of the chemical quality status of water bodies but also for the rehabilitation of their ecological status. In the light of these recommendations, it becomes essential to use biota to assess not only the chemical status of water bodies through contaminant load analysis, but also their ecological status, in what must be an integrated, multidisciplinary approach.

Three fundamental questions arise if this kind of methodology is to be invoked to monitor the health of water bodies: (a) Which of the countless species inhabiting any given aquatic ecosystem should be used as a source of environmental information? (b) Which assessment strategy should be employed: the analysis of entire organisms or specific tissue discrimination? (c) Which supplementary methodologies can be applied to complement the information supplied by biota analysis so as to facilitate a more rigorous evaluation of the state of the ecosystem?

This chapter presents an overview of the different approaches and methodologies used for gathering environmental information from biota analysis. The choice of species, the analysis of individual whole organisms or selected tissues, and the study of the partition of contaminants within organisms to subcellular methods such as stress-related enzymatic studies and metal-binding proteins [metallothioneins (MTs) or phytochelatins (PCs)] will be discussed as a way of responding to these three major issues. Of the multitude of contaminants present in aquatic environments, we shall focus principally on metals; these are regarded as one of the main and most toxic groups of environmental contaminants, and very liable to bioaccumulate. But even though many of the aspects and methodologies are discussed in the context of metals, they are equally applicable to other classes of contaminant.

Case studies will be presented, essentially from an extensively studied system (the Ria de Aveiro coastal lagoon, west coast of Portugal), where a well-defined anthropogenic mercury contamination gradient has justified numerous studies on the bioaccumulation of contaminants and its effects.

6.2   CHOICE OF SPECIES

The use of bioindicators and biomonitors for gathering environmental information is currently widespread, although some abuse of these terms is customary. The terms *bioindicator* and *biomonitor* are commonly but incorrectly taken to be synonymous: a *bioindicator* provides qualitative information on the quality of the environment, whereas a *biomonitor* supplies quantitative data on environmental contamination.[14] This section focuses on the latter, since the trace metal content of biota has been commonly used in biomonitoring programs of metal pollution in the marine environment and is considered to provide a time-integrated measure of metal bioavailability.[1,15,16]

In a review on the subject, Rainbow and Phillips[15] enumerate the desirable properties for a suitable biomonitor species. According to these authors, a suitable biomonitor should be sessile, abundant, and easy to identify and sample. They also consider resistance and tolerance to stress and environmental variations to be important features, as is the absence of regulation in the accumulation process. However, taking into account the different pathways by which organisms are exposed to contaminants,[15,17,18] species-specific traits, and the specificity of any given contaminant, the suitability of different species as biomonitors may differ. The solution, according to Rainbow and Phillips,[15] would be to use not one biomonitor species, but a suite of biomonitors that would reflect the overall bioavailability of contaminants in the various compartments of an ecosystem (dissolved and particulate fractions of the water column and sediments). Also, since some contaminants are transferred from prey to predator via biomagnification processes, it would be useful to choose a set of biomonitors involving different trophic levels.

6.2.1 Primary Producers

Macroalgae have generally been considered efficient biomonitors of dissolved metal sources,[2,15,19–23] but there appear to be two major limitations to their use. Some genera, especially green macroalgae from the genus *Ulva*, show marked morphological polymorphism, which makes taxonomic identification problematic. In view of the variability in contaminant accumulation between even closely related species, some bias may result from using species from this genus for biomonitoring studies.[19] The other crucial limitation of macroalgae stems from their surfaces becoming contaminated by epiphytes and particulate matter, which may give rise to overestimations of contaminant accumulation.[24] The observed positive correlations with sediment contamination rather than with dissolved fraction contaminant levels[25,26] may in part reflect surface contamination by highly contaminated fine particulates. The washing techniques to cope with this limitation are not standardized,[19,24] and the use of correction factors (for instance, the Al content, an indicator of lithogenic particles) does not cover the contribution of adherent epiphytes to the overall contaminant accumulation. Species-specific traits should also be taken into account when choosing an algal species, since free-floating algae (mainly green macroalgae like *Ulva*) will be more mobile and, unlike the more sessile strains, may not reflect the environmental contamination at the site of collection.[23]

Lastly, the contaminant to be monitored may also influence the choice of algal biomonitor, since some species may be more suitable for specific contaminants. Coelho et al.[23] studied the response of dominant species of macroalgae (*Fucus vesiculosus*, *Gracilaria verrucosa*, and *Enteromorpha intestinalis*) to a mercury contamination gradient in a temperate estuary. While for total mercury the three species showed a similar pattern, the organic mercury fraction (methylmercury and other organic mercury compounds) was species-specific and displayed opposing trends with distance to contamination source (Figure 6.1); their suitability as biomonitors for this contaminant is thus questionable.

Fewer data are available on contaminant accumulation by sea grasses than for macroalgae. Nevertheless, sea grasses of several genera have been selected as biomonitors, with special focus on the genera *Posidonia* and *Zostera*.[19] In contrast to algae, sea grasses may reflect both dissolved contaminant concentrations and sediment contaminant loads, since they may accumulate these substances through both the root system and the aerial parts. This was observed in a study by Lafabrie et al.[27] of the seagrass *Posidonia oceanica* in the Mediterranean, where metals accumulated preferably in the blades, which suggested uptake from the water column. In another study, the same species was found to reflect contamination in both the water column (Cd and Pb) and the sediment (Co, Cr, Hg, and Ni).[28] Such differential uptake routes may lead to uncertainty about the source of contamination responsible for tissue contamination.

Salt marsh plants, in turn, reflect mainly sediment contaminant loads rather than the dissolved fraction, given their uptake route via the root system. Some research has been performed on the genera *Spartina* and *Phragmites*,[29–33] which accumulate metals essentially in the underground

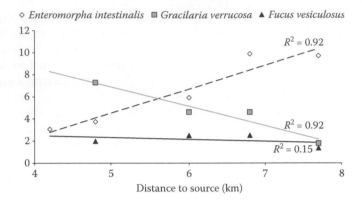

FIGURE 6.1 Organic mercury fraction (%) in three macroalgae species along a mercury gradient in a temperate estuary. (After Coelho et al. 2005. *Estuar. Coast. Shelf Sci.* 65: 492–500. With permission.)

biomass and restrict their translocation to the photosynthetic parts. Roots would therefore be a potential biomonitor of sediment contaminants, which is not always the case since salt marsh plants actively modify the environment around the roots, by oxygen transportation and its release in the root system, altering the chemical equilibrium and contaminant bioavailability. Furthermore, many salt marsh plants exhibit marked seasonality, which can lead to increased bias and possibly mask correlations with environmental contamination. The use of sea grasses and salt marsh plants to gather environmental information will be discussed later in the chapter.

6.2.2 Suspension Feeders

To assess the bioavailability of contaminants in the water column (in both dissolved and suspended particulates), the usual and most reliable option is to use suspension feeders, the best known of which are mussels. The Mussel Watch Program in the United States has been monitoring contaminant concentrations in mussels and oysters since 1986,[34–36] and similar programs have been implemented in other areas of the globe.[37–39] These take advantage of the worldwide distribution of the genus *Mytilus* and its favorable characteristics for biomonitoring purposes, such as high tolerance to contamination and salinity fluctuations, sessile existence, and ease of collection.[15,34] The principal limitation regarding the use of mussels (the genus *Mytilus* in temperate waters and the closely related genus *Perna* in tropical areas) is the coexistence of closely related species, taxonomically difficult to identify, which may bring the validity of data intercomparisons into question,[15] as was discussed by Szefer et al.[40] Despite this shortcoming, such an approach provides invaluable information about the ecological status of aquatic ecosystems, and can supply long-term databases from which the temporal trends of contaminant bioavailability at any given location can be inferred, provided that the same species is sampled.[34–36,39] To minimize physiology-associated bias, certain sampling guidelines should be followed. The sampling season ought to be invariable and outside the reproductive periods, preferably in winter, since many lipophilic compounds may be released during spawning owing to the high lipid content of eggs and sperm.[34] Also, sampling should focus on a predetermined size range, given the many years' lifespan of mussels, in order to reduce disparities in exposure time to contaminants, which could artificially mask any existing environmental trend. Finally, the sampling site should remain unchanged in order to ensure comparable exposure conditions to contaminants.[34]

A variation on this theme is the transplantation of individuals to desired monitoring sites, an active biomonitoring method. In this approach, however, some of the inherent natural variability is reduced by ensuring comparable biological samples.[41–44] Moreover, exposure time is controlled, the organisms are statistically similar, and the method is independent of the natural occurrence of the species selected; on the other hand, the experimental design and logistics are more complex and subject to disturbances during exposure.[41]

6.2.3 Sediment Dwellers

To assess contaminant bioavailability in sediments a different type of biomonitor is required, preferably a sediment-dwelling deposit feeder that will respond to contaminant levels in freshly deposited superficial sediments.[15] A substantial amount of research has been conducted on numerous benthic bivalve species, of which *Scrobicularia plana* and *Macoma balthica* are by general consent regarded as the most reliable biomonitors, not only in field monitoring studies,[41,45–47] but also in laboratory bioavailability and toxicity testing studies.[48–52] One major drawback with these organisms, however, is their more restricted geographical distribution compared to that of mussels and oysters, which reduces their usefulness for global monitoring programs.[15] Also, as they are sediment-dwelling species, sampling is more labor-intensive.[50]

Another limitation was pointed out by Coelho et al.[53] on the basis of the available literature on *S. plana*, he noted the lack of criteria homogeneity between different studies using the same species. Since each researcher often uses just one single size class,[47,51] or does not even make reference to the size class studied,[45,46,54] intercomparison of studies is virtually impossible. Coelho et al.[53] stress the importance of having some knowledge of the lifespan and the annual bioaccumulation patterns of each species: results can then be extrapolated from one specific size class to another, and meaningful comparison between different studies becomes possible. Their study of the bioaccumulation pattern of *S. plana* suggested that this species accumulates mercury linearly during its lifespan and, moreover, that the ratio of annual bioaccumulation rates to the levels of this contaminant in suspended particulate matter (SPM) is consistent. If confirmed, this could prove to be a good predictor of accumulation rates for *S. plana* (Table 6.1). Since the bioaccumulation of mercury appears to follow a linear model, its annual accumulation rates could be predicted for any given size class, and study comparison would then be feasible.[53]

Polychaetes, which are generally deposit-feeding detritivores, are another group frequently considered for sediment biomonitoring purposes. The genus *Hediste* has been widely used,[55–61] especially the species *Hediste diversicolor*. Some caution is necessary in their use, however, since the taxonomy of polychaetes is complex, and the coexistence of related species is common. Moreover, *H. diversicolor* is known to regulate the tissue concentrations of several trace elements;[15] it also exhibits a wide range of feeding strategies,[62] which may prevent a correct assessment of the source

---

TABLE 6.1
Mercury Concentrations in SPM (mg kg^{-1}, Mean Values ± Standard Deviation), Annual Bioaccumulation Rates in *S. plana* (Calculated from the Slope of Regression Lines, mg kg^{-1} y^{-1}), and the Ratio between Annual Accumulation Rates and SPM Hg Concentrations

Station	SPM [Hg] (mg kg^{-1})	Annual Bioaccumulation Rates	Accumulation Rate/SPM
A1	25.8 ± 0.4	0.258	0.01
A2	20.1 ± 2.6	0.254	0.013
A4	6.5 ± 0.2	0.064	0.01
A5	8.9 ± 0.5	0.035	0.004
A11	1.0 ± 0.1	0.009	0.009
A15	1.2 ± 0.5	−0.002	−0.002

Source: After Coelho et al. 2006. *Estuar. Coast. Shelf Sci.* 69: 629–635. With permission.

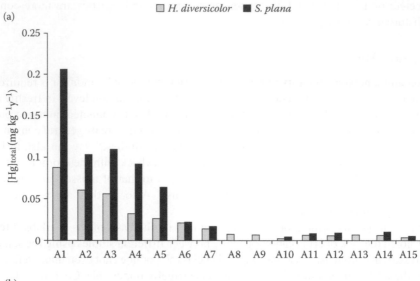

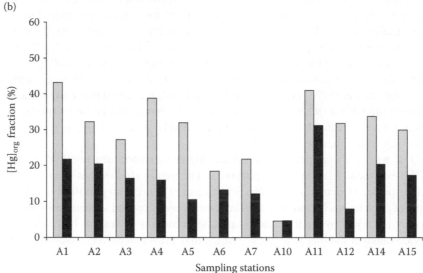

FIGURE 6.2   (a) Annual mercury accumulation (mg kg^{-1} y^{-1}) and (b) organic mercury fraction (%) in *S. plana* and *H. diversicolor*. (After Coelho et al. 2008. *Estuar. Coast. Shelf Sci.* 78: 516–523. With permission.)

and extent of contamination. In a recent comparative study on mercury accumulation by *H. diversi-color* and *S. plana*, Coelho et al.[61] found that the clam accumulated more mercury than the worm, both on an absolute and on an annual level, which may indicate regulation of mercury uptake by the polychaete (Figure 6.2a). In the same study, the organic mercury fraction of *H. diversicolor* was found to be consistently higher than that of *S. plana* (Figure 6.2b), and the worm's omnivore diet was considered at least partially responsible for these results. Therefore, the use of tellinid bivalves for sediment biomonitoring seems preferable to that of polychaete worms.

6.2.4 PELAGIC SPECIES

An extensive literature exists on contaminant accumulation in fish with a wide range of dietary strategies.[63–68]

The major limitation associated with the use of pelagic species for monitoring, particularly in field surveys, is the increased mobility of these larger organisms when compared to species normally used as biomonitors, that is, bivalves. Especially in point source contamination situations, there is no certainty whether organisms have been subject to the same level of contamination and for the same time periods, as would be in the case of sessile species.[65,69] As tidal and seasonal migrations within aquatic systems are common,[63,65] increased bias may result; this implies that fish may not accurately reflect environmental contamination at the site of collection. A solution to minimize this limitation could be to focus on immature individuals, which commonly have a restricted geographical range and would thus better reflect local contaminant stress. However, such an approach calls for caution, since other sources of uncertainty may emerge from the use of juveniles, such as dietary changes with growth. Transplantation experiments involving the caging of individuals could also cope with this limitation, although in these situations the risk of stress-related bias is not negligible.

Nonetheless, the use of pelagic species to monitor aquatic environments is not without its merits. Two main goals are usually pursued when analyzing fish contaminant loads: the determination of the health risk to humans and the use of fish as environmental indicators of aquatic ecosystem quality.[70] Not surprisingly, since fish are a major part of the human diet, most research focuses on the health perspective and selects edible species, as contaminants are mostly quantified in muscular tissue.[64,67] This anthropocentric perspective assesses mainly the risks associated with consumption of fish[70–72] and does not provide a great deal of environmental information. In contrast, studies directed toward lifespan accumulation patterns and dietary effects of accumulation[66,73] may supply invaluable information on the bioaccumulation, biomagnification, and toxicity of contaminants within aquatic food webs. Given the high trophic position of fish, piscivorous species in particular may reflect the contamination and toxicity risks to top predators with diets similar to those of aquatic mammals and birds.

6.3   ASSESSMENT STRATEGIES

The choice of assessment strategy will depend mainly on the purpose of the study. Most monitoring programs rely on whole-body contaminant analysis as a preferential methodology. It is especially effective for simple structured, homogenous organisms such as algae, and also small invertebrates. Sample preparation and manipulation is minimal, reducing contamination hazards, and is cost effective, given the rather unspecific nature of the process. Mussel Watch programs worldwide[36–39] have used this methodology to monitor temporal trends in contaminant bioavailability since 1986.[34] However, while whole-body analysis can provide information about contaminant bioavailability and bioaccumulation patterns, the origin and pathways of such accumulation are not always easy to infer. Since organisms may accumulate contaminants from water, sediments, or diet,[15,17,18] studying their distribution and accumulation within organisms may provide an insight to specific bioaccumulation pathways. This information is essential for understanding the relative contribution of the various environmental compartments in accumulation and toxicity processes as well as for defining appropriate quality guidelines for water and sediments.[17] A disadvantage of this kind of approach, however, is the multiplication of samples for each organ selected for analysis, involving more laboratory hours and costs for biomonitoring programs. In addition, the reduction of sample mass may prevent the use of individual organisms and impose the need to use composite samples, which will somewhat impair data interpretation. Lastly, dissection and tissue separation require expert skills to avoid sample contamination, which when working with small organisms will represent an increased risk.[41]

Several studies have focused on clarifying the pathways of contaminant uptake.[17,74–76] These will depend not only on the specific aspects of a contaminant, but will also be species-specific and influenced by environmental variables; interpretational difficulties will inevitably ensue. For this reason, laboratory assays under controlled conditions are commonly used to assess the relative importance of each pathway in the overall contaminant accumulation in aquatic biota, and also the effect of altering environmental variables. Such studies can prove to be useful tools not only when interpreting field data, but also when choosing species to use for monitoring specific contaminants or compartments.

6.3.1 ACCUMULATION AND PARTITIONING OF CONTAMINANTS IN PLANTS

The use of macrophytes, that is, macroalgae, sea grasses, and salt marsh plant species, to monitor the environmental quality of aquatic environments has been discussed widely, and in general, research in this area follows a similar strategy: contaminant loads are analyzed in different organs or tissues of the plant and in the surrounding environment.[29,31,33,77,78] The calculation of the concentration factors (CF = $[M]_{plant}/[M]_{environment}$, in which M is the metal concentration) allows the determination of the quantitative proportions in which a given metal occurs in a macrophyte tissue or organ and in the surrounding environment. While macroalgae absorb nutrients and contaminants from the dissolved fraction, salt marsh plants incorporate them primarily through the root system, and some sea grasses through a combination of both processes. Most plants restrict the translocation of contaminants to photosynthetic tissues, accumulating them mainly in the underground biomass[30,32]; this makes plant roots candidates for sediment monitoring purposes. Still, some authors have also observed consistent ratios between the metal concentration in the photosynthetic tissues of sea grasses and that in the sediment (*biosediment factor*).[28] This implies that different species demonstrate contrasting accumulation and translocation patterns (species-specific behavior) and that different contaminants will be accumulated differently within the same species;[29,31,33,77,79] the choice of biomonitor therefore requires caution. Some studies have reported seasonally dependent variations in metal concentrations in macrophytes.[29,77,80] Thus, a plant's annual cycle should also be taken into account, since during the growing season, the metals may be subject to a dilution effect. In addition, there may be spatial variation within and between systems in the distribution of pollutants as a consequence of the physical and chemical characteristics of the environment (e.g., hydrology, sedimentation rates, sediment composition, temperature, salinity, and pH).

Differential metal concentrations in tissues or organs, as well as the growth rates and production of sea grasses[78] and salt marsh plants,[81] have been used to calculate the potential cycling/turnover of metals within a system and/or the annual export of contaminants from an estuarine environment to adjacent coastal waters. Similar studies have been performed for macroalgae[23] and have provided valuable information on contaminant transport and bioavailability processes within and between aquatic ecosystems, since contaminants associated with decaying plant biomasses will become bioavailable through herbivory or the detritivore food web.

One final advantage associated with analyzing contaminant levels in different tissues or organs of aquatic plants is linked with the ongoing search for plant species suitable for phytoremediation/phytoextraction, which aims to remove pollutants from the environment by repeatedly harvesting the plant biomass,[14,82] and phytostabilization, the purpose of which is to restrict the bioavailability of metals in that the plant's roots, by accumulating metals, become effective sinks.[77,83] A strong candidate for phytoremediation should be a hyperaccumulator of pollutants in its aboveground biomass, that is, a plant capable of removing significant amounts of contaminant from the environment.[84] Separate analyses of the different tissues or organs will therefore be essential in the search for plants appropriate to restoration efforts.[84]

6.3.2 DIFFERENTIAL TISSUE ANALYSIS IN ANIMALS

Differential tissue analysis is not frequent in studies involving small invertebrates, such as bivalves, polychaetes, and gastropods, since, as discussed earlier, their diminutive size poses challenging methodological problems. Nonetheless, examples of the usefulness of this approach can be found in the literature. One study compared the specific tissue distribution of cadmium and two forms of mercury in the clam *Corbicula fluminea* when exposed to contaminated water or sediments;[85] the metal distributions in five selected organs were reported to display strong specificities, in accordance with their different contamination modalities. In a study on two other bivalve species, the spiny and Pacific scallops, specific tissue analysis enabled Nørum et al.[86] to assess the effects of species, gender, and reproductive state on the differential accumulation of several metals. On the basis of these findings, the authors were able to suggest some strategies for improving biomonitoring programs, such as gender separation or equal/constant sex composition when using pooled samples.

For a large number of contaminants, however, most research has made use of larger organisms, mainly decapod crustaceans,[87–90] cephalopods,[74] and fish.[75,76,91]

Despite mobility being referred as a source of bias in a field study on the pattern and pathways of mercury lifespan bioaccumulation in *Carcinus maenas*[90] (Figure 6.3), the use of differential tissue analysis still managed to discriminate two different pathways for the accumulation of this metal. In areas of low contamination, diet was considered to be the main source of mercury; this was reflected by higher levels in the internal organs (muscle and hepatopancreas) than in the gills, and higher organic mercury fractions (80–90%) in muscle tissue, suggesting uptake from organic mercury-rich dietary items. In highly contaminated areas, however, environmental exposure was found to be predominant, with lower organic fractions in all tissues and higher concentrations in gills. Similar findings had previously been reported by Laporte et al.,[87] who stated that mercury accumulation in the gills was associated with a higher intake of inorganic mercury, as the gills are in contact mostly with the dissolved and particulate species in water.

The same arguments are valid for the widespread tissue discrimination in fish contaminant analyses. Alquezar et al.[75,76] used a similar approach to assess metal accumulation and uptake pathways in the tissues of *Tetractenos glaber*, and also noticed gender differential accumulation. Studies involving fish, however, are usually approached from a rather anthropocentric standpoint, given that in general only edible tissues are analyzed, despite other tissues being more sensitive and indicative of contamination.

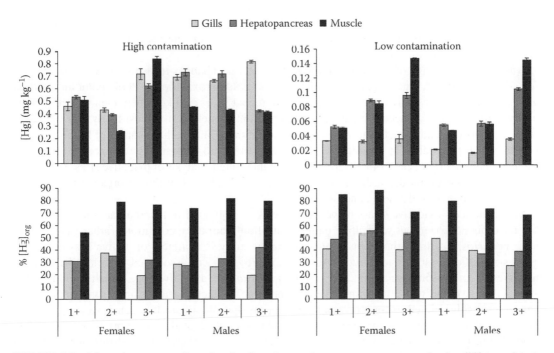

FIGURE 6.3 Mercury concentrations (mg kg⁻¹) and organic mercury percentages in the different tissues and genders of 1+, 2+, and 3+ year old *Carcinus maenas*. (After Coelho et al. 2008. *Mar. Pollut. Bull.* 56: 1104–1110. With permission.)

6.4   SUPPLEMENTARY METHODOLOGIES

Ecological or environmental risk assessment (ERA) is defined as the procedure by which the likely or actual adverse effects of pollutants and other anthropogenic activities on ecosystems and their components are estimated with a known degree of certainty using scientific methodologies.[92] Despite the usefulness (extensively discussed above) of bioaccumulation markers for establishing connections between external levels of exposure and internal levels of tissue contamination, some limitations of this approach must be underlined in the light of the previous concept of ERA. Even if applied in an organ-specific perspective, bioaccumulation markers have failed with respect to hazard identification and effect assessment, since they are incapable of directly indicating the possible development of harmful effects or the actual induction of damage. To fill this gap, therefore, and thus to enhance the efficacy of monitoring programs, the assessment of adaptive responses or early adverse effects on biota is strongly recommended as a complementary tool.

As a general rule, organisms have developed protective mechanisms that increase their resistance to contaminants, including metals. Biochemical in nature, these mechanisms include the induction of MTs as well as the synthesis and activation of antioxidant defenses. Therefore, as biochemical markers they play a key role, acting as early-warning signals, the detection of which can prevent adverse effects at higher hierarchical levels.[93]

6.4.1   Oxidative Stress

There is a close relationship between environmental exposure to contaminants and the generation in the organism of intracellular reactive oxygen species (ROS), such as superoxide ($O_2^{\bullet-}$) and hydroxyl (HO$^{\bullet}$) radicals, and hydrogen peroxide (H_2O_2). To cope with the overproduction of potentially damaging ROS, eukaryotic organisms can increase the levels of protective antioxidant enzymes— catalase (CAT), glutathione peroxidase (GPx), and superoxide dismutase (SOD)—as well as nonenzymatic free radical scavengers such as reduced glutathione (GSH). When the balance between oxidants and antioxidants is disturbed in favor of the former, the organism is considered to be in a state of "oxidative stress." Under these conditions, reactive oxygen intermediates may damage DNA and membrane lipids, and affect the function of cellular proteins.

Among a wide range of contaminant-induced molecular changes measured in aquatic species, oxidative stress [DNA damage, protein oxidation, and lipid peroxidation (LPO)] and/or antioxidant responses have been rapidly gaining recognition in recent years as a key phenomenon, being commonly employed as nonspecific biomarkers.[94] In relation to metals, it is most likely that oxidative stress is a sensitive endpoint, because mitochondria, the major intracellular source of ROS, are common targets for this class of chemicals.[95] Although metals are recognized as important oxidative agents in aquatic organisms from different taxa, the integrated analysis of oxidative stress (oxidative damage and antioxidant defense) parameters and metal loads in key tissues/organs has not been extensively explored under field conditions.

Metal-induced oxidative stress responses have been neglected in macroalgae; the few studies that have been carried out were done under controlled laboratory conditions rather than in real field scenarios. To the authors' knowledge, a single study is available on native seaweeds:[96] A species-specific resistance to metals (As, Cu, Cd, Pb, and Zn) associated with a GSH content increase is reported in *Fucus* spp., *Rhizoclonium tortuosum*, and *Ulva* spp., which enabled these species to thrive in highly contaminated environments. The lack of studies on this group of organisms is probably related to the prevalent assumption that plants are less sensitive to chemicals than aquatic animal species.[97] However, according to Nimptsch et al.,[98] susceptibility to toxicants depends both on the species used and on the contaminants rather than on differences between plants and animals.

Indigenous mussels (*Mytilus galloprovincialis*) of the Saronikos Gulf in Greece were used for monitoring heavy metal pollution.[99] The results indicated the induction of CAT and SOD activity as well as an increase in LPO in the gills and mantle as a result of cellular oxyradical generation associated with metal body burdens.

A similar approach was applied in biomonitoring studies using fish. Changes to the antioxidant enzymes, glutathione system, and LPO induction reflected the presence of heavy metals (Zn, Cu, Cd, As, and Pb) in different tissues of environmentally exposed *Clarias gariepinus*.[100] In a study carried out along an estuarine area impacted by mercury discharges (Ria de Aveiro, Portugal), oxidative stress responses were assessed in fish liver and related to total mercury concentrations in *Liza aurata* seasonally collected and caged in the study area.[101] Despite the significant increase in hepatic mercury load, no evidence of peroxidative damage was detected in wild or caged fish, an observation attributed to the effectiveness of antioxidant defenses, that is, GSH and CAT activity.

Although the integrated approach adopted in the previously mentioned studies provides a more representative picture of the systems under study, there are numerous examples in the literature of studies investigating oxidative stress responses without the complementary assessment of the tissue contaminant load. The significance of such isolated analysis may be compromised by certain factors capable of confounding the interpretation of the results. For instance, false-negative results can arise from the presence of antioxidant enzyme inhibitors in the environment; in the presence of mixtures of chemicals, therefore, antioxidant responses reflect the balance between inducers and inhibitors.[102] In addition, other environmental parameters unrelated to pollution, such as temperature, salinity, and oxygen, can cause important changes in these biochemical responses.[103–105] Therefore, basic information about the influence of these factors on the biomarker to be used in different species is required in order to ensure accuracy of measurement.[105]

Summarizing, to minimize uncertainties and avoid misinterpretations, the combined approach (bioaccumulation markers/oxidative stress responses) is recommended. Moreover, this integrated approach may provide insight into the potential mechanisms of contaminant effects, so that cause–effect relationships can be established, something that is particularly relevant given the presence of complex mixtures.

6.4.2 METALLOTHIONEINS

The majority of biomonitoring studies involves analyzing the bulk contaminant concentration in the whole body or selected tissues of organisms. However, many marine organisms (both vertebrate and invertebrate) possess defense mechanisms to cope with contamination, through immobilization and accumulation in specific organs, cells, or proteins;[106] accumulated contaminants can then be stored, metabolized, or excreted. In view of this, the validity of bulk chemical analyses as a measure of contaminant impact and toxicity must be questioned.

In the specific case of heavy metal pollution, particular attention has been given to MTs. These are heat-resistant, nonenzymatic proteins with a high cysteine content and low molecular weight, whose major function is considered to be the regulation and metabolism of essential metals such as Cu and Zn.[72,107] In addition to this regulatory function, MTs are thought to play a role in metal detoxification. An excess of intracellular free metal ions, whether essential or not, can have damaging effects, impairing several vital cellular functions. Generally, MT synthesis is assumed to be induced under conditions of elevated metal concentration, providing more binding sites for metal ions and limiting possible damage. Generally, MT expression increases with the elevation of tissue concentrations of MT-inducing metals (Ag, Cd, Cu, Hg, etc.), thus reflecting metal bioavailability in the environment.[108] The induction of MT is, therefore, a potentially powerful biochemical indicator of response to metal contamination.[107,109]

Research on the subject has encountered some inconsistencies, giving rise to a few questions that may cripple their usefulness as an environmental biomonitoring tool. In polluted environments, animals are generally exposed to a mixture of different metals, and it is generally impossible to attribute MT induction to one element or another. On the other hand, several papers have reported the absence of MT induction in invertebrates exposed to metals, or even a decrease in MT concentrations following exposure to metals.

In a review of this theme, Amiard et al.[107] raised three main issues regarding the use of MT as a biomarker of metal pollution that need clarification: its dependency on specific metals, on biological characteristics (species, population, and organ), and on the period and dose of exposure.

The first issue was highlighted by Barka et al.,[110] among others, who tested MT induction in the copepod *Tigriopus brevicornis* by a suite of metals and reported differences in the induction capacities of selected metals. Bebianno and Machado,[111] on the other hand, found MT levels to correlate positively with Cd and Cu concentrations but not with Zn in *Mytilus galloprovincialis*.

The dependency of MT induction on biological characteristics is more evident in invertebrate species, whereas in vertebrates metal detoxification is carried out mainly through binding to MTs. Metal sequestration in invertebrates occurs through various processes, such as immobilization as inorganic precipitates, and accumulation as insoluble lipofuschin pigments in tertiary lysosomes and non-MT metal-binding proteins.[106,112] Differences between vertebrate and invertebrate species with regard to MT induction patterns are therefore common. Even among closely related species, such as the two decapod crustaceans *Carcinus maenas* and *Pachygrapsus marmoratus*, Legras et al.[113] found significant differences in metal accumulation and MT induction, suggesting species-specific traits in the role of MT in metal detoxification. In the oyster *Crassostrea gigas*, a seasonal study[114,115] demonstrated that MT concentrations in the digestive gland were only occasionally correlated with accumulated metal concentrations, whereas in the gills, such correlations were observed over the major part of the year. However, even in the gills, seasonal variations in MT concentrations were high enough to conceal intersite differences when annual means were calculated for all the individuals analyzed monthly.

In considering the period and dose of exposure, Mouneyrac et al.[116] reported that populations of the amphipod *Orchestia gammarellus* from estuaries with contrasting metal contamination failed to demonstrate differences in MT induction, since MT body concentrations did not increase upon exposure to raised availabilities of Cu, Zn, or Cd. From their findings, these authors concluded that MT-like proteins were unsuitable as potential biomarkers. In addition, a number of studies have indicated that MTs are also induced by nonmetals, for example, organic aromatic compounds.[117,118] This has raised doubts regarding the use of this parameter as a specific biomarker for metal exposure.

Despite all these inconsistencies and uncertainties, however, MT induction is a valuable methodology, although not alone but as part of a suite of biomonitoring strategies, provided that caution is exercised and the experimental design optimized. The choice of species is crucial: While some authors[106] recommend the use of fish in view of the various competing mechanisms of metal sequestration in invertebrates, the excessive mobility of such pelagic species is a major limitation. Bivalves are therefore most probably the best candidates for biomonitoring programs involving MT concentrations as biomarkers,[107] for the same reasons that made them ideal as contaminant load biomonitors. The same authors also emphasize the importance of making a good choice of organ in which to measure MT concentrations when these proteins are used as a biomarker, since protein induction is usually stronger in the gills and digestive glands. Finally, confounding factors such as salinity,[119] seasonality,[112,115] sex and reproductive state,[113,120] and size[72] should be taken into account and efforts made to minimize the associated bias.

A very similar approach using primary producers is based on an analogous defense mechanism—the production of PCs. In the presence of excess metals, PCs are formed, which effectively capture metals. A major advantage of PCs over MTs is their specificity for metals, since no other environmental factors are known to induce PC accumulation; the activation of PCs is considered to ensue from a direct "sensing" of metal excess.[121] An elevated PC level in macroalgae and sea grasses has also been used as a specific biomarker of heavy metal bioavailability and stress.[122] PCs do indeed seem to be representative of the level of perturbation of the medium and of the health of the organisms, and they could therefore be used for the early detection of changes in water quality.[122]

6.5 CONCLUSIONS

In a general perspective, the use of biota as a source of environmental information provides invaluable data on the health status of aquatic ecosystems, with numerous advantages over sediment or dissolved contaminant analyses. Biomonitoring programs and scientific research must rely on careful experimental designs, in which three main issues should be addressed: Firstly, the meticulous selection of species to monitor, which will depend on the desired environmental compartments to be assessed and the target contaminant. No universal biomonitor exists, hence a suite of species is advisable, each reflecting specific trophic levels and thus different compartments of the ecosystem. Secondly, the assessment strategy should be adjusted according to the selected species, since tissue discrimination will provide additional information on accumulation pathways and specific toxicity mechanisms; but this procedure is more labor-intensive and time-consuming, not to mention the difficulties associated with sampling mass in smaller organisms. Finally, the use of biomarkers such as oxidative stress indicators or MT induction may provide useful, quantitative information on the effects of contaminants, given that whole-body analyses fail to indicate what portion of the contaminant body burden is actually reactive and thus responsible for adverse effects in biota.

REFERENCES

1. Rainbow, P.S. 1995. Biomonitoring of heavy metal availability in the marine environment. *Mar. Pollut. Bull.* 31: 183–192.
2. Leal, M.C.F., M.T. Vasconcelos, I. Sousa-Pinto, and J.P.S. Cabral. 1997. Biomonitoring with benthic macroalgae and direct assay of heavy metals in seawater of the Oporto coast (Northwest Portugal). *Mar. Pollut. Bull.* 34: 1006–1015.
3. Leermakers, M., S. Galleti, N. De Galan, Brion, and W. Baeyens. 2001. Mercury in the Southern North Sea and Scheldt estuary. *Mar. Chem.* 75: 229–248.
4. Ramalhosa, E., P. Monterroso, S. Abreu, E. Pereira, C. Vale, and A. Duarte. 2001. Storage and export of mercury from a contaminated bay (Ria de Aveiro, Portugal). *Wetlands Ecol. Manage.* 9: 311–316.
5. Lawson, N.M., R.P. Mason, and J.-M. Laporte. 2001. The fate and transport of mercury, methylmercury, and other trace metals in Chesapeake Bay tributaries. *Water Res.* 35: 501–515.
6. Luoma, S.N. 1996. The developing framework of marine ecotoxicology: Pollutants as a variable in marine ecosystems? *J. Exp. Mar. Biol. Ecol.* 200: 29–55.
7. Wang, W.-X., I. Stupakoff, C. Gagnon, and N.S. Fisher. 1998. Bioavailability of inorganic and methylmercury to a marine deposit-feeding polychaete. *Environ. Sci. Technol.* 32: 2564–2571.
8. Farkas, A., C. Erratico, and L. Viganò. 2007. Assessment of the environmental significance of heavy metal pollution in surficial sediments of the River Po. *Chemosphere* 68: 761–768.
9. Heim, S., J. Schwarzbauer, A. Kronimus, R. Littke, C. Woda, and A. Mangini. 2004. Geochronology of anthropogenic pollutants in riparian wetland sediments of the Lippe River (Germany). *Org. Geochem.* 35: 1409–1425.
10. Cantwell, M.G., J.W. King, R.M. Burgess, and P.G. Appleby. 2007. Reconstruction of contaminant trends in a salt wedge estuary with sediment cores dated using a multiple proxy approach. *Mar. Environ. Res.* 64: 225–246.
11. Evenset, A., G.N. Christensen, J. Carroll, et al. 2007. Historical trends in persistent organic pollutants and metals recorded in sediment from Lake Ellasjøen, Bjørnøya, Norwegian Arctic. *Environ. Pollut.* 146: 196–205.
12. Larner, B.L., A.J. Seen, A.S. Palmer, and I. Snape. 2007. A study of metal and metalloid contaminant availability in Antarctic marine sediments. *Chemosphere* 67: 1967–1974.
13. Zhong, H. and W.-X. Wang. 2006. Metal–solid interactions controlling the bioavailability of mercury from sediments to clams and sipunculans. *Environ. Sci. Technol.* 40: 3794–3799.
14. Mertens, J., S. Luyssaert, and K. Verheyen. 2005. Use and abuse of trace metal concentrations in plant tissue for biomonitoring and phytoextraction. *Environ. Pollut.* 138: 1–4.
15. Rainbow, P.S. and D.J.H. Phillips. 1993. Cosmopolitan biomonitors of trace metals. *Mar. Pollut. Bull.* 26: 593–601.
16. Saiz-Salinas, J.I., J.M. Ruiz, and G. Frances-Zubillaga. 1996. Heavy metal levels in intertidal sediments and biota from the Bidasoa estuary. *Mar. Pollut. Bull.* 32: 69–71.

17. Wang, W.-X. and N.S. Fisher. 1999. Delineating metal accumulation pathways for marine invertebrates. *Sci. Total Environ.* 237/238: 459–472.

18. Rainbow, P.S. 2002. Trace metal concentrations in aquatic invertebrates: Why and so what? *Environ. Pollut.* 120: 497–507.

19. Phillips, D.J.H. 1994. Macrophytes as biomonitors of trace metals. In: K.J.M. Kramer (ed.), *Biomonitoring of Coastal Waters and Estuaries*, pp. 85–103. Boca Raton, FL: CRC Press.

20. Conti, M.E. and G. Cecchetti. 2003. A biomonitoring study: Trace metals in algae and molluscs from Tyrrhenian coastal areas. *Environ. Res.* 93: 99–112.

21. Gosavi, K., J. Sammut, S. Gifford, and J. Jankowski. 2004. Macroalgal biomonitors of trace metal contamination in acid sulfate soil aquaculture ponds. *Sci. Total Environ.* 324: 25–39.

22. Runcie, J.W. and M.J. Riddle. 2004. Metal concentrations in macroalgae from East Antarctica. *Mar. Pollut. Bull.* 49: 1109–1126.

23. Coelho, J.P., M.E. Pereira, A. Duarte, and M.A. Pardal. 2005. Macroalgae response to a mercury contamination gradient in a temperate coastal lagoon (Ria de Aveiro, Portugal). *Estuar. Coast. Shelf Sci.* 65: 492–500.

24. Nassar, C.A.G., L.Y. Salgado, Y. Yoneshigue-Valentin, and G.M. Amado Filho. 2003. The effect of iron-ore particles on the metal content of the brown alga *Padina gymnospora* (Espírito Santo Bay, Brazil). *Environ. Pollut.* 123: 301–305.

25. Haritonidis, S. and P. Malea. 1999. Bioaccumulation of metals by the green alga *Ulva rigida* from Thermaikos Gulf, Greece. *Environ. Pollut.* 104: 365–372.

26. Melville, F. and A. Pulkownik. 2007. Investigation of mangrove macroalgae as biomonitors of estuarine metal contamination. *Sci. Total Environ.* 387: 301–309.

27. Lafabrie, C., C. Pergent-Martini, and G. Pergent. 2008. Metal contamination of *Posidonia oceanica* meadows along the Corsican coastline (Mediterranean). *Environ. Pollut.* 151: 262–268.

28. Lafabrie, C., G. Pergent, R. Kantin, C. Pergent-Martini, and J.-L. Gonzalez. 2007. Trace metals assessment in water, sediment, mussel and seagrass species—validation of the use of *Posidonia oceanica* as a metal biomonitor. *Chemosphere* 68: 2033–2039.

29. Caçador, I., C. Vale, and F. Catarino. 2000. Seasonal variation of Zn, Pb, Cu and Cd concentrations in the root ± sediment system of *Spartina maritima* and *Halimione portulacoides* from Tagus estuary salt marshes. *Mar. Environ. Res.* 49: 279–290.

30. Weis, P., L. Windham, D.J. Burke, and J.S. Weis. 2002. Release into the environment of metals by two vascular salt marsh plants. *Mar. Environ. Res.* 54: 325–329.

31. Fitzgerald, E.J., J.M. Caffrey, S.T. Nesaratnam, and P. McLoughlin. 2003. Copper and lead concentrations in salt marsh plants on the Suir Estuary, Ireland. *Environ. Pollut.* 123: 67–74.

32. Windham, L., J.S. Weis, and P. Weis. 2003. Uptake and distribution of metals in two dominant salt marsh macrophytes, *Spartina alterniflora* (cordgrass) and *Phragmites australis* (common reed). *Estuar. Coast. Shelf Sci.* 56: 63–72.

33. Quan, W.M., J.D. Han, A.L. Shen, et al. 2007. Uptake and distribution of N, P and heavy metals in three dominant salt marsh macrophytes from Yangtze River estuary, China. *Mar. Environ. Res.* 64: 21–37.

34. O'Connor, T.P., A.Y. Cantillo, and G.G. Lauenstein. 1994. Monitoring of temporal trends by the NOAA NS&T Mussel Watch Project. In: K.J.M. Kramer (ed.), *Biomonitoring of Coastal Waters and Estuaries*, pp. 29–50. Boca Raton, FL: CRC Press.

35. O'Connor, T.P. 2002. National distribution of chemical concentrations in mussels and oysters in the USA. *Mar. Environ. Res.* 53: 117–143.

36. O'Connor, T.P. and G.G. Lauenstein. 2006. Trends in chemical concentrations in mussels and oysters collected along the US coast: Update to 2003. *Mar. Environ. Res.* 62: 261–285.

37. Sericano, J.L., T.L. Wade, T.J. Jackson, et al. 1995. Trace organic contamination in the Americas: An overview of the US national status & trends and the International 'Mussel Watch' Programmes. *Mar. Pollut. Bull.* 31: 214–225.

38. Villeneuve, J.-P., F.P. Carvalho, S.W. Fowler, and C. Cattini. 1999. Levels and trends of PCBs, chlorinated pesticides and petroleum hydrocarbons in mussels from the NW Mediterranean coast: Comparison of concentrations in 1973–1974 and 1988–1989. *Sci. Total Environ.* 237/238: 57–65.

39. Monirith, I., D. Ueno, S. Takahashi, et al. 2003. Asia-Pacific mussel watch: Monitoring contamination of persistent organochlorine compounds in coastal waters of Asian countries. *Mar. Pollut. Bull.* 46: 281–300.

40. Szefer, P., S.W. Fowler, K. Ikuta, et al. 2006. A comparative assessment of heavy metal accumulation in soft parts and byssus of mussels from subarctic, temperate, subtropical and tropical marine environments. *Environ. Pollut.* 139: 70–78.

41. De Kock, W.C. and K.J.M. Kramer. 1994. Active biomonitoring (ABM) by translocation of bivalve molluscs. In: K.J.M. Kramer (ed.), *Biomonitoring of Coastal Waters and Estuaries*, pp. 51–85. Boca Raton, FL: CRC Press.

42. Mikac, N., Z. Kwokal, D. Martincic, and M. Branica, 1996. Uptake of mercury species by transplanted mussels *Mytilus galloprovincialis* under estuarine conditions (Krka river estuary). *Sci. Total Environ.* 184: 173–182.

43. Abbe, G.R., G.F. Riedel, and J.G. Sanders. 2000. Factors that influence the accumulation of copper and cadmium by transplanted eastern oysters (*Crassostrea virginica*) in the Patuxent River, Maryland. *Mar. Environ. Res.* 49: 377–396.

44. Olivier, F., M. Ridd, and D. Klumpp. 2002. The use of transplanted cultured tropical oysters (*Saccostrea commercialis*) to monitor Cd levels in North Queensland coastal waters (Australia). *Mar. Pollut. Bull.* 44: 1051–1062.

45. Ruiz, J.M. and J.I. Saiz-Salinas. 2000. Extreme variation in the concentration of trace metals in sediments and bivalves from the Bilbao estuary (Spain) caused by the 1989–90 drought. *Mar. Environ. Res.* 49: 307–317.

46. Ridgway, J., N. Breward, W.J. Langston, R. Lister, J.G. Rees, and S.M. Rowlatt. 2003. Distinguishing between natural and anthropogenic sources of metals entering the Irish Sea. *Appl. Geochem.* 18: 283–309.

47. Cheggour, M., A. Chafik, N.S. Fisher, and S. Benbrahim. 2005. Metal concentrations in sediments and clams in four Moroccan estuaries. *Mar. Environ. Res.* 59: 119–137.

48. Boisson, F., M.G.J. Hartl, S.W. Fowler, and C. Amiard-Triquet. 1998. Influence of chronic exposure to silver and mercury in the field on the bioaccumulation potential of the bivalve *Macoma balthica*. *Mar. Environ. Res.* 45: 325–340.

49. Stecko, J.R.P. and L.I. Bendell-Young. 2000. Uptake of [109]Cd from sediments by the bivalves *Macoma balthica* and *Protothaca staminea*. *Aquat. Toxicol.* 47: 147–159.

50. Byrne, P.A. and J. O'Halloran. 2001. The role of bivalve molluscs as tools in estuarine sediment toxicity testing: A review. *Hydrobiologia* 465: 209–217.

51. García-Luque, E., T.A. Delvalls, C. Casado-Martínez, J.M. Forja, and A. Gómez-Parra. 2004. Simulating a heavy metal spill under estuarine conditions: Effects on the clam *Scrobicularia plana*. *Mar. Environ. Res.* 58: 671–674.

52. Riba, I., M.C. Casado-Martínez, J. Blasco, and T.A. Delvalls. 2004. Bioavailability of heavy metals bound to sediments affected by a mining spill using *Solea senegalensis* and *Scrobicularia plana*. *Mar. Environ. Res.* 58: 395–399.

53. Coelho, J.P., M. Rosa, E. Pereira, A. Duarte, and M.A. Pardal. 2006. Pattern and annual rates of *Scrobicularia plana* mercury bioaccumulation in a human induced mercury gradient (Ria de Aveiro, Portugal). *Estuar. Coast. Shelf Sci.* 69: 629–635.

54. Pérez, E., J. Blasco, and M. Solé. 2004. Biomarker responses to pollution in two invertebrate species: *Scrobicularia plana* and *Nereis diversicolor* from the Cádiz Bay (SW Spain). *Mar. Environ. Res.* 58: 275–279.

55. Hylland, K., M. Sköld, J.S. Gunnarsson, and J. Skei. 1997. Interactions between Eutrophication and contaminants. IV. Effects on sediment-dwelling organisms. *Mar. Pollut. Bull.* 33: 90–99.

56. Muhaya, B.B.M., M. Leermakers, and W. Baeyens. 1997. Total mercury and methylmercury in sediments and in the polychaete *Nereis diversicolor* at Groot Buitenschoor (Scheldt estuary, Belgium). *Water Air Soil Pollut.* 94: 109–123.

57. Baeyens, W., C. Meuleman, B. Muhaya, and M. Leermakers. 1998. Behaviour and speciation of mercury in the Scheldt estuary (water, sediments and benthic organisms). *Hydrobiologia* 366: 63–79.

58. Bernds, D., D. Wübben, and G.-P. Zauke. 1998. Bioaccumulation of trace metals in polychaetes from the German Wadden Sea: Evaluation and verification of toxicokinetic models. *Chemosphere* 37: 2573–2587.

59. Ruus, A., M. Schaanning, S. Øxnevad, and K. Hylland. 2005. Experimental results on bioaccumulation of metals and organic contaminants from marine sediments. *Aquat. Toxicol.* 72: 273–292.

60. Durou, C., L. Poirier, J.-C. Amiard, et al. 2007. Biomonitoring in a clean and a multi-contaminated estuary based on biomarkers and chemical analyses in the endobenthic worm *Nereis diversicolor*. *Environ. Pollut.* 148: 445–458.

61. Coelho, J.P., M. Nunes, M. Dolbeth, M.E. Pereira, A.C. Duarte and M.A. Pardal. 2008. The role of two sediment dwelling invertebrates on the mercury transfer from sediments to the estuarine trophic web. *Estuar. Coast. Shelf Sci.* 78: 516–523.

62. Scaps, P. 2002. A review of the biology, ecology and potential use of the common ragworm *Hediste diversicolor* (O.F. Müller) (Annelida: Polychaeta). *Hydrobiologia* 470: 203–218.

63. Abreu, S.N., E. Pereira, C. Vale, and A.C. Duarte. 2000. Accumulation of mercury in sea bass from a contaminated lagoon (Ria de Aveiro, Portugal). *Mar. Pollut. Bull.* 40: 293–297.

64. Alonso, D., P. Pineda, J. Olivero, H. González, and N. Campos. 2000. Mercury levels in muscle of two fish species and sediments from the Cartagena Bay and the Ciénaga Grande de Santa Marta, Colombia. *Environ. Pollut.* 109: 157–163.

65. Usero, J., C. Izquierdo, J. Morillo, and I. Gracia. 2003. Heavy metals in fish (*Solea vulgaris*, *Anguilla anguilla* and *Liza aurata*) from salt marshes on the southern Atlantic coast of Spain. *Environ. Int.* 29: 949–956.

66. Henry, F., R. Amara, L. Courcot, D. Lacouture, and M.-L. Bertho. 2004. Heavy metals in four fish species from the French coast of the Eastern English Channel and Southern Bight of the North Sea. *Environ. Int.* 30: 675–683.

67. Marcovecchio, J.E. 2004. The use of *Micropogonias furnieri* and *Mugil liza* as bioindicators of heavy metals pollution in La Plata river estuary, Argentina. *Sci. Total Environ.* 323: 219–226.

68. Meador, J.P., D.W. Ernest, and A.N. Kagley. 2005. A comparison of the non-essential elements cadmium, mercury, and lead found in fish and sediment from Alaska and California. *Sci. Total Environ.* 339: 189–205.

69. Pereira, M.E., S.N. Abreu, J.P. Coelho, et al. 2006. Seasonal fluctuations of tissue mercury contents in the European shore crab *Carcinus maenas* from low and high contamination areas (Ria de Aveiro, Portugal). *Mar. Pollut. Bull.* 52: 1450–1457.

70. Davis, J.A., M.D. May, B.K. Greenfield, et al. 2002. Contaminant concentrations in sport fish from San Francisco Bay, 1997. *Mar. Pollut. Bull.* 44: 1117–1129.

71. Kehrig, H.A., O. Malm, and I. Moreira. 1998. Mercury in a widely consumed fish *Micropogonias furnieri* (Demarest, 1823) from four main Brazilian estuaries. *Sci. Total Environ.* 213: 263–271.

72. Bebianno, M.J., C. Santos, J. Canário, N. Gouveia, D. Sena-Carvalho, and C. Vale. 2007. Hg and metallothionein-like proteins in the black scabbardfish *Aphanopus carbo*. *Food Chem. Toxicol.* 45: 1443–1452.

73. Penedo de Pinho, A., J.R.D. Guimarães, A.S. Martins, P.A.S. Costa, G. Olavo, and J. Valentin. 2002. Total mercury in muscle tissue of five shark species from Brazilian offshore waters: Effects of feeding habit, sex, and length. *Environ. Res. Section A* 89: 250–258.

74. Danis, B., P. Bustamante, O. Cotret, J.L. Teyssié, S.W. Fowler, and M. Warnau. 2005. Bioaccumulation of PCBs in the cuttlefish *Sepia officinalis* from seawater, sediment and food pathways. *Environ. Pollut.* 134: 113–122.

75. Alquezar, R., S.J. Markich, and D.J. Booth. 2006. Metal accumulation in the smooth toadfish, *Tetractenos glaber*, in estuaries around Sydney, Australia. *Environ. Pollut.* 142: 123–131.

76. Alquezar, R., S.J. Markich, and J.R. Twining. 2008. Comparative accumulation of ^{109}Cd and ^{75}Se from water and food by an estuarine fish (*Tetractenos glaber*). *J. Environ. Radioact.* 99: 167–180.

77. Almeida, C.M.R., A.P. Mucha, and M.T.S.D. Vasconcelos. 2006. Comparison of the role of the sea club-rush *Scirpus maritimus* and the sea rush *Juncus maritimus* in terms of concentration, speciation and bioaccumulation of metals in the estuarine sediment. *Environ. Pollut.* 142: 151–159.

78. Kaldy, J.E. 2006. Carbon, nitrogen, phosphorus and heavy metal budgets: How large is the eelgrass (*Zostera marina* L.) sink in a temperate estuary? *Mar. Pollut. Bull.* 52: 332–356.

79. Sousa A.I., I. Caçador, A.I. Lillebø, and M. Pardal. 2008. Heavy metal accumulation in *Halimione portulacoides*: Intra- and extra-cellular metal binding sites. *Chemosphere* 70: 850–857.

80. Żbikowski, R., P. Szefer, and A. Latała. 2006. Distribution and relationships between selected chemical elements in green alga *Enteromorpha* sp. from the southern Baltic. *Environ. Pollut.* 143: 435–448.

81. Válega, M., A.I. Lillebø, I. Caçador, M.E. Pereira, A.C. Duarte, and M.A Pardal. 2008. Mercury mobility in a salt marsh colonised by *Halimione portulacoides*. *Chemosphere* 73: 1224–1229.

82. Riddle, S.G., H.H. Tran, J.G. Dewitt, and J.C. Andrews. 2002. Field, laboratory, and X-ray absorption spectroscopic studies of mercury accumulation by water hyacinths. *Environ. Sci. Technol.* 36: 1965–1970.

83. Reboreda, R. and I. Caçador. 2007. Halophyte vegetation influences in salt marsh retention capacity for heavy metals. *Environ. Pollut.* 146: 147–154.

84. Weis, J.S. and P. Weis. 2004. Metal uptake, transport and release by wetland plants: Implications for phytoremediation and restoration. *Environ. Int.* 30: 685–700.

85. Inza, B., F. Ribeyre, R. Maury-Brachet, and A. Boudou. 1997. Tissue distribution of inorganic mercury, methylmercury and cadmium in the Asiatic clam (*Corbicula fluminea*) in relation to the contamination levels of the water column and sediment. *Chemosphere* 35: 2817–2836.

86. Nørum, U., V.W.-M. Lai, and W.R. Cullen. 2005. Trace element distribution during the reproductive cycle of female and male spiny and Pacific scallops, with implications for biomonitoring. *Mar. Pollut. Bull.* 50: 175–184.

87. Laporte, J.M., F. Ribeyre, J.P. Truchot, and A. Boudou. 1997. Combined effects of water pH and salinity on the bioaccumulation of inorganic mercury and methylmercury in the shore crab *Carcinus maenas*. *Mar. Pollut. Bull.* 34: 880–893.

88. Laporte, J.M., S. Andres, and R.P. Mason. 2002. Effect of ligands and other metals on the uptake of mercury and methylmercury across the gills and the intestine of the blue crab (*Callinectes sapidus*). *Comp. Biochem. Physiol. Part C* 131: 185–196.

89. Bodin, N., A. Abarnou, A.-M. Le Guellec, V. Loizeau, and X. Philippon. 2007. Organochlorinated contaminants in decapod crustaceans from the coasts of Brittany and Normandy (France). *Chemosphere* 67: S36–S47.

90. Coelho, J.P., A.T. Reis, S. Ventura, M.E. Pereira, A.C. Duarte, and M.A. Pardal. 2008. Pattern and pathways for mercury lifespan bioaccumulation in *Carcinus maenas*. *Mar. Pollut. Bull.* 56: 1104–1110.

91. Klinck, J.S., W.W. Green, R.S. Mirza, et al. 2007. Branchial cadmium and copper binding and intestinal cadmium uptake in wild yellow perch (*Perca flavescens*) from clean and metal-contaminated lakes. *Aquat. Toxicol.* 84: 198–207.

92. Depledge, M.H. and M.C. Fossi. 1994. The role of biomarkers in environmental assessment (2). *Ecotoxicology* 3: 161–172.

93. van der Oost, R., J. Beyer, and N.P.E. Vermeulen. 2003. Fish bioaccumulation and biomarkers in environmental risk assessment: A review. *Environ. Toxicol. Pharmacol.* 13: 57–149.

94. Monterroso, P., S.N. Abreu, E. Pereira, C. Vale, and A.C. Duarte. 2003. Estimation of Cu, Cd and Hg transported by plankton from a contaminated area (Ria de Aveiro). *Acta Oecol.* 24: S351–S357.

95. Wang, G. and B.A. Fowler. 2008. Roles of biomarkers in evaluating interactions among mixtures of lead, cadmium and arsenic. *Toxicol. Appl. Pharmacol.* 233: 92–99.

96. Pawlik-Skowrońska, B., J. Pirszel, and M.T. Brown. 2007. Concentrations of phytochelatins and glutathione found in natural assemblages of seaweeds depend on species and metal concentrations of the habitat. *Aquat. Toxicol.* 83: 190–199.

97. Mohan, B.S. and B.B. Hosetti. 1999. Review: Aquatic plants for toxicity assessment. *Environ. Res. Section A* 81: 259–274.

98. Nimptsch, J., D.A. Wunderlin, A. Dollar, and S. Pflugmacher. 2005. Antioxidant and biotransformation enzymes in *Myriophyllum quitense* as biomarkers of heavy metal exposure and eutrophication in Suquía River basin (Córdoba, Argentina). *Chemosphere* 61: 147–157.

99. Vlahogianni, T., M. Dassenakis, M.J. Scoullos, and A. Valavanidis. 2007. Integrated use of biomarkers (superoxide dismutase, catalase and lipid peroxidation) in mussels *Mytilus galloprovincialis* for assessing heavy metals' pollution in coastal areas from the Saronikos Gulf of Greece. *Mar. Pollut. Bull.* 54: 1361–1371.

100. Farombi, E.O., O.A. Adelowo, and Y.R. Ajimoko. 2007. Biomarkers of oxidative stress and heavy metal levels as indicators of environmental pollution in African cat fish (*Clarias gariepinus*) from Nigeria Ogun River. *Int. J. Environ. Res. Public Health* 4: 158–165.

101. Guilherme, S., M. Válega, M.E. Pereira, M.A. Santos, and M. Pacheco. 2008. Antioxidant and biotransformation responses in *Liza aurata* under environmental mercury exposure—relationship with mercury accumulation and implications for public health. *Mar. Pollut. Bull.* 56: 845–859.

102. Ahmad, I., M. Pacheco, and M.A. Santos. 2006. *Anguilla anguilla* L. oxidative stress biomarkers: An *in situ* study of freshwater wetland ecosystem (Pateira de Fermentelos, Portugal). *Chemosphere* 65: 952–962.

103. Malek, R.L., H. Sajadi, J. Abraham, M.A. Grundy, and G.S. Gerhard. 2004. The effects of temperature reduction on gene expression and oxidative stress in skeletal muscle from adult zebra fish. *Comp. Biochem. Physiol. Part C* 138: 363–373.

104. Olsvik, P.A., T. Kristensen, R. Waagbo, K.E. Tollefsen, B.O. Rosseland, and H. Toften. 2006. Effects of hypo- and hyperoxia on transcription levels of five stress genes and the glutathione system in liver of Atlantic cod *Gadus morhua*. *J. Exp. Biol.* 209: 2893–2901.

105. Almeida, E.A., C.D. Bainy, A.P. Loureiro, et al. 2007. Oxidative stress in *Perna perna* and other bivalves as indicators of environmental stress in the Brazilian marine environment: Antioxidants, lipid peroxidation and DNA damage. *Comp. Biochem. Physiol. Part A* 146: 588–600.

106. George, S.G. and P.-E. Olsson. 1994. Metallothioneins as indicators of trace metal pollution. In: K.J.M. Kramer (ed.), *Biomonitoring of Coastal Waters and Estuaries*, pp. 151–179. Boca Raton, FL: CRC Press.

107. Amiard, J.-C., C. Amiard-Triquet, S. Barka, J. Pellerin, and P.S. Rainbow. 2006. Metallothioneins in aquatic invertebrates: Their role in metal detoxification and their use as biomarkers. *Aquat. Toxicol.* 76: 160–202.

108. Monserrat, J.M., P.E. Martínez, L.A. Geracitano, et al. 2007. Pollution biomarkers in estuarine animals: Critical review and new perspectives. *Comp. Biochem. Physiol. Part C* 146: 221–234.
109. Langston, W.S., B.S. Chesman, G.R. Burt, N.D. Pope, and J. McEvoy. 2002. Metallothionein in liver of eels *Anguilla anguilla* from the Thames Estuary: An indicator of environmental quality? *Mar. Environ. Res.* 53: 263–293.
110. Barka, S., J.-F. Pavillon, and J.-C. Amiard. 2001. Influence of different essential and non-essential metals on MTLP levels in the Copepod *Tigriopus brevicornis. Comp. Biochem. Physiol. Part C* 128: 479–493.
111. Bebianno, M.J. and L.M. Machado. 1997. Concentrations of metals and metallothioneins in *Mytilus galloprovincialis* along the South Coast of Portugal. *Mar. Pollut. Bull.* 34: 666–671.
112. Geffard, A., C. Amiard-Triquet, and J.-C. Amiard. 2005. Do seasonal changes affect metallothionein induction by metals in mussels, *Mytilus edulis*? *Ecotoxicol. Environ. Saf.* 61: 209–220.
113. Legras, S., S. Mouneyrac, J.-C. Amiard, C. Amiard-Triquet, and P.S. Rainbow. 2000. Changes in metallothionein concentrations in response to variation in natural factors (salinity, sex, weight) and metal contamination in crabs from a metal-rich estuary. *J. Exp. Mar. Biol. Ecol.* 246: 259–279.
114. Geffard, A., C. Amiard-Triquet, J.-C. Amiard, and C. Mouneyrac. 2001. Temporal variations of metallothionein and metal concentrations in the digestive gland of oysters (*Crassostrea gigas*) from a clean and a metal-rich site. *Biomarkers* 6: 91–107.
115. Geffard, A., J.C. Amiard, and C. Amiard-Triquet. 2002. Use of metallothionein in gills from oysters (*Crassostrea gigas*) as a biomarker: Seasonal and intersite fluctuations. *Biomarkers* 7: 123–137.
116. Mouneyrac, C., J.C. Amiard, C. Amiard-Triquet, A. Cottier, P.S. Rainbow, and B.D. Smith. 2002. Partitioning of accumulated trace metals in the talitrid amphipod crustacean *Orchestia gammarellus*: A cautionary tale on the use of metallothionein-like proteins as biomarkers. *Aquat. Toxicol.* 57: 225–242.
117. Pedrajas, J.R., J. Peinado, and J. LopezBarea. 1995. Oxidative stress in fish exposed to model xenobiotics. Oxidatively modified forms of Cu, Zn superoxide dismutase as potential biomarkers. *Chem. Biol. Interact.* 98: 267–282.
118. Kling, P., L.J. Erkell, P. Kille, and P.E. Olsson. 1996. Metallothionein induction in rainbow trout gonadal (RTG-2) cells during free radical exposure. *Mar. Environ. Res.* 42: 33–36.
119. Leung, K.M.Y., J. Svavarsson, M. Crane, and D. Morritt. 2002. Influence of static and fluctuating salinity on cadmium uptake and metallothionein expression by the dogwhelk *Nucella lapillus* (L.). *J. Exp. Biol. Ecol.* 274: 175–189.
120. Mouneyrac, C., C. Amiard-Triquet, J.C. Amiard, and P.S. Rainbow. 2001. Comparison of metallothionein concentrations and tissue distribution of trace metals in crabs (*Pachygrapsus marmoratus*) from a metal-rich estuary, in and out of the reproductive season. *Comp. Biochem. Physiol. Part C* 129: 193–209.
121. Clemens, S. 2006. Toxic metal accumulation, responses to exposure and mechanisms of tolerance in plants. *Biochimie* 88: 1707–1719.
122. Ferrat L., C. Pergent-Martini, and M. Roméo. 2003. Assessment of the use of biomarkers in aquatic plants for the evaluation of environmental quality: Application to sea grasses. *Aquat. Toxicol.* 65: 187–204.

7 Speciation Analytics in Aquatic Ecosystems

A. de Brauwere, Y. Gao, S. De Galan, W. Baeyens, M. Elskens, and M. Leermakers

CONTENTS

7.1 Introduction 121
7.2 Speciation of Dissolved Cr, Fe, Mn, and As 122
    7.2.1 *In Situ* Speciation in the Aquatic System 123
    7.2.2 Sampling 125
    7.2.3 Sample Preservation and Storage 126
    7.2.4 Electrochemical Speciation 126
    7.2.5 Hyphenated Methods 127
        7.2.5.1 Exchange Columns (Trap and Elute) 127
        7.2.5.2 Chromatography 127
        7.2.5.3 Chemiluminescence and Colorimetric Reactions 128
    7.2.6 Other Methods 128
7.3 Speciation of Dissolved Hg 129
    7.3.1 Sample Handling and Storage 129
    7.3.2 Analytical Methods for Hg(0), DMHg, Hg-R, and Hg-T 130
    7.3.3 Analytical Methods for MMHg 130
        7.3.3.1 Extraction Procedures 130
        7.3.3.2 Gas Chromatographic Separation Methods 130
        7.3.3.3 Derivatization and Validation 130
        7.3.3.4 GC Improvements 131
        7.3.3.5 Liquid Chromatographic Separation Methods 131
        7.3.3.6 Detection Methods 132
    7.3.4 Speciation Modeling of Hg 132
Acknowledgment 132
References 132

7.1 INTRODUCTION

The speciation of redox-sensitive elements (e.g., Cr, Fe, and Mn), redox-sensitive elements that also form organometallic compounds (e.g., As), and redox-sensitive elements that also form organometallic and volatile compounds (e.g., Hg) in aquatic systems is discussed in this chapter. Because the scope of this subject is so large, only the dissolved phase in the aquatic system will be considered.

There are many reasons why chemists are not satisfied with the assessment of total trace metal levels in aquatic systems, but want to go further and gather information about the speciation of these metals. There are, for example, differences in toxicity [e.g., Cr(III) versus Cr(VI), As(III) versus

TABLE 7.1

Concentration Levels of Trace Metal Species in Various Aquatic Systems

Species	Groundwater	River or Lake Water	Coastal Water	Open Seawater
Cr-T	0.20–67 nM[1,2,3]	0.50–100 nM[4]		0.10–16 nM[4]
Fe-T	0.60–180 µM[5]	4–120 µM[6,7]	5–400 nM[8]	0.050–2 nM[8]
Mn-T	21–25 µM[1]	0.050–18 µM[6,9]	0.10–49 µM[7]	0.50–3 nM[10]
As-T	0.13–1.33 nM (unpolluted)[11]	0.15 µM (mean)		0.13–0.27 nM[11]
	1.30–67 µM (polluted)[11]	Up to 1.70 µM[12]		
Hg-T		0.40–16 pM[13]	0.60–7.1 pM[13]	
Hg-R		0.50–10 pM[13]	0.40–1.8 pM[13]	0.70–2.0 pM[13]
Hg(0)		0.10–0.67 pM[13]	0.10–0.80 pM[13]	0.05–0.89 pM[13]
MMHg		0.050–3 pM[13]	0.075–0.94 pM[13]	

As(V), and Hg(II) versus MMHg (monomethyl-Hg)], solubility [e.g., Fe(II) versus Fe(III) and Mn(II) versus Mn(IV)], the difference in volatility [Hg(0) versus Hg(II)], and bioavailability (e.g., labile-bound metal complex/strongly bound metal complex). It is therefore a challenge to identify and quantify the major species of elements such as Cr, Fe, Mn, As, and Hg in aquatic systems. Table 7.1 gives an overview of the levels of common trace metals in various aquatic systems.

In general, speciation of trace metal species involves the following steps:

1. Sampling
2. Sample pretreatment/preservation/storage
3. Extraction/derivatization/preconcentration of some (all) species
4. Species separation
5. Species detection

Each of these steps can modify the natural speciation distribution, and so the goal is to minimize their number. These steps will now be discussed for two groups of trace metals: Cr, Fe, Mn, and As (Section 7.2), and Hg (Section 7.3).

7.2 SPECIATION OF DISSOLVED Cr, Fe, Mn, AND As

The major oxidation states of chromium in natural waters are III and VI. The ratio of their concentrations is highly variable depending on the specific physicochemical (e.g., redox, pH, etc.) conditions of the water column. Chromium(III) is the most stable oxidation state of this metal and an element essential for human health, whereas chromium(VI) is reported as being a possible human carcinogen and mutagen. Cr(VI) compounds are more soluble, mobile, and bioavailable than Cr(III) species.[4,14] The presence of these two forms and their ratio depend on the pH,[15] redox potential and reactions (oxygen concentration, presence of appropriate reducers, and photochemical redox transformation), and mediators acting as ligands or catalysts.[16–18] It has been reported that Cr(VI) compounds are about 100 times more toxic than Cr(III) compounds owing to their high oxidation potential and ready passage through biological membranes.[19]

Fe and Mn appear to be increasingly important for photosynthetic carbon fixation by marine phytoplankton[20,21] and thus also in the process of the earth's warming up. In fact, the contemporary ocean is the largest sink of carbon dioxide, scavenging 45 gigatons of carbon per annum from the atmosphere, of which 11 gigatons are exported[22] to the ocean interior. To sustain this C flux through marine ecosystems, essential elements such as Fe and Mn must be supplied in a ratio reflecting the composition of marine phytoplankton species. In most surface waters of the oceans, the concentrations of the essential trace elements are extremely low, especially during bloom periods. Moreover, they are present in different forms: in the dissolved phase they can be labile, strongly bound to organic ligands, or colloidal, this last form also being regarded as a fraction of the dissolved phase. Not all forms are suited for uptake by phytoplankton: in fact, only the free ions and the labile-bound complexes are.

Nowadays, not only Fe but other trace metals as well, for example, Mn, Co, or Cu, are thought to limit primary production. It is thus a real challenge for oceanographers not just to assess correctly the very low levels of Fe and Mn in the oceans but also to carry out the speciation of these elements (total dissolved concentrations are at the nM level, labile forms <nM level). Fe occurs in two oxidation states in natural aquatic systems: Fe(II), which is readily soluble, and Fe(III), which is almost insoluble. However, both Fe ions can form diverse complexes with organic ligands with different labilities and solubilities, and colloidal particles, which are also considered part of the dissolved phase. Manganese also exists in two oxidation states in aquatic systems: soluble Mn(II) and insoluble Mn(IV); both are present in a dynamic cycle in seawater. The nonlabile Mn pool consists of oxidized Mn(IV) species, but these can be photochemically reduced and thus solubilized.[23]

Arsenic occurs in several oxidation states (+III, +V, 0, and −III), and a variety of inorganic and organic As forms are widespread in the environment. Potentially toxic to humans, animals, and plants, arsenic is present in the environment as a result of both natural and anthropogenic processes. Chronic and acute arsenic poisoning due to exposure to elevated concentrations has been reported worldwide, the worst arsenic disaster being the contamination of groundwater (arsenic levels >0.67 μM) in West Bengal and Bangladesh, where groundwater is the main source of drinking water.[24] Extensive toxicity studies of As have shown that different forms exhibit different toxicities, as is the case with many environmental pollutants.[25] Inorganic As species are more toxic than methylated compounds, arsenobetaine (AB), arsenocholine (AsC$^+$), or arsenosugars. With the exception of the tetramethylarsonium ion, acute toxicity generally decreases with an increasing degree of methylation. Indeed, some organoarsenicals, such as AB, which is commonly found in seafood, and AsC$^+$ are considered to be nontoxic toward living organisms, while the inorganic As species arsenite [As(III)] and arsenate [As(V)] have been identified as being the most toxic.[25–27] Only four As species [As(III), As(V), MMA or monomethylarsonic acid, and DMA or dimethylarsinic acid] are mainly present and studied in the water column, because the arsenosugars, AB and AsC$^+$ (the larger organoarsenicals) produced in and by aquatic organisms are not observed in the aquatic phase. The distribution of the most common arsenic species in natural waters depends mainly on the redox potential and the pH conditions.[28] Under oxidizing conditions (i.e., surface waters) the predominant species is As(V), whereas under mildly reducing conditions (e.g., anoxic groundwaters) As(III) is the thermodynamically stable form.[28] According to the literature, the MMA and the DMA fraction found in estuaries is highly variable depending on salinity, turbidity, temperature, and phytoplankton activity. For example, in the Humber and the Thames Estuary in winter, no methylated species were found; in midsummer, however, their levels ranged from 0% to 12% in the Humber, depending on the above-mentioned variables,[29] with DMA concentrations from 0.27 to 2.7 nM, and made up about 8% of the dissolved arsenic concentration in the Thames.[30] In winter, we were unable to find any methylated As species in the Zenne River (Belgium).[27]

7.2.1 IN SITU SPECIATION IN THE AQUATIC SYSTEM

Speciation is best carried out directly in the aquatic system, without sampling. This has been possible since the development of Diffusive Equilibrium in Thin Films (DET) and Diffusive Gradient in Thin Films (DGT) probes.[31] The DET probe consists of a very thin gel layer that is immersed in the aquatic system and allowed to equilibrate with the bulk solution. The concentration of solutes in the gel is similar to that in the bulk solution for all solutes that can diffuse through the pore openings of the gel (some gels have open pores >5 nm and some gels have restricted pores <1 nm). The DGT

TABLE 7.2
LODs for Trace Metal Species in Aquatic Samples

Speciation Procedure for Cr	LOD	Speciation Procedure for Fe	LOD	Speciation Procedure for Mn	LOD	Speciation Procedure for As	LOD
Solid extraction (GFAAS)	0.035–0.19 nM (Cr(III))[14]	Resin-based column chromatography (GFAAS)	0.020 nM[35]			Solid extraction (GFAAS)	0.50–1.5 nM (As(III))[43]
						Solid extraction (ICPMS)	0.10 nM (As(V))[43]
Ion exchange (ICPMS)	0.096–0.13 (Cr(III))[14] 0.23–0.31 (Cr(VI))[14]					IC-HG-GFAAS	50 nM (As(III))[43]
						HG-AFS	0.3–53 nM (As(III))[43]
						HG-GF-ICPMS	0.030 nM (As(V))[43]
Selective coprecipitation (FAAS)	0.0077–0.012 nM (Cr(III) and Cr(VI))[14]						
Bidirectional eletrostacking (AAS)	0.12 nM (Cr(III))[14] 0.096 nM (Cr(VI))[14]	Sequential injection analysis (UV-VIS)	1.8 µM (Fe(III))[36] 2.7 µM (Fe(II))[36]	Sequential injection analysis (UV-VIS)	89 nM (Mn(II))[41] 143 nM (Mn-T)[41]	HPLC-ICPMS	0.30–0.80 nM (As(III))[43]
Capillary reaction (chemiluminescence)	0.60 pM (Cr(III))[32] 8.0 pM (Cr(VI))[32]	Selective reagent (chemiluminescence)	0.10 M (Fe(II))[37] 0.20 M (Fe(III))[37] 0.021 nM (Fe-T)[38]	Chelation (UV)	0.39 nM[42]		
Electrochemistry (CSV)	0.10 nM (Cr(III) and Cr(VI))[33] 0.080 nM (Cr(VI))[14]		0.010–0.11 nM[39,40]		4.0 nM (Mn(II))[9]		0.060–40 nM (As(III))[43]
DET-ICPMS	70 nM (Cr-T)[34]		4.0 µM (Fe-T)[34]		17 nM (Mn-T)[34]		2.0 nM (As-T)[34]
DGT-ICPMS	1.8 nM (unpublished data)[a]		19 nM[34,a]		1.4 nM[34,a]		0.14 nM (unpublished data)[a]

a   These values are blank values. LODs for DGT depend in fact on the exposure time.

technique is not a static (equilibrium) sampling method. Based on the mass transport control of solutes from the bulk solution to a backup resin, it makes use of two hydrogel layers: a polyacrylamide gel is the diffusive layer, which is backed up by a second thin film gel layer containing a resin, generally a chelex cation exchanger. With the DGT probe, it is thus possible to preconcentrate the solutes by increasing the exposure time in the solution. The limits of detection (LODs) obtained with the DET and DGT techniques are shown in Table 7.2.

With a DGT device, Cr(III) can be bound to the chelex resin because of its cationic nature, whereas Cr(VI) is not bound to the resin (it has an anionic nature) but is present in the diffusive gel layer (as in a DET probe), reaching equilibrium with Cr(VI) in the aquatic system. Hence, Cr(VI) can be measured in the diffusive layer and Cr(III) in the resin layer.[44] For Mn the same procedure can be adopted. The oxidized Mn(IV) species form colloids or even larger particles and will not be sampled by the DGT probe, whereas Mn(II) species are free or labile complexes. For Fe speciation, DGTs with open pores and with restricted pores are often used. Since in aquatic systems, Fe(III) is present mostly as a ligand complex or in colloidal form, the restrictive pore size excludes these forms and makes only Fe(II) species available to the restrictive DGT,[45] whereas the open-pore DGT allows the passage of Fe(II) and small and labile Fe(III) complexes. In the case of arsenic speciation, As(III) and As(V) diffuse through the diffusive gel layer of the DGT, but only As(III) is immobilized on the chelating resin layer; As(V) remains in the diffusive layer as an anionic compound.

The species separation obtained in this way is stable and no longer changes. In the laboratory, the solutes scavenged by the DGT resin or present in the diffusive gel layer can be solubilized and measured with a sensitive technique such as graphite furnace atomic absorption spectrometry (GF-AAS) or inductively coupled plasma mass spectrometry (ICP-MS), attaining detection limits of around 0.5 ppt. Laser ablation inductively coupled plasma mass spectrometry (LA-ICP-MS), which used to be applied to generate trace metal flux distributions in shells or sediments, is now utilized directly on the dry resin gels from DGT; with this technique a greater precision is achieved than with the classical ICP-MS.[46]

7.2.2 Sampling

For analyses not using the *in situ* DET and DGT techniques, classic sampling procedures are needed. Sample collection and storage are key aspects of the overall analysis methodology; hence, there is a need to standardize protocols for trace metal sampling in aquatic systems. Low-density polyethylene (LDPE), high-density polyethylene (HDPE), fluorinated HDPE, and Teflon [polytetrafluoroethylene (PTFE) or fluorinated ethylene propylene (FEP)] are traditional sampling materials (see e.g., Achterberg et al.[8]). A strict cleaning protocol needs to be followed. According to Achterberg et al.,[8] the containers should first be immersed in 5% detergent for a week, then copiously rinsed with Milli-Q water, then immersed again in 6 M analytical grade HCl or HNO_3 for two weeks, and stored double bagged until use. Immediately before sampling they should be rinsed three times with Milli-Q, and then rinsed three times with the sample before being filled finally.[8] Our samples of surface water (North Sea, Scheldt River, and Zenne River) were collected manually from a rubber dinghy by submerging the precleaned sampling bottles approximately 20 cm beneath the water surface. Arm-length gloves were worn during sampling. The dinghy moved gently against the current during sampling and was positioned approximately 100 m upcurrent of the research vessel. When sampling from a rubber dinghy is not possible because of adverse weather conditions or when subsurface samples (10 m depth or more) are required, NOEX (Technicap, France) (Go-Flo type) sampling bottles and plastic-coated messengers, which are also thoroughly precleaned, are used.[47] In addition, a Kevlar cable is mounted on the oceanographic winch. Filtration in the field is performed as soon as possible in a clean lab container or on a clean air bench close to the sampling spot. The filtration apparatus consists of an FEP separating funnel onto which a Teflon filter holder is connected. Filtration is performed under pressure using N_2 gas.

7.2.3 Sample Preservation and Storage

Although sample preservation is to be avoided, sometimes there are no other options. For chromium sampling, the water sample was acidified at pH 2 and no Cr(III) was lost to the wall of the pre-cleaned polyethylene (PE) or the polypropylene (PP) bottle for more than one month.[48,49] While it is clear that Cr(III) is stable for a long time in an acidified sample, it was reported that acidification of coastal seawater resulted in the rapid reduction of Cr(VI) to Cr(III).[48] On the other hand, a stable Cr(VI) concentration could be ensured at a nearly neutral pH, especially under a CO_2 blanket.[48,50]

Traditionally, in order to prevent oxidation of unstable Fe(II) species, the water sample has to be filtered immediately after sampling (filtration of the sample in the field needs to be carried out under completely oxygen-free conditions) and stabilized: stabilization depends on the subsequent analytical method.[51] Even when all sample treatment protocols are rigorously applied, Fe(II) is so easily oxidized that the initial speciation can be distorted simply by contact of the sample with air. Mn(II) oxidizes much more slowly than Fe(II): this reaction is about 10^7 times slower than that of Fe(II) at pH 8 and 25°C,[52] reducing the risk of error during the speciation procedure. After filtration, only Mn(II) and Mn(IV) colloids remain in the sample. Filtered samples, mostly acidified, are commonly stored in precleaned Teflon bottles at 4°C.

For As, contamination will occur only rarely as long as standard procedures for trace elements are followed. The preservation of samples is more likely to be one of the troublesome steps in As speciation analyses. Events like changes in oxidation state, changes induced by microbial activity, or losses by volatilization or adsorption have to be avoided.[25] It has been observed that aqueous samples intended for total As determination did not sustain any losses during storage when kept in acid-washed glass, PTFE, or PE containers.[53] As far as storage for As-speciation experiments is concerned, little information is available on appropriate storage conditions for As. An overview of the influence of critical factors for species stability (pH, temperature, light, and container material) and of procedures for the preservation of the integrity of species is given by Ariza et al.[54] The recommended procedures are freezing, cooling, acidification, sterilization, deaeration, addition of ascorbic acid, and/or storage in the dark, but there is no general agreement on these procedures and some reports are conflicting. For samples in which bacteria may exist naturally, storage at low temperatures or even freeze-drying[55,56] is required to prevent biological activity from modifying the nature of the sample.

For aqueous samples, time and temperature studies report that, at higher concentrations (0.27 μM), immediate storage of filtered [0.45 μm polycarbonate filters (nucleopore)] and acidified [to 1% with HCl (supra-pure)] natural waters at about 5°C can preserve As(III) and As(V) concentrations for about 30 days.[57] It is advisable that samples with lower As concentrations be kept in the dark at 4°C.[58]

7.2.4 Electrochemical Speciation

Cathodic stripping voltammetry (CSV) and adsorptive cathodic stripping voltammetry (AdCSV) allowed Cr speciation by direct determination of Cr(VI) in the presence of predominant Cr(III) levels with a detection limit (for LODs, see Table 7.2) of 0.08 nM for Cr(VI).[59] Gledhill et al.[60] used CSV to assess dissolved and total Fe concentrations in the North Sea, whereas Boye et al.[61] used CSV to obtain Fe speciation in the northeast Atlantic Ocean. A much greater sensitivity for Fe speciation and lower LODs (for LODs, see Table 7.2) was obtained by CSV using an adsorptive and competing ligand.[39] CSV was also used by a research group from Liverpool to determine dissolved and particulate Mn in the water column.[62,63] Carbon film electrodes were used by Filipe and Brett[9] to determine trace levels of Mn(II) in pore water samples; the detection limit was very low (Table 7.2).

Electrochemical methods for arsenic determination were initially based on polarography with a dropping mercury electrode. More recent methods, based on anodic stripping voltammetry (ASV), anodic stripping chronopotentiometry (SC), and CSV, rely almost exclusively on the detection of As(III), since As(V) is detected with difficulty because of its perceived electro-inactivity.

These methods use either a gold- or a mercury-based electrode.[64,65] Despite past problems with determining inorganic arsenic species, Salaün et al.[65] showed that As(III) can be determined by ASV using a gold microwire electrode at any pH, including the neutral pH typical of natural waters, whereas As(V) requires acidification to pH 1. Detection limits with this microelectrode are 0.2 nM As(III) at pH 8 and 0.3 nM combined arsenic (III + V) at pH 1 with a 30-s deposition time (Table 7.2). Additionally, copper is codetermined with this technique.

7.2.5 HYPHENATED METHODS

7.2.5.1 Exchange Columns (Trap and Elute)

Cox and Mcleod[66] passed their water samples through activated alumina microcolumns in the field, isolating and retaining both Cr(III) and Cr(VI) species. The microcolumns were then returned to the laboratory and inserted into a flow injection inductively coupled plasma-emission spectrometry (FI-ICP-ES) system for elution and quantification (the lowest results reported are around 40 nM). The pretreatment of the microcolumns and the FI-ICP-ES method was, however, complicated and time-consuming. Recently, Dogutan et al.[67] and Latif et al.[68] preconcentrated the Cr species on an exchange column and first eluted one species, and subsequently both of them.

Special resin-based columns can perform iron speciation from water samples.[51,69] Resin-based column chromatography procedures are attractive for several reasons. Mini- and microcolumns packed with ion exchange and adsorbing resins are effective for separating and preconcentrating simple cationic species of Fe(II) and Fe(III) or their complexes with different chromogenic reagents. Flow injection systems with resin-based columns make the procedures fast and simple to operate and allow automation. Automation of sample handling and analysis will especially minimize the risk of contamination and enhance the repeatability.

A promising solid-phase extraction adsorbent for metals and more particularly As is nanometer TiO_2 material and immobilized nanometer TiO_2.[70,71] Both As(III) and As(V), or As(III) alone, were quantitatively absorbed on immobilized nanometer TiO_2 depending on the pH (for LODs, see Table 7.2).

7.2.5.2 Chromatography

The most widely used separation techniques include high-performance liquid chromatography (HPLC), ion chromatography (IC), and capillary electrophoresis (CE), and all can be used in combination with other treatments such as hydride generation (HG), coprecipitation, and voltammetry (see e.g., Liang and Liu,[71] Ronkart et al.,[72] Sounderajan et al.,[73] and Hu et al.[74]). For As, the most common of these methods is liquid chromatography combined with HG: this makes use of the ability of As species to form volatile hydrides when reacting with $NaBH_4$. A novel separation method is capillary microextraction (CME) with an ordered meso-porous Al_2O_3 coating. This method can be used to simultaneously separate inorganic As(III)/As(V) and Cr(III)/Cr(VI).[74] In many cases we are only interested in inorganic As speciation: this involves the determination of the total As content and the content of one of the two species, the other one being obtained by subtraction.

The most popular analytical technique for the speciation of the predominant As species in an aquatic system [As(III), As(V), MMA, and DMA] is HPLC-HG-AFS; the technique used in our laboratory was described by Baeyens et al.[27] The optimization of HPLC and HG-AFS coupling provided chromatograms of the four "anionic" arsenicals, similar to those in Figure 7.1, for a mixed standard of 6.7 nM of each of the compounds. The chromatographic conditions are as follows: column [Hamilton PRP-X100 (250 × 4.1 mm; 10 μm)]; mobile phase [KH_2PO_4/K_2HPO_4 buffer; 20 mM, pH 6.0 (HCl)]; flow rate (1 mL min^{-1}); and injection volume (200 μL). For the HG system, the Ar gas flow rates are the same as for the total As determinations but they are modified for HCl (1.5 M; 1 mL min^{-1}) and $NaBH_4$ [2.5% (m/v); 1 mL min^{-1}].

The detection limits (Table 7.2) are As(III) (0.40 nM), DMA (0.57 nM), MMA (0.55 nM), and As(V) (0.89 nM). Precision and accuracy on the lowest sensitivity scale were calculated from 10

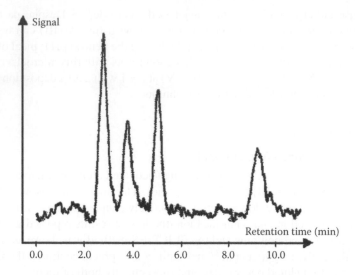

FIGURE 7.1 Chromatogram of a 0.5 µg As L^{-1} mixed standard of As(III), As(V), MMA, and DMA, using HPLC-HG-AFS. Order of elution: As(III)—DMA—MMA—As(V).

consecutive measurements of an artificial standard containing 5.3 nM of each of the four compounds of interest. The relative standard deviation varied between 7% and 13% whereas the recovery rate was within 92–98%, except for As(V), which was 81%. Accuracy measurements could not be made from measurements of reference material, since no liquid reference material, which is certified for the separate As compounds, was available.

7.2.5.3 Chemiluminescence and Colorimetric Reactions

According to the review by Marques et al.,[75] the most frequent pretreatment used for chromium speciation is complex formation. Extraction processes are frequently used after complex formation to extract the complexes formed prior to UV-VIS detection. Recently, the simultaneous determination of Cr(III) and Cr(VI) using an in-capillary reaction, CE separation, and chemiluminescence detection was reported with LODs (Table 7.2) for Cr(III) and Cr(VI) of 0.6 and 8 pM, respectively.[32]

Spectrophotometric techniques combined with flow injection analysis (FIA) and on-line preconcentration can meet the required detection limits for natural Fe concentrations in aquatic systems (Table 7.2) by also using very specific and sensitive ligands, such as ferrozine [3-(2-bipyridyl)-5,6-bis(4-phenylsulfonic acid)-1,2,4-triazine], that selectively bind Fe(II). Determining Fe(II) as well as the total Fe after on-line reduction of Fe(III) to Fe(II) with ascorbic acid allows a kind of speciation.[37] A drawback is that the selective complexing agents can shift the iron redox speciation in the sample. For example, several researchers have reported a tendency for ferrozine to reduce Fe(III) to Fe(II) under certain conditions.[76] Most ferrozine methods involve sample acidification, which may also promote reduction of Fe(III) in the sample. Fe(II) is a transient species in most seawater environments and is rapidly oxidized to Fe(III); therefore, unacidified samples are required in order to maintain redox integrity.[8] An alternative is to couple FIA with a chemiluminescence reaction.[77,78]

7.2.6 OTHER METHODS

In the 1980s, an analytical technique was developed for the study of chromium speciation in natural waters based on the atomization of electrodeposited species on graphite tubes.[79] Two independent automated platforms consisting of an ultraviolet (UV) on-line unit and a chelation/preconcentration/

matrix elimination module were specifically developed to process trace element samples, including Mn, on-site to avoid sample storage prior to ICP-MS analysis. This method has been used to determine total, labile, and organically bound dissolved Mn concentration[42] (for LODs, see Table 7.2).

7.3 SPECIATION OF DISSOLVED Hg

Although all forms of Hg are poisonous, the ecological and human health effects of mercury are generally related to environmental transformations of inorganic mercury to the toxic and biomagnification-prone compound monomethylmercury (MMHg). Significant improvements in instrumentation have been made in recent years, allowing reliable measurements of all Hg species, but traditional problems related to contamination, nonquantitative recoveries, and to questions about the possibility of artifact formation and transformations of methylmercury during the sample preparation and separation steps require the rigorous execution of validated analytical protocols. The importance of Hg speciation studies has been highlighted in various review articles published in the last 10 years by several authors.[80–84]

The relevant speciation methods depend on the nature of the sample (e.g., freshwater, seawater, and anoxic water) and the concentration level. In general, ambient Hg levels in natural aquatic systems, especially open ocean waters, are very low (for ppt–ppq levels, see Table 7.1) and require appropriate precleaning, sampling, and storage procedures. In terms of speciation of organomercury species, extraction is a very subtle step because (1) the whole species content may not be released and (2) artifacts can occur, so that some organomercury species may be destroyed or formed (interspecies exchange). Often, the extraction step for mercury speciation is applied in combination with a cleanup/preconcentration step such as distillation, solvent extraction, or headspace. Artifacts during the extraction-cleanup phase have been specifically studied with the latter methods. By using isotope-labeled compounds it is possible to study interspecies exchange.

Precleaning of material, sampling, and filtration in the field are not very different from the procedures used for the other trace metals described above, except that for Hg borosilicate glass bottles can also be used, and that samples collected for volatile, metallic mercury Hg(0) and dimethylmercury (DMHg) species are not filtered. When filtration cannot be carried out in the field, samples should be kept unpreserved, cold, and in the dark. More specific information about our techniques can be found in Baeyens[80] and Leermakers et al.[47,84]

7.3.1 SAMPLE HANDLING AND STORAGE

The most volatile forms present in water are Hg(0) and DMHg. They should be removed from the samples immediately after collection by purging and trapping on gold (for total gaseous Hg) and Carbotrap or Tenax (for DMHg). When purging and trapping in the field is not possible, samples should be collected in completely full glass bottles with Teflon-lined caps, as these species are lost rapidly ($t_{1/2}$ = 10–20 h) from Teflon and PE bottles.[85] Because acids can accelerate the oxidation of volatile species, these samples should be stored, refrigerated and unacidified, and processed within 1–2 days.

After filtration, samples for reactive Hg (Hg-R) and total dissolved Hg (Hg-T) were acidified with 0.5% HCl.[47] BrCl is also often used to preserve samples intended for Hg-T determination. The acidification of samples to be used for Hg-R determination was not recommended by Parker and Bloom,[85] especially those with high levels of dissolved organic carbon (DOC). DOC may coagulate after acidification of the solution, with concomitant adsorption and precipitation of Hg-R. Labile Hg (Hg-R) appears to be relatively stable (days to weeks) in filtered, unpreserved samples.

Samples designated for the determination of MMHg were stored deep-frozen and unpreserved.[47] Alternatively, Parker and Bloom[85] suggested storing freshwater and seawater samples for MMHg in the refrigerator and in the dark after the addition of 0.4% HCl or 0.2% H_2SO_4. Sulfuric acid was recommended for seawater to minimize Cl⁻ levels, which interfere when the distillation extraction method is used.

7.3.2 ANALYTICAL METHODS FOR Hg(0), DMHg, Hg-R, AND Hg-T

Hg-T, Hg-R, DMHg, and Hg(0) were determined by cold vapor atomic fluorescence spectrometry (CVAFS) using an Au-amalgamation preconcentration step.[86] Hg(0) was purged from the sample on a gold column in the field; DMHg on a Carbotrap column. Both were transferred to an analytical Au column in the laboratory and determined with CVAFS. Hg-R was measured using $SnCl_2$ as a reducing agent; Hg-T was analyzed by BrCl oxidation and reduction with $NH_2OH.HCl$ prior to reduction with $SnCl_2$.[47,87] The detection limits were 0.025 pM for Hg(0) and DMHg, 0.075 pM for Hg-R, and 0.25 pM for Hg-T (see Table 7.2). The detection limit for Hg-T in a water sample without any pretreatment is much higher (10 pM) with ICP-MS, but ICP-MS is often used in combination with a cold vapor module or an Au column.

7.3.3 ANALYTICAL METHODS FOR MMHg

7.3.3.1 Extraction Procedures

Total dissolved MMHg can be analyzed by aqueous-phase ethylation after separating MMHg from the interfering chloride matrix by extraction with methylene chloride.[88] For a 200-mL sample a detection limit of 0.075 pM is achieved. An alternative method for the simultaneous extraction of Hg(II) and MMHg in natural waters at fM levels is to extract both into toluene as dithiozonates after acidification of the water sample, followed by back extraction into an aqueous solution of Na_2S and removal of H_2S by purging with N_2.[89]

Nagase et al.[90] and Horvat et al.[91] proposed vapor distillation in a stream of air or nitrogen at 150°C for the nonchromatographic separation of inorganic Hg and MMHg. In combination with the ethylation technique, Carbotrap or Tenax preconcentration, GC separation, and AFS detection,[88,92] this quickly became the method of choice for the extraction of MMHg because of its high efficiency (practically 100% recoveries of MMHg), the elimination of inorganic Hg in the extract, and the formation of clean aqueous extracts that eliminate interferences in the ethylation step. However, investigations in the mid-1990s showed that the distillation procedure used to separate methylmercury from both water and sediment samples artificially generates MMHg in the presence of natural organic substances. A special issue of *Chemosphere* was published in 1999,[93] summarizing the state-of-the-art regarding the artifact formation of MMHg during derivatization and analysis (during separation owing to the presence of the silanizing agent), and also during sample storage.

7.3.3.2 Gas Chromatographic Separation Methods

Apart from the above-mentioned problems associated with the extraction of organomercurials, difficulties were also encountered with the chromatography of organomercury halides. The different packed and capillary columns used were reviewed by Baeyens.[80] In order to prevent ion exchange and adsorption processes on the column (which cause undesirable effects such as tailing, changing of the retention time, and a decrease in peak areas/heights), passivation of the packing material with Hg(II) chloride in benzene (or toluene) is needed. Moreover, the more common GC detectors may lack the selectivity required for use in the speciation of Hg in environmental samples. For instance, electron capture detection (ECD) was commonly used for methylmercury speciation in environmental samples, but its unselective response required laborious cleanup processes of the extract in the organic phase.

7.3.3.3 Derivatization and Validation

To overcome these problems, alternative methods involving precolumn derivatization of Hg species have been developed. These nonpolar derivatives can then be separated on nonpolar packed[88,94] or capillary columns.[95] Iodation with acetic acid,[94,96] hydration with $NaBH_4$,[97,98] aqueous-phase ethylation with sodium tetraethylborate ($NaBEt_4$),[88] and derivatization with a Grignard reagent (ethylation, butylation, propylation, etc.)[95] are the most commonly used methods.

Aqueous-phase ethylation, room-temperature precollection, and separation by GC with CV-AFS detection have become the most frequently used techniques in laboratories involved in studies of the biogeochemical cycle of mercury. Like elemental Hg and DMHg, the ethylated species are volatile and can therefore be purged from solution at room temperature and collected on sorbents such as Carbotrap or Tenax. After thermal release, all the mercury compounds (natural or derivatized) are separated by cryogenic,[88] isothermal,[99] or temperature-programmed GC.[100] Instead of being collected on Carbotrap or Tenax, the ethylated compounds may be injected directly into the GC column by headspace injection[100,101] or cryotrapped on a fused silica column and desorbed by flash heating.[102,103] As the Hg species are eluted from the column, they are thermally decomposed in a pyrolytic column (900°C) before being measured by an Hg-specific detector (e.g., CV-AFS, CV-AAS, QF-AAS, MIP-AES, and ICP-MS). It should be mentioned that the ethylation procedure cannot be used for the determination of other organomercurials; moreover, it is not clear whether ethylmercury compounds were originally present in the sample. Therefore, the usefulness of other derivatization agents has been investigated. Sodium tetrapropylborate ($NaBPr_4$) was proposed by De Smaele et al.,[104] and sodium tetraphenylborate ($NaBPh_4$) by Abuin et al.[105] and Grinberg et al.[106] Sodium borohydride may also be used to form volatile methylmercury hydride, which is then quantified by gas chromatography in line with a Fourier transform infrared spectrophotometer.[107]

If derivatization of the native species is carried out, derivatization yields should also be assessed. In aqueous samples these yields are relatively easy to assess when a derivatized standard similar to the derivatized organomercury compound is available. Use of the standard addition method allows the derivatization yield to be determined.

7.3.3.4   GC Improvements

Several techniques have been used to overcome the problem of low column loadings on capillary columns. Capillary columns have also been used after preconcentration of alkyl derivatives on a wide-bore fused silica column[103] or by solid-phase microextraction (SPME).[106]

Multicapillary GC (MCGC) [919 capillaries, 1 m*40 μm id coated with 0.2 μm SE 30 stationary phase (Alltech)] coupled to ICP-MS[103] allows column loadings and carrier gas flow rates to approach those of packed columns. The basic and unique features are the high speed of separation at large sample injection volumes with an exceptionally high range of volumetric velocities of the carrier gas at which the column retains its high efficiency. This makes plasma source detection ideally suited for MCGC, leading to a coupled technique with a tremendous potential for separation analysis.

Solid-phase microextraction capillary gas chromatography (SPME-GC) is also an interesting preconcentration method. After derivatization with tetraethylborate, tetrapropylborate, or tetraphenylborate, the ethylated compounds are extracted by SPME on a silica fiber coated with polydimethylsiloxane (PDMS). SPME can be performed either in the aqueous phase or in the headspace. After SPME extraction, species are thermally desorbed, separated by GC, and analyzed.[106]

7.3.3.5   Liquid Chromatographic Separation Methods

Applications of HPLC for Hg speciation studies have been reviewed by Harrington.[83] Practically all HPLC methods for Hg speciation reported in the literature are based on reversed-phase separations, involving the use of a silica-bonded phase column and a mobile phase containing an organic modifier, a chelating or an ion pair reagent, and in some cases, a pH buffer.

The interface to couple HPLC columns with the atomizer can be very simple, with a direct connection from the exit of the column to the nebulizer of the AAS or plasma detector. Unfortunately, the efficiency of the nebulizer is very low (1–3%), which limits the sensitivity. A general way to circumvent this lack of sensitivity is postcolumn derivatization to form cold Hg vapor. However, the generation of a cold vapor from organomercury species requires an extra step: their conversion to

Hg(II). This conversion is usually on-line, but in an effort to analyze low levels of mercury species, some workers have developed on- and off-line sample preconcentration methods.[108-110]

Besides reversed-phase HPLC, IC has also been used to separate Hg species.[111,112] IC enables the direct separation of more polar and ionic species, so that sample pretreatment can be simplified. The coupling of IC with CV-ICP-MS allows very low detection limits to be obtained.[112]

7.3.3.6   Detection Methods

The development of a commercial, relatively inexpensive, extremely sensitive, and selective CV-AFS instrumentation in the late 1980s and 1990s[113] made this the most popular detector in laboratories working on the biogeochemical cycling of Hg. In recent years, the use of ICP-MS in speciation analysis has increased tremendously. Besides its high sensitivity and selectivity, ICP-MS offers the opportunity to perform speciated isotope dilution mass spectrometry (SID-MS).[114] Not only is this technique highly accurate and precise, the isotopically enriched isotopes can also be used as tracers to check for species transformations and extraction recoveries. However, to determine the low levels of Hg in natural aquatic systems that are quantifiable with a gold column and a CV-AFS instrument, ICP-MS is often coupled to either a cold vapor generation module or an Au column.

7.3.4   Speciation Modeling of Hg

The extent of complexation of dissolved mercury in estuarine waters will vary markedly with the nature and concentration of the inorganic and organic ligands as well as their respective stability constants. The following equations describe the relations between the different species:

1. $Hg_{(total)} = Hg_{(labile)} + Hg_{(nonlabile)}$
2. $Hg_{(labile)} = HgL_{inorg} + Hg_{(free)}^{2+} + Hg(0)$
3. $HgL_{inorg} = \Sigma Hg_{(free)}^{2+}(L_{inorg(free)})$
4. $Hg_{(nonlabile)} = HgL_{org} + MeHg$

The major species can be calculated using known values of equilibrium constants and the concentrations of the ligands $Hg_{(total)}$, $Hg_{(labile)}$, and MeHg:

5. $\beta_{HgLn} = [HgL_n]/([Hg_{(free)}^{2+}][L]^n)$,

where β_{HgLn} is the conditional stability constant of the complex HgL_n, and [HgLn], [Hg^{2+}], and [L] are the concentrations of the complex, the free mercuric ion, and the free ligand, respectively.

A correct speciation involves a multitude of chemical equilibria, as any metal can form a complex with any ligand.

Using the above equations, model simulations of the various Hg species in the Scheldt estuary were carried out using the TK-Solver program.[115] A conditional stability constant of 10^{19} was estimated for Hg–humic acid interactions in the Scheldt.

ACKNOWLEDGMENT

The authors gratefully acknowledge the financial support from the Interuniversity Attraction Poles Programme—Belgian State—Belgian Science Policy (TIMOTHY—P6/13). Anouk de Brauwere is a postdoctoral researcher of the Flanders Research Foundation (FWO).

REFERENCES

1. Cheng, K.P. and J.J. Jiao. 2008. Metal concentrations and mobility in marine sediment and groundwater in coastal reclamation areas: A case study in Shenzhen, China. *Environ. Pollut.* 151: 576–584.
2. Farnham, I.M., K.H. Johannesson, A.K. Singh, V.F. Hodge, and K.J. Stetzenbach. 2003. Factor analytical approaches for evaluating groundwater trace element chemistry data. *Anal. Chim. Acta* 490: 123–138.

3. Galindo, G., C. Sainato, C. Dapen, et al. 2007. Surface and groundwater quality in the northeastern region of Buenos Aires Province, Argentina. *J. S. Amer. Earth Sci.* 23: 336–345.

4. Kotas, J. and Z. Stasicka. 2000. Chromium occurrence in the environment and methods of its speciation. *Environ. Pollut.* 107: 263–283.

5. Windom, H.L., W.S. Moore, L.F.H. Niencheski, and R.A. Jahnke. 2006. Submarine groundwater discharge: A large, previously unrecognized source of dissolved iron to the South Atlantic Ocean. *Mar. Chem.* 102: 252–266.

6. Huang, X., M. Sillanpa, B. Duo, and E.T. Gjessing. 2008. Water quality in the Tibetan Plateau: Metal contents of four selected rivers. *Environ. Pollut.* 156: 270–277.

7. Cidu, R. and R. Biddau. 2007. Transport of trace elements under different seasonal conditions: Effects on the quality of river water in a Mediterranean area. *Appl. Geochem.* 22: 2777–2794.

8. Achterberg, E.P., T.W. Holland, A.R. Bowie, R.F.C. Mantoura, and P.J. Worsfold. 2001. Determination of iron in seawater—a review. *Anal. Chim. Acta* 442: 1–14.

9. Filipe, O.M.S. and C.M.A. Brett. 2003. Cathodic stripping voltammetry of trace Mn(II) at carbon film electrodes. *Talanta* 61: 643–650.

10. De Jong, J.T.M., M. Boye, M.D. Gelado-Caballero, et al. 2007. Inputs of iron, manganese and aluminium to surface waters of the Northeast Atlantic Ocean and the European continental shelf. *Mar. Chem.* 107: 120–142.

11. Environmental Health Criteria 224. 2001. *Arsenic and Arsenic Compounds*, 2nd edition, p. 248. Geneva: World Health Organization.

12. Pichler, T., R. Brinkmann, and G.I. Scarzella. 1998. Arsenic abundance and variation in golf course lakes. *Sci. Total Environ.* 394: 313–320.

13. Baeyens, W. 1998. Evolution of trace metal concentrations in the Scheldt estuary (1978–1995). A comparison with estuarine and ocean levels. *Hydrobiologia* 366: 157–167.

14. Gomez V. and M.P. Callao. 2006. Chromium determination and speciation since 2000. *Trends Anal. Chem.* 25: 1006–1015.

15. Campanella, L. 1996. Problems of speciation of elements in natural water: The case of chromium and selenium. In: S. Caroli (ed.), *Element Speciation in Bioinorganic Chemistry*, pp. 419–444. New York: Wiley-Interscience.

16. Cranston, R.E. and J.W. Murray. 1980. Chromium species in the Columbia River and estuary. *Limnol. Oceanogr.* 25: 1104–1112.

17. Pettine, M., F.J. Millero, and T. La Noce. 1991. Chromium(III) interactions in seawater through its oxidation kinetics. *Mar. Chem.* 34: 29–46.

18. Kieber, R.J. and G.R. Helz. 1992. Indirect photoreduction of aqueous chromium (VI). *Environ. Sci. Technol.* 26: 307–312.

19. Barnowski, C., N. Jakubowski, D. Stuewer, and J.A.C. Broekaert. 1997. Speciation of chromium by direct coupling of ion exchange chromatography with inductively coupled plasma mass spectrometry. *J. Anal. At. Spectrom.* 12: 1155–1161.

20. Morel, F.M.M. and N.M. Price. 2003. The biogeochemical cycles of trace metals in the oceans. *Science* 300: 944–947.

21. Baeyens, W., M. De Gieter, M. Leermakers, and I. Windal. 2006. Speciation in environmental samples. In: L. Nollet (ed.), *Chromatographic Analyses of the Environment*, Chapter 20, pp. 743–778. Boca Raton, FL: CRC Press-Taylor & Francis Group.

22. Laws, E.A., P.G. Falkowski, W.O. Jr. Smith, H. Ducklow, and J.J. McCarthy. 2000. Temperature effects on export production in the Open Ocean. *Global Biogeochem. Cycles* 14(4): 1231–1246.

23. Sunda, W.G. and S.A. Huntsman. 1988. Effect of sunlight on redox cycles of manganese in the southwestern Sargasso Sea. *Deep-Sea Res.* 35: 1297–1317.

24. Kinnibrugh, D.G. and P.L. Smedley. 2001. *Arsenic Contamination of Groundwater in Bangladesh*, Vol. 2: Final Report; BGS Technical Report WC/00/19. Keyworth: British Geological Survey.

25. Leermakers, M., W. Baeyens, M. De Gieter, et al. 2006. Toxic arsenic compounds in environmental samples: Speciation and validation. *Trends Anal. Chem.* 25(1): 1–10.

26. Caruso, J.A. and M. Montes-Bayon. 2003. Elemental speciation studies—new directions for trace metal analysis. *Ecotoxicol. Environ. Saf.* 56: 148–163.

27. Baeyens, W., A. de Brauwere, N. Brion, M. De Gieter, and M. Leermakers. 2007. Arsenic speciation in the River Zenne, Belgium. *Sci. Total Environ.* 384: 409–419.

28. Cullen, W.R. and K.J. Reimer. 1989. Arsenic speciation in the environment. *Chem. Rev.* 41: 2146–2157.

29. Kitts, H.J., G.E. Millward, A.W. Morris, and L. Ebdon, 1994. Arsenic biogeochemistry in the Humber Estuary, U.K. *Estuar. Coast. Shelf Sci.* 39(2): 157–172.

30. Millward, G.E., H.J. Kitts, L. Ebdon, J.I. Allen, and A.W. Morris. 1997. Arsenic species in the Humber Plume, U.K. *Cont. Shelf Res.* 17(4): 435–454.

31. Davison, W. and H. Zhang 1994. *In situ* speciation measurements of trace components in natural waters using thin-film gels. *Nature* 367: 546–548.

32. Yang, W.P., Z.J. Zhang, and W. Deng. 2003. Speciation of chromium by in-capillary reaction and capillary electrophoresis with chemiluminescence detection. *J. Chromatogr. A* 1014: 203–214.

33. Li, Y.J. and H.B. Xue. 2001. Determination of Cr(III) and Cr(VI) species in natural waters by catalytic cathodic stripping voltammetry. *Anal. Chim. Acta* 448: 121–134.

34. Leermakers, M., Y. Gao, C. Gabeille, et al. 2005a. Determination of high resolution pore water profiles of trace metals in sediments of the Rupel River (Belgium) using DET (diffusive equilibrium in thin films) and DGT (diffusive gradients in thin films) techniques. *Water Air Soil Pollut.* 166: 265–286.

35. Lohan, M.C., A.M. Aguilar-Islas, R.P. Franks, and K.W. Bruland. 2005. Determination of iron and copper in seawater at pH 1.7 with a new commercially available chelating resin, NTA Superflow. *Anal. Chim. Acta* 530: 121–129.

36. Mulaudzi, L.V., J.F. Van Staden, and R.I. Stefan. 2002. On-line determination of iron(II) and iron(III) using a spectrophotometric sequential injection system. *Anal. Chim. Acta* 467: 35–49.

37. Blain, S. and P. Treguer. 1995. Iron(II) and iron(III) determination in sea water at the nanomolar level with selective on-line preconcentration and spectrophotometric determination. *Anal. Chim. Acta* 305: 425–432.

38. De Jong, J.T.M., J. Den Das, U. Bathmann, et al. 1998. Dissolved iron at subnanomolar levels in the Southern Ocean as determined by ship-board analysis. *Anal. Chim. Acta* 377: 113–124.

39. Van den Berg, C.M.G. 2006. Chemical speciation of iron in seawater by cathodic stripping voltammetry with dihydroxynaphthalene. *Anal. Chem.* 78: 156–163.

40. Riso, R.D., B. Pernet-Coudrier, M. Waeles, and P. Le Corre. 2007. Dissolved iron analysis in estuarine and coastal waters by using a modified adsorptive stripping chronopotentiometric (SCP) method. *Anal. Chim. Acta* 598: 235–241.

41. Staden, J.F., L.V. Mulaudzi, and R.I. Stefan. 2003. Speciation of Mn(II) and Mn(VII) by on-line spectrophotometric sequential injection analysis. *Anal. Chim. Acta* 499: 129–137.

42. Point, D., G. Bareille, H. Pinaly, C. Belin, and O.F.X. Donard. 2007. Multielemental speciation of trace elements in estuarine waters with automated on-site UV photolysis and resin chelation coupled to inductively coupled plasma mass spectrometry. *Talanta* 72: 1207–1216.

43. Huang, D.Q., O. Nekrassova, and R.G. Compton. 2004. Analytical methods for inorganic arsenic in water: A review. *Talanta* 64: 269–277.

44. Ernstberger, H., H. Zhang, and W. Davison. 2002. Determination of chromium speciation in natural systems using DGT. *Anal. Bioanal. Chem.* 373: 873–879.

45. Zhang, H. and W. Davison. 1999. Diffusional characteristics of hydrogels used in DGT and DET techniques. *Anal. Chim. Acta* 398: 329–340.

46. Warnken, K.W., H. Zhang, and W. Davison. 2004. Analysis of polyacrylamide gels for trace metals using diffusive gradients in thin films and laser ablation inductively coupled plasma mass spectrometry. *Anal. Chem.* 76: 6077–6084.

47. Leermakers, M., S. Galletti, S. De Galan, N. Brion, and W. Baeyens. 2001. Mercury in the southern North Sea and Scheldt estuary. *Mar. Chem.* 75: 229–248.

48. Sirinawin, W. and S. Westerlund. 1997. Analysis and storage of sample for chromium determination in seawater. *Anal. Chim. Acta* 356: 35–40.

49. Zhang, N., J.S. Suleiman, M. He, and B. Hu. 2008. Chromium (III)-imprinted silica gel for speciation analysis of chromium in environmental water samples with ICP-MS detection. *Talanta* 75: 536–543.

50. Dyg, S., R. Cornelis, B. Griepink, and P. Verbeeck. 1994. Development and interlaboratory testing of aqueous and lyophilised Cr(III) and Cr(VI) reference materials. *Anal. Chim. Acta* 286: 297–308.

51. Ozturk, M. and N. Bizsel. 2003. Iron speciation and biogeochemistry in different nearshore waters. *Mar. Chem.* 83: 145–156.

52. Morgan, J.J. 2000. Manganese in natural waters and earth's crust: Its availability to organisms. *Met. Ions Biol. Syst.* 37: 1–34.

53. Cheam, V. and H. Agemian. 1980. Preservation of inorganic arsenic species at microgram levels in water samples. *Analyst* 105: 737–743.

54. Ariza, J.L.G., E. Morales, D. Sanchez-Rodas, and I. Giraldez. 2000. Stability of chemical species in environmental matrices. *Trends Anal. Chem.* 19: 200–209.

55. Thurow, K., A. Koch, N. Stoll, and C.A. Haney. 2002. General approaches to the analysis of arsenic containing warfare agents. In: R.R. McGuire and J.C. Compton (eds), *Environmental Aspects of Converting CW Facilities to Peaceful Purposes*, pp. 123–138. Dordrecht, The Netherlands: Kluwer Academic Publishers.

56. McSheehy, S., J. Szpunar, R. Morabito, and Ph. Quevauviller. 2003. The speciation of arsenic in biological tissues and the certification of reference materials for quality control. *Trends Anal. Chem.* 22: 191–209.

57. Hall, G.E.M., J.C. Pelchat, and G. Gauthier. 1999. Stability of inorganic arsenic(III) and arsenic(V) in water samples. *J. Anal. At. Spectrom.* 14: 205–213.

58. Lagarde, F., M.B. Amran, M.J.F. Leroy, et al. 1999. Improvement scheme for the determination of arsenic species in mussel and fish tissues. *Fresenius J. Anal. Chem.* 363: 5–11.

59. Bobrowski, A., B. Bas, J. Dominik, et al. 2004. Chromium speciation study in polluted waters using catalytic adsorptive stripping voltammetry and tangential flow filtration. *Talanta* 63: 1003–1012.

60. Gledhill, M., C.M.G. van den Berg, R.F. Nolting, and K.R. Timmermans. 1998. Variability in the speciation of iron in the northern North Sea. *Mar. Chem.* 59: 283–300.

61. Boye, M., A. Aldrich, C.M.G. van den Berg, et al. 2006. The chemical speciation of iron in the north-east Atlantic Ocean. *Deep-Sea Res.* 53: 667–683.

62. Achterberg, E.P., C.M.G. van den Berg, M. Boussemart, and W. Davison. 1997. Speciation and cycling of trace metals in Esthwaite Water: A productive English lake with seasonal deep-water anoxia. *Geochim. Cosmochim. Acta* 61: 5233–5253.

63. Hlawatsch, S., T. Neumann, C.M.G. van den Berg, M. Kersten, J. Harff, and E. Suess. 2002. Fast-growing, shallow-water ferro-manganese nodules from the western Baltic Sea: Origin and modes of trace element incorporation. *Mar. Geol.* 182: 373–387.

64. Munoz, E. and S. Palmero. 2005. Analysis and speciation of arsenic by stripping potentiometry: A review. *Talanta* 65: 613–620.

65. Salaün, P., B. Planer-Friedrich, and C.M.G. van den Berg. 2007. Inorganic arsenic speciation in water and seawater by anodic stripping voltammetry with a gold microelectrode. *Anal. Chim. Acta* 585: 312–322.

66. Cox, A.G. and C. Mcleod. 1992. Field sampling technique for speciation of inorganic chromium in rivers. *Mikrochim. Acta* 109: 161–164.

67. Dogutan, M., H. Filik, and I. Tor. 2003. Preconcentration and speciation of chromium using a melamine based polymeric sequestering succinic acid resin: Its application for Cr(VI) and Cr(III) determination in wastewater. *Talanta* 59: 1053–1060.

68. Latif, E., A.K. Aslihan, and S. Mustafa. 2008. Solid phase extraction method for the determination of iron, lead and chromium by atomic absorption spectrometry using Amberite. *J. Hazard. Mater.* 153: 454–461.

69. Pohl, P. and B. Prusisz. 2006. Redox speciation of iron in waters by resin-based column chromatography. *Trends Anal. Chem.* 25: 909–916.

70. Liang, P., T.Q. Shi, and J. Li. 2004. Nanometer-size titanium dioxide separation/preconcentration and FAAS determination of trace Zn and Cd in water sample. *Int. J. Environ. Anal. Chem.* 84: 315–321.

71. Liang, P. and R. Liu. 2007. Speciation analysis of inorganic arsenic in water samples by immobilized nanometer titanium dioxide separation and graphite furnace atomic absorption spectrometric determination. *Anal. Chim. Acta* 602: 32–36.

72. Ronkart, S.N., V. Laurent, P. Carbonnelle, N. Mabon, A. Copin, and J.-P. Barthélemy. 2007. Speciation of five arsenic species (arsenite, arsenate, MMAA, DMAA and AsBet) in different kind of water by HPLC-ICP-MS. *Chemosphere* 66: 738–745.

73. Sounderajan, S., A.C. Udas, and B. Venkataramani. 2007. Characterization of arsenic(V) and arsenic(III) in water samples using ammonium molybdate and estimation by graphite furnace atomic absorption spectroscopy. *J. Hazard. Mater.* 149: 238–242.

74. Hu, W., F. Zheng, and B. Hu. 2008. Simultaneous separation and speciation of inorganic As(III)/As(V) and Cr(III)/Cr(VI) in natural waters utilizing capillary microextraction on ordered mesoporous Al_2O_3 prior to their on-line determination by ICP-MS. *J. Hazard. Mater.* 151: 58–64.

75. Marques, M.J., A. Salvador, A. Morales Rubio, and M. De la Guardia. 2000. Chromium speciation in liquid matrices: A survey of the literature. *Fresenius J. Anal. Chem.* 367: 601–613.

76. Hudson, R.J.M., D.T. Covault, and F.M.M. Morel. 1992. Investigating of iron coordination and redox reactions in seawater using ^{59}Fe radiometry and ion-pair solvent extraction of amphiphilic iron complexes. *Mar. Chem.* 38: 209–235.

77. Powell, R.T., D.W. King, and W.M. Landing. 1995. Iron distributions in surface waters of the south Atlantic. *Mar. Chem.* 50: 13–20.

78. Bowie, A.R., E.P. Achterberg, R.F.C. Mantoura, and P.J. Worsfold. 1998. Determination of sub-nanomolar levels of iron in seawater using flow injection with chemiluminescence detection. *Anal. Chim. Acta* 361: 189–200.

79. Batley, G.E. and J.P. Matousek. 1980. Determination of chromium speciation in natural waters by electrodeposition on graphite tubes for electrothermal atomization. *Anal. Chem.* 52: 1570–1574.

80. Baeyens, W. 1992. Speciation of mercury in different compartments of the environment. *Trends Anal. Chem.* 11: 245–254.

81. Quevauviller, Ph., O.F.X. Donard, E.A. Maier, and B. Griepink. 1992. Improvements of speciation analyses in environmental matrices. *Mikrochim. Acta* 109: 169–190.

82. Sanchez Uria, J.E. and A. Sanz-Medel. 1998. Inorganic and methylmercury speciation in environmental samples. *Talanta* 47: 509–524.

83. Harrington, C.F. 2000. The speciation of mercury and organomercury compounds by using high-performance liquid chromatography. *Trends Anal. Chem.* 19: 167–179.

84. Leermakers, M., Ph. Quevauviller, M. Horvat, and W. Baeyens. 2005b. Mercury in environmental samples: Speciation, artefacts and validation. *Trends Anal. Chem.* 24: 383–393.

85. Parker, J.L. and N.S. Bloom. 2005. Preservation and storage techniques for low-level aqueous mercury speciation. *Sci. Total Environ.* 337(1–3): 253–263.

86. Gill, G.A. and W.F. Fitzgerald. 1987. Mercury in surface waters of the open ocean. *Global Biochem. Cycles* 1(3): 199–212.

87. Bloom, N.S. and E.A. Crecelius. 1983. Determination of mercury in seawater at sub-nanogram per liter levels. *Mar. Chem.* 14(1): 49–59.

88. Bloom, N. 1989. Determination of pictogram levels of methylmercury by aqueous phase ethylation, followed by cryogenic gas chromatography with gold vapour atomic fluorescence detection. *Can. J. Fish. Aquat. Sci.* 46: 1131–1140.

89. Logar, M., M. Horvat, H. Akagi, and B. Pihlar. 2002. Simultaneous determination of inorganic mercury and methylmercury compounds in natural waters. *Anal. Bioanal. Chem.* 374: 1015–1021.

90. Nagase, H., Y. Ose, T. Sato, and T. Ishikawa. 1982. Methylation of mercury by humic substances in an aquatic environment. *Sci. Total Environ.* 25: 133–142.

91. Horvat, M., K. May, M. Stoeppler, and A.R. Byrne. 1988. Comparative studies of methylmercury determination in biological and environmental samples. *Appl. Organomet. Chem.* 2: 515–524.

92. Horvat, M., N.S. Bloom, and L. Liang. 1993. Comparison of distillation with other current isolation methods for the determination of methyl mercury compounds in low level environmental samples: Part 1. Sediments. *Anal. Chim. Acta* 281: 135–152.

93. Falter, R. (ed.) 1999. *Sources of Error in Methylmercury Determination during Sample Preparation, Derivatisation and Detection.* Germany: Wiesbaden-Schierstein, May 27–29, 1998; *Chemosphere* 39: 1037–1224.

94. Decadt, G., W. Baeyens, D. Bradley, and L. Goeyens. 1985. Determination of methylmercury in biological samples by semiautomated headspace analysis. *Anal. Chem.* 57: 2788–2791.

95. Bulska, E., D.C. Daxter, and W. Frech. 1991. Capillary column gas chromatography for mercury speciation. *Anal. Chim. Acta* 249: 545–554.

96. Lansens, P. and W. Baeyens. 1990. Improvement of the semiautomated headspace analysis method for the determination of methylmercury in biological samples. *Anal. Chim. Acta* 228: 93–99.

97. Craig, P.J., D. Mennie, M.I. Needham, O.F.X. Donard, and F. Martin. 1993. Mass spectroscopic nuclear magnetic resonance evidence confirming the existence of methyl mercury hydride. *J. Organomet. Chem.* 447: 5–8.

98. Tseng, C.M., A. De Diego, F.M. Martin, D. Amouroux, and O.F.X. Donard. 1997. Rapid determination of inorganic mercury and methylmercury in biological references materials by hydride generation, cryo-focusing, atomic absorption spectrometry after open focused microwave-assisted alkaline digestion. *J. Anal. At. Spectrom.* 12: 743–750.

99. Liang, L., M. Horvat, and N.S. Bloom. 1993. An improved speciation method for mercury by GC/CVAFS after aqueous phase ethylation and room temperature precollection. *Talanta* 41: 371–379.

100. Leermakers, M., H.L. Nguyen, S. Kurunczi, B. Vanneste, S. Galletti, and W. Baeyens. 2003. Determination of methylmercury in environmental samples using static headspace gas chromatograph and atomic fluorescence detection after aqueous phase ethylation. *Anal. Bioanal. Chem.* 377: 327–333.

101. Baeyens, W., M. Leermakers, R. Molina, L. Holsbeek, and C. Joiris. 1999. Investigation of headspace and solvent extraction methods for the determination of dimethyl- and monomethylmercury in environmental matrices. *Chemosphere* 39(7): 1107–1117.

102. Tseng, C.M., A. De Diego, H. Pinaly, D. Amouroux, and O.F.X. Donard. 1998. Cryofocusing coupled to atomic absorption spectrometry for rapid and simple mercury speciation in environmental matrices. *J. Anal. At. Spectrom.* 13: 755–764.

103. Lobinski, R., I.R. Pereiro, H. Chassaigne, A. Wasik, and J. Szpunar, 1998. Elemental speciation and coupled techniques—towards faster and reliable analyses. *J. Anal. At. Spectrom.* 13: 859–867.

104. De Smaele, T., L. Moens, R. Dams, P. Sandra, J. Van de Eycken, and J. Vandyck. 1998. Sodium tetra(*n*-propyl)borate: A novel aqueous *in situ* derivatization reagent for the simultaneous determination of organomercury, lead- and tin-compounds with capillary gas chromatography-inductively coupled plasma mass spectrometry. *J. Chromatogr. A* 793: 99–106.

105. Abuin, M., A.M. Carro, and R.A. Lorenzo. 2000. Experimental design of a microwave-assisted extraction-derivatization method for the analysis of methylmercury. *J. Chromatogr. A* 889: 185–193.

106. Grinberg, P., R.C. Campos, Z. Mester, and R.E. Sturgeon. 2003. A comparison of alkyl derivatization methods for speciation of mercury based on solid phase microextraction gas chromatography with furnace atomization plasma emission spectrometry detection. *J. Anal. At. Spectrom.* 18: 902–909.

107. Filipelli, M., F. Baldi, and J.H. Weber. 1992. Methylmercury determination as volatile methylmercury hydride by purge and trap gas chromatography in line with Fourier transform infrared spectroscopy. *Environ. Sci. Technol.* 26: 1457–1460.

108. Munaf, E., H. Haraguchi, D. Ishii, T. Takeuchi, and M. Goto. 1990. Speciation of mercury compounds in waste water by microcolumn liquid chromatography using a preconcentration column with cold-vapour atomic absorption spectrometric detection. *Anal. Chim. Acta* 235: 399–404.

109. Wu, J.C.G. 1991. Interfacing HPLC and cold-vapor AA with on-line preconcentration for mercury speciation. *Spectrosc. Lett.* 24: 681–697.

110. Falter, R. and H.F. Scholer. 1995. Determination of mercury species in natural waters at pictogram level with on-line RP C18 preconcentration and HPLC-UV-PCO-CVAAS. *Fresenius J. Anal. Chem.* 353: 34–38.

111. Schlegel, D., J. Mattusch, and K. Dittrich. 1994. Speciation of arsenic and selenium compounds by ion chromatography with inductively coupled plasma atomic emission spectrometry detection using the hydride technique. *J. Chromatogr. A* 683: 261–267.

112. Qiang, T., W. Johnson, and B. Buckley. 2003. Mercury speciation analysis in soil samples by ion chromatography, post-column cold vapor generation and inductively coupled plasma mass spectrometry. *J. Anal. At. Spectrom.* 18: 696–701.

113. Bloom, N.S. and W.F. Fitzgerald. 1988. Determination of volatile mercury species at the picogram level by low temperature gas chromatography with cold-vapor atomic fluorescence detection. *Anal. Chim. Acta* 208: 151–161.

114. Hintelmann, H., R. Falter, R. Ilgen, and R.D. Evans. 1997. Determination of artifactual formation of monomethylmercury (CH_3Hg^+) in environmental samples using stable Hg^{2+} isotopes with ICP-MS detection: Calculation of contents applying species specific isotope addition. *Fresenius J. Anal. Chem.* 358: 363–370.

115. Leermakers, M., C. Meuleman, and W. Baeyens. 1995. Mercury speciation in the Scheldt estuary. *Water Air Soil Pollut.* 80: 641–652.

8 Immunochemical Analytical Methods for Monitoring the Aquatic Environment

Javier Adrian, Fátima Fernández, Alejandro Muriano,
Raquel Obregón, Javier Ramón, Nuria Tort,
and M.-Pilar Marco

CONTENTS

8.1 Introduction 140
8.2 Immunochemical Determination of Industrial Contaminants 142
    8.2.1 Polycyclic Aromatic Hydrocarbons 142
    8.2.2 Surfactants 145
    8.2.3 Organohalogenated Compounds 147
    8.2.4 Heavy Metals and Metalloids 149
    8.2.5 Other Industrial Pollutants: Bisphenol A 150
8.3 Immunochemical Methods for Pesticides 151
    8.3.1 Insecticides 152
    8.3.2 Herbicides and Plant Growth Regulators 154
8.4 Immunochemical Determinations of Pharmaceutical and
Personal Care Products 158
    8.4.1 Antibiotics 158
        8.4.1.1 Sulfonamides 160
        8.4.1.2 Fluoroquinolones 161
        8.4.1.3 Amphenicols 162
        8.4.1.4 Tetracyclines 163
        8.4.1.5 ß-Lactams 164
        8.4.1.6 Macrolides 165
        8.4.1.7 Other Drugs 165
    8.4.2 Steroid Hormones 166
        8.4.2.1 Estrogens 166
        8.4.2.2 Androgens 168
        8.4.2.3 Gestagens 169
        8.4.2.4 Corticosteroids 169
8.5 General Summary 169
Acknowledgments 170
References 170

8.1   INTRODUCTION

In recent decades, efforts to better understand the occurrence, fate, and environmental effects of anthropogenic chemicals have focused largely on industrial compounds and agricultural pesticides. This emphasis has been in response to the increasing production of these chemicals, including the so-called emerging contaminants, such as pharmaceutical active ingredients and personal care products, their concentrated use, the potential risk of persistence, and their sometimes unknown acute and chronic toxic effects (see Table 8.1 for the most important groups of emerging pollutants).[1]

Pollutants can enter the environment in a great many ways.[2] Some compounds, such as pesticides, are deliberately released during agricultural applications; others, like industrial by-products, get into our water and air resources during regulated and unregulated industrial discharges. On the other hand, pharmaceuticals and biogenic hormones are normally discharged into the environment through wastewater treatment plants (WWTPs), which are often not designed to remove them from the effluent. From there on, the transport, fate, and possible adverse consequences of these pollutants on human health and the ecosystem are frequently unknown or at best not clearly understood.[3] Potential concerns include reproductive impairment,[4–6] increased incidence of cancer,[7] development of antibiotic-resistant bacteria,[8] or the potentially elevated toxicity of chemical mixtures due to synergistic effects;[9] these concerns demand extensive investigation of all possible situations.

The aim of regulations and regulatory methods to assess and control the impact of these substances in the aquatic environment is to protect the ecosystem and public health while keeping watch on their contamination levels and potential negative effects. In addition to the REACH (Registration, Evaluation, Authorization, and Restriction of Chemicals) law (EC 1907/2006) regarding chemicals and their safe use, which came into force on June 1, 2007, there are specific regulations for protecting health and ensuring the good quality of all water resources, such as the Drinking Water Directive (DWD, the Council Directive 98/83/EC), the Bathing Water Directive (2006/7/EC), or the Urban Waste Water Directives (91/271/EEC and 98/15/EC). Moreover, the intention of the Water Framework

---

TABLE 8.1
Some of the Most Important Contaminants Needed to be Monitored

Industrial Contamintants	Pesticides	Pharmaceuticals
Surfactants and metabolites	*Insecticides*	*Antibiotics*
Nonanionic	Organochlorines	Sulfonamides
Anionic	Organophosphorus	Fluoroquinolones
Cationic		Amphenicols
		Tetracyclines
		ß-Lactams
		Macrolides
Organohalogenated compounds	*Herbicides/plant growth regulators*	*Other drugs*
Polychlorinated	Triazines	Diclofenac
Chlorophenols	Phenylurea compounds	Indomethacin
Dioxins	Chloroacetanilides	Nitrofurantion
Polybrominated biphenyl ethers	Sulfonylureas	Paclitaxel
Bromophenols		Spectinomycin
Heavy metals and metalloids		*Steroid hormones*
As, Cd, Cr, Cu, Pb, Hg, Ag, Se, Zn, etc.		Estrogens
		Androgens
		Gestagens
		Corticosteroids
Industrial additives and others		
Phthalate esters		
Bisphenol A		

Directive (WFD, 2000/60/EC) is to provide an overall framework for a cleaner and safer aquatic ecosystem, particularly with regard to surface freshwater and groundwater bodies (i.e., lakes, streams, rivers, estuaries, coastal waters, etc.). Thus, in line with this directive, the Marine Strategy Directive (2008/56/EC) aims to achieve a good environmental status of the EU's seawaters by 2021. Similarly, a River Basin Management Plan (RBMP) is being developed for implementation in each river basin district; each such plan must include knowledge of the particular pressures and impacts of human activities on the river basin in question, as well as protection programs, controls, and remediation measures. The first RBPM is scheduled to be published at the end of 2009; in parallel with it, the new Groundwater Directive (2006/118/EC) establishes a regime that sets underground water quality standards (QSs) and introduces measures to prevent or limit inputs of pollutants into groundwater.

The aim of all these directives is to maintain water quality in all its aspects, including chemical pollution, on which this chapter will focus. Community policy regarding dangerous or hazardous substances in European waters was introduced almost three decades ago by Council Directive 76/464/EEC. This Directive is now integrated in the WFD (codified as 2006/11/EC). The main strategy against the pollution of surface waters involves the identification of substances and development of control measures. More than 100,000 existing chemicals have been identified in the European Inventory of Existing Commercial Chemical Substances (EINECS). Data collection on the potential adverse effects of these chemicals has led to 141 of them being classified as priority substances. The latest list of priority substances presenting a significant risk to or via the aquatic environment was published in Council Decision No 2455/2001/EC. The list identifies a further 33 substances or groups of substances that have been shown to be of major concern. Moreover, in July 2006, the EC adopted proposals for a new Directive to protect surface water from pollution by setting up Environmental Quality Standards (EQS, concentrations of pollutants which should not be exceeded) and measures to monitor pollution (COM(2006)397 final). Although for surface waters the WFD aims to ensure at least a minimum chemical quality, particularly in relation to very toxic substances, the case regarding groundwater is somewhat different: the assumption is that groundwater should not be polluted at all. Therefore, the WFD strategy includes a prohibition on direct discharges to groundwater and, to cover indirect discharges, requires groundwater bodies to be monitored so that any change in chemical composition can be detected.

In order to achieve these objectives, more efficient analytical techniques need to be developed. This is so that the public health can be protected by ensuring that levels of pollutants remain below EQS concentrations and by providing, on a continuous basis, information on the fate of existing chemicals. A variety of multiresidue analytical procedures capable of analyzing an important number of chemicals[10–14] in a single run have been reported in recent years. Most of them combine the high resolution of chromatographic methods [gas chromatography (GC), high performance liquid chromatography (HPLC), and ultra performance liquid chromatography (UPLC)] with the excellent detectability of sophisticated detectors like those based on mass spectrometry. In spite of this, these procedures lack the requisite sample throughput for a really effective assessment of the health of aquatic environments. The main limitations are due to the sequential nature of these analytical methods as well as the need to clean up and preconcentrate samples prior to analysis. In this context, immunochemical methods, which are based on the affinity of an antibody for an antigen (Figure 8.1), should be looked at as an interesting alternative for a number of reasons.

Firstly, measurements are performed by default in aqueous media, which makes these methods excellent tools for the analysis of aqueous samples. Secondly, owing to high specificity and detectability of certain substances or groups of substances, it is often possible to analyze them directly in the same matrix without having to purify, extract, and preconcentrate the sample. It is also possible to develop a variety of immunochemical configurations targeted to particular analytical requirements. Thus, a great number of immunochemical methods for the on-site analysis of environmental pollutants have been reported or are commercially available.[15–19] On the other hand, simultaneous analysis of many samples is possible by incorporating the immunoreagents on formats using 96- or 384-well microplates and developing formats with high sample-throughput capacities.[20–22] Selective sample treatments based on immunosorbents yield very clean aqueous extracts; such techniques can

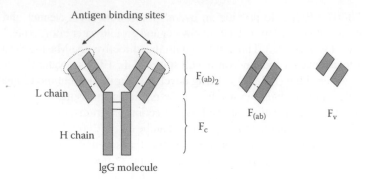

FIGURE 8.1   Basic H_2L_2 structure of the G immunoglobulins (IgG). It is formed by two pairs of polypeptide chains interlinked by disulfide bonds. The F_c fragment is the constant region and is involved in immune regulation, whereas the $F_{(ab)}$ (antibody binding fraction) fragment is the region that contains the variable fraction (F_v) with specific binding sites that allow interaction with Ag.

then be applied in tandem with some other type of immunochemical or chromatographic analysis.[23–26] Recent years have witnessed interesting advances in the incorporation of selective immunoreagents into electronic devices, which emit optical or electrical signals in response to the presence of an analyte. Immunosensors have made it possible to develop integrated devices capable of handling data and of taking automated remedial action when the analyte is present.[3,27–33] Since the reader will find extensive information in recent reviews on the fundamentals[34–38] of these immunochemical analytical methods and examples of their application in environmental analysis,[29,30,33,39–44] we will now concentrate our attention on the applicability of these methods to the determination of some of the most important pollutants in aquatic environments—specifically, marine and freshwater ecosystems and their biota. The chapter does not pretend to be an exhaustive review of what has been reported in the literature; rather, its aim is to provide the reader with representative examples of immunochemical analytical technologies that have either been applied to the aquatic environment or exhibit great potential in this respect. Examples of the several groups of chemicals have been selected according to their toxicological risk, relevance, and regular use or production.

8.2   IMMUNOCHEMICAL DETERMINATION OF INDUSTRIAL CONTAMINANTS

Worldwide industrial activity generates vast amounts of chemical residues such as metals, polycyclic aromatic hydrocarbons (PAHs), polyhalogenated biphenylethers, and surfactants. One of the main concerns regarding the contamination caused by such activity is the enormous amount of unknown substances generated as by-products during the manufacture of other chemicals. The highly toxic dioxins, for example, are released into the environment largely as the unwanted by-products of industrial processes. In addition, there is the risk of major accidents, like the Seveso accident in 1976, with dramatic consequences for the population. Immunochemical methods for following up and monitoring the emissions of some of these pollutants into the aquatic environment have been reported.

8.2.1   POLYCYCLIC AROMATIC HYDROCARBONS

PAHs are chemical compounds consisting of fused aromatic rings without any heteroatoms or substituents. They are formed mainly from the incomplete combustion of organic carbon-containing fuels such as wood, coal, diesel fuel, fat, or tobacco. PAHs can be found airborne in the gaseous phase or adsorbed to particles,[45] in aqueous matrices like groundwater, wastewater, drinking water, or river water,[46] and even adsorbed to solids in sediments or soils.[47] The U.S. Environmental Protection Agency (EPA) includes 16 PAHs in the list of priority pollutants in wastewater to be

TABLE 8.2
Immunochemical Techniques Developed for the Detection of PAHs

Target Analyte	Technique	Matrix	Sensitivity		Reference
			LOD	IC_{50}	
Benzo(a)pyrene[a]	ELISA	Water and sediments	0.7 µg L^{-1b}	20.5 µg L^{-1b}	50
	ELISA	Tap, lake, and river water	24 µg L^{-1b}	65 µg L^{-1b}	51
	ELISA[c] (SDI, Newark, USA)	Soil		3.2 µg L^{-1b}	54
	ELISA[c] (SDI, Alton, UK)	Sediment	5.5 µg kg^{-1}	255 µg kg^{-1}	55
PAHs	ELISA[c] (Ohmicron, PA, USA)	River water		0.1–37 µg L^{-1b}	57
	Immunosorbent-LC	Surface water	0.002 µg L^{-1}		48
	Immunoaffinity and GC-MS	Corals	25 µg kg^{-1}		62
Phenanthrene[a]	Amperometric immunosensor	River water	5.0 µg L^{-1}	18.0 µg L^{-1}	58
		Tap water	6.3 µg L^{-1}	26.0 µg L^{-1}	
	Amperometric immunosensor	Sea, river, and tap water	1.4 µg L^{-1b}	29.3 µg L^{-1b}	59

[a] Indicator of total PAHs.
[b] LODs and IC_{50} calculated in buffer solution.
[c] Commercial kit.

monitored[48] because of their carcinogenic, mutagenic, and toxic properties.[7] The European Council Directive 98/83/EC concerning the quality of water intended for human consumption (WIHC) established a limit of 0.01 µg L^{-1} for benzo(a)pyrene, the lowest limit set for any individual chemical parameter in this directive.

As can be seen in Table 8.2, several immunochemical techniques, principally enzyme-linked immunosorbent assays (ELISAs), have been developed and applied to environmental samples such as water, soil, and sediments with very good limits of detection (LOD).[49–51]

EPA has included commercially available ELISA kits in its list of official methods (Method 4035)[52] capable of detecting PAH concentrations >1 mg kg^{-1} in soil samples.[53–56] The usual ELISA configurations for the analysis of small molecules are shown in Figure 8.2.

Most of the antibodies described were produced against one particular PAH congener, but because of their structural similarities they are in fact able to detect several PAHs. For this reason, different assays employ distinct target analytes as indicators of total PAH content, for example, phenanthrene,[53,56] benzo(a)pyrene,[50,51,54,55] or pyrene,[49] When real samples are immunoassayed, PAH levels have often been overestimated due to cross-reactivity with other congeners in the same sample.[50,54,57] In contrast, the underestimated PAH levels reported following the analysis of sediments[55] or contaminated soil samples[56] can be explained by the low extraction efficiency of the methods recommended in most commercial kits, where samples are merely shaken manually in the presence of an organic solvent. Nevertheless, the possibility must also be entertained that the sample contains PAHs with a low response factor to the particular antibody employed in the kit.[49,53] Amperometric immunosensors employing electrodes directly printed with an immobilized competitor of the target analyte have also been investigated to detect phenanthrene in environmental samples. Following this approach, Fahnrich et al.[58] achieved detection limits of 5 µg L^{-1} in river water and 6 µg L^{-1} in tap water samples. Moore et al.[59] applied the same transducing principle to

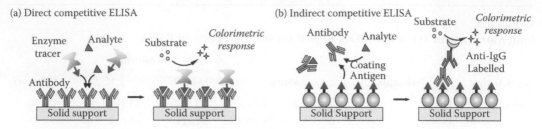

(a) Direct competitive ELISA   (b) Indirect competitive ELISA

FIGURE 8.2 Scheme of the ELISA formats most frequently used for the analysis of low-molecular-weight analytes. (a) Direct competitive ELISA: the Ab is coated on the surface and a competition is established between the analyte and the enzyme tracer. After the washing step, a substrate is added to produce a chromogen product that is easily quantified. (b) Indirect competitive ELISA: a coating antigen is immobilized on the solid support while the specific IgG and the analyte are in solution during the competition step. After the removal of unbound reagents, a secondary IgG labeled with the enzyme (IgG-enzyme), which specifically recognizes the Ab, is added. Finally, after another washing step, the amount bound is also quantified by the addition of the substrate solution.

analyze phenanthrene in sea, river, and tap water samples, obtaining an LOD of 1.4 μg L^{-1}. These biosensors have great potential in field assays, where simple, quick detection systems are required: unlike the microplate ELISA methods, the measurements do not involve many operational steps. Biosensors based on surface plasmon resonance (SPR) have also been reported, although they have not often been used to analyze PAHs in real environmental samples. Figure 8.3 illustrates both the SPR principle and the schematic representations of a typical sensogram obtained with this technique. Gobi et al.[60] used this methodology to analyze benzo(a)pyrene in buffer with promising results (LOD = 0.1 μg L^{-1}). Liu et al.[61] report the use of an immunosensor based on piezoelectric signal transduction for the analysis of benzo(a)pyrene, pyrene, and naphthalene compounds; when measuring buffered samples, this device can achieve an LOD in the nM range.

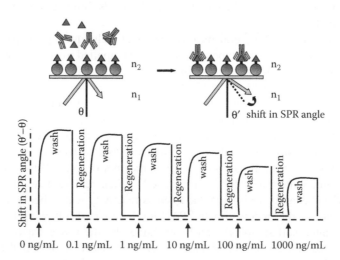

0 ng/mL   0.1 ng/mL   1 ng/mL   10 ng/mL   100 ng/mL   1000 ng/mL

FIGURE 8.3 Sensogram generated by SPR principle for the constant concentration of antigen and antibody, and the varying concentration of analyte. Surface plasmons are excited by the light energy at a critical angle (θ) causing an oscillation and the generation of an evanescent wave. Under this condition a decrease in the reflected light intensity is observed. The angle θ depends on the dielectric medium close to the metal surface and is therefore strongly affected by molecules directly adsorbed on the metal surface. This principle allows the direct detection of the interaction between the analyte and the antibody.

As an alternative, immunoaffinity procedures have been developed to selectively extract PAHs from environmental samples. Thomas and Li[62] demonstrated the greater efficiency of immunoaffinity methods in comparison with conventional extraction procedures. Bouzige et al.[48] prepared an immunosorbent for use in an on-line analytical procedure, followed by liquid chromatography coupled to fluorescence detection, to monitor surface water samples. The sensitivity of the fluorescence detection in combination with the selectivity of the immunosorbent (IS) extraction enabled PAH compounds to be detected at levels between 2 and 10 ng L^{-1}.

8.2.2 SURFACTANTS

Surfactants are very common water pollutants, mainly because of their extensive use in detergent formulations that are directly discharged into the environment via wastewater.[63] They are usually organic compounds described as being amphiphilic, that is, containing both hydrophobic and hydrophilic groups. The most commonly accepted and scientifically sound classification of a surfactant is based on the charge present in the hydrophilic portion of the molecule after its dissociation in aqueous solution. Anionic surfactants are based mainly on sulfate, sulfonate, or carboxylate compounds: examples include linear alkylbenzene sulfonates (LAS), fatty alcohol sulfates (FAS), alkyl sulfonates (AS), alkyl ether sulfates (AES), and sodium dodecyl sulfate (SDS).[64] Commercial LAS, which represent more than 40% of all surfactants, consist of a mixture of at least 20 compounds, including isomers.[65] Residues of LAS, as well as their metabolites, the sulfophenyl carboxylates (SPCs), have been found in marine sediments[66] and in surface waters from the low μg L^{-1} up to the 500 μg L^{-1} range.[67–69] On the one hand, the toxicity of these compounds is low, but on the other, they may assist the permeation of other pollutants into aquatic biota.[70] Nonionic surfactants, such as alkylphenol ethoxylates (APEs) and alkyl ethoxylates (AEs), do not ionize in aqueous solution; they are extensively used in detergent formulations or as stabilizers in plastics. The main components of APEs are isomers of nonylphenol ethoxylate (NPE) and to a lesser extent, compounds related to octylphenol ethoxylate (OPE). Alkylphenols (APs), APE metabolites, and their halogenated derivatives are much more persistent in the aquatic environment and thus give reason for concern. Nonylphenol (NP) and octylphenol (OP) are listed as surface water priority substances; they are endocrine disruptors with powerful estrogenic effects.[71,72] Accordingly, the OSPAR commission (from the Oslo and Paris conventions),[73] whose brief it is to protect the marine environment of the North-East Atlantic, decided to phase out the use of APEs by the year 2000.[74] Nevertheless, monitoring of APEs will still be necessary in the coming years. Finally, the class of cationic surfactants includes nitrogen compounds such as fatty amine and quaternary ammonium salts (QAS), in which the hydrophobic groups are attached to positively charged nitrogen. These surfactants, which also have a biocide effect, are in general more expensive than the anionic ones and therefore less often used.[64] Their poor solubility and tendency to adsorb to solids or to form complexes with anionic substances reduce the risk to the aquatic environment. The application of immunoassays to the analysis of surfactants started in 1982 with the determination of Triton X, one of the best known APEs, with ELISA achieving detection limits in the μg L^{-1} range.[75] Since then, numerous immunochemical techniques have been developed, mainly to detect anionic and nonionic surfactants in the environment (Table 8.3).

With regard to anionic surfactants, Farré et al.[76] and Ramón-Azcón et al.[77] reported an immunoassay for LAS determination in wastewater samples (LOD = 2 μg L^{-1}). The assay also identified, with a high level of cross-reactivity, the long SPC chain that is formed after LAS degradation. Analyses can be performed directly in wastewater merely by diluting the samples 10-fold to eliminate matrix interferences. Similarly, Estevez and co-workers recently reported an immunoassay for short alkyl chain SPCs, which are the final degradation products of LAS.[78] Because these products are highly polar, it is quite difficult to analyze SPCs if an extraction/preconcentration step is necessary. In contrast, immunochemical methods allow the direct determination of these chemical markers in aqueous samples. Thus, Zhang et al.[79] developed a sequential injection analysis (SIA) combined

TABLE 8.3

Immunochemical Techniques Developed for the Detection of Surfactants

Target Analyte	Technique	Matrix	Sensitivity		Reference
			LOD	IC_{50}	
AEs	ELISA (tube assay)	Tap and river water	$2\,\mu g\,L^{-1a}$	$12\,\mu g\,L^{-1a}$	89
	ELISA (plate assay)	Tap and river water	$20\,\mu g\,L^{-1}$	$71\,\mu g\,L^{-1}$	89
APEs	ELISA	River water	$16\,\mu g\,L^{-1a}$	$79\,\mu g\,L^{-1a}$	63
	ELISA	Polluted and tap water	$10\,\mu g\,L^{-1a}$	$246\,\mu g\,L^{-1a}$	72
	Sequential injection CL assay	River water	$10\,\mu g\,L^{-1}$	$30\,\mu g\,L^{-1a}$	86
APs	ELISA	Surface river water	$\mu g\,L^{-1}$ range		87
	SPR	Marine products, shellfish	$10\,\mu g\,kg^{-1}$		90
LASs	ELISA	Wastewater	$1.8\,\mu g\,L^{-1a}$	$28,1\,\mu g\,L^{-1a}$	76
	IS fluorescence detection assay	Waste- and groundwater	$7\,\mu g\,L^{-1}$		81
	FPIA	Waste- and groundwater	$30\,\mu g\,L^{-1}$		394

[a] LODs and IC_{50} calculated in buffer solution.

with a chemiluminescence detector and neodymium magnet to perform magneto-immunoassay experiments for the analysis of LAS; they achieved an LOD of $25\,\mu g\,L^{-1}$. Moreover, Sánchez-Martínez et al.[80] developed a very sensitive fluorescence polarization immunoassay (FPIA) for analyzing LAS in wastewater and groundwater samples. Previously, the same authors had developed an immunoaffinity chromatography procedure followed by fluorescence detection to analyze LAS in tap water, groundwater, and wastewater samples.[81] Recoveries were between 86% and 111% with dodecylbenzenesulfonate (LDS) as analyte (LOD = $7\,\mu g\,L^{-1}$).

The immunochemical determination of APEs and AP has been undertaken by several research groups[63,72,82,83] (for further details, see the review by Estevez-Alberola and Marco[84]), although the detectability achieved is not good enough to analyze NP at the EQS value established for all surface waters ($0.33\,\mu g\,L^{-1}$).[85] An LOD of $10\,\mu g\,L^{-1}$ has recently been reported for a chemiluminescent immunoassay of APE.[86] Previously, a highly sensitive and reproducible ELISA method had been described for NP, but the LOD obtained ($2.3\,\mu g\,L^{-1}$) was again still insufficient.[82] This is why it is often necessary to include a preconcentration step prior to immunochemical analysis. The precision and accuracy of most of these ELISAs have therefore been evaluated by measuring these surfactants in spiked river samples, following a solid-phase extraction (SPE) procedure; recoveries have been good: 85–118%.[63,87] The validation studies carried out by several groups have also shown good correlation with chromatographic methods,[72,76,88] although some samples analyzed were clearly overestimated. The accuracy of the APE immunoassay mentioned above[86] was also successfully evaluated by measuring spiked river samples. Usually, a wide cross-reactivity pattern is observed, in which APEs with a distinct number of ethoxylate units are detected in addition to AP and certain carboxylate metabolites. A highly selective ELISA for AEs has also been developed and evaluated by measuring different spiked water matrices, such as distilled, tap, and river water samples, with recoveries reported to be in the 75–134% range.[89] Furthermore, an SPR Biacore sensor was applied by Samsonova et al.[90] to detect NPs in different shellfish matrices like mussels, oysters, cockles, and scallops.

The detection limits obtained for all the aquatic biota samples were around $10\,\mu g\,kg^{-1}$,[90] which is good enough considering the QS estimated for aquatic biota [10 mg NP kg^{-1} food (wet weight)].[91] Several amperometric biosensors have been described, but none which might be suitable for aquatic environment monitoring. Rose et al.,[92] for example, analyzed APE and AP in buffer using a capillary immunoassay with subsequent amperometric detection, and Evtugyn et al.[93] developed another amperometric immunosensor for the analysis of NP with an LOD of 10 µg L^{-1}.

Several ELISA methods for determining cationic surfactants have been reported although we have not found any examples of their application to real environmental samples. The detectability obtainable by these methods is very good: an LOD of 0.04 µg L^{-1} has been reported for benzyldomethyldodecylammonium chloride (BDD12AC), a component of benzalkonium chloride (BAK).[94]

8.2.3 ORGANOHALOGENATED COMPOUNDS

The toxicity, bioaccumulative potential, and ecological impact of organohalogenated substances such as polychlorinated biphenyls (PCBs), polychlorinated dibenzofurans (PCDFs), polychlorinated dibenzo-*para*-dioxins (PCDDs), or polybrominated diphenylethers (PBDEs) have been extensively reviewed.[95–98] All are referred to as persistent organic pollutants (POPs), that is, chemical substances that remain in the environment, bioaccumulate through the food chain, and pose a risk to human health and the environment. The international community is calling for action to reduce and then eliminate the production or formation of these substances and to monitor their emission. In this case, the detectability obtainable by analytical methods should be very low, since the limits established for these residues are in the ng per liter range.

PCBs, which have been commonly used as lubricants, immersion oils, or fire retardants, are normally formed by the chlorination of biphenyl in the presence of an FeCl$_3$ catalyst. PCB production was banned in the 1970s because of the high toxicity of most PCB congeners and mixtures. The number and the location of chlorine atoms in a specific PCB congener determine its physicochemical properties, environmental pathways, and toxicity. Isomers with a higher chlorine content bind preferentially to organic matter present in the solid phase; consequently, they are not easily degraded and are also poorly leached from sediments by water.[99] On the other hand, some PCBs are more mobile, hence their tendency to volatilize, and therefore to circulate, in gaseous form, through different environmental compartments.[100] PCBs are classified as a probable human carcinogen by EPA, which has established a maximum contaminant level goal of zero and a maximum contaminant level and practical quantification limit of 0.5 µg L^{-1} in drinking water.[101] In the USA, regulatory limits for soil remediation vary according to state and site, but in general are 5 or 10 µg g^{-1} for industrial restricted access areas and 1 or 2 µg g^{-1} for residential access areas.[102] For this reason, EPA has established method 4020, a procedure for screening soils and nonaqueous waste liquids, to determine when total PCBs are present at concentrations above 5, 10, or 50 mg kg^{-1}.[103] Table 8.4 shows other results regarding the analysis of organohalogenated compounds in environmental matrices. A commercially available ELISA test (PCB RaPID Assay®) has also been used to analyze these compounds; an LOD of 0.6 µg g^{-1} was obtained in mussel tissues.[104] Lawruk et al. applied the same kit but with super paramagnetic particles as the solid support to analyze soil and water samples; they obtained detection limits of 0.2 µg L^{-1} and 500 µg kg^{-1}, respectively.[105] Another commercial ELISA kit (EnBioTec Laboratories) was evaluated for PCB 118 determination in retail fish samples (LOD = 0.05 µg L^{-1}).[106]

Now let us take a brief look at other approaches. Zhao et al.[107] developed an optical immunosensor consisting of a quartz crystal fiber coated with partially purified polyclonal antibodies to detect PCBs in soil and water samples. The optical signal was generated by the fluorescence produced when the 2,4,5-trichlorophenoxybutyrate fluorescein conjugate binds to the previously coated antibody.[107] Electrochemical immunosensing strategies using carbon-based screen-printed electrodes as transducers in a direct competitive immunoassay have also been applied in the analysis of PCBs in marine sediment extracts[108,109] (LOD = <1 µg L^{-1}). Pribyl et al.[110] developed a piezoelectric quartz crystal (PQC) immunosensor for the *in situ* determination of different PCB congeners in soil toluene extracts without any additional purification step. Also, a high-performance immunochromatographic (HPIAC) procedure has been successfully used as a cleanup method to isolate PCBs from water samples.[111]

TABLE 8.4

Immunochemical Techniques Developed for the Detection of Organohalogenated Compounds

Target Analyte	Technique	Matrix (Pretreatment)	Sensitivity LOD	Sensitivity IC_{50}	Reference
PCBs	PQC immunosensor	Soil samples		6 µg L^{-1}	110
118	ELISA (EnBioTec Labs)	Fish muscle tissue		0.05 µg L^{-1a}	106
Aroclor® 1248	Immunomagnetic	Marine sediment extracts	24 µg L^{-1}	0.4 µg L^{-1}	108,109
Aroclor® 1242	amperometric		8 µg L^{-1}	0.5 µg L^{-1}	
Aroclor® 1016	immunosensor		94 µg L^{-1}	0.8 µg L^{-1}	
Aroclor® 1254	ELISA^a (PCB Assay®)	Mussel tissue		600 µg kg^{-1}	104
	MAG particle IA	Water		0.2 µg L^{-1}	105
	(PCB RaPID Assay® kit)	Soil		500 µg L^{-1}	
Aroclor® 1242	Fiber optic immunosensor	Soil, river, and bay water		10 µg L^{-1}	107
PCDDs (TMDD)	ELISA	Soil	0.123 µg kg^{-1}	0.028 µg kg^{-1}	121

^{a} Commercial kit.

Other immunochemical techniques, not applied to environmental samples as yet, could be interesting for the future analysis of these residues.[112,113]

Immunochemical methods have also been reported for PCDDs and PCDFs. These substances are unintentionally formed during combustion processes and in the synthesis of chlorine gas and other chemicals used in the bleaching procedures of the pulp or paper industry.[114,115] PCDDs have 75 positional congeners with different grades of toxicity, 2,3,7,8-tetrachlorodibenzo-*p*-dioxin (TCDD) being the most toxic. The extremely low water solubility of these compounds,[116] which is approximately 1000-fold lower than that of PCBs and PAHs, has significant downstream effects on the development of immunoassays because they are applied in aqueous media. EPA has established an official immunoassay method (4025) based on a commercially available ELISA kit that uses polyclonal antibodies for the analysis of these compounds in soil samples at 0.5 µg kg^{-1} levels.[117] Harrison and Carlson developed both a tube test and a microplate assay using one of Stanker's monoclonal antibodies[118,119] and obtained detection limits of 167 and 50 pg L^{-1} for TCDDs in soil samples.[120] The use of accelerated solvent extraction (ASE) followed by ELISA for the rapid screening of dioxin-contaminated soils has been reported recently.[121] An immunoaffinity chromatography method for the purification of polychlorinated dibenzo-*p*-dioxins and furans from biological samples was explored with the aim of simplifying the cleanup procedure and thereby reducing the time and cost of analysis.[122] A study of the effect of organic solvents on the development of an ELISA has also been reported; an IC_{50} of 0.24 µg L^{-1} for TCDD was obtained.[123] Additionally, a piezoelectric immunosensor system was developed for the rapid detection of PCDDs primarily in buffer. In this case, the antibodies deposited in the quartz crystal resonator were able to quantitatively detect concentrations between 0.01 and 1.30 µg L^{-1}.[124]

Finally, PBDEs—mainly three commercial mixtures known as Penta-BDE, Octa-BDE, and Deca-BDE—are still widely used as flame retardants in products such as polymers, resins, electronic devices, building materials, textiles, and the polyurethane foam padding used in furniture and carpets. The intensive production and use of these compounds has made them ubiquitous in the environment and in biota.[125,126] EPA is working with industry, governments, and environmental and public health groups to research and better understand the potential health risks posed by these substances.[127] The European Commission is also aware of these risks to the environment and public health and has established EQS in the low ppt level. Thus, for Penta-BDE, the annual average (AA) EQS is 0.0005 µg L^{-1} for inland surface waters and 0.0002 µg L^{-1} for other surface waters. There have been a few attempts to develop immunochemical methods for polybrominated flame retardant compounds. An ELISA kit for the analysis of PBDE is commercially available from Abraxis LLC. Based on the use of magnetic particles, this kit is addressed to the BDE-47 and BDE-99 congeners that compose the Penta-BDE formulation; LOD for the 47th congener is <25 ng L^{-1}. Shelver et al.[128] have reported an ELISA with an IC$_{50}$ value of 28 µg L^{-1} for BDE-47 in buffer.

8.2.4 HEAVY METALS AND METALLOIDS

Heavy metals are also considered dangerous and persistent environmental contaminants. At least 20 are known to be toxic in some way and fully half of these, including As, Cd, Cr, Cu, Pb, Hg, Ni, Ag, Se, or Zn, are released into the environment in sufficient quantities to constitute a risk to human health. Metals bind easily to soils or sediments, and in this form they are relatively nontoxic; but changes in the weather or medium pH in combination with other environmental factors can mobilize them, thereby increasing their availability and effective toxicity. For this reason, sites contaminated with heavy metals must be monitored regularly. By way of example, mercury exceeds the 1 mg kg^{-1} action level, established by the U.S. Food and Drug Administration (FDA), in many marine and freshwater fish samples. For mercuric chloride used in medicine, the minimum risk level for Hg exposure is 0.007 mg kg^{-1} per day.[129] EPA has therefore certified an immunoassay (Method 4500) that provides a screening procedure for the determination of mercury in soils at concentrations up to 0.5 mg kg^{-1}. On the other hand, the Lead and Copper Rule (LCR), introduced by the EPA in 1991, established an action level of 0.015 mg L^{-1} for lead and 1.3 mg L^{-1} for copper based on the 90 percentile level of tap water samples.[130] Another method (4510), again proposed by EPA, also determines lead in water and soil by means of an immunoassay.[131] Nowadays, many of the heavy metals mentioned above are also regulated by the European Union; in water for human consumption, the permitted levels of chromium and mercury are 50 and 1 µg L^{-1}, respectively.[132] Moreover, cadmium and its compounds are among the 33 priority substances with an EQS of 0.08 µg L^{-1} in water listed in the proposal for a Directive [COM(2006)398 final] presented by the European Commission.

Basically, there are two ways of producing antibodies against heavy metals. Firstly, when the immunogen consists of a heavy metal bound to a chelator like ethylenediaminetetraacetic acid (EDTA), the antibodies raised do not recognize the metal itself but identify the entire structure.[133–135] On the other hand, if the antibodies are produced directly against the heavy metal attached to a suitable immunogen,[136,137] the free metal can be recognized instead of the cage-like chelate structure. Wylie et al.[137] developed highly specific antibodies for mercury using a glutathione complex, which is the basis of the only commercially available metal ion immunoassay (BiMelyze® Mercury Immunoassay). Alternatively, Barbas et al.[138] reported the isolation of recombinant antibody fragments that preferentially recognize certain metals complexed to iminodiacetic acid. Some examples of immunochemical techniques applied to detect heavy metals in environmental samples are given in Table 8.5.

TABLE 8.5
Immunochemical Techniques Developed for the Detection of Heavy Metals

| Target Analyte | Technique | Matrix | Sensitivity | | Reference |
			LOD	IC$_{50}$	
Cd(II)	ELISA	Environmnetal water samples	7 µg L^{-1}	—	134
Hg(II)	ELISA	Water (EPA samples)	0.5 µg L^{-1}	—	137
	ELISA[a] (BiMelyze®)	Scallop tissue extract	100 µg kg^{-1}	—	395
Pb(II)	FPIA	Soil	20 ng kg^{-1}	—	140

[a] Commercial kit.

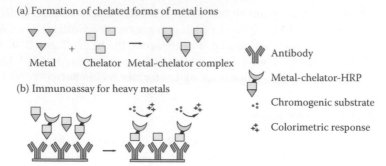

(a) Formation of chelated forms of metal ions

Metal   +   Chelator   Metal-chelator complex

Y   Antibody

Metal-chelator-HRP

(b) Immunoassay for heavy metals

Chromogenic substrate

Colorimetric response

FIGURE 8.4   Scheme of the ELISA format most commonly used for the analysis of heavy metals, where antibodies recognize chelated forms of metal ions.

ELISAs for detecting cadmium(II), nickel(II), lead(II), and mercury(II) in water samples have also been reported.[134,139] The most common ELISA format used for the analysis of these compounds can be seen in Figure 8.4.

Similarly, an FPIA used polyclonal antibodies raised against the lead(II)-EDTA chelate to detect the metal in soils, solid waste leachates, airborne dust, and drinking water samples.[140] Moreover, the Kin ExA™ 3000 automated immunoassay instrument was adapted to analyze Cd(II), Co(II), Pb(II), and U(VI) metals in groundwater samples.[141] On the other hand, monoclonal antibodies, raised against the Cd-EDTA complex, have been used to develop an immunochromatography (IC) procedure for the quick testing of Cd in food (LOD = 0.3 μg kg^{-1}).[135] The development and validation of a one-step immunoassay for the determination of Cd(II) in human serum with an LOD of 0.24 μg L^{-1} has also been described,[133] as has the optimization and validation of an immunoassay that measures soluble indium at 0.005 μg L^{-1} in buffer.[142] An alternative way of detecting the presence of heavy metals is to use molecular biomarkers such as the diagnostic and prognostic tools used in marine pollution monitoring. Metallothioneins (MTs) are synthesized by toxic metals such as Cd, Hg, and Cu by chelation through cysteine residues and act as biomarkers of metal exposure in both vertebrates and invertebrates. These biomarkers are used with a range of molecular approaches to evaluate the exposure of various sentinel marine organisms, for example, mussels, clams, oysters, snails, and fishes, to metal contaminants.

The demonstration that MTs from a wide variety of fish species are recognized by an antiserum raised against one piscine MT has enabled the development of immunotechniques based on ELISA[143] and radioimmunoassay (RIA) procedures[144] for the quantification of these compounds. A competitive solid-phase assay based on dissociation-enhanced lanthanide fluoroimmuno-detection (DELFIA) of anti-MT monoclonal antibody bound to a solid phase has been reported.[145] An electrochemical determination of MTs by square wave cathodic stripping voltammetry has also been developed and optimized.[146]

8.2.5   OTHER INDUSTRIAL POLLUTANTS: BISPHENOL A

Among the emerging pollutants of industrial origin, Bisphenol A [2,2 bis(4-hydroxydiphenyl)propane] (BPA) has special relevance since it was one of the first chemicals discovered to mimic estrogens as endocrine disrupters.[147] This compound was first reported by Dianin in 1891.[148] BPA is produced in large quantities worldwide, mainly for the preparation of polycarbonates, epoxy resins, and unsaturated polyester-styrene resins.[149] The final products are used in many ways, such as coatings on cans, powder paints, additives in thermal paper, in dental composite fillings, and even as antioxidants in plasticizers or polymerization inhibitors in polyvinyl chloride (PVC). To a minor extent, BPA is also used as precursor for flame retardants such as tetrabromobisphenol A or tetrabromobisphenol-S-bis(2,3-dibromopropyl) ether.[150] This substance can enter the environment

TABLE 8.6
Immunochemical Techniques Developed for the Detection of Bisphenol A

Target Analyte	Technique	Matrix	Sensitivity		Reference
			LOD (μg L^{-1})	IC$_{50}$ (μg L^{-1})	
Bisphenol A	TIRF immunosensor	Water	0.005[a]	—	163
	SPR immunosensor	MilliQ water	0.014	0.86	162
		Groundwater	0.168	5.12	
		River water	0.292	8.38	

[a] LODs and IC$_{50}$ calculated in buffer solution.

via the effluent from the factories producing it because it is not completely removed during waste-water treatment.[151,152] Several studies have demonstrated that BPA released to ground or surface water may be strongly adsorbed to soil or sediments.[153–155] Several ELISA applications have been developed to determine BPA in environmental and industrial waste samples:[156,157] Zhao et al.[158] obtained an LOD of 0.1 μg L^{-1} in real water samples. Recently, immunosensors have appeared on the market to complement conventional immunoassays for the analysis of this compound (Table 8.6). Optical SPR immunosensors have been reported to analyze BPA in buffer,[159–161] while a fully automated sensor called River ANAlyzer (RIANA), based on a combination of fluorescent labels and the evanescent wave principle, achieved an LOD of 14 ng L^{-1} in natural water samples.[162,163] Moreover, an impedimetric immunosensor, based on label-free direct detection of BPA with a quartz crystal microbalance, has been reported to obtain a detection limit of approximately 0.3 μg L^{-1} in human serum.[164]

Park et al. demonstrated the effectiveness of piezoelectric immunosensors as a valuable alternative screening method for BAP environmental monitoring, achieving an LOD of 0.1 μg L^{-1}, although so far, only in buffer.[165] Further immunoaffinity chromatographic methods have been developed with the aim of improving the analytical procedure of BPA in biological fluid samples[166] and in canned food.[167] The same approach could be used to shorten cleanup steps, as well as the cost and time of analysis of other environmental samples.

8.3 IMMUNOCHEMICAL METHODS FOR PESTICIDES

Pesticides are chemical substances used for preventing or limiting the damage caused by pests. Thus, unlike other groups of chemicals, pesticides are intentionally released into the environment. Moreover, there is a high risk of these chemicals turning up in the food chain: foodstuffs may become contaminated during agricultural production, processing, packaging, and storage. Owing to the sheer volume of their usage, coupled with their universal distribution, environmental persistence, and toxicological properties, pesticides are considered a major public health hazard. Agricultural pesticides may lead to contamination of surface and groundwaters by drift, runoff, drainage, and leaching.[168] Surface water contamination may have ecotoxicological effects on aquatic flora and fauna as well as on human health.[169,170] In the aquatic ecosystem, there is a continuous interchange of these compounds between the land, sediment, sediment–water interface, interstitial waters, aquatic organisms, and air–water interface. The distribution of pesticides between water and biotic materials can affect their dynamics in the ecosystem. Thus, their mobility, possible transformation, and biomagnification constitute a real threat to human health, wildlife, and the entire environment. This situation is reflected by the number of analyses on the influence of pesticides on particular aquatic ecosystems,[171–179] and also by the presence of these biocides on most governmental priority lists of compounds that should be monitored. Intensive research has been carried out during the last 20 years aiming to develop analytical immunochemical technologies with improved capabilities regarding detectability and sample-throughput capabilities for pesticide analysis in environmental matrices.[41,180–185] In the following, we will take a brief look at some of the most important of these methodologies and their application to the aquatic environment; the reader will find more detailed information in recent reviews.[25,41,43]

8.3.1 Insecticides

Used in agriculture to combat insect pests since the 1940s, insecticides include several chemical families that constitute a serious environmental risk. Thus, organochlorine (OC) or organophosphorus (OP) insecticides, the first generation of pesticides, are known to be highly persistent in the environment, unlike the pyrethroids that break down quickly in direct sunlight, usually just a few days after application. Nowadays, the use of OC has been banned in most developing countries because of concerns about their environmental impact and human health effects; nevertheless, their residues are still present in many environmental and biological compartments. The most important immunochemical techniques developed to analyze these compounds are listed in Table 8.7.

In a recent study, we reported on how the general population is still exposed to these substances, as evidenced by the excretion of chlorophenols and bromophenols in urine.[186] The study used antibodies developed for trichlorophenol,[187–189] an insecticide used as a wood and textile preservative, to extract these analytes from urine and to analyze them by combining immunosorbent cartridges[190] and an ELISA on a 96 setup, in such a way that 96 samples could be immunoextracted and analyzed in parallel. The results obtained from the immunochemical analyses were validated by gas chromatography-mass spectrometry (GC-MS), showing excellent correlation.[186] One of the best known OC pesticides is dichloro-diphenyl-trichloroethane (DDT), and many research groups have attempted to develop antibodies to detect this compound.[191,192] Beasley et al.,[193] in 1998, were the first to apply a DDT ELISA to environmental samples (LOD = 0.3 µg L^{-1}). Later, Amitarani et al.[194] reported on another ELISA where DDT was detected in river water samples at levels close to 1 µg L^{-1}. An FPIA for the detection of DDT and its isomers in drinking water was developed by Eremin et al.,[195] who achieved LODs of 12 µg L^{-1} and 30 ng L^{-1}, respectively. Several immunochemical analytical methods have also been developed for chlorinated cyclodienes (CCDs), such as endosulfan, heptachlor, chlordane aldrin, endrin, and dieldrin, since the pioneering work of Langone and Van Vunakis,[196] who designed a RIA for dieldrin and aldrin in 1975. Manclus et al.[197] produced monoclonal antibodies against endosulfan (α/β), which recognized almost all structurally related cyclodiene insecticides with good detectability. On the other hand, Stanker et al.[198] adapted a commercially available ELISA kit for the analysis of endosulfan in environmental water samples with very good results. Lee and Kennedy[199] produced polyclonal antibodies to develop an ELISA to detect endosulfan in runoff water and soil extracts with an LOD of 0.2 µg L^{-1}. A direct ELISA was also developed for screening aldrin, dieldrin, and endrin compounds in tap and Nile river water samples: LODs of 5 and 10 µg L^{-1} were obtained for aldrin and dieldrin, respectively.[200] Finally, a fiber optic immunosensor was developed, which can detect most of the cyclodiene congeners at ppb levels in soil extracts and environmental water samples, using rabbit polyclonal antibodies raised against the chlorendic caproic acid hapten.[201]

Several qualitative and quantitative immunochemical methods and their application to the analysis of environmental samples have been described for OP insecticides, a family that includes widely used pesticides such as azinphos-ethyl/methyl, dichlorvos, fenitrothion or fenthion, malathion, mevinphos, and parathion. Mercader and Montoya[202] produced monoclonal antibodies against azinphos-methyl and developed an ELISA that was used for the analysis of water samples from different sources, reaching detectability levels near 0.05 µg L^{-1}. Watanabe et al.[203] reported the production of polyclonal antibodies and ELISA procedures to analyze fenitrothion in river, tap, and mineral water (LOD = 0.3 µg L^{-1}). Banks et al.[204] produced polyclonal antibodies against dichlorvos, an organophosphate insecticide used for stored grain, which also cross-reacts with fenitrothion. Nishi et al.[205] reported the first immunoassay for malathion. Residues of this insecticide have

TABLE 8.7
Immunochemical Techniques Developed for the Detection of Insecticides

| Target Analyte | Technique | Matrix | Sensitivity | | Reference |
			LOD	IC$_{50}$	
Aldrin	ELISA	Tap and river water	5 µg L^{-1}		200
	Optical immunosensor	Soil and water samples		5 µg L^{-1}	201
Azinphos-ethyl/ methyl	ELISA	Well, tap, channel, cistern, and drinking water	0.05 µg L^{-1}	0.33 µg L^{-1}	202
Carbaryl	ELISA	Cucumber and strawberry		0.13 µg L^{-1}	396
	SPR	Ground, river, and tap water	0.86 µg L^{-1}	3.97 µg L^{-1}	397
Carbofuran	Magneto-ELISA	Water and soil	5 µg L^{-1}	0.056 µg L^{-1}	213
	Planar array evanescent immunosensor	Ground and river water	0.1 µg L^{-1}		222
	SPR immunosensor	Drinking water	0.03 µg L^{-1}	1.06 µg L^{-1}	223
Deltamethrin	ELISA	River sample	1.1 µg L^{-1}	17.5 µg L^{-1}	398
Endosulfan	ELISA[a]	Environmental water		2 µg well^{-1}	198
	ELISA	Soil and runoff water	0.2 µg L^{-1}		399
Esfenvalerate	ELISA	Tap and river water		30 µg L^{-1}	400
	Fluorescence immunoassay	River water	0.04 µg L^{-1}	0.8 µg L^{-1}	233
Fenitrothion	ELISA	Fruit extracts	40 ng well^{-1}	297 ng well^{-1}	401
	ELISA	River and tap water	0.3 µg L^{-1}	6 µg L^{-1}	402
	ELISA	Rice extracts	3 µg L^{-1}	14 µg L^{-1}	403
Fenthion	Dipstick immunoassay	Food samples	0.5 µg L^{-1}	15 µg L^{-1}	404
Flucythrinate	ELISA	River water and soil	10, 0.2 mg L^{-1}		405
Malathion	Sol-gel immunosorbent	Surface water	0.50 µg L^{-1}	0.10 µg L^{-1}	406
Mevinphos	ELISA	Buffer	52 ng well^{-1}	3700 µg L^{-1}	407
Parathion	ELISA	Buffer		600 µg L^{-1}	407
Parathion-methyl	Commercial kit	Water		0.3 µg L^{-1}	408
Triazophos	ELISA	Buffer	0.11 µg L^{-1}	5.51 µg L^{-1}	409
	ELISA	Buffer	0.10 µg L^{-1}	0.65 µg L^{-1}	410

[a] Commercial kit.

been detected in ground and surface water at levels up to 6.1 µg L^{-1}.[206,207] Brun's group[208] recently developed an assay with an LOD of approximately 0.1 µg L^{-1} that was successfully applied to the analysis of river and groundwater samples. In contrast, monoclonal antibodies have been produced to develop an ELISA for the analysis of parathion and parathion-methyl compounds in water and milk samples at levels around 1 µg L^{-1}.[209] Similarly, a commercially available ELISA kit (EnviroGard Parathion Plate Kit) has been validated for application in water samples: LODs of 0.03 and 0.05 µg L^{-1} were obtained for parathion and parathion-methyl, respectively.

Since their commercial introduction in the early 1960s, N-methylcarbamate pesticides (carbaryl, carbofuran (CF), methiocarb, etc.) have been used worldwide as substitutes for OCs because of their excellent efficiency as insecticides and nematicides, their relatively low mammalian toxicities in many cases, and their low bioaccumulation potentials. In recent years, several ELISAs have become commercially available or have been developed to determine these pesticides in water samples,[210–215] as well as in fruits and vegetables.[214,216–219] For example, the performance of two ELISA formats (microtiter plates and magnetic particles) were compared with the EPA method 531.1 (liquid chromatography-postcolumn derivatization-fluorescence detection, LC-PCR-FD) for the determination of carbaryl in groundwater samples of the Campo de Nijar aquifer (Almeria, Spain).[212] A close correspondence was found for the results obtained when spiked and well water samples were split for analysis by ELISA and by LC-PCR-FD, but the absence of a matrix effect and the high throughput capability of the ELISA formats pointed to the superiority of these immuno-chemical methods for screening purposes. The presence of CF during a year in lake, well, and irri-gation ditch water in an agricultural area south of Milan has been evaluated using a fluorescent immunoassay with a time-resolved revelation system. Results show that CF peaked at around 87 ng mL^{-1} in September and October.[220] Another interesting approach is the homogenous immu-noassay developed for CF by the same group. In this case, the determination used liposomes and a mastoparan (Mast)-hapten conjugate as cytolitic agent. Dipicolinic acid (DPA) was used as fluores-cent chelating agent. Liposome lysis was proportional to the standard concentrations in a dynamic range between 10 pg and 10 ng. The assay was applied to the analysis of tap water and environmen-tal water samples taken from the same agricultural area, with recoveries of between 90% and 105%.[221] Automated methods and immunosensors have also been reported.[215,220,222–227] Mauriz et al.[226] described the application of a commercial optical sensor system based on SPR detection to the direct analysis of carbaryl in different environmental water samples without any sample pre-treatment. Detection limits obtained for ground, river, and tap water were 1.3, 1.2, and 0.9 μg L^{-1}, respectively, whereas the IC$_{50}$ values obtained were in the 4.0–4.6 μg L^{-1} range.

Immunochemical analytical methods have also been developed for pyrethroid insecticides. Lee et al.[228,229] developed an immunoassay for analyzing pyrethroids of the second group that was applied to detect deltamethrin and bifenthrin in water and soil samples as well as deltamethrin in wheat grain. Watanabe et al.[230] and Mak's group[231] developed a class-specific immunoassay for the first and second group of pyrethroids respectively, obtaining very good detection limits. A competitive ELISA has recently been developed for the detection of the pyrethroid insecticide cyhalothrin, giving an LOD of 4.7 μg L^{-1}; it was evaluated using fortified tap water, well water, and wastewater samples with recoveries between 80% and 114%.[232] Another interesting example is the fluorescence-quenching competitive immunoassay in microdroplets reported for the sensitive detection of the pyrethroid insecticide esfenvalerate using laser-induced fluorescence from a rhodamine hapten conjugate. The competitive immunoreaction was performed in microdroplets generated by a vibrating orifice aerosol generator system with a 10 μm diameter orifice. The fluorescence emitted from the droplets was detected by a 1/8 inch imaging spectrograph with a 512 × 512 thermoelectrically cooled, charged-coupled device. A very small mass of analyte could be detected with this method; thus, monitoring of a picoliter droplet sample enabled detection down to ~0.1 nM. Matrix interferences were negligible when this technique was applied to the analysis of river water samples.[233] Sasaki et al.[234] developed a novel SPR biosensor chip by using a plasma-polymerized ethylene diamine film over a gold layer sputtered onto glass. Antietofenprox antibody was immobilized on the glass surface using glutaral-dehyde, and the response of the SPR biosensor was compared to that of a commercial chip. The result was not so different from that obtained with the commercial chip, but the fact that the plasma polym-erized membrane is optically homogeneous might have helped to produce a higher response.

8.3.2 HERBICIDES AND PLANT GROWTH REGULATORS

Herbicides are used to get rid of unwanted plants like weeds, brush, unproductive trees, and other vegetation that may deprive crops and other "useful" plants of nutrients. Numerous immunological techniques for the analysis of triazines, such as atrazine, propazine, simazine, ametryn, and cyanazine, have been developed recently (Table 8.8).[185,235–237] Owing to their environmental persistence and their

TABLE 8.8
Immunochemical Techniques Developed for the Detection of Herbicides

Target Analyte	Technique	Matrix	Sensitivity		Reference
			LOD	IC_{50}	
2,4-D	Optical immunosensor	River and lake water	0.1 μg L^{-1}		248
	Dipstick immunoassay	Pond water	0.5 μg L^{-1}	6 μg L^{-1}	249
2,4,5-T	ELISA	Soil and water samples		11.6 μg L^{-1}	246
Acetochlor	Polarization fluoroimmunoassay	Drinking water	9 μg L^{-1}		271
Atrazine	ELISA	Soil, ground, and water	1 ng L^{-1}	20 ng L^{-1}	238
	ELISA	River, lake, and tap water		5 μg L^{-1}	411
	ELISA	Creek and drinking water	0.2 μg L^{-1}		412
	ELISA[a]	River, estuarine, and sea water	0.05 μg L^{-1}	0.3 μg L^{-1}	239
	FIIA	Estuarine and sea water	75 ng L^{-1}	470 ng L^{-1}	413
	ELISA		60 ng L^{-1}	9 ng L^{-1}	
	Amperometric immunosensor	Drinking water	6 ng L^{-1}	0.17 μg L^{-1}	240
	SPR immunosensor	Well, river, and tap water	26 ng L^{-1}	0.18 μg L^{-1}	243
Isoproturon	Optical immunosensor	River and estuarine water	0.14 μg L^{-1}	1.65 μg L^{-1}	255
Metsulfuron-methyl	ELISA	Drinking water	40 ng L^{-1}	1.4 μg L^{-1}	274
Propanil	RIANA	Drinking water	0.6 ng L^{-1}	52 μg L^{-1}	163
Simazine	ELISA	Ground and tap water	50 ng L^{-1}	0.1 μg L^{-1}	414
	ELISA	Lake, rain, and mineral water	0.01 μg L^{-1}	0.07 μg L^{-1}	415
	ELISA	Ground and well water		2.03 μg L^{-1}	416
	ELISA	Tap water Putah creek water Bay water		0.75 μg L^{-1}	417
	Magneto-ELISA	Distilled water	0.349 μg L^{-1}	1.76 μg L^{-1}	418
		Groundwater	0.402 μg L^{-1}	2.09 μg L^{-1}	
		Estuarine water	0.416 μg L^{-1}	2.10 μg L^{-1}	
	FIIA	Drinking water	0.02 μg L^{-1}		419
	Optical immunosensor	Drinking water	0.026 μg L^{-1}		420
	Sol-gel immunosorbent	Surface water	0.25 μg L^{-1}	0.05 μg L^{-1}	406
	m-ISLMA_1	Surface water	1 × 10^{-5} μg L^{-1}	0.25 μg L^{-1}	421
	m-ISLMA_2		2 × 10^{-2} μg L^{-1}	1 × 10^{-4} μg L^{-1}	
	m-ELISA		1 × 10^{-1} μg L^{-1}	13.8 μg L^{-1}	
	μ-IA	Mineral water	0.2 ng L^{-1}		244
	μ-ISLMA		0.1 ng L^{-1}		
Trifluralin	ELISA	Surface water	0.85 μg L^{-1}	5.78 μg L^{-1}	422
	ELISA (OWLS)	Surface water	0.8 μg L^{-1}	2.87 μg L^{-1}	278
Linuron	RIANA	River and MilliQ water	0.01 μg L^{-1}	1.03 μg L^{-1}	255

water solubility (33 mg L^{-1}), triazines are distributed mainly in groundwaters and surface waters. For this reason, a large number of reported immunoassays have focused on the analysis of natural water samples.

Wittmann and Hock[238] measured atrazine in drinking and groundwater samples, reaching detection limits close to 1 ng L^{-1} without using any preconcentration or cleanup step. The presence of atrazine has also been analyzed in estuarine and seawater samples using other immunological techniques like magneto ELISA and FIIA with very low detection limits—close to 50 ng L^{-1}.[239] Several electrochemical immunosensors have been described for atrazine detection in food and aquatic matrices, like the one presented by Zacco et al.,[240] who developed an amperometric immunosensor based on modified magnetic particles with antibodies that are captured by a graphite-epoxy magneto composite, also used as the transducer for the electrical immunosensing; the LOD for drinking water samples was 6 ng L^{-1}. Figure 8.5 shows a schematic representation of the biosensor.

Several authors[185,237,241] developed an impedimetric immunosensor based on interdigitated electrodes without the use of any label; this system achieved detection limits of 0.04 and 0.19 µg L^{-1} in buffer and wine samples, respectively. Following up the same idea, but exploring a conductometric transduction system, Valera et al.[242] produced an immunosensor using antibodies labeled with gold nanoparticles; the LOD in buffer was 0.1 µg L^{-1}. Both systems should be easily adaptable for the analysis of environmental samples. Recently, Farré et al.[243] developed another immunosensor based on the SPR principle to analyze atrazine in well, river, and drinking water; detection limits were approximately 26 ng L^{-1} in all cases. A micro-immune-supported liquid membrane assay (µ-ISLMA) based on chemiluminescent detection has been developed to detect simazine in a single miniaturized cartridge system.[244] This chapter also discusses the influence of using different SAMs and different kinds of antibodies (polyclonal, affinity purified polyclonal, and monoclonal) on extraction parameters and assay sensitivity. LODs obtained for mineral water samples were at the ng L^{-1} level. Tschmelak et al.[163,245] applied the RIANA biosensor to detect propanil, a selective postemergent herbicide, in water samples without any pretreatment (LOD = 0.6 ng L^{-1}).

Chlorophenoxy acid herbicides are also widely used to control broadleaf weeds and grass plants. Several immunoassays have been reported for 2,4-dichlorophenoxyacetic acid (2,4-D) and 2,4,5-trichlorophenoxyacetic acid (2,4,5-T).[246,247] Several immunosensors have been described using a transducing principle similar to the RIANA system already described in this chapter. Thus, Meusel et al.[248] reported the use of monoclonal antibodies in a sensor chip to analyze river and lake water samples, obtaining detection limits of 0.1 µg L^{-1}. Moreover, monoclonal antibodies, produced by Cuong et al.,[249] were used in a dipstick immunoassay format to analyze pond water samples. When applied to the 2,4-D compound, this semiquantitative method yielded for an IC$_{50}$ of 6 µg L^{-1} and an LOD of 0.5 µg L^{-1}.

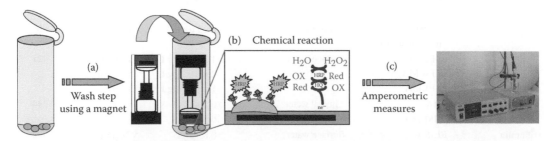

FIGURE 8.5 Schematic representation of an electrochemical magneto immunosensing strategy for the detection of low-molecular-weight compounds. After the immunoreaction, the antibody-modified magnetic beads are captured by the m-GEC electrode. Chemical reactions occurring at the m-GEC surface polarized at −0.150 V (versus Ag/AgCl) upon the addition of H$_2$O$_2$ in the presence of mediator (hydroquinone) are recorded. (From Zacco, E. et al. 2006. *Anal. Chem.* 78: 1780–1788. With permission.)

Immunochemical methods have also been reported for the analysis of phenylurea herbicides in different matrices, including food and environmental samples.[250-254] Thus, recombinant antibodies have been applied to the analysis of the phenylurea herbicide diuron with very good detectability (IC_{50} = 2 and 12 µg L^{-1} in the indirect and direct ELISA formats, respectively).[251] Similarly, isoproturon has been analyzed in soil extracts using an ELISA.[254] An ultrasensitive time-resolved fluoroimmunoassay (TR-FIA) for diuron in water samples has been recently reported. This assay was performed using the diuron-specific polyclonal antibody raised in sheep; rabbit antisheep IgG was used as fluorescent marker, conjugated with a chelating molecule complexed with Eu^{3+}. Even though the sensitivity of the lanthanide chelate was up to 10 times better than in other techniques, this level was 20 ng L^{-1} below the European Community limits. Water samples collected monthly from an agricultural area showed that peak diuron concentrations were 65 pg mL^{-1} in ditch water samples in June and 180 pg mL^{-1} in lake water samples in September.[220] On the subject of immunosensors, it is worth mentioning the work by Mallat et al.,[255] who again applied the RIANA system to monitor isoproturon, diuron, and linuron in Ebro delta waters (Tarragona, Spain) (LOD = 0.01 µg L^{-1}). A flow-through fluoroimmunosensor has also been developed for isoproturon in well water with a detectability in the µg L^{-1} range.[256] The use of antibodies in SPE methods against phenylurea herbicides has been investigated by immobilizing the antibodies on different solid supports[257-259] or encapsulating them in sol-gel matrices.[260] These immunosorbents have been applied as both a cleanup and a preconcentration step of these herbicides from ground,[252] drinking,[259] and surface[258,259] water samples, using on- and off-line procedures. Thus, with the immunosorbent conveniently packed in a C18 column coupled to a liquid chromatography (LC) system, about 10 phenylureas were monitored from the Seine River.[259] The class-selectivity profile demonstrated by these immunosorbents makes them useful for multiresidue analysis procedures of this particular family of herbicides.

Several immunoanalytical techniques have been developed for the analysis of chloroacetanilides, another important family of herbicides, such as alachlor,[261,262] metolachlor,[263-265] and their metabolites[266] in various matrices. Immunoassays for detecting butachlor and acetochlor have received less attention than the above-mentioned analogs, although some immunochemical developments have also been reported.[267-269] An electrochemical immunosensor[270] and a fluororimmunoassay[271] have also been described for acetochlor. Interesting are the interlaboratory collaborative field experiments[264] carried out to compare solid-phase extraction-gas chromatography (SPE-GC), solid-phase microextraction-gas chromatography (SPME-GC), and ELISA tests for the analysis of metolachlor. Runoff water samples were collected during the first rain event following herbicide application and analyzed using different methods. Larger metolachlor concentrations were found in surface runoff (1.4–54.9 µg L^{-1}) than in tile drainage (0.01–8.5 µg L^{-1}). The results demonstrated that although ELISA overestimated the concentration of this chloroacetanilide herbicide, correlation with the chromatographic methods was very good. An amperometric immunosensor for acetochlor detection[270] based on screen-printed electrodes has been reported, although the detectability achieved was not sufficient for the direct analysis of drinking water. The LOD described were around 25 and 60 µg L^{-1} for drinking and surface water, respectively.

Several ELISAs have been developed for the analysis of sulfonylurea herbicides like chlorsulfuron,[272] triasulfuron,[273] and metsulfuron methyl.[274] Schlaeppi et al.,[273] for example, developed an immunoassay using monoclonal antibodies for the analysis of fortified soil samples. The sensitivity of the assay, after an optimized extraction procedure, was 0.1 pg kg^{-1}. Eremin et al.[195] developed a FPIA for chlorsulfuron detection in MilliQ water samples; the LOD obtained in 50 µL of sample was 10 µg L^{-1}. Another example of a fluoroimmunoassay was the one developed by Wang et al.[275] consisting of TR-FIA method for bensulfuron-methyl based on fluorescence resonance energy transfer (FRET) from a Tb^{3+} fluorescent chelate to an organic dye, Cy3 or Cy3.5; this method achieved a detection limit of 2.1 µg L^{-1}. The same author[276] developed a new immunoassay method by using graphite furnace atomic absorption spectrometry with an EDTA-Cd^{2+} chelate as the label; bensulfuron-methyl was analyzed using this technique (LOD = 0.95 µg L^{-1}). Dzantiev et al.[277] developed an electrochemical immunosensor for analyzing chlorsulfuron herbicide in just 15 min. The working range for the quantitative detection of chlorsulfuron was from 0.01 to 1 μg L^{-1}. Finally, Szekacs et al.[278] developed a highly sensitive immunosensor using optical waveguide lightmode spectroscopy (OWLS) to detect trifluralin, a selective pre-emergence herbicide, with an IC$_{50}$ of 1.0 μg L^{-1}. The principle is based on the precise measurement of the resonance angle of polarized laser light, diffracted by grating and coupled onto a thin waveguide.

8.4   IMMUNOCHEMICAL DETERMINATIONS OF PHARMACEUTICAL AND PERSONAL CARE PRODUCTS

The term "pharmaceutical and personal care products (PPCPs)" refers to any product used for personal health or cosmetic reasons or used in agriculture to enhance the growth or health of livestock;[279] it comprises a diverse collection of thousands of chemical substances.[40] The overall pharmaceutical production in Europe, Japan, and the United States amounted to USD373 billion in 2005.[280] PPCPs have probably been present in the environment and water for as long as humans have been using them. While an important number of these substances enter the environment directly from industry, treated and untreated domestic sewage containing excreted PPCPs and their metabolites following human use is a major source of these compounds in the environment.[39] Nowadays, sewage systems and municipal WWTPs are still not equipped for the complete removal of PPCPs or other unregulated contaminants.[281] In addition to the framework provided by the Water Directive mentioned in the introduction, Directive 2001/82/EC regulates the requirements for the ecotoxicity testing of pharmaceuticals. With the advances in technologies that have improved the ability to detect, control, and quantify these chemicals, we can now begin to identify what effects, if any, these chemicals have on human and environmental health.

8.4.1   ANTIBIOTICS

Antibiotics are chemical substances extremely active at low doses that kill or slow the growth of bacteria. Since their discovery, antimicrobials have been an essential part of modern human and veterinary medicine as well as in aquaculture or even in plants for the treatment of infectious diseases produced by bacteria. In the last decade, the general misuse of antibiotics as growth promoters or for prophylactic purposes[282] has become a decisive factor favoring the increase of bacterial resistance. This risk situation may spread from animals to humans through the food chain[8] but may also have a crucial impact on the ecosystem itself by producing adverse effects in animals and plants. At present, WWTP effluents and confined animal feeding operations (AFOs) represent the prime sources of antibiotics entering the environment:[281,283] the greatest percentage of antibiotics are excreted after consumption, and thousands of tonnes of them reach the terrestrial and aquatic environment every year.[1] Governmental agencies have therefore set limitations on the levels of residues in accordance with available toxicological data by laying down specific regulations or proposals that are to complement the restrictions already in place with respect to animal foods destined for human consumption. Antibiotics are classified into several families, for example, penicillins, fluoroquinolones (FQs), sulfonamides (SAs), tetracyclines (TCs), macrolides, and chloramphenicols (CAPs). In general, all these compounds are quite resistant to biodegradation since they were designed to demonstrate a certain metabolic stability during their pharmacological action; they are likely to remain in the environment in unchanged form or as persistently active metabolites.[284] In this context, different techniques based on wholly divergent principles have been developed to deal with the problems relating to antibiotic residues.[285–287] Most of them, like growth inhibition tests, take advantage of their antibacterial activity. Others, such as chromatographic methods, are highly specific and sensitive but require extensive sample preparation, sophisticated and therefore expensive equipment, and skilled laboratory staff. On the other hand, immunochemical techniques can be excellent tools for assessing antibiotic contamination in different environmental matrices as a result of their excellent detectability, specificity, and throughput capacities.[44,287] Table 8.9 summarizes some of these techniques reported for the detection of several families of antibiotics.

TABLE 8.9
Immunochemical Techniques Developed for the Detection of Antibiotics

Target Analyte	Technique	Matrix	Sensitivity		Reference
			LOD ($\mu g\ L^{-1}$)	IC_{50} ($\mu g\ L^{-1}$)	
Sulfonamides	ELISA	Fish muscle	<100[a]		303
	Charm II RIA test	Drinking water sources	0.05		310
	RIANA	Water samples	0.01	100	311
Sulfamethazine	RIA	Lagoon and river samples	5.00		309
	SPIE with MALDI-TOF	Drinking water	0.10		313
	MS	Soil	1.00		
		Manure	1.00		
Sulfamethizole	AWACSS	River water samples	0.02		312
Fluoroquinolones	ELISA	Shrimp tissues	~4.00		323
	ELISA	Fish and shrimp sample	0.70[b]	<10[c]	322
Ciprofloxacin	ELISA	Milk, chicken, and pork		0.32[a]	324
Enrofloxacin	RIA	Lagoon and river samples	5.00		309
Tetracyclines	Charm II RIA test	Hog lagoon Surface water Groundwater	1.00	1–20[d]	348
	Charm II RIA test	Drinking water sources	0.05		310
Tetracycline	ELISA[e] (IDS Corporation)	Surface and groundwater	0.20		345
	ELISA[e] (Ridascreen)		0.10		
Tetracycline	ELISA[e] (R-Biopharm	Manure samples from	0.38[a]	1.02[a]	346
Anhydrotetracycline	GmH)	hog lagoons and cattle	0.25[a]	5.40[a]	
Chlortetracycline		feedlots	0.01[a]	0.21[a]	
Anhydrochlor-TC			0.01[a]	6.92[a]	
Oxytetracycline			0.05[a]	0.97[a]	
Chloramphenicol	ELISA[e] (5091CAP1p, EDiagnostica)	Shrimp tissue	0.10 0.13[b]	0.22[c]	334
	SPR Biosensor (Biacore Q)	Prawn samples	1.00[a] 0.04[b]	0.07[d]	336
	Membrane-based CL sensor	Shrimp samples	3.23		339
Chloramphenicol	TR-FIA	Shrimp samples	0.05		423
CAP succinate			0.10		
Thiamphenicol	SPR Biosensor	Shrimp tissue	0.5[b]	0.13[a]	337
Florefenicol	(Biacore Q, sensor chip		0.2[b]	0.47[a]	
Florefenicol amine	CM5)		250[b]	887[a]	
Chloramphenicol			0.1[b]	1.26[a]	
Chloramphenicol	SPR Biosensor	Prawn	0.04[b]	0.07[c]	338
CAP glucuronide	(Biacore Q and Qflex® Kit)			76% CR	
Penicillin G	RIA	Lagoon and river samples	1.00		309
Penicillin G	Fluoro immunoassay	Wastewater	2.40	30.0	354
Amoxicillin		Sewage water	~5.00	58.0	
Tylosin	ELISA (IDS Corporation)	Surface and groundwater	0.20		345
	ELISA (Ridascreen)		0.10		
Erythromycin	RIA	Lagoon and river samples	10.0		309

[a] LODs and IC_{50} calculated in buffer solution.
[b] General decision limit.
[c] Detection capacity.
[d] Linear range.
[e] Commercial kit.

A significant number of immunochemical methods for antibiotic residue analysis with narrow or broad specificity, or even possessing multianalyte capabilities, have been described, and some of them are commercially available.[39,40,287,288] We recently reported a microplate-based ELISA method that can detect 25 antibiotics from the ß-lactam (BL), SA, and FQ families.[22] Moreover, the assay based on a dipstick platform is becoming the new simple, rapid and easy-to-use sensing device for on-site measurements. The biological recognition elements normally used in this user-friendly technology are the receptors;[289–292] applications using antibodies have appeared in recent years[254] but have so far been implemented only in the analysis of food samples, not yet in environmental samples.

8.4.1.1 Sulfonamides

SAs are an important group of broad-spectrum synthetic bacteriostatic antibiotics, whose chemical structure contains a 4-aminobenzensulfonamide functional group with different heterocycles attached to the N1-position of the SA bridge. This antibiotic family is widely used in animal husbandry in most European countries.[293,294] The pharmacokinetic profile of SAs ensures that they are quickly eliminated from the organism (40–90%), usually as the parent compound or as bioactive metabolites.[295] As with many other pharmaceuticals, SAs are fairly water-soluble, polar compounds that ionize depending on the pH of the matrix. In addition to hydrophobic partitioning, these compounds can absorb to soils via cation exchange, cation bridging, surface complexes, and hydrogen bonding.[296] Hence, SAs will persist in the environment and, because of their relatively high mobility, will enter groundwater and be transported to aquifers and surface waters;[284] relevant methodologies for monitoring environmental samples are therefore necessary. Besides the large number of chromatographic methods reported,[294,297–300] several immunochemical techniques have been developed for the analysis of water samples. Several ELISAs with broad[301–304] or narrower selectivity profiles[305–308] within the SA family have been applied to the analysis of these residues in various food matrices in compliance with legislation. Campagnolo et al.[309] studied the presence of different antimicrobials in wastewater samples from pig and poultry farms using a commercial radioimmunoassay (Charm II RIA). Prior to the analysis, samples were simply filtered through a 0.45 µm glass fiber filter; an LOD of 5 µg L^{-1} was obtained for sulfamethazine. In order to achieve lower detection limits, Yang and Carlson[310] coupled SPE cartridges to the same RIA test as a preconcentration technique. This method was optimized to detect SA and TC (see below) compounds in water samples from rivers and the influent/effluent of a WWTP. The detection limit for sulfamethazine was 0.05 µg L^{-1} using the SPE/RIA method; quantification of sulfamethoxazole, sulfadimethoxazone, and sulfathiazole was also possible. Initially developed for biochemical studies, RIA has the disadvantage of handling and producing radioactive residues, so their use should be avoided whenever possible. On the other hand, the already-mentioned RIANA immunosensor was used to detect SAs in drinking, ground, and surface water samples.[311] With this biosensor and a mixture of antibodies, it was possible to achieve detection limits <10 ng L^{-1}, limits of quantification (LOQ) <100 ng L^{-1}, and IC$_{50}$ values between 0.5 and 5 µg L^{-1} for five SAs without sample pretreatment. The automated water analyser computer supported system (AWACSS) instrument represents a development of the RIANA sensor in that the multianalyte analysis capability has been expanded, theoretically permitting simultaneous measurements of up to 30 analytes from the groups of modern pesticides, endocrine disrupting compounds, and pharmaceuticals. With this system, Tschmelak et al.[184,312] achieved an LOD of <0.02 µg L^{-1} for sulfamethoxazole in river water samples. On the other hand, Grant et al.[313] described a method for detecting residues of sulfamethazine and its major metabolite N_4-acetylsulfamethazine in water, aqueous suspensions of soil, and composted manure samples, using solid-phase immunoextraction (SPIE) coupled with MALDI-TOF MS. The LODs for both compounds in all kinds of samples were <1 µg L^{-1}. No further immunochemical methods for the direct detection of SAs in environmental samples were found; nonetheless, the application to environmental water samples of those currently applied in complex biological matrices[305,314–316] is predicted to be straightforward. Several novel immunosensing strategies for detecting SAs have been developed by our group. Zacco et al.[317] immobilized class-specific anti-SA antibodies to magnetic

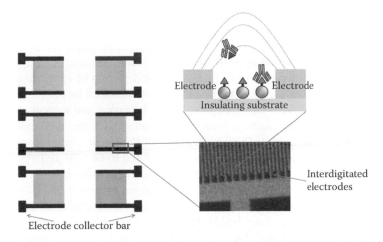

Electrode

Electrode

Insulating substrate

Interdigitated electrodes

Electrode collector bar

FIGURE 8.6 Schematic representation of a new transducer for biosensor application based on a three-dimensional IDEA. Binding of molecules to the chemically modified and biofunctionalized transducer surface induces important conductivity changes between the electrodes, which can be monitored. (From Ramón-Azcón, J. et al. 2008. *Biosens. Bioelectron.* 23: 1367–1373. With permission.)

particles to be captured, after the immunochemical reaction, by a magneto sensor made of graphite-epoxy composite (m-GEC) that is also used as the transducer for the electrochemical detection. The LOD obtained for sulfapyridine in milk was 1.4 μg L^{-1}. Another example, using the same immunoreagents, is presented by Bratov et al.[318]: a new transducer for biosensor applications based on a three-dimensional interdigitated electrode array (IDEA) with electrode digits, separated by an insulating barrier. The binding of molecules to the chemically modified surface of the transducer induces important changes in conductivity between the electrodes; impedance measurements with this immunosensor detected sulfapyridine with an IC$_{50}$ of 5.6 μg L^{-1} in buffer. As can be seen in Figure 8.6, using this strategy, it was possible to place the immunoreaction where most of the electric field is, instead of using just a small percentage.

Finally, class-selective immunoreagents for SA detection were implemented in a waveguide interrogated optical system (WIOS). The label-free sensor, developed by the Swiss Center for Electronics and Microtechnology (CSEM), is based on the evanescent wave principle, where changes in the refractive index of the modified chip surface are detected by scanning the resonance condition at which a light wave is coupled in the waveguide through a conveniently designed grating.[319] Monitoring of the resonance wavelength allows real-time monitoring of the binding of nonlabeled molecules to the waveguide grating surface, previously modified with the immunoreagents by means of a photopolymerizable dextran layer. The LOD obtained with this methodology for sulfapyridine in milk was 0.5 μg L^{-1}.

8.4.1.2 Fluoroquinolones

FQs are a synthetic class of antibiotics widely used for both prevention and treatment of various diseases in animal husbandry and aquaculture, as well as in humans. The environmental concern regarding FQs is evinced not only by their potential to promote antibiotic resistance, but also by their unfavorable ecotoxicity profile.[320] FQs are excreted as parent compounds, as conjugates, or as oxidation, hydroxylation, dealkylation, or decarboxylation products. FQs bind strongly to topsoils, thereby reducing the threat of surface water and groundwater contamination; this implies, however, that the terrestrial environment is a further relevant exposure pathway.[40] The strong binding of FQs to soils and sediments delays their biodegradation and explains their persistence in the environment. Wastewater treatment eliminates 79–87% of FQs before their arrival in rivers; adverse effects on the aquatic habitats of surface waters are thus rather unlikely. On the other hand, these compounds are also susceptible to photodegradation in water: this involves the oxidation, dealkylation, and cleavage of the piperazine ring.[321] Fluoroquinolone residues in marine products are an important analytical target because they are regarded as good indicators of environmental quality. In this context, Huet et al.[322] reported the development of an ELISA for the detection of 15 fluoroquinolones in fish and shrimps, as well as in other samples (kidney, eggs, and muscle). The pretreatment required for the analysis of the marine products involved sample centrifugation and solvent extraction prior to 10-fold dilution. The assay was characterized in accordance with the recommendations of the European Commission (Commission Decision 2002/657/EC) by calculating the general decision limit (CCα) and detection capacity (CCβ). CCα was calculated to be 0.70 μg L^{-1}, whereas CCβ for most of these compounds was <10 μg L^{-1}, except in the case of sarafloxacin, oxolinic acid, flumequine, and cinoxacin, the detection capacities of which did not exceed 4, 25, 100, and 200 μg L^{-1}, respectively. An ELISA using monoclonal antibodies with a broad specificity for fluoroquinolone antibiotics (12 FQ congeners) was described by Wang et al.[323] for chicken, honey, egg, and shrimp samples: IC$_{50}$ in buffer varied from 2.1 μg L^{-1} (norfloxacin) to 4.4 μg L^{-1} (lomefloxacin). Shrimp samples were fortified at different levels (50, 100, and 200 μg L^{-1}), separately with fluoroquinolones such as enrofloxacin, ciprofloxacin, norfloxacin, ofloxacin, flumequine, and danofloxacin; recoveries were between 63% and 90%. The last example found of an immunoassay applied to the analysis of fluoroquinolones in environmental samples is the already-mentioned immunosensor developed by Campagnolo et al.;[309] these authors achieved an LOD for enrofloxacin of 5.0 μg L^{-1}. Other immunoassays performed to detect FQs in complex biological samples[322,324–326] should a priori be easily adaptable for monitoring these substances in environmental water samples. The polyclonal antibodies developed and evaluated by Pinacho et al.[327] have been implemented in different immunochemical techniques to analyze a wide range of fluoroquinolone congeners. The same authors developed an ELISA capable of analyzing milk samples after a very simple dilution step, obtaining detection limits for most important fluoroquinolones of <0.4 μg L^{-1}.[328] Other uses of these immunoreagents have focused on electrochemical devices.[329,330] One example is an amperometric immunosensor that follows the same format as the one described in the sulfonamide section;[317] with this instrument LODs of 5.3 ng L^{-1} for ciprofloxacin in whole milk were obtained.[331] Impedance spectroscopy combined with immunosensor technology has been used to detect ciprofloxacin at 10 ng L^{-1} levels in buffer.[332] In this approach, the sensor electrode was based on the immobilization of the antibodies by chemical binding onto a poly(pyrrole-N-hydroxysuccinimide [NHS]) film electrogenerated on a solid gold substrate. The final immunoreaction triggers a signal via impedance spectroscopy measurements. Again, the application of these new analytical approaches to environmental samples should be straightforward.

8.4.1.3 Amphenicols

CAP, a bacteriostatic antimicrobial originally derived from the bacterium *Streptomyces Venezuelae*, was the first antibiotic to be manufactured synthetically on a large scale. Although CAP is effective against a wide variety of microorganisms, its use has been banned in the EU since 1994 because of certain toxicological side effect problems such as aplastic anemia, brown marrow suppression, or the so-called gray baby syndrome.[39] For this reason, a zero tolerance was established for the presence of these residues in any kind of animal products. On the other hand, this antibiotic has been widely used in the last 10 years by many low-income Asian countries for aquaculture disease treatment, because of its exceedingly low price. Although the use of this antibiotic in animal production has recently been prohibited in these countries too, CAP residues have been detected in marine products intended for the EU market.[333] According to several European Commission Decisions (2001/699/EC, 2001/705/EC, 2002/249/EC, 2002/250/EC, and 2002/251/EC), certain fishery and aquaculture products imported for human consumption must be subjected to a test in order to ensure the absence of CAP residues.[334] Thus, the main efforts have focused on the study of CAP residues in marine food products to control the problems mentioned above. Impens et al.[334] described the use of a commercial ELISA kit (5091CAP1p) to detect CAP in shrimp tissue after organic/aqueous extraction, obtaining an LOD of 0.1 μg L^{-1}. The method was revalidated according to Commission Decision 2002/657/EC,[335] which has been more commonly used for chromatographic techniques; CCα and CCβ values of 0.13 and 0.22 μg L^{-1}, respectively, were obtained. On the other hand, a commercial SPR immunosensor (Biacore Q) has been used by several authors to detect CAP residues in different kinds of matrices.[336–338] For example, Ferguson et al.[338] reported the implementation of a commercial detection kit (Qflex) in the cited biosensor to accurately determine CAP residues in milk, poultry muscle, honey, and prawn. CCα and CCβ values for prawn samples, after a tedious pretreatment, were 0.04 and 0.07 μg L^{-1}, respectively, while the glucuronide form of CAP cross-reacted 76% in this matrix. Using the same biosensor, Ashwin et al.[336] obtained similar parameters for CAP detection in prawn samples but with a simpler sample pretreatment procedure. Furthermore, the immunoreagents developed by Dumont et al.[337] were implemented in the same SPR sensor as described above for the simultaneous residue detection of several fenicol antibiotic congeners in shrimps from a single sample extract. The IC$_{50}$ values obtained for thiamphenicol (TAP), florefenicol (FF), and CAP were 0.13, 0.47, and 1.26 μg L^{-1}, respectively. CCβ values were also estimated for each compound in this study (TAP: 0.13, FF: 0.47, and CAP: 1.26 μg L^{-1}). In a different context, Park and Kim described the development of a membrane-based chemiluminescent immunosensor for the analysis of very low levels of CAP residues in different samples of animal food for human consumption, such as pork, beef, chicken, milk, and shrimps.[339] The shrimp samples were simply filtered through Whatman paper to avoid undesirable matrix effects (LOD $\simeq$ 3 μg L^{-1}). Alternatively, a large number of immunoassay screening methods for CAP detection in foods (e.g., milk, eggs, and meat) and other related complex matrices have been reported in the literature.[339–341] Despite the use of immunochemical methods to analyze residues in marine biota, we have not found examples of their application to analyze CAP residues in environmental samples, although these methodologies should be readily adaptable to the analysis of these types of matrices.

8.4.1.4 Tetracyclines

TCs are an important group of broad-spectrum antibiotics used against *Gram*-negative and *Gram*-positive microorganisms in modern human and veterinary medicine practice for both prevention and treatment of diseases, as well as additives in animal foodstuffs to promote growth in concentrated animal feeding operations (CAFOs). As with most types of antibiotics, only small portions of the tetracyclines administered are actually metabolized or absorbed in the body, and most of the drug is eliminated in feces and urine in unchanged form.[342] Normally, tetracyclines are not found at high levels in the environment: because of their chelation properties, they readily precipitate in the presence of divalent cations (i.e., Ca^{2+}, Mg^{2+}, or Zn^{2+}) and are accumulated in sewage sludge or sediments.[343] On the other hand, tetracycline residues have also been detected in many surface water resources that receive discharges from municipal WWTPs and agricultural runoff.[2,344] Besides the demonstrated persistence of TCs in agricultural soils that have received manure containing antibiotics, the biodegradation of these compounds to even more toxic substances must activate new strategies to improve their control and the efficiency of their removal in wastewater plants. To this end, a commercially available ELISA kit (RIDASCREEN®), commonly used for detecting tetracycline residues in meat and milk samples, was easily adapted for the ultratrace analysis of surface and groundwaters.[345] The assay was found to be highly sensitive to tetracycline and chlortetracycline with detection limits of 0.1 μg L^{-1} in lake waters, runoff samples, and soil saturation extracts. Furthermore, Aga et al.[346] evaluated another commercial ELISA kit (R-Biopharm GmbH) for investigating the occurrence and fate of tetracyclines in the environment. In this case, the potential use of class generic antibodies led to the multiple recognition of several TCs such as tetracycline, chlortetracycline, and oxytetracycline, and also their epimers and corresponding dehydration by-products with IC$_{50}$ values from 0.2 to 6.9 μg L^{-1}. Subsequently, the same immunochemical detection kit was evaluated by measuring the presence of tetracyclines in samples from different manured soil surface layers (0–5 cm).[347] Only trace amounts (<1 µg L^{-1}) of oxytetracycline were recorded in these samples and none was detected in water samples from field lysimeters; tetracyclines thus have a low mobility in soil, as suggested before.[342] The Charm II RIA method was applied to environmental samples but with the focus on tetracycline detection.[348] This Charm II RIA, previously developed as a screening tool for detecting tetracycline residues in serum, urine, milk, and tissues, was adapted for the analysis of water samples by Meyer et al.[348] who achieved an LOD of 1 µg L^{-1} and a semiquantitative analytical range of 1–20 µg L^{-1}. In this study, liquid waste samples were obtained from several hog lagoons, and the surface and groundwater samples were from areas given over to intensive poultry production; the analytical results were well correlated with those acquired by means of liquid chromatography-mass spectrometry (LC-MS) techniques. The same RIA technique, again applied by Campagnolo et al.[309] to different aqueous environmental samples, was able to detect chlortetracycline at sensitivity levels of 1 µg L^{-1}. Yang and Carlson[310] used SPE as a preconcentration technique in conjunction with Charm II RIA to obtain lower detection limits for tetracycline measurements in water matrices. In this case, detection limits of 0.05 µg L^{-1} for tetracycline, oxytetracycline, and chlortetracycline were obtained in the analysis of different wastewater samples. Other immunochemical methods developed to analyze manifold tetracycline residues, mainly in honey, milk, and animal tissues intended for human consumption,[349] could be adapted to analyze water samples.

8.4.1.5   ß-Lactams

The BL group is one of the most important families of antibiotics used in veterinary medicine for the treatment of septicemia, urinary infections, and pulmonary infections. The presence of penicillin residues in food of animal origin, such as milk or meat, can have the same drawbacks as other antibiotics: unfavorable microbiological effects in the dairy industry, possible hypersensitivity reactions in consumers, and antibiotic resistance.[350] On the other hand, their persistence in environmental samples should be very low, mainly because of the chemically unstable BL ring, which is highly sensitive to pH, heat, and ß-lactamase enzymes.[351] Some authors therefore point out the absence of this kind of antibiotic residue in water samples, but aim to detect their degradation products in order to evaluate possible future environmental risks. Several immunochemical techniques, based on different detection principles, have thus been developed to detect BL compounds in food samples of animal origin.[309,352–354] Many of these technologies are applied to the analysis of milk samples, because this antibiotic family is the most frequently used for the treatment of mastitis in dairy cows. Gaudin et al.[352] applied the Biacore SPR sensor, described previously for chloramphenicol detection, to detect ampicillin in milk samples using commercial monoclonal antibodies. Samples were pretreated to facilitate the opening of the BL ring; final detection limits of 5.9 and 12.5 µg L^{-1} for ampicillin in buffer and in milk, respectively, were obtained. This immunoassay revealed high cross-reaction values for other BL antibiotics such as penicillin G and M. The same biosensor was also used by Gustavsson et al.[353] to assay the activity of a carboxypeptidase and antibodies against the enzymatic product generated in milk samples. Detection limits for penicillin G were approximately 1 µg L^{-1}, and seven BL compounds were detected below their maximum residue limits (MRLs). It can be assumed that, as in the case of the antibiotic analyses mentioned earlier, application of immunoassays originally developed for other biological samples to environmental water samples should produce even fewer matrix effects. Benito-Pena et al.[354] prepared polyclonal antibodies to develop an automated flow-through fluoroimmunosensor for the analysis of penicillin antibiotics in wastewater samples from influent and effluent sewage water; LOD and IC$_{50}$ values for penicillin G and amoxicillin in buffer were 2.4, 5.0, and 30, 58 µg L^{-1}, respectively. This immunosensor was applied to the analysis of both compounds in wastewater samples passed through 0.45 µm glass fiber filters; the technique was validated by chromatography. Moreover, as in the case of SAs, fluoroquinolone, tetracycline, and chloramphenicol compounds, Campagnolo et al.[309] measured BLs in water samples taken from the vicinity of a farm; they obtained a detection limit of 2 µg L^{-1} for penicillin G.

8.4.1.6 Macrolides

Macrolide antibiotics, such as tylosin, roxithromycin, and erythromycin, are an important group of pharmaceuticals used in human and veterinary medical practice. Their activity stems from the presence of a large macrocyclic lactone ring containing 14, 15, or 16 atoms, with deoxy sugars, usually cladinose and desosamine, linked via glycosidic bonds. After application, a certain fraction of these macrolides is metabolized to inactive compounds, but a significant amount is excreted as active metabolites.[355] Most macrolide structures enter the environment via animal manure, which limits their mobility and bioactivity. In any case, control of these residuals in the environment is necessary so as to avoid future negative impacts on public health. Kumar et al.[315] reported on the detection of tylosin, used extensively in pig production for both growth promotion and therapeutic purposes, using two commercial ELISAs for surface and groundwater samples. Samples were diluted twice in buffer prior to their analysis; an LOD of 0.2 μg L^{-1} was obtained. Several antibiotic growth promoters, including tylosin, were analyzed in ground feed samples using a multianalyte ELISA after a cleanup step on OASIS® HLB cartridges by Situ et al.[356] Polyclonal antibodies were developed for this purpose. With this method, LOD and CCß values for five banned substances in animal feeds were respectively 0.28 and 0.30 mg kg^{-1} for bacitracin, 1.02 and 1.50 mg kg^{-1} for olaquindox, 0.21 and 0.60 mg kg^{-1} for spiramycin/tylosin, and 0.09 and 0.20 mg kg^{-1} for virginiamycin. Campagnolo et al.[309] measured erythromycin macrolides along with five other antibiotics, achieving RIA detection limits of around 10 μg L^{-1}.

8.4.1.7 Other Drugs

This section deals with immunochemical methods developed to determine drugs that do not belong to any of the most common antibiotic family groups described above, but because of their importance and general use require to be considered, too. Table 8.10 shows a few examples of ELISAs for the analysis of these compounds in environmental samples.

A highly sensitive and specific ELISA for the determination, in different types of water samples, of diclofenac, a commonly used nonsteroidal anti-inflammatory drug (NSAID), has been developed by Deng et al.[357] This analyte belongs to the most frequently detected, pharmaceutically active compounds in the water cycle. The immunoassay was able to measure tap water samples directly— respective LOD and IC$_{50}$ values were 6 and 60 ng L^{-1}. On the other hand, surface water samples required fivefold dilution and the wastewater samples 10-fold dilution in buffer to be analyzed correctly; the LODs were then 20 and 60 ng L^{-1}, respectively. Recently, the development and validation of a highly sensitive and specific ELISA for the detection of pharmaceutical indomethacin in

TABLE 8.10
Immunochemical Techniques Developed for the Detection of Other Drugs

Target Analyte	Technique	Matrix	Sensitivity		Reference
			LOD (μg L^{-1})	IC$_{50}$ (μg L^{-1})	
Diclofenac	ELISA	Tap water	6×10^{-3}	60×10^{-3}	357
		Surface water	19×10^{-3}		
		Wastewater	60×10^{-3}		
Indomethacin	ELISA	Tap water	0.01	<0.25	358
(acemetacin 92% CR)		Driking water	0.01	<0.25	
		Surface water	0.01	<0.25	
		Wastewater	0.10	<2.50	
Nitrofurantoin	ELISA	Animal fed water	0.20	3.20	359

water samples from different sites in the Chengdu area was presented by Huo et al.[358] This commonly used compound is also included in the NSAID group. Although indomethacin is considered stable in the environment, its long-term presence in aquatic systems may increase chronic toxicity and more insidious effects, like endocrine disruption, growth inhibition, and cytotoxicity, in aquatic animals. The study measured tap water and drinking water samples directly (LOD = 0.01 µg L^{-1}); the same LOD was obtained for surface water samples after these had been filtered through a 0.45 µm nylon cartridge. Wastewater samples required a 10-fold dilution step prior to analysis with the immunoassay (LOD = 0.1 µg L^{-1}). In all cases, around 90% of acemetacin cross-reacted. Liu et al.[359] prepared polyclonal antibodies for the immunochemical detection of nitrofurantoin residues in water samples. Nitrofurans are a group of synthetic broad-spectrum antibiotics frequently employed in animal production to treat and prevent gastrointestinal infections caused by *Escherichia coli* and *Salmonella*. They are also used as growth promoters in pig, poultry, and fish production. Using the relevant ELISA, LOD and IC$_{50}$ values of 0.20 and 3.20 µg L^{-1}, respectively, were obtained in drinking water fed to animals. A fluorescence-based continuous-flow immunosensor for the sensitive, precise, accurate, and fast determination of paclitaxel was developed by Sheikh and Mulchandani.[360] A natural product, this compound is known to be one of the most active anticancer agents approved by FDA for application in clinical oncology practice. The assay is based on the displacement and detection downstream of rhodamine-labeled paclitaxel by a flow-through spectrofluorometer, as a result of the competition with paclitaxel introduced as a pulse into the stream of carrier buffer flowing through the system. The detection limit found in buffer and human plasma samples was around 4 µg L^{-1}. Finally, a fluorescence immunoassay to detect spectinomycin, which is used as an oral treatment to control bacterial enteritis in pigs and to prevent and control losses due to chronic respiratory disease in chickens, was developed by Medina et al.[361] The antibodies and secondary immunoreagents implemented in the assay enabled an LOD of approximately 5 µg L^{-1} in buffer to be obtained.

8.4.2 STEROID HORMONES

Steroid hormones are a group of biologically active compounds controlling human body functions related mainly to the endocrine and immune systems. Synthesized from cholesterol, they have a cyclopenta-o-perhydrophenanthrene ring in common.[40] Mammalian steroid hormones, which are secreted by the adrenal cortex, testicles, ovary, and placenta, can be classified into different groups, such as estrogens, gestagens, androgens, and glucocorticoids, depending on the intracellular receptor to which they bind in order to become active.[6] Apart from the endogenous hormones, many synthetic steroids have been produced for their high bioactivity. Thus, the consumption of natural and synthetic steroids in human medicine and animal farming has increased steadily in recent decades. On the other hand, humans and animals excrete hormone steroids from their bodies, which readily enter the aquatic environment through sewage discharge and animal waste disposal, mainly via effluents from WWTPs.[4,6] Once in waterways, they may adsorb to solid particles, like bed sediments or soils, where steroids may persist for long periods.[362,363] The increasing number of steroid hormones in the environment may interfere with the normal functioning of endocrine systems, thus affecting reproduction and development in wildlife.[363] Apart from the standard chromatographic techniques, many examples can be found in the literature of the immunochemical determination of steroid residues, mainly in biological samples, but also in environmental matrices.[364,365] Table 8.11 shows some examples of the immunological methods described for these compounds.

8.4.2.1 Estrogens

Estradiol is one of the main female sexual hormones; it is also the structural backbone for the engineering of some synthetic estrogens, such as ethynyl estradiol or mestranol, used in human hormone treatments. Both natural and synthetic estrogens are classified as endocrine disrupting chemicals (EDCs).[6,362] Many of these substances and their metabolites end up in the environment where

TABLE 8.11
Immunochemical Techniques Developed for the Detection of Steroid Hormones

Target Analyte	Technique	Matrix	Sensitivity		Reference
			LOD	IC$_{50}$	
Estradiol	ELISA[a] (Abraxis, USA)	Urban wastewater River and groundwater	0.05 µg L^{-1}		370
	SPEoptical sensor	Seawater (sewage plants) Seawater	1.5 µg L^{-1} 0.16 µg L^{-1}		372
	SPE-CLEIA	Tap and wastewater	1.5 µg L^{-1}		369
	TIRF	Wastewater	0.16 µg L^{-1}	1.84 µg L^{-1}	371
	ETIA	Wastewater	0.85 µg L^{-1}	1.2 µg L^{-1}	371
Estriol	ELISA[a] (Abraxis, USA)	Urban wastewater River and groundwater	0.05 µg L^{-1}		370
Estrone	ELISA[a] (Abraxis, USA)	Urban wastewater River and groundwater	0.05 µg L^{-1}		370
	TIRF	River water Groundwater	0.08 µg L^{-1} 0.08 µg L^{-1}	0.53 µg L^{-1} 0.56 µg L^{-1}	424
	TIRF	Wastewater	0.01 µg L^{-1}	0.51 µg L^{-1}	371
	ETIA	Wastewater	0.50 µg L^{-1}	0.81 µg L^{-1}	371
Ethynylestradiol	ELISA[a] (Abraxis, USA)	Urban wastewater River and groundwater	0.05 µg L^{-1}		370
	TIRF	Wastewater	0.07 µg L^{-1}	1.07 µg L^{-1}	371
	ETIA	Wastewater	0.01 µg L^{-1}	2.70 µg L^{-1}	371
Noresthindrone	EIA	River and potable water	10 ng L^{-1}		386
Progesterone	RIA	River and potable water	5 ng L^{-1}		386
	CLEIA	River and potable water	15 pg per tube		385
	TIRF	MilliQ water	0.96 ng L^{-1}		364
Testosterone	RIANA	Drinking water River water	0.2 ng L^{-1} 0.2 ng L^{-1}		375

Notes: SPE, solid-phase extraction.
[a] Commercial kit.

they can have adverse effects on wildlife organisms even at very low concentrations.[366–368] Thus, environmental monitoring programs on estrogens call for analytical techniques capable of achieving very low detection limits. With regard to immunochemical methods, there are several commercial tests on the market, addressed mainly to food residue analysis. Nevertheless, application to the analysis of aquatic ecosystems should be easy to implement. Zhao et al.[369] developed a chemiluminescence enzyme immunoassay (CLEIA) for the determination of 17β-estradiol in wastewater samples; the working linear range obtained was from 2.5 to 1600 ng L^{-1}, with a detection limit of 1.5 ng L^{-1}. Recoveries of spiked tap water and wastewater samples at 0, 2.5, 10, and 50 ng L^{-1} were in the 80–110% range. Results were compared with the commercially available radioimmunoassay kit (MARCA): a good correlation ($R^2 = 0.997$) was obtained. Farré et al.[76] evaluated four different commercially available ELISAs for the rapid screening of estrogens in different water matrices, including natural and spiked samples from urban wastewater, river water, and groundwater from the vicinity of Barcelona. ELISA kits[370] were configured to measure 40 samples per plate with sufficient sensitivity, high cross-reactivity with other congeners, and reproducibility. All the samples extracted by SPE yielded recoveries from 79% to 86%; assay validation was carried out by comparison with high performance liquid chromatography tandem mass spectrometry (HPLC-MS/MS) using a triple quadrupole (QqQ) instrument. Coille et al.[371] described the use of two fluorescence immunochemical methods to detect different estrogenic compounds in synthetic wastewater. In the first one, the immunosensor is based on the internal reflection fluorescence (TIRF) principle (LOD for estradiol = 0.16 µg L^{-1} and for estrone and ethynylestradiol = 0.07 µg L^{-1}). The other method is an energy transfer immunoassay (ETIA), in which the specific antibody is labeled to a donor fluorescent dye, while the antigen is coupled to an acceptor dye via a bovine serum albumin (BSA) molecule. The fluorescence is quenched as a consequence of the biorecognition reaction. In this case, the respective detection limits obtained were 0.5, 0.85, and 0.01 µg L^{-1} for estrone, estradiol, and ethylestradiol. Recovery rates for both techniques when measuring spiked wastewater samples were between 70% and 112%. Recently, Zhang et al.[372] have developed a sensitive and simple immunoassay based on the SPR technique for monitoring 17β-estradiol. Previous to their analysis, seawater samples were hydrolyzed with HCl/methanol solution and preconcentrated using C18 SPE; recovery values were approximately 92%. Subsequent studies showed that the precision and repeatability of the SPR assay were good; cross-reactivity with other estrogens was very low.

8.4.2.2 Androgens

The main applications of immunochemical techniques for androgens are in the doping control of athletes, forensic chemistry, farm animals for human consumption, and food analysis.[373,374] However, there have been a few applications in the environmental field for natural or synthetic androgens. For example, Barel-Cohen et al.[4] monitored natural steroids in sewage and fishpond effluents, finding levels of testosterone between 2.1 and 7.8 ng L^{-1} at different collecting points along a river. A TIRF immunosensor has been developed for reliable sub-ng L^{-1} detection of testosterone in aquatic environmental matrices without sample pretreatment. Thus, direct analysis of spiked lab water, drinking water, and river water samples gave an LOD of 0.2 ng L^{-1} with recovery rates between 70% and 120%.[375] This sensor system was therefore shown to be a suitable warning tool in environmental analysis, in addition to the standard analytical methods. On the other hand, several ELISA procedures have been developed for the analysis of testosterone and related compounds in biological samples.[365,376] In the case of performance-enhancing anabolic steroids, such as stanozolol, nandrolone, and the recently designed steroid known as tetrahydrogestrinone, immunochemical assays have also been performed on equine and human urine samples during doping controls.[377–379] Additionally, powerful techniques combining multiresidue immunoaffinity chromatography with GC-MS or ELISAs are available for the simultaneous identification and semiquantification of various androgens in samples of urine and feces.[39,378,379] Figure 8.7 shows a scheme of a multi-immunoaffinity chromatography (multi-IAC) procedure.

Immunosensors have made a great contribution in the field of androgenic steroid detection, giving detection limits comparable to those obtained with standard ELISA procedures. Several electrochemical immunosensors have been developed for detecting testosterone, methyltestosterone,

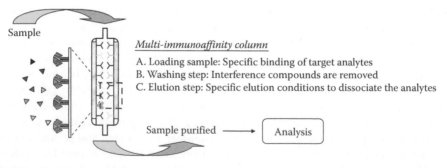

Sample

Multi-immunoaffinity column

A. Loading sample: Specific binding of target analytes
B. Washing step: Interference compounds are removed
C. Elution step: Specific elution conditions to dissociate the analytes

Sample purified ⟶ Analysis

FIGURE 8.7 Schematic sequential step procedure for chromatographic determination after multi-IAC purification.

19-nortestosterone, boldenone, and methylboldenone in spiked bovine urine using screen-printed electrodes.[380–382] Once again, the techniques already used to analyze these compounds in complex biological matrices can be adapted for the analysis of environmental samples.

8.4.2.3 Gestagens

Gestagens are hormones that produce similar effects to those of endogenous progesterone, which is excreted from the ovary to act as a balancer in the menstrual cycle, pregnancy, and embryogenesis. As with estrogens, most of the immunochemical techniques described in the literature have been developed to analyze these compounds in biological matrices.[383–385] On the other hand, Aherne et al. described a RIA for progesterone and norethindrone detection in water samples, obtaining LODs of 6 and 17 ng L^{-1}, respectively.[386] This group also reported that norethindrone underwent 28% biodegradation in an activated sludge system in 6 h and was completely degraded in one day, which is the time required to purify river water for drinking purposes. Furthermore, Käppel et al.[364] developed an immunosensor based on TIRF detection to analyze progesterone in spiked MilliQ water. The assay was optimized to obtain results in 5 min with an LOD of 0.96 ng L^{-1}. All these results show fairly well the usefulness of immunochemical techniques for the determination of micro- and nanogram per liter quantities of gestagens in aqueous samples.[387]

8.4.2.4 Corticosteroids

Endogenous corticosteroids are produced by the adrenal glands in response to stressors such as exercise, illness, and starvation.[5] Synthetic cortisone derivatives were synthesized in the late 1940s for therapeutic purposes. Lately, these products have found their way into the world of sports because of their anti-inflammatory properties, but they are now on the list of substances banned by the International Olympic Committee (IOC). Moreover, corticosteroids like dexamethasone are used not only in veterinary practice for the treatment of respiratory and gastrointestinal disorders, but also as illegal growth promoters in animal feedstuffs. To control this undesirable situation, efficient screening procedures, based mainly on ELISA methods, have been described for the analysis of most important corticosteroids in biological matrices,[388–390] but they have not yet been applied to environmental samples. Pujos[390] reported the analysis of 18 human corticosteroids, both endogenous and synthetic, in spiked urine samples. The samples required a pretreatment based on simple 1/50 dilution. ELISA is a suitable technique for the systematic detection of corticosteroids by the food and agriculture industries in many different sample matrices;[391–393] their implementation in the analysis of aquatic environmental samples seems appropriate.

8.5 GENERAL SUMMARY

In recent decades, immunochemical techniques have been widely demonstrated to be an interesting alternative to the more conventional analytical methodologies in many areas, but additional work is still necessary to completely adapt them to the analysis of environmental contaminants. On the other hand, considering that the analysis of very complex biological samples with these methods has been successful, the prospects for their application to the analysis of water and soil samples seem highly promising. In this relatively new situation, where data on the occurrence, risk assessment, and environmental toxicity of most of these emerging pollutants are not available, collaboration and interchange of expertise between analytical and immunochemists are needed to achieve this objective. The benefits accruing from these methods (i.e., high sensitivity, selectivity, cost-effectiveness, high sample processing capabilities *vis-à-vis* target analytes) are nowadays available for assessing risk and protecting public health from the adverse effects of these types of pollutants. From now on, research efforts should focus on the development of multianalyte immunochemical systems, in which more than one compound or group of compounds can be detected simultaneously, and on the design of new analytical user-friendly devices (i.e., immunosensors) for continuous or on-site measurements. Technical development should be accompanied by some officially organized efforts to find ways of validating screening immunoassay techniques and recognizing them as practicable routine methods in environmental monitoring laboratories. For the time being, directives are issued, regulations enacted, and conferences held to ensure water quality, protect water resources, and ensure the good health of the entire environment.

ACKNOWLEDGMENTS

This work has been supported by the Ministry of Science and Technology (Contract numbers AGL2005-07700-C06-01, NAN2004-09195-C04-04, and NAN2004-09415-C05-02). The AMR group is a Group de Recerca de la Generalitat de Catalunya and has support from the Departament d'Universitats, Recerca i Societat de la Informació de la Generalitat de Catalunya (expedient 2005SGR 00207).

REFERENCES

1. Xia, K., A. Bhandari, K. Das, et al. 2005. Occurrence and fate of pharmaceuticals and personal care products (PPCPs) in biosolids. *J. Environ. Qual.* 34: 91–104.
2. Kolpin, D.W., E.T. Furlong, M.T. Meyer, et al. 2002. Pharmaceuticals, hormones, and other organic wastewater contaminants in U.S. streams, 1999–2000: A national reconnaissance. *Environ. Sci. Technol.* 36: 1202–1211.
3. Rodriguez-Mozaz, S., M.J.L. de Alda, and D. Barceló. 2006. Biosensors as useful tools for environmental analysis and monitoring. *Anal. Bioanal. Chem.* 386: 1025–1041.
4. Barel-Cohen, K., L.S. Shore, M. Shemesh, et al. 2006. Monitoring of natural and synthetic hormone in polluted river. *J. Environ. Manage.* 78: 16–23.
5. Nozaki, O. 2001. Steroid analysis for medical diagnosis. *J. Chromatogr. A* 935: 267–278.
6. Ying, G.-G., R.S. Kookana, and Y.-J. Ru. 2002. Occurrence and fate of hormone steroids in the environment. *Environ. Int.* 28: 545–551.
7. IARC (International Agency for research on cancer). 1985. Polynuclear Aromatic compounds: Bituminous, coal tar and derived products, shale oils and soots. *Monographs on the Evaluation of the Carcinogenic Risks of Chemicals to Humans*, 35: 4.
8. Wegener, H.C. 2003. Antibiotics in animal feed and their role in resistance development. *Curr. Opin. Microbiol.* 6: 439–445.
9. Sumpter, J.P. and S. Jobling. 1995. Vitellogenesis as a biomarker for estrogenic contamination of the aquatic environment. *Environ. Health Perspect.* 103: 174–178.
10. Gowik, P., B. Jülicher, and S. Uhlig. 1998. Multi-residue method for non-steroidal anti-inflammatory drugs in plasma using high-performance liquid chromatography-photodiode-array detection: Method description and comprehensive in-house validation. *J. Chromatogr. B* 716: 221–232.
11. Cohen, E., R.J. Maxwell, and D.J. Donoghue. 1999. Automated multi-residue isolation of fluoroquinolone antimicrobials from fortified and incurred chicken liver using on-line microdialysis and high-performance liquid chromatography with programmable fluorescence detection. *J. Chromatogr. B* 724: 137–145.
12. Koole, A., J.-P. Franke, and R.A. de Zeeuw. 1999. Multi-residue analysis of anabolics in calf urine using high-performance liquid chromatography with diode-array detection. *J. Chromatogr. B* 724: 41–51.
13. Petrovic, M., M. Gros, and D. Barceló. 2006. Multi-residue analysis of pharmaceuticals in wastewater by ultra-performance liquid chromatography-quadrupole-time-of-flight mass spectrometry. *J. Chromatogr. A* 1124: 68–81.
14. Kruve, A., A. Künnapas, K. Herodes, et al. 2008. Matrix effects in pesticide multi-residue analysis by liquid chromatography-mass spectrometry. *J. Chromatogr. A* 1187: 58–66.
15. Gerlach, R.W., M.S. Gustin, and J.M. Van Emon. 2001. On-site mercury analysis of soil at hazardous waste sites: by immunoassay and ASV. *Appl. Geochem.* 16: 281–290.
16. Shan, G., W.R. Leeman, S.J. Gee, et al. 2001. Highly sensitive dioxin immunoassay and its application to soil and biota samples. *Anal. Chim. Acta* 444: 169–178.
17. Bogdanovic, J., M. Koets, I. Sander, et al. 2006. Rapid detection of fungal [alpha]-amylase in the work environment with a lateral flow immunoassay. *J. Allergy Clin. Immunol.* 118: 1157–1163.
18. Kim, S.J., K.V. Gobi, R. Harada, et al. 2006. Miniaturized portable surface plasmon resonance immunosensor applicable for on-site detection of low-molecular-weight analytes. *Sens. Actuat. B: Chem.* 115: 349–356.

19. Centi, S., E. Silva, S. Laschi, et al. 2007. Polychlorinated biphenyls (PCBs) detection in milk samples by an electrochemical magneto-immunosensor (EMI) coupled to solid-phase extraction (SPE) and disposable low-density arrays. *Anal. Chim. Acta* 594: 9–16.

20. Zhao, L.X. and J.M. Lin. 2005. Development of a micro-plate magnetic chemiluminescence enzyme immunoassay (MMCLEIA) for rapid- and high-throughput analysis of 17 beta-estradiol in water samples. *J. Biotechnol.* 118: 177–186.

21. Nichkova, M., M. Germani, and M.P. Marco. 2008. Immunochemical analysis of 2,4,6-Tribromophenol for assessment of wood contamination. *J. Agric. Food Chem.* 56: 29–34.

22. Adrian, J., D.G. Pinacho, B. Granier, et al. 2008. A multianalyte ELISA for immunochemical screening of sulfonamide, fluoroquinolone and ß-lactam antibiotics in milk samples using class-selective bioreceptors. *Anal. Bioanal. Chem.* 391: 1703–1712.

23. Li, Z., S. Wang, N. Alice Lee, et al. 2004. Development of a solid-phase extraction-enzyme-linked immunosorbent assay method for the determination of estrone in water. *Anal. Chim. Acta* 503: 171–177.

24. Situ, C., E. Grutters, P. van Wichen, et al. 2006. A collaborative trial to evaluate the performance of a multi-antibiotic enzyme-linked immunosorbent assay for screening five banned antimicrobial growth promoters in animal feeding stuffs. *Anal. Chim. Acta* 561: 62–68.

25. Farré, M., M. Kuster, R. Brix, et al. 2007. Comparative study of an estradiol enzyme-linked immunosorbent assay kit, liquid chromatography-tandem mass spectrometry, and ultra performance liquid chromatography-quadrupole time of flight mass spectrometry for part-per-trillion analysis of estrogens in water samples. *J. Chromatogr. A* 1160: 166–175.

26. Jeon, M., J. Kim, K.-J. Paeng, et al. 2008. Biotin-avidin mediated competitive enzyme-linked immunosorbent assay to detect residues of tetracyclines in milk. *Microchem. J.* 88: 26–31.

27. Badihi-Mossberg, M., V. Buchner, and J. Rishpon. 2007. Electrochemical biosensors for pollutants in the environment. *Electroanalytical* 19: 2015–2028.

28. Daniels, J.S. and N. Pourmand. 2007. Label-free impedance biosensors: Opportunities and challenges. *Electroanalytical* 19: 1239–1257.

29. Gonzalez-Martinez, M.A., R. Puchades, and A. Maquieira. 2007. Optical immunosensors for environmental monitoring: How far have we come? *Anal. Bioanal. Chem.* 387: 205–218.

30. Marco, M.-P. and D. Barceló. 2000. Immunosensors for environmental analysis. In: D. Barceló (ed.), *Sample Handling and Trace Analysis of Pollutants Techniques, Applications and Quality Assurance*, pp. 1075–1103. Amsterdam: Elsevier.

31. Merkoci, A. 2007. Electrochemical biosensing with nanoparticles. *FEBS J.* 274: 310–316.

32. Rodriguez-Mozaz, S., M.P. Marco, M.J.L. de Alda, et al. 2004. Biosensors for environmental monitoring of endocrine disruptors: A review article. *Anal. Bioanal. Chem.* 378: 588–598.

33. Shankaran, D.R., K.V. Gobi, and N. Miura. 2007. Recent advancements in surface plasmon resonance immunosensors for detection of small molecules of biomedical, food and environmental interest. *Sens. Actuat. B: Chem.* 121: 158–177.

34. Grego, S., J.R. McDaniel, and B.R. Stoner. 2008. Wavelength interrogation of grating-based optical biosensors in the input coupler configuration. *Sens. Actuat. B: Chem.* 131: 347–355.

35. Leung, A., P.M. Shankar, and R. Mutharasan. 2007. A review of fiber-optic biosensors. *Sens. Actuat. B: Chem.* 125: 688–703.

36. Li, X.-M., X.-Y. Yang, and S.-S. Zhang. 2008. Electrochemical enzyme immunoassay using model labels. *Trends Anal. Chem.* 27: 543–553.

37. Pingarrón, J.M., P. Yáñez-Sedeño, and A. González-Cortés. 2008. Gold nanoparticle-based electrochemical biosensors. *Electrochim. Acta* 53: 5848–5866.

38. Pumera, M., S. Sánchez, I. Ichinose, et al. 2007. Electrochemical nanobiosensors. *Sens. Actuat. B: Chem.* 123: 1195–1205.

39. Salvador, J.-P., J. Adrian, R. Galve, et al. 2007. Application of bioassays/biosensors for the analysis of pharmaceuticals in environmental samples. In: M. Petrovic and D. Barceló (eds), *Analysis, Fate and Removal of Pharmaceuticals in the Water Cycle*, pp. 279–334. Amsterdam: Elsevier.

40. Estévez, M.C., H. Font, M. Nichkova, et al. 2005. Immunochemical determination of pharmaceuticals and personal care products as emerging pollutants. In: D. Barceló (ed.), *Emerging Organic Pollutants in Waste Waters and Sludge*, pp. 181–244. Berlin: Springer.

41. Jiang, X., D. Li, X. Xu, et al. 2008. Immunosensors for detection of pesticide residues. *Biosens. Bioelectron.* 23: 1577–1587.

42. Oubiña, A., B. Ballesteros, P. Bou, et al. 2000. Immunoassays for environmental analysis. In: D. Barceló (ed.), *Sample Handling and Trace Analysis of Pollutants Techniques, Applications and Quality Assurance*, pp. 289–340. Amsterdam: Elsevier.

43. Picó, Y., M. Fernández, M.J. Ruiz, et al. 2007. Current trends in solid-phase-based extraction techniques for the determination of pesticides in food and environment. *J. Biochem. Biophys. Methods* 70: 117–131.

44. Farré, M., L. Kantiani, and D. Barcelo. 2007. Advances in immunochemical technologies for analysis of organic pollutants in the environment. *Trends Anal. Chem.* 26: 1100–1112.

45. Schnelle-Kreis, J., I. Gebefugi, G. Welzl, et al. 2001. Occurrence of particle-associated polycyclic aromatic compounds in ambient air of the city of Munich. *Atmos. Environ.* 35(Suppl.): 71–81.

46. Negrao, M.R. and M.F. Alpendurada. 1998. Solvent-free method for the determination of polynuclear aromatic hydrocarbons in waste water by solid-phase microextraction-high-performance liquid chromatography with photodiode-array detection. *J. Chromatogr. A* 823: 211–218.

47. Tuhackova, J., T. Cajthaml, K. Novak, et al. 2001. Hydrocarbon deposition and soil microflora as affected by highway traffic. *Environ. Pollut.* 113: 255–262.

48. Bouzige, M., V. Pichon, and M.C. Hennion. 1998. On-line coupling of immunosorbent and liquid chromatographic analysis for the selective extraction and determination of polycyclic aromatic hydrocarbons in water samples at the ng l^{-1} level. *J. Chromatogr. A* 823: 197–210.

49. Knopp, D., M. Seifert, V. Vaananen, et al. 2000. Determination of polycyclic aromatic hydrocarbons in contaminated water and soil samples by immunological and chromatographic methods. *Environ. Sci. Technol.* 34: 2035–2041.

50. Li, K., L.A. Woodward, A.E. Karu, et al. 2000. Immunochemical detection of polycyclic aromatic hydrocarbons and 1-hydroxypyrene in water and sediment samples. *Anal. Chim. Acta* 419: 1–8.

51. Matschulat, D., A.P. Deng, R. Niessner, et al. 2005. Development of a highly sensitive monoclonal antibody based ELISA for detection of benzo[a] pyrene in potable water. *Analyst.* 130: 1078–1086.

52. U.S. EPA (SW-846) (Environmental Protection Agency). 1995. (ed.). *Method 4035, Soil Screening for Polynuclear Aromatic Hydrocarbons by Immunoassay*, 3rd edition. *Test Methods for Evaluating Solid Waste, Physical/Chemical Methods*, Vols. I and II, pp. 1–10. Washington DC.

53. Bowadt, S., L. Mazeas, D.J. Miller, et al. 1997. Field-portable determination of polychlorinated biphenyls and polynuclear aromatic hydrocarbons in soil using supercritical fluid extraction. *J. Chromatogr. A* 785: 205–217.

54. Chuang, J.C., J.M. Van Emon, Y.-L. Chou, et al. 2003. Comparison of immunoassay and gas chromatography-mass spectrometry for measurement of polycyclic aromatic hydrocarbons in contaminated soil. *Anal. Chim. Acta* 486: 31–39.

55. Fillmann, G., M.C. Bicego, A. Zamboni, et al. 2007. Validation of immunoassay methods to determine hydrocarbon contamination a in estuarine sediments. *J. Brazil. Chem. Soc.* 18: 774–781.

56. Nording, M., K. Frech, Y. Persson, et al. 2006. On the semi-quantification of polycyclic aromatic hydrocarbons in contaminated soil by an enzyme-linked immunosorbent assay kit. *Anal. Chim. Acta* 555: 107–113.

57. Barceló, D., A. Oubina, J.S. Salau, et al. 1998. Determination of PAHs in river water samples by ELISA. *Anal. Chim. Acta* 376: 49–53.

58. Fahnrich, K.A., M. Pravda, and G.G. Guilbault. 2003. Disposable amperometric immunosensor for the detection of polycyclic aromatic hydrocarbons (PAHs) using screen-printed electrodes. *Biosens. Bioelectron.* 18: 73–82.

59. Moore, E.J., M.P. Kreuzer, M. Pravda, et al. 2004. Development of a rapid single-drop analysis biosensor for screening of phenanthrene in water samples. *Electroanalytical* 16: 1653–1659.

60. Gobi, K.V., S.J. Kim, H. Tanaka, et al. 2007. Novel surface plasmon resonance (SPR) immunosensor based on monomolecular layer of physically-adsorbed ovalbumin conjugate for detection of 2,4-dichlorophenoxyacetic acid and atomic force microscopy study. *Sens. Actuat. B: Chem.* 123: 583–593.

61. Liu, M., Q.X. Li, and G.A. Rechnitz. 1999. Flow injection immunosensing of polycyclic aromatic hydrocarbons with a quartz crystal microbalance. *Anal. Chim. Acta* 387: 29–38.

62. Thomas, S.D. and Q.X. Li. 2000. Immunoaffinity chromatography for analysis of polycyclic aromatic hydrocarbons in corals. *Environ. Sci. Technol.* 34: 2649–2654.

63. Goda, Y., A. Kobayashi, S. Fujimoto, et al. 2004. Development of enzyme-linked immunosorbent assay for detection of alkylphenol polyethoxylates and their biodegradation products. *Water Res.* 38: 4323–4330.

64. Salaguer, J.-L. 2002. Surfactants, types and uses. Teaching Aid in Surfactant Science & Engineering. p. 49.

65. Lunar, L., S. Rubio, and D. Perez-Bendito. 2004. Differentiation and quantification of linear alkyl benzenesulfonate isomers by liquid chromatography-ion-trap mass spectrometry. *J. Chromatogr. A* 1031: 17–25.

66. Lara-Martin, P. 2005. Determination and distribution of alkyl, ethoxysulfates and linear alkylbenzene sulfonates in coastal marine sediments from the Bay of Cadiz (Southwest of Spain). *Environ. Toxicol. Chem.* 24: 2196–2202.

67. Waters, J. and T.C.J. Feijte. 1995. AIS+/CESIO+ Environmental surfactant monitoring programme: Outcome of five national pilot studies on linear alkylbenzene sulphonate (LAS). *Chemosphere* 30: 1939–1956.

68. Marcomini, A. 2000. Behavior of anionic and nonionic surfactants and their persistent metabolites in the Venice lagoon, Italy. *Environ. Toxicol. Chem.* 19: 2000–2007.

69. Álvarez-Muñoz, D., P.A. Lara-Martín, J. Blasco, et al. 2007. Presence, biotransformation and effects of sulfophenylcarboxylic acids in the benthic fish *Solea senegalensis. Environ. Int.* 33: 565–570.

70. Nomura, Y., K. Ikebukuro, K. Yokoyama, et al. 1998. Application of a linear alkylbenzene sulfonate bio-sensor to river water monitoring. *Biosens. Bioelectron.* 13: 1047–1053.

71. Ahel, M., C. Schaffner, and W. Giger. 1996. Behaviour of alkylphenol polyethoxylate surfactants in the aquatic environmentIII. Occurrence and elimination of their persistent metabolites during infiltration of river water to groundwater. *Water Res.* 30: 37–46.

72. Mart'ianov, A.A., B.B. Dzantiev, A.V. Zherdev, et al. 2005. Immunoenzyme assay of nonylphenol: Study of selectivity and detection of alkylphenolic non-ionic surfactants in water samples. *Talanta* 65: 367–374.

73. OSPAR Commission for the Protection of the Marine Environment of the North-East Atlantic. 2008. Available at http://www.ospar.org/.

74. de Voogt, P., K. de Beer, and F. van der Wielen. 1997. Determination of alkylphenol ethoxylates in indus-trial and environmental samples. *Trends Anal. Chem.* 16: 584–595.

75. Wie, S.I. and B.D. Hammock. 1982. The use of enzyme-linked immunosorbent assays (ELISA) for the determination of Triton X nonionic detergents. *Anal. Biochem.* 125: 168–176.

76. Farré, M., J. Ramon, R. Galve, et al. 2006. Evaluation of a newly developed enzyme-linked immunosor-bent assay for determination of linear alkyl benzenesulfonates in wastewater treatment plants. *Environ. Sci. Technol.* 40: 5064–5070.

77. Ramón-Azcón, J., R. Galve, F. Sanchez-Baeza, et al. 2006. Development of an enzyme-linked immuno-sorbent assay for the determination of the linear alkylbenzene sulfonates and long-chain sulfophenyl carboxylates using antibodies generated by pseudoheterologous immunization. *Anal. Chem.* 78: 71–81.

78. Estevez, M.C., R. Galve, F. Sanchez-Baeza, et al. 2008. Disulfide symmetric dimers as stable pre-hapten forms for bioconjugation: A strategy to prepare immunoreagents for the detection of sulfophenyl car-boxylate residues in environmental samples. *Chem. Eur. J.* 14: 1906–1917.

79. Zhang, R., K. Hirakawa, D. Seto, et al. 2005. Sequential injection chemiluminescence immunoassay for anionic surfactants using magnetic microbeads immobilized with an antibody. *Talanta* 68: 231–238.

80. Sánchez-Martínez, M., M. Aguilar-Caballos, S. Eremin, et al. 2006. Long-wavelength fluorimetry as an indirect detection system in immunoaffinity chromatography: Application to environmental analysis. *Anal. Bioanal.Chem.* 386: 1489–1495.

81. Sanchez-Martinez, M.L., M.P. Aguilar-Caballos, S.A. Eremin, et al. 2005. Determination of linear alky-lbenzenesulfonates in water samples by immunoaffinity chromatography with fluorescence detection. *Anal. Chim. Acta* 553: 93–98.

82. Estevez, M.C., M. Kreuzer, F. Sanchez-Baeza, et al. 2006. Analysis of nonylphenol: Advances and improvements in the immunochemical determination using antibodies raised against the technical mix-ture and hydrophilic immunoreagents. *Environ. Sci. Technol.* 40: 559–568.

83. Mart'ianov, A.A., A.V. Zherdev, S.A. Eremin, et al. 2004. Preparation of antibodies and development of enzyme-linked immunosorbent assay for nonylphenol. *Int. J. Environ. Anal. Chem.* 84: 965–978.

84. Estevez-Alberola, M.C. and M.P. Marco. 2004. Immunochemical determination of xenobiotics with endocrine disrupting effects. *Anal. Bioanal. Chem.* 378: 563–575.

85. Commission of the European Communities. 2000. Proposal for a directive of the European Parliament and of the Council on Environmental Quality Standards in the field of water policy and amending Directive 2000/60/EC.

86. Zhang, R., H. Nakajima, N. Soh, et al. 2007. Sequential injection chemiluminescence immunoassay for nonionic surfactants by using magnetic microbeads. *Anal. Chim. Acta* 600: 105–113.

87. Cespedes, R., K. Skryjova, M. Rakova, et al. 2006. Validation of an enzyme-linked immunosorbent assay (ELISA) for the determination of 4-nonylphenol and octylphenol in surface water samples by LC-ESI-MS. *Talanta* 70: 745–751.

88. Matsunaga, T., F. Ueki, K. Obata, et al. 2003. Fully automated immunoassay system of endocrine dis-rupting chemicals using monoclonal antibodies chemically conjugated to bacterial magnetic particles. *Anal. Chim. Acta* 475: 75–83.

89. Goda, Y., M. Hirobe, A. Kobayashi, et al. 2005. Production of a monoclonal antibody and development of enzyme-linked immunosorbent assay for alkyl ethoxylates. *Anal. Chim. Acta* 528: 47–54.

90. Samsonova, J.V., N.A. Uskova, A.N. Andresyuk, et al. 2004. Biacore biosensor immunoassay for 4-nonylphenols: Assay optimization and applicability for shellfish analysis. *Chemosphere* 57: 975–985.

91. CIRCA (Communication Information Resource Centre Administration) 2008. Implementing the Water Framework Directive, European Commission. Available at http://circa.europa.eu/Public/irc/env/wfd/information.

92. Rose, A., C. Nistor, J. Emneus, et al. 2002. GDH biosensor based off-line capillary immunoassay for alkylphenols and their ethoxylates. *Biosens. Bioelectron.* 17: 1033–1043.

93. Evtugyn, G.A., S.A. Eremin, R.P. Shaljamova, et al. 2006. Amperometric immunosensor for nonylphenol determination based on peroxidase indicating reaction. *Biosens. Bioelectron.* 22: 56–62.

94. Bull, J.P., A.N. Serreqi, T. Chen, et al. 1998. Development of an immunoassay for a quaternary ammonium compound, benzyldimethyldodecylammonium chloride. *Water Res.* 32: 3621–3630.

95. Safe, S. 1990. Polychlorinated-biphenyls (PCBs), dibenzo-para-dioxins (PCDDS), dibenzofurans (PCDFS), and related-compounds—environmental and mechanistic considerations which support the development of toxic equivalency factors (TEFS). *Crit. Rev. Toxicol.* 21: 51–88.

96. Alcock, R.E., P.A. Behnisch, K.C. Jones, et al. 1998. Dioxin-like PCBs in the environment—human exposure and the significance of sources. *Chemosphere* 37: 1457–1472.

97. Birnbaum, L.S., D.F. Staskal, and J.J. Diliberto. 2003. Health effects of polybrominated dibenzo-*p*-dioxins (PBDDs) and dibenzofurans (PBDFs). *Environ. Int.* 29: 855–860.

98. Srogi, K. 2008. Levels and congener distributions of PCDDs, PCDFs and dioxin-like PCBs in environmental and human samples: A review. *Environ. Chem. Lett.* 6: 1–28.

99. Harner, T. and M. Shoeib. 2002. Measurements of octanol-air partition coefficients (KOA) for polybrominated diphenyl ethers (PBDEs): Predicting partitioning in the environment. *J. Chem. Eng. Data* 47: 228–232.

100. Gdaniec-Pietryka, M., L. Wolska, and J. Namiesnik. 2007. Physical speciation of polychlorinated biphenyls in the aquatic environment. *Trends Anal. Chem.* 26: 1005–1012.

101. U.S. EPA (Environmental Protection Agency). 1995. National Primary Drinking Water Regulations: CFR 40: Parts 141–143.

102. U.S. EPA (Environmental Protection Agency). 2008. Polychlorinated Biphenyls (PCBs) Manufacturing, Processing, Distribution in Commerce, and Use Prohibitions: CFR 40: Part 761.

103. U.S. EPA (Environmental Protection Agency). 1996. Method 4020-Screening for Polychlorinated Biphenyls by Immunoassay.

104. Fillmann, G., T.S. Galloway, R.C. Sanger, et al. 2002. Relative performance of immunochemical (enzyme-linked immunosorbent assay) and gas chromatography-electron-capture detection techniques to quantify polychlorinated biphenyls in mussel tissues. *Anal. Chim. Acta* 461: 75–84.

105. Lawruk, T.S., C.E. Lachman, S.W. Jourdan, et al. 1996. Quantitative determination of PCBs in soil and water by a magnetic particle-based immunoassay. *Environ. Sci. Technol.* 30: 695–700.

106. Tsutsumi, T., Y. Amakura, A. Okuyama, et al. 2006. Application of an ELISA for PCB 118 to the screening of dioxin-like PCBs in retail fish. *Chemosphere* 65: 467–473.

107. Zhao, C.Q., N.A. Anis, K.R. Rogers, et al. 1995. Fiber optic immunosensor for polychlorinated biphenyls. *J. Agric. Food Chem.* 43: 2308–2315.

108. Centi, S., S. Laschi, M. Franek, et al. 2005. A disposable immunomagnetic electrochemical sensor based on functionalised magnetic beads and carbon-based screen-printed electrodes (SPCEs) for the detection of polychlorinated biphenyls (PCBs). *Anal. Chim. Acta* 538: 205–212.

109. Centi, S., S. Laschi, and M. Mascini. 2007. Improvement of analytical performances of a disposable electrochemical immunosensor by using magnetic beads. *Talanta* 73: 394–399.

110. Pribyl, J., M. Hepel, and P. Skládal. 2006. Piezoelectric immunosensors for polychlorinated biphenyls operating in aqueous and organic phases. *Sens. Actuat. B: Chem. Spec. Issue* 113: 900–910.

111. Concejero, M.A., R. Galve, B. Herradon, et al. 2001. Feasibility of high-performance immunochromatography as an isolation method for PCBs and other dioxin-like compounds. *Anal. Chem.* 73: 3119–3125.

112. Endo, T., A. Okuyama, Y. Matsubara, et al. 2005. Fluorescence-based assay with enzyme amplification on a micro-flow immunosensor chip for monitoring coplanar polychlorinated biphenyls. *Anal. Chim. Acta* 531: 7–13.

113. Shimomura, M., Y. Nomura, W. Zhang, et al. 2001. Simple and rapid detection method using surface plasmon resonance for dioxins, polychlorinated biphenyls and atrazine. *Anal. Chim. Acta* 434: 223–230.

114. Bumb, R.R., W.B. Crummett, S.S. Cutie, et al. 1980. Trace chemistries of fire: A source of chlorinated dioxins. *Sci. Total Environ.* 210: 385–390.

115. Rappe, C., L.-O. Kjeller, S.-E. Kulp, et al. 1991. Levels, profile and pattern of PCDDs and PCDFs in samples related to the production and use of chlorine. *Chemosphere* 23: 1629–1636.

116. Schroy, J.M., F.D. Hileman, and S.C. Cheng. 1985. Physical/chemical properties of 2,3,7,8-TCDD. *Chemosphere* 14: 877–880.

117. U.S. EPA (Environmental Protection Agency). 2002. Method 4025-screening for polychlorinated dibenzodioxins and polychlorinated dibenzofurans (PCDD/Fs) by immunoassay. Available at http://www.epa.gov/SW-846/pdfs/4025.pdf.

118. Harrison, R.O. and R.E. Carlson. 1997. An immunoassay for TEQ screening of dioxin/furan samples: Current status of assay and applications development. *Chemosphere* 34: 915–928.

119. Stanker, L.H., B. Watkins, N. Rogers, et al. 1987. Monoclonal antibodies for dioxin: Antibody characterization and assay development. *Toxicology* 45: 229–243.

120. Watkins, B.E., L.H. Stanker, and M. Vanderlaan. 1989. An immunoassay for chlorinated dioxins in soils. *Chemosphere* 19: 267–270.

121. Nording, M., M. Nichkova, E. Spinnel, et al. 2006. Rapid screening of dioxin-contaminated soil by accelerated solvent extraction/purification followed by immunochemical detection. *Anal. Bioanal. Chem.* 385: 357–366.

122. Huwe, J.K., W.L. Shelver, L. Stanker, et al. 2001. On the isolation of polychlorinated dibenzo-*p*-dioxins and furans from serum samples using immunoaffinity chromatography prior to high-resolution gas chromatography-mass spectrometry. *J. Chromatogr. B* 757: 285–293.

123. Sugawara, Y., S.J. Gee, J.R. Sanborn, et al. 1998. Development of a Highly Sensitive enzyme-linked immunosorbent assay based on polyclonal antibodies for the detection of polychlorinated dibenzo-*p*-dioxins. *Anal. Chem.* 70: 1092–1099.

124. Zhou, X.C. and L. Cao. 2001. High sensitivity microgravimetric biosensor for qualitative and quantitative diagnostic detection of polychlorinated dibenzo-*p*-dioxins. *Analyst* 126: 71–78.

125. de Wit, C.A. 2002. An overview of brominated flame retardants in the environment. *Chemosphere* 46: 583–624.

126. Hites, R.A. 2004. Polybrominated diphenyl ethers in the environment and in people: A meta-analysis of concentrations. *Environ. Sci. Technol.* 38: 945–956.

127. U.S. EPA (Environmental Protection Agency). 2008. Polybrominated diphenylethers (PBDEs). Available at http://www.epa.gov/oppt/pbde/.

128. Shelver, W.L., Y.-S. Keum, H.-J. Kim, et al. 2005. Hapten syntheses and antibody generation for the development of a polybrominated flame retardant ELISA. *J. Agric. Food Chem.* 53: 3840–3847.

129. ATSDR (Agency for Toxic Substances and Disease Registry). 2006. Mercury. Available at http://www.atsdr.cdc.gov/cabs/mercury/#risk.

130. U.S. EPA (Environmental Protection Agency). 2005. Lead and copper rule: A quick reference guide for schools and child care facilities that are regulated under the safe drinking water act. Available at http://www.epa.gov/safewater/schools/pdfs/lead/qrg_lcr_schools.pdf.

131. U.S. EPA (Environmental Protection Agency). 2003. Index to EPA test methods. Available at http://www.epa.gov/region1/info/testmethods/pdfs/testmeth.pdf.

132. European Union Commission. 1998. Council Directive 98/83/EC of 3 November 1998 on the quality of water intended for human consumption. OJ 1998 L 330, p. 32.

133. Darwish, I.A. and D.A. Blake. 2002. Development and validation of a one-step immunoassay for determination of cadmium in human serum. *Anal. Chem.* 74: 52–58.

134. Khosraviani, M., A.R. Pavlov, G.C. Flowers, et al. 1998. Detection of heavy metals by immunoassay: Optimization and validation of a rapid, portable assay for ionic cadmium. *Environ. Sci. Technol.* 32: 137–142.

135. Sasaki, K., K. Tawarada, H. Okuhata, et al. 2006. Development of MAb-based immunochromatographic assay for cadmium from biological samples. *Proceedings of the 28th Annual International Conference of the IEEE Engineering in Medicine and Biology Society*, Vol. 1–15, pp. 1499–1502.

136. Marx, A. and B. Hock. 2000. Monoclonal antibody-based enzyme immunoassay for mercury(II) determination. *Methods: Companion Methods Enzymology* 22: 49–52.

137. Wylie, D.E., L.D. Carlson, R. Carlson, et al. 1991. Detection of mercuric ions in water by ELISA with a mercury-specific antibody. *Anal. Biochem.* 194: 381–387.

138. Barbas, C.F., J.S. Rosenblum, and R.A. Lerner. 1993. Direct selection of antibodies that coordinate metals from semisynthetic combinatorial libraries. *Proceedings of the National Academy of Sciences*, Vol. 90, pp. 6385–6389.

139. Blake, D.A., R.C. Blake II, M. Khosraviani, et al. 1998. Immunoassays for metal ions. *Anal. Chim. Acta* 376: 13–19.

140. Johnson, D.K., S.M. Combs, J.D. Parsen, et al. 2002. Lead analysis by anti-chelate fluorescence polarization immunoassay. *Environ. Sci. Technol.* 36: 1042–1047.

141. Blake, D.A., R.M. Jones, R.C. Blake, et al. 2001. Antibody-based sensors for heavy metal ions. *Biosens. Bioelectron.* 16: 799–809.

142. Chakrabarti, P., F.M. Hatcher, R.C. Blake, et al. 1994. Enzyme immunoassay to determine heavy metals using antibodies to specific metal-EDTA complexes: Optimization and validation of an immunoassay for soluble indium. *Anal. Biochem.* 217: 70–75.

143. Norey, C.G., W.E. Lees, B.M. Darke, et al. 1990. Immunological distinction between piscine and mammalian metallothioneins. *Comp. Biochem. Physiol. B Biochem. Mol. Biol.* 95: 597–601.

144. Hogstrand, C., P.-E. Olsson, and C. Haux. 1989. A radioimmunoassay for metallothionein in fish. *Mar. Environ. Res.* 28: 183–186.

145. Butcher, H., W. Kennette, O. Collins, et al. 2003. A sensitive time-resolved fluorescent immunoassay for metallothionein protein. *J. Immunol. Methods* 272: 247–256.

146. El Hourch, M., A. Dudoit, and J.-C. Amiard. 2003. Optimization of new voltammetric method for the determination of metallothionein. *Electrochim. Acta* 48: 4083–4088.

147. Krishnan, A.V., P. Stathis, S.F. Permuth, et al. 1993. Bisphenol-A: An estrogenic substance is released from polycarbonate flasks during autoclaving. *Endocrinology* 132: 2279–2286.

148. Szemiea, M., M. Gryta, and E. Grzywa. 2000. Study of the separation of Dianin's compound formed in the bisphenol-A synthesis. *J. Inclus. Phenom. Macrocycl. Chem.* 37: 59–66.

149. Deng, H., G. Tang, and W. Pan 2007. Method for producing epoxy using bisphenol A and epoxychloropropane, China Patent, Editor.

150. Kong, Q., Q. Dong, and Z. Shang. 2006. Production process of tetrabromobisphenol-S-bis (2,3-dibromopropyl) ether. China Patent, Editor.

151. Korner, W., U. Bolz, W. Süßmuth, et al. 2000. Input/output balance of estrogenic active compounds in a major municipal sewage plant in Germany. *Chemosphere* 40: 1131–1142.

152. Staples, C.A., P.B. Dome, G.M. Klecka, et al. 1998. A review of the environmental fate, effects, and exposures of bisphenol A. *Chemosphere* 36: 2149–2173.

153. Bolz, U., H. Hagenmaier, and W. Korner. 2001. Phenolic xenoestrogens in surface water, sediments, and sewage sludge from Baden-Wurttemberg, south-west Germany. *Environ. Pollut.* 115: 291–301.

154. Fromme, H., T. Kuchler, T. Otto, et al. 2002. Occurrence of phthalates and bisphenol A and F in the environment. *Water Res.* 36: 1429–1438.

155. Heemken, O.P., H. Reincke, B. Stachel, et al. 2001. The occurrence of xenoestrogens in the Elbe River and the North Sea. *Chemosphere* 45: 245–259.

156. Goda, Y., A. Kobayashi, K. Fukuda, et al. 2000. Development of the ELISAs for detection of hormone-disrupting chemicals. *Water Sci. Technol.* 42: 81–88.

157. Hirobe, M., Y. Goda, Y. Okayasu, et al. 2006. The use of enzyme-linked immunosorbent assays (ELISA) for the determination of pollutants in environmental and industrial wastes. *Water Sci. Technol.* 54: 1–9.

158. Zhao, M.P., Y.Z. Li, Z.Q. Guo, et al. 2002. A new competitive enzyme-linked immunosorbent assay (ELISA) for determination of estrogenic bisphenols. *Talanta* 57: 1205–1210.

159. Marchesini, G.R., K. Koopal, E. Meulenberg, et al. 2007. Spreeta-based biosensor assays for endocrine disruptors. *Biosens. Bioelectron.* 22: 1908–1915.

160. Matsumoto, K., T. Sakai, A. Torimaru, et al. 2005. A surface plasmon resonance-based immunosensor for sensitive detection of bisphenol A. *J. Fac. Agric. Kyushu Univ.* 50: 625–634.

161. Soh, N., T. Watanabe, Y. Asano, et al. 2003. Indirect competitive immunoassay for bisphenol A, based on a surface plasmon resonance sensor. *Sens. Mater.* 15: 423–438.

162. Rodriguez-Mozaz, S., M.L. de Alda, and D. Barceló. 2005. Analysis of bisphenol A in natural waters by means of an optical immunosensor. *Water Res.* 39: 5071–5079.

163. Tschmelak, J., G. Proll, and G. Gauglitz. 2004. Verification of performance with the automated direct optical TIRF immunosensor (River Analyser) in single and multi-analyte assays with real water samples. *Biosens. Bioelectron.* 20: 743–752.

164. Rahman, M.A., M.J.A. Shiddiky, J.S. Park, et al. 2007. An impedimetric immunosensor for the label-free detection of bisphenol A. *Biosens. Bioelectron.* 22: 2464–2470.

165. Park, J.W., S. Kurosawa, H. Aizawa, et al. 2006. Piezoelectric immunosensor for bisphenol A based on signal enhancing step with 2-methacrolyloxyethyl phosphorylcholine polymeric nanoparticle. *Analyst* 131: 155–162.

166. Zhao, M.P., Y. Liu, Y.Z. Li, et al. 2003. Development and characterization of an immunoaffinity column for the selective extraction of bisphenol A from serum samples. *J. Chromatogr. B* 783: 401–410.

167. Braunrath, R., D. Podlipna, S. Padlesak, et al. 2005. Determination of bisphenol A in canned foods by immunoaffinity chromatography, HPLC, and fluorescence detection. *J. Agric. Food Chem.* 53: 8911–8917.
168. Hyer, K.E., G.M. Hornberger, and J.S. Herman. 2001. Processes controlling the episodic streamwater transport of atrazine and other agrichemicals in an agricultural watershed. *J. Hydrol.* 254: 47–66.
169. Holland, J. and P. Sinclair. 2004. Environmental fate of pesticides and the consequences for residues in food and drinking water. In: D. Hamilton and S. Crossley (eds), *Pesticide Residues in Food and Drinking Water*, pp. 27–62. New York: Wiley.
170. Miyamoto, J., N. Mikami, and Y. Takimoto. 1990. The fate of pesticides in aquatic ecosystems. In: D.H. Hutson and T.R. Roberts (eds), *Progress in Pesticide Biochemistry and Toxicology*, pp. 123–147. New York: Wiley.
171. Aguilar, C., I. Ferrer, F. Borrull, et al. 1999. Monitoring of pesticides in river water based on samples previously stored in polymeric cartridges followed by on-line solid-phase extraction-liquid chromatography-diode array detection and confirmation by atmospheric pressure chemical ionization mass spectrometry. *Anal. Chim. Acta* 386: 237–248.
172. Barceló, D. 1991. Applications of gas chromatographymass spectrometry in monitoring environmentally important compounds. *Trends Anal. Chem.* 10: 323–329.
173. Fernandez-Alba, A.R., A. Aguera, M. Contreras, et al. 1998. Comparison of various sample handling and analytical procedures for the monitoring of pesticides and metabolites in ground waters. *J. Chromatogr. A* 823: 35–47.
174. Irace-Guigand, S., J.J. Aaron, P. Scribe, et al. 2004. A comparison of the environmental impact of pesticide multiresidues and their occurrence in river waters surveyed by liquid chromatography coupled in tandem with UV diode array detection and mass spectrometry. *Chemosphere* 55: 973–981.
175. Lacorte, S., J.J. Vreuls, J.S. Salau, et al. 1998. Monitoring of pesticides in river water using fully automated on-line solid-phase extraction and liquid chromatography with diode array detection with a novel filtration device. *J. Chromatogr. A* 795: 71–82.
176. Mallat, E., D. Barceló, C. Barzen, et al. 2001. Immunosensors for pesticide determination in natural waters. *Trends Anal. Chem.* 20: 124–132.
177. Najdek, M. and D. Bazulic. 1988. Chlorinated hydrocarbons in mussels and some benthic organisms from the northern Adriatic Sea. *Mar. Pollut. Bull.* 19: 37–38.
178. Perez-Ruzafa, A., S. Navarro, A. Barba, et al. 2000. Presence of pesticides throughout trophic compartments of the food web in the Mar Menor Lagoon (SE Spain). *Mar. Pollut. Bull.* 40: 140–151.
179. Rivera, J., J. Caixach, and M. De Torres. 1986. Fate of atrazine and trifluralin from an industrial waste dumping at the Llobregat river. Presence in fish, raw and finished water. *Int. J. Anal. Chem.* 24: 183–191.
180. Barzen, C., A. Brecht, and G. Gauglitz. 2002. Optical multiple-analyte immunosensor for water pollution control. *Biosens. Bioelectron.* 17: 289–295.
181. Delaunay-Bertoncini, N. and M.C. Hennion. 2004. Immunoaffinity solid-phase extraction for pharmaceutical and biomedical trace-analysiscoupling with HPLC and CE-perspectives. *J. Pharmac. Biomed. Anal.* 34: 717–736.
182. Hennion, M.-C. and V. Pichon. 2003. Immuno-based sample preparation for trace analysis. *J. Chromatogr. A* 1000: 29–52.
183. Rodriguez-Mozaz, S., S. Reder, M. Lopez de Alda, et al. 2004. Simultaneous multi-analyte determination of estrone, isoproturon and atrazine in natural waters by the RIver ANAlyser (RIANA), an optical immunosensor. *Biosens. Bioelectron.* 19: 633–640.
184. Tschmelak, J., G. Proll, and G. Gauglitz. 2005. Optical biosensor for pharmaceuticals, antibiotics, hormones, endocrine disrupting chemicals and pesticides in water: Assay optimization process for estrone as example. *Talanta* 65: 313–323.
185. Valera, E., J. Ramón-Azcón, Á. Rodríguez, et al. 2007. Impedimetric immunosensor for atrazine detection using interdigitated [mu]-electrodes (ID[mu]E's). *Sens. Actuat. B: Chem.* 125: 526–537.
186. Nichkova, M. and M.-P. Marco. 2006. Biomonitoring human exposure to organohalogenated substances by measuring urinary chlorophenols using a High-Throughput Screening (HTS) immunochemical method. *Environ. Sci. Technol.* 40: 2469–2477.
187. Galve, R., F. Camps, F. Sanchez-Baeza, et al. 2000. Development of an immunochemical technique for the analysis of trichlorophenols using theoretical models. *Anal. Chem.* 72: 2237–2246.
188. Nichkova, M., R. Galve, and M.-P. Marco. 2002. Biological monitoring of 2,4,5-trichlorophenol (I): Preparation of antibodies and development of an immunoassay using theoretical models. *Chem. Res. Toxicol.* 15: 1360–1370.

189. Galve, R., F. Sánchez-Baeza, F. Camps, et al. 2002. Indirect competitive immunoassay for trichlorophenol: Rational evaluation of the competitor heterology. *Anal. Chim. Acta* 452: 191–206.

190. Nichkova, M. and M.P. Marco. 2005. Development and evaluation of C18 and immunosorbent solid-phase extraction methods prior to immunochemical analysis of chlorophenols in human urine. *Anal. Chim. Acta* 533: 67–82.

191. Centero, E.R., W.J. Johnson, and A.H. Sehon. 1970. Antibodies to two common pesticides, DDT and malathion. *Int. Arch. Allergy Appl. Immunol.* 37: 1–13.

192. Haas, G.J. and E.J. Guardia. 1968. Production of antibodies against insecticide-protein conjugates. *Proc. Soc. Exp. Biol. Med.* 129: 546–551.

193. Beasley, H.L., T. Phongkham, M.H. Daunt, et al. 1998. Development of a panel of immunoassays for monitoring DDT, its metabolites, and analogues in food and environmental matrices. *J. Agric. Food Chem.* 46: 3339–3352.

194. Amitarani, B.E., A. Pasha, P. Gowda, et al. 2002. Comparison of ELISA and GC methods to detect DDT residues in water samples. *Indian J. Biotechnol.* 1: 292–297.

195. Eremin, S.A., I.A. Ryabova, J.N. Yakovleva, et al. 2002. Development of a rapid, specific fluorescence polarization immunoassay for the herbicide chlorsulfuron. *Anal. Chim. Acta* 468: 229–236.

196. Langone, J.J. and H. Van Vunakis. 1975. Radioimmunoassay for dieldrin and aldrin. *Res. Commun. Chem. Pathol. Pharmacol.* 10: 163–171.

197. Manclus, J.J., A. Abad, M.Y. Lebedev, et al. 2004. Development of a monoclonal immunoassay selective for chlorinated cyclodiene insecticides. *J. Agric. Food Chem.* 52: 2776–2784.

198. Stanker, L.H., M. Vanderlaan, and B.E. Watkins. 1994. Monoclonal antibodies to cyclodiene insecticides and method for detecting the same. U.S. Patent, Editor.

199. Lee, N.A. and I.R. Kennedy. 2001. Environmental monitoring of pesticides by immunoanalytical techniques: Validation, current status, and future perspectives. *J. AOAC Int.* 84: 1393–1406.

200. Ragab, A.A., A.M.A. Ibrahim, and C.J. Smith. 1997. Quantification by ELISA of aldrin/dieldrin in river Nile water and tap water samples collected in Egypt. *Food Agric. Immunol.* 9: 51–55.

201. Brummel, K.E., J. Wright, and M.E. Eldefrawi. 1997. Fiber optic biosensor for cyclodiene insecticides. *J. Agric. Food Chem.* 45: 3292–3298.

202. Mercader, J.V. and A. Montoya. 1999. Development of monoclonal ELISAs for azinphos-methyl. II. Assay optimization and water sample analysis. *J. Agric. Food Chem.* 47: 1285–1293.

203. Watanabe, E., K. Baba, H. Eun, et al. 2006. Evaluation of performance of a commercial monoclonal antibody-based fenitrothion immunoassay and application to residual analysis in fruit samples. *J. Food Prot.* 69: 191–198.

204. Banks, J.N., M.Q. Chaudhry, W.A. Matthews, et al. 1998. Production and characterisation of polyclonal antibodies to the common moiety of some organophosphorus pesticides and development of a generic type ELISA. *Food Agric. Immunol.* 10: 349–361.

205. Nishi, K., Y. Imajuku, M. Nakata, et al. 2003. Preparation and characterization of monoclonal and recombinant antibodies specific to the insecticide malathion. *J. Pestic. Sci.* 28: 301–309.

206. U.S. EPA (Environmental Protection Agency). 1992. Pesticides in groundwater databaseA compilation of monitoring studies: 1971–1991. EPA 734-12-92-001. Office of Prevention, Pesticides and Toxic Substances.

207. ATSDR (Agency for Toxic Substances and Disease Registry). 2000. Toxicological profile information sheets-malathions-draft for public comment, Atlanta, GA.

208. Brun, E.M., M. Garces-Garcia, M.J. Banuls, et al. 2005. Evaluation of a novel malathion immunoassay for groundwater and surface water analysis. *Environ. Sci. Technol.* 39: 2786–2794.

209. Ibrahim, A.M.A., M.A. Morsy, M.M. Hewedi, et al. 1994. Monoclonal antibody based ELISA for the detection of ethyl parathion. *Food Agric. Immunol.* 6: 23–30.

210. Abad, A. and A. Montoya. 1997. Development of an enzyme-linked immunosorbent assay to carbaryl. 2. Assay optimization and application to the analysis of water samples. *J. Agric. Food Chem.* 45: 1495–1501.

211. Itak, J.A., E.G. Olson, J.R. Fleeker, et al. 1993. Validation of a Paramagnetic Particle-Based Elisa for the quantitative-determination of carbaryl in water. *Bull. Environ. Contam. Toxicol.* 51: 260–267.

212. Marco, M.-P., S. Chiron, J. Gascón, et al. 1995. Validation of two immunoassay methods for environmental monitoring of carbaryl and 1-naphthol in ground water samples. *Anal. Chim. Acta* 311: 319–329.

213. Jourdan, S.W., A.M. Scutellaro, J.R. Fleeker, et al. 1995. Determination of carbofuran in water and soil by a Rapid Magnetic Particle-Based Elisa. *J. Agric. Food Chem.* 43: 2784–2788.

214. Abad, A., M.J. Moreno, R. Pelegrí, et al. 1999. Determination of carbaryl, carbofuran and methiocarb in cucumbers and strawberries by monoclonal enzyme immunoassays and high-performance liquid chromatography with fluorescence detection: An analytical comparison. *J. Chromatogr. A* 833: 3–12.

215. Yang, G. and S. Kang. 2008. SPR-based antibody-antigen interaction for real time analysis of carbamate pesticide residues. *Food Sci. Biotechnol.* 17: 15–19.
216. Abad, A., M.J. Moreno, R. Pelegri, et al. 2001. Monoclonal enzyme immunoassay for the analysis of carbaryl in fruits and vegetables without sample cleanup. *J. Agric. Food Chem.* 49: 1707–1712.
217. Nunes, G.S., M.P. Marco, M. Farré, et al. 1999. Direct application of an enzyme-linked immunosorbent assay method for carbaryl determination in fruits and vegetables. Comparison with a liquid chromatography-postcolumn reaction fluorescence detection method. *Anal. Chim. Acta* 387: 245–253.
218. Abad, A., M.J. Moreno, and A. Montoya. 1997. A monoclonal immunoassay for carbofuran and its application to the analysis of fruit juices. *Anal. Chim. Acta* 347: 103–110.
219. Abad, A. and A. Montoya. 1995. Application of a monoclonal antibody-based elisa to the determination of carbaryl in apple and grape juices. *Anal. Chim. Acta* 311: 365–370.
220. Bacigalupo, M.A. and G. Meroni. 2007. Quantitative determination of diuron in ground and surface water by time-resolved fluoroimmunoassay: Seasonal variations of diuron, carbofuran, and paraquat in an agricultural area. *J. Agric. Food Chem.* 55: 3823–3828.
221. Bacigalupo, M.A., G. Meroni, and R. Longhi. 2006. Determination of carbofuran in water by homogeneous immunoassay using selectively conjugate mastoparan and terbium/dipicolinic acid fluorescent complex. *Talanta* 69: 1106–1111.
222. Xing, W.-L., G.-R. Ou, Z.-H. Jiang, et al. 2000. Detection of multiple herbicide residues using a planar array evanescent field immunosensor. *Anal. Chem.* 33: 1071–1078.
223. Mauriz, E., A. Calle, J.J. Manclus, et al. 2007. Multi-analyte SPR immunoassays for environmental biosensing of pesticides. *Anal. Bioanal. Chem.* 387: 1449–1458.
224. Gonzalez-Martinez, M.A., S. Morais, R. Puchades, et al. 1997. Monoclonal antibody-based flow-through immunosensor for analysis of carbaryl. *Anal. Chem.* 69: 2812–2818.
225. Gonzalez-Martinez, M.A., S. Morais, R. Puchades, et al. 1997. Development of an automated controlled-pore glass flow-through immunosensor for carbaryl. *Anal. Chim. Acta* 347: 199–205.
226. Mauriz, E., A. Calle, A. Montoya, et al. 2006. Determination of environmental organic pollutants with a portable optical immunosensor. *Talanta* 69: 359–364.
227. Penalva, J., J.A. Gabaldon, A. Maquieira, et al. 2000. Determination of carbaryl in vegetables using an immunosensor working in organic media. *Food Agric. Immunol.* 12: 101–114.
228. Lee, N., D.P. McAdam, and J.H. Skerritt. 1998. Development of immunoassays for type II synthetic pyrethroids. 1. Hapten design and application to heterologous and homologous assays. *J. Agric. Food Chem.* 46: 520–534.
229. Lee, N., H.L. Beasley, and J.H. Skerritt. 1998. Development of immunoassays for type II synthetic pyrethroids. 2. Assay specificity and application to water, soil, and grain. *J. Agric. Food Chem.* 46: 535–546.
230. Watanabe, T., G. Shan, D.W. Stoutamire, et al. 2001. Development of a class-specific immunoassay for the type I pyrethroid insecticides. *Anal. Chim. Acta* 444: 119–129.
231. Mak, S.K., G. Shan, H.J. Lee, et al. 2005. Development of a class selective immunoassay for the type II pyrethroid insecticides. *Anal. Chim. Acta* 534: 109–120.
232. Gao, H.B., Y. Ling, T. Xu, et al. 2006. Development of an enzyme-linked immunosorbent assay for the pyrethroid insecticide cyhalothrin. *J. Agric. Food Chem.* 54: 5284–5291.
233. Feng, J., G.M. Shan, B.D. Hammock, et al. 2003. Fluorescence quenching competitive immunoassay in micro droplets. *Biosens. Bioelectron.* 18: 1055–1063.
234. Sasaki, S., E. Kai, H. Miyachi, et al. 1998. Direct determination of etofenprox using surface plasmon resonance. *Anal. Chim. Acta* 363: 229–233.
235. Ballesteros, B., D. Barceló, A. Dankwardt, et al. 2003. Evaluation of a field test kit for triazine herbicides (SensoScreen® TR500) as an alarm system of pesticide water samples contamination. *Anal. Chim. Acta* 475: 105–115.
236. Gonzalez-Martinez, M.A., S. Morais, R. Puchades, et al. 1998. Enzyme immunoassay for atrazine. Comparison of three immobilization supports. *Fresenius J. Anal. Chem.* 361: 179–184.
237. Ramón-Azcón, J., E. Valera, Á. Rodrígucz, et al. 2008. An impedimetric immunosensor based on inter-digitated microelectrodes (IDμE) for the determination of atrazine residues in food samples. *Biosens. Bioelectron.* 23: 1367–1373.
238. Wittmann, C. and B. Hock. 1989. Improved enzyme immunoassay for the analysis of triazines in water samples. *Food Agric. Immunol.* 1: 211–224.
239. Gascon, J., A. Oubina, I. Ferrer, et al. 1996. Performance of two immunoassays for the determination of atrazine in sea water samples as compared with on-line solid phase extraction-liquid chromatography-diode array detection. *Anal. Chim. Acta* 330: 41–51.

240. Zacco, E., M.I. Pividori, and S. Alegret. 2006. Electrochemical magnetoimmunosensing strategy for the detection of pesticides residues. *Anal. Chem.* 78: 1780–1788.

241. Rodríguez, Á., E. Valera, J. Ramón-Azcón, et al. 2008. Single frequency impedimetric immunosensor for atrazine detection. *Sens. Actuat. B: Chem.* 129: 921–928.

242. Valera, E., J. Ramón-Azcón, F.J. Sanchez, et al. 2008. Conductimetric immunosensor for atrazine detection based on antibodies labelled with gold nanoparticles. *Sens. Actuat. B: Chem.* 134: 95–103.

243. Farré, M., E. Martínez, J. Ramón, et al. 2007. Part per trillion determination of atrazine in natural water samples by surface plasmon resonance immunosensor. *Anal. Bioanal. Chem.* 388: 207–214.

244. Tudorache, M. and J. Emneus. 2006. Micro-immuno supported liquid membrane assay (mu-ISLMA). *Biosens. Bioelectron.* 21: 1513–1520.

245. Tschmelak, J., G. Proll, J. Riedt, et al. 2005. Biosensors for unattended, cost-effective and continuous monitoring of environmental pollution: Automated water analyser computer supported system (AWACSS) and river analyser (RIANA). *Int. J. Environ. Anal. Chem.* 85: 837–852.

246. Morais, S., P. Casino, M.L. Marin, et al. 2002. Assessment of enzyme-linked immunosorbent assay for the determination of 2,4,5-TP in water and soil. *Anal. Bioanal. Chem.* 374: 262–268.

247. Sanchez, F.G., A.N. Diaz, A.F.G. Diaz, et al. 1999. Quantification of 2,4,5-trichlorophenoxyacetic acid by fluorescence enzyme-linked immunosorbent assay with secondary antibody. *Anal. Chim. Acta* 378: 219–224.

248. Meusel, M., D. Trau, A. Katerkamp, et al. 1998. New ways in bioanalysisone-way optical sensor chip for environmental analysis. *Sens. Actuat. B: Chem.* 51: 249–255.

249. Cuong, N.V., T.T. Bachmann, and R.D. Schmid. 1999. Development of a dipstick immunoassay for quantitative determination of 2,4-dichlorophenoxyacetic acid in water, fruit and urine samples. *Fresenius J. Anal. Chem.* 364: 584–589.

250. Ben Rejeb, S., N. Fischer Durand, A. Martel, et al. 1998. Development of anti-phenylurea antibody purification techniques for improved environmental applications. *Anal. Chim. Acta* 376: 41–48.

251. Scholthof, K.B.G., Z. Guisheng, and A.E. Karu. 1997. Derivation and properties of recombinant fab antibodies to the phenylurea herbicide diuron. *J. Agric. Food Chem.* 45: 1509–1517.

252. Ferrer, I., V. Pichon, M.C. Hennion, et al. 1997. Automated sample preparation with extraction columns by means of anti-isoproturon immunosorbents for the determination of phenylurea herbicides in water followed by liquid chromatography-diode array detection and liquid chromatography-atmospheric pressure chemical ionization mass spectrometry. *J. Chromatogr. A* 777: 91–98.

253. Pichon, V., L. Chen, M.C. Hennion, et al. 1995. Preparation and evaluation of immunosorbents for selective trace enrichment of phenylurea and triazine herbicides in environmental waters. *Anal. Chem.* 67: 2451–2460.

254. Liégeois, E., Y. Dehon, B. de Brabant, et al. 1992. ELISA test, a new method to detect and quantify isoproturon in soil. *Sci. Total Environ.* 123–124: 17–28.

255. Mallat, E., C. Barzen, R. Abuknesha, et al. 2001. Part per trillion level determination of isoproturon in certified and estuarine water samples with a direct optical immunosensor. *Anal. Chim. Acta* 426: 209–216.

256. Pulido-Tofiño, P., J.M. Barrero-Moreno, and M.C. Pérez-Conde. 2000. Flow-through fluoroimmunosensor for isoproturon determination in agricultural foodstuff: Evaluation of antibody immobilization on solid support. *Anal. Chim. Acta* 417: 85–94.

257. Lawrence, J.F., C. Ménard, M.-C. Hennion, et al. 1996. Use of immunoaffinity chromatography as a simplified cleanup technique for the liquid chromatographic determination of phenylurea herbicides in plant material. *J. Chromatogr. A* 732: 277–281.

258. Pichon, V., L. Chen, N. Durand, et al. 1996. Selective trace enrichment on immunosorbents for the multiresidue analysis of phenylurea and triazine pesticides. *J. Chromatogr. A* 725: 107–119.

259. Pichon, V., L. Chen, and M.C. Hennion. 1995. On-line preconcentration and liquid chromatographic analysis of phenylurea pesticides in environmental water using a silica-based immunosorbent. *Anal. Chim. Acta* 311: 429–436.

260. Pulido-Tofiño, P., J.M. Barrero-Moreno, and M.C. Pérez-Conde. 2001. Sol-gel glass doped with isoproturon antibody as selective support for the development of a flow-through fluoroimmunosensor. *Anal. Chim. Acta* 429: 337–345.

261. Lawruk, T.S., C.S. Hottenstein, D.P. Herzog, et al. 1992. Quantification of alachlor in water by a novel magnetic particle-based ELISA. *Bull. Environ. Contam. Toxicol.* 48: 643–650.

262. Feng, P.C.C., S.J. Wratten, S.R. Horton, et al. 1990. Development of an enzyme-linked immunosorbent-assay for alachlor and its application to the analysis of environmental water samples. *J. Agric. Food Chem.* 38: 159–163.

263. Schraer, S.M., D.R. Shaw, M. Boyette, et al. 2000. Comparison of enzyme-linked immunosorbent assay and gas chromatography procedures for the detection of cyanazine and metolachlor in surface water samples. *J. Agric. Food Chem.* 48: 5881–5886.

264. Gaynor, J.D., D.A. Cancilla, G.R.B. Webster, et al. 1996. Comparative solid phase extraction, solid phase microextraction, and immunoassay analyses of metolachlor in surface runoff and tile drainage. *J. Agric. Food Chem.* 44: 2736–2741.

265. Schmitt, A., V. Hingst, L. Erdinger, et al. 1992. Aspects in the development of an enzyme immunosorbent-assay for the detection of the herbicide metolachlor in water samples. *Zentral. Hyg. Umweltmed.* 193: 272–286.

266. Tessier, D.M. and J. Marshall Clark. 1998. An enzyme immunoassay for mutagenic metabolites of the herbicide alachlor. *Anal. Chim. Acta* 376: 103–112.

267. Yakovleva, J., A.V. Zherdev, V.A. Popova, et al. 2003. Production of antibodies and development of enzyme-linked immunosorbent assays for the herbicide butachlor. *Anal. Chim. Acta* 491: 1–13.

268. Guo, Y., J. Chen, N. Wang, et al. 2002. Preparation and application of polyclonal antibody to butachlor. *Beijing Daxue Xuebao Ziran Kexue Ban/Acta Scientiarum Naturalium Universitatis Pekinensis* 38: 447.

269. Hegedus, G., V. Krikunova, I. Belai, et al. 2002. An enzyme-linked immunosorbent assay (ELISA) for the detection of acetochlor. *Int. J. Environ. Anal. Chem.* 82: 879–891.

270. Solna, R., P. Skládal, and S.A. Eremin. 2003. Development of a disposable electrochemical immunosensor for detection of the herbicide acetochlor. *Int. J. Environ. Anal. Chem.* 83: 609–620.

271. Yakovleva, J.N., A.I. Lobanova, O.A. Panchenko, et al. 2002. Production of antibodies and development of specific polarization fluoroimmunoassay for acetochlor. *Int. J. Environ. Anal. Chem.* 82: 851–863.

272. Kelley, M.M., E.W. Zahnow, W.C. Petersen, et al. 1985. Chlorsulfuron determination in soil extracts by enzyme immunoassay. *J. Agric. Food Chem.* 33: 962–965.

273. Schlaeppi, J.M.A., W. Meyer, and K.A. Ramsteiner. 1992. Determination of triasulfuron in soil by monoclonal antibody-based enzyme immunoassay. *J. Agric. Food Chem.* 40: 1093–1098.

274. Simon, E., D. Knopp, P.B. Carrasco, et al. 1998. Development of an enzyme immunoassay for metsulfuron-methyl. *Food Agric. Immunol.* 10: 105–120.

275. Wang, G., J. Yuan, K. Matsumoto, et al. 2001. Homogeneous time-resolved fluoroimmunoassay of bensulfuron-methyl by using terbium fluorescence energy transfer. *Talanta* 55: 1119–1125.

276. Wang, G., J. Yuan, B. Gong, et al. 2001. Immunoassay by graphite furnace atomic absorption spectrometry using a metal chelate as a label. *Anal. Chim. Acta* 448: 165–172.

277. Dzantiev, B.B., E.V. Yazynina, A.V. Zherdev, et al. 2004. Determination of the herbicide chlorsulfuron by amperometric sensor based on separation-free bienzyme immunoassay. *Sens. Actuat. B: Chem.* 98: 254–261.

278. Szekacs, A., N. Trummer, N. Adanyi, et al. 2003. Development of a non-labeled immunosensor for the herbicide trifluralin via optical waveguide lightmode spectroscopic detection. *Anal. Chim. Acta* 487: 31–42.

279. U.S. EPA (Environmental Protection Agency). 2008. Pharmaceuticals and personal care products (PPCPs). Available at http://www.epa.gov/ppcp/.

280. OECD, EFPIA. 2005. Pharmaceutical associations of the European countries, VFA. The Pharmaceutical Production Worldwide Report. Available at http://www.vfa.de/en/articles/index-en.html.

281. Batt, A.L., I.B. Bruce, and D.S. Aga. 2006. Evaluating the vulnerability of surface waters to antibiotic contamination from varying wastewater treatment plant discharges. *Environ. Pollut.* 142: 295–302.

282. Kelly, L., D.L. Smith, E.L. Snary, et al. 2004. Animal growth promoters: To ban or not to ban? A risk assessment approach. *Int. J. Antimicrob. Agents* 24: 205–212.

283. Centner, T.J. and T.A. Feitshans. 2006. Regulating manure application discharges from concentrated animal feeding operations in the United States. *Environ. Pollut.* 141: 571–573.

284. Halling-Sorensen, B., S. Nors Nielsen, P.F. Lanzky, et al. 1998. Occurrence, fate and effects of pharmaceutical substances in the environment—A review. *Chemosphere* 36: 357–393.

285. Adanyi, N., M. Varadi, N. Kim, et al. 2006. Development of new immunosensors for determination of contaminants in food. *Curr. Appl. Phys.* 6: 279–286.

286. Fatta, D., A. Achilleos, A. Nikolaou, et al. 2007. Analytical methods for tracing pharmaceutical residues in water and wastewater. *Trends Anal. Chem.* 26: 515–533.

287. Buchberger, W.W. 2007. Novel analytical procedures for screening of drug residues in water, waste water, sediment and sludge. *Anal. Chim. Acta* 593: 129–139.

288. Hock, B. 2002. Immunochemical analysis of water pollutants. *Acta Hydrochim. Hydrobiol.* 29: 375–390.

289. Gustavsson, E. and A. Sternesjö. 2004. Biosensor analysis of β-lactams in milk: Comparison with microbiological, immunological, and receptor-based screening methods. *J. AOAC Int.* 87: 614–620.

290. Link, N., W. Weber, and M. Fussenegger. 2007. A novel generic dipstick-based technology for rapid and precise detection of tetracycline, streptogramin and macrolide antibiotics in food samples. *J. Biotechnol.* 128: 668–680.

291. Reybroeck, W., S. Ooghe, H.D. Brabander, et al. 2007. Validation of the tetrasensor honey test kit for the screening of tetracyclines in honey. *J. Agric. Food Chem.* 55: 8359–8366.

292. Salter, R.S., D. Legg, N. Ossanna, et al. 2001. Charm safe-level β-lactam test for amoxicillin, ampicillin, ceftiofur, cephapirin, and penicillin G in raw commingled milk. *J. AOAC Int.* 84: 29–36.

293. Wang, X., K. Li, D. Shi, et al. 2007. Development of an Immunochromatographic lateral-flow test strip for rapid detection of sulfonamides in eggs and chicken muscles. *J. Agric. Food Chem.* 55: 2072–2078.

294. Raich-Montiu, J., J. Folch, R. Compano, et al. 2007. Analysis of trace levels of sulfonamides in surface water and soil samples by liquid chromatography-fluorescence. *J. Chromatogr. A* 1172: 186–193.

295. Tolls, J. 2001. Sorption of veterinary pharmaceuticals in soils: A review. *Environ. Sci. Technol.* 35: 3397–3406.

296. Thiele-Bruhn, S., T. Seibicke, H.R. Schulten, et al. 2004. Sorption of sulfonamide pharmaceutical antibiotics on whole soils and particle-size fractions. *J. Environ. Qual.* 33: 1331–1342.

297. Li, J.-D., Y.-Q. Cai, Y.-L. Shi, et al. 2007. Determination of sulfonamide compounds in sewage and river by mixed hemimicelles solid-phase extraction prior to liquid chromatography-spectrophotometry. *J. Chromatogr. A* 1139: 178–184.

298. Malintan, N.T. and M.A. Mohd. 2006. Determination of sulfonamides in selected Malaysian swine wastewater by high-performance liquid chromatography. *J. Chromatogr. A* 1127: 154–160.

299. Nieto, A., F. Borrull, R.M. Marce, et al. 2007. Selective extraction of sulfonamides, macrolides and other pharmaceuticals from sewage sludge by pressurized liquid extraction. *J. Chromatogr. A* 1174: 125–131.

300. Riediker, S., J.M. Diserens, and R.H. Stadler. 2001. Analysis of β-lactam antibiotics in incurred raw milk by rapid test methods and liquid chromatography coupled with electrospray ionization tandem mass spectrometry. *J. Agric. Food Chem.* 49: 4171–4176.

301. Spinks, C.A., G.M. Wyatt, S. Everest, et al. 2002. Atypical antibody specificity: Advancing the development of a generic assay for sulphonamides using heterologous ELISA. *J. Sci. Food Agric.* 82: 428–434.

302. Franek, M., I. Diblikova, I. Cernoch, et al. 2006. Broad-specificity immunoassays for sulfonamide detection: Immunochemical strategy for generic antibodies and competitors. *Anal. Chem.* 78: 1559–1567.

303. Zhang, H.Y., L. Wang, Y. Zhang, et al. 2007. Development of an enzyme-linked immunosorbent assay for seven sulfonamide residues and investigation of matrix effects from different food samples. *J. Agric. Food Chem.* 55: 2079–2084.

304. Adrian, J., H. Font, F. Sanchez-Baeza, et al. 2008. Preparation of polyclonal antibodies for the generic determination of sulfonamide antibiotics and development of an enzyme-linked immunosorbent assay (ELISA) for milk analysis. *J. Agric. Food Chem.* 57: 385–394.

305. Cliquet, P., E. Cox, W. Haasnoot, et al. 2003. Extraction procedure for sulfachloropyridazine in porcine tissues and detection in a sulfonamide-specific enzyme-linked immunosorbent assay (ELISA). *Anal. Chim. Acta* 494: 21–28.

306. Font, H., J. Adrian, R. Galve, et al. 2008. Immunochemical assays for direct sulfonamide antibiotic detection in milk and hair samples using antibody derivatized magnetic nanoparticles. *J. Agric. Food Chem.* 56: 736–743.

307. Muldoon, M.T., C.K. Holtzapple, S.S. Deshpande, et al. 2000. Development of a monoclonal antibody-based cELISA for the analysis of sulfadimethoxine. 1. Development and characterization of monoclonal antibodies and molecular modeling studies of antibody recognition. *J. Agric. Food Chem.* 48: 537–544.

308. Pastor-Navarro, N., C. Garcia-Rover, A. Maquieira, et al. 2004. Specific polyclonal-based immunoassays for sulfathiazole. *Anal. Bioanal. Chem.* 379: 1088–1099.

309. Campagnolo, E.R., K.R. Johnson, A. Karpati, et al. 2002. Antimicrobial residues in animal waste and water resources proximal to large-scale swine and poultry feeding operations. *Sci. Total Environ.* 299: 89–95.

310. Yang, S. and K. Carlson. 2004. Routine monitoring of antibiotics in water and wastewater with a radioimmunoassay technique. *Water Res.* 38: 3155–3166.

311. Tschmelak, J., M. Kumpf, G. Proll, et al. 2004. Biosensor for seven sulphonamides in drinking, ground, and surface water with difficult matrices. *Anal. Lett.* 37: 1701–1718.

312. Tschmelak, J., G. Proll, J. Riedt, et al. 2005. Automated water analyser computer supported system (AWACSS): Part II: Intelligent, remote-controlled, cost-effective, on-line, water-monitoring measurement system. *Biosens. Bioelectron.* 20: 1509–1519.

313. Grant, G.A., S.L. Frison, and P. Sporns. 2003. A sensitive method for detection of sulfamethazine and N4-acetylsulfamethazine residues in environmental samples using solid phase immunoextraction coupled with MALDI-TOF MS. *J. Agric. Food Chem.* 51: 5367–5375.

314. Korpimaki, T., E.C. Brockmann, O. Kuronen, et al. 2004. Engineering of a broad specificity antibody for simultaneous detection of 13 sulfonamides at the maximum residue level. *J. Agric. Food Chem.* 52: 40–47.

315. Pastor-Navarro, N., E. Gallego-Iglesias, A. Maquieira, et al. 2007. Development of a group-specific immunoassay for sulfonamides: Application to bee honey analysis. *Talanta* 71: 923–933.

316. Wang, X., K. Li, D. Shi, et al. 2007. Development and validation of an immunochromatographic assay for rapid detection of sulfadiazine in eggs and chickens. *J. Chromatogr. B* 847: 289–295.

317. Zacco, E., J. Adrian, R. Galve, et al. 2007. Electrochemical magneto immunosensing of antibiotic residues in milk. *Biosens. Bioelectron.* 22: 2184–2191.

318. Bratov, A., J. Ramon, N. Abramova, et al. 2008. Three-dimensional interdigitated electrode array as a transducer for label-free biosensors. *Biosens. Bioelectron.* 24: 729–735.

319. Adrian, J., S. Pasche, J-M. Diserens, et al. 2009. Waveguide interrogated optical immunosensor (WIOS) for detection of sulfonamide antibiotics in milk. *Biosens. Bioelectron.* submitted.

320. Golet, E.M., A.C. Alder, and W. Giger. 2002. Environmental exposure and risk assessment of fluoroquinolone antibacterial agents in wastewater and river water of the Glatt Valley Watershed, Switzerland. *Environ. Sci. Technol.* 36: 3645–3651.

321. Sukul, P. and M. Spiteller. 2007. Fluoroquinolone antibiotics in the environment. *Rev. Environ. Contam. Toxicol.* 191: 131–162.

322. Huet, A.C., C. Charlier, S.A. Tittlemier, et al. 2006. Simultaneous determination of (fluoro)quinolone antibiotics in kidney, marine products, eggs, and muscle by enzyme-linked immunosorbent assay (ELISA). *J. Agric. Food Chem.* 54: 2822–2827.

323. Wang, Z., Y. Zhu, S. Ding, et al. 2007. Development of a monoclonal antibody-based broad-specificity ELISA for fluoroquinolone antibiotics in foods and molecular modeling studies of cross-reactive compounds. *Anal. Chem.* 79: 4471–4483.

324. Duan, J. and Z. Yuan. 2001. Development of an indirect competitive ELISA for ciprofloxacin residues in food animal edible tissues. *J. Agric. Food Chem.* 49: 1087–1089.

325. Holtzapple, C.K., S.A. Buckley, and L.H. Stanker. 1999. Immunosorbents coupled on-line with liquid chromatography for the determination of fluoroquinolones in chicken liver. *J. Agric. Food Chem.* 47: 2963–2968.

326. VanCoillie, E., J. DeBlock, and W. Reybroeck. 2004. Development of an indirect competitive ELISA for flumequine residues in raw milk using chicken egg yolk antibodies. *J. Agric. Food Chem.* 52: 4975–4978.

327. Pinacho, D.G., F. Sanchez-Baeza, and M.P. Marco. 2009. Development of a class selective indirect competitive enzyme-linked immunosorbent assay (ELISA) for detection of fluoroquinolone antibiotics. *J. Agric. Food Chem.* submitted.

328. Pinacho, D.G., F. Fernández, F. Sanchez-Baeza, et al. 2009. An immunochemical high-throughput screening method for fluoroquinolone antibiotics in milk samples. *Anal. Chim. Acta* submitted.

329. Garifallou, G.Z., G. Tsekenis, F. Davis, et al. 2007. Labeless immunosensor assay for fluoroquinolone antibiotics based upon an AC impedance protocol. *Anal. Lett.* 40: 1412–1422.

330. Pinacho, D.G., K. Gorgy, S. Cosnier, et al. 2008. Electrogeneration of polymer films functionalized by fluoroquinolone models for the development of antibiotic immunosensor. *ITBM-RBM* 29: 181–186.

331. Pinacho, D.G., F. Sanchez-Baeza, M.P. Marco, et al. 2009. Development of an amperometric magneto immunosensor for detection of fluoroquinolone antibiotics. *J. Agric. Food Chem.* submitted.

332. Ionescu, R.E., N. Jaffrezic-Renault, L. Bouffier, et al. 2007. Impedimetric immunosensor for the specific label free detection of ciprofloxacin antibiotic. *Biosens. Bioelectron.* 23: 549–555.

333. Huys, G., K. Bartie, M. Cnockaert, et al. 2007. Biodiversity of chloramphenicol-resistant mesophilic heterotrophs from southeast Asian aquaculture environments. *Res. Microbiol.* 158: 228–235.

334. Impens, S., W. Reybroeck, J. Vercammen, et al. 2003. Screening and confirmation of chloramphenicol in shrimp tissue using ELISA in combination with GC-MS2 and LC-MS2. *Anal. Chim. Acta* 483: 153–163.

335. Commission of the European Communities. Decision 2002/657/EC of 12 August 2002 implementing Council Directive 96/23/EC concerning the performance of analytical methods and the interpretation of results (2002/657/EC) L221/8.

336. Ashwin, H.M., S.L. Stead, J.C. Taylor, et al. 2005. Development and validation of screening and confirmatory methods for the detection of chloramphenicol and chloramphenicol glucuronide using SPR biosensor and liquid chromatography-tandem mass spectrometry. *Anal. Chim. Acta* 529: 103–108.

337. Dumont, V., A.C. Huet, I. Traynor, et al. 2006. A surface plasmon resonance biosensor assay for the simultaneous determination of thiamphenicol, florefenicol, florefenicol amine and chloramphenicol residues in shrimps. *Anal. Chim. Acta* 567: 179–183.

338. Ferguson, J., A. Baxter, P. Young, et al. 2005. Detection of chloramphenicol and chloramphenicol glucuronide residues in poultry muscle, honey, prawn and milk using a surface plasmon resonance biosensor and Qflex(R) kit chloramphenicol. *Anal. Chim. Acta* 529: 109–113.

339. Park, I.-S. and N. Kim. 2006. Development of a chemiluminescent immunosensor for chloramphenicol. *Anal. Chim. Acta* 578: 19–24.

340. Scortichini, G., L. Annunziata, M.N. Haouet, et al. 2005. ELISA qualitative screening of chloramphenicol in muscle, eggs, honey and milk: Method validation according to the Commission Decision 2002/657/EC criteria. *Anal. Chim. Acta* 535: 43–48.

341. Zhang, S., Z. Zhang, W. Shi, et al. 2006. Development of a chemiluminescent ELISA for determining chloramphenicol in chicken muscle. *J. Agric. Food Chem.* 54: 5718–5722.

342. Kim, S., P. Eichhorn, J.N. Jensen, et al. 2005. Removal of antibiotics in wastewater: Effect of hydraulic and solid retention times on the fate of tetracycline in the activated sludge process. *Environ. Sci. Technol.* 39: 5816–5823.

343. Nelson, M., W. Hillen, and R.A. Greenwald. 2001. *Tetracyclines in Biology, Chemistry and Medicine.* Birkhäuser: Springer.

344. Hirsch, R., T. Ternes, K. Haberer, et al. 1999. Occurrence of antibiotics in the aquatic environment. *Sci. Total Environ.* 225: 109–118.

345. Kumar, K., A. Thompson, A.K. Singh, et al. 2004. Enzyme-linked immunosorbent assay for ultratrace determination of antibiotics in aqueous samples. *J. Environ. Qual.* 33: 250–256.

346. Aga, D.S., R. Goldfish, and P. Kulshrestha. 2003. Application of ELISA in determining the fate of tetracyclines in land-applied livestock wastes. *Analyst* 128: 658–662.

347. Aga, D.S., S. O'Connor, S. Ensley, et al. 2005. Determination of the persistence of tetracycline antibiotics and their degradates in manure-amended soil using enzyme-linked immunosorbent assay and liquid chromatography-mass spectrometry. *J. Agric. Food Chem.* 53: 7165–7171.

348. Meyer, M.T., J.E. Bumgarner, J.L. Varns, et al. 2000. Use of radioimmunoassay as a screen for antibiotics in confined animal feeding operations and confirmation by liquid chromatography/mass spectrometry. *Sci. Total Environ.* 248: 181–187.

349. Pastor-Navarro, N., S. Morais, A. Maquieira, et al. 2007. Synthesis of haptens and development of a sensitive immunoassay for tetracycline residues: Application to honey samples. *Anal. Chim. Acta* 594: 211–218.

350. Beausse, J. 2004. Selected drugs in solid matrices: A review of environmental determination, occurrence and properties of principal substances. *Trends Anal. Chem.* 23: 753–761.

351. Li, D., M. Yang, J. Hu, et al. 2008. Determination of penicillin G and its degradation products in a penicillin production wastewater treatment plant and the receiving river. *Water Res.* 42: 307–317.

352. Gaudin, V., J. Fontaine, and P. Maris. 2001. Screening of penicillin residues in milk by a surface plasmon resonance-based biosensor assay: Comparison of chemical and enzymatic sample pre-treatment. *Anal. Chim. Acta* 436: 191–198.

353. Gustavsson, E., J. Degelaen, P. Bjurling, et al. 2004. Determination of β-lactams in milk using a surface plasmon resonance-based biosensor. *J. Agric. Food Chem.* 52: 2791–2796.

354. Benito-Pena, E., M.C. Moreno-Bondi, G. Orellana, et al. 2005. Development of a novel and automated fluorescent immunoassay for the analysis of β-lactam antibiotics. *J. Agric. Food Chem.* 53: 6635–6642.

355. Yang, S. and K.H. Carlson. 2004. Solid-phase extraction-high-performance liquid chromatography-ion trap mass spectrometry for analysis of trace concentrations of macrolide antibiotics in natural and waste water matrices. *J. Chromatogr. A* 1038: 141–155.

356. Situ, C. and C.T. Elliott. 2005. Simultaneous and rapid detection of five banned antibiotic growth promoters by immunoassay. *Anal. Chim. Acta* 529: 89–96.

357. Deng, A., M. Himmelsbach, Q.Z. Zhu, et al. 2003. Residue analysis of the pharmaceutical diclofenac in different water types using ELISA and GC-MS. *Environ. Sci. Technol.* 37: 3422–3429.

358. Huo, S.-M., H. Yang, and A.-P. Deng. 2007. Development and validation of a highly sensitive ELISA for the determination of pharmaceutical indomethacin in water samples. *Talanta* 73: 380–386.

359. Liu, W., C. Zhao, Y. Zhang, et al. 2007. Preparation of polyclonal antibodies to a derivative of 1-aminohydantoin (AHD) and development of an indirect competitive ELISA for the detection of nitrofurantoin residue in water. *J. Agric. Food Chem.* 55: 6829–6834.

360. Sheikh, S.H. and A. Mulchandani. 2001. Continuous-flow fluoro-immunosensor for paclitaxel measurement. *Biosens. Bioelectron.* 16: 647–652.

361. Medina, M.B. 2004. Development of a fluorescent latex immunoassay for detection of a spectinomycin antibiotic. *J. Agric. Food Chem.* 52: 3231–3236.

362. Kuster, M., M. Jose Lopez de Alda, and D. Barcelo. 2004. Analysis and distribution of estrogens and progestogens in sewage sludge, soils and sediments. *Trends Anal. Chem.* 23: 790–798.

363. Desbrow, C., E.J. Routledge, G.C. Brighty, et al. 1998. Identification of estrogenic chemicals in STW effluent. 1. Chemical fractionation and in vitro biological screening. *Environ. Sci. Technol.* 32: 1549–1558.

364. Käppel, N.D., F. Proll, and G. Gauglitz. 2007. Development of a TIRF-based biosensor for sensitive detection of progesterone in bovine milk. *Biosens. Bioelectron.* 22: 2295–2300.

365. Lu, H., G. Conneely, M.A. Crowe, et al. 2006. Screening for testosterone, methyltestosterone, 19-nortestosterone residues and their metabolites in bovine urine with enzyme-linked immunosorbent assay (ELISA). *Anal. Chim. Acta* 570: 116–123.

366. Hansen, P.D., H. Dizer, B. Hock, et al. 1998. Vitellogenin—a biomarker for endocrine disruptors. *Trends Anal. Chem.* 17: 448–451.

367. Pelissero, C., G. Flouriot, J.L. Foucher, et al. 1993. Vitellogenin synthesis in cultured hepatocytes; an in vitro test for the estrogenic potency of chemicals. *J. Steroid Biochem. Mol. Biol.* 44: 263–272.

368. Purdom, C.E., P.A. Hardiman, V.V.J. Bye, et al. 1994. Estrogenic effects of effluents from sewage treatment works. *Chem. Ecology* 8: 275–285.

369. Zhao, L., J.-M. Lin, Z. Li, et al. 2006. Development of a highly sensitive, second antibody format chemiluminescence enzyme immunoassay for the determination of 17[beta]-estradiol in wastewater. *Anal. Chim. Acta* 558: 290–295.

370. Farré, M. 2006. Evaluation of commercial immunoassays for the detection of estrogens in water by comparison with high-performance liquid chromatography tandem mass spectrometry HPLC–MS/MS (QqQ). *Anal. Bioanal. Chem.* 385: 1001.

371. Coille, I., S. Reder, S. Bucher, et al. 2002. Comparison of two fluorescence immunoassay methods for the detection of endocrine disrupting chemicals in water. *Biomol. Eng.* 18: 273–280.

372. Zhang, W.-W., Y.-C. Chen, Z.-F. Luo, et al. 2007. Analysis of 17[beta]-estradiol from sewage in coastal marine environment by surface plasmon resonance technique. *Chem. Res. Chinese Univers.* 23: 404–407.

373. Kazlauskas, R. 2000. Drugs in sports: Analytical trends. *Ther. Drug Monit.* 22: 103–109.

374. Ueki, M. and M. Okano. 1999. Doping with naturally occurring steroids. *J. Toxicol. Toxin. Rev.* 18: 177–195.

375. Tschmelak, J., M. Kumpf, N. Kappel, et al. 2006. Total internal reflectance fluorescence (TIRF) biosensor for environmental monitoring of testosterone with commercially available immunochemistry: Antibody characterization, assay development and real sample measurements. *Talanta* 69: 343–350.

376. Degand, G., P. Schmitz, and G. Maghuin-Rogister. 1989. Enzyme immunoassay screening procedure for the synthetic anabolic estrogens and androgens diethylstilbestrol, nortestosterone, methyltestosterone and trenbolone in bovine urine. *J. Chromatogr. B* 489: 235–243.

377. Salvador, J.P., F. Sanchez-Baeza, and M.P. Marco 2007. Preparation of antibodies for the designer steroid tetrahydrogestrinone and development of an Enzyme-Linked Immunosorbent Assay for human urine analysis. *Anal. Chem.* 79: 3734–3740.

378. Salvador, J.P., F. Sanchez-Baeza, and M.P. Marco. 2008. Simultaneous immunochemical detection of stanozolol and the main human metabolite, 3'-hydroxy-stanozolol, in urine and serum samples. *Anal. Biochem.* 376: 221–228.

379. Tang, P.W., D.L. Crone, C.S. Chu, et al. 1993. Measuring the nandrolone threshold ratio by enzyme-linked immunosorbent assay for 5[alpha]-estrane-3[beta],17[alpha]-diol. *Anal. Chim. Acta* 275: 139–146.

380. Conneely, G., M. Aherne, H. Lu, et al. 2007. Electrochemical immunosensors for the detection of 19 nortestosterone and methyltestosterone in bovine urine. *Sens. Actuat. B: Chem.* 121: 103–112.

381. Conneely, G., M. Aherne, H. Lu, et al. 2007. Development of an immunosensor for the detection of testosterone in bovine urine. *Anal. Chim. Acta* 583: 153–160.

382. Lu, H., G. Conneely, M. Pravda, et al. 2006. Screening of boldenone and methylboldenone in bovine urine using disposable electrochemical immunosensors. *Steroids* 71: 760–767.

383. Corrie, J.E.T., W.A. Ratcliffe, and J.S. Macpherson. 1981. Generally applicable 125 iodine-based radioimmunoassays for plasma progesterone. *Steroids* 38: 709–717.

384. Elder, P.A., K.H.J. Yeo, J.G. Lewis, et al. 1987. An enzyme-linked immunosorbent assay (ELISA) for plasma progesterone: Immobilised antigen approach. *Clin. Chim. Acta* 162: 199–206.

385. Kohen, F., J.B. Kim, H.R. Lindner, et al. 1981. Development of a solid-phase chemiluminescence immunoassay for plasma progesterone. *Steroids* 38: 73–88.

386. Aherne, G.W., J. English, and V. Marks. 1985. The role of immunoassay in the analysis of microcontaminants in water samples. *Ecotoxicol. Environ. Saf.* 9: 79–83.

387. Carralero, V., A. Gonzalez-Cortes, P. Yanez-Sedeno, et al. 2007. Nanostructured progesterone immunosensor using a tyrosinase-colloidal gold-graphite-Teflon biosensor as amperometric transducer. *Anal. Chim. Acta* 596: 86–91.

388. Roberts, C.J. and L.S. Jackson. 1995. Development of an ELISA using a universal method of enzyme-labelling drug-specific antibodies Part I: Detection of dexamethasone in equine urine. *J. Immunol. Methods* 181: 157–166.

389. Anfossi, L., C. Tozzi, C. Giovannoli, et al. 2002. Development of a non-competitive immunoassay for cortisol and its application to the analysis of saliva. *Anal. Chim. Acta* 468: 315–321.

390. Pujos, E. 2005. Comparison of the analysis of corticosteroids using different techniques. *Anal. Bioanal. Chem.* 381: 244–254.

391. Brambilla, G., M. Fiori, and E. Pierdominici. 1998. A possible correlation between the blood leukocyte formula and the use of glucocorticoids as growth promoters in beef cattle. *Veterin. Res. Commun.* 22: 457–465.

392. Delahaut, P., P. Jacquemin, Y. Colemonts, et al. 1997. Quantitative determination of several synthetic corticosteriods by gas chromatography-mass spectrometry after purification by immunoaffinity chromatography. *J. Chromatogr. B* 696: 203–215.

393. Brunn, H. 1994. Identification and quantification of dexamethasone and related xenobiotic corticosteroids in cattle urine with ELISA and HPLC/ELISA. *Archiv Lebensmittelhyg* 45: 96.

394. Sanchez-Martinez, M.L., M.P. Aguilar-Caballos, S.A. Eremin, et al. 2007. Long-wavelength fluorescence polarization immunoassay for surfactant determination. *Talanta* 72: 243–248.

395. Carlson, L., B. Holmquist, R. Ladd, et al. 1996. Immunoassay for mercury in seafood and animal tissues. In: R.C. Beier and L.H. Stanker (eds), *Immunoassays for Residue Analysis*, pp. 388–394. Oxford: Oxford University Press.

396. Abad, A., M.J. Moreno, and A. Montoya. 1999. Development of monoclonal antibody-based immunoassays to the *N*-methylcarbamate pesticide carbofuran. *J. Agric. Food Chem.* 47: 2475–2485.

397. Mauriz, E., A. Calle, J.J. Manclus, et al. 2006. Single and multi-analyte surface plasmon resonance assays for simultaneous detection of cholinesterase inhibiting pesticides. *Sens. Actuat. B: Chem.* 118: 399–407.

398. Lee, H.J., G. Shan, T. Watanabe, et al. 2002. Enzyme-linked immunosorbent assay for the pyrethroid deltamethrin. *J. Agric. Food Chem.* 50: 5526–5532.

399. Lee, N.J., H.L. Beasley, S.W.L. Kimber, et al. 1997. Application of immunoassays to studies of the environmental fate of endosulfan. *J. Agric. Food Chem.* 45: 4147–4155.

400. Shan, G.M., D.W. Stoutamire, I. Wengatz, et al. 1999. Development of an immunoassay for the pyrethroid insecticide esfenvalerate. *J. Agric. Food Chem.* 47: 2145–2155.

401. Garrett, S.D., D.J.A. Appleford, G.M. Wyatt, et al. 1997. Production of a recombinant anti-parathion antibody (scFv); stability in methanolic food extracts and comparison to an anti-parathion monoclonal antibody. *J. Agric. Food Chem.* 45: 4183–4189.

402. Watanabe, E., Y. Kanzaki, H. Tokumoto, et al. 2002. Enzyme-linked immunosorbent assay based on a polyclonal antibody for the detection of the insecticide fenitrothion. Evaluation of antiserum and application to the analysis of water samples. *J. Agric. Food Chem.* 50: 53–58.

403. Kim, Y.J., Y.A. Kim, Y.T. Lee, et al. 2007. Enzyme-linked immunosorbent assays for the insecticide fenitrothion—influence of hapten conformation and sample matrix on assay performance. *Anal. Chim. Acta* 591: 183–190.

404. Cho, Y.A., Y.J. Kim, B.D. Hammock, et al. 2003. Development of a microtiter plate ELISA and a dipstick ELISA for the determination of the organophosphorus insecticide fenthion. *J. Agric. Food Chem.* 51: 7854–7860.

405. Nakata, M., A. Fukushima, and H. Ohkawa. 2001. A monoclonal antibody-based ELISA for the analysis of the insecticide flucythrinate in environmental and crop samples. *Pest. Manage. Sci.* 57: 269–277.

406. Vera-Avila, L.E., J.C. Vazquez-Lira, M.G. De Llasera, et al. 2005. Sol-gel immunosorbents doped with polyclonal antibodies for the selective extraction of malathion and triazines from aqueous samples. *Environ. Sci. Technol.* 39: 5421–5426.

407. Alcocer, M.J.C., P.P. Dillon, B.M. Manning, et al. 2000. Use of phosphonic acid as a generic hapten in the production of broad specificity anti-organophosphate pesticide antibody. *J. Agric. Food Chem.* 48: 2228–2233.

408. Hennion, M.-C. and D. Barceló. 1998. Strengths and limitations of immunoassays for effective and efficient use for pesticide analysis in water samples: A review. *Anal. Chim. Acta* 362: 3–34.

409. Liu, S.H., L. Wang, and L.H. Wei. 2005. Studies on the immunoassay for triazophos. *Chin. J. Anal. Chem.* 33: 1697–1700.
410. Gui, W.J., R.Y. Jin, Z.L. Chen, et al. 2006. Hapten synthesis for enzyme-linked immunoassay of the insecticide triazophos. *Anal. Biochem.* 357: 9–14.
411. Karu, A.E., R.O. Harrison, D.J. Schmidt, et al. 1991. Monoclonal immunoassay of triazine herbicides—development and implementation. *ACS Symp. Ser.* 451: 59–77.
412. Bushway, R.J., L.B. Perkins, L. Fukal, et al. 1991. Comparison of enzyme-linked-immunosorbent-assay and high-performance liquid-chromatography for the analysis of atrazine in water from Czechoslovakia. *Archiv. Environ. Contam. Toxicol.* 21: 365–370.
413. Gascon, J., A. Oubina, B. Ballesteros, et al. 1997. Development of a highly sensitive enzyme-linked immunosorbent assay for atrazine—performance evaluation by flow injection immunoassay. *Anal. Chim. Acta* 347: 149–162.
414. Wortberg, M., M.H. Goodrow, S.J. Gee, et al. 1996. Immunoassay for simazine and atrazine with low cross-reactivity for propazine. *J. Agric. Food Chem.* 44: 2210–2219.
415. Winklmair, M., M.G. Weller, J. Mangler, et al. 1997. Development of a highly sensitive enzyme-immunoassay for the determination of triazine herbicides. *Fresenius J. Anal. Chem.* 358: 614–622.
416. Brena, B.M., L. Arellano, C. Rufo, et al. 2005. ELISA as an affordable methodology for monitoring groundwater contamination by pesticides in low-income countries. *Environ. Sci. Technol.* 39: 3896–3903.
417. Schneider, P. and B.D. Hammock. 1992. Influence of the ELISA format and the hapten enzyme conjugate on the sensitivity of an immunoassay for S-triazine herbicides using monoclonal-antibodies. *J. Agric. Food Chem.* 40: 525–530.
418. Gascon, J., E. Martinez, and D. Barcelo. 1995. Determination of atrazine and alachlor in natural-waters by a rapid-magnetic particle-based elisa—influence of common cross-reactants—deethylatrazine, deisopropylatrazine, simazine and metolachlor. *Anal. Chim. Acta* 311: 357–364.
419. Kramer, P. and R. Schmid. 1991. Flow-Injection Immunoanalysis (FIIA)—a new immunoassay format for the determination of pesticides in water. *Biosens. Bioelectron.* 6: 239–243.
420. Klotz, A., A. Brecht, C. Barzen, et al. 1998. Immunofluorescence sensor for water analysis. *Sens. Actuat. B: Chem.* 51: 181–187.
421. Tudorache, M., M. Co, H. Lifgren, et al. 2005. Ultrasensitive magnetic particle-based immunosupported liquid membrane assay. *Anal. Chem.* 77: 7156–7162.
422. Hegedus, G., I. Belai, and A. Szekacs. 2000. Development of an enzyme-linked immunosorbent assay (ELISA) for the herbicide trifluralin. *Anal. Chim. Acta* 421: 121–133.
423. Shen, J., Z. Zhang, Y. Yao, et al. 2006. A monoclonal antibody-based time-resolved fluoroimmunoassay for chloramphenicol in shrimp and chicken muscle. *Anal. Chim. Acta* 575: 262–266.
424. Rodriguez-Mozaz, S., M.J.L. de Alda, and D. Barcelo. 2006. Fast and simultaneous monitoring of organic pollutants in a drinking water treatment plant by a multi-analyte biosensor followed by LC-MS validation. *Talanta* 69: 377–384.

9 Application of Biotests

Lidia Wolska, Agnieszka Kochanowska, and Jacek Namieśnik

CONTENTS

9.1 Introduction 189
9.2 Chemical Monitoring in Assessing the Extent of Environmental Pollution 190
9.3 Importance of Toxicity Tests 192
    9.3.1 Toxkit Tests 195
9.4 Integrated System of Water Pollution Assessment 200
9.5 Legal Regulations Applying to the Use of Bioassays in Environmental Monitoring 201
9.6 Ecotoxicological Classification of Environmental Samples 201
9.7 Bioassay Application in Environmental Monitoring: Some Case Studies 210
    9.7.1 Identification of Toxic Compounds 210
    9.7.2 Identification of Hot Spots (Sites with a Very High Level of Pollution) 211
    9.7.3 Ranking of Problems in Polluted Areas Managed by Specific Authorities 212
    9.7.4 Revision of Monitoring Parameters 214
9.8 Conclusions 216
References 216

9.1 INTRODUCTION

Rapid developments of new technologies, progressive urbanization, and consumer lifestyles cause adverse and sometimes even irreversible changes in the environment. Air, water, and soil are exposed to large-scale pollution, originating mostly from anthropogenic sources. The pollution present in the abiotic part of the environment is subject to the following processes:

- Transport, which causes it to occur in places distant from emission sources.
- Physical and chemical changes (photochemical and biochemical) responsible for the production of new compounds (secondary pollution), and a decrease in the concentration of primary pollution.

From the abiotic part of the environment, chemical compounds permeate into plant, animal, and human organisms. Living organisms function in an integrated network of connections between themselves and their surroundings. A stable exchange of matter, energy, and information takes place between the elements of the networks, and the correct functioning of all the elements is possible only in a state of mutual dynamic equilibrium, that is, homeostasis. Changes in the environment disrupt this equilibrium and if far-reaching, often cause irreversible damage to individual species or entire ecosystems.

At the beginning of the 1970s, increasing environmental awareness regarding the influence of toxic substances resulted in the emergence of many nongovernmental organizations; in cooperation with government departments they established systems for monitoring and reducing pollution.

In the early days of these systems, monitoring was based mostly on chemical parameters, both total and individual. General criteria for the selection of these parameters resulted from the then state of knowledge about the presence of toxic compounds in the environment and included the toxicity of substances and the scale of their occurrence in the environment. Along with an increasing recognition of the network relationships occurring in ecosystems and the identification of new environmental toxins, the list of monitored substances gradually grew. At present, according to the Water Directive, some 170 substances need to be monitored in aquatic environments (including 33 priority compounds), and new substances are waiting to be added to the list.[1]

Gathering information about the environment is significantly hampered by

- The presence of thousands of substances (with well-known or unknown toxic properties), which have not been detected by monitoring.
- The bioavailability of toxic substances, which is not identical to their total content in individual compartments of the environment.
- Synergetic effects, the estimation of which has too great a risk of error in such a complex system as environmental samples.

Information based on chemical monitoring records alone is difficult to convert into specific knowledge about any real threat to living organisms in a given compartment of the environment caused by the presence in it of toxic substances. This is a consequence of a basic fault in the present system, namely, the impossibility of indicating the potential toxic effects on the investigated ecosystem of a diverse (with regard to the kinds of pollutants and their concentrations) mixture of compounds. The question then arises: How is it possible to accurately quantify the threat associated with such a complex mixture of pollutants present in the environment?

9.2 CHEMICAL MONITORING IN ASSESSING THE EXTENT OF ENVIRONMENTAL POLLUTION

At present, the monitoring of aquatic environments across the European Union (EU) is administered by the Water Framework Directive (WFD) and its daughter directives, which oblige EU member states to improve the status of the waters in their river basins and maintain them in "good status" by 2015.[1] Within the framework of a "management by river basin" system, the member states establish surveillance and operational monitoring. Surveillance monitoring is conducted to obtain information on long-term changes in the natural conditions prevalent in a given area, or changes resulting from widespread anthropogenic activity. When there is a risk of the environmental objectives for the waters in a given river basin not being fulfilled, operational monitoring is applied. In some cases, it is necessary to invoke investigative monitoring, the aim then being to determine why the waters in a particular area do not meet environmental objectives. Each of the three kinds of monitoring requires appropriate tools to obtain significant and reliable information for effective water management.

Although the WFD does not include recommendations concerning specific monitoring methods, it is self-evident that their selection should allow for costs and the applicability of a given method. The achievement of WFD objectives, however, also depends on the availability of suitable tools and technologies.

According to Annex V of the WFD, surface water monitoring should include the following indicative parameters of water quality elements[1]:

- Biological
- Hydromorphological
- Chemical and physicochemical

The hydromorphological quality elements include the dynamics of water flow, the size of the investigated water district, and the structure and composition of the ground. The biological quality elements include the species composition and population sizes of organisms living in a given aquatic ecosystem (flora, benthic invertebrates, and ichthyofauna). The changes in the composition and numbers of organisms ensuing from environmental pollution (biological, chemical, or physical) appear after a certain time lag (often after many months or years) and therefore relate to the existing effects of pollution. As such, this information cannot result in either an efficient and effective program for future monitoring or preventive actions within operational monitoring.

Annex V in the WFD describes the framework for groundwater monitoring; it specifies monitoring of the quantitative status (by measuring the groundwater level) and chemical status.

Until quite recently, chemical monitoring relied exclusively on determining the concentrations of certain chemical compounds (selected as indicators of chemical environmental pollution) in water samples, sediments, or soils using classical analytical methods. From a theoretical point of view, the best use of the appropriate analytical methods would be to provide a full analytical characterization of the environment, that is, to determine the concentrations of all known and unknown pollutants in each of its compartments. However, it is doubtful whether such a task is possible or even relevant, bearing in mind

- The number of compounds which should be determined
- The diversity of concentrations (mostly trace and microtrace)
- The fluctuations of pollutant concentrations over time and in a given area
- The complex composition of sample matrices
- The complicated and therefore time-consuming and labor-intensive procedures of preparing samples for analysis
- The additional stress on the environment resulting from the use of chemical reagents, primarily organic solvents used in sample preparation
- The additional costs involved in the purchase of high-purity reagents, and the necessity to use an excess of reagents
- The problems with obtaining suitable reference materials necessary for the validation of analytical methods and the calibration of measuring and control instruments

Classical analytical methods also have other limitations. Measurement data cannot be a source of information about possible interactions of toxic substances because[2]

- Toxic effects are summed (additive synergism)
- The overall toxic effect is significantly greater than that resulting from the simple summation of the effects of the individual components (hyperadditive synergism)
- The effects of chemical compounds may weaken and even cancel each other out (antagonism)

In practice, the classical analytical methods used to assess the degree of environmental pollution are intended for the determination of only a limited number of chemical compounds (or groups of compounds), that is, those whose presence in the environment and permitted levels of concentration are regulated for environmental protection. Present legal regulations do not take the following into account[3]:

- Newly synthesized compounds
- Unidentified pollutants, due to the imperfection of analytical procedures and of monitoring and measuring instruments

The impact of pollutants (mostly from anthropogenic sources) on ecosystems is increasing rapidly. Unfortunately, knowledge about their harmful influence on living organisms is accumulating at a considerably slower pace. At present, List I in Council Directive 76/464/EEC,[4] which includes especially dangerous pollutants, defined as permanent, toxic, and subject to bioaccumulation, contains 33 priority substances. This list is open; with time and increasing knowledge, new and dangerous compounds, specified by the European Committee, the Council, and member states, are added to it. The requirements concerning the discharge of pollutants specified in List I and the standards of environmental water quality with respect to admissible levels of these substances have been defined at EU level. Where pollution poses a lesser threat but nonetheless restricts or hinders the use of water resources to a significant extent (List II), suitable standards of quality are established at member state level, although they have been introduced for only 25 out of 139 substances.

The inconveniences associated with analyzing a complex mixture of pollutants can be avoided by determining the total parameters in a given compartment of the environment. Chemical oxygen demand (COD) and biological oxygen demand (BOD) or total carbon concentration (TC) and total organic carbon (TOC), which measure the total organic matter or the entire content of a given element in the pollutants present in an investigated sample, can be successfully applied in environmental analytics.[5,6] Unfortunately, however, the information obtained in this manner is still difficult to convert into knowledge about the toxicity of the environmental compartment in question *vis-à-vis* its living organisms.[7] The diverse bioavailability of the forms in which chemical compounds can occur in the environment makes it virtually impossible to make a reliable risk assessment of potential ecotoxicological effects based on total chemical parameters.[8]

Bearing in mind the aforementioned problems, one can see that analysis of environmental samples based on the determination of every chemical compound and/or the selection of total parameters may supply only part of the knowledge necessary for assessing the toxic impact of chemicals on living organisms. At present, chemical monitoring is only to a limited extent capable of identifying and determining the quantity of compounds with a possible toxic action. Moreover, the complex interactions taking place between pollutants as well as their different bioavailabilities significantly restrict the ability of such systems to assess the quality of environmental compartments.

These limitations of the current system of assessing aquatic environment quality indicate that further research and newer, more reliable tools are needed. Such tools introduced into analytical practice would enable fresh information to be obtained. This information would then complement the data obtained from chemical monitoring and would enable the real risk from the presence of a mixture of diverse pollutants in the environment to be adequately assessed.

9.3 IMPORTANCE OF TOXICITY TESTS

Ecotoxicity studies have become increasingly important in pollution assessment; an effective tool and valuable complement to the information obtained from chemical monitoring, they are being implemented more and more frequently. However, until quite recently, toxicity tests served exclusively to determine and compare the toxicity of individual chemical substances, for example, during research conducted in order to authorize the sale of a specific pesticide. In this instance, results in the form of estimated toxicity coefficients are used for the registration, licensing, and enactment of legal regulations for chemical substances. On the basis of toxicity tests, one can also rate the influence of an environmental sample (waters, sediments, or soils), which is a mixture of many chemical substances, on the health status of living organisms. At present, it is usual to assess the effective concentration EC50 in research on environmental samples, although this value is still often identified with the toxic influence of a single chemical compound.

Therefore, toxicity tests, or in other words bioassays, can supply information on the total load of an investigated sample in a diverse (in terms of type and quantity) mixture of pollutants, which allows for the possibility of their interactions.[8–12] Bioassays are based on the use of particularly sensitive species (bioindicators), which are characterized by their quick reaction to changes in their environment. This results from their relatively low ability to maintain a stable state of equilibrium, that is, from their narrow range of tolerance to specific toxic factors. Such organisms show a special ability to accumulate pollutants.[13] Hence, they can work as so-called Biological Early Warning

Systems (BEWS), delivering the first signals about toxicity in an environmental compartment and simultaneously indicating the need for further, more detailed, analyses using classical methods.[14–17]

A bioassay may therefore be understood as an analytical method based on the observation of a response at a given level (or levels) of biological organization, resulting from homeostasis disorders induced by changes occurring in a given compartment of the environment (i.e., the occurrence of pollution).[18,19] The aim of a bioassay is to prove the presence of toxic substances in the environment and/or demonstrate their harmfulness through the quantitative assessment of a given substance's impact on living organisms (based on a comparison with a control sample). Measurement of toxicity is thus an example of relative measurement, so common in classical chemical analysis.

Bioassays used in analytical practice can be classified according to the type of bioindicators used in a given toxicity test. The organisms most frequently used are bacteria, plants, and animals; detailed information on their application in bioassays to assess environmental pollution is given in a review study.[20] In this chapter, we present tables for the use of selected bioindicators in toxicity tests.

Basic information about the degree of pollution in a given environmental element is delivered by bioassays done using a single plant or animal species (single species tests) representing a specific trophic state. The tests are conducted according to standardized procedures and under controlled laboratory conditions that are optimal for the tested organism.[21–26] Obviously, there is no ideal bioindicator. The selection of a suitable organism for toxicity testing depends on the type of information required, the state, physical and chemical properties of the analyzed sample (its origin), the type of investigated substances or mixture of chemical substances, and the sensitivity of the examined species. Each bioindicator species displays a different sensitivity to different groups of pollutants. This limitation should be considered during the planning and interpretation of bioindicator-based findings on the degree of environmental pollution. A bioassay where only one bioindicator species is used to determine the toxicity reflects only the sensitivity of the tested species.

Such a procedure carries the risk of underestimating the toxicity of investigated substances with regard to an entire ecosystem. It is important that toxicity tests be conducted simultaneously with several bioindicators, that is, with a battery of bioassays characterized by different sensitivities and representing different trophic levels.[19] Such an approach is often applied in research on environmental samples, which are usually complex mixtures of compounds with unknown physicochemical properties.

Single species tests and tests using bioassay batteries are called lower-tier tests.[19] But in order to carry out a more stringent examination of the complex interactions between potentially toxic chemical compounds and organisms inhabiting specific ecosystems, experiments are sometimes carried out in microcosms and mesocosms; the literature describes these as higher-tier tests.[19]

An important requirement for toxicity tests is their reproducibility and repeatability. In conformation with the principles of good laboratory practice (GLP),[27] it is recommended that they be performed according to standard procedures and guidelines prepared by world standardization organizations such as the OECD, ISO, or CEN. Table 9.1 presents a list of selected ISO standards and OECD guidelines with regard to the modality of toxicity tests.

Many European states have already taken note of the benefits of ecotests, this is reflected in national regulations. Also, the European Commission, in the provisions of the WFD, requires EU member states to have achieved good ecological status of surface waters and groundwaters by 2015, and suggests the use of ecotests as one way of reaching this objective.[1] Article 16 of the WFD, which deals with protection strategies against water pollution, stipulates that the order in which activities are undertaken with regard to priority pollutants should be based on the degree of the related threat to the aquatic environment or, through the aquatic environment, to human. Such an assessment can be conducted using a method introduced in Council Regulation 793/93 (on the evaluation and control of the risks of existing substances).[28] This assessment refers exclusively to toxicity in aquatic and human environments (exposure through the mediation of water). Additionally, Annex V in the WFD contains recommendations concerning the establishment of environmental

TABLE 9.1
Standards (ISO) and Guidelines (OECD) Determining the Manner of Performing Toxicity Tests Using Selected Bioindicators[29–32]

Number of the Standard/ Research Method——Assay Guidelines	Bioindicator	Type/Application of the Assay
1	2	3
ISO 10712:1995 PN-EN ISO 10712:2001	Bacteria (*Pseudomonas putida*)	Growth inhibition (testing the inhibition of cell proliferation)
ISO 11348-1:1998 PN-EN ISO 11348-1:2002	Marine luminescent bacteria (*V. fischeri*)	Bioluminescence inhibition of bacteria in water samples (method using freshly prepared bacteria)
ISO 11348-2:1998 PN-EN ISO 11348-2:2002		Bioluminescence inhibition of bacteria in water samples (method using dried bacteria)
ISO 11348-3:1998 PN-EN ISO 11348-3:2002		Bioluminescence inhibition of bacteria in water samples (method using lyophilized bacteria)
ISO 15522:1999		Growth inhibition of microorganisms in activated sludge
ISO 8692:2004 PN-EN 8692:2005 OECD 201 (modification of guidelines, passed in 1984)	Algae——chlorophytes (*Scenedesmus subspicatus, Selenastrum capricornutum, Chlorella vulgaris*)	Growth inhibition of freshwater algae
ISO 10253:2006 PN-EN ISO 10253:2002	Algae—diatoms (*Skelotonema costatum, Phaeodactylum tricornutum*)	Growth inhibition of marine algae
OECD 221(new guidelines, 2000)	Duckweed (*Lemna minor*) Duckweed (*Lemna gibba*)	Growth inhibition
OECD 208 (original guidelines, passed in 1984)	Land plants	Influence on growth
OECD 208A (project of guideline modifications 2000)		Germination and seedling growth
OECD 208B (project of guideline modifications 2000)		Vegetative abilities
OECD 202 (guideline modifications, passed in 1984) ISO 6341:1996 PN-EN ISO 6341:2002 OECD 211 (original guidelines, passed in 1998) ISO 10706:2000	Freshwater crustacean—*Daphnia* (*Daphnia magna*)	Mobility inhibition Influence on reproduction
ISO 14669:1999	Marine crustaceans (*Acartia tonsa, Tisbe battagliai, Nitocra spinipes*)	Acute toxicity
OECD 218 (original guidelines, passed in 2000)	Chironomidae (*Chironomus tentans, Chironomus riparius*)	Sediment toxicity
OECD 219 (original guidelines, passed in 2000)		Water toxicity

TABLE 9.1   (continued)

Number of the Standard/ Research Method—Assay Guidelines	Bioindicator	Type/Application of the Assay
1	2	3
OECD 207 (original guidelines, passed in 1984)	Redworm (*Eisenia fetida*)	Acute toxicity using artificial soil substrate
ISO 11268-1:1993 PN-ISO 11268-1:1997		Influence on reproduction
ISO 11268-2:1998 PN-ISO 11268-2:2001		
OECD 220 (project of guideline modifications 2000)	Potworm (*Enchytraeus* sp.)	Influence on reproduction
OECD 203 (guidelines passed in 1992)	Fish Zebra fish (*Brachydanio rerio*)	Acute toxicity
ISO 7346-1:1996 PN-EN ISO 7346-1:2002		Acute toxicity (static method)
ISO 7346-2:1996 PN-EN ISO 7346-2:2002		Acute toxicity (semistatic method)
ISO 7346-3:1996 PN-EN ISO 7346-3:2002		Acute toxicity (flow-through method)
OECD 212 (original guidelines 1998)	Fish Zebra fish (*Brachydanio rerio*),	Toxicity in embryonic stage (short-term test)
OECD 210 (original guidelines 1992)	Rainbow trout (*Oncorhynchus mykiss*), Fathead minnow (*Pimephales promelas*)	Toxicity in early life (fry)
OECD 204 (original guidelines 1992)	Zebra fish (*Danio rerio*)	Influence on growth (long-term test)
ISO 10229:1994	Rainbow trout (*Oncorhynchus mykiss*)	

quality standards with regard to priority pollutants. If possible, countries should obtain information on the acute and chronic toxicities of these pollutants, both for the "basic set" of species for a given type of water,[1] including

- Algae and/or macrophytes
- Daphnia or organisms representative of saline waters
- Fish

and also other aquatic species for which such data are available.

9.3.1   TOXKIT TESTS

The tests most frequently used in analytical practice in many countries are conventional toxicity tests, often recommended by national and international standard organizations to assess the toxicity of

- Freshwaters using, inter alia, algae (*Chlorella vulgaris*), daphnia (*Daphnia magna* Straus), rainbow trout (*Oncorhynchus mykiss*), zebra fish (*Brachydanio rerio* Hamilton–Buchanan), guppies (*Lebistes reticulatus* Peters), and *Gammarus varsoviensis* (Jażdż.)

- Saline waters using algae (*Skeletonema costatum* and *Phaeodactylum tricornutum*), shrimps (*Crangon crangon*), Pacific Ocean oyster larvae (*Crassostrea gigas*), young turbots (*Rhombus maximus*), and flatfish (*Pleuronectes platessa*)
- Soils using redworm (*Eisenia fetida*)

Despite the increasing popularity of such tests in laboratories worldwide, their deficiencies and limitations should be borne in mind. They demand continuous cultivation of test organisms, which considerably increases research costs. At the same time, because of changes occurring in the populations of bioindicator organisms, considerable differences in sensitivity can arise between test organisms originating from different laboratories.[19] Moreover, the results of a conventional test are obtained over a period ranging from 24 h to several days, which means that this type of test is practically useless for a pollution event that requires immediate action.

The need to perform environmental studies on an increasing scale in order to obtain the most comprehensive information on the state of the environment and its processes has led to an increase in the importance and range of rapid miniaturized bioassays, variously known as microbioassays, alternative tests, or second-generation assays.[33] The bioassays utilize unicellular or small multicellular organisms that exhibit a specific response on contact with a liquid sample.

Bioluminescent bacteria are exceptionally sensitive, and even trace quantities of toxins in water reduce the amount of light they emit. In 1979, this phenomenon was used to assess the degree of pollution in water samples, the intensity of the bioluminescent light of these bacteria being measured.[34] A year later, the American company AZUR Environmental (formerly Microbics Corporation) manufactured the first specialist equipment—Microtox®—for carrying out analyses of environmental samples.

Nowadays, Microtox is the most popular bioassay available that uses bioluminescent bacteria as its active element. Analysts find it a useful tool for assessing pollution in different compartments of the environment, possessing as it does both the advantages of bioindicator techniques and the precision of classic instrumental analysis. Here are some examples of its application:

- Monitoring of surface water quality[35-38]
- Monitoring of groundwater[39,40]
- Determination of soil and sediment toxicity[39-49]
- Investigation of leachates from landfill sites[39,50-52]
- Initial determination of the toxicity of sewage delivered to a sewage treatment plant[39,49,53-59]
- Monitoring the stages of the treatment of municipal, industrial, and pesticide sewage[7,39,49,54,60-64]
- Monitoring of treated sewage at each stage of purification, and of water before and after water dumping[39,45]
- Monitoring the treatment stages of water for drinking purposes[39]

During the 30-year period of Microtox usage, 1500 potentially toxic substances have been tested and an immense number of environmental samples examined; the results have spawned several hundred research papers.[65] The sensitivity of bacteria to the presence of simple organic substances is similar to that displayed by daphnia and fish. However, there was greater differentiation with respect to the toxicity of larger molecules (pesticides and pharmaceuticals) and complex effluents, for example, from the pharmaceutical and chemical industries. The German DIN standards and international ISO standards recommend the use of bacteria that have been freeze-dried or have originated from continuous cultivation conducted in a laboratory. *Vibrio fischeri* bacteria are available on the market in freeze-dried form. After hydration, their cell walls are slightly

TABLE 9.2
Types of Information Obtained Based on Toxicity Assays Conducted for Bioluminescent
***V. fischeri* Bacteria[66]**

Type of Assay	Type of Obtained Information
Screening test	Results of the assay answer the question whether a sample is toxic or not. To this end, an undiluted sample is subjected to analysis. If a decrease in the intensity of bacterial luminescence is smaller than the established threshold value, then the sample need not be subjected to further research.
Basic test	Results of the assay answer the question about the toxicity of a sample. The sample for which a decrease in the intensity of bacterial luminescence exceeds an established threshold value is subjected to detailed analysis in a series of dilutions and determination of toxicity index, e.g., EC50.

damaged, which enables easier absorption of toxic substances. Their sensitivity to organic compounds is several times greater than bacteria derived from a fresh culture.[67] Thanks to the use of standard bacterium strains (NRRLB-11177), the quality of which is guaranteed by the manufacturer; the repeatability of results is very high.

Research on toxicity using bioluminescent bacteria can be conducted in two ways. Table 9.2 lists the types of information obtainable, depending on the form of the assay.

The Microtox system requires neither specialist training nor experience in work with bioindicators: the results of an assay are read automatically. As opposed to other bioassays, staffs are not required to have received a specifically biological education. The results of toxicity determinations for a whole series of samples examined at the same time are obtained within 30 min. This enables quick and effective action to be taken, which is especially important in monitoring drinking water, in assessing the efficiency of sewage treatments, and in testing the toxicity of surface waters above and below sewage outfalls.

The Microtox test can be used in field studies and in mobile laboratories.[7,8,54] This is especially important in unexpected and rapidly developing environmental threats, when continuous inspection of pollution levels is necessary.[68] Since the test is instantly applicable, and because the results are obtained within an hour of the sample being delivered to the laboratory, this type of assay has become very popular worldwide. But despite the aforementioned advantages, this system is not perfect. The most important shortcomings are

- Tests can be carried out only on samples that are colorless (although color correction is possible), clear, and of low viscosity.
- The formation of complex compounds with chloride ions in the case of samples polluted by heavy metals can lead to erroneous results (a solution of sodium chloride is added to the sample to obtain the appropriate conditions for bacteria).[69]
- The need for an additional device to control pH in the samples under examination (the optimum pH range is 6–8).
- The required presence of sodium ions in the analyzed sample, which regulates the stages of bacterial bioluminescence.[69]

Assays based on the application of bioluminescent bacteria are attracting increasing interest; in 1991, their use was regulated by the standard DIN 38412 (part 34). Although this standard was developed with regard to the control of harmful substances occurring in sewage, it can be applied to any type of water—from drinking water to water passing through a landfill.

The assays are available worldwide in the form of ready-made kits, which enable the toxicity in samples to be estimated in a short time. Work on the idea and development of a methodology for microbioassays (commonly known as Toxkits) using microorganisms not requiring continuous cultivation was pioneered by a team of scientists led by Professor Guido Persoone of Ghent University, Belgium.[70–72] Toxkits contain microbioassays equipped with all the accessories (including test organisms) necessary for easy, quick, sensitive, and repeatable toxicity assays.[73] Bioindicator organisms (from standard cultures) are provided to a laboratory in cryptobiotic form, that is, in a state of physiological rest (dormancy), for example:

- Rotifer cysts
- Crustacean eggs
- Algae—in the form of cells immobilized in a suitable medium and prevented from proliferating by a special solution

Organisms in such forms can be kept refrigerated for several months; when necessary, they are quickly prepared ("on demand") for an assay. Prior to an assay, the cysts or eggs are placed in water. Under the influence of strong light, these forms begin to grow and after 18–96 h (depending on the organism) a batch of young organisms is generated, ready to be used as biological components for a suitable assay.[74,75]

The test organisms most often used in toxicity microbioassays are *Daphnia magna*, *Daphni pulex*, and *Ceriodaphnia dubia*.[76] These crustaceans, like rotifers and protozoans, are typical organisms whose natural environments are aquatic ecosystems. Their eggs, included in the kit, are protected by chitinous capsules (Lat. *ephippium*) and can be kept for a long time without any loss of properties.

Microorganisms lie at the lowest level of the trophic pyramid, which is why any adverse changes occurring in them may directly or indirectly affect organisms at higher trophic levels, and consequently the state of the entire ecosystem. Because of their large surface area and the immediate contact of a cellular membrane with the investigated medium, microorganisms are more sensitive to toxic substances than invertebrates or fish.[33]

Toxicity is typically a function of exposure duration, which is why long-term assays have special importance in ecotoxicology. However, using long-lived species is troublesome. Microorganisms characterized by the short life span of a single generation provide a convenient means of determining the effect of chronic exposure to a toxic substance. The growing interest in this approach to environmental pollution studies is also due to other factors, such as[73]

- The elimination of continuous cultivation
- The low costs of conducting a single sample analysis
- The possibility of testing several samples simultaneously
- The short response duration
- The small sample volume of a sample
- The small space occupied in the laboratory, and also the possibility of conducting research in the field (*in situ*)

Because of the numerous benefits of Toxkit assays, they can be used for sample testing even by small analytical laboratories and local monitoring stations for routine toxicity assessments of environmental samples as well as in emergencies. The use of standard organisms means that assays can be standardized and that repeatable results can be obtained in different laboratories. Another, not unimportant aspect is that conducting toxicity assays on microorganisms does not require the consent of an ethical committee.[77] Table 9.3 provides information on microbioassays commercially available in kit form (Toxkit).[73]

TABLE 9.3
List of Commercially Available Microbioassays

Toxkit 1	Taxon 2	Test Species 3	Test Duration 4	Type of Test 5	Recommended by 6
		Freshwater Environment			
ALGALTOXKIT F™	Algae (chlorophytes)	Selenastrum capricornutum (Raphidocelis subcapitata or Pseudokirchneriella subcapitata)	72 h	Chronic toxicity (growth inhibition)	OECD, ISO
DAPHTOXKIT F™ magna	Crustaceans (daphniae)	Daphnia magna	24–48 h	Acute toxicity	OECD, ISO
DAPHTOXKIT F™ pulex	Crustaceans (daphniae)	Daphnia pulex	24–48 h	Acute toxicity	OECD
CERIODAPHTOXKIT F™	Crustaceans (daphniae)	Ceriodaphnia dubia	24 h	Acute toxicity	US EPA
THAMNOTOXKIT F™	Crustaceans (fairy shrimp)	Thamnocephalus platyurus	24 h	Acute toxicity	—
RAPIDTOXKIT™			30–60 min	Acute toxicity for larvae	—
ROTOXKIT F™	Rotifers	Brachionus calyciflorus	24 h	Acute toxicity	ASTM
ROTOXKIT F™ short-chronic			48 h	Short-term chronic toxicity (reproduction)	AFNOR
PROTOXKIT F™	Protozoans (ciliates)	Tetrahymena thermophila	24 h	Chronic toxicity (growth inhibition)	OECD
OSTRACODTOXKIT F™	Crustaceans (ostracods)	Heterocypris incongruens	6 days	Chronic toxicity (lethality/growth inhibition)	—
PHYTOTOXKIT™	Plants (monocotyledons and dicotyledons)	1. Sorghum saccharatum 2. Lepidium sativum 3. Sinapis alba	72 h	Chronic toxicity (inhibition of germination and root growth)	—
		Estuarine and Marine Environment			
MARINE ALGALTOXKIT™	Algae (diatoms)	Phaeodactylum tricornutum	72 h	Chronic toxicity (growth inhibition)	ISO
ROTOXKIT M™	Rotifers	Brachionus plicatilis	24–48 h	Acute toxicity	ASTM
ARTOXKIT M™	Crustaceans (fairy shrimp)	Artemia franciscana (previously Artemia salina)	24–48 h	Acute toxicity	—

9.4   INTEGRATED SYSTEM OF WATER POLLUTION ASSESSMENT

The stipulations of the WFD, concerning the obligation to carry out monitoring, provide legal regulations for 172 priority pollutants.[1,4] However, it has been estimated that there may be some 100,000 substances (in different and varying concentrations) in the aquatic environment. This means that the vast majority (>99%) of pollutants are not covered by chemical monitoring.[78]

The results obtained from toxicity assays are thus a valuable complement to the information derived from physicochemical and chemical studies. This simultaneous use of monitoring and ecotoxicity assays to assess the quality of the environment is called an integrated approach or integrated tool; recent years have seen a rise in its popularity among analysts. This interest is engendered by the significant information that can be obtained with this approach: information relating to the complex interactions occurring between pollutants and the different bioavailability of the forms in which these pollutants can occur.

The main sources of environmental water pollution are industrial effluents, municipal sewage, and run-off waters. Somewhat less significant is the influence of landfill site effluents and sewage sludge. Because of the pollutant load they contain, all these "waters" must comply with rigid requirements before entering the environment so as to prevent deterioration in its quality. Information on the potential risks posed by the release of these pollutants into aquatic ecosystems is a good basis for determining

- The discharge of effluents and landfill waters into the environment.
- The effectiveness of treating effluents and landfill waters.
- The manner of storage or use of sludge.

The United States is the leader as regards the integrated assessment of the quality of effluents introduced into aquatic environments. In 1984, the United States Environmental Protection Agency (US EPA) introduced the notion of Whole Effluent Toxicity (WET). WET assays may involve the following samples:[79]

- Municipal run-off waters (precipitation)
- Municipal (household) sewage
- Mine effluents (from the drainage of mines)
- Run-off waters from fields (containing pesticides and loaded with nutrients as a result of the application of fertilizers)
- All types of industrial effluents
- Landfill leachates (water drained from the bottom of a landfill)

as well as other mixtures with complex compositions that could enter the environment but whose exact composition and toxicity is unknown.

Depending on the country, the approaches to ecoassay use in assessing effluent quality vary and are described by different names.[80] In Australia and New Zealand, and also in Great Britain, the notion of direct toxicity assessment (DTA) began to be used at the beginning of the 1990s in relation to the immediate assessment of toxicity, both in effluent samples and in samples of waters into which the effluents are discharged. In European countries and in the United States WET is used, but this additionally includes tests for the bioaccumulation, biodegradation, and persistence of pollutants in the environment. The notion of Whole Effluent Assessment (WEA) has become especially common in Europe since 2000 following the Oslo-Paris Commission agreement (OSPAR).[81] In a nutshell, WEA combines tests of acute and chronic toxicity, bioaccumulation, mutagenicity, and persistence.[82] Sometimes, it also includes measurements of total parameters, such as BOD, COD, or TOC.[80] In the future this type of assessment may also include the concentrations of endocrine disruptors, whose presence can lead to disorders in hormonal and immunological systems.[82] In Germany, the term ICE is used (Integrated Control of Effluents) and in the Netherlands WEER (Whole Effluent Environmental Risk) is used.

In different compartments of the environment, sediments accumulate mainly heavy metals and persistent organic pollutants (POPs), which reach aquatic ecosystems from different sources. Metal ions do not remain dissolved in water for long: they are liable to be precipitated as a result of oxygenation, the formation of different compounds (carbonates and sulfates), or sorption on mineral surfaces and the organic fraction of sediments.[83]

The bioavailability of pollutants adsorbed on the surface of sediments is very important in toxicity assessment. Chemical analysis enables the quantitative determination of toxic substances in sediment samples or those present in aqueous phases, for example, in pore water. However, the solubility, mobility and, in later stages, the availability to benthic organisms of pollutants such as metals and organic pollutants are all also influenced by a number of other parameters, for example, the pH of the environment, TOC concentration, or the granulometric distribution of sediments.[42] The pH value is often considered to be the main factor influencing the mobility of metal ions.[52] As numerous reports have shown, the total concentration of metals in a sample cannot be treated as a measure of its toxicity. Only the form in which a given metal appears, that is, dissolved (ionic), or as a compound with organic or inorganic substances, determines the bioavailability of the metal and its toxicity toward living organisms. This is why it is so significant to combine the two types of assay: toxicity assays and chemical analysis.

Table 9.4 presents a list of studies on the pollution levels of effluents, landfill leachates, sludge, and sediments, undertaken using an integrated approach to pollution assessment, that is, the simultaneous use of chemical analysis and bioassays.

9.5   LEGAL REGULATIONS APPLYING TO THE USE OF BIOASSAYS IN ENVIRONMENTAL MONITORING

The integrated approach to environmental pollution assessment and the practical use of such testing in environmental management is developing intensively in Western Europe (Germany, Belgium, the Netherlands, Italy, Sweden, and Norway), Canada, the United States, Australia, and New Zealand, and also in some postcommunist countries (Lithuania, Estonia, and Slovakia).[96] Table 9.5 presents information on legal regulations (on national and regional levels), valid in the countries of Northern America and Europe, applying to the measurement of toxicity in environmental samples.

9.6   ECOTOXICOLOGICAL CLASSIFICATION OF ENVIRONMENTAL SAMPLES

The results of analytical determinations (including ecotoxicological assays) obtained during monitoring measurements are usually converted into information intelligible to nonspecialists in the field of analytics. Classification systems are one way of presenting monitoring databases in a nontechnical fashion.

The results of toxicity assays enable the ecotoxicological quality of environmental samples to be assessed on the basis of the value (expressed as a percentage) of the observed effect of toxic activity, for example, bioluminescence inhibition, algal growth inhibition, and lethalities of crustaceans, as well as estimated toxicity indices such as $L(E)C_{20}$, $L(E)C_{50}$, or toxicity unit (TU).

The principles of ecotoxicological quality classification based on the TU index are included in the 2002 recommendations of the Helsinki Commission (HELCOM). The classification applies to samples of treated effluents discharged to waters from industrial plants manufacturing chemicals,[98] textiles,[99] and pesticides.[100] HELCOM recommends testing the acute toxicity of effluent samples using two of the four suggested indicator organisms (Table 9.6).

TABLE 9.4
Examples of Studies on the Pollution Caused by Effluents, Landfill Leachates, Sludge, and Sediments Based on the Simultaneous Use of Chemical Analytics and Toxicity Assays

Matrix	Spectrum of Compounds and Parameters Tested for	Bioassays	Notes	References
1	2	3	4	5
Effluents				
Industrial effluents	Organic compounds: chloro- and nitrophenols, nonionic surfactants, linear alkylated benzene sulfonates, benzene and naphthalene sulfonates, estradiol, ethinyl estradiol	*V. fischeri (ToxAlert® 10, ToxAlert® 100)*	There were certain correlations between the results of the chemical analyses and bioassays; to a large extent, however, the composition of the samples remained unknown Distinct correlations were found between the inhibition of bioluminescence and compound content, but only in the case of nonylphenol, nonylphenol carboxylate, nonylphenol ethoxylate, and chlorophenols	84
Municipal and industrial effluents	BZT_7, COD N_{NH4}, N_{tot}, P_{tot} General suspensions Heavy metals: Cd, Cr, Cu, Mn, Ni, Pb, Zn	*Selenastrum capricornutum (Algaltoxkit F™)* *Nitellopsis obtusa (Charatox)* *Daphnia magna (Daphtoxkit F™ magna)* *Thamnocephalus platyurus (Thamnotoxkit F™)* *Tetrahymena thermophila (Protoxkit F™)* *V. fischeri (Microtox®)*	Significant correlations were found between the indices of chemical pollution and toxicity (chemical pollutants explain about 70% of the toxicity variability) Conformity between chemical parameters and toxicity indices was greatest when the tests included information obtained from all the tests in a battery (and not just the toxicity indices of the most sensitive assay)	85
Sludge				
Sludge (municipal and industrial)	PCB, PAH TOC Heavy metals: As, Cd, Cr, Cu, Pb, Mn, Zn	*V. fischeri (LUMIStox®)*	Significant correlations were found between EC_{20} values and the concentrations of almost all PAH compounds, PCB-118, Pb, and Zn	8
Sludge (municipal and industrial)	PAH	*V. fischeri (ToxAlert® 100)*	There were significant correlations between the total content of PAH compounds and the indices assessed—TII50 (15 min) and TII50 (30 min) The high phenanthrene content had a significant influence on the toxicity of the samples	86

Landfill Leachates

Sample	Parameters	Biotests	Description	Ref.
Municipal landfill leachates	Organic compounds: BTEX; propyl benzene derivatives, bicyclic compounds, naphthalene, chlorinated aliphatic compounds, phenols, pesticides, and phthalates	V. fischeri (Biotox®); Selenastrum capricornutum (conventional assay); Salmonella typhimurium (UmuC)	The observed toxicity to both bacteria and algae was due to the presence of nonvolatile organic compounds, mainly naphthalene and 4-chloro-m-cresol (volatiles were lost during sample preparation) Assays using bacteria were generally more sensitive than those using algae. In certain cases, however, the reverse situation prevailed; a battery of bioassays was then applied The worst polluted sample (according to the number of compounds identified) was also the most toxic. The concentration levels at which pollutants occurred in the samples had no direct influence on their toxicity	51
	pH COD N_{NH4}, N_{tot}, P_{tot}, sulfates General suspensions, dissolved substances Heavy metals: Cd, Cr, Cu, Mn, Ni, Pb, Zn, Fe	Selenastrum capricornutum (conventional assay)	The concentration of metal ions in a landfill leachate sample depended on the pH and the content of organic compounds; the latter may have bound metal ions in complexes, thus decreasing their bioavailability (a high organic matter content causes a reduction in toxicity) The total concentration of metals in a sample provides no reliable information on its toxicity Bioassays provide information on the chemical speciation of metals and their bioavailability	52
Leachates from the oil-shale industry and from spent shale dumps	Phenols Sulfates	V. fischeri (Biotox®, Microtox®) Daphnia pulex (Daphtoxkit F™ pulex) Brachionus calyciflorus (Rotoxkit F™)	Bioassays based on bacteria proved to be the most sensitive, while those based on rotifers were the least sensitive. At the same time, bacterial bioassays were deemed the most suitable for screening tests The results of the bacterial bioassays indicated that the toxicity of the samples was caused mainly (75%) by phenols (p-cresol, 3,4-dimethylphenol, and phenol), and to a lesser degree by sulfates (about 25%)	87

Sediments

Sample	Parameters	Biotests	Description	Ref.
Marine sediments	Heavy metals: Cd, Pb, Cu, Cr, Zn, Mn, Ni, Fe TOC	V. fischeri (LUMIStox®)	No correlation was observed between the toxicity and the TOC or concentration of individual heavy metals (taking account of their concentration in a bioavailable fraction). Therefore, there may have been interactions between individual components of the sediments and other pollutants	88
	Heavy metals: Hg, Pb, Cu, Ni, Zn, Mn, Fe, Cd TOC Humus compound content Sulfur and nitrogen content	V. fischeri (Microtox® Basic Solid-phase Test, Microtox® Acute Toxicity Basic Test)	The observed toxicity to bioluminescent bacteria was strictly correlated with the content of elemental sulfur and sulfides Owing to the low levels of metals in the bioavailable fraction, there was no correlation between their content and the observed toxic effect	89
Freshwater sediments	PAH TOC	V. fischeri Daphnia magna	No distinct linear correlation was found between toxicity and the presence of the organic compounds investigated	90

continued

TABLE 9.4 (continued)

Matrix	Spectrum of compounds and Parameters Tested for	Bioassays	Notes	References
1	2	3	4	5
Sediments taken from sewers	PCB PAH Heavy metals: Hg, Cu, Ni, Co, Cd, Pb, Zn, Fe, Cr Sulfur	V. fischeri (Microtox® Basic Solid-phase Test, Microtox® Basic Test—assay with organic extracts)	An immediate assay of sediment samples enabled the toxicity due to the presence of inorganic pollutants (metals and sulfates) to be assessed. The assay of organic extracts obtained from the samples enabled the toxicity due to the PAH or PCB content to be determined	40
	Elemental sulfur	V. fischeri (Microtox® Basic Solid-Phase Test)	No elemental sulfur S_8 was detected in groundwater samples The bioavailable fraction (responsible for the inhibition of bacterial bioluminescence) constituted only 3–8% of the total content of elemental sulfur in the sample	91
Port sediments	Heavy metals PAH PCB	V. fischeri (Microtox® Solid-Phase Test) Corophium volutator Acartia tonsa Skeletonema costatum	No distinct correlation was found between toxicity and the presence of the compounds identified in the samples	92
	Heavy metals PAH PCB	Daphnia magna Selenastrum capricornutum Pimephales promelas Hyaella azteca	A connection was found between the bioavailability of metals and their toxicity Chemical analyses revealed a high content of organic compounds (mainly PAHs), which were probably responsible for the observed toxic effect	93
Marine sediments	PAH TOC	V. fischeri (LUMIStox®) Lemna minor	The toxicity was probably associated with the presence of aliphatic compounds and elemental sulfur The difficulty in finding a correlation between the results of chemical analyses and ecotoxicological assays may have been due to antagonistic and synergic effects	94
Fluvial sediments	HCH DDT PAH Tributyltin Saturated, unsaturated, and aliphatic carbohydrates Biphenyls Elemental sulfur Aromatic esters	V. fischeri (LUMIStox®) Daphnia magna (conventional assay) Scenedesmus vacuolatus (conventional assay)	Extract samples were divided into fractions for a toxicity assessment and the identification of toxic compounds The high toxicity of the acetone extract samples obtained from the sediments to V. fischeri was due to the presence of elemental sulfur The toxicity due to the total toxicity of individual fractions was significantly greater than that calculated for the complete extract The toxicity due to the total toxicity of individual subfractions significantly exceeded calculated on the basis of the examination of an individual fraction	95

TABLE 9.5
Legal Regulations (on National and Regional Scales) in Various Countries of North America and Europe Applicable to the Measurement of Toxicity[97]

Country	Legal Regulations
United States of America	Studies of toxicity in the aquatic environment are performed to assess potential threats and risks related to the discharge of municipal and industrial effluents to waters. The discharge of industrial effluents is regulated by US EPA in the program of permissions issued by the National Pollutant Discharge Elimination System——NPDES, which was founded by the Clean Water Act (CWA). The toxicity monitoring, required by the permission, is a program for supervising and maintaining the quality of waters in which toxicity assays are performed at regular intervals (once in 3–4 months) so as to establish whether life in a given aquatic environment is safe from any risk related to pollution discharge. Bioindicators and the methods applied are described in US EPA documents, for example: • US EPA 1991a: Methods for measuring the acute toxicity of effluents to aquatic organisms EPA-600/4-90-027 • US EPA 1991b: Short-term methods for estimating the chronic toxicity of effluents and receiving waters to freshwater organisms EPA-600/4-91/002 Standard methods for assessing the toxicity of waters and effluents also include chapters on toxicity tests. Similarly, the American Society for Testing and Materials——(ASTM) has published standard procedures for toxicity determination.
Canada	Federal law requires the combination of acute lethality and sublethal toxicity tests for effluents from paper mills and metallurgical plants. Discharge of effluents from other industries is regulated by various local administrations Legal regulations are based on the following tests: 1. Luminescence inhibition test with bacteria 2. Algal growth inhibition test 3. Test of acute toxicity to *Daphnia magna* or *Ceriodaphnia dubia* 4. Test of acute toxicity to fish
United Kingdom	A set of methods is being developed for a DTA of effluent toxicity, which is to meet the requirements of SNIFER (Scotland and Northern Ireland Forum for Environmental Research). The set includes the following toxicity tests: – Algal growth inhibition test – *Daphnia magna* immobilization test – Juvenile fish lethality test
France	The decree of October 28, 1975, stipulates that the discharge of effluents is subjected to taxation, the level of which depends on the number of "TU," determined by a test of acute toxicity to *Daphnia magna*. The tax is collected by "Agences de Bassin"
Germany	The different Länder in Germany require the following tests in the control of effluents: 1. Luminescence inhibition test with bacteria 2. Algal growth inhibition test 3. Acute toxicity test with *Daphnia magna* 4. Acute toxicity test with fish (soon to be replaced by a test on fish eggs) 5. Genotoxicity test by the Umu assay 6. Toxicity test on higher plants (*Lemna*)
Austria	Toxicity tests of waters are required to assess potential threats connected with the discharge of municipal and industrial effluents to receiving waters. Effluents from water treatment plants and receiving waters are periodically monitored and regulated by the Austrian authorities.

continued

TABLE 9.5 (continued)

Country	Legal Regulations
	Austrian (ÖNORM) and German recommendations (DIN) use the following four standard toxicity tests: – Luminescence inhibition test with bacteria – Acute toxicity test with *Daphnia* – Acute toxicity test with fish – Algal growth inhibition test In order to determine the threshold of safety for all the toxicity tests for effluents and receiving waters, a G value is used. The G values depend on the type of emission: G_L (bacteria), G_D (Daphnia) G_F (fish), and G_A (algae). The G values for industrial effluents, and discharges from paper mills, textile plants, tanneries, chemical plants, and detergent production plants are $G_F = 2$; G_L and $G_D = 4$; and $G_A = 8$. Effluents from the pharmaceutical industry and pesticide production plants are limited by the legal requirement of $G_A = 16$. For municipal effluents, only the test on fish is required, and the value G_F needs to be smaller than 2. The frequency of toxicity tests is once in five years, but individual regulations may be more restrictive.
Spain	Toxicity testing is required by the authorities responsible for river basins as a complement to chemical analysis. In their Order No. 10/1993 of October 26, 1993, the regional authorities of Madrid require the toxicity of effluents to be tested using the following assays: – Luminescence inhibition test with bacteria – Algae growth inhibition test – Acute toxicity test with *Daphnia magna* – Respiration inhibition test on sludges – Acute toxicity test with rotifers – Acute toxicity test with *Thamnocephalus*
Portugal	Three microbioassays (Toxkits) were accredited by the Portuguese Institute for Standardization in 2001: – Daphtoxkit F magna microbiotest (acute toxicity tests with *Daphnia magna*) – The Thamnotoxkit F microbiotest (acute toxicity tests with *Thamnocephalus platyurus*) – The Artoxkit M microbiotest (acute toxicity tests with *Artemia salina/franciscana*) The assays are to be included in the national law as official ecotoxicological parameters.
Norway	The State Pollution Control Authority requires ecotoxicological characterization of industrial effluents in combination with the renewal of permissions for the discharge of effluents.
Sweden	Although Swedish legislation does not require determination of the toxicity of effluents, it nevertheless indicates that those who might affect the natural environment have to prove that their actions do not adversely affecting it. As a result, many industrial effluents are in practice examined with regard to their toxicity to selected bioindicator organisms.
Italy	Since 1999 toxicity assays for effluents are required by law (order D.L. 152/99). The following tests are recommended: – Acute toxicity test on *Daphnia magna* or *Ceriodaphnia dubia* – Algal growth inhibition test on algae *Selenastrum capricornutum* – Luminescence inhibition test on bacteria – Acute toxicity test on *Artemia salina* (for saline discharges)
Denmark	Industrial effluents discharged directly to receiving waters have to be tested for toxicity within the framework of discharge consents. The applied toxicity assays conform with international recommendations, that is: – Algal growth inhibition test – Acute toxicity test with *Daphnia* – Acute toxicity test with fish

continued

TABLE 9.5 (continued)

Country	Legal Regulations
Belgium	**Flanders**
	The Flemish Environmental Agency may order industrial plants to perform toxicity assays, so plants obtain permission to discharge their effluents. At present, many plants are asked, at regular intervals (a few times a year) to perform toxicological tests on their effluents. The effluents are initially subjected to the following assays:
	– Luminescence inhibition test with bacteria – Growth inhibition test with algae – Acute toxicity test with *Daphnia magna* – Acute toxicity test with fish
	Subsequent effluent analyses are performed on those organisms that have proved to be the most sensitive in the battery of tests.
	Wallonia
	The system is basically similar to the Flemish one, but it is still in preparation.
Greece	There are no legal regulations on the use of ecotoxicological assays in environmental monitoring, but there are many ongoing research projects which include ecotoxicological assays, mainly concerning effluents.
Czech Republic	There are no legal regulations on the use of ecotoxicological assays in environmental monitoring.
Poland	There are no legal regulations on the use of ecotoxicological assays in environmental monitoring.

For example, the use of toxicity assays with *V. fischeri* bacteria and *Daphnia magna* crustaceans, for which the TU index is 8, gives a No Observed Effect Concentration (NOEC) of 12.5% (Equation 9.1):

$$TU = \frac{100}{NOEC}. \tag{9.1}$$

Effluents should therefore be of such a quality that at a dilution of 1:7 (12.5%) acute toxicity is not observed in *V. fischeri* and *Daphnia magna*. If NOEC < 12.5%, then the ecotoxicological quality of the investigated effluents is poor and they cannot be released to surface waters.

In 2003, Professor Persoone et al.[101] developed a classification of acute toxicity levels in natural waters and effluents discharged to these waters based on two systems:

• A hazard classification system for natural waters
• A toxicity classification system for wastes discharged into the aquatic environment

TABLE 9.6
List of Toxicity Tests and Critical Values of the TU Index According to the Recommendations by the HELCOM[100]

Bioindicator	Test Duration (h)	TU Index Value
Fish	96	2
Daphnia crustaceans	48	8
Algae	72	16
V. fischeri bacteria	0.5	8

TABLE 9.7
Hazard Classification System for Natural Waters According to Persoone[101]

Class	PE Value	Hazard
I	<20%	No acute hazard
II	$20\% \leq PE < 50\%$	Slight acute hazard
III	$50\% \leq PE < 100\%$	Acute hazard
IV	100% in at least one test	High acute hazard
V	100% in all tests	Very high acute hazard

Each of which is in turn based on two criteria:

- A ranking in five acute hazard classes
- A weight score for each hazard class

Natural waters are classified according to the percentage effect (PE) obtained with each of the microbiotests. The water is ranked into one of five classes on the basis of the highest toxic response shown by at least one of the tests applied. Table 9.7 shows the classification system for natural waters.

The toxicity of effluents discharged into the aquatic environment is also classified using the numerical value of the PE index obtained in a test on an undiluted sample. However, for samples in which PE > 50%, additional assays are conducted in which increasing dilutions of the examined samples are tested. The $L(E)C_{50}$ values obtained are converted into TU. Depending on the numerical values of the TU index, effluent samples are classified according to the criteria listed in Table 9.8.

The advantage of these systems is the possibility of estimating the weight score for each hazard class to indicate the quantitative importance (weight) of the toxicity in that class. The class weight score is calculated according to the following equation:

$$\text{class weight score} = \frac{\sum \text{all test scores}}{n}, \tag{9.2}$$

where n is the number of tests performed.

The percentage value of the class weight score is calculated in the following way:

$$\text{class weight score in \%} = \frac{\text{class score}}{\text{maximum class weight score}} \times 100. \tag{9.3}$$

TABLE 9.8
Toxicity Classification System for Effluents Discharged into the Aquatic Environment According to Persoone[101]

Class	TU Value	Hazard
I	<0.4	No acute hazard
II	$0.4 \leq TU < 1$	Slight acute hazard
III	$1 \leq TU < 10$	Acute hazard
IV	$10 \leq TU < 100$	High acute hazard
V	≥ 100	Very high acute hazard

TABLE 9.9
Ecotoxicity of Surface Waters Based on the CF[36]

Risk for Aqueous Environment	CF Value Corresponding the EC_{50} Value for *Daphnia magna*	CF Value Corresponding the EC_{50} Value with *V. fischeri*
High	≤10	≤40
Medium	≤20	≤80
Low	>20	>80

This takes into account the number of tests performed on various bioindicators (the greater the number of tests performed, the more reliable the assessment of a given sample) and the variability in the toxicity estimated using various bioindicators (plants and animals).

In their study of surface waters in the Po River basin (Italy), Galassi et al. introduced a classification of ecotoxicological quality based on the concentration factor (CF).[36] The pollutants present in a sample were enriched by solid-phase extraction (SPE). This procedure enabled the samples to be differentiated ecotoxicologically in that the presence of only those toxic compounds extractable with a particular solvent were determined. Ecotoxicity was assessed according to the criteria set out in Table 9.9, the most toxic effect being observed with *V. fischeri* during a 30-min test or *Daphnia magna* during a 48-h test.

The German Federal Institute of Hydrology (Bundesanstalt für Gewässerkunde—BfG) has developed a system of ecotoxicological classification of sediments based on the numerical pT value (*potentia toxicologiae*).[102] This value is the negative binary logarithm of the dilution coefficient of a sample in which no acute toxicity is recorded. In other words, the numerical pT value shows how many times a given sample should be diluted (in a 1:2 ratio) for it to cease being toxic. The pT value determined for the most sensitive bioindicator among all the organisms in the battery of bioassays gives the toxicity class of a given sediment. All the toxicity tests applied, and also all the liquid phases examined—pore water[103] as well as aquatic or organic elutriates—are equivalent to the criteria established by Krebs (Table 9.10). For example, if the first pT value for which no toxic effect is observed is equal to 5, then the sediment examined belongs to toxicity class V.

The ecotoxicological quality of sediments can also be assessed using the classification developed by the ARGE-Elbe project.[104] It allocates sediment samples to one of five ecotoxicological quality classes on the basis of the recorded PE (Table 9.11).

TABLE 9.10
Ecotoxicological Classification of Sediments Based on the pT Value[102]

Highest Dilution Level without Effect	Dilution Factor	pT Value	Toxicity Classes	Classification of Sediments with Regard to Environmental Management	
				Color Coding	Three-Level Assessment System
Undiluted Sample	2^0	0	0	0	Unproblematic
1:2	2^{-1}	1	I	I	
1:4	2^{-2}	2	II	II	
1:8	2^{-3}	3	III	III	Critical
1:16	2^{-4}	4	IV	IV	
1:32	2^{-5}	5	V	V	Hazardous
≤(1:64)	$≤2^{-6}$	≥6	VI	VI	

TABLE 9.11
Ecotoxicological Classification of Sediments Developed in the ARGE–Elbe Project[104,105]

Toxicity Class	PE Value	Ecological Status (with Reference to WFD)	Color Coding
I	≤15%	Very good	Blue
II	>15% PE ≤ 30%	Good	Green
III	>30% PE ≤ 50%	Moderate	Yellow
IV	>50% PE ≤ 70%	Weak	Orange
V	>70%	Bad	Red

In recent years, more complex systems for classifying the toxicity of aquatic environmental samples have been developed, for example:

- Potential ecotoxic effects probe (PEEP)[106–108]
- Potentially affected fraction (PAF)[109,110]
- Sediment toxicity (SEDTOX)[111]

Also, a sediment quality system has been developed that combines chemical analyses with biotic indices in the so-called TRIAD (Sediment Quality Triad-integrated assessments of sediment quality based on measures of chemistry, toxicity, and benthos).[112]

9.7 BIOASSAY APPLICATION IN ENVIRONMENTAL MONITORING: SOME CASE STUDIES

The information contained in the results of an ecotoxicity study has a special significance with respect to chemical analysis. Integrating chemical and ecotoxicological studies offer the same advantages as environment quality estimation and enables

- Samples to be screened for further chemical monitoring tests and/or tests for identifying toxic substances
- Pollution hot spots to be identified and inventorized
- Monitoring stations to be identified where chemical parameters have not been correctly defined
- The problems occurring in a given area to be prioritized
- The real risks resulting from the bioavailability and mobility of pollutants to be determined

9.7.1 IDENTIFICATION OF TOXIC COMPOUNDS

Detecting and identifying toxic compounds in environmental samples (compounds with unknown structures and properties) require the use of time-consuming, costly methods to isolate them from the matrix, then the application of complex techniques to separate the compounds present in an extract, and finally the determination of their structure (identification). Applying such a procedure to all samples collected from a selected area is very expensive. Samples therefore need to be selected, for example, with the aid of the results of ecotoxicological tests: samples with the determined toxicity may contain toxic compounds.

A procedure of this kind was used in the *International Odra Project—IOP* (1997–2001).[113] Screening tests of water samples on *V. fischeri* bacteria showed that samples taken from two measuring stations (the town of Brzeg Dolny and the confluence of the Mała Panew River with the Odra River, Poland) were toxic toward the bacteria. Chemical analysis for the detection of organic compounds (volatiles and those extracted with dichloromethane) identified trichloroethylene (274 $\mu g\ L^{-1}$) and tetrachloroethylene (1.6 $\mu g\ L^{-1}$) in a sample from Brzeg Dolny. Since the EC_{50} of trichloroethylene is 176 $\mu g\ L^{-1}$, it is highly probable that it was this compound that was mainly responsible for the poor surface water quality at Brzeg Dolny. No significant quantities of organic compounds were detected in the sample from the Mała Panew River. However, as the report by the Provincial Environmental Protection Inspectorate in Opole suggests (1997), the waters of the Mała Panew River were polluted with Zn and Pb compounds, the levels of which exceeded several times the permissible concentrations of these metals in surface waters. So, it was probably the high heavy metal content that was responsible for the observed toxicity of the water in the Mała Panew River.

In a project to examine the influence of selected landfill sites in the province of Pomerania (Pomorze) in northern Poland on the ecotoxicity of groundwaters,[114] ecotoxicological tests were performed on groundwater samples taken from monitoring piezometers located around seven of the largest and oldest (without insulation) landfill sites. The project included an acute toxicity test with *V. fischeri* bacteria as well as acute and chronic toxicity tests with *Daphnia magna* crustaceans. The groundwater samples from one of the piezometers located at a landfill site that also received waste from pharmaceutical companies were highly toxic; but the reason for this was not explained by any of the monitored chemical parameters. Additional chemical tests identified the presence of chlorobenzene, aniline, and dibutyl phthalate. The concentration of aniline was high (approximately 0.5 mg L^{-1}).

9.7.2 Identification of Hot Spots (Sites with a Very High Level of Pollution)

From the point of view of environmental management, it is very important to identify sites characterized by significant changes (compared with natural ecosystems) caused by the presence of primary or secondary pollutants. To identify the threats due to these changes, the range of influence of pollutants, the distribution of pollution intensity, and the directions and dynamics of changes need to be determined. The results of physicochemical studies are not a suitable tool with which to achieve this aim for the following reasons:

- The limited information potential of studies of physicochemical parameters (there is a finite number of determined parameters)
- The frequent lack of knowledge about secondary pollution
- Problems with identifying the effects due to the synergistic interactions of pollutants
- The different bioavailabilities of the toxic substances determined

Ecotoxicological studies appear to be the most informative and cost-effective tools for the identification of pollution hot spots.

The EUROCAT project[115] involved the determination of acute toxicity in the waters of the Gulf of Gdańsk with *V. fischeri*. Sea water samples were collected in four series and did not show any toxic effects, but the water samples collected in the final reaches of the Vistula River, near the bridge at Kiezmark (some 12 km upstream of the river mouth) and in the area where the river water merges with the sea showed a 30–60% decrease in luminescence. Considering the direction in which the gulf waters circulate (new water flows in from the northwest) and the fact that the Vistula River receives pollution from its entire catchment area, the results are not surprising. They show that the waters of the river are polluted with toxic substances; at the river mouth, after mixing with the waters of the gulf, there is a slight decrease in their toxicity. As a result of the northwesterly inflow of water into the gulf, the polluted river waters are directed to the east.

In a pilot project on the monitoring and assessment of the quality of the waters of the Bug River where it forms the eastern border of Poland, tests were carried out to assess the acute and chronic toxicity of water samples taken from the river, sediment samples, and treated effluents reaching surface waters from the water treatment plants located in the river basin.[116] Using a much simplified classification system, resembling that used in the ARGE-Elbe project (a classification based on the PE), the quality of waters, sediments, and effluents were determined. Comparison of the results of ecotoxicological tests done on surface waters and effluents reveals a distinct link between the very low ecotoxicological quality of the effluents and the poor quality of surface waters near the point of discharge.

9.7.3 Ranking of Problems in Polluted Areas Managed by Specific Authorities

Solving environmental problems occurring in a given area usually requires considerable financial and human resources, not to mention suitable instruments and infrastructure. The availability of these resources is limited, hence it is necessary to rank the importance of problems and create corresponding lists of importance. The results of ecotoxicological studies may provide important and sometimes even key information permitting such sequences to be established. The examples given below illustrate well this potential of ecotoxicological studies. They assist in the choice of the most appropriate environmental decisions and help lower the costs of these decisions.

Using a battery of tests consisting of five bioindicators (rotifers—*Brachionus calyciflorus*, bacteria—*V. fischeri*, crustaceans—*Thamnocephalus platyurus* and *Daphnia magna*, and duckweed—*Lemna minor*), ecotoxicological tests were conducted on effluents released into surface waters from water treatment plants located in the Bug River basin. Table 9.12 presents the results of toxicity determinations for the aforementioned organisms and the ecotoxicological quality of effluents according to the toxicity classes recommended by the HELCOM (TU)[98–100] and toxicity classes using the pT value to assess the toxicity of the effluent.[102,117] Comparison of the results obtained with these two systems of ecotoxicological classification (TU and pT) shows significant differences in the quality of the same samples; this is mainly because two of the bioindicators used in the project (*Daphnia magna* and *V. fischeri*), recommended the HELCOM system, are less sensitive than *Brachionus calyciflorus* and *Thamnocephalus platyurus*, used in the pT value system. Four or five of the water treatment plants participating in the project require further testing so that an explanation for the poor ecotoxicological quality of effluents can be found.

In the project mentioned above, to assess the influence of selected landfill sites in the province of Pomerania (Poland) on the ecotoxicity of groundwaters,[118] ecotoxicological tests were carried out on groundwater samples collected from monitoring piezometers around seven of the largest and oldest (without insulation) landfill sites. The study used the acute toxicity test for *V. fischeri* and the acute and chronic toxicity tests for *Daphnia magna*. The quality of groundwater was examined using three systems of ecotoxicological classification HELCOM (TU), Krebs (pT), and Persoone's system.[101] The groundwaters were also classified on the basis of physicochemical parameters, according to recommendations issued by the Polish Ministry of Environment.[118]

Comparison of the ecotoxicological quality of the examined samples with the physicochemical quality reveals distinct differences. Table 9.13 presents examples of two quite different situations, which arose during this comparison:

- Piezometer P4: poor physicochemical quality (class 5) but good ecotoxicological quality.
- Piezometer P6: very poor ecotoxicological quality of waters and poor physicochemical quality (class V).

In the case of piezometer P4, the poor physiochemical quality was the result of admissible values for parameters such as turbidity, color, iron, and chloride being exceeded; but the ecological hazards posed by these excessive values are insignificant or nonexistent.

The water monitored by piezometer P6 was assigned to quality class V (admissible values for the following parameters were exceeded: ammonia nitrogen, phenols, COD_{Cr}, COD_{Mn}, chloride, and total content of solutes); at the same time, however, the ecotoxicological quality of these waters was stated to be very poor. The slight excess of permissible values for the physiochemical markers does

TABLE 9.12
Assessment of Toxicity from Water Treatment Plants According to the Krebs (pT) and HELCOM (TU) Classifications

Effluents	*Brachionus calyciflorus* Acute Toxicity (24 h)		*Thamnocephalus platyurus* Acute Toxicity (24 h)		*Daphnia magna* Acute Toxicity (48 h)		*V. fischeri* Acute Toxicity (30 min)	
	TU[a]	pT	TU[a]	pT	TU[a]	pT	TU[a]	pT
(October 2001)								
1. PGKiM Tomaszów Lubelski	64	6	32	5	4	2	3	1
2. Piezometer no. 1[b]	64	6	256	>6	64	6	5	2
3. PGKiM Hrubieszów	64	6	128	>6	—	0	6	3
4. MPGK Chełm	128	>6	128	>6	—	0	3	1
5. NZPS Orchówek	64	6	—	0	8	3	3	1
6. PGKiM Terespol	128	>6	256	>6	16	4	3	1
7. PWiK Biała Podlaska	64	6	64	6	8	3	3	1
8. PUiK Łuków	—	0	8	3	—	0	3	1
9. PUiK Sokołów Podlaski	64	6	256	>6	128	>6	3	1
10. PWiK Siedlce	8	3	16	4	8	3	3	1
11. PWiK Wyszków	8	3	8	3	8	3	3	1
(May 2003)								
1. PWiK Wyszków					—	0	8	3

HELCOM classification (TU)

Krebs classification (pT)

—Meets criteria
—Does not meet criteria
—Unproblematic
—Critical
—Hazardous

[a] TU, toxicity unit.
[b] Dump leachate near Tomaszów Lubelski.

TABLE 9.13

Ecotoxicological and Physicochemical Assessment of Undergroundwater Sample Collected from Piezometers[118]

Piezometer	HELCOM		Krebs (pT)		Persoone					Chemical Classification
	V. fischeri	D. magna	V. fischeri	D. magna	Wastewaters		Undergroundwaters			Exceeded Parameters
1	2	3	4	5	6	7	8	9	10	11
P4 (10.11.2003)	Good	Good	I	0	III	50	II	50	V	Turbidity, color, iron, and chloride
P6 (07.12.2003)	Low	Low	III	VI	V	75	IV	83	V	Ammonia nitrogen, phenols, COD_{Cr}, COD_{Mn}, chloride, total content of solutes

not explain such a poor ecotoxicological quality in the analyzed water samples. One problem did emerge from the results of the ecotoxicological studies, however, the solution to which required time-consuming and costly chemical analyses—chlorobenzene, aniline, and dibutyl phthalates were found to be present.

This example shows how helpful information obtained from an ecotoxicological study can be in undertaking important decisions (related to the need to manage with limited financial resources).

9.7.4 Revision of Monitoring Parameters

The assumption underlying the monitoring models used so far is that testing covers a certain range of parameters specified by legal regulations. Local discharges of other substances (not specified in the regulations) as well as secondary pollution effects remain "invisible" to the monitoring model and are not perceived as real environmental threats. The current provisions of the Water Directive (investigative monitoring) allow for further monitoring parameters to be added to the existing list. Toxicity studies can be particularly useful in this situation.

A strict relationship should exist between toxicity (the average value of the indicated toxicity parameters in the case of the organism analyzed) and the monitoring parameter of the chemical load of the sediment sample (the total concentration of the indicated parameter in relation to the average value of this parameter across all the samples analyzed) (Figure 9.1). A condition of the appearance of such a relationship is that the spectrum of these physiochemical parameters mirrors the factors that actually pollute the environmental compartments under scrutiny and indicate toxicity in relation to biota.

Monitoring of sediments from the Bug River basin indicates that they are not chemically polluted.[119] Unfortunately, however, the relevant analyses did not take organic compounds into consideration; only the following were analyzed:

- Heavy metals (Pb, Zn, Cr, Cu, Cg, Hg, Ni, and Co)
- Macroelements (Fe, Mg, Al, and Mn)
- Cations and anions (NH_4^+, Cl^-, NO_3^-, NO_2^-, and SO_4^{2-})

Examination of the relation between the toxicity parameters and the chemical load of the samples allows three separate groups of monitoring stations to be distinguished:

- Three stations (Bug-Włodawa, Bug-Terespol, and Krzna-Neple), where the ecotoxicity was higher than expected owing to the high level of chemical pollution.

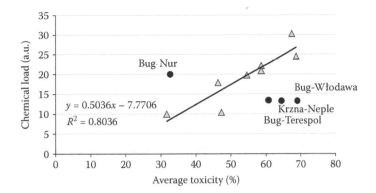

FIGURE 9.1 Relationship between sediment toxicity in the Bug River basin and the level of chemical pollution.

- One station (Bug-Nur), where chemical loads were high but the toxicity low.
- A cluster of stations (gray triangles in Figure 9.1), which indicates a linear relationship between the parameters determined and the toxicity of the sample (coefficient of determination $R^2 = 0.8036$).

The high level of chemical pollution in sediments at the Bug-Nur station was due to the Mg and Zn contents being twice as high as the average for the analyzed samples. Such a pollutant composition did not increase the toxicity of the sample.

The particularly high toxicity of samples collected at the Bug-Włodawa, Bug-Terespol, and Krzna-Neple stations suggests that substances other than the ones monitored had an influence on the level of pollution in those samples. The monitoring parameters applied at these stations should be revised in order to make

- An inventory of pollutants released into the water near these points
- An independent attempt at identifying substances responsible for such a high toxicity in the analyzed samples

The objective of another project was to evaluate the level of pollution in water and sediments in Lake Turawskie, a storage reservoir built in the 1930s on the Mała Panew River in southwestern Poland, and then to search for a correlation between the analytical chemical results and the toxicity parameters estimated from the application of biotests. Polluted waters as well as large amounts of polluted sediments enter the reservoir from the Mała Panew and its tributary, the Libawa. Industrial activities in the river basin include silver, zinc, and lead processing plants, steel and glass manufacture, and the production of cellulose and chemicals. Agriculture presents a further potential threat (e.g., fertilizer and pesticide run-off), as does the use of the reservoir's banks for recreational purposes.

Chemometric studies indicated a lack of correlation between individual chemical parameters and estimated toxicity parameters. A significant aspect of the toxicity effect is the bioavailability form in which a pollutant is present in the analyzed sample. Efforts were made to find a link between the toxicity of a sediment sample and the mobility of the heavy metal forms it contained, but no relationship could be found between the determined toxicity and the potential toxicity, the latter calculated on the basis of the load of mobile forms of metals. A relationship was found, however, between the determined toxicity effects using the crustacean *Heterocypris incongruens* and the potential toxicity resulting from total metal loads. These results indicate that forms of heavy metals that are insoluble in water may nonetheless be available to *H. incongruens*.

The Lake Turawskie analytical project confirmed the increased threat posed to the reservoir's ecosystem by the unfortunate siting of a dump for postproduction sediments from the "Mała Panew" works in Ozimek (left bank, near the river's point of entry to the lake). A more detailed explanation of how this dump will affect the environmental condition of Lake Turawskie will, however, require further study.

9.8   CONCLUSIONS

Integration of chemical monitoring based on the measurement of each individual indicator of environmental pollution, including toxicity parameters, will yield fuller information regarding the state of an environmental compartment.

Knowledge of all the possible biological effects that a given combination of pollutants will have in an ecosystem can be the basis for taking more accurately targeted administrative decisions and managing the environment more effectively.

The possibility of including ecotoxicological studies in the monitoring of environmental pollution should therefore be considered. Beforehand, however, an appropriate classification of environmental samples will need to be prepared, a suitable ecotest chosen, and implementation tests conducted.

The use of chemometric methods will cut the costs of environmental monitoring; in the future this can be carried out on the basis of an optimal number of indispensable parameters to be determined, without any loss of significant information on environmental pollution.

A clear correlation between the results of chemical analysis and biotests is lacking, which precludes the separate application of these two types of tests. Ecotoxicological tests do, however, provide additional information on the state of the environment. They indicate the need for further, more detailed analytical studies, the aim of which should be to identify in samples those compounds not yet covered by current chemical monitoring programs, and whose presence in the environment is not yet controlled by any legal regulations on environmental protection.

Polish and EU legislations suggest in directly that toxicity could be applied to assess the quality of the environment. However, there is a lack of knowledge and motivation enabling the routine application of biotests in environmental monitoring. Further analytical studies are therefore necessary before toxicity parameters can be included in environment quality assessment systems.

REFERENCES

1. Directive 2000/60/EU of the European Parliament and of the Council of October 23, 2000, establishing a framework for community action in the field of water policy.
2. Sieńczuk, W. 1999. *Toksykologia*. Warszawa: PZWL.
3. Klamer, H.J.C., P.E.G. Leonards, M.H. Lamoree, L.A. Villerius, J.E. Åkerman, and J.F. Bakker. 2005. A chemical and toxicological profile of Dutch North Sea surface sediments. *Chemosphere* 58: 1579–1587.
4. Council Directive 76/464/EEC of May 4, 1976, on pollution caused by certain dangerous substances discharged into the aquatic environment of the community.
5. Baena, J.R. and M. Valcárcel. 2003. Total indices in analytical sciences. *Trends Anal. Chem.* 22: 641–646.
6. Namieśnik, J. and T. Górecki. 2002. Application of total parameters in environmental analytics. *Am. Lab.* 34: 18–21.
7. Castillo, M. and D. Barceló. 1999. Identification of polar toxicants in industrial wastewaters using toxicity-based fractionation with liquid chromatography/mass spectrometry. *Anal. Chem.* 71: 3769–3776.
8. Mantis, I., D. Voutsa, and C. Samara. 2005. Assessment of the environmental hazard from municipal and industrial wastewater treatment sludge by employing chemical and biological methods. *Ecotoxicol. Environ. Saf.* 62: 397–407.
9. Bulich, A.A., M.W. Greene, and D.L. Isenberg. 1981. Reliability of the bacterial luminescence bioassay for the determination of toxicity of pure compounds and complex effluents. In: D.R. Branson and K.L. Dickson (eds), *Aquatic Toxicology and Hazard Assessment: Fourth Conference*, pp. 338–347. Baltimore, USA: ASTM STP737.

10. Nohava, M., W.R. Vogel, and H. Gaugitisch. 1995. Evaluation of the luminescent bacteria bioassay for the estimation of the toxicological potential of effluent water samples—comparison with data from chemical analyses. *Environ. Int.* 21: 33–37.

11. Kiebling, M. and M. Rayner-Brandes. 1998. ToxAlert systems for the toxicity testing of environmental samples with luminescent bacteria. *GIT Lab. J.* 4: 254–255.

12. Wang, C., Y. Wang, F. Kiefer, A. Yedeler, Z. Wang, and A. Kettrup. 2003. Ecotoxicological and chemical characterization of selected treatment process effluents of municipal sewage treatment plant. *Ecotoxicol. Environ. Saf.* 56: 211–217.

13. Wardencki, W. 2004. Ogólna charakterystyka metod biologicznych wykorzystywanych w ocenie zanieczyszczeń środowiska. In: W. Wardencki (ed.), *Bioanalityka w ocenie zanieczyszczeń środowiska*, pp. 9–22. Gdańsk: CEEAM.

14. Van der Schalie, W.H., T.R. Shedd, P.L. Knechtges, and M.W. Widder. 2001. Using higher organisms in biological early warning systems for real-time toxicity detection. *Biosens. Bioelectron.* 16: 457–465.

15. Tahedl, H. and D.P. Häder. 2001. Automated biomonitoring using real time movement analysis of *Euglena gracilis*. *Ecotoxicol. Environ. Saf.* 48: 161–169.

16. Cho, J. Ch., K.J. Park, H.S. Ihm, et al. 2004. A novel continuous toxicity test system using a luminously modified freshwater bacterium. *Biosens. Bioelectron.* 20: 338–344.

17. Kirkpatrick, A.J., A. Gerhardt, J.T. Dick, P. Laming, and J.A. Berges. 2006. Suitability of *Crangonyx pseudogracilis* (Crustacea: Amphipoda) as an early warning indicator in the multispecies freshwater biomonitor. *Environ. Sci. Pollut. Res. Int.* 13: 242–250.

18. Markert, B.A., A.M. Breure, and H.G. Zechmeister. 2003. Bioindication/biomonitoring of the environment. In: B.A. Markert, A.M. Breure, and H.G. Zechmeister (eds), *Biomarkers and Biomonitors. Principles, Concepts and Applications*, pp.17–55. Oxford: Elsevier.

19. Ratte, H.T., M. Hammers-Writz, and M. Cleuvers. 2003. Ecotoxicity testing. In: B.A. Markert, A.M. Breure, and H.G. Zechmeister (eds), *Biomarkers and Biomonitors. Principles, Concepts and Applications*, pp. 221–250. Oxford: Elsevier.

20. Kuczyńska, A., L. Wolska, and J. Namieśnik. 2005. Application of biotests in environmental research. *Crit. Rev. Anal. Chem.* 35: 135–154.

21. Sakai, M. 2001. Chronic toxicity tests with *Daphnia magna* for examination of river water quality. *J. Environ. Sci. Health B* 36: 67–74.

22. Hund, K. 1997. Algal growth inhibition test—feasibility and limitations for soil assessment. *Chemosphere* 35: 1069–1082.

23. Okamura, H., M. Piao, I. Aoyama, M. Sudo, T. Okubo, and M. Nakamura. 2002. Algal growth inhibition by river water pollutants in the agricultural area around Lake Biwa, Japan. *Environ. Pollut.* 117: 411–419.

24. Ricking, M., E. Beckman, and A. Svenson. 2002. Polycyclic aromatic compounds and Microtox® Acute Toxicity in contaminated sediments in Sweden. *J. Soil Sediments* 2: 129–136.

25. Baudrimont, M., S. Andrès, J. Metivaud, et al. 1999. Field transplantation of the freshwater bivalve *Corbicula fluminea* along a polymetallic contamination gradient (river Lot, France): II. Metallothionein response to metal exposure. *Environ. Toxicol. Chem.* 18: 2472–2477.

26. Cope, W.G., J.G. Wiener, M.T. Steingraeber, and G.J. Atchison. 1994. Cadmium, metal-binding proteins, and growth in bluegill (*Lepomis macrochirus*) exposed to contaminated sediments from the upper Mississippi River basin. *Can. J. Fish. Aquat. Sci.* 51: 1356–1367.

27. OECD Series on Principles of Good Laboratory Practice and Compliance Monitoring. Number 1: *OECD Principles on Good Laboratory Practice (as revised in 1997)*. Chemicals Group and Management Committee. ENV/MC/CHEM(98)17.

28. Council Regulation (EEC) No. 793/93 of March 23, 1993, on the evaluation and control of the risks of existing substances.

29. http://www.oecd.org.

30. http://www.iso.org.

31. http://www.pkn.pl.

32. http://www.citib.hg.pl/citib/normy/13.060.70.htm.

33. Rojíčková-Padrtová, R., B. Maršálek, and I. Holoubek.1998. Evaluation of alternative and standard toxicity assays for screening of environmental samples: Selection of an optimal test battery. *Chemosphere* 37: 495–507.

34. Bulich, A.A. 1979. Use of luminescent bacteria for determining toxicity in aquatic environments. In: L.L. Markings and R.A. Kimerle (eds), *Aquatic Toxicology*, pp. 98–106. Philadelphia, PA: ASTM STP 667.

35. Boluda, R., J.F. Quintanilla, J.A. Bonilla, E. Sáez, and M. Gamón. 2002. Application of the *Microtox®* test and pollution indices to the study of water toxicity in the Albufera Natural Park (Valencia, Spain). *Chemosphere* 46: 355–369.

36. Galassi, S., L. Guzzella, and V. Croce. 2004. Screening organic micropollutants in surface waters by SPE extraction and ecotoxicological testing. *Chemosphere* 54: 1619–1624.

37. Phyu, Y.L., M.St.J. Warne, and R.P. Lim. 2005. Effect of river water, sediment and time on the toxicity and bioavailability of molinate to the marine bacterium *Vibrio fischeri* (Microtox). *Water Res.* 39: 2738–2746.

38. Bihari, N., M. Fafandel, B. Hamer, and B. Kralj-Bilen. 2006. PAH content, toxicity and genotoxicity of coastal marine sediments from the Rovinj area, Northern Adriatic, Croatia. *Sci. Total Environ.* 366: 602–611.

39. Steinberg, S.M., E.J. Poziomek, W.H. Engelmann, and K.R. Rogers. 1995. A review of environmental applications of bioluminescence measurements. *Chemosphere* 30: 2155–2197.

40. Salizzato, M., B. Pavoni, A. Volpi Ghirardini, and P.F. Ghetti. 1998. Sediment toxicity measured using *Vibrio fischeri* as related to the concentrations of organic (PCBs, PAHs) and inorganic (metals, sulphur) pollutants. *Chemosphere* 36: 2949–2968.

41. Brohon B. and R. Gourdon. 2000. Influence of soil microbial activity level on the determination of contaminated soil toxicity using Lumistox and MetPlate bioassays. *Soil Biol. Biochem.* 32: 853–857.

42. Guzzela, L. 1998. Comparison of test procedures for sediment toxicity evaluation with *Vibrio fischeri* bacteria. *Chemosphere* 37: 2895–2909.

43. Brohon B., C. Delome, and R. Gourdon. 2001. Complementarity of bioassays and microbial activity measurements for the evaluation of hydrocarbon-contaminated soils quality. *Soil Biol. Chem.* 33: 883–891.

44. Lau, S.S. and L.M. Chu. 1999. Contaminant release from sediments in a coastal wetland. *Water Res.* 33: 909–918.

45. Stefess, G.C. 1998. Monitoring of environmental effects and process performance during biological treatment of sediment from the petroleum harbour in Amsterdam. *Water Sci. Technol* 37: 395–402.

46. Bispo, A., M.J. Jourdain, and M., Jauzein. 1999. Toxicity and genotoxicity of industrial soils polluted by polycyclic aromatic hydrocarbons (PAHs). *Org. Geochem.* 30: 947–952.

47. Ho, K.T., R.M. Burgess, M.C. Pelletier, et al. 2002. An overview of toxicant identification in sediments and dredged material. *Mar. Pollut. Bull.* 44: 286–293.

48. Nendza, M. 2002. Inventory of marine biotest methods for the evaluation of dredged material and sediments. *Chemosphere* 48: 865–883.

49. Reemtsma, T., A. Putschew, and M. Jekel. 1999. Industrial wastewater analysis: A toxicity-directed approach. *Waste Manag.* 19: 181–188.

50. Benfenati, E., D. Barcelo, I. Johnson, and S. Galassi. 2003. Emerging organic contaminants in leachates from industrial waste landfills and industrial effluent. *Trends Anal. Chem.* 22: 757–765.

51. Baun, A., A. Ledin, L.A. Reitzel, P.L. Bjerg, and T.H. Christensen. 2004. Xenobiotic organic compounds in leachates from ten Danish MSW landfills—chemical analysis and toxicity tests. *Water Res.* 38: 3845–3858.

52. Magdaleno, A. and E. De Rosa. 2000. Chemical composition and toxicity of waste dump leachates using *Selenastrum capricornutum* Printz (Chlorococcales, Chlorophyta) *Environ. Toxicol.* 15: 76–80.

53. LeBlond, J.B. and L.K. Duffy. 2001. Toxicity assessment of total dissolved solids in effluent of Alaskan mines using 22-h chronic Microtox® and *Selenastrum capricornatum* assays. *Sci. Total Environ.* 271: 49–59.

54. Farré, M., M.J. García, L. Tirapu, A. Ginebreda, and D. Barceló. 2001. Wastewater toxicity screening of non-ionic surfactants by Toxalert® and Microtox® bioluminescence inhibition assays. *Anal. Chim. Acta* 427: 181–189.

55. Klinkow, N. and M. Jekel. 1999. Use of toxicity-directed fractionation procedures for the localization and identification of toxicants in wastewater and environmental samples—a review. *Vom Wasser* 93: 325–348.

56. Reemtsma, T., A. Putschew, and M. Jekel. 1999. Application of toxicity directed analysis to industrial wastewaters. *Vom Wasser* 92: 243–255.

57. Fiehn, O., L. Vigelahn, G. Kalnowski, T. Reemtsma, and M. Jekel. 1997. Toxicity-directed fractionation of tannery wastewater using solid-phase extraction and luminescence inhibition in Microtiter plates. *Acta Hydrochim. Hydrobiol.* 25: 11–16.

58. Reemtsma, T., O. Fiehn, and M. Jekel. 1999. A modified method for the analysis of organics in industrial wastewater as directed by their toxicity to *Vibrio fischeri*. *Fresenius J. Anal. Chem.* 363: 771–776.

59. Sherry, J., B. Scott, and B. Dutka. 1997. Use of various acute, sublethal and early life-stage tests to evaluate the toxicity of refinery effluents. *Environ. Toxicol. Chem.* 16: 2249–2257.

60. Jacobs, M.W., J.A. Coates, J.J. Delfino, G. Bitton, W.M. Davis, and K.L. Garcia. 1993. Comparison of sediment extract Microtox toxicity with semi-volatile organic priority pollutant concentrations. *Arch. Environ. Contam. Toxicol.* 24: 461–468.

61. Asami, M., N. Suzuki, and J. Nakanishi. 1999. Aquatic toxicity emission from Tokyo: Wastewater measured using marine luminescent bacterium, *Photobacterium phosphoreum. Water Sci. Technol.* 33: 121–128.

62. Trevizo, C. and N. Nirmalakhandan. 1999. Prediction of microbial toxicity of industrial organic chemicals. *Water Sci. Technol* 39: 63–69.

63. Parados, M., C. Benninghoff, C. Guequen, and J. Dobrowolski. 1999. Acute toxicity assessment of Polish (waste)water with a microplate-based *Hydra attenuata* assay: A comparison with the Microtox® test. *Sci. Total Environ.* 244: 141–148.

64. Gupta, G. and M. Karuppiah. 1996. Toxicity study of a Chesapeake Bay tributary—Wicomico River. *Chemosphere* 32: 1193–1215.

65. Kaiser, K.L.E. and V.S. Palabrica. 1991. *Photobacterium phosphoreum* toxicity data index. *Water Pollut. Res. J. Can.* 26: 361–431.

66. Operating manual: *ToxAlert® 100, Version 0.09.* 1989. Merck Darmstadt, Darmstadt, Germany.

67. Workshop materials "Microcystins and other aqueous pollutants", 12–13.05.2004, Tresta, Poland.

68. Nałęcz-Jawęcki, G. and J. Sawicki. 1996. Microtox—szybki system bioindykacyjny do oceny toksyczności wód i ścieków. *Gaz, Woda i Technika Sanitarna* 2: 47–51.

69. Campbell, M., G. Bitton, B. Koopman, and J.J. Delfino. 1992. Preliminary comparison of sediment extraction procedures and exchange solvents for hydrophobic compounds based on inhibition of bioluminescence. *Environ. Toxicol. Water Qual.* 7: 329–338.

70. Persoone, G. 1992. Cyst-based toxicity tests: I. A promising new tool for rapid and cost-effective toxicity screening of chemicals and effluents. *Zeitschr. für Angew. Zool.* 79: 17–36.

71. Persoone, G. 1992. Cyst-based toxicity tests: VI. Toxkits and Fluotox tests as cost-effective tools for routine toxicity screening. In: K.G. Steinhauser and P.D. Hansen (eds), *Biologische Testverfahren*, pp. 563–576. Stuttgart: Gustav Fischer Verlag.

72. Calleja M.C. and G. Persoone. 1992. Cyst-based toxicity tests: IV. The potential of ecotoxicological tests for the prediction of acute toxicity in man as evaluated on the first ten chemicals of the MEIC programme. *ATLA* 20: 395–405.

73. http://www.microbiotests.be.

74. Standard operational procedure: *Daphtoxkit F*™ magna. *Crustacean Toxicity Screening Test for Freshwater.* MicroBioTests Inc., Nazareth, Belgium.

75. Standard operational procedure: *Ostracodtoxkit F. Chronic "Direct Contact" Toxicity Test for Freshwater Sediments.* MicroBioTests Inc., Nazareth, Belgium.

76. Mitchell, J.A.K., J.E. Burgess, and R.M. Stuetz. 2002. Developments in ecotoxicity testing. *Rev. Environ. Sci. Biotech.* 1: 169–198.

77. Sosak-Świderska, B. 2002. Mikrobiotesty w badaniach wody. *Bioskop* 2: 9–10.

78. Van der Oost, R. 2005. Bioassays and biomarkers: The missing link in the EU Water Framework Directive. In: *CIPAC International Symposium.* the Netherlands: Utrecht.

79. Van Dam, R.A. and J.C. Chapman. 2001. Direct toxicity assessment. *Australas. J. Ecotoxicol.* 7: 175–198.

80. Power, E.A. and R.S. Boumphrey. 2004. International trends in bioassay use for effluent management. *Ecotoxicology* 13: 377–398.

81. *Point and Diffuse Sources: OSPAR Background Document Concerning the Elaboration of Programmes and Measures Relating to Whole Effluent Assessment.* OSPAR Commission, London, UK, 2000.

82. Tonkes, M., H. Pols, H. Warner, and V. Bakker. 1998. *Whole-effluent Assessment.* RIZA Report 98.034. 1-20. Lelystad (Netherlands): Institute for Inland Water Management and Waste Water Treatment (RIZA).

83. Poreda, A. 2006. Akumulacja jonów metali ciężkich przez mikroorganizmy. *Laboratorium* 8: 20–23.

84. Farré, M., G. Klöter, M. Petrovic, M.C. Alonso, M.J.L. de Alda, and D. Barceló. 2002. Identification of toxic compounds in wastewater treatment plants during a field experiment. *Anal. Chim Acta* 456: 19–30.

85. Manusadžianas, L., L. Balkelytè, K. Sadauskas, I. Blinova, L. Põllumaa, and A. Kahru. 2003. Ecotoxicological study of Lithuanian and Estonian wastewaters: Selection of the biotests, and correspondence between toxicity and chemical-based indices. *Aquat. Toxicol.* 63: 27–41.

86. Pérez, S., M. Farré, M.J. García, and D. Barceló. 2001. Occurrence of polycyclic aromatic hydrocarbons in sewage sludge and their contribution to its toxicity in the *ToxAlert* 100 bioassay. *Chemosphere* 45: 705–712.

87. Kahru, A., R. Reiman, and A. Rätsep. 1998. The efficiency of different phenol-degrading bacteria and activated sludges in detoxification of phenolic leachates. *Chemosphere* 37: 301–318.

88. Zabetoglou, K., D. Voutsa, and C. Samara. 2002. Toxicity and heavy metal contamination of surficial sediments from the Bay of Thessaloniki (Northwestern Aegean Sea) Greece. *Chemosphere* 49: 17–26.

89. Calace, N., S. Ciardullo, B.M. Petronio, et al. 2005. Influence of chemical parameters (heavy metals, organic matter, sulphur and nitrogen) on toxicity of sediments from the Mar Piccolo (Taranto, Ionian Sea, Italy). *Microchem. J.* 79: 243–248.

90. Hyotylainen, T. and A. Oikari. 1999. Correlation between concentration in urine and in blood of cadmium and lead among women in Asia. *Sci. Total Environ.* 243: 97–107.

91. Ricking, M., H. Neumann-Hensel, J. Schwarzbauer, and A. Svenson. 2004. Toxicity of octameric elemental sulphur in aquatic sediments. *Environ. Chem. Lett.* 2: 109–112.

92. Pedersen, F., E. Bjørnestad, H.V. Andersen, J. Kjøholt, and C. Poll. 1998. Characterization of sediments from Copenhagen Harbour by use of biotests. *Water Sci. Technol* 37: 233–240.

93. McCarthy, L.H., R.L. Thomas, and C.I. Mayfield. 2004. Assessing the toxicity of chemically fractionated Hamilton Harbour (Lake Ontario) sediment using selected aquatic organisms. *Lakes Reser.: Res. Manag.* 9: 89–102.

94. Olajire, A.A., R. Altenburger, E. Küster, and W. Brack. 2005. Chemical and ecotoxicological assessment of polycyclic aromatic hydrocarbon—contaminated sediments of the Niger Delta, Southern Nigeria. *Sci. Total Environ.* 340: 123–136.

95. Brack, W., R. Altenburger, U. Ensenbach, M. Möder, H. Segner, and G. Schüürmann. 1999. Bioassay-directed identification of organic toxicants in river sediment in the industrial region of Bitterfeld (Germany)—a contribution to hazard assessment. *Arch. Environ. Contam. Toxicol.* 37: 164–174.

96. Vondráček, J., M. Machala, K. Minksová, et al. 2001. Monitoring river sediments contaminated predominantly with polyaromatic hydrocarbons by chemical and *in vitro* bioassay techniques. *Environ. Toxicol. Chem.* 20: 1499–1506.

97. Based on material available from the company Tigret Sp. z o.o.

98. HELCOM Recommendation 23.11.2002. *Requirements for discharging of waste water from the chemical industry.* 61–66. Helsinki (Finland): Adopted March 6, having regard to Article 20, Paragraph 1 b) of the Helsinki Convention. Annex 13.

99. HELCOM Recommendation 23.12.2002. *Reduction of discharges and emissions from production of textiles.* 67–72. Helsinki (Finland): Adopted March 6, having regard to Article 20, Paragraph 1 b) of the Helsinki Convention. Annex 14.

100. HELCOM Recommendation 23.10.2002 *Reduction of discharges and emissions from production and formulation of pesticides.* 56–60. Helsinki (Finland): Adopted March 6, having regard to Article 20, Paragraph 1 b) of the Helsinki Convention. Annex 12.

101. Persoone, G., B. Marsalek, I. Blinova, et al. 2003. A practical and user-friendly toxicity classification system with microbiotests for natural waters and wastewaters. *Environ. Toxicol.* 18: 395–402.

102. Krebs, F. 1998. The pT-value as a classification index in aquatic toxicology. *GIT Fachzeitschrift für das Laboratorium* 32: 293–296.

103. Commission Decision of May 24, 2006, on the national provisions notified by the Czech Republic under Article 95(4) of the EC Treaty concerning the maximum admissible content of cadmium in fertilisers (notified under document number C(2006) 2036).

104. Reincke, H., U. Schulte-Oehlmann, M. Duft, B. Markert, J. Oehlmann, and B. Stachel. 2001. Biologisches Effektmonitoring an Sedimenten der Elbe mit *Potamopyrgus antipodarum* und *Hinia* (*Nassarius*) *rericulata* (Gastropoda: Prosobranchia). ARGE-Elbe.

105. Grote, M., R. Altenburger, W. Brack, et al. 2005. Ecotoxicological profiling of transect River Elbe sediments. *Acta Hydrochim. Hydrobiol.* 33: 555–569.

106. Costan, G., N. Bermingham, C. Blaise, and J.F. Ferard. 1993. Potential ecotoxic effects probe (PEEP): A novel index to assess and compare the toxic potential of industrial effluents. *Environ. Toxicol. Water Qual.* 8: 115–140.

107. Blaise, C. and J.F. Ferard. 2005. Effluent assessment with the PEEP (potential ecotoxic effects probe) index. In: C. Blaise and J.F. Ferard. (eds), *Small-scale Freshwater Toxicity Investigations*, Vol. 2, pp. 69–87. New York: Springer.

108. Kusui, T. 2002. Japanese application of bioassays for environmental management. *Sci. World J.* 2: 537–541.

109. Roghair, C.J., J. Struijs, and D. de Zwart. 1997. *Measurement of Toxic Potency in Freshwaters in the Netherlands. Part A. Methods.* RIVM Report 607504 004. the Netherlands: National Institute of Public Health and Environment.

110. Struijs, J., R. Ritsema, v.d. Kamp, and D. de Zwart. 2000. *A Pilot of New Monitoring Techniques*. RIVM report 607200 003. The Netherlands: National Institute of Public Health and Environment.
111. Bombardier, M. and N. Bermingham. 1999. The SED-TOX index: Toxicity-directed management tool to assess and rank sediments based on their hazard concept and application. *Environ. Toxicol. Chem.* 18: 685–688.
112. Long, E.R. and P.M.A. Chapman. 1985. Sediment quality triad: Measures of sediment contamination, toxicity and infaunal community composition in Puget Sound. *Mar. Pollut. Bull.* 16: 405–415.
113. Wolska, L. and Ż. Polkowska. 2001. Bacterial luminescence test screening of highly polluted areas in the Odra River. *Bull. Environ. Contam. Toxicol.* 67: 52–58.
114. Wolska, L., A. Kuczyśnka, and J. Namieśnik. 2006. Quality of groundwater contaminated by leachates from seven Polish landfills— chemical and ecotoxicological classifications. *Toxicol. Environ. Chem.* 88: 501–513.
115. European Catchments. Catchment changes and their impact on the coast—EUROCAT. EVK1-CT-2000-00044.
116. Wolska, L., M. Michalska, and M. Bartoszewicz. 2003. Toxicity of ecosystem. In: J. Dojlido, W. Kowalczewski, R. Miłaszewski, and J. Ostrowski (eds), *Bug River, Water and Natural Resources*, pp. 352–361. Warszawa: IMGW, Wyż. Szk. Ekol. i Zarz.
117. Directives for the Management of Dredged Material from Waters within the Jurisdiction of the German Federal Waterways and Shipping Administration—HABAB-WSV 2000, HABAK-WSV 1999.
118. Decree of the Minister of the Environment, dated February 11, 2004, concerning the classification of the conditions of surface and underground waters, methods of monitoring and interpreting results and presenting the conditions of these waters.
119. Reports on Environmental Impact Assessment (EIA). Available at http://www.wios.lublin.pl/ (in Polish).

10 Total Parameters as a Tool for the Evaluation of the Load of Xenobiotics in the Environment

Tadeusz Górecki and Heba Shaaban El-Hussieny Mohamed

CONTENTS

10.1 Introduction 223
10.2 Biochemical Oxygen Demand, Chemical Oxygen Demand,
and Total Organic Carbon 224
10.3 Total Parameters in Environmental Analysis 227
10.4 Conclusions 233
References 233

10.1 INTRODUCTION

One of the unintended consequences of industrialization and urbanization is the global distribution of numerous chemicals throughout the atmosphere, hydrosphere, and lithosphere. Many of these compounds are xenobiotics, that is, they are foreign to living organisms.[1] In the vast majority of cases, xenobiotics are considered to be environmental pollutants. They can be hazardous both to various ecosystems and to humans. One of the main tasks of environmental analytical chemistry is the analysis of those compounds in the environment.

Environmental analytics has been the driving force for the development of many new or improved analytical methods. The growing concern regarding the potential harmful effects of long-term exposure to very low concentrations of some xenobiotics, especially those with carcinogenic properties, has resulted in an on-going demand for the determination of such compounds at ever decreasing levels and/or in increasingly complex matrices. This has led to the introduction of a number of new methods and instrumental techniques into analytical practice.[2] One prominent example of such techniques is comprehensive two-dimensional gas chromatography (GC × GC).[3,4] In this technique, samples are subjected to chromatographic separation in two columns coated with stationary phases of different selectivities. The columns are connected through a special interface (modulator), whose role is to collect the eluate from the first column and periodically inject it into the second, much shorter column. In the ideal case, analyte separation in the second column is completed before the next injection takes place. The cycle is repeated throughout the entire run. Compared to conventional, one-dimensional gas chromatography (1D-GC), GC × GC offers vastly improved separation power. Analytes that cannot be separated using any single column can often be baseline resolved using this technique. The number of analytes that can be resolved in a single GC × GC run is much higher than in 1D-GC because the peak capacity of a GC × GC system is to a first approximation the product of the peak capacities of the individual dimensions. In addition, GC × GC with thermal modulation offers significantly better sensitivity compared to 1D-GC owing to (a) band compression prior to reinjection into the second dimension, and (b) chromatographic separation of the first dimension column bleed from the analyte peaks in the second column. Taking these advantages into account, it should come as no surprise that GC × GC has found numerous applications in environmental analysis.[5]

Environmental analysis and monitoring are the foundation of all environmental science. Although they cannot solve any environmental problems by themselves, they do supply information on the condition of the environment, the effectiveness of remedial and preventive actions, and the impact of various technologies on the environment.[6] Depending on the goals of the analysis and the method(s) used, environmental analysis can produce information of varying degrees of detail. In the simplest case scenario, only the elemental composition of the sample is determined, such as is often the case in inorganic analysis. Methods such as atomic absorption spectroscopy (AAS), inductively coupled plasma optical emission spectroscopy (ICP-OES), or inductively coupled plasma mass spectrometry (ICP-MS) yield exclusively this type of information (unless the final determination is preceded by analyte separation, as in the coupling of gas chromatography with atomic emission detection—GC-AED). At the other end of this spectrum is full speciation.[7] According to the IUPAC definition, speciation analysis is a process leading to the identification and determination of the various forms of occurrence of a given element in a sample. There are two types of speciation analysis: physical speciation (the forms in which a chemical compound occurs) and chemical speciation (the identification and determination of all chemicals containing a given element). Numerous papers contain information on the particular types of chemical speciation (e.g., screening, distribution, group, chiral, or individual speciation).

Taking into account the diversity of chemicals that can pollute a given environmental compartment (e.g., air, water, or soil), the separation, identification, and determination of all pollutants can be a daunting task with respect to both the complexity and the cost of the analysis. When full speciation is neither desirable nor necessary, and the information delivered by elemental analysis is insufficient, the third option is to determine a suitable total (or summary) parameter(s) describing the total content of a given element in all the pollutants or in a particular subgroup of pollutants.[6] In some cases, the use of suitable total parameters could significantly reduce the number of necessary determinations, thereby allowing more efficient assessment of the degree of pollution. Both approaches (speciation analysis and determination of total parameters) should be considered complementary—the value of a suitable total parameter could be used to decide whether full speciation analysis is in fact necessary.

10.2 BIOCHEMICAL OXYGEN DEMAND, CHEMICAL OXYGEN DEMAND, AND TOTAL ORGANIC CARBON

Biochemical oxygen demand (BOD) is arguably the oldest total parameter used for the characterization of water quality. It was introduced in the first decade of the twentieth century as a test for the organic pollution of rivers. BOD is the amount of oxygen in mg L^{-1} required for the oxidation of the organic matter contained in water by biological action under standardized test conditions (usually a temperature of 20°C and an incubation time of 5 days).[8,9] The test is often used to evaluate the efficiency of wastewater treatment processes.

In the classical method of BOD determination, dissolved oxygen (DO) is determined in the sample using titration before and after the incubation period. The sample is introduced to a flask and diluted to a predetermined volume with distilled water. The flask is shaken to make sure that the water is saturated with oxygen. The pH of the sample is adjusted if necessary, and the covered flask is stored for the duration of the incubation period away from light. A blank sample is prepared in the same way. BOD is calculated as the difference between the initial DO content and the content after the incubation period, taking into account the dilution factor. When the typical incubation period of 5 days is used, the parameter determined is denoted BOD_5. The limit of detection of this method is ~1 mg L^{-1}.

Determination of BOD is based on the measurement of DO consumed in the process of biodegradation of organic matter in a certain period of time. It is a universal parameter, the determination of which does not require expensive equipment. On the other hand, only compounds that undergo biodegradation contribute to BOD—refractory organic compounds cannot be detected by this technique. Reproducibility of this measurement is usually quite poor, and the method is very time-consuming. A number of alternative approaches have been proposed to overcome this last limitation, including respirometric methods, headspace BOD determination, and direct measurement of absorbance.[8] Another interesting alternative is the use of microbial BOD sensors.[10] Such sensors consist of an immobilized microbial film sandwiched between two gas-permeable membranes. The response of the sensor is related to the change in the concentration of DO caused by biodegradation of the organic matter in the sample by the biofilm. A physical transducer is used to monitor this process. BOD biosensors measure short-term BOD, which is not necessarily identical to the conventional BOD_5. Consequently, correlations between the two parameters need to be examined. Due to their short response times, BOD biosensors are particularly useful for control purposes during aerobic treatment of wastewaters.

An alternative parameter that can be used to characterize the organic load of water is chemical oxygen demand (COD). COD is the amount of oxygen required to oxidize the organic matter present in the sample using chemical methods.[8] In a COD assay, a known excess of a strong oxidizing agent (typically potassium dichromate under acidic conditions) is added to the sample and incubated with it for a period of time. Excess oxidant is then determined, usually by titration. COD determination is much faster than BOD determination—it can usually be completed in a few hours. In addition, the method is simpler and more reproducible than BOD determination. The values of COD and BOD for a given sample are usually correlated, although the exact correlations may vary widely from one sample to the other. COD values are usually greater than the corresponding BOD values because COD is related to the concentration of all oxidizable chemicals, whereas BOD is determined by the concentration of biodegradable chemicals only.

The limitations of both BOD and COD determination can be overcome by direct determination of the organic carbon (OC) content of the sample. The corresponding parameter is called total organic carbon (TOC). It is defined as the amount of carbon covalently bonded in organic compounds in a water sample.[8] The analytical value of TOC was recognized since 1931. A large number of papers dealing with the instrumental and methodological aspects of TOC determination in samples of various kinds have been published since then. TOC is a convenient measure of the overall contamination of water with organic compounds. It can also be a very useful measure of the efficiency of water and wastewater treatment.

In both TOC and dissolved organic carbon (DOC) determinations, OC in the water sample is oxidized to produce carbon dioxide (CO_2), which is then measured by a detection system. Inorganic carbon (IC) is removed prior to the analysis by acidifying the sample. Alternatively, TOC can be determined indirectly through the measurement of total carbon (TC) and IC. TOC in the indirect method is calculated as the difference between the two.

There are two main approaches to the oxidation of OC in water samples to CO_2: combustion in an oxidizing gas and UV-promoted or heat-catalyzed chemical oxidation. Other approaches are sometimes used, but are much less widespread.[11] Carbon dioxide, which is released from the oxidized sample, can be detected in several ways, including conductivity detection, nondispersive infrared (NDIR) detection, or conversion to methane and measurement with a flame ionization detector (FID).[12,13] The limits of detection in TOC determination can be as low as 1 μg L^{-1}, and the dynamic range can span many orders of magnitude. The precision of the method is usually very good, and the analysis can be completed in a few minutes. Another advantage is the very small amount of sample required—from 10 to 2000 μL.

The TC content of a sample can be subdivided into many fractions depending on the particular needs. The classification is often based on purely operational parameters, for example, the method used to release the carbon from the sample (examples include purgeable organic carbon—POC or acid-released organic carbon—AROC). Table 10.1 lists examples of the different fractions of TC that can be determined in liquid and solid samples. This particular classification is based on the determination of carbon content, but similar classifications could be prepared for other elements present in the pollutants.

The TOC content can also be determined in air. However, the parameter used in this case is called total hydrocarbons (TH) rather than TOC. One significant difference between the OC in air and in other matrices is that the atmosphere contains a nearly constant background concentration of methane (~1.7 µg g^{-1} v/v),[14] derived mostly from natural sources. Thus, any TH measurement will

TABLE 10.1
TC Fractions in Liquid and Solid Samples[11]

Parameter	Symbol	Description
Organic carbon	OC	
Inorganic carbon	IC	
Liquid matrix		
Total carbon	TC	All carbon present in any particle and compound TC = TIC + TOC
Total inorganic carbon	TIC	All IC present in the form of carbonate, bicarbonate, and dissolved CO_2. Quantity depending on pH, temperature, and partial pressure of CO_2 TIC = DIC + PIC
Dissolved inorganic carbon	DIC	
Particulate inorganic carbon	PIC	Suspended particle material
Total organic carbon	TOC	All carbon from all organic sources covalently bound TOC = DOC + POC or TOC = NPOC + VOC
Dissolved organic carbon	DOC	All organic species that are soluble or pass through a filter of 0.45 µm DOC = VOC + NPDOC
Particulate organic carbon	POC	Suspended particles, moieties that are kept back by a 0.2–10 µm filter
Volatile organic carbon	VOC	Low boiling (<100°C), low-molecular-weight compounds
Purgeable organic carbon	POC	OC released by sparging. POC and VOC are often used interchangeably
Acid-released organic carbon	AROC	OC released by acid treatment
Nonpurgeable organic carbon	NPOC	Not removed by sparging NPOC = NPDOC + POC
Nonvolatile organic carbon	NVOC	
Nonpurgeable dissolved organic carbon	NPDOC	
Nonvolatile dissolved organic carbon	NVDOC	
Solid matrix		
Total carbon	TC	All in solid form TC = TIC + TOC
Total inorganic carbon	TIC	
Total organic carbon	TOC	
Volatile organic carbon	VOC	
Nonvolatile organic carbon	NVOC	
Acid-soluble organic carbon	ASOC	Might be lost during separation of the spent acid (up to 45%). Increases roughly with the percentage of $CaCO_3$ in the sample
Acid-insoluble organic carbon	AIOC	TOC = AIOC + ASOC
Oxidizable carbon	OXC	Easily oxidizable OC, not stabilized in organic–mineral complexes
Soil organic matter	SOM	Organic materials that accompany soil particles through a 2 mm sieve

TABLE 10.2

Parameters and Techniques Used for the Determination of TH in Air[6]

Parameter	Measurement Technique	Comments
TH	Air is supplied directly to an FID	Detector signal is not always proportional to the number of carbon atoms (e.g., due to the presence of heteroatoms). Methane concentration in the air is much higher than the total concentration of the remaining hydrocarbons, which makes it difficult to monitor changes in the levels of the latter
	CO and CO_2 are removed from the air, which is then directed through a suitable catalyst to a nondispersive infrared detector (NDIR)	Detector signal is proportional to the number of carbon atoms. Sensitivity is poor
	After removing CO and CO_2, air is directed to an FID through a suitable catalyst (to oxidize organic compounds to CO_2) and a methanizer (metallic nickel heated to 400°C, facilitating conversion of CO_2 to CH_4 in the presence of hydrogen)	Very good sensitivity
TNMHC	Air is directed to an FID either directly or through a catalyst capable of selective oxidation of nonmethane hydrocarbons. TNMHC is determined as the difference between the two signals ($TH–CH_4$)	Finding a suitable catalyst is a problem
	A chromatographic column separates CO, CO_2, and CH_4 from other sample components. The three gases are detected by an FID after passing through a methanizer. The remaining organic compounds trapped at the column head are backflushed to the detector as a single peak	Chromatographic determinations are discontinuous by nature
	Air is passed through a cryotrap, where all organic compounds except CH_4 are selectively trapped. The analytes are injected to the detector by rapid heating of the trap	Ice formation inside the trap is a serious problem
	Air passes through a catalytic reactor, where CO is selectively oxidized to CO_2, a CO_2 trap, and another catalytic reactor, where organic compounds (except CH_4) are selectively oxidized. The CO_2 formed (equivalent to TNMHC) is trapped using molecular sieves. Its amount can be determined using various techniques following thermal desorption	The procedure is very complex and error-prone due to its multistage character

include methane in addition to other hydrocarbons. A different parameter, total nonmethane hydrocarbons (TNMHC), can be used to eliminate this background contribution of methane. Table 10.2[6] contains a brief description of the techniques used for determining TH and TNMHC in air.

10.3 TOTAL PARAMETERS IN ENVIRONMENTAL ANALYSIS

Although detailed speciation analysis is gaining significance in environmental analytics, a large number of total parameters are still being used for the more general characterization of environmental samples. For example, the US Environmental Protection Agency (US EPA) specifies methods for the determination of the following total parameters among the 9000 series subset of methods for method SW-846[15]: Total and Amenable Cyanide (methods 9010C and 9012B), Total Organic

Halides (TOX, methods 9020B and 9022), Purgeable Organic Halides (POX, method 9021), Extractable Organic Halides (EOX, method 9023), Acid-Soluble and Acid-Insoluble Sulfides (methods 9030B and 9034), Extractable Sulfides (method 9031), Sulfates (methods 9035, 9036 and 9038), Total Organic Carbon (method 9060A), Phenolics (methods 9065, 9066 and 9067), n-Hexane Extractable Material (HEM, method 9071B), Total Recoverable Petroleum Hydrocarbons (method 9074), Total Chlorine in Petroleum Products (methods 9075, 9076, and 9077), and Total Coliform (methods 9131 and 9132). All these methods are based on different principles, but a feature they have in common is that the total content of a given element or species is determined rather than the individual chemicals (or organisms in the case of Total Coliform) contributing to the parameter. Some of those parameters are subsets of other parameters (e.g., POX in the case of TOX), which allows for a somewhat more detailed characterization of the samples under study. This differentiation is typically based on the physical properties of the contributing analytes and the corresponding methods used to recover them from the samples. For example, TOX characterizes the total content of organic halides in water. It detects all organic halides containing chlorine, bromine, and iodine that are adsorbed by granular activated carbon under the conditions of the method (fluorine-containing species are not determined).[16] In this method, a sample of water free of nondissolved solids is passed through a column containing 40 mg of activated carbon. The column is washed to remove any trapped inorganic halides and the activated carbon is then combusted to convert the adsorbed organohalides to hydrogen halides (HX), which are trapped and titrated electrolytically using a microcoulometric detector. POX, on the other hand, allows the determination of volatile organic halides only. In this method, volatile organic halides are purged into a pyrolysis furnace using a stream of CO_2. The HX formed during the pyrolysis of volatile analytes containing halogen atoms is trapped and titrated electrolytically with a microcoulometric detector.[17] Finally, EOX is used to determine organic halides in solids. While this parameter is clearly aimed at determining the total content of organic halides in solids, its name reflects the fact that one can never be sure whether all the relevant analytes could be recovered from the sample using liquid extraction. In EOX determination, an aliquot of a solid sample is extracted with ethyl acetate by sonication to isolate organic halides. A 25 µL aliquot of the extract is introduced into a pyrolysis furnace using a stream of CO_2/O_2, and the HX pyrolysis product is determined by microcoulometric titration.[18]

The number of different total (summary) parameters that can be used in environmental analysis and monitoring is very large. For illustration, Table 10.3[6] lists examples of various parameters used by researchers to characterize gaseous, liquid, and solid samples, published between 1978 and 2000.

Table 10.3 is followed by a brief literature review covering mostly research published in the last decade. The review is not intended to be comprehensive—its main goal is to illustrate that total parameters are still valuable tools in environmental analytics, in spite of the rapid development of ever more powerful chemical speciation methods and techniques.

Total volatile organic carbon (TVOC) was determined in indoor air using an adsorption/combustion-type gas sensor.[103] The TVOC concentration was obtained from the detector output based on the different responses obtained for two different adsorption periods of the sensor; the output increased with increasing concentration of toluene as a typical VOC. An adsorption/combustion-type TVOC gas sensor with a low heat capacity employing $Pd/\gamma-Al_2O_3$ as a sensing material fabricated on a silicon substrate was developed.[104] The sensor was driven by low–high pulse heating at temperatures of 200°C and 400°C. Total volatile organic compounds were determined in ambient air using gas chromatography-mass spectrometry (GC-MS) following active collection on multisorbent tubes and thermal desorption.[105] Unidentified analytes were quantified as toluene equivalents.

Total sulfur (TS), acid-volatile sulfide (AVS), Cr-reducible sulfur (CRS), and extractable sulfate in sediments were determined by coulometric titrimetry,[106] a method that yielded improved data quality and increased laboratory throughput.

Total mercury, inorganic mercury, and methylmercury in water were determined using a simple and ultrasensitive method, based on cold vapor generation (CVG), coupled to atomic fluorescence

TABLE 10.3

Examples of Total Parameters Used in Environmental Analysis and Monitoring[6]

State of Aggregation	Sample Origin	Total Parameter		
		Name	Acronym	Reference
Gaseous samples	Atmospheric air	Total hydrocarbons	TH	19
	Indoor air	Total volatile organic compounds	TVOC	20
	Atmospheric aerosols	Organic carbon	OC	21
		Elemental carbon	EC	
	Suspended particulate matter	Elemental carbon	EC	
		Total soluble organic carbon	TSOC	22
		Total insoluble organic carbon	TSIC	
	Suspended particulate matter	Total carbon	TC	23
	Exhaust gases	Total organic halogens	TOX	24
		Volatile total organic halogens	VTOX	
	Exhaust gas from waste incinerator	Extractable organic halogens	EOX	25
	Landfill gas	Total sulfur	TS	26
Liquid samples	Liquid samples	Chemical oxygen demand	COD	27
	Liquid samples	Adsorbable organic sulfur	AOS	28–30
	Aqueous samples	Adsorbable organically bound elements	AOE	31
	Aqueous samples	Total organic carbon	TOC	32–37
	Aqueous samples	Purgeable organic halogens	POX	38
	Ultrapure water	Total organic carbon	TOC	39,40
	Geothermal waters	Total mercury		41
	Aqueous samples	Adsorbable organic halogen	AOX	31,42,43
	Water	Suspended organic carbon (particulate organic carbon)	SOC (POC)	44
	Water	Total dissolved nitrogen	TDN	45
		Dissolved organic nitrogen	DON	
	Water	Assimilable organic carbon	AOC	46
	Aqueous solutions	Dissolved organic sulfur	DOS	47
	Groundwater	Total organic halogen	TOX	48,49
	Groundwater	Dissolved organic carbon	DOC	50–54
	Groundwater	Purgeable organic carbon	POC	55
		Volatile halogenated hydrocarbons	VHH	
	Surface water	Total sulfur	TS	56
		Organically bound sulfur	OBS	
	River water	Dissolved organic carbon	DOC	57–59
	River water	Dissolved organic carbon	DOC	60
		Suspended organic carbon (particulate organic carbon)	SOC (POC)	
		Dissolved organic nitrogen	DON	
		Suspended organic nitrogen	SON	
	River water	Adsorbable organic halogens	AOX	61,62
	River water	Dissolved total carbohydrates	DTCH	63
		Dissolved free monosaccharides	DFMS	
		Dissolved organic carbon	DOC	
		Biodegradable dissolved organic carbon	BDOC	
		Humic substances	HS	

continued

TABLE 10.3   (continued)

State of Aggregation	Sample Origin	Total Parameter		
		Name	Acronym	Reference
	River and lake water	Total organic carbon	TOC	64–67
		Dissolved organic carbon	DOC	
		Suspended organic carbon	SOC	
	Seawater	Total nitrogen	TN	68
	Seawater	Dissolved organic nitrogen	DON	69
	Seawater	Dissolved inorganic carbon	DIC	70
	Seawater	Dissolved organic carbon	DOC	71–75
	Wastewater	Total organic carbon	TOC	76
	Wastewater	Adsorbable organic halogens	AOX	77,78
	Wastewater	Dissolved organic carbon	DOC	79
	Wastewater	Adsorbable organic halogens	AOX	80,81
	Wastewater and process water	Total organic carbon	TOC	82
		Total nitrogen	TN	
	Hospital wastewater	Total organic halogens	TOX	83
		Extractable organic halogens	EOX	
		Purgeable organic halogens	POX	
	Surface water and seawater	Total petroleum hydrocarbons	TPH	84
Solid samples	Soil	Organic chlorine	Cl_{org}	85.86
	Soil and sediments	Water-soluble organic carbon	WSOC	87
	Sediments	Chemical oxygen demand	COD	88
		Total nitrogen	TN	
		Total phosphorous	TP	
		Difference on ignition	DOI	
	Sediments	Organic carbon	C_{org}	89
		Total nitrogen	N_{tot}	
	Marine sediments	Difference on ignition	DOI	90
	Marine sediments	Total organic carbon	TOC	91
	Marine sediments	Dissolved organic carbon	DOC	92
	Marine sediments and biota	Extractable organic halogens	EOX	93,94
	Lake sediments	Extractable organic halogens	EOX	95,96
	Lake sediments	Adsorbable organic halogen	AOX	97
	Sewage sludge	Total organic carbon	TOC	98
	Sewage sludge	Total organic carbon	TOC	99
		Chemical oxygen demand	COD	
		Biochemical oxygen demand	BOD	
	Electronic waste	Organic carbon	OC	100
		Total organic halogens	TOX	
	Incineration residue	Elemental carbon	EC	101
		Total organic carbon	TOC	
	Particulates from waste incineration	Extractable organic halogens	EOX	102

spectrometry (AFS).[107] In the presence of UV irradiation, all the mercury (MeHg + Hg_2^+) in a sample solution was reduced to Hg(0) by $SnCl_2$; in the absence of UV irradiation, only Hg_2^+ species could be determined. Total Hg and inorganic Hg were also determined by a novel method based on the on-line coupling of high-intensity focused ultrasound (HIFU) with a sequential injection/flow injection analysis (SIA/FIA) system.[108] This method provided high throughput, automation, and low reagent consumption.

Total mercury was determined in wastewater by cold-vapor atomic absorption spectrometry in an alkaline medium using sodium hypochlorite solution.[109] This technique was simple and rapid because no digestion was required for the determination. The time required for the complete procedure was only about 5 min. Inorganic and organic Hg were also determined in wastewaters using cold-vapor atomic absorption spectrometry and pyrolysis atomic absorption spectrometry.[110] Sample digestion was not required; inorganic Hg was directly determined by the cold vapor method, and total Hg was determined by the pyrolysis method. The value of total organic Hg was obtained as the difference of the two values. Total inorganic mercury was determined together with organomercury species in sediments using solid-phase microextraction and multicapillary GC hyphenated to inductively coupled plasma-time-of-flight mass spectrometry.[111] Headspace solid-phase microextraction with a carboxen/polydimethylsyloxane fiber was used for the extraction/preconcentration of mercury species after derivatization with sodium tetraethylborate and subsequent volatilization. Total mercury in sewage sludge was determined using solid sampling Zeeman atomic absorption spectrometry,[112] where a specially designed furnace was used and atomization of mercury was performed at a constant temperature in the 900–1000°C range. The method allowed the measurement of very low Hg concentrations without extraction or preconcentration procedures.

Humic substances (HS) were determined in water using catalytic cathodic stripping voltammetry (CSV).[113] This method was based on the adsorptive properties of iron–HS complexes on the mercury drop electrode at natural pH. HS were also determined in water using a method based on the binding of a dye, Toluidine Blue (TB), to HS molecules to produce a dye–HS complex, which caused a decrease in absorbance at 630 nm.[114] The method was rapid, sensitive, and practicable.

Total phosphorus in water was determined using FIA,[115] on-line microwave digestion and flow injection spectrophotometry,[116] and digital imaging colorimetry.[117] In the last method, the interaction of potassium dihydrogen phosphate with ammonium molybdate, potassium antimonyl tartrate, and ascorbic acid in acidic media resulted in a blue complex. With increasing potassium dihydrogen phosphate concentration, the color of the solution and the RGB value of digital imaging increased. An electrochemical method based on using a nano-TiO_2-$K_2S_2O_8$ film for photocatalytic oxidation with PMo_{12} film-modified electrode was also used for the determination of total phosphorus in water.[118]

Total nitrogen was determined in water using high-temperature oxidation and chemiluminescence,[119] as well as microwave digestion-UV spectrophotometry.[120] Total nitrogen in solid waste was determined by modified Kjeldahl–Nessler reagent colorimetrically and modified Kjeldahl titration with hydrochloric acid.[121] No significant differences were found between the two methods in terms of accuracy and precision.

Extractable organic halides (EOX; also known as extractable organohalogens) together with extractable organochlorinated (EOCl), extractable organobrominated (EOBr), and extractable organoiodinated (EOI) compounds were determined in air by instrumental neutron activation analysis (INAA) combined with organic solvent extraction.[122] EOX in sediments were assayed coulometrically; samples were extracted with *n*-hexane in a Soxhlet apparatus.[123] EOX was determined in contaminated soils and sediments using two mixed solvents—toluene/*n*-hexane and acetone/*n*-hexane—as extractants.[124]

COD was determined in water using photoelectro-synergistic catalysis (PEC) with FIA[125] and an on-line monitoring system based on the photoelectrochemical degradation principle. The latter method employed a specially designed thin-layer photoelectrochemical cell that incorporated a highly effective nanoparticulate TiO_2 photoanode;[126,127] it allowed direct quantitation of electron transfer at the TiO_2 nanoporous film electrode during exhaustive photoelectrocatalytic degradation of organic matter in a thin-layer photoelectrochemical cell. The method was environmentally friendly, robust, and rapid, and did not require the use of a standard for calibration. COD was also determined in water by using a sensitive spectrophotometric method based on the use of a nano-TiO_2-$KMnO_4$ photocatalytic oxidation system[128] and by using an electrochemical sensor with an electrode-surface grinding unit,[129] where the oxidizing action of Cu on an organic species was used as the basis of the COD measurement.

BOD was determined in wastewater using a microbial sensor based on an organic–inorganic hybrid material for immobilization of the biofilm. The biosensor response to the sample exhibited good reproducibility, long-term stability, and required only 10 min for each measurement.[130] Near-infrared spectroscopy was used for the rapid determination of COD and BOD in wastewater.[131]

Reference filters for the analysis of elemental carbon (EC) and OC in aerosol particles were produced by a spray-drying method and a carbon aerosol sampling system.[132] Submicrometer carbon particles were produced by nebulizing a carbon black hydrosol and a potassium hydrogen phthalate solution. The TC concentration measured at three different locations on the filter showed that the carbon particles were uniformly distributed on the filter. EC and OC were also determined in snow and ice using a two-step heating GC system.[133] OC and EC were transformed into CO_2 in a stream of oxygen at 340°C and 650°C, respectively. The resulting CO_2 was accumulated in two molecular sieve traps, and then introduced into a gas chromatograph by heating the traps to 200°C in a helium stream.

Assimilable organic carbon (AOC) was determined in water using flow-cytometric enumeration and a natural microbial consortium as inoculum.[134] Two bacterial species were used for the measurement of AOC in water, based on their respective 16S rDNA sequences. The AOC content in 41 water samples was determined with these two sets by quantitative real-time polymerase chain reaction (qRT-PCR).[135]

Total organic halogens (TOX), including TOCl, TOBr, and TOI, were determined in drinking water using pyrolysis and off-line ion chromatography.[136] TOX and total volatile organic halogen (TVOX) were determined in exhaust gases using a method based on the adsorption of gaseous organohalogen compounds by a special-grade activated carbon.[137] TOX was defined as the organo-halogen compounds collected both in the water drain and in the activated carbon columns, whereas TVOX was defined as only the compounds collected by the activated C columns. The carbon particles were combusted in an electric furnace at 900°C. The HCl formed from the halogenated organics was determined by electrochemical titration. TOX determined by this method was proposed as an alternative index of dioxins in flue gas.

TOX, EOX, AOX, and POX were determined in hospital waste sludge[138] treated with 400 µg g^{-1} of hypochlorite. Ethanol is a solvent commonly used for extracting organic halides from sludge, but its extraction efficiency proved to be poor.

Water-soluble organic carbon (WSOC) was characterized in atmospheric aerosols using solid-state ^{13}C nuclear magnetic resonance (NMR) spectroscopy[139] and anion-exchange chromatography.[140] An instrument for the on-line measurement of WSOC was described, in which a particle-into-liquid sampler impacted ambient particles grown to large water droplets onto a plate, then washed them into a flow of purified water. The resulting liquid was filtered and the carbon content quantified by a TOC analyzer.[141] TOC was determined in the neutral compound (NC), mono- and diacid (MDA), and polyacid (PA) fractions of WSOC in atmospheric aerosol particles using anion-exchange high-performance liquid chromatography.[142] The method enabled direct TOC analysis of the eluted fractions without any pretreatment.

Dissolved inorganic carbon (DIC) was determined in seawater using a method in which the seawater was acidified with 10% H_3PO_4; the evolved CO_2 gas was absorbed by NaOH solution and titrated against HCl with the end points detected using phenolphthalein and a mixture of bromocresol green and methyl red.[143] DIC and its isotope composition were determined in water using a gas chromatograph coupled to an isotope ratio mass spectrometer (GCIRMS). The instrument was capable of analyzing some 50 water samples per day.[144] DIC and DOC were measured by sequential injection spectrophotometry with on-line UV photo-oxidation.[145] Direct measurement of DIC in water was made possible by reagent-free ion chromatography (RF-IC)[146] with an electrolytically generated hydroxide eluent. All inorganic forms of carbon were converted into carbonate, which was detected as a single chromatographic peak using conductivity detection.

DOC was determined in water using an automated segmented flow analyzer,[147] in which dissolved organic matter was converted to CO_2 by UV-persulfate oxidation. The CO_2 formed induced a change in pH that altered the color intensity of a phenolphthalein solution. The change was measured automatically by colorimetry. DOC and total dissolved nitrogen (TDN) were determined in water using a coupled high-temperature combustion TOC-nitrogen chemiluminescence detection system; this allowed the simultaneous determination of DOC and TDN in the same sample using a single injection and provided low detection limits and excellent linear ranges for both DOC and TDN.[148]

TC and total nitrogen were determined in soil using near-infrared spectroscopy.[149] Thermal combustion combined with ion chromatography was used to measure TC in air particulate matter.[150] OC, IC, and TC were determined in soil using a dual temperature combustion method, which enabled all three parameters to be obtained in a single run; the variability of this method was significantly reduced.[151] DIC, OC, and TC were determined in natural waters using a combination of two methods: RF-IC and inductively coupled plasma atomic emission spectrometry (ICP-AES).[152] DIC was measured in untreated samples using RF-IC and by in-line mixing with 0.1 M HNO_3 to enhance CO_2 removal in the nebulizer, followed by ICP-AES analysis. Total dissolved carbon (TDC) was measured by in-line mixing with 0.1 M NaOH followed by ICP-AES analysis. DOC was obtained as the difference between DIC and TDC. Only nonvolatile organic carbon could be detected with the method.

The TS content was determined in waste activated sludge using microwave digestion followed by ICP-AES[153] and by vacuum combustion extraction-quadrupole mass spectrometry (VCE-QMS).[154] TS was determined in landfill gases by oxidative combustion followed by detection of the resulting SO_2 with a quartz piezoelectric crystal microbalance (QCM).[155]

Dissolved organic sulfur was determined in aqueous solutions after isolation by solid-phase extraction on macroporous resins and reversed-phase sorbents.[156] The sulfur in the extracts was determined by pyrohydrogenolysis of the extract in a heated quartz tube (1100°C) in a hydrogen atmosphere followed by flame photometric detection.

10.4 CONCLUSIONS

Determination of total parameters is an example of activities classified as group speciation. Taking into account the sheer number and diversity of different chemicals that may be present in environmental samples, determination of total parameters could be an attractive alternative to full chemical speciation, especially for the initial screening of samples. Samples for which the value of a given total parameter exceeds prescribed limits can then be further analyzed using more advanced methods. Overall, group speciation can save both time and money in the characterization of environmental samples.

REFERENCES

1. IUPAC Compendium of Chemical Terminology, Electronic version. http://goldbook.iupac.org (2006), created by M. Nic, J. Jirat, B. Kosata; updates compiled by A. Jenkins, doi:10.1351/goldbook.
2. Namieśnik, J. 2000. Trends in environmental analytics and monitoring. *Crit. Rev. Anal. Chem.* 30: 221–269.
3. Górecki, T., J. Harynuk, and O. Panić. 2003. Comprehensive two-dimensional gas chromatography (GC × GC). In: J. Namieśnik, W. Chrzanowski, and P. Zmijewska (eds), *New Horizons and Challenges in Environmental Analysis and Monitoring*, pp. 61–83. Gdańsk (Poland): Centre of Excellence in Environmental Analysis and Monitoring, Gdańsk University of Technology.
4. Górecki, T., O. Panić, and N. Oldridge. 2006. Recent advances in comprehensive two-dimensional gas chromatography (GC × GC). *J. Liquid Chromatogr. Rel. Technol.* 29: 1077–1104.
5. Panić, O. and T. Górecki. 2006. Comprehensive two-dimensional gas chromatography (GC × GC) in environmental analysis and monitoring. *Anal. Bioanal. Chem.* 386: 1013–1023.
6. Namieśnik, J. and T. Górecki. 2002. Application of total parameters in environmental analytics. *Am. Environ. Lab.* 34: 18–21.

7. Kot, A. and J. Namies´nik. 2000. The role of speciation in analytical chemistry. *Trends Anal. Chem.* 19: 69–79.

8. Domini, C.E., L. Vidal, and A. Canals. 2007. Main parameters and assays involved with organic pollution of water. In: L.M.L. Nollet (ed.), *Handbook of Water Analysis*, 2nd edition, pp. 337–366. Boca Raton, FL: CRC Press LLC.

9. McNaught, A.D. and A. Wilkinson (ed.), 1997. *Compendium of Chemical Terminology* (2nd edition), Blackwell Scientific Publications, Oxford, UK.

10. Liu, J. and B. Mattiasson. 2002. Microbial BOD sensors for wastewater analysis. *Water Res.* 36: 3786–3802.

11. Bisutti, I., I. Hilke, and M. Raessler. 2004. Determination of total organic carbon—an overview of current methods. *Trends Anal. Chem.* 23: 716–726.

12. EPA method 9060A. 2004. Available at http://epa.gov/sw-846/pdfs/9060a.pdf.

13. EPA method 415.3. 2005. Available at http://www.epa.gov/microbes/m_415_3Rev1_1.pdf.

14. Wayne, R.P. 2000. *Chemistry of Atmospheres*, 3rd edition. Oxford, NY: Oxford University Press.

15. EPA method SW-846. 2007. Available at http://www.epa.gov/sw-846/main.htm.

16. EPA method 9020B. 1994. Available at http://www.epa.gov/sw-846/pdfs/9020b.pdf.

17. EPA method 9021. 1992. Available at http://www.epa.gov/sw-846/pdfs/9021.pdf.

18. EPA method 9023. 1996. Available at http://www.epa.gov/sw-846/pdfs/9023.pdf.

19. Bella, D., G. Cavazzutti, M. Petrucci, and C. Rossi. 1976. Inquinamento atmosferico da sostanze organiche e loro determinazione come idrocarburi totali. *Boll. Chim. Lab. Prov.* 27: 107–132.

20. Molhave, L. 1999. The TVOC concept. In: T. Salthammer (ed.), *Organic Indoor Air Pollutants. Occurrence, Measurement, Evaluation*, pp. 305–318. Wiley-VCH, Weinheim, Germany.

21. Birch, M.E. 1998. Analysis of carbonaceous aerosols: Interlaboratory comparison. *Analyst* 123: 851–857.

22. Iwatsuki, M., T. Kyotani, and K. Matsubara. 1998. Fractional determination of elemental carbon and total soluble and insoluble organic compounds in airborne particulate matter by thermal analysis combined with extraction and heavy liquid separation. *Anal. Sci.* 14: 321–326.

23. Fung, Y.S. and K.L. Dao. 1998. Determination of total carbon in air particulate matter by thermal combustion—ion chromatography. *Int. J. Environ. Anal. Chem.* 69: 125–139.

24. Kawamoto, K. 1999. TOX as a novel alternative index of dioxins in flue gas. *Organohalogen Compd.* 40: 157–160.

25. Akimoto, Y. and Y. Inouye. 1999. Behaviour of extractable organic halogens in a municipal solid waste incinerator and the relationship with that of polychlorinated dibenzo-*p*-dioxins and polychlorinated dibenzofurans. *J. Health Sci.* 45: 256–261.

26. Rocha, T.A.P., A.B.P. Oliveira, and A.C. Armando. 2000. Determination of total sulfur in landfill gases using a quartz crystal microbalance. *Int. J. Environ. Anal. Chem.* 75: 121–126.

27. Chen, J.-M., T.-C. Pan, and C.-W. Huang. 1994. A modified sealed oven-UV method of COD determination. *Jpn. J. Toxicol. Environ. Health* 40: 338–343.

28. Kupka, H.J., H. Stremming, and M. Spitaler. 1989. AOS—Analyse organischer Schwefelverbindungen in Wasser. *LaborPraxis* 13(4): 270–278.

29. Wilke, D., M. Geisler, and M. Hahn. 2000. H. Matschiner, Summarische Bestimmung von organischen Schwefelverbindungen. *CLB—Chemie in Labor und Biotechnik* 51: 13–17.

30. Randt, C. and R. Altenbeck. 1997. Problems in the development of a method for the determination of the total amount of adsorbable organic sulfur compounds (AOS) in water/wastewater. *Vom Wasser* 88: 217–225.

31. Lehnert, H., T. Twiehaus, D. Rieping, W. Buscher, and K. Cammann. 1998. Thermal desorption and atomic emission spectrometric determination of adsorbable organically bound elements for water analysis. *Analyst* 123: 637–640.

32. Burkhardt, M.R., R.W. Brenton, J.A. Kammer, V.K. Jha, P.G. O'Mara-Lopez, and M.T. Woodworth. 1999. Improved method for the determination of non-purgeable suspended organic carbon in natural water by silver filter filtration, wet chemical oxidation, and infrared spectrometry. *Water Resour. Res.* 35: 329–334.

33. Espinoza, L.H., D. Lucas, D. Littlejohn, and S. Kyauk. 1999. Total organic carbon content in aqueous samples determined by near-IR spectroscopy. *Appl. Spectrosc.* 53: 103–107.

34. Winnett, W.K. and M.P. Murphy. 1994. A novel sample introduction technique for combustion total organic carbon analysis in aqueous materials. *Talanta* 41: 1627–1630.

35. Carniel, A., C. Del Bianco, and N. Zanin. 1998. Determinazione della sostanza organica nelle acque: Correlazione tra metodo Kuebel (ossidabilita al Permanganato) e TOC (Total Organic Carbon). *Boll. Chim. Igien.* 49: 223–226.

36. MacCraith, B., K.T.V. Grattan, D. Connolly, R. Briggs, W.J.O. Boyle, and M. Avis. 1993. Cross comparison of techniques for the monitoring of total organic carbon (TOC) in water sources and supplies. *Water Sci. Technol.* 28: 457–463.

37. van Leeuwen, J., M. Drikas, D. Bursill, and B. Nicholson. 1997. Contamination of samples for DOC Analysis. *Water* 24(1): 12.

38. Zuercher, F. 1981. Simultaneous determination of total purgeable organo-chlorine, bromine- and fluorine-compounds in water by ion-chromatography. In: A. Bjorseth and G. Angeletti (eds), *Analysis of Organic Micropollutants in Water*, pp. 272–276. Dordrecht, Holland: D. Riedel Publishing Company.

39. Melanson, P. and M. Retzik. 1995. A second-generation TOC analysis system for high purity water systems. *Ultrapure Water* 12(3): 76–79.

40. Cohen, N. 1994. TOC as a replacement for the oxidizable substances test. *Ultrapure Water* 11(3): 48–49.

41. Sakamoto, H., J. Taniyama, and N. Yonehara. 1997. Determination of ultra-trace amounts of total mercury by gold amalgamation-cold vapor AAS in geothermal water samples by using ozone as pretreatment agent. *Anal. Sci.* 13: 771–775.

42. Pasturenzi, M., M. Bianchi, and H. Muntau. 1997. Measurement of AOX for the screening of halogenated organic compounds in water. *Ann. Chim.* 87: 611–625.

43. Koschuh, B., M. Montes, J.F. Camuña, R. Pereiro, and A. Sanz-Medel. 1998. Total organochloride and organobromide determinations in aqueous samples by microwave induced plasma-optical emission spectrometry. *Microchim. Acta* 129: 217–223.

44. Anderson, D.J., T.B. Bloem, and J.V. Higgins. 1998. Sub-sampling technique for the determination of particulate-phase organic carbon in water. *J. Gt. Lakes Res.* 24: 838–844.

45. Bronk, D.A., M.W. Lomas, P.M. Glibert, K.J. Schukert, and M.P. Sanderson. 2000. Total dissolved nitrogen analysis: Comparisons between the persulfate, UV and high temperature oxidation methods. *Mar. Chem.* 69: 163–178.

46. Escobar, I.C. and A.A. Randall. 2000. Sample storage impact on the assimilable organic carbon (AOC) bioassay. *Water Res.* 34: 1680–1686.

47. Binde, F. and H.H. Rüttinger. 1997. Isolation and determination of dissolved organic sulfur compounds— development of the organic group parameter DOS. *Fresenius J. Anal. Chem.* 357: 411–415.

48. Gron, C. and H.P. Dybdahl. 1996. Determination of total organic halogens (TOX); bias from a non-halogenated organic compound. *Environ. Int.* 22: 325–329.

49. Sebastiano, G., V. Rebizzi, E. Bellelli, et al. 1995. Total halogenated organic (TOX) measurement in ground waters treated with hypochlorite and chlorine dioxide. Initial results of an experiment carried out in Emilia-Romagna (Italy). *Water Res.* 29: 1207–1209.

50. Cappelli, F., P.D. Goulden, J. Lawrence, and D.J. MacGregor. 1978. Determination of the adsorption efficiency of the "organics carbon adsorbable" standard method by dissolved organic carbon analysis. *J. Environ. Sci. Health A* 13: 167–176.

51. Zwiener, C. and F.H. Frimmel. 1998. Application of headspace GC/MS screening and general parameters for the analysis of polycyclic aromatic hydrocarbons in groundwater samples. *Fresenius J. Anal. Chem.* 360: 820–823.

52. Schiff, S.L., R. Aravena, S.E. Trumbore, M.J. Hinton, R. Elgood, and P.J. Dillon. 1997. Export of DOC from forested catchments on the Precambrian Shield of Central Ontario: Clues from ^{13}C and [14]C. *Biogeochem.* 36: 43–65.

53. Mitra, S. and R.M. Dickhut. 1999. Three-phase modelling of polycyclic aromatic hydrocarbon association with pore-water-dissolved organic carbon. *Environ. Toxicol. Chem.* 18: 1144–1148.

54. Vogl, J. and K.G. Heumann. 1998. Development of an ICP-IDMS method for dissolved organic carbon determinations and its application to chromatographic fractions of heavy metal complexes with humic substances. *Anal. Chem.* 70: 2038–2043.

55. Fürhacker, M. 1998. Messung von POX und/oder LHKW im Grundwasser als Beurteilungsgrundlage der Belastungssituation. *Österreichische Wasser-und Abfallwirtschaft* 50: 128–136.

56. Crowther, J., Fr.B. Lo, M.W. Rawlings, and B. Wright. 1995. Determination of organically bound sulfur in swamp and terrestrial waters by continuous flow oxidation and ion chromatography. *Environ. Sci. Technol.* 29: 849–855.

57. Meyer, J.L., J. Bruce Wallace, and S.L. Eggert. 2000. Leaf litter as a source of dissolved organic carbon in streams. *Ecosystems* 1: 240–249.

58. Tao, S. 1998. Spatial and temporal variation in DOC in the Yichun River, China. *Water Res.* 32: 2205–2210.

59. Küchler, I.L., N. Miekeley, and B.R. Forsberg. 1994. Molecular mass distributions of dissolved organic carbon and associated metals in waters from Rio Negro and Rio Solimôes. *Sci. Total Environ.* 156: 207–216.

60. Rostad, C.E., J.A. Leenheer, and S.R. Daniel. 1997. Organic carbon and nitrogen content associated with colloids and suspended particulates from the Mississippi River and some of its tributaries. *Environ. Sci. Technol.* 31: 3218–3225.

61. Pasturenzi, M., M. Bianchi, F. Pelusio, and H. Muntau. 1997. Indagine conoscitiva sul contenuto di AOX nelle acque del fiume Olona. *Boll. Chim. Igien.* 48: 85–90.

62. Luitjens, M., W. Schwanenberg, and H. Kupka. 1992. Säulenmethode zur AOX-Bestimmung. *LaborPraxis* 16(1): 37–41.

63. Gremm, Th.J. and L.A. Kaplan. 1998. Dissolved carbohydrate concentration, composition, and bioavailability to microbial heterotrophs in stream water. *Acta Hydrochim. Hydrobiol.* 26: 167–171.

64. Barałkiewicz, D., M. Kraska, and J. Siepak. 1996. The content of DOC, POC and TOC in Lobelian Lakes. *Pol. J. Environ. Stud.* 5: 17–22.

65. Barałkiewicz, D. and J. Siepak. 1994. The contents and variability of TOC, POC and DOC concentration in natural waters. *Pol. J. Environ. Stud.* 3: 15–18.

66. Siepak, J. 1999. Total organic carbon (TOC) as a sum parameter of water pollution in selected Polish rivers (Vistula, Odra, and Warta). *Acta Hydrochim. Hydrobiol.* 27: 282–285.

67. Barałkiewicz, D. and J. Siepak. 1995. Levels and seasonal variation of dissolved and suspended organic carbon in Góreckie Lake (in Polish). *Morena* 3: 73–78.

68. Rossi, G. and M. Savarese. 1997. On-line determination of total nitrogen in natural sea water samples by alkaline persulphate oxidation. *Mar. Pollut. Bull.* 35: 174–175.

69. Doval, M.D., F. Fraga, and F.F. Perez. 1997. Determination of dissolved organic nitrogen in seawater using Kjeldahl digestion after inorganic nitrogen removal. *Oceanologica Acta* 20: 713–720.

70. Ji, L. and X.K. Lu. 1997. Study on determination of DIC in seawater by coulometric method. *Chin. J. Oceanol. Limnol.* 15: 357–362.

71. Wang, J.T., Z.B. Zhang, and L.S. Liu. 1997. Determination of dissolved organic carbon in seawater using UV/persulphate method and HTCO method. *Chin. J. Oceanol. Limnol.* 15: 25–31.

72. Borsheim, K.Y., S.M. Myklestad, and J. Sneli. 1999. Monthly profiles of DOC, mono- and polysaccharides at two locations in the Trondheimsfjord (Norway) during two years. *Mar. Chem.* 63: 255–272.

73. Thingstad, T.F., Å. Hagström, and F. Rassoulzadegan. 1997. Accumulation of degradable DOC in surface waters: Is it caused by a malfunctioning microbial loop? *Limnol. Oceanogr.* 42: 398–404.

74. McKenna, J.H. and P.H. Doering. 1995. Measurement of dissolved organic carbon by wet chemical oxidation with persulfate: Influence of chloride concentration and reagent volume. *Mar. Chem.* 48: 109–114.

75. Moran, S.B., M.A. Charette, S.M. Pike, and C.A. Wicklund. 1999. Differences in seawater particulate organic carbon concentration samples collected using small-and large-volume methods: The importance of DOC adsorption to the filter blank. *Mar. Chem.* 67: 33–42.

76. Lahl, U. and B. Zeschmar-Lahl (Oyten). 1998. TOC-Grenzwerte (Eluat) für das nachsorgearme Deponieren von Restabfällen. *Korrespondenz-Abwasser* 45: 1321–1329.

77. Ferguson, J.F. 1994. Anaerobic and aerobic treatment for AOX removal. *Water Sci. Technol.* 29: 149–162.

78. Bornhardt, C., J.E. Drewes, and M. Jekel. 1997. Removal of organic halogens (AOX) from municipal wastewater by powdered activated carbon (PAC)/activated sludge (AS) treatment. *Water Sci. Technol.* 35: 147–153.

79. Fiehn, O., G. Wegener, J. Jochimsen, and M. Jekel. 1998. Analysis of the ozonation of 2-mercaptobenzothiazole in water and tannery wastewater using sum parameters, liquid- and gas chromatography and capillary electrophoresis. *Water Res.* 32: 1075–1084.

80. Schröder, H.Fr. 1998. Characterization and monitoring of persistent toxic organics in the aquatic environment. *Water Sci. Technol.* 38: 151–158.

81. Rudolph, J. 1994. AOX-Elimination und AOX-Produktion bei der Oxidation mit H_2O_2/Fe^{++} und $H_2O_2/$ UV. *Korrespondenz-Abwasser* 41: 1794–1801.

82. Fabinski, W., A. Grunewald, B. Hielscher, and Ch. Wolff. 1993. Die kontinuierliche On-Line-Messung des organischen Kohlenstoffs und des gebundenen Stickstoffs in Ab—und Prozeßwässern mit dem Tocas. *GWF—Wasser/Abwasser* 134: 613–619.

83. Tsai, C.T., C.T. Kuo, and S.T. Lin. 1999. Analysis of organic halides in hospital waste sludge disinfected using sodium hypochlorite (NaOCl). *Water Res.* 33: 778–784.

84. Hutcheson, M.S., D. Pederson, N.D. Anastas, J. Fitzgerald, and D. Silverman. 1996. Beyond TPH. Health based evaluation of petroleum hydrocarbons exposures. *Regul. Toxicol. Pharmacol.* 25: 85–101.

85. Öberg, G. 1998. Chloride and organic chlorine in soil. *Acta Hydrochim. Hydrobiol.* 26: 137–144.

86. Öberg, G. and C. Gron. 1998. Sources of organic halogens in spruce forest soil. *Environ. Sci. Technol.* 32: 1573–1579.

87. Tao, S. and B. Lin. 2000. Water soluble organic carbon and its measurement in soil and sediment. *Water Res.* 34: 1751–1755.

88. Zink-Nielsen, I. 1977. Intercalibration of methods for chemical analysis of sediments; results from inter-calibrations of methods for determining loss on ignition, COD, total nitrogen, total phosphorus, and heavy metals in sediments. *Vatten* 33: 14–20.

89. Lohse, L., R.T. Kloosterhuis, H.C. de Stigter, W. Helder, W. Van Raaphorst, and T.C.E. van Weering. 2000. Carbonate removal by acidification causes loss of nitrogenous compounds in continental margin sediments. *Mar. Chem.* 69: 193–201.

90. Luczak, Ch., M.-A. Janquin, and A. Kupka. 1997. Simple standard procedure for the routine determination of organic matter in marine sediment. *Hydrobiologia* 345: 87–94.

91. Goni, M., K.C. Ruttenberg, and T.I. Eglinton. 1997. Sources and contribution of terrigenous organic carbon to surface sediments in the Gulf of Mexico. *Nature* 389: 275–278.

92. Burdige, D.J. and K.G. Gardner. 1998. Molecular weight distribution of dissolved organic carbon in marine sediment pore waters. *Mar. Chem.* 62: 45–64.

93. Kawano, M., S. Kitamura, J. Falandysz, and R. Tatsukawa. 1994. Occurrence of extractable organic halogens (EOX) in Polish marine sediments. *Proceedings of the 19th Conferencein Baltic Oceanographers*, Sopot, pp. 725–731.

94. Kannan, K., M. Kawano, Y. Kashima, M. Matsui, and J.P. Giesy. 1999. Extractable organohalogens (EOX) in sediment and biota collected at an estuarine marsh near a former chloralkali facility. *Environ. Sci. Technol.* 33: 1004–1008.

95. Suominen, K.P., C. Wittmann, M.A. Kähkönen, and M.S. Salkinoja-Salonen. 1998. Organic halogen, heavy metals and biological activities in pristine and pulp mill recipient lake sediments. *Water Sci. Technol.* 37: 79–86.

96. Kähkönen, M.A., K.P. Suominen, P.K.G. Manninen, and M.S. Salkinoja-Salonen. 1998. 100 years of sediment accumulation history of organic halogens and heavy metals in recipient and nonrecipient lakes of pulping industry in Finland. *Environ. Sci. Technol.* 32: 1741–1746.

97. Suominen, K.P., T. Jaakkola, E. Elomaa, R. Hakulinen, and M.S. Salkinoja-Salonen. 1997. Sediment accumulation of organic halogens in pristine forest lakes. *Environ. Sci. Pollut. Res.* 4: 21–30.

98. Heron, G., M.J. Barcelona, M.L. Andersen, and T.H. Christensen. 1997. Determination of nonvolatile organic carbon in aquifer solids after carbonate removal by sulfurous acid. *Ground Water* 35: 6–11.

99. El-Rehaili, A.M. 1994. Implications of activated sludge kinetics based on total or soluble BOD, COD and TOC. *Environ. Technol.* 15: 1161–1172.

100. Stachel, B., G. Behringer, and U. Schacht. 1994. Determination of organohalogen compounds and organic carbon as sum parameter in electronic waste. *Fresenius J. Anal. Chem.* 350: 375–378.

101. Rubli, S., E. Medilanski, and H. Belevi. 2000. Characterization of total organic carbon in solid residues provides insight into sludge incineration processes. *Environ. Sci. Technol.* 34: 1772–1777.

102. Kawano, M., M. Ueda, M. Matsui, Y. Kashima, M. Matsuda, and T. Wakimoto. 1998. Extractable organic halogens (EOX: Cl, Br, and I), polychlorinated naphthalenes and polychlorinated dibenzo-*p*-dioxins and dibenzofurans in ashes from incinerators located in Japan. *Organohalogen Compd.* 36: 221–224.

103. Shoda, T., N. Takahashi, and T. Sasahara. 2007. A concentration monitor using an adsorption/combustion-type gas sensor for total volatile organic compounds in indoor air. *Chem. Senses* 23(Suppl. B): 28–30.

104. Sasahara,T., H. Kato, A. Saito, M. Nishimura, and M. Egashira. 2007. Development of a ppb-level sensor based on catalytic combustion for total volatile organic compounds in indoor air. *Sens. Actuat. B* B126: 536–543.

105. Massold, E., C. Bahr, T. Salthammer, and S.K. Brown. 2005. Determination of VOC and TVOC in air using thermal desorption GC-MS—practical implications for test chamber experiments. *Chromatographia* 62: 75–85.

106. Wilkin, R.T. and K.J. Bischoff. 2006. Coulometric determination of total sulfur and reduced inorganic sulfur fractions in environmental samples. *Talanta* 70: 766–773.

107. Li, H.,Y. Zhang, C. Zheng, L. Wu, Y. Lv, and X. Hou. 2006. UV irradiation controlled cold vapor generation using $SnCl_2$ as reductant for mercury speciation. *Anal. Sci.* 22: 1361–1365.

108. Fernandez, C., A.C.L. Conceicao, R. Rial-Otero, C. Vaz, and J.L. Capelo. 2006. Sequential flow injection analysis system on-line coupled to high intensity focused ultrasound: Green methodology for trace analysis applications as demonstrated for the determination of inorganic and total mercury in waters and urine by cold vapor atomic absorption spectrometry. *Anal. Chem.* 78: 2494–2499.

109. Kagaya, S., Y. Kuroda, Y. Serikawa, and K. Hasegawa. 2004. Rapid determination of total mercury in treated waste water by cold vapor atomic absorption spectrometry in alkaline medium with sodium hypochlorite solution. *Talanta* 64: 554–557.

110. Ou, H., Y. Zhang, Q. Wu, J. Fang, and Y. Shu. 2004. Rapid determination of trace inorganic and total organic mercury in waste water by cold vapor atomic absorption spectrometry and pyrolysis atomic absorption spectrometry. *Fenxi Ceshi Xuebao* 23: 68–70.

111. Jitaru, P. and F.C. Adams. 2004. Speciation analysis of mercury by solid-phase microextraction and multicapillary gas chromatography hyphenated to inductively coupled plasma-time-of-flight-mass spectrometry. *J. Chromatogr. A* 1055: 197–207.

112. Jitaru, P. and F.C. Adams. 2004. Speciation analysis of mercury by solid-phase microextraction and multicapillary gas chromatography hyphenated to inductively coupled plasma-time-of-flight-mass spectrometry. *J. Chromatogr. A* 1055: 197–207.

113. Laglera, L.M., G. Battaglia, and M.G. van den Berg. 2007. Constant determination of humic substances in natural waters by cathodic stripping voltammetry of their complexes with iron. *Anal. Chim. Acta* 599: 58–66.

114. Sheng, G., M. Zhang, and H. Yu. 2007. A rapid quantitative method for humic substances determination in natural waters. *Anal. Chim. Acta* 592: 162–167.

115. Lin, X. 2006. Determination of total phosphorus in water by flow injection analysis instrument. *Fujian Fenxi Ceshi* 15: 42–44.

116. Su, L., H. Zhang, Q. Wang, and H. Xie. 2007. On-line microwave digestion for the determination of total phosphorus in circulating water by flow injection spectrophotometry. *Gongye Shuichuli* 27: 80–82.

117. Yang, C., X. Sun, B. Liu, and H. Lian. 2007. Determination of total phosphorus in water sample by digital imaging colorimetry. *Fenxi Huaxue* 35: 850–853.

118. Ai, S., B. Zhang, X. Qu, X. Zou, and L. Jin. 2006. Determination of total phosphorus in water based on nanometer titanium dioxide film photocatalytic oxidation with electrochemical method. *Fenxi Huaxue* 34: 1068–1072.

119. Yang, C., R. Wu, P. Zhu, and X. Yang. 2007. Determination of total nitrogen in water using high temperature oxidation and chemiluminescence method. *Fenxi Huaxue* 35: 529–531.

120. Zhang, F. and J. He. 2006. Determination of total nitrogen in water by microwave digestion UV spectrophotometry. *Fenxi Ceshi Xuebao* 25: 112–114.

121. Zhang, Y., L. Chen, J. Yuan, Y. Yuan, P. Ji, and P. Ji. 2007. Determination of total nitrogen in solid waste by two methods. *Wuran Fangzhi Jishu.* 20: 92–95.

122. Xu, D., Q. Tian, and Z. Chai. 2006. Determination of extractable organohalogens in the atmosphere by instrumental neutron activation analysis. *J. Radioanal. Nucl. Chem.* 270: 5–8.

123. Niemirycz, E., A. Kaczmarczyk, and J. Blazejowski. 2005. Extractable organic halogens (EOX) in sediments from selected Polish rivers and lakes—a measure of the quality of the inland water environment. *Chemosphere* 61: 92–97.

124. Ren, L., Q. Zhou, Y. Song, and P. Li. 2002. An analytical method for determination of extractable organic halogens in contaminated soils and sediments. *Huanjing Kexue Xuebao* 22: 701–705.

125. Li, J., L. Zheng, L. Li, G. Shi, Y. Xian, and L. Jin. 2007. Photoelectro-synergistic catalysis combined with a FIA system application on determination of chemical oxygen demand. *Talanta* 72: 1752–1756.

126. Zhang, S., D. Jiang, and H. Zhao. 2006. Development of chemical oxygen demand on-line monitoring system based on a photoelectrochemical degradation principle. *Environ. Sci. Technol.* 40: 2363–2368.

127. Zhao, H., D. Jiang, S. Zhang, K. Catterall, and R. John. 2004. Development of a direct photoelectrochemical method for determination of chemical oxygen demand. *Anal. Chem.* 76: 155–160.

128. Chen, Y., Y. Wu, L. Zhu, et al. 2006. Rapid determination of chemical oxygen demand using a nano-TiO_2-$KMnO_4$ photocatalytic oxidation system. *Fenxi Huaxue* 34: 1595–1598.

129. Jeong, B.G., S.M. Yoon, C.H. Choi, et al. 2007. Performance of an electrochemical COD (chemical oxygen demand) sensor with an electrode-surface grinding unit. *J. Environ. Monit.* 9: 1352–1357.

130. Liu, C., J. Qu, J. Jia, L. Qi, and S. Dong. 2005. Online measurement of BOD using a microbial sensor based on organic–inorganic hybrid material. *Fenxi Huaxue* 33: 609–613.

131. He, J., X. Yang, L. Wang, and J. Pan. 2007. Rapid determination of chemical oxygen demand (COD), biochemical oxygen demand (BOD5) and pH in wastewater using near-infrared spectroscopy. *Huanjing Kexue Xuebao* 27: 2105–2108.

132. Lee, H.M., K. Okuyama, A. Mizohata, T. Kim, and H. Koyama. 2007. Fabrication of reference filter for measurements of EC (elemental carbon) and OC (organic carbon) in aerosol particles. *Aerosol Sci. Technol.* 41: 284–294.

133. Xu, B., T. Yao, X. Liu, and N. Wang. 2006. Elemental and organic carbon measurements with a two-step heating-gas chromatography system in snow samples from the Tibetan Plateau. *Ann. Glaciol.* 43: 257–262.

134. Hammes, F.A. and T. Egli. 2005. New method for assimilable organic carbon determination using flow-cytometric enumeration and a natural microbial consortium as inoculum. *Environ. Sci. Technol.* 39: 3289–3294.

135. Zhang, T., X.L. Qin, and H.H.P. Fang. 2007. Use of P-17 and NOX specific primer sets for assimilable organic carbon (AOC) measurements. *Water Sci. Technol: Water Supply* 7: 157–163.

136. Hua, G. and D.A. Reckhow. 2006. Determination of TOCl, TOBr and TOI in drinking water by pyrolysis and off-line ion chromatography. *Anal. Bioanal. Chem.* 384: 495–504.

137. Kawamoto, K. 1999. TOX as a novel alternative index of dioxins in flue gas. *Organohalogen Compd.* 40: 157–160.

138. Tsai, C.T., C.T. Kuo, and S.T. Lin. 1998. Analysis of organic halides in hospital waste sludge disinfected using sodium hypochlorite (NaOCl). *Water Res.* 33: 778–784.

139. Sannigrahi, P., A.P. Sullivan, R.J. Weber, and E.D. Ingall. 2006. Characterization of water-soluble organic carbon in urban atmospheric aerosols using solid-state 13C NMR spectroscopy. *Environ. Sci. Technol.* 40: 666–672.

140. Chang, H., P. Herckes, J.L. Collett, and L. Jeffrey, Jr. 2004. On the use of anion exchange chromatography for the characterization of water soluble organic carbon. *Geophys. Res. Lett.* 32: L01810/1–L01810/4.

141. Sullivan, A.P., R.J. Weber, A.L. Clements, J.R. Turner, M.S. Bae, and J.J. Schauer. 2004. A method for on-line measurement of water-soluble organic carbon in ambient aerosol particles: Results from an urban site. *Geophys. Res. Lett.* 31: L13105/1–L13105/4.

142. Mancinelli, V., M. Rinaldi, E. Finessi, et al. 2007. An anion-exchange high-performance liquid chromatography method coupled to total organic carbon determination for the analysis of water-soluble organic aerosols. *J. Chromatogr. A* 1149: 385–389.

143. Song, J., X. Li, N. Li, X. Gao, H. Yuan, and T. Zhan. 2004. Simple and accurate method for determining accurately dissolved inorganic carbon in seawaters. *Fenxi Huaxue* 32: 1689–1692.

144. Assayag, N., K. Rive, M. Ader, D. Jezequel, and P. Agrinier. 2006. Improved method for isotopic and quantitative analysis of dissolved inorganic carbon in natural water samples. *Rapid Commun. Mass Spectrom.* 20: 2243–2251.

145. Tue-Ngeun, O., R.C. Sandford, J. Jakmunee, K. Grudpan, I.D. McKelvie, and P.J. Worsfold. 2005. Determination of dissolved inorganic carbon (DIC) and dissolved organic carbon (DOC) in freshwaters by sequential injection spectrophotometry with on-line UV photo-oxidation. *Anal. Chim. Acta* 554: 17–24.

146. Polesello, S., G. Tartari, P. Giacomotti, R. Mosello, and S. Cavalli. 2006. Determination of total dissolved inorganic carbon in freshwaters by reagent-free ion chromatography. *J. Chromatogr. A* 1118, 56–61.

147. Duarte, A.C. 2006. The assembling and application of an automated segmented flow analyzer for the determination of dissolved organic carbon based on UV-persulfate oxidation. *Anal. Lett.* 39: 1979–1992.

148. Pan, X., R. Sanders, A.D. Tappin, P.J. Worsfold, and E.P. Achterberg. 2005. Simultaneous determination of dissolved organic carbon and total dissolved nitrogen on a coupled high-temperature combustion total organic carbon-nitrogen chemiluminescence detection (HTC TOC-NCD) system. *J. Autom. Meth. Manage. Chem.* 4: 240–246.

149. Barthes, B.G., D. Brunet, H. Ferrer, J. Chotte, and C. Feller. 2006. Determination of total carbon and nitrogen content in a range of tropical soils using near infrared spectroscopy: Influence of replication and sample grinding and drying. *J. Near Infrared Spectrosc.* 14: 341–348.

150. Fung, Y.S. and K.L. Dao. 1998. Determination of total carbon in air particulate matters by thermal combustion—ion chromatography, *Int. J. Environ. Anal. Chem.* 69: 125–139.

151. Bisutti, I., I. Hilke, J. Schumacher, and M. Raessler. 2007. A novel single-run dual temperature combustion (SRDTC) method for the determination of organic, inorganic and total carbon in soil samples. *Talanta* 71: 521–528.

152. Stefansson, A., I. Gunnarsson, and N. Giroud. 2007. New methods for the direct determination of dissolved inorganic, organic and total carbon in natural waters by reagent-free ion chromatography and inductively coupled plasma atomic emission spectrometry. *Anal. Chim. Acta* 582: 69–74.

153. Dewil, R., J. Baeyens, J. Roelandt, and M. Peereman. 2006. The analysis of the total sulfur content of wastewater treatment sludge by ICP-OES. *Environ. Eng. Sci.* 23: 904–907.

154. Sayi, Y.S., P.S. Shankaran, C.S. Yadav, G.C. Chhapru, K.L. Ramakumar, and V. Venugopal. 2002. Determination of total sulfur by vacuum combustion extraction-quadrupole mass spectrometry (VCE-QMS). *Proc. Natl. Acad. Sci. India Sect. A* 72: 241–248.
155. Rocha, T.A.P., J.A.B.P. Oliveira, and A.C. Duarte. 1999. Determination of total sulfur in landfill gases using a quartz crystal microbalance. *Int. J. Environ. Anal. Chem.* 75: 121–126.
156. Binde, F. and H.H. Ruttinger. 1997. Isolation and determination of dissolved organic sulfur compounds. Development of the organic group parameter DOS. *Fresenius J. Anal. Chem.* 357: 411–415.

11 Determination of Radionuclides in the Aquatic Environment

Bogdan Skwarzec

CONTENTS

11.1 Introduction 242
11.2 Determination of Radionuclides in the Aquatic Environment 242
    11.2.1 Primordial Radionuclide: ^{40}K 242
    11.2.2 Neutron Activation Products: ^{55}Fe and ^{63}Ni 242
        11.2.2.1 Determination of ^{55}Fe and ^{63}Ni 243
    11.2.3 Fission Products: ^{90}Sr and ^{137}Cs 246
        11.2.3.1 Determination of ^{137}Cs 246
        11.2.3.2 Determination of ^{90}Sr 247
    11.2.4 ^{226}Ra in Water 248
    11.2.5 ^{222}Rn in Water 249
    11.2.6 Natural and Transuranic Alpha Radionuclides: ^{210}Po, ^{234}U,
        ^{235}U, ^{238}U, ^{238}Pu, $^{239+240}Pu$, and ^{241}Pu 249
        11.2.6.1 Radiochemical Methods for Determination of
                Polonium, Radiolead, Uranium, and Plutonium
                in Aquatic Samples 249
        11.2.6.2 Coprecipitation of Polonium, Radiolead, Uranium,
                and Plutonium with Manganese Dioxide in
                Natural Waters 249
        11.2.6.3 Mineralization of Suspended Matter 250
        11.2.6.4 Mineralization of Sediment 251
        11.2.6.5 Mineralization of Biota 251
        11.2.6.6 Separation and Determination of Polonium
                ^{210}Po and Lead ^{210}Pb 251
        11.2.6.7 Separation, Purification, and Electrolysis of Plutonium 252
        11.2.6.8 Separation, Purification, and Electrolysis of Uranium 252
        11.2.6.9 Measurement of Radionuclide Activity 253
        11.2.6.10 Activity Determination of Polonium, Radiolead,
                Uranium, and Plutonium Radionuclides 253
        11.2.6.11 Determination of ^{241}Pu Activity 256
References 256

11.1 INTRODUCTION

The radionuclides present in the environment are classified as being either of natural or of anthropogenic origin[1]:

Naturally occurring radionuclides
a. Radionuclides of terrestrial origin (e.g., ^{40}K and ^{87}Rb)
b. Cosmogenic radionuclides (e.g., ^{3}H, ^{14}C, ^{32}Si, and ^{36}Cl)
c. Primary radionuclides: these long-lived radionuclides have been ubiquitous on the Earth ever since its formation, that is, approximately 4.5×10^9 years ago. The radionuclides ^{238}U, ^{232}Th, and ^{235}U are the respective parent members of the uranium, thorium, and actino-uranium radioactive decay series

Anthropogenic radionuclides
a. Neutron activation products (e.g., ^{22}Na, ^{54}Mn, ^{55}Fe, ^{60}Co, ^{63}Ni, ^{64}Cu, ^{65}Zn, ^{110m}Ag, ^{124}Sb, and ^{125}Sb)
b. ^{235}U and ^{239}Pu fission radionuclides (e.g., ^{90}Sr, ^{95}Zr, ^{131}I, ^{132}I, ^{132}Te, ^{137}Cs, and ^{144}Ce)
c. Transuranic elements (e.g., ^{237}Np, ^{238}Pu, ^{239}Pu, ^{240}Pu, ^{241}Pu, ^{241}Am, and ^{243}Am)

Because of their long half-life, type of decay, and strong radiotoxicity, the most important radionuclides in the aquatic environment are ^{40}K (primordial radionuclide), ^{55}Fe and ^{63}Ni (neutron activation products), ^{90}Sr and ^{137}Cs (fission products), ^{210}Po, ^{210}Pb, ^{234}U, ^{235}U, and ^{238}U (naturally occurring alpha and beta emitters), and ^{238}Pu, ^{239}Pu, ^{240}Pu, and ^{241}Pu (artificial plutonium radionuclides).[1]

11.2 DETERMINATION OF RADIONUCLIDES IN THE AQUATIC ENVIRONMENT

11.2.1 PRIMORDIAL RADIONUCLIDE: ^{40}K

The isotope ^{40}K can be analyzed in natural water samples with the Cherenkov counting technique.[2,3] Because of the lack of a suitable radiotracer for K and the similarity between the chemistries of rubidium and potassium, ^{86}Rb can be used as a tracer for K.[4] Also, thermal ionization mass spectrometry (TIMS) has been used to determine ^{40}K in environmental samples. The interference of mass 40 can be solved by double spiking with ^{43}Ca/^{48}Ca; the procedure for the routine high-precision isotope analysis of the K–Ca system will then be free of Ca fractionations.[5]

11.2.2 NEUTRON ACTIVATION PRODUCTS: ^{55}Fe AND ^{63}Ni

A number of artificial radionuclides are produced as a result of activation during nuclear weapons tests, the operation of reprocessing plants and reactors in nuclear power stations, and in nuclear studies. Novel radioanalytical techniques have enabled activation products such as ^{22}Na, ^{51}Cr, ^{54}Mn, ^{65}Zn, ^{110m}Ag, and ^{124}Sb to be detected in the environment.[6,7]

Stainless steel contains iron and nickel—important materials in nuclear power reactors and possible constituents of the materials used to construct nuclear test devices or their supporting structures.[8,9] During nuclear weapons tests, stable Fe and Ni isotopes are neutron activated, giving rise to radioactive Fe and Ni along with fission products. In nuclear power plants, moreover, stable Fe and Ni isotopes are released from stainless steel through corrosion, become activated, and are transported to different parts of the reactor system.

Neutron activation of the stable isotopes of iron produces two radioactive isotopes—^{55}Fe and ^{59}Fe. ^{55}Fe (half-life = 2.685 years), which is a beta (electron-capture) emitter and decays to the stable ^{55}Mn isotope, is the more important isotope.[10]

The production of radioactive nickel isotopes through neutron activation yields ^{59}Ni, ^{63}Ni, and ^{65}Ni. Of these three, ^{63}Ni is important, because the activity ratio of $^{59}Ni/^{63}Ni$ is only 0.01.[6] The ^{63}Ni isotope is a beta-particle emitter with a half-life of 100.1 years, decaying to the stable ^{63}Cu isotope.[10]

11.2.2.1 Determination of ^{55}Fe and ^{63}Ni

The radioanalytical methods of determining ^{55}Fe and ^{63}Ni were worked out by Holm et al.[6] and Skwarzec et al.[11] Figures 11.1 and 11.2 illustrate the procedures for the radiochemical analysis of ^{55}Fe and ^{63}Ni in aquatic environmental samples.

The radiochemical yield is determined by the atomic absorption spectrometry (AAS) of stable Fe and Ni before and after electrodeposition. The activities of ^{55}Fe and ^{63}Ni are measured using an anticoincidence-shielded windowless low-level beta-particle gas-flow counter operating in the Geiger–Müller region. The gas-flow counter is a four-channel counter from the Risø National Laboratory, Denmark, using argon (99%) and isobutane (1%) as the counting gas.[11]

11.2.2.1.1 Separation and Determination of ^{55}Fe

Figure 11.1 presents a scheme for the radioanalytical determination of ^{55}Fe in water, biota, and sediment samples.[11] This procedure is based on the separation of Fe from other metals (especially Cd, Cs, Cu, Ni, and Zn) on an anion exchange resin. The iron is then purified by coprecipitation with cupferron (the ammonium salt of nitrosophenylhydroxylamine).

11.2.2.1.2 Coprecipitation of ^{55}Fe in Natural Water with Iron Hydroxide

11.2.2.1.2.1 Procedure 1
About 5–10 L of natural water is passed through a preweighed membrane filter of 0.45 μm pore diameter. The water is adjusted to pH ≈ 9 with ammonia, and about 1 g of iron ($FeCl_3$ solution) is added as radiochemical yield determinant. Nitrogen is bubbled for about 2 h to ensure good mixing and isotope exchange. The iron hydroxide $Fe(OH)_3$ deposit is allowed to settle overnight; the overlying liquid phase is sucked off and the precipitate is collected by decantation. The precipitate containing ^{55}Fe is dissolved in 20 mL 9 M HCl and the solution is passed through a column containing Dowex 1 × 4 (100–200 mesh) anion exchange resin. The ^{55}Fe is then separated, purified, and electroplated according to procedure 2.

11.2.2.1.3 Mineralization, Separation, and Electrolysis of ^{55}Fe in Aquatic Sediment and Biota Samples

11.2.2.1.3.1 Procedure 2
10 mL perhydrol (30% H_2O_2) is added to 10 g of a sediment or biota sample ashed for 12 h at 550°C and heated until the perhydrol is completely decomposed. After evaporation, 20 mL aqua regia is added to the dry residue; the sample is then heated for 1 h and evaporated to dryness. The dry residue is dissolved in 15 mL 9 M HCl and the solution is passed through a 0.8 μm pore diameter Sartorius membrane filter. The deposit retained on the filter is leached with 15 mL 9 M HCl and filtered again. The residue is added to the supernatant from the first filtration. 1 mL of the solution is diluted with distilled water and Fe determined by AAS. The remainder of the solution is passed through a column (100 × 10 mm²) containing Dowex 1 × 4 (100–200 mesh) anion exchange resin, after which the column is flushed with 50 mL 9 M HCl.

The eluate and flushing solution (in 9 M HCl) are discarded. The Fe adsorbed on the resin is eluted using 30 mL 0.5 M HCl. 10 mg cupferron per 1 mg of iron is added to the solution, which is then stored overnight at temperatures between 0°C and 4°C. The resulting brown precipitate [$(Cup)_3Fe$] is removed with a Sartorius membrane filter, and the cupferrate is decomposed with 50 mL aqua regia and 10 mL 65% HNO_3. After evaporation, the dry residue is dissolved in 10 mL 4 M HCl. 1 mL of this solution is taken for the AAS measurement of stable Fe. The remainder of the solution is transferred to a plating cell and a "pinch" of ascorbic acid (reducing Fe^{3+} to Fe^{2+}), NH_4OH (pH = 9), and 5 mL 0.6 M $(NH_4)_2SO_4$ (aq) are added. The Fe is electroplated on a polished copper disc (diameter = 17 mm) for 2 h with a current of 0.3 A. After electrodeposition, a known fraction of the electrolyte is again taken for an AAS measurement of stable iron.

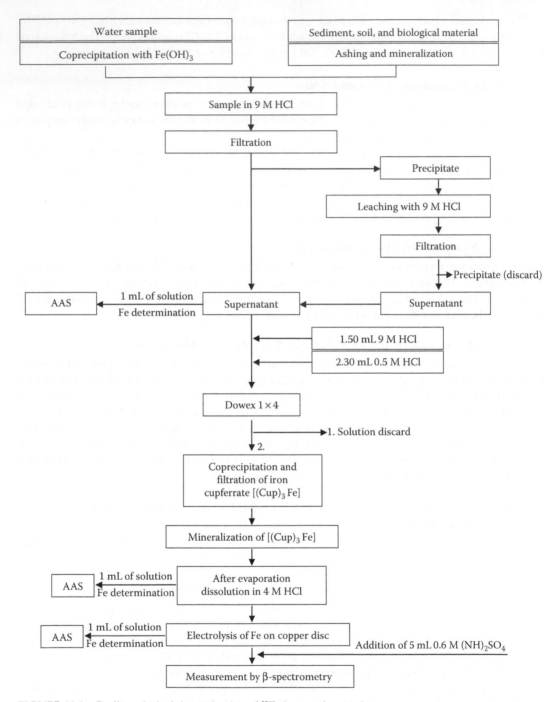

FIGURE 11.1   Radioanalytical determination of ^{55}Fe in aquatic samples.

11.2.2.1.4   Separation and Determination of ^{63}Ni

Figure 11.2 shows a scheme for the radioanalytical determination of ^{63}Ni in aquatic samples.[11] The method for determining nickel activity is based on the separation of this element from other radionuclides, particularly ^{55}Fe. To separate ^{63}Ni, the stable dimethylglyoxime (DMG) complex (DMG)$_2$Ni is formed in ammonia and extracted with chloroform.

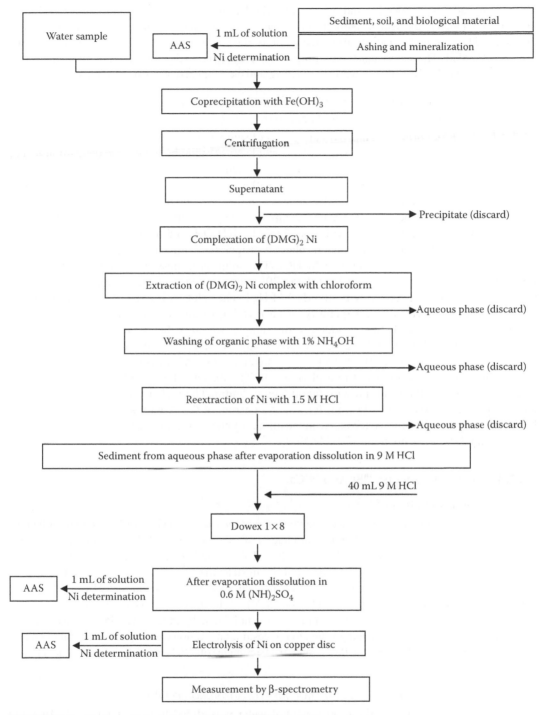

FIGURE 11.2 Radioanalytical determination of ^{63}Ni in aquatic samples.

11.2.2.1.5 Separation of ^{63}Ni in Natural Water from other Radionuclides by Coprecipitation with Hydroxide

11.2.2.1.5.1 Procedure 3 About 20–30 L of natural water is passed through a preweighed membrane filter of 0.45 μm diameter. The water is adjusted to pH $\cong$ 9 with NH$_4$OH, and about

200 mg of stable Ni is added as radiochemical yield determinant. Next, approximately 10 mL 1 M $FeCl_3$ (aq) is added to the sample. Nitrogen is bubbled for about 2 h to ensure good mixing and isotope exchange. The iron hydroxide $Fe(OH)_3$ precipitate is allowed to settle overnight and, after decantation, is discarded. The supernatant containing ^{63}Ni is evaporated to dryness and the residue dissolved in 30 mL 1 M HCl. ^{63}Ni is then separated, purified, and electroplated according to procedure 4.

11.2.2.1.6   Mineralization, Separation, and Electrolysis of ^{63}Ni in Aquatic Sediment and Biota Samples

11.2.2.1.6.1   Procedure 4   Approximately 50 g of sample (dry sediment or biological material) is ashed for 12 h at 550°C together with 200 mg stable Ni as radiochemical yield determinant. The sample is then leached with 50 mL aqua regia, the sediment centrifuged, and the supernatant evaporated to dryness. The residue is dissolved in 30 mL 1 M HCl and the hydroxides precipitated with NH_4OH (pH $\geq$ 9) leaving Ni in the supernatant. After centrifugation, 15 mL 1% DMG in ethanol is added and the pink $(DMG)_2Ni$ complex formed is extracted with 3 × 30 mL chloroform. The organic phase is rinsed with 2 × 25 mL 1% NH_4OH, and the Ni is backextracted with 2 × 25 mL 1.5 M HCl. The acidic aqueous phase is evaporated with a few drops of 65% HNO_3 and the dry residue dissolved in 10 mL 9 M HCl. The solution is passed through a column (100 mm × 10 mm) containing Dowex 1 × 8 (100–200 mesh) anion exchange resin, and the nickel adsorbed on the resin is eluted with 40 mL 9 M HCl. The solution is evaporated to dryness and the residue treated with a small quantity of 65% HNO_3 to destroy any organic matter. After evaporation, 1 mL conc. H_2SO_4 is added to the sample, which is then heated until white fumes appear. After the addition of 5 mL 0.6 M $(NH_4)_2SO_4$, the sample is adjusted to pH $\geq$ 9 and the volume adjusted to 10 mL. 1 mL of electrolyte is taken for an AAS measurement of stable Ni. The rest of the solution is transferred to a plating cell. The distance between the electrodes can be adjusted by moving the platinum wire (anode) up to 5 mm. The nickel is electroplated on a polished copper disc (diameter = 17 mm) for 2 h at a current of 0.2 A with a few drops of NH_4OH being added every 30 min. After electrodeposition, a known fraction of the electrolyte is again taken for an AAS measurement of stable Ni.

11.2.3   Fission Products: ^{90}Sr and ^{137}Cs

11.2.3.1   Determination of ^{137}Cs

The artificial ^{137}Cs radionuclide is one of the most important long-lived (T = 30.17 years) fission products and a common contaminant. It emits β-radiation of two energies—1176 keV (6%) and 514 keV (94%)—exciting a 2.55 min isomeric level of $^{137}Ba*$. This isomeric level de-excites itself by the emission of a single γ-ray of 661.66 keV. In the equilibrium state, the activities of ^{137}Cs and $^{137}Ba*$ are the same.[10] ^{137}Cs activity can be determined directly using beta spectrometry or indirectly by measuring the $^{137}Ba*$ activity with gamma spectrometry (E = 662 keV).[12,13] In aquatic samples, ^{137}Cs determination is based on its adsorption on AMP (ammonium phosphomolybdate hydrate, $(NH_4)_3PO_4 \cdot 12MoO_3 \cdot 3H_2O$) in water samples, the separation and purification of cesium on a cation exchange resin, coprecipitation of cesium hexachloroplatinate Cs_2PtCl_6, and measurement of ^{137}Cs activity in a low-level flow beta counter.[13]

11.2.3.1.1   Separation of Cesium and Strontium by Adsorption on AMP

11.2.3.1.1.1   Procedure 1   About 30 L natural water is acidified to pH = 1 with 6 M HCl and approximately 50 mL 7.5 mM CsCl (aq) and 50 mL 0.15 M $SrCl_2$ (aq) (both chemical recovery tracers) are added. After mixing, approximately 10 g AMP is added and nitrogen gas bubbled for about 0.5 h to ensure good mixing. The AMP precipitate with cesium is allowed to settle overnight; the overlying liquid phase containing strontium is collected for ^{90}Sr determination; the AMP precipitate is filtered, and then rinsed with 0.05 M HCl. Next, the AMP sediment is dissolved in 20 mL EDTA (sodium ethylenediaminetetraacetate) and NaOH solution (32 g NaOH + 20 g EDTA L^{-1}). After filtration, the solution is passed through a column ($Ø = 10$ mm; length $= 120$ mm) containing Dowex cation exchange resin. The column is flushed with 50 mL H_2O, after which the cations (Na^+, K^+, and Rb^+) adsorbed on the resin are eluted with 350 mL 0.3 M HCl. Finally, the cesium (Cs^+) adsorbed on the resin is eluted with 120 mL 3 M HCl. The eluate is evaporated and the residue treated with a few drops of conc. HCl and conc. HNO_3 to destroy any remaining organic matter and ammonia (NH_4^+). Next, a few drops of 6 M HCl and 5 mL distilled H_2O are added to the dry residue; the resulting solution is cooled, and then added to 1 mL 0.22 mM of aqueous hexachloroplatinic acid (H_2PtCl_6). The cesium hexachloroplatinate (Cs_2PtCl_6) precipitate is filtered, dried, and weighed. The ^{137}Cs activity in the precipitate is measured with a low-level flow beta counter.

11.2.3.1.2   Calculation of ^{137}Cs Activity

The calibration coefficient η between the mass of Cs_2PtCl_6 sediment and ^{137}Cs activity is calculated from

$$\eta = \frac{A}{CPM_{100\%}}, \tag{11.1}$$

where η is the calibration coefficient (Bq count^{-1} min^{-1}) (usual range 0.03–0.06), A is the activity of ^{137}Cs standard, and $CPM_{100\%}$ is the number of counts calculated for 100% cesium recovery.

The ^{137}Cs activity in the samples is calculated from

$$C = \frac{CPM_\eta \cdot \eta \cdot 100\%}{m \cdot Y}, \tag{11.2}$$

where C is the ^{137}Cs concentration (Bq L^{-1}), m is the volume or sample mass (L), η is the calibration coefficient (Bq min^{-1} count^{-1}), CPM_η is the number of counts during ^{137}Cs measurement with the beta counter (without background), and Y is the recovery (%).

The standard deviation (SD) for activity of ^{137}Cs is calculated from

$$SD = \sqrt{\frac{CPM_b}{t_p} + \frac{CPM_t}{t_t}}, \tag{11.3}$$

where CPM_b is the number of counts including background during sample measurement with the beta counter, CPM_t is the number of background counts with the beta counter, t_p is the sample counting time, and t_t is the background counting time.

11.2.3.2   Determination of ^{90}Sr

^{90}Sr is one of the most hazardous and dangerous radioactive isotopes. It is a pure beta emitter ($E_{max} = 546$ keV) and decays to another pure beta emitter, ^{90}Y ($E_{max} = 2283.9$ keV).[10] The radiochemical methods for determining ^{90}Sr in aquatic samples (water, sediment, and biota) are based on the adsorption of radiostrontium on AMP in water samples, mineralization of sediment and biota, and sorption on Sr resin.[14-16]

11.2.3.2.1   Procedure 2

Before analysis, ^{85}Sr is added to each sample as a chemical recovery tracer. 250 g oxalic acid $C_2H_2O_4$ is added to the liquid phase (supernatant) containing strontium from the water samples (see procedure 1); the solution is stirred for 10 min, after which conc. NH_4OH is added to bring the solution to pH $= 7$. After decantation of the supernatant, the oxalate precipitate is rinsed with distilled water, and then dissolved in 50 mL of a mixture of 3 M HNO_3 and 0.01 M $C_2H_2O_4$. Samples of aquatic organisms and sediments, after mineralization, are heated under watch-glass covers to boiling point and kept simmering for about 4 h. The Sr present in the sample is leached into the solution. After cooling, the sample is filtered and the solid residue discarded. The solution is evaporated to dryness and redissolved in 50 mL of the 3 M HNO_3 + 0.01 M $C_2H_2O_4$ mixture. Next, strontium is separated on a chromatographic column filled with Sr resin. This column is conditioned with 50 mL of the 3 M HNO_3 + 0.01 M $C_2H_2O_4$ mixture. The sample solution is then poured into the column, followed by 20 mL of the 3 M HNO_3 + 0.01 M $C_2H_2O_4$ mixture as a flushing solution. Under such conditions the majority of the sample matrix passes straight through the column, whereas Sr is strongly retained. Strontium (also ^{90}Sr) is then eluted from the column with 50 mL distilled water. This fraction may also contain traces of ^{210}Pb. This nuclide and its daughter products ^{210}Bi and ^{210}Po decay by emitting beta and alpha particles, respectively; so if they are present in the sample, they may obscure the ^{90}Sr signal during liquid scintillation counting. Therefore, an additional purification step has to be introduced to the procedure. The strontium fraction is evaporated to dryness with the addition of 10 mg Pb carrier in $Pb(NO_3)_2$ solution (1 mL). The evaporation is necessary to remove possible traces of nitric acid that may still be remaining in the sample after its elution from the column. The sample is then redissolved in 30 mL distilled water containing a few drops of CH_3COOH (1:1) and 3 mL of NH_4I (5 g/100 mL) solution is added. The yellow PbI_2 precipitate is then dissolved by first heating the sample and then recrystallizing it by cooling the beaker in a cold water bath. The solution is then suction-filtered through a paper filter. The solid residue is discarded and the solution containing Sr evaporated to dryness with the addition of conc. HNO_3 to remove excess iodine. After evaporation, the dry residue is dissolved in 1 mL 1 M HNO_3 and transferred to a plastic liquid scintillation vial. The beaker is rinsed twice with 1 mL distilled water, which is then combined with the sample. This is then measured with a low-background gamma spectrometer equipped with an HPGe detector to determine the ^{85}Sr activity in order to determine the chemical recovery. After 14 days, the sample containing ^{90}Sr in equilibrium with its daughter ^{90}Y is measured with a 1414-003 Wallac Guardian liquid scintillation counter.

11.2.3.2.2   Calculation of ^{90}Sr Activity

After measurement with the beta counter the ^{90}Sr activity is calculated from

$$A = \frac{29.55 \cdot N \cdot 100\%}{2 \cdot t \cdot \text{eff} \cdot Y \cdot m}, \tag{11.4}$$

where A is the ^{90}Sr activity (Bq L^{-1}), N is the number of counts, t is the counting time (s), Y is the recovery of ^{85}Sr tracer (%) from gamma measurement, m is the volume or mass of sample (L or g), 29.55 is the proportionality factor between number of counts in ^{90}Y and ^{90}Sr–^{90}Y spectral energy, 2 is the value of the activity of either ^{90}Sr or ^{90}Y—these two activities are in equilibrium, and eff is the effective factor for beta radiation (usually from 0.90 to 1.00).

$$\text{eff} = \frac{A_2}{A},$$

where A_2 is the ^{90}Y activity in sample and A is the real ^{90}Y activity.

The detection limit of ^{90}Sr is calculated from[17]

$$L_d = 2.86 + 4.78\sqrt{(B + 1.36)}, \tag{11.5}$$

where B is the number of background counts.

11.2.4 ^{226}Ra IN WATER

^{226}Ra (half-life = 1602 years) is a naturally occurring radioisotope of the ^{238}U decay series. Earth, marine, and environmental scientists often require analysis of ^{226}Ra in natural water because of public health concerns[18] and because it has proved to be a useful tracer of geochemical processes, particularly in the aquatic environment.[19] The measurement of radium in natural, public, and drinking water supplies has become a matter of interest since it is one of the most hazardous elements with respect to internal radiation exposure[20]; indeed, the enforcement of regulations has made the analysis of ^{226}Ra and ^{228}Ra in ground water very common. Natural waters are by far the most frequent sample matrices assayed for radium by liquid scintillation methods,[21,22] Cherenkov counting of its daughter nuclides[23] and alpha measurement, and mass spectrometric analysis (ICP-MS).[24]

11.2.5 ^{222}Rn in Water

^{222}Rn (half-life = 3.8 days) is an inert noble gas and the immediate daughter nuclide of ^{226}Ra. Zikovsky and Roireau[25] have developed simple methods for the measurement of radon in water using a proportional counter. The method is based on purging radon from water with argon, which is bubbled through the water sample and then directed to the counting tube. The argon picks up the radon that was dissolved in the water. A gas purification system removes moisture and oxygen. A high voltage is set for the alpha plateau; this gives a very low background of <0.2 cpm (counts per minute), a counting efficiency of 25%, and thus a detection limit of 0.02 Bq L^{-1}. This limit compares favorably with that of other methods developed for the determination of radon in water, such as liquid scintillation,[26] Cherenkov counting,[27] or luminescence analysis.[28]

11.2.6 Natural and Transuranic Alpha Radionuclides: ^{210}Po, ^{234}U, ^{235}U, ^{238}U, ^{238}Pu, $^{239+240}$Pu, and ^{241}Pu

A number of natural and artificial alpha radionuclides are used, or could be used, as indicators for studying geochemical and biological processes in the natural aquatic environment. The concentrations of these radionuclides in natural components are very low. Thus, high-quality analytical procedures are needed for the measurement of radionuclides in environmental samples. Until now a large number of radioanalytical methods for either a single radionuclide or a limited number of radionuclides have been described in the literature. However, only a few methods are available for multiradionuclide determination.[29–32]

11.2.6.1 Radiochemical Methods for Determination of Polonium, Radiolead, Uranium, and Plutonium in Aquatic Samples

The radiochemical procedure for the simultaneous determination of natural (^{210}Po, ^{210}Pb, ^{234}U, and ^{238}U) and artificial (^{238}Pu, $^{239+240}$Pu, and ^{241}Pu) isotopes in aquatic samples (water, sediments, and biological material) is based on the coprecipitation of radionuclides with manganese dioxide in natural water, the mineralization of sediment and biota samples, and the sequential separation and purification of radionuclides on anion exchange resins. The separated elements are electrodeposited on silver (polonium) or steel discs (uranium and plutonium), and their activities measured by alpha spectrometry with low-level-activity silicon detectors. The radiochemical analysis of aquatic samples (water, biota, and sediments) presented in Figure 11.3 enables polonium, radiolead, uranium, and plutonium to be determined, and ensures high recoveries as well.[33,34]

11.2.6.2 Coprecipitation of Polonium, Radiolead, Uranium, and Plutonium with Manganese Dioxide in Natural Waters

11.2.6.2.1 Procedure 1

About 20 L (analysis of polonium and uranium) or 100 L (analysis of plutonium) of natural water is passed through a preweighed membrane filter (with a pore diameter of 0.45 μm). The water is acidified to pH = 1 with 65% HNO$_3$, and about 50 mBq of each of the tracers ^{209}Po, ^{232}U, and ^{242}Pu are added. Nitrogen gas is bubbled for about 3 h to ensure good mixing and isotope exchange.

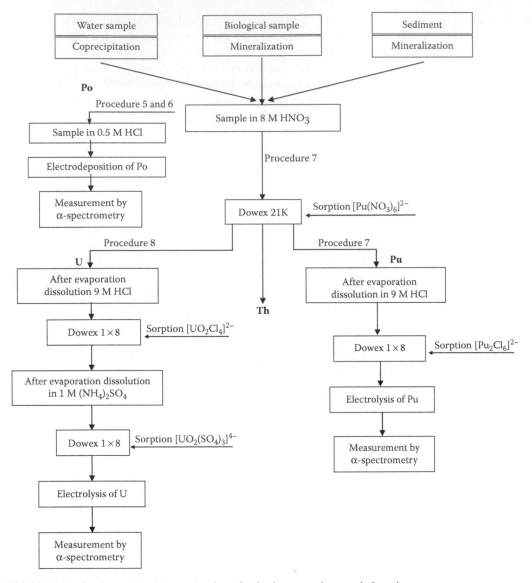

FIGURE 11.3 Radioanalytical determination of polonium, uranium, and plutonium.

10 mL 0.2 M KMnO$_4$ is then added to the water sample, which is brought to pH = 9 with conc. NH$_4$OH. Next, 10 mL 0.3 M MnCl$_2$ is added, and nitrogen continues to be bubbled for another hour. The MnO$_2$ is allowed to settle overnight; the overlying liquid phase is sucked off and the precipitate collected by centrifugation. The MnO$_2$ precipitate is dissolved in 100 mL 1% H$_2$O$_2$ solution in 1.2 M HCl and the new solution evaporated to dryness. The dry residue is dissolved in 50 mL 8 M HNO$_3$.

11.2.6.3 Mineralization of Suspended Matter

The mineralization of suspended matter involves the decontamination of membrane filters with sediment using a mixture of hydrochloric and hydrofluoric acids, and the subsequent removal of silicon as volatile SiF$_4$.[33]

11.2.6.3.1 Procedure 2

Wet digestion is carried out with 0.5 mL 40% HF and 5 mL 6 M HCl in a polytetrafluoroethane (PTFE) autoclave in which the filter with the suspended matter has been placed. The autoclave is then put in a drying oven at 120°C for 2 h. After cooling, the autoclave is centrifuged and the solution evaporated to dryness on a hot plate. 1 mL 30% H_2O_2 and 0.5 mL 60% $HClO_4$ are added and the mixture is heated until the H_2O_2 is completely decomposed. The closed PTFE vessel is then kept at 180–190°C for 2 h. After cooling and centrifuging, the $HClO_4$ is evaporated until white fumes cease to evolve and the digestive residue is dissolved in 5 mL 8 M HNO_3.

11.2.6.4 Mineralization of Sediment

The method used to mineralize sediment depends on its chemical composition. In his research work, the author applied different chemical mineralization procedures to sandy, slimy, and loamy sediments.[34,35] If the origin and nature of the sediment samples were unknown, procedure 3 was used.

11.2.6.4.1 Procedure 3

About 1 g of dry sediment is placed in a PTFE evaporating dish, 5 mL 30% H_2O_2 are added, and the mixture is slowly heated until the H_2O_2 has decomposed completely. Then 5 mL 65% HNO_3 is added and the mixture is heated until the HNO_3 has decomposed entirely. After evaporation, 5 mL 30% HCl and 5 mL 40% HF are added to the dry residue and the mixture is heated for 3 h. After evaporation, 5 mL 70% $HClO_4$ is added to the dry residue and the solution is heated for 2 h. The perchloric acid is evaporated until white fumes cease to evolve, and the digestive residue is dissolved in 20 mL 8 M HNO_3.

11.2.6.5 Mineralization of Biota

11.2.6.5.1 Procedure 4

About 1 g of dry (polonium analysis) or ashed biota (uranium and plutonium analysis) is placed in a conical flask, 10 mL 65% HNO_3 is added, and the mixture is slowly warmed until the HNO_3 has decomposed completely. This step is performed three times. After evaporation, the digestive residue is dissolved in 10 mL 8 M HNO_3.[34]

11.2.6.6 Separation and Determination of Polonium ²¹⁰Po and Lead ²¹⁰Pb

11.2.6.6.1 Procedure 5

The aquatic samples in 8 M HNO_3 are evaporated and the dry residue is dissolved in 20 mL 0.5 M HCl. After the addition of approximately 50 mg of ascorbic acid (reduction of Fe^{3+} to Fe^{2+}), the solution is transferred to PTFE vessels equipped with a silver sheet bottom. Polonium is electrodeposited at 90°C for 4 h.[34,36]

The direct measurement of lead ²¹⁰Pb activity in aquatic samples is difficult in view of the low energy of the β particles emitted. The activity of lead ²¹⁰Pb is calculated indirectly by measuring the activity of polonium ²¹⁰Po resulting from the decay of lead ²¹⁰Pb.[34,37]

11.2.6.6.2 Procedure 6

After electrodeposition of ²¹⁰Po, the solution is evaporated and the dry residue dissolved in 10 mL 10 M HCl. The solution is passed through a column (Ø = 10 mm; length = 100 mm) containing Dowex 1 × 8 (100–200 mesh) anion exchange resin. The remainder of the ²¹⁰Po is adsorbed on the resin, whereas lead ²¹⁰Pb passes through the column. The column is flushed with 40 mL 10 M HCl. The ²¹⁰Pb fraction is evaporated to dryness and the residue treated with a little HNO_3 to destroy any remaining organic matter. After an interval of several months, the dry residue is dissolved in 20 mL 0.5 M HCl. ²⁰⁹Po tracer is added and the ²¹⁰Po (formed from ²¹⁰Pb) is subsequently electrodeposited on a silver disc.

11.2.6.7 Separation, Purification, and Electrolysis of Plutonium

The plutonium(IV) in the 8 M HNO_3 + 10 M HCl solution comprises the anion complexes $[Pu(NO_3)_6]^{2-}$ and $[PuCl_6]^{2-}$, which are adsorbed on the anion exchange resin, whereas the Pu(III) occurs as the Pu^{3+} cation. Reduction of the adsorbed Pu(IV) anion complexes by ammonium iodine causes their decomposition to Pu(III).[34,38]

11.2.6.7.1 Procedure 7

Thirty percent H_2O_2 (1 mL perhydrol for each 100 mL of samples) is added to the sample solution in 8 M HNO_3 and heated until the perhydrol has completely decomposed. After cooling, a "pinch" of $NaNO_2$ is added to stabilize the Pu(IV), and the solution is passed through a column ($\emptyset$ = 10 mm; length = 80 mm) containing Dowex 21 K (50–100 mesh) anion exchange resin. The column is flushed with 90 mL 8 M HNO_3. The eluate is collected in a beaker and stored for uranium analysis (procedure 8). The thorium, neptunium, and americium adsorbed on the resin are eluted with 100 mL 10 M HCl, whereas Pu(IV), after reduction, is eluted with 70 mL 1 M NH_4I in 10 M HCl. The eluate containing the plutonium is evaporated to dryness and the residue treated with a little aqua regia to destroy any organic matter and ammonium iodide. The dry residue is dissolved in 5 mL 10 M HCl and passed through a column ($\emptyset$ = 10 mm; length = 100 mm) containing Dowex 1 × 4 (100–200 mesh) anion exchange resin in order to separate the plutonium from the thorium and uranium. The remaining thorium is eluted with 50 mL 10 M HCl, whereas the remainder of the uranium is eluted with 150 mL 8 M HNO_3. After reduction, the Pu(IV) is eluted with 70 mL 1 M NH_4I in 10 M HCl. The eluate is evaporated and the residue treated with a small quantity of aqua regia to destroy any organic matter and ammonium iodide. The dry residue is dissolved in 0.2 mL conc. H_2SO_4 and warmed for about 10 min. After cooling, 5 mL 0.01 M oxalic acid in 1 M $(NH_4)_2SO_4$, two drops of 0.5 M DTPA-NH_4 (the ammonium salt of diethylamine triamine pentaacetic acid), and one drop of 1% tropeolin 00 (the sodium salt 4-[(4-anilinophenyl)azo]benzenesulfonic acid) pH indicator are added. This solution is transferred to a plating cell, and NH_4OH is added to bring its pH to approximately 2 (straw-yellow color). The distance between the electrodes can be adjusted by moving the platinum wire (anode) up to 5 mm. The plutonium is electroplated on a polished stainless steel disc for 2 h with a current of 1.0 A (approximately 0.63 A per 1 cm^2 of cathode). Following electrolysis, 0.1 mL conc. NH_4OH is added and the current switched off. The plutonium disc is removed from the plating cell and rinsed with acetone.[34,39–42]

11.2.6.8 Separation, Purification, and Electrolysis of Uranium

Uranium U(VI) (also Fe, Co, Cu, Zn, and Cd) in 10 M HCl solution is present in the form of the complex uranyl anion $UO_2Cl_4^{2-}$, which is adsorbed on an anion exchange resin.[43,44] Separation and purification of uranium from other elements is possible in sulfuric acid solution. When the H_2SO_4 (aq) concentration is >0.01 M, uranium exists in the anionic forms $UO_2(SO_4)_2^{2-}$ and $UO_2(SO_4)_3^{4-}$. In contrast to uranium, other elements (Fe, Co, Cu, Zn, and Cd) do not form anionic complexes in sulfuric acid solution.[34]

11.2.6.8.1 Procedure 8

After evaporation, the uranium eluate (procedure 7) is dissolved in 10 mL 9 M HCl and the solution is passed through a column ($\emptyset$ = 10 mm; length = 100 mm) containing Dowex 1 × 8 (100–200 mesh) anion exchange resin. The column is flushed with 60 mL 9 M HCl. The U, Fe, Co, and Cu adsorbed on the resin are eluted with 60 mL 0.5 M HNO_3 and the solution is evaporated to dryness. The residue is then dissolved in 5 mL 1 M $(NH_4)_2SO_4$ (pH = 1.5) and warmed for about 10 min; after cooling, the solution is passed through a column containing Dowex 1 × 8 (100–200 mesh) resin in the sulfate form. The column is flushed with 60 mL 1 M $(NH_4)_2SO_4$ (pH = 1.5) solution. The uranium adsorbed on the resin is eluted with 50 mL 0.5 M HCl. The solution is evaporated to dryness and the residue treated with a little aqua regia to destroy any organic matter. The uranium fraction is dissolved in 5 mL 0.75 M $(NH_4)_2SO_4$ (pH = 2) and the solution transferred to a plating cell. Uranium is electroplated on a polished stainless steel disc at a current of 1.0 A for 1.5 h. After electrolysis, approximately 0.5 mL of conc. NH_4OH is added and the current switched off. On removal, the uranium disc is rinsed with acetone.

11.2.6.9   Measurement of Radionuclide Activity

The activities of ^{210}Po, ^{234}U, ^{238}U, ^{238}Pu, and $^{239+240}Pu$ radionuclides are measured using alpha spectrometry and a surface barrier detector with an active surface of 300–450 mm^2 (ORTEC, USA) placed in a vacuum chamber connected to a 1024 multichannel analyzer (Canberra-Packard). Measuring the activity of a single preparation of polonium and uranium takes 2–4 days, and that of plutonium takes 5–10 days, depending on the activity of the sample. Figures 11.4 through 11.6 present typical spectra for the alpha measurements of polonium, uranium, and plutonium.[33,34]

The minimum detectable activities (MDAs) of polonium, uranium, and plutonium radionuclides are calculated from

$$MDA = \frac{L_d}{t_p \cdot e \cdot Y},$$ (11.6)

where L_d is the detection limit of radionuclide activity proposed by Hurtgen et al.[17] (Equation 11.5), t_p is the counting time (s), e—detector efficiency (usually 0.29–0.35), and Y is the recovery.

The MDAs of ^{210}Po, ^{238}U, and $^{239+240}Pu$ is between 0.10 and 0.15 mBq.

11.2.6.10   Activity Determination of Polonium, Radiolead, Uranium, and Plutonium Radionuclides

The activities of ^{210}Po, ^{234}U, ^{235}U, ^{238}U, ^{238}Pu, and $^{239+240}Pu$ are calculated using

$$A_i = \frac{I_i}{e_i \cdot Y_i} \pm SD_i,$$ (11.7)

where i is ^{210}Po, ^{234}U, ^{235}U, ^{238}U, ^{238}Pu, or $^{239+240}Pu$; A_i is the activity (Bq); I_i is the count rate of the sample (without background) defined as the ratio N_i/t_p; N_i is the ^{210}Po, ^{234}U, ^{235}U, ^{238}U, ^{238}Pu, or

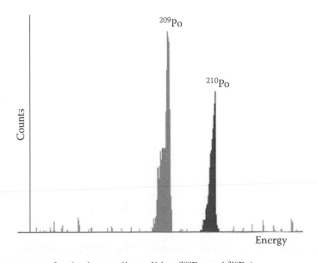

FIGURE 11.4   Alpha spectra of polonium radionuclides (^{209}Po and ^{210}Po).

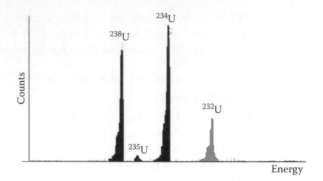

FIGURE 11.5   Alpha spectra of uranium radionuclides (^{232}U, ^{234}U, ^{235}U, and ^{238}U).

$^{239+240}$Pu count (without background); t_p is the ^{210}Po, ^{234}U, ^{235}U, ^{238}U, ^{238}Pu, or $^{239+240}$Pu counting time [s]; e_i is the detector efficiency; Y_i is the recovery of ^{210}Po, ^{234}U, ^{235}U, ^{238}U, ^{238}Pu, or $^{239+240}$Pu; and SD_i is the SD of sample activity.

The activity of ^{210}Po in aquatic samples has to be calculated on the basis of the electrodeposition time of the polonium on a silver disc using

$$A_0 = A_t \exp\left(\frac{t \cdot \ln 2}{T}\right) = A_t \exp(0.00502 \cdot t), \tag{11.8}$$

where A_0 is the ^{210}Po activity at the time of electrodeposition on the silver disc [Bq], A_t is the ^{210}Po activity at the time of counting (Bq), t is the time interval between electrodeposition and the ^{210}Po count (days), and T is the ^{210}Po half-life (138.4 days).

The recoveries of ^{210}Po, ^{234}U, ^{235}U, ^{238}U, ^{238}Pu, or $^{239+240}$Pu are calculated from

$$Y_i = \frac{I_i}{e_i \cdot s}, \tag{11.9}$$

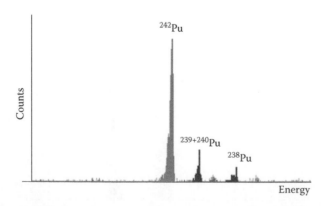

FIGURE 11.6   Alpha spectra of plutonium radionuclides (^{238}Pu, $^{239+240}$Pu, and ^{242}Pu).

where Y_i is the recovery; I_i is the count rate of recovery indicator (^{209}Po, ^{232}U, and ^{242}Pu) defined as the ratio N_i/t_p; s is the ^{209}Po, ^{232}U, or ^{242}Pu activity added before radiochemical analysis (Bq); and e_i is the detector efficiency.

The SD for the activity of polonium (also of uranium and plutonium) is calculated from

$$SD = \frac{A_0}{A_t} \frac{\sqrt{(I_p/t_p + I_t/t_t)}}{Y \cdot e}, \tag{11.10}$$

where A_0 is the activity of ^{210}Po at the time of electrodeposition (Bq), A_t is the activity of ^{210}Po at the time of counting (Bq), I_p is the sample count rate (with background), I_t is the background count rate, t_p is the sample counting time (s), t_t is the background counting time (s), Y is the recovery, and e is the detector efficiency.

If the background count is very low, and during the analysis time $A_0 = A_t$ (for uranium and plutonium radionuclides), then

$$SD = \frac{\sqrt{N_p}}{Y \cdot e \cdot t_p}, \tag{11.11}$$

where N_p is the number of counts of uranium (^{234}U, ^{235}U, and ^{238}U) or plutonium (^{238}Pu and $^{239+240}$Pu) radionuclides.

The radiolead ^{210}Pb activity is estimated on the basis of ^{210}Po in growth after the lead fraction has been purified and stored for several months (up to two years) (procedure 6). The ^{210}Pb activity at the time of sample collection is calculated from

$$A_0(^{210}Pb) = \left[\frac{A_2(^{210}Po)}{1 - \exp\left[-k(t_2 - t_1)\right]} \right], \tag{11.12}$$

where $A_0(^{210}Pb)$ is the activity of ^{210}Pb at the time of sample collection, $A_2(^{210}Po)$ is the activity of ^{210}Po originating from ^{210}Pb decay following the second electrodeposition, t_1 is the time interval between collection and the first ^{210}Po count, t_2 is the time interval between collection and the second ^{210}Po count, and k is the ^{210}Po decay constant.

The uranium concentration is estimated on the basis of ^{238}U activity from

$$1 \text{ Bq } ^{238}U = 81.6 \text{ μg U.} \tag{11.13}$$

The impact of the Chernobyl plutonium fraction in the aquatic samples is calculated using the following formula[33].

$$F_{ch} = \frac{R_{obs} - R_n}{R_{ch} - R_n} = \frac{R_{ch} - 0.04}{0.56}, \tag{11.14}$$

where R_{obs} is the ^{238}Pu/$^{239+240}$Pu activity ratio in the aquatic sample, R_n is the ^{238}Pu/$^{239+240}$Pu activity ratio in the global atmospheric fallout (0.04), and R_{ch} is the ^{238}Pu/$^{239+240}$Pu activity ratio in the Chernobyl accident (0.60).

11.2.6.11   Determination of ^{241}Pu Activity

^{241}Pu is a low-energy beta emitter with an E_{max} of 21 keV and a half-life of 14.4 years.[10] ^{241}Pu is the only significant beta-emitting transuranium nuclide in low-level waste from nuclear power plants. The quantitation of ^{241}Pu in low-level waste and environmental samples is of interest because it is a precursor of other transuranium radionuclides with longer half-lives, greater environmental mobility, and greater radiotoxicity. As a beta emitter ^{241}Pu has a relatively low radiotoxicity; however, it decays to ^{241}Am (half-life = 432 years), which is a highly radiotoxic, alpha emitter. The two predominant sources of environmental contamination are the fallout from past nuclear weapons tests and discharges from nuclear fuel cycle operations (reprocessing, in particular) in the form of gaseous emissions or liquid effluents.[45] ^{241}Pu can be determined directly by measurement in a beta proportional counting system[46] and by liquid scintillation counting (samples with a relatively high content of ^{241}Pu), and indirectly by alpha spectrometric measurements of its daughter radionuclide ^{241}Am.[47,48] Measurement based on the growth of ^{241}Am can be carried out only after a long growth period (between 4 and 20 years). Even after four years the activity ratio ^{241}Am/^{241}Pu is only 1/166. The plutonium alpha spectra obtained have to be compared with the same spectra obtained 4–20 years earlier. This then enables the ^{241}Pu content to be estimated on the basis of the increase in the 5.49 MeV peak of ^{241}Am, which takes into account the ^{238}Pu present in environmental samples from the Chernobyl accident. ^{241}Pu activity is calculated from

$$A_{Pu_0} = 31.3074 \cdot \frac{A_{241_{Am}} \cdot e^{+\lambda_{Am} \cdot t}}{1 - e^{-\lambda_{Pu} \cdot t}}, \tag{11.15}$$

where A_{Pu_0} is the ^{241}Pu activity at the time of sampling, $A_{241_{Am}}$ is the ^{241}Am activity measured after 4–20 years, λ_{Pu} is the decay constant of ^{241}Pu (0.050217 year^{-1}), λ_{Am} is the decay constant of ^{241}Am (0.001604 year^{-1}), 31.3074 is the $\lambda_{Pu}/\lambda_{Am}$ ratio, and t is the time from sampling to the measurement of ^{241}Am (4–20 years).

REFERENCES

1. Skwarzec, B. 2000. *Radiochemia środowiska i ochrona radiologiczna.* Gdańsk: Wydawnictwo DJ s.c.
2. Pullen, B.P. 1986. Cherenkov counting of ^{40}K in KCl using a liquid scintillation spectrometer. *J. Chem. Educ.* 63: 971–976.
3. Rao, D.D., V. Sudheendren, A. Baburajan, S. Chandramouli, A.G. Hegde, and U.C. Mishra. 1986. Measurement of high energy gross beta and ^{40}K by Cherenkov counting in liquid scintillation analyzer. *J. Radioanal. Nucl. Chem.* 243: 651–655.
4. Noor, A., M. Zakir, B. Rasyid, Maming, and M.F. L'Annaunziata. 1996. Cherenkov and liquid scintillation analysis of the triple label ^{86}Rb-^{35}S-^{33}P. *Appl. Radiat. Isot.* 47: 659–668.
5. Fletcher, I.R., A.L. Maggi, K.J. Rosmanand, and N.J. McNaughton. 1997. Isotopic abundances of K and Ca using a wide-dispersion multi-collector mass spectrometer and low fractionation ionization techniques. *Int. J. Mass Spectrom.* 163: 1–17.
6. Holm, E., B. Oregoni, D. Vas, H. Pettersson, J. Rioseco, and U. Nilsson. 1990. Nickel-63: Radiochemical separation and measurement with an ion implanted silicon detector. *J. Radioanal. Nucl. Chem.* 138: 111–118.
7. Holm, E. 1994. Source and distribution of anthropogenic radionuclides in marine environment. In: E. Holm (ed.), *Radioecology,* pp. 63–83. Singapore, New Jersey, London, Hong Kong: World Scientific.
8. Holm, E., P. Roos, and B. Skwarzec. 1992. Nickel-63 in Baltic Fish and sediments. *Appl. Radiat. Isot.* 43: 371–376.
9. Skwarzec, B., E. Holm, P. Roos, and J. Pempkowiak. 1994. Nickel-63 in Baltic fish and sediments. *Appl. Radiat. Isot.* 54: 609–611.
10. Browne, E. and F.B. Firestone. 1986. *Table of Radioactive Isotopes.*New York: Wiley.
11. Skwarzec, B., E. Holm, and D. Strumińska. 2001. Radioanalytical determination of ^{55}Fe and ^{63}Ni in the environmental samples. *Chem. Anal. (Warsaw)* 46: 23–30.

12. Folsom, T.R. and C. Sreekumaran. 1970. *Some reference methods for determining radioactive and natural cesium for marine studies*. Technical reports series no. 118. Reference Methods for Marine Radioactivity Studies, International Atomic Energy Agency, Vienna.

13. CLOR. 1978. *Procedure for radiochemical and chemical analysis of environmental and biological samples*. Raport no. CLOR-110/D, Centralne Laboratorium Ochrony Radiologicznej, Warszawa.

14. Horwitz, E.P., R. Chiarizia, and M.L. Dietz. 1992. A novel strontium-selective extraction chromatographic resin. *Solvent Extr. Ion Exch.* 10: 313–336.

15. Fliss, M., W. Botsch, J. Handl, and R. Michel. 1998. A fast method for the determination of strontium-89 and strontium-90 in environmental samples and its application to the analysis of strontium-90 in Ukrainian soils. *Radiochim. Acta* 83: 81–92.

16. Gaca, P. 2004. Nagromadzenie i ocena skażenia środowiska przyrodniczego Polski izotopem ^{90}Sr. Praca doktorska, Wydział Chemii Uniwersytetu Gdańskiego.

17. Hurtgen, C., S. Jerome, and M. Woods. 2000. Revisiting Currie—how low can you go? *Appl. Radiat. Isot.* 53: 45–50.

18. Aupiais, J., C. Fayolle, P. Gilbert, and N. Dacheux. 1998. Determination of Ra-226 in mineral drinking water by alpha liquid scintillation with rejection of beta emitters. *Anal. Chem.* 70: 2353–2359.

19. Burnett, W.C. and W.-C. Tai. 1992. Determination of radium in natural waters by alpha liquid scintillation. *Anal. Chem.* 64: 1691–1697.

20. Higuchi, H., M. Uesugi, K. Satoh, and N. Ohashi, 1984. Determination of radium in water by liquid scintillation counting after preconcentration with ion exchange resin. *Anal. Chem.* 56: 761–763.

21. Higuchi, H. 1981. Analytical methods for radium in environmental samples. *Radioisotopes* 30: 618–627.

22. Saarinen, L. and J. Suksi. 1992. Determination of uranium series radionuclides Pa-231 and Ra-226 using liquid scintillation counting (LSC). In: *Report on the Nuclear Waste Commission of Finnish Power Companies*. Technical Report YJT-92-20, Helsinki, Finland.

23. Blackburn, R. and M.S. Al-Masri. 1992. Determination of radium-226 in aqueous samples using liquid scintillation counting. *Analyst* 117: 1949–1951.

24. Becker, J.K. and H.-J. Dietze. 1998. Determination of long-lived radionuclides by double-focusing sector field ICP mass spectrometry. *Adv. Mass Spectrom.* 14: 681–689.

25. Zikovsky, L. and N. Roireau. 1990. Determination of radon in water by argon purging and alpha counting with a proportional counter. *Appl. Radiat. Isot.* 41: 679–681.

26. Salonen, L. 1993. Measurement of low levels of ^{222}Rn in water with different commercial liquid scintillation counters and pulse shape analysis. In: J.E. Noakes, F. Schoenhofer, and H.A. Polach (eds), *Advances in Liquid Scintillation Spectrometry 1992*, pp. 361–371. Tucson: Radiocarbon Publishers, University of Arizona.

27. Blackburn, R. and M.S. Al-Masri. 1993. Determination of radon-222 and radium-226 in water samples by Cherenkov counting. *Analyst* 118: 873–876.

28. Homma, Y., Y. Murase, and M. Takiue. 1987. Determination of ^{222}Rn by air luminescence method. *J. Radioanal. Nucl. Chem. Lett.* 119: 457–465.

29. Bowen, V.T., J.S. Olsen, C.L. Osterberg, and J. Revera. 1971. *Radioactivity in the Marine Environment*. Washington, DC: National Academy of Sciences.

30. Koide, M. and K.W. Bruland, 1975. The electrodeposition and determination of radium by isotopic dilution in sea water and sediment simultaneously with other natural radionuclides. *Anal. Chem.* 75: 1–19.

31. Wong, K.M., G.S. Brown, and V.E. Noshkin. 1978. A rapid procedure for plutonium separation in large volume of fresh and saline water by manganese dioxide coprecipitation. *J. Radioanal. Chem.* 42: 7–15.

32. Jaakkola, T. 1994. Radiochemical separation. In: E. Holm (ed.), *Radioecology*, pp. 233–253. Singapore, New Jersey, London, Hong Kong: World Scientific.

33. Skwarzec, B. 1995. *Polon, uran i pluton w ekosystemie południowego Bałtyku*. Rozprawy i monografie, 6. Instytut Oceanologii PAN, Sopot.

34. Skwarzec, B. 1997. Radiochemical methods for the determination of polonium, radiolead, uranium and plutonium in environmental samples. *Chem. Anal. (Warsaw)* 42: 107–115.

35. Szefer, P. and B. Skwarzec. 1988. Distribution and possible sources of some elements in the sediment cores of the southern Baltic. *Mar. Chem.* 23: 109–129.

36. Flynn, W.W. 1968. The determination of low levels of polonium-210 in environmental materials. *Anal. Chim. Acta* 43: 221–227.

37. Takizawa, Y., L. Zhao, M. Yamamoto, T. Abe, and K. Ueno. 1990. Determination of ^{210}Pb and ^{210}Po in human tissues of Japanese. *J. Radioanal. Nucl. Chem.* 138: 145–152.

38. Holm, E. 1977. Plutonium isotopes in the environment. PhD dissertation, University of Lund, Sweden, LUNF D6 NFRA-1005.

39. Chu, Y. 1971. Plutonium determination in soil by leaching and ion exchange. *Anal. Chem.* 43: 449–452.

40. Talvitie, N.A. 1972. Electrodeposition of actinides for alpha-spectrometric determination. *Anal. Chem.* 44: 280–283.

41. Kressin, I.K. 1977. Electrodeposition of Pu and Am from high resolution of alpha spectrometry. *Anal. Chem.* 49: 842–845.

42. Purphal, W.K., T.D. Filer, and G.J. McNabb. 1984. Electrodeposition of actinides in a mixed oxalate-chloride electrolyte. *Anal. Chem.* 56: 113–116.

43. Strelow, F.W. and E. Bohme. 1967. Anion exchange and selectivity scale for elements in surface acid media with a strongly basic resin. *Anal. Chem.* 39: 595–599.

44. Sing, N.P. and W. Wrenn. 1973. Determination of alpha-emitting uranium isotopes in soft tissue by solvent extraction and alpha-spectrometry. *Talanta* 30: 271–274.

45. Pimpl, M. 1992. Increasing the sensitivity of [241]Pu determination for emission and immission control of nuclear installations by aid liquid scintillation counting. *J. Radioanal. Nucl. Chem.* 161: 429–435.

46. Rosner, G., H. Hotzl, and R. Winkler. 1992. Determination of [241]Pu by low level beta proportional counting, application to Chernobyl fallout samples and comparison with the [241]Am build-up method. *J. Radioanal. Nucl. Chem.* 163: 225–233.

47. Struminśka, D.I., B. Skwarzec, and M. Mazurek-Pawlukowska. 2005. Plutonium isotopes [238]Pu, [239+240]Pu and [241]Pu in Baltic Sea ecosystem. *Nukleonika* 50: S45–S48.

48. Strumińska, D.I. and B. Skwarzec. 2006. [241]Pu concentration in southern Baltic Sea ecosystem. *J. Radioanal. Nucl. Chem.* 268: 59–63.

12 Analytical Techniques for the Determination of Inorganic Constituents

Jorge Moreda-Piñeiro and Antonio Moreda-Piñeiro

CONTENTS

12.1 Introduction 260
    12.1.1 Characteristics of Inorganic Constituents 260
    12.1.2 Classification of Inorganic Constituents 260
12.2 Analytical Techniques 261
    12.2.1 Gravimetric Measurements 262
    12.2.2 Titrimetric Measurements 262
    12.2.3 Spectrophotometric Techniques Based on Molecular Absorption
            Radiation: Ultraviolet-Visible Spectrophotometry 263
            12.2.3.1 Instrumentation 264
    12.2.4 Atomic Spectrometry 265
            12.2.4.1 Flame Atomic Absorption Spectrometry 266
            12.2.4.2 Electrothermal Atomic Absorption Spectrometry 268
            12.2.4.3 High-Resolution Continuous Source
                   Atomic Absorption Spectrometry 269
            12.2.4.4 Atomic Emission Spectrometry 270
            12.2.4.5 Inductively Coupled Plasma-Optical Emission Spectrometry 270
            12.2.4.6 Atomic Fluorescence Spectrometry 271
    12.2.5 Mass Spectrometric Techniques 272
            12.2.5.1 Inductively Coupled Plasma-Mass Spectrometry 272
    12.2.6 Chemical Vapor Generation-Atomic Spectrometry (CVG-AS) 273
            12.2.6.1 Instrumentation 274
    12.2.7 Electrochemical Techniques 275
            12.2.7.1 Anodic Stripping Voltammetry 275
            12.2.7.2 Potentiometric Sensors: ISEs and Gas-Permeable
                   Membrane Sensors 276
    12.2.8 Separation Techniques 278
            12.2.8.1 Ion Chromatography 278
            12.2.8.2 Capillary Electrophoresis 280
    12.2.9 Automatic Analyzers and Monitoring 281
12.3 Determination of Inorganic Constituents 282
Acronyms and Abbreviations 295
References 296

12.1 INTRODUCTION

12.1.1 CHARACTERISTICS OF INORGANIC CONSTITUENTS

Inorganic constituents are present in major, minor, and trace concentrations in aquatic environments as a result of weathering, atmospheric, and biogeochemical processes, and also as a consequence of industrial pollution. Inorganic constituents in aquatic ecosystems are a group of chemical species that are characterized by

1. A wide variety of chemical structures, from simple monoatomic ions and molecular ions to complex molecules.
2. Different roles in aquatic ecosystems, from acting as nutrients for living organisms (nitrogen and phosphorus compounds) to exerting toxic effects on such organisms (arsenic and mercury).
3. A variety of aggregations, for example, dissolved inorganic constituents coexisting with different dissolved gases such as O_2, NO_2, or CO_2, or inorganic constituents (such as phosphorus compounds and metals) are bonded to particulate matter suspended in water—the latter is more common.
4. A wide range of concentrations ranging from ultratrace levels of around ng L^{-1} (some heavy metals in sea water) to concentrations in the mg L^{-1} and % (m/v) ranges (major ions and nutrients).

In the absence of any human impact, the relative proportions and rates of dissolution of substances in natural waters are highly variable, depending on the local geological, climatic, and geographical conditions. Ions such as Cl^- and Na^+ (and, to a lesser extent, SO_4^{2-} and Mg^{2+}) are the major components of sea water. The major constituents of world river waters are Ca^{2+} and HCO_3^- (from weathering of $CaCO_3$). In river waters, several ratios of SO_4^{2-}, HCO_3^-, K^+, Mg^{2+}, and Ca^{2+} to Cl^- are usually greater than those in sea water.[1] Since ground water often occurs in association with geological materials containing soluble minerals, higher concentrations of dissolved salts are normally expected in ground water than in surface water. Ground water usually contains high concentrations of ions such as Ca^{2+}, HCO_3^-, Na^+, Mg^{2+}, SO_4^{2-}, and Cl^- (from 1 to 1000 mg L^{-1}),[2] and some toxic elements (As and Se). In general, trace elements such as As, Ba, Cd, Cr, Co, Cu, Mn, Ni, Pb, Sr, and Zn are present at lower concentrations in sea water than in surface water. This is what justifies the use of analytical techniques of very different sensitivities, which enable ions and dissolved gases to be determined at very different concentration levels.

12.1.2 CLASSIFICATION OF INORGANIC CONSTITUENTS

The inorganic constituents of aquatic ecosystems can be classified as follows:

1. Nutrients—these are elements essential to the metabolism of living organisms. Nitrogen, phosphorus, and silicon are the most important and commonly studied nutrients in aquatic ecosystems.
   a. Nitrogen compounds are essential for living organisms as they are important constituents of proteins. In the aquatic environment, inorganic nitrogen occurs in a range of oxidation states: nitrate (NO_3^-), nitrite (NO_2^-), the ammonium ion (NH_4^+), and molecular nitrogen (N_2).
   b. Phosphorus compounds occur in aquatic systems in both particulate and dissolved species. The speciation of different phosphate compounds (PO_4^{3-}, $H_2PO_4^-$, and HPO_4^{2-}) in water depends on the ambient pH. Artificial increases in phosphate concentrations as a result of human activities are the main cause of eutrophication of rivers and lakes.

c. Silicon is also an essential element for aquatic plants such as diatoms. Si is present in water in dissolved (silicic acid), suspended, and colloidal forms.

2. Trace elements—elements such as Al, As, Be, Cd, Cr, Cu, Fe, Hg, Mn, Mo, Ni, Pb, Sb, Se, Tl, V, and Zn are present in the aquatic environment as a result of the weathering of rocks and soils (surface and ground waters) and of industrial waste water discharges and mining activities. The toxicity of trace elements in water depends on their concentration and oxidation state (Cr or As).

3. Major ions and inorganic nonmetallic substances—these constitute a heterogeneous group of species including major ions, halides, cyanide, and sulfur species (Na^+, K^+, Ca^{2+}, Mg^{2+}, CO_3^{2-}, HCO_3^-, Cl^-, F^-, SO_4^{2-}, SO_3^-, S^{2-}, and CN^-) as well as dissolved gases (O_2, N_2, CO_2, and NO_2).

   a. Major ions, halides, cyanide, and sulfur species: sodium (Na^+) and chloride (Cl^-) are two of the most abundant constituents of natural fresh and sea water. Aquatic environments contain high concentrations of Na^+ and Cl^- and also Ca^{2+}, Mg^+, CO_3^{2-}, F^-, SO_4^{2-}, SO_3^-, and S^{2-}. This is because their salts are extremely soluble in water and are readily dissolved from rocks and minerals (sedimentary rocks, fluorapatite, gypsum, pyrite, etc.) as a result of weathering and surface runoff. Ions such as Ca^{2+}, Mg^+, CO_3^{2-}, and HCO_3^- are responsible for the hardness and alkalinity of water. K^+ and CN^- are found in low concentrations in natural waters since potassium-containing rocks are relatively resistant to weathering, and cyanide is present in the environment as a result of industrial discharges. Certain chlorine compounds (hypochlorite and chloramines) are present as a consequence of the chlorination of drinking water and waste water treatment.

   b. Dissolved gases such as O_2 and N_2 are the major dissolved gases in natural waters; inorganic carbon, in the form of dissolved CO_2, together with N_2O and Cl_2 from the disinfection of drinking and waste waters are also present in trace concentrations.

12.2 ANALYTICAL TECHNIQUES

In view of the wide variety and concentrations of inorganic constituents in aquatic environments, there are a great number of analytical techniques used to determine these compounds. They include

1. Gravimetric measurements for silica and sulfate.
2. Titrimetric measurements: alkalinity titration for carbonate and bicarbonate ion determinations, argentometric and potentiometric titrations for determining chloride, and iodometric titration for sulfite, chlorine, and dissolved oxygen.
3. Spectrophotometric techniques based on molecular absorption radiation for determining nutrients (NO_3^-, NO_2^-, NH_4^+, N_2, phosphorus, and silicon) as well as chlorine, fluoride, cyanide, sulfate, and sulfide.
4. Spectrophotometric techniques based on the dispersion of radiation, for example, the nephelometric determination of sulfate.
5. Spectrometric techniques based on atomic absorption or the emission of radiation: flame atomic absorption spectrometry (FAAS), electrothermal atomic absorption spectrometry (ETAAS), inductively coupled plasma-optical emission spectrometry (ICP-OES), inductively coupled plasma-mass spectrometry (ICP-MS), and cold vapor (CV)/hydride generation (HG), mainly for trace and ultratrace metal determinations.
6. Electrochemical techniques: anodic stripping voltammetry (ASV) and cathodic stripping voltammetry (CSV) for determining trace elements, and potentiometric sensors for determining dissolved gases (CO_2, NO_2, SO_2, NH_3, H_2S, HCN, and HF) as well as chloride, fluoride, cyanide, and sulfide.
7. Separation techniques such as ion chromatography (IC) and capillary electrophoresis (CE) for determining major and minor ions.

12.2.1 Gravimetric Measurements

Gravimetric methods are based on the quantitative measurement of an analyte by weighing a pure and insoluble compound from the analyte. Analytical balances, which provide accurate and precise data, are used for this purpose. Detailed information on gravimetric methods and their applications can be found in the literature.[3]

In summary, a typical gravimetric procedure to determine an unknown concentration of an analyte in solution is as follows:

1. Quantitative and selective precipitation of the analyte from solution, for example,

$$Ag^+(aq) + Cl^-(aq) \rightarrow AgCl(s);$$

$$Ca^{2+}(aq) + C_2O_4^{2-}(aq) \rightarrow CaC_2O_4(s);$$

$$Ba^{2+}(aq) + SO_4^{2-}(aq) \rightarrow BaSO_4(s).$$

The composition of the insoluble compound (precipitate) obtained from the analyte must be known and stable. Poorly soluble substances may form colloidal suspensions (particle diameters from 10^{-7} to 10^{-4} cm). The formation of a colloidal suspension can be minimized or prevented by carrying out the precipitation from a dilute solution of the analyte, at a temperature close to the boiling point of water and with constant stirring. The relative supersaturation affects the particle size and is expressed as $Q - S/S$, where Q is the instantaneous concentration of the added species and S is the equilibrium solubility of the compound that precipitates. Particle size seems to be inversely proportional to relative supersaturation. The electric double layer formed during precipitation keeps the colloidal precipitate particles from coming into contact with each other, thus preventing further coagulation.

There are two ways of bringing the particles closer together and of increasing the probability of coagulation: (a) Heating increases overall thermal motion, affecting the mobility of adsorbed ions and colloidal precipitate particles. The final effect is that there are collisions of particles, which cause the particle size to increase as a result of coagulation. (b) Increasing the electrolyte concentration in the solution leads to a decrease in the mean radius of the electric double layer and encourages further coagulation. A procedure known as high-temperature precipitate digestion is usually carried out to enhance coagulation, and hence increase the particle size of the precipitate. It must be mentioned that other processes such as peptization, surface adsorption, mixed crystal formation, occlusion, and mechanical entrapment can occur during the precipitation procedure, causing the results to be positively or negatively erroneous.

2. Isolation of the precipitate by filtering and rinsing. Having been isolated, the precipitate must be thoroughly rinsed in order to remove contaminants; great care must be taken to ensure that no precipitate is lost during either filtration or rinsing.
3. Drying the solid in an oven to remove the solvent: the precipitate must be heated (at different temperatures) until a stable, dry state is reached. Reliable results are founded on a thorough knowledge of the precipitate's properties.
4. Weighing of the solid on an analytical balance.
5. Calculation of the analyte concentration in the original solution from the weight of the precipitate.

12.2.2 Titrimetric Measurements

Titrations represent a comprehensive set of procedures useful for performing quantitative determinations in analytical chemistry. Detailed information on titrimetric methods and their applications can be found in the literature.[4]

Titration involves measuring the volume of a standardized solution containing a known concentration of a reagent (titrant) that reacts quantitatively with the analyte. Titrations can be done by adding the standardized solution from a burette to a sample solution until the reaction between the titrant and the analyte is complete. The volume of the titrant used to carry out the titration is determined by the difference between the initial and final readings of the burette. The equivalence point of a titration is reached when the amount of titrant is chemically equivalent to the amount of analyte in the sample; at this point, the number of gram equivalents of titrant and analyte are equal. However, the equivalence point is a theoretical point that cannot be determined experimentally. We can only estimate this point by observing any physical change that accompanies the equivalence condition. We must ensure that the difference between the equivalence point volume and the end point volume is minimal; this difference is known as the titration error. Sometimes, an indicator is added to the sample solution in order to elicit an observable physical change (end point) at or near the equivalence point. Typical changes of an indicator are the appearance or disappearance of color, color changes, and the appearance or disappearance of turbidity. Instruments are commonly used to detect end points; they respond to a certain property of the solution (e.g., pH for potentiometric titrations; conductivity, conductometric titrations; and radiation absorption, photometric titrations) that typically changes during the titration. End point detection by instrumental methods provides, among other benefits, greater accuracy and precision, because the error inherent in the visual observance of indicator change is avoided. In addition, these end point detection methods are more appropriate for samples where turbidity or color is involved. Potentiometric titrations are useful when a change in potential occurs during titration. Thus, acid–base titrations can be monitored with a pH meter, precipitation, and complexometric titrations with ion-selective electrodes (ISEs). The equivalence point is determined mathematically: it is the point of inflection on a plot of potential versus the volume of titrant added. The equivalence point can be better visualized if the first or second derivative ($\Delta E/\Delta V$ or $\Delta^2 E/\Delta^2 V$, respectively) is plotted against the volume of titrant added.

12.2.3 SPECTROPHOTOMETRIC TECHNIQUES BASED ON MOLECULAR ABSORPTION RADIATION: ULTRAVIOLET-VISIBLE SPECTROPHOTOMETRY

Spectrophotometry is based on the simple relationship between the molecular absorption of ultraviolet-visible (UV-VIS) radiation by a solution and the concentration of a colored species in such a solution. For these groups of techniques involving matter-radiation interaction, electromagnetic radiation should be regarded as a stream of discrete packets behaving as individual particles called photons. The energy of a photon is proportional to the frequency of the radiation and is given by the Planck equation: $E = h\nu = hc/\lambda$. Detailed information on UV-VIS spectrophotometry can be found in the monographs by Heinz-Helmut[5] and Harris and Bashford.[6]

The absorption of radiation is the process by which a chemical species selectively attenuates (or selectively decreases the intensity of) certain frequencies of electromagnetic radiation. According to quantum theory, each particle (atom, ion, and molecule) has a unique set of energy states. At room temperature, most particles are in the ground state. When a photon interacts with a particle, it is likely to be absorbed if its energy matches any of the energy jumps quantified for this particle. Under these conditions, the energy of the photon is transferred to the particle, whose valence electrons are promoted to a higher energy state, called an excited state ($M + h\nu \rightarrow M^*$). After a short time (10^{-6} to 10^{-9} s), the excited species relaxes to the ground state; during this process the excess energy is transferred to adjacent particles. Relaxation is manifested by the photochemical decomposition of the excited species (M^*) and the formation of a new species, or by the re-emission of the energy in the form of fluorescent or phosphorescent radiation. The absorption characteristics of a species can be described by an absorption spectrum showing the beam radiation attenuation versus the wavelength, frequency, or wave number. Each line is attributed to the transition of electrons from one of the many states of vibrational and rotational energy associated with ground state to the excited electronic state. Because of the large number of possible vibrational and rotational states, their energies differ little among themselves, and therefore the number of lines contained in a typical band is very high. Molecular spectrophotometry makes use of the transitions located in the ultraviolet (UV) (10–400 nm), visible (VIS) (400–1000 nm), and infrared (IR) (0.78–1000 μm) regions of the electromagnetic spectrum.

Two important terms are used in absorption spectrophotometry. One is the transmittance T, defined as the ratio (%) of the intensity of the radiation beam emerging from the solution I to that of the incident beam I_0 : $T = I/I_0$. The other is the absorbance, A, defined as the negative logarithm of the transmittance T : $A = -\log_{10} T = \log I_0/I$. Unlike the transmittance, the absorbance of a solution increases with increasing beam attenuation. The functional relationship between the analytical signal measurement (absorbance) and the analytical parameter of interest (concentration) is known as the Bouguer–Lambert–Beer law and is expressed as $A = \log (I_0/I) = abc$, where A is the absorbance of the solution, a is the proportionality constant called the absorptivity, b is the radiation path length within a sample, and c is the concentration of the solution. When the concentration is expressed in mol L^{-1}, absorptivity is represented by e and is called the molar absorptivity. This linear relationship is a generalization, however, as there are deviations from the direct proportionality between absorbance and concentration for a fixed path length. Some of these deviations are technical deviations from the law when concentrations higher than 0.01 M are measured. Others are consequences of the experimental measurements, for example, the use of polychromatic and dispersive radiation (instrumental deviations), or they may derive from analyte association or dissociation with the solvent or chemical changes to give products with different absorption properties (chemical deviations).

Because the sample must be placed in a cell (sample container), there is an interaction between the radiation and the cell walls, which produces a loss of power at each interface as a result of reflections and absorptions. In order to prevent or minimize these effects, the power of the transmitted beam is usually compared with the same radiation beam that passes through a reference cell containing only the solvent. Therefore, the measured absorbance is defined by the equation: $A = \log (I_{solvent}/I_{sample}) = \log (I_0/I)$, where $I_{solvent}$ and I_{sample} are the intensities of the beams emerging from the solvent and sample cell, respectively.

12.2.3.1 Instrumentation

Instruments used for transmittance or absorbance measurements consist of five basic elements:

1. A light source—a tungsten filament; hydrogen, mercury, or deuterium lamps; a xenon discharge lamp—that provides an intense, stable, and constant radiation.
2. A wavelength selector—interference and absorption filters or a monochromator (prisms and diffraction gratings)—to isolate the desired emission line.
3. A sample cell—normally parallelepiped in shape with a standard length of 1 cm and made of glass for the VIS region or quartz (or fused silica) for the UV region. The cell has an opening for inserting the sample and a stopper to prevent evaporation.
4. A detector—phototubes, photomultiplier tubes, photovoltaic cells, or photodiodes, which convert radiant energy into a measurable signal.
5. A data processing unit—an electronic device that amplifies, filters, and performs mathematical processes on the signal.

These components are assembled in different ways to produce several instrument designs. Here we consider two general types of spectroscopic instruments:

1. The photometer: A simple instrument that uses absorption or interference filters for wavelength isolation and a photoelectric device to measure the radiant power.
2. The spectrophotometer: A specialized device capable of recording the various components of complex radiation.

These instruments exist in three configurations:

1. Single-beam instruments: These consist of a radiation source, a monochromator, and two cells for the reference and the sample solutions, which are alternately inserted in the light path; also a detector, an amplifier, and a reading device. These instruments require a stable voltage source to prevent errors arising from variations in the beam intensity. Also, differences between cells (mainly irregularities in the walls) are not easily compensated for.
2. Dual-beam instruments: In these the incident light beam is split by a rotating mirror-chopper into two separate beams, one of which passes through the sample and the other passes through the reference. The alternating beams reaching the detector thus permit a simple mathematical treatment of signals (signal modulation). This design is routinely used and leads to good results, since it minimizes drift of the radiation source and the amplifier.
3. Multichannel instruments: These are equipped with a photodiode array detection system. The radiation from a tungsten or deuterium lamp is focused on the sample or solvent cell, and then passes to a diffracting grating. The scattered radiation arrives at the diode array, which simultaneously detects and analyzes various wavelengths.

Specific applications of spectrophotometry are turbidimetry and nephelometry; both techniques are based on the dispersion of radiation from particles containing the target analyte. Measurement of the incident beam attenuation due to the presence of particles is the basis of the turbidimetry approach. In contrast, measurement of the light scattered (dispersed) by such particles at 90° to the incident beam is the basis of nephelometry. The amount of dispersed radiation depends on the density, size, shape, and refractive index of such particles. The effects of all these factors are covered by the Rayleigh law, which establishes the amount of radiation reflected toward the detector. They also depend on the wavelength: $I/I_0 = K(Nv^2/\lambda^4) \sin \alpha^2$, where I and I_0 are the respective intensities of the dispersed and incident beams, α is the angle measured relative to the excitation light, N is the number of particles of volume v, and λ is the wavelength of the light.

12.2.4 Atomic Spectrometry

Atomic spectrometry techniques make use of the absorption or emission of radiation (optical emission or fluorescence emission) by atoms or ions. Therefore, the physicochemical principles underlying this huge group of techniques (based on the interaction between electromagnetic radiation and matter) are similar to those of UV-VIS or IR absorption spectrometry for molecules. However, there is an important difference between atomic spectrometry and molecular spectrometry: in the former the sample must first be atomized. In fact, the ways in which elemental atoms or ions (atomization) are obtained constitute the foundations of the various atomic spectrometry techniques in use. The atomization mechanisms in atomic spectrometry can be split into two main groups: thermal atomization (atomization by heating at very high temperatures) and chemical atomization (atomization through a chemical reaction that converts the aqueous analyte into an atomic vapor).

Flames (temperatures from 1700°C to 3100°C) and plasmas (temperatures ranging between 4000°C and 6000°C) are very efficient means of atomization; they are used in FAAS, flame atomic emission spectrometry (FAES), ICP-OES, and ICP-MS. High temperatures (up to 3000°C) can also be obtained if an intense electrical current is set up between two electrical contacts. This type of atomization by electrical heating (electrothermal atomization) is the basis of ETAAS.

For some elements such as Hg, atoms (atomic vapor) can also be obtained by chemical reaction. The ions in solution (Hg^{2+}) can be efficiently reduced to Hg, which is a vapor at room temperature. Similarly, other elements with high electronegativities (i.e., electronegativities close to that of hydrogen) can be efficiently converted into vapors by a reduction reaction similar to that used for mercury. This is done with the so-called covalent hydride-forming elements (As, Bi, Pb, Sb, Se, Sn, and Te), which are converted into gaseous hydrides at room temperature. These hydrides are then heated (electrically in most cases) and decomposed into an atomic vapor. In general, both CV and HG are included in a more global concept—chemical vapor generation (CVG)—which embraces not only classical reduction reactions but also other reactions that generate volatile compounds. Following CVG, measurements are mostly carried out by ICP-OES (cold vapor-inductively coupled plasma-optical emission spectrometry, CV-ICP-OES; hydride generation-inductively coupled plasma-optical emission spectrometry, HG-ICP-OES; or chemical vapor generation-inductively coupled plasma-optical emission spectrometry, CVG-ICP-OES), but also by atomic absorption spectrometry (AAS) (cold vapor-atomic absorption spectrometry, CV-AAS; hydride generation-atomic absorption spectrometry, HG-AAS; or chemical vapor generation-atomic absorption spectrometry, CVG-AAS), atomic fluorescence spectrometry (AFS) (cold vapor-atomic fluorescence spectrometry, CV-AFS; hydride generation-atomic fluorescence spectrometry, HG-AFS; or chemical vapor generation-atomic fluorescence spectrometry, CVG-AFS), atomic emission spectrometry (AES), or ICP-MS (cold vapor-inductively coupled plasma-mass spectrometry, CV-ICP-MS; hydride generation-inductively coupled plasma-mass spectrometry; HG-ICP-MS; or chemical vapor generation-inductively coupled plasma-mass spectrometry, CVG-ICP-MS).

Atomization sources can also be classified according to the technique of introducing the sample into the atomizer. Sample introduction is continuous when the sample is aspirated during fixed flow and noncontinuous when a discrete volume of sample is introduced into the atomizer. The first system (flame- and plasma-based atomizers) supplies a constant atomic signal; in the second one (electrothermal atomizers) the atomic signal reaches a maximum value and then drops to zero.

12.2.4.1 Flame Atomic Absorption Spectrometry

Because of its inherent simplicity and low capital cost, FAAS is one of the most commonly used atomic techniques in the analytical laboratory for the determination of inorganic compounds in waters. A more detailed description of FAAS instrumentation will be found elsewhere.[7]

12.2.4.1.1 Instrumentation

12.2.4.1.1.1 Source of Radiation The radiation source for FAAS instrumentation is quite similar to that of other AAS techniques, such as ETAAS or CVG-AAS (CV-AAS and HG-AAS). The one most commonly applied is the line source (LS), which generates a characteristic narrow-line emission of a selected element. There are two principal LSs for AAS: the hollow cathode lamp (HCL) and the electrodeless discharge lamp (EDL).[8]

An HCL consists of a tungsten rod (anode) and a hollow cylindrical cathode lined with the target element. Both electrodes are contained within a glass envelope filled with an inert gas (Ar or Ne) at low pressures (1–5 Torr). The characteristic radiation is obtained when a potential of approximately 300–500 V is applied between the two electrodes, which causes the inert gas contained in the lamp to ionize, and the ions (cations) and electrons to migrate to the cathode and anode, respectively. On application of a high voltage, the bombardment of the cations on the inner surface of the cathode causes metal atoms to sputter out of the cathode cup, producing an atomic vapor. Further collisions excite these metal atoms, which, on returning to the ground state, produce an intense and characteristic radiation.

A typical EDL consists of a hermetically sealed quartz envelope containing an inert gas (Ar) at very low pressure and the element or salt of the target element. In order to ionize the inert gas, microwave radiation (approximately 100 MHz) or, as is usually the case, radio frequency (RF) radiation (from 100 kHz to 100 MHz) is applied. Commercially available RF EDLs have a built-in starter, run at 27 MHz, which provides a high voltage spark to ionize the filler gas to initiate the discharge.

Although most of the radiation sources for AAS are LSs, the great advances in detector technology, especially the development of solid-state array detectors and charge-coupled devices (CCDs), have led to the successful application of continuous sources (CSs) for AAS. A modern CS is based on a conventional xenon short-arc lamp that has been optimized to run in the so-called hot-spot mode.[9] This discharge mode requires the appearance of a small plasma spot close to the cathode surface. In order to obtain this plasma, new CSs have a short interelectrode distance and operate at high xenon pressure. In the same way, new materials and geometries for both anode and cathode rods have been carefully optimized.

12.2.4.1.1.2   Flame Atomizer   The energy supplied by a flame for AAS is used both to atomize the sample and to maintain the atomic vapor within the light path of the spectrometer. In this sense the flame acts as an atomization cell in which the atoms interact with the radiation from the lamp or in which the excited atoms deactivate by emitting the characteristic radiation. The premixed laminar flame, in which the fuel and the oxidant gases are mixed in an expansion chamber prior to entering the burner, is the most commonly used atomization cell.[7,8] There are two different flames depending on the nature of the oxidant gas: the air–acetylene flame and the nitrous oxide–acetylene flame. The air–acetylene flame (with a slot length of 100 mm) is the one most commonly used and reaches temperatures from 2000°C to 2500°C. But such temperatures are insufficient for atomizing refractory elements or those forming thermally stable oxides or carbides. In such cases, the nitrous oxide–acetylene flame (with a slot length of 50 mm), which attains temperatures of approximately 3000°C, is preferred.

12.2.4.1.1.3   Nebulizers   The liquid sample is introduced into the flame as a fine aerosol, which is generated by a nebulizer/expansion chamber arrangement. Pneumatic nebulizers, especially the pneumatic concentric nebulizer, are the most widely used in FAAS instruments. Once the aerosol is generated, it passes through the expansion (spray) chamber, where large droplets collect on its walls and drain away. A number of baffles are placed in the spray chamber to ensure that only the smallest droplets reach the flame. The efficiency of aerosol generation with the combination of a pneumatic nebulizer and an expansion chamber is around 10–15%; but the sensitivity of this arrangement is poor. In order to obtain a more concentrated aerosol, an impact bead may be placed in the path of the initial aerosol inside the spray chamber.[8] The primary aerosol generated by the nebulizer impacts on the bead and secondary fragmentation takes place; this increases the number of fine drops that can reach the flame and improves the efficiency of nebulization. The impact bead is normally made of glass, but there are other designs based on ceramics.

The development of new, highly efficient nebulizers, described in detail in the section on ICP-OES (Section 12.2.4.4.1), has meant that a more concentrated aerosol and a more sensitive FAAS determination is achievable. Similarly, the use of slotted tube atom traps (STATs) and water-cooled atom traps (WCAT)[10,11]—the latter have undergone modification in recent years[12]—enhances sensitivity with regard to volatile elements like Cd and Pb because of the long residence time of these atoms in the tube.

12.2.4.1.1.4   Background Correction   Analyte absorption can be affected by background absorption. Possibly nonspecific, this latter type of absorption is attributed to light scattering on particulates formed by recombination of the sample matrix at cold spots. Background absorption can also be a broad molecular absorption signal caused by radicals or molecules vaporized in the atomizer. Therefore, in order to obtain the net absorbance of the analyte atoms, the incident radiation scattered or absorbed must be subtracted from the total measured absorbance. This operation is known as background correction, and there are different ways of carrying it out.

The background correction system usually used is based on the nonspecific continuous radiation emitted by an additional source; for example, a deuterium arc lamp or a hydrogen arc lamp for correction below 400 nm, or a tungsten halogen lamp for the VIS region.[7] The level of atomic absorption due to the deuterium lamp is negligible, but the level of nonspecific absorption is the same. Therefore, the signal recorded with the LSs can be subtracted from that recorded with the CS, and the background absorption can be removed. This method for background correction is inexpensive, but background signals larger than 0.5 units cannot be compensated for. In addition, there are other drawbacks associated with the relatively high noise and overcorrection, which occurs when emission lines from other elements in the sample lie close to the characteristic emission line of the target element.

Other background correction systems include the Zeeman effect and the Smith–Hieftje background correction. A detailed description of the operational principles of these methods is beyond the scope of this chapter and the required information can be found in the relevant literature.[7,13] The advantages of these methods over deuterium lamps are that high background signals (up to 2.0 units) and structured backgrounds can easily be corrected for.

12.2.4.2 Electrothermal Atomic Absorption Spectrometry

The atomic absorption spectrometer for ETAAS is the same as that used for FAAS/FAES except that a graphite furnace atomizer replaces the flame/burner arrangement in the light path of the spectrometer. The graphite furnace consists of a graphite tube with graphite electrodes clamped at either end and held axially in line with the light source. The furnace is heated by a low voltage (10 V) and high current (up to 500 A) via water-cooled contacts at each end and the whole atomizer is purged with an inert gas (Ar or N_2).

12.2.4.2.1 Instrumentation

12.2.4.2.1.1 Atomizer   Sample introduction in ETAAS is discrete, with small volumes of between 5 and 100 μL being placed on the inner surface of a graphite tube through a small opening, often with the aid of an autosampler. The discrete volume can also be deposited directly on to the inner walls of the tube (wall atomization) or on to a small platform (L'vov platform) located within the graphite tube (platform atomization). A conventional graphite tube for longitudinal heating consists of cylinders 3–5 cm in length and 3–8 mm in diameter. As the electrical current is applied at the ends of the tube, there is a temperature gradient along the graphite tube: the central portion is several hundred degrees hotter than the ends (nonisothermal conditions). This can lead to condensation of the analyte or recombination with other species at the cooler ends of the tube. As L'vov himself suggested, the sample can be deposited on a small graphite platform (L'vov platform) only loosely connected to the tube walls.[14] Thus, although the graphite tube is directly heated by the electric current, the platform is heated by radiation and convection from the tube walls. There is therefore a time lag between the heating of the tube and that of the platform, and atomization occurs only when the surrounding gas is relatively hot and the whole operation is taking place isothermally.

In addition to the L'vov platform technology, the tube in some modern instruments is heated transversely, that is, from the sides. These atomizers are commonly known as transversal heated graphite atomizers (THGAs). This heating method prevents a temperature gradient from occurring along the tube and offers other advantages over longitudinal heating, such as more efficient atomization, less tailing of the absorption signal, and a decrease in the memory effect.

The use of isothermal operation (L'vov platform and/or transversal heating) is one of the characteristics of the stabilized temperature platform furnace (STPF) concept, which is the basis of most ETAAS applications.[15] These recommendations include (1) isothermal operation; (2) the use of a matrix modifier; (3) measurement of an integrated absorbance signal rather than peak height; (4) rapid heating during atomization (maximum power heating); (5) fast electronics to follow the transient signal; and (6) the use of the Zeeman effect background correction system.

Apart from atomization in the isothermal operation mode, there is another characteristic derived from the STPF concept that is worth commenting on, namely, matrix modification.

12.2.4.2.1.2 Matrix Modification   Matrix modification, also called chemical modification,[16] can be defined as a process aiming to separate the analyte from the matrix, therefore facilitating interference-free determinations. This process consists of the addition of a reagent (modifier) or a combination of reagents, which react with the analyte or with the matrix, thus permitting selective volatilization and, consequently, the separation of analyte from the matrix at some point of the graphite furnace temperature program. A matrix or chemical modifier acts in two main ways. Firstly, it removes concomitants (matrix) by reducing matrix volatility. In this case, the chemical modifier reacts with the matrix and the product of this reaction is more volatile, so it may be lost at low temperatures (i.e., during the charring step). Classical examples of this behavior are exhibited by compounds such as ammonium nitrate and nitric acid, which volatilize chloride ions by forming volatile ammonium chloride, or by gas matrix modifiers such as oxygen or synthetic air, which are used to combust a sample with a high organic matrix content.

The second mode of action of a modifier is direct reaction with the analyte to convert it into a phase with greater thermal stability, that is, to reduce analyte volatility. In this way, the charring stage can be carried out at higher temperatures, allowing a more efficient removal of the matrix but without the loss of analyte. Examples of this type of matrix modifier include transition metal ions (mainly Pd), which form thermally stable intermetallic compounds with analytes, and magnesium nitrate, which thermally decomposes to magnesium oxide, and in the process traps analyte atoms in its crystalline matrix; it is thermally stable until 1100°C. In fact, the most frequently reported mixture for matrix modification consists of $Pd(NO_3)_2$ and $Mg(NO_3)_2$, proposed by Schlemmer and Welz as a universal chemical modifier.[17]

Some chemical modifiers behave in other ways—they decrease analyte volatility or concomitant volatility—so that concomitants (matrix) are volatilized during the cleaning step. Examples of the first behavior are certain organic acids, such as ascorbic or citric acids, which react with volatile elements, thereby diminishing their volatilities. An example of the second type of behavior is the use of ammonium molybdate, which reacts with phosphate ions to form the highly refractory ammonium molybdophosphate.

The use of organic reducing agents in combination with Pd has extended the usefulness of Pd as a chemical modifier for volatile elements.[18] The formation of thermally stable intermetallic compounds with Pd is temperature-dependent and occurs at temperatures close to 1000°C. At these temperatures volatile elements may be lost before Pd can react with them. The combination of Pd and a reducing agent guarantees the reduction of Pd at an early stage of the temperature program; for determining volatile elements this mixture is extremely convenient.[18]

Finally, the use of permanent chemical modifiers must be mentioned. Such chemical modification involves coating the graphite tube with a noble metal such as Ir, Pt, W, or Zr. These modifiers behave in much the same way as aqueous Pd in that thermally stable intermetallic compounds are formed on the hot inner surfaces of the coated graphite tube.

12.2.4.3   High-Resolution Continuous Source Atomic Absorption Spectrometry

New developments in solid-state array detectors and CCDs, as well as powerful, specially designed echelle spectrometers and improvements in CSs, have led to a fresh concept for AAS, which allows the simultaneous determination of several elements based on atomic absorption measurements.[9]

12.2.4.3.1   Instrumentation

There is a commercially available instrument for HR-CS AAS in which a flame, a graphite furnace, or a CVG system are used to carry out atomization. The instrument has a double monochromator with a prism premonochromator and a high resolution echelle monochromator, which allows a wavelength from 189 to 900 nm to be used in a sequential measurement mode.[19]

The echelle monochromator and the prism are arranged in similar Littrow mountings coupled with two small folding mirrors. The double-pass mode in the Littrow prism increases the angular dispersion for minimum prism size and the autocollimation mode of the 76° echelle grating results in maximum dispersion and resolving power. The radiation from the CS is focused through the atomizer on to the spectrometer entrance slit with the aid of two elliptical mirrors. The collimated beam having been refracted by the prism, a small segment of the low-dispersed continuum spectrum passes through the entrance slit of the high-resolution echelle monochromator. The rotation of the prism ensures that the selected spectral interval, within the analytical line of interest and its neighborhood, is transmitted.

A detector for HR-CS AAS must have a large dynamic range, which means that most of the application area of the detector must be free of shot-noise. This parameter is calculated by the ratio between the saturation capacity and the square of the readout noise. Detectors with large dynamic ranges are needed because analyte measurements as well as background absorption arrive at the detector simultaneously. CCDs are therefore the most favored CS AAS detectors because of their saturation capacities between 600,000 and 800,000 electrons per pixel and readout noises of 5–30 electrons. This leads to a shot-noise-limited dynamic range of 600–800, the results of which are adequate for AAS requirements. The CCD in commercially available instruments has 576 pixels, which make it possible to make highly resolved and truly simultaneous observations of a spectral range of almost 1 nm.

12.2.4.4   Atomic Emission Spectrometry

When samples are atomized at very high temperatures, a significant proportion of atoms or ions may be excited by the absorption of thermal energy. On returning to the ground state, these atoms or ions emit radiation of the element-characteristic wavelength, the intensity of such emission being proportional to the concentration of atoms. Although many types of excitation sources are available, flames and plasmas are nowadays the most important ones for AES; they are the basis of FAES and (ICP-OES). Formally, the instrumentation for FAES is similar to that for FAAS, except for the external radiation source, which is not required. The instrumentation for ICP will be discussed in the following section.

12.2.4.5   Inductively Coupled Plasma-Optical Emission Spectrometry

Because of its versatility and productivity, ICP-OES is one of the most useful techniques in instrumental element analysis. The multielement determination capacity of this technique enables it to deal with the basic workload in many routine laboratories. Complete information on all aspects relating to ICP-OES can be found in a few monographs.[20–22]

12.2.4.5.1   Instrumentation

The basis of ICP is plasma, a very hot ionized gas, sometimes referred to as the fourth state of aggregation. This energetic plasma is sustained in a quartz torch placed in a RF oscillating magnetic field. The very high temperature in the plasma causes the complete breakdown of the sample, and the atoms and ions are excited to emit their characteristic radiations. Argon is usually chosen as the plasma gas because of its inertness, optical transparency to the UV-VIS spectrum, moderately low thermal conductivity, and high first ionization energy.[8]

12.2.4.5.1.1   Plasma Torch   The torch consists of three concentric cylinders: the outer tube is the channel for the coolant gas (outer, coolant, or plasma gas), the intermediate one is the channel for the auxiliary gas, and the inner one the channel for the sample carrier gas (nebulizer gas). The plasma gas flows tangentially into the torch along the outer tube until shortly before it reaches the plasma. The gas cools the torch but also maintains the plasma; flow rates between 10 and 20 L min^{-1} are typical. The intermediate tube is approximately 16 mm in diameter and serves to force the coolant gas to flow tangentially along the outer tube; it also enables another gas (the auxiliary gas) to be introduced. Typical auxiliary gas flow rates are between 0 and 2 L min^{-1}. After nebulization, the sample is introduced as an aerosol through the inner tube, also called the injector. This tube is usually made of quartz or aluminum oxide. Nebulizer flows are normally low (from 0.6 to 1.0 L min^{-1}). Modern ICP-OES instruments allow both radial and axial viewing (the dual view concept); in them, the plasma lies horizontally and the optics is set up for axial viewing.

12.2.4.5.1.2   RF Generators   In order to ignite the plasma a high frequency electrical field is applied. Two types of designs for RF generators can be used: crystal-controlled and free-running generators. In the former, an oscillator circuit incorporating a crystal oscillating at a fixed frequency is responsible for inducing the electrical field. In the latter, changes in power loading are compensated by slight shifts in the frequency of the oscillation circuit in order to bring the whole circuit back into resonance.

12.2.4.5.1.3 Nebulizers and Spray Chambers The nebulizer converts the sample liquid into an aerosol. Unlike FAAS/FAES, where solution uptake is by free aspiration, the solution to be nebulized in ICP is usually moved by a peristaltic pump.

There are three different types of nebulizers: pneumatic nebulizers, glass frit (fritted disc) nebulizers, and ultrasonic nebulizers.[23] The pneumatic concentric nebulizer, commonly used in FAAS/FAES, and the cross-flow and Babington nebulizers are the most commonly reported pneumatic nebulizers in ICP-OES. Low-flow nebulizers and micronebulizers deliver a higher mass of analyte to the plasma but have low limits of detection.[24,25] There are new developments in low-flow nebulizers, such as the parallel path nebulizer, and especially the dual nebulizer sample introduction system or multimode sample introduction system,[26] which permits direct pneumatic nebulization and/or introduction of vapors (hydrides). Direct injection nebulizers (Vulkan direct nebulizers and direct injection high efficiency nebulizers, DIHN) are becoming common, because they transport nearly 100% of the sample to the plasma, thereby increasing sensitivity.[8] However, DIHNs require very low-flow rates of sample uptake, typically <100 µL min^{-1}; with samples containing suspended particulate matter; however, the quality of results may be below the desired level, because this matter can block the instrument.

An aerosol generated by nebulization is directed through a spray chamber (nebulizer chamber). This is usually made of glass, quartz, or inert polymers (Ryton or several fluorine-based polymers), which prevents large aerosol droplets from reaching the plasma. The classical Scott chamber design has been superseded by the cyclonic chamber, which has a 50% better sensitivity.

12.2.4.5.1.4 Optics and Detectors Optical components can be arranged in two principal types of instruments: those that measure all wavelengths at the same time (simultaneous spectrometers), and those in which one wavelength is measured after another (sequential spectrometers). The classical optical mounts based on the Paschen–Runge or Czerny–Turner mounts, which used to be the basis of many simultaneous and some sequential spectrometers, have now been superseded by the echelle mount. In such a mount very good resolution is obtained with a mechanically ruled grating, which typically has only 50–100 grooves per mm. Both a prism and a grating are used as cross-dispersing media, the former to sort the orders of the VIS range, and the latter to perform this task for the UV range. In such configurations, modern ICP-OES instruments use solid-state detectors, mainly charge injection devices (CIDs) and CCDs.

12.2.4.6 Atomic Fluorescence Spectrometry

AFS is based on the absorption of radiation of a certain frequency (the energy transition from the outermost electronic orbitals to a higher energy state) and the subsequent deactivation of the excited atoms with the release of radiation. The most useful type of fluorescence, resonance fluorescence, involves a fluorescence emission radiation of the same wavelength as that used for excitation. Because of the inherent sensitivity of the fluorescence emission process, AFS is one of the most sensitive atomic techniques. All the benefits of AFS are enhanced when this spectrometric technique is used in combination with vapor generation methods, especially for covalent-hydride-forming elements.

12.2.4.6.1 Instrumentation

12.2.4.6.1.1 Source of Radiation As the sensitivity in AFS is directly proportional to the source intensity, intense LSs are needed. Typical HCLs are insufficient to guarantee a high intensity of excitation radiation; the previously described EDLs, however, can do so. High-intensity hollow cathode lamps (HI-HCL), first designed by Sullivan and Walsh, are commonly used in modern atomic fluorescence spectrometers.[27] These are based on HCL, but the atomization and excitation processes, which in earlier lamps occurred over the same cathode, are now kept separate by the use of an additional cathode. In the operation of these lamps, a low current (between 10 and 50 mA) is passed to atomize the element deposited on the cathode; then, a high current (200–500 mA) is supplied by the auxiliary cathode to excite the atoms. With atomization and excitation now taking place over two separate electrodes, the intensity of the lamp is increased by a factor of 20–100. Ongoing developments of HI-HCL have been based on this initial design. Lasers likewise provide high-intensity monochromatic radiation to saturate atomic transition. For laser-excited AFS (LEAFS) or laser-induced atomic fluorescence spectrometry (LIF), the laser must be capable of generating wavelengths throughout the VIS and UV regions in order to excite as many elements as possible. In addition, since atomic fluorescence is mainly resonance fluorescence, the use of an ICP, fed with a solution containing elements at high concentrations, provides an alternative quasi-monochromatic radiation source. CSs such as high-pressure xenon-arc lamps can also be used as AFS sources because of the high intensity of the radiation they emit. The use of these lamps eliminates the need to change the source for each element, but the downside is that there are problems with scattering.

12.2.4.6.1.2 Atomizers Flames have usually been used as atomizers for AFS, although plasmas and electrothermal atomizers have been also suggested.[28] Typical air–acetylene and nitrous oxide–acetylene flames, as well as hydrogen–argon–air flames, have been used as atomizers for AFS. However, there have been problems with high background signals; attempted solutions of this problem have included the separation of the flame with a quartz tube (torch), or with a shear inert gas (argon), which prevents the entrance of air. Despite the use of these approaches, however, all such flames have a high background emission. Flames in which an inert gas (argon or nitrogen) is burned in hydrogen have low background emissions; but as the temperature of these flames is also low, they are unsuitable for work involving refractory elements. However, the temperature in these flames, commonly known as diffusion flames, is quite sufficient to atomize hydrides, so they are normally used in AFS instruments with HG systems for sample introduction.

12.2.5 Mass Spectrometric Techniques

Mass spectrometric techniques are based on the measurement or counting of ions produced at high temperatures. An ion can be identified on the basis of its mass-to-charge ratio (m/z), characteristic of a certain isotope. In addition, quantification is based on the dependence between the number of ions and the concentration of a given isotope in the sample. Mass spectrometers consist of an ion source, a mass analyzer, and an ion detector. The ion source is typically the basis for the different types of mass spectrometric techniques. Plasmas are the most common ion sources for Mass spectrometric elemental determinations, and it is mass spectrometry (MS) using this ion source that will now be described. Complete details of this technique can be found in published monographs.[29,30]

12.2.5.1 Inductively Coupled Plasma-Mass Spectrometry

12.2.5.1.1 Instrumentation

Besides the sample introduction system (a peristaltic pump as in ICP-OES), the basic components for ICP-MS are a horizontally configured (axial configuration) argon plasma torch, which allows the plasma gases to be sampled by the ion analyzer via a differential pumping unit, and the ion analyzer itself, which is a quadrupole/magnetic sector mass spectrometer. The interface (ion sampling interface) between the plasma and the mass analyzer consists of a series of differentially pumped vacuum chambers held at consecutively lower pressures. This interface is needed to extract ions from the hot plasma, which is at atmospheric pressure, into a mass spectrometer at very low pressure (approximately 10^{-9} atm). The plasma is therefore aligned axially with the tip of a water-cooled, nickel or copper sampling cone with a narrow orifice (1 mm); behind this sampling cone the pressure is reduced to 2×10^{-3} atm by means of a vacuum pump. Plasma gases and analyte ions pass through another metal cone aligned axially with the sampling cone (orifice diameter 0.7 mm—the skimmer cone) to an intermediate vacuum chamber held at a pressure of $<10^{-7}$ atm. At this instant, the ions form a beam that can be focused on to the entrance of the mass analyzer by means of a series of ion lenses.

12.2.5.1.1.1　Mass Analyzers　Once in the mass analyzer, the ions are separated in accordance with their different mass-to-charge ratios (*m/z*). There are three types of mass analyzers: quadrupole, double-focusing magnetic sector, and time of flight (TOF). Quadrupoles consist of four cylindrical metal rods arranged in parallel and operating as electrodes. A combination of RF and direct current (DC) voltages are applied to each pair of opposite rods across a temperature gap of 180°C. Depending on the RF/DC ratio, the electric field allows ions to pass in a certain *m/z* range. By varying the RF/DC ratio the quadrupole scans the *m/z* range and allows ions of consecutively higher *m/z* ratio to pass to the detector.

The double-focusing magnetic sector mass analyzer consists of two devices to focus the ion beam: an electric sector analyzer and a magnetic sector analyzer. Ions from the source are accelerated through the entrance slit of the electric sector, which acts as an energy filter. After a narrow band of ions with certain kinetic energies has been selected, the ion beam is focused on to the magnetic sector, where the ions are deflected in accordance with the *m/z* ratio (a high degree of deflection for ions with high *m/z* ratios). In the same way as for quadrupoles, a mass spectrum is obtained by scanning the magnetic field and allowing ions of consecutively higher *m/z* ratio to pass the exit slit of the magnetic sector in the direction of the detector.

TOF mass analyzers are based on bombardment by a pulse of electrons or photons to periodically produce positive ions. The pulses have frequencies between 10 and 50 kHz. The generated ions are then accelerated by an electric sector (voltages from 10^3 to 10^4 V) at the same frequency as the ionizing bombardment but with a certain gap. The accelerated ions pass to a 1 m long analyzer rod, which is not subjected to an electrical or magnetic field. As all the ions have the same kinetic energy, their velocities along the analyzer rod must be inversely proportional to the *m/z* ratio. In this way, those ions with lower *m/z* ratios reach the detector first. The times to reach the detector (the TOF) are between 1 and 30 μs.

12.2.5.1.1.2　Ion Detectors　The channel electron multiplier is the usual ion detector for MS. This consists of a curved glass tube (1 mm i.d.), the inner surface of which is coated with a resistive material; the end of the tube is flared. When the instrument is operating in the pulse counting mode (the most sensitive mode), ions are attracted into the funnel opening by a high applied voltage (up to −3500 V). When these ions collide with the inner coating, a significant number of secondary electrons are ejected from the resistive surface and accelerated down the tube. They then collide with the inner walls of the tube and cause further electrons to be ejected from that surface. As a result, an exponential cascade of electrons is produced, which eventually reaches saturation point and results in a large electron pulse (a gain of 10^7–10^8 over the original collision). The second operation mode is the analogous mode; this works in a similar way to the pulse counting mode, but lower voltages (between −500 and −1500 V) are applied and the multiplier does not become saturated. The pulses therefore vary in size, and there is a gain of only 10^3–10^4.

ICP-MS is an attractive technique for multielement determinations. In addition, it allows isotope ratio measurements[31] and isotope dilution analysis[30] to be carried out. The concepts related to these approaches are, however, beyond the scope of this chapter; the reader will find full details in the literature.[30,31]

12.2.6　Chemical Vapor Generation-Atomic Spectrometry (CVG-AS)

As mentioned at the beginning of this chapter, besides thermal energy to produce atoms, there are chemical methods for obtaining atomic or molecular vapors that are readily atomizable. The atoms generated from these vapors interact with electromagnetic radiation, and the resulting atomic absorption or fluorescence phenomena can be monitored; or else they can be excited and their typical emission radiation used for measurement purposes. In this context, Dědina and Tsalev[32] have written an excellent monograph on HG and AAS.

Besides mercury CV and covalent hydrides obtained by chemical reduction, there are other chemical routes that lead to CVG, for example, alkylation, halide generation, metal carbonyl generation, and chelate or oxide generation.[33] The classical, CV/HG techniques are thus now known as CVG techniques—this concept includes all possible chemical reactions for generating a chemical vapor (atomic or molecular). According to Sturgeon et al.,[33] there has been a resurgence of interest in CVG because the number of elements that can be determined following CVG has increased in recent years. Therefore, to the well-known CVG of mercury, arsenic, selenium, antimony, lead, bismuth, tin, germanium, and tellurium, new elements have been added to the chemical vapor "family," among them cadmium, chromium, manganese, nickel, copper, silver, and gold.[34,35] In addition, fresh developments involving new chelating agents such as thiourea and 8-hydroxyquinoline, UV-assisted CVG approaches, and trapping strategies like solid-phase microextraction (SPME) or single droplet microextraction will ensure that there will be alternatives to the classical CV/HG methodologies.

Current CV and HG procedures are based on a reduction reaction using sodium borohydride.[36] The decomposition of this reagent at pH < 1.0 is complete within a few microseconds, and the resulting nascent hydrogen reduces analytes to hydrides. The Marsh reaction (most often Zn/HCl), originally proposed for covalent hydrides, and the reaction with tin (II) chloride to obtain mercury CV have been superseded by sodium borohydride/HCl.[37] One of the drawbacks of this chemical reduction is that the reduction reaction and the vapor released depend on the valence state of the element. The hydrides of arsenic and antimony respond less vigorously when these elements are in their highest valence state (+5), but the efficiency of HG is close to 100% for the +3 state. Similarly, generating the hydrides of selenium and tellurium is more efficient when these elements are in the +4 state. In such cases, the elements have to be prereduced, usually with potassium iodide, potassium bromide, or L-cysteine; alternatively, the HCl concentration is increased, or heat is applied. For lead, HG efficiency is poor when this element is in the +2 state (the most common valence state of lead). Lead therefore has to be oxidized to the +4 state with hydrogen peroxide prior to reduction; this leads to a substantial improvement in HG.[8] When an element is present in the form of different organic species (e.g., arsenic as monomethyl arsenic acid, dimethyl arsenic acid; antimony as trimethyl antimony), the HG process must guarantee that all species are converted into the corresponding hydride. This can be achieved by acid digestion of the water sample or by using reaction solutions, such as thioglycolic acid for arsenic, in which the HG efficiency is independent of the chemical form.

Electrochemical hydride generation (EcHG) deserves attention. In addition to the first batch approaches, new EcHG developments based on continuous flow (CF) or flow injection (FI) systems have also been proposed. An electrochemical reduction of hydride-forming elements in solution is achieved in special electrolytic cells, avoiding the use of a reducing agent (sodium borohydride). Lower blanks and lower limits of detection are inherent in EcHG.[32,38]

12.2.6.1 Instrumentation

CVG (CV/HG) methods are based on direct transfer modes, mainly continuous mode—CF or FI—and batch mode.[32] If prereduction or oxidation steps are needed, auxiliary lines can be inserted into flow modes to perform these stages automatically before chemical reduction. CF modes deliver the sample and acid to the mixing coil separately, although in many cases the sample can be acidified off-line. FI systems use an injection valve to inject a discrete volume of sample into the carrier (acid) flow. The acidified sample (CF) or the carrier with the injected sample (FI) merges with the reducing solution flow upstream of the reaction coil, where hydride formation takes place. A purging gas (argon) is introduced either upstream or downstream of the reaction coil. Argon introduction upstream is more convenient because of the stripping effect of the purging gas on the hydride vapor released. In batch mode, the acidified sample is transported to a stirred glass cell containing the reducing solution or to a plastic vessel without stirring but with an entrance line through which the reducing solution is pumped. After the chemical reduction, the liberated vapor is flushed with the inert gas into the atomizer.

Regardless of the chemical vapor transfer mode, CVG uses a gas–liquid separator to separate the chemical vapor from the liquid reagents prior to its introduction into the atomizer. There are several designs of gas–liquid separators, but they can be classified into three basic types: hydrostatic separators, forced outlet separators, and membrane separators. A detailed description of gas–liquid separators will be found in specialized monographs.[32]

12.2.6.1.1   Atomizers

The vapor can be atomized in inert gas-hydrogen diffusion flames, in narrow-bore quartz tubes electrically heated or heated over an air–acetylene flame, and in plasmas. Additionally, the atomizer can act as a vapor preconcentration medium just before atomizing. This is what happens in graphite furnace atomizers (in situ trapping) or on silver or gold wires for direct amalgamation of mercury.

Diffusion flames are often argon–hydrogen flames, which are preferable when using atomic fluorescence detectors. These atomizers have several disadvantages, however, such as vapor dilution into the flame gases and background absorption of the flame. Heated quartz tube atomizers are normally T-shaped tubes with the crossbar-tube aligned in the optical path of the spectrometer. These tubes employ either an electrical resistance device or an air–acetylene flame for heating the crossbar-tube of the atomizer; they were the first type to become commercially available. Vapors have also been introduced to plasma-based instrumentation. The sensitivity of the vapor-forming elements increases markedly in comparison with conventional nebulization because of the improved efficiency of transport (close to 100%).

Since the first application of a graphite furnace for hydride atomization in 1980, electrothermal atomization has been used for both preconcentration and atomization of trapped vapors.[39] The *in situ* trapping approach uses commercial graphite furnaces as the trapping medium and trapping temperatures from room temperature to around 600°C. Coated graphite tubes, mainly iridium-coated graphite tubes, have been shown to be excellent trapping media for covalent hydrides. The interface used to transfer the vapor from the gas–liquid separator to the graphite furnace is a quartz capillary inserted into the autosampler.[40] In modern instrumentation, the capillary is inserted into the graphite furnace when vapor generation starts; on its completion, the capillary is automatically removed from the graphite furnace and atomization can commence. In direct amalgamation, mercury is collected on a silver or gold wire, from which it is released by heating. Both the *in situ* trapping graphite furnace and gold amalgamation enable very sensitive determinations of vapor-forming elements.

12.2.7   Electrochemical Techniques

12.2.7.1   Anodic Stripping Voltammetry

Voltammetry involves microelectrolytic techniques in which the working electrode potential is forced by external instrumentation to follow a known potential–time function, and the resultant current–potential and current–time curves are analyzed to obtain information about the solution composition. They operate under conditions in which the polarization of the working electrode is at a maximum, which maximizes the variation of the intensity with potential. The working electrode is an ideally polarizable electrode (i.e., the electrode exhibits a large change in potential when an infinitesimally small current is passed through it), such as the dropping mercury electrode (DME). Its surface area is kept as small as possible in order to maintain a high concentration of active species in the adjacent area, and therefore a high level of polarization. A wide variety of voltammetry techniques are available: stripping voltammetry and the voltammetry oxygen sensor are the ones widely applied in the determination of trace metals and dissolved oxygen, respectively. Detailed information on voltammetry methods is given in the monographs by Bard and Faulkner,[41] Wang,[42] and Brainina and Neyman.[43]

Stripping voltammetry is based on the electrochemical deposition or accumulation of analyte at a voltammmetric electrode (anode or cathode) of constant surface during controlled-potential electrolysis. After a certain delay (30 s), the analyte is stripped or dissolved from the electrode by a voltammetric technique or chemical reaction. To give an example, in the first step, the increase in potential causes copper to be electrochemically deposited on the cathode by the reaction: $Cu^{2+} + 2e^- \rightarrow Cu$; in the second step, when the potential has fallen sufficiently, the deposited copper redissolves as a result of the reverse reaction: $Cu \rightarrow Cu^{2+} + 2e^-$. A plot of current intensity versus potential gives a peak. If several species are present, there is a peak for each one on the voltammogram (current–potential curve). The potential for redissolving the deposited metal can be used for qualitative purposes, the peak height or area for quantification. ASV is employed for the determination of amalgamate-forming metals (metallic ions) and CSV for determining species that form water-insoluble salts with mercury on the electrode surface.

12.2.7.1.1 Instrumentation

The instrument consists of three electrodes: (1) a working electrode that should be easily polarizable, usually a microelectrode based on mercury DME; (2) an auxiliary electrode, usually a platinum wire or a droplet of mercury; (3) a reference electrode, usually a calomel electrode (Hg/Hg_2Cl_2 and KCl). The DME is a micro working electrode, on which the mercury droplet is formed at the end of a glass capillary (length 10–20 cm; i.d. 0.05 mm). The mercury droplets are of a highly reproducible diameter and have a lifetime from 2 to 6 s. The advantages of the DME are the following: (1) the constant renewal of the electrode surface; (2) the charge-transfer overvoltage of the hydrogen ions present in the aqueous solvent is high on an Hg drop surface; and (3) Hg form amalgams with many metals, thereby lowering their reduction potential. To carry out the measurements, an excess of a strong electrolyte (KCl) is added to the sample; the ionic strength and conductivity thus remain constant. Because oxygen is easily reduced, the dissolved oxygen must be removed: this is done by bubbling an inert gas (N_2) for a few minutes and then keeping a stream of nitrogen on the surface to prevent oxygen redissolution.

12.2.7.1.1.1 Voltammetry Sensor: Voltammetry Oxygen Electrode (Clark Electrode)   The Clark electrode is a voltammetry sensor capable of measuring the current generated in a redox reaction (a reaction in which dissolved oxygen is involved), which is proportional to the concentration of the dissolved oxygen. This device allows the determination of dissolved oxygen in aqueous samples (fresh water, sea water, blood, sewage, and industrial effluents). It is fitted with a membrane separating the working electrode from the sample solution, thus allowing oxygen to migrate through the membrane. The Clark electrode consists of a platinum disc cathode maintained at a potential of approximately −0.6 V with respect to the annular silver anode surrounding it. A thin (approximately 20 μm) gas-permeable membrane [made from polytetrafluoroethylene (PTFE) or polyethylene] is held in tension across the end of this assemblage such that there is a thin film (~10 μm) between the membrane and the anode and cathode immersed in a buffered KCl electrolyte solution. When the sensor is immersed in water containing dissolved oxygen, molecular oxygen diffuses through the membrane and the internal electrolyte film, and the following electrode processes occur:

Cathode reaction: $O_2 + 4H_3O^+ + 4e^- \leftrightarrow 6H_2O$
Anode reaction: $Ag + Cl^- \rightarrow AgCl(s) + e^-$

The diffusion current that flows between the electrodes is proportional to the oxygen concentration.

12.2.7.2 Potentiometric Sensors: ISEs and Gas-Permeable Membrane Sensors

Potentiometric sensors are based on a membrane that separates the sample solution of a reference solution contained within the electrode. The membranes are permeable to particular types of ions (ISEs) or gases (gas-permeable membrane sensors). These electrodes generate a potential that is proportional to the concentration of a single analyte. This proportionality is expressed by an equation similar to the Nernst equation. Thus, several dissolved ions or certain dissolved gases (CO_2, NO_2, SO_2, NH_3, H_2S, HCN, and HF) can be measured with high selectivity and sensitivity. These devices are commonly employed in monitoring studies. Detailed information on ISEs can be found in the monographs by Bard and Faulkner[41] and Evans.[44]

12.2.7.2.1 Instrumentation

Potentiometric sensors consist of two electrodes: one is the indicator or membrane electrode (the ion-specific electrode surrounded by a thin film of an intermediate electrolyte solution and enclosed in an ion-permeable membrane), and the other is the reference electrode. Both electrodes are immersed in the sample solution. The indicator electrode consists of an internal reference electrode with an internal filling solution and a membrane located at one end. The membrane separates two liquids, the internal filling solution and the sample solution. The liquid that fills the membrane electrode is a solution containing the ion to be analyzed in the sample, and the concentration of which remains constant. The external sample solution also contains those same ions, but their concentration is unknown. The membrane allows these ions (and only these ions) to migrate through the membrane from both sides. The speed of migration depends on the concentrations of such ions on either side of the membrane. This creates a charge imbalance between the two sides, which is reflected by the emergence of a potential. Since the potential due to the reference electrode is constant, signal variations are attributed only to the potential dependent on the ion concentration in the sample solution. All these considerations, along with the implementation of the relevant equations, lead to the conclusion that the potential measured by a potentiometer connected to the two electrodes is the sum of a constant value E_K plus a value depending on the concentration of the ion in the sample solution, according to the following Nernstian expression:

$$E_{measured} = E_k + \left(\frac{RT}{nF}\right)\ln C, \tag{12.1}$$

where C is the analyte concentration, the value of E_K encompasses all potential constants and can be determined after calibration, R and F are constants (8.314 JK^{-1} mol^{-1} and $96,485$ C mol^{-1}, respectively), and T is the temperature expressed in K.

Membranes are made of glass or a glass of a particular compound and certain insoluble polymers filled with certain liquids. Each membrane electrode is designed for the specific measurement of a particular ion. The fluoride (F^-) ISE, for example, contains an Ag/AgCl electrode immersed in a solution of NaCl and NaF and a membrane glass made from LaF_3. ISEs have also been designed for the determination of Cl^-, Br^-, I^-, CN^-, SCN^-, NO_3^-, S^{2-}, Ag^+, Cd^{2+}, Ca^{2+}, Pb^{2+}, and so on. Table 12.1

TABLE 12.1
ISE Glass Membranes

Target Analyte	Membrane
I^-	LaF_3
Cl^-	AgCl
Br^-	AgBr
I^-	AgI
CN^-	AgI
SCN^-	AgSCN
Ag^+/S^{2-}	Ag_2S
Cu^{2+}	Ag_2S (CuS)
Cd^{2+}	Ag_2S (CdS)
Pb^{2+}	Ag_2S (PbS)

TABLE 12.2
Types of Potentiometric Dissolved Gas Sensors Based on ISEs

Target Analyte	Equilibrium in the Internal Solution	Sensing Electrode
CO_2	$CO_2 + 2H_2O \leftrightarrow HCO_3^- + H_3O^+$	pH glass
NO_2	$2NO_2 + 3H_2O \leftrightarrow NO_2^- + NO_3^- + 2H_3O^+$	pH glass
SO_2	$SO_3 + 2H_2O \leftrightarrow HSO_3^- + H_3O^+$	pH glass
NH_3	$NH_3 + H_2O \leftrightarrow NH_4^+ + OH^-$	pH glass
H_2S	$H_2S + 2H_2O \leftrightarrow S^{2-} + 2H_3O^+$	Ag_2S (S^{2-}) ISE
HCN	$HCN + H_2O \leftrightarrow CN^- + H_3O^+$	Ag_2S (Ag^+) ISE
HF	$HF + H_2O \leftrightarrow F^- + H_3O^+$	LaF_3 ISE

lists the membranes used in some ISE electrodes. Finally, protons (H_3O^+) can be detected by this technique, which permits pH measurements. There are also potentiometric sensors that detect molecules, especially gases and certain biochemical molecules.

Gas-permeable membrane sensors usually incorporate a conventional ISE surrounded by a thin film of an intermediate electrolyte solution and enclosed in a gas-permeable membrane. An internal reference electrode is usually included; so the sensor represents a complete electrochemical cell. Gas-permeable membranes are usually made of polymeric materials and are from 10 to 100 μm thick. Highly selective, gas-permeable membranes allow the gas to migrate through the membrane and then to react with the reagent contained within. The reaction usually generates or consumes protons, the numbers of protons generated or consumed being proportional to the concentration of dissolved gas. There are applications for gases such as CO_2, NO_2, NH_3, HCN, HF, H_2S, and SO_3 in water. Gas-permeable membrane sensors are remarkably selective and sensitive devices; the most commonly used ones are shown in Table 12.2.

12.2.8 SEPARATION TECHNIQUES

IC and CE are the most commonly applied separation techniques for the determination of inorganic ions.

12.2.8.1 Ion Chromatography

IC is a liquid chromatography subclass in which analyte separation is based on ion-exchange mechanisms. Analytes interact with both the stationary phase (ion-exchange resin) and the mobile phase. The full details of liquid chromatography, especially IC, will be found in the literature.[45,46]

IC is based on the ion-exchange equilibrium between the ions in solution and those on the surface of a solid—an ion-exchange resin (the stationary phase). The resins are polymers (typically, a styrene-divinylbenzene copolymer) that contain chemical groups capable of capturing an anion or cation from the solution and at the same time releasing OH^- or H^+ ions, respectively. The ion-exchange resin can be either a strong or a weak cation/anion exchanger. The exchange capacity of the resin also depends on the number of ionic groups that it carries. There are cation exchange resins (containing sulfonic groups, $-SO_3^-$ H^+ or carboxylic groups, $-COO^-$ H^+) and anion exchange resins (containing $-NR_3^+$ OH^- groups, where R = H or CH_3). These resins are deposited on nonporous microparticles of glass or polymers, or on porous silica. When an ion-exchange resin with sulfonic acid groups is in contact with an aqueous solution containing cations, M^{x+}, the ion-exchange equilibria can be described as follows:

$$x RSO_3^- \ H^+ \text{ (solid)} + M^{x+} \text{ (solution)} \leftrightarrow (RSO_3^-)_x \ M^{x+} \text{ (solid)} + x H^+ \text{ (solution)},$$

where $RSO_3^- H^+$ represents one of the many sulfonic acid groups bonded to the polymer. In a similar way, an anion exchange resin interacts with anions, A^{x-}, according to the following equilibrium:

$$xRN(CH_3)_3^+ \cdot OH^- (solid) + A^{x-}(solution) \leftrightarrow [Rh(Ch_3)_{3+}]_x A^{x-} + xOH^-.$$

The mobile phases are usually aqueous solutions that contain a competing ion carrying the same charge as the analytes to be separated. These competing ions displace the analyte ion from the stationary phase and finally elute the analytes from the column. During the separation, the analyte ions compete with the ions in the eluent for the charged sites on the stationary phase. The separation is based on an adsorption–desorption process between the analyte and the ionic groups of the stationary phase. The most crucial parameters for the separation are the nature of the resin, the pH, and the ionic strength of the eluent.

12.2.8.1.1 Instrumentation

Since IC is a type of high-performance liquid chromatography (HPLC) in which the stationary phase is an ion-exchange resin, an IC instrument is similar to a typical HPLC. It consists of (1) a pump unit; (2) solvent reservoirs; (3) an injector; (4) a chromatographic column; and (5) a detector. The principle of operation is simple. The pump pushes the eluent (mobile phase) through the column at a certain flow rate. For injecting the sample, the eluent passes through the injector and transfers the sample to the column. The sample components are separated in the column, each of which is then successively detected by the detector. The pumps (usually reciprocating pumps and displacement pumps) produce pressures up to 400 bar (40 MPa) covering a large range of flow rates (0.1–10 mL min^{-1}) with a high degree of precision and no pulsation. Solvents should be filtered before use to remove suspended particles. Similarly, dissolved gases must also be removed by out-gassing with helium or nitrogen or by processing in ultrasonic baths. The solvents used as the mobile phase are stored in glass or stainless steel reservoirs. The instrument enables operation with a constant solvent composition (isocratic elution) or with a variable solvent composition (gradient elution).

The sample injection system allows the injection of sample volumes from 1 to 500 μL. To prevent depressurization of the system, samples are injected through special six-way valves to which a sample loop has been attached. The sample is injected directly into the loop by means of a microliter syringe (filling position) while the eluent flows to the column. The eluent flow is then directed via the sample loop to the column when the valve is switched to the injection position.

The column length varies from 3 to 30 cm, the inner diameter from 1 to 10 mm. Columns are made of very resistant materials to withstand high pressures (<40 MPa): Most commonly, a stainless steel or heavy-walled glass tube is inserted into a metal tube. Columns are packed with a solid material of particle size <10 μm (spherical and nonporous microglass beads or polymer particles). A guard column is usually placed before the column to remove suspended particles from solvents, as well as certain sample constituents that could irreversibly bind to the stationary phase.

Conductivity and photometric detectors (detection of ionic conductivity or the color of the mobile phase, respectively) are used as IC detectors. The low sensitivity is the main disadvantage of the conductivity detector, because the conductivity of the mobile phase (buffered solution) is strong and tends to mask that due to analytes. This detection problem has been solved by coupling a suppressor column or an exchange membrane to the analytical column. These devices convert mobile phase ions (e.g., carbonate, bicarbonate, and hydroxide) into their weakly conducting conjugate acids, thereby decreasing the conductivity. Since the analyte ions do not react in the suppressor, their conductivity can be detected at very low levels. After a certain time, the suppressor column has to be regenerated—this is the main shortcoming of this approach. The suppressor column is unnecessary if a direct or indirect photometric detector is used. Indirect photometric detection uses a UV-VIS absorbing substance that provides a constant background absorption. Ions eluted from the column displace the absorbing substance and are measured indirectly. Direct photometric detection uses a complex-forming agent that is mixed with the eluent. This remains colorless during the flow of the pure mobile phase. However, metal-complex formation takes place when metal ions are present in the eluent, and the intense color of such complexes can be distinguished from the colorless background.

12.2.8.2 Capillary Electrophoresis

Separations by electrophoresis are based on the differential migration of solutes in an electrical field. In CE, electrophoresis is performed in narrow-bore capillaries, typically of 25–75 μm i.d., which are usually filled with a buffer solution. There are several advantages accruing from the use of capillaries, related mainly to the detrimental effects of Joule heating. The high electrical resistance of the capillary enables the application of very powerful electrical fields (100–500 Vcm^{-1}) with only minimal heat generation. The use of strong electrical fields results in short analysis times together with high efficiency and resolution. In addition, several separation modes with different separation mechanisms and selectivities are available, sample volume requirements are minimal (1–10 nL), and in-capillary detection is possible. It may also be possible to apply CE to quantitative analysis, and to automate the technique. Detailed descriptions of CE and its principal applications can be found in the monographs by Morteza[47] and Shintani and Polonsky.[48]

Separation by electrophoresis is based on solute velocity differences in an electrical field. The velocity of an ion, v, can be given by $v = \mu_e E$, where additionally μ_e is the electrophoretic mobility and E is the applied electrical field. The electrical field is a function of the applied voltage and the capillary length. For a given ion and medium, the mobility is a constant characteristic of that ion. It is determined by the electric force exerted on the ion, balanced by its frictional drag through the medium, that is, $\mu_e \propto F_E/F_F$, where F_E is the electrical force and F_F is the frictional force. The electrical force is given by $F_E = qE$ and the frictional force (for a spherical ion) by $F_F = -6\pi\eta rv$, where q is the ionic charge, η is the viscosity of the solution, r is the ionic radius, and v is the ionic velocity. During electrophoresis, a steady state defined by the balance of these forces is attained. At this point the forces are equal but opposite, and $qE = 6\pi\eta rv$. Since $v = \mu_e E$, the mobility can be described by physical parameters such as $\mu_e = q/6\pi\eta r$. From this last equation it is evident that the mobilities of small, highly charged species are high, whereas those of large, minimally charged species are low.

Another fundamental aspect of CE separation is electro-osmotic flow (EOF). EOF is the bulk flow of liquid in the capillary and is a consequence of the surface charge on the inner capillary walls. EOF results from the effect of the applied electric field on the solution double-layer at the wall. It governs the amount of time that solutes remain in the capillary by the superposition of flow on to soluble mobility. This can have the effect of altering the required capillary length, but it does not affect selectivity. The magnitude of EOF can be expressed in terms of velocity or mobility by $V_{EOF} = (\varepsilon\zeta/\eta) E$ or $\mu_{EOF} = (\varepsilon\zeta/\eta)$, where V_{EOF} is the velocity, μ_{EOF} is the EOF "mobility," ε is a dielectric constant, and ζ is the zeta potential. The zeta potential is determined essentially by the surface charge on the capillary wall. Since this charge is strongly pH dependent, the magnitude of EOF varies with the pH. The zeta potential is also dependent on the ionic strength of the buffer solution, as described by the double-layer theory. A unique feature of EOF in the capillary is the flat profile of the flow. Since the driving force of the flow is uniformly distributed along the capillary (i.e., at the walls), there is no pressure drop within the capillary, and the flow is nearly uniform throughout. This is in contrast to the pressure generated by an external pump, which yields a laminar or parabolic flow owing to the shear force at the wall.

12.2.8.2.1 Instrumentation

One key feature of CE is the overall simplicity of the instrumentation. Briefly, the ends of a narrow-bore, fused silica capillary (25–75 μm i.d., 350–400 μm o.d., and 10–100 cm in length) are placed in buffer reservoirs. The content of the reservoirs is identical to that within the capillary. The reservoirs also contain the electrodes used to make electrical contact between the high voltage power supply and capillary. The sample is loaded into the capillary as follows: one of the reservoirs (usually at the anode) is replaced by the sample reservoir and either an electric field (electrokinetic injection) or external pressure (hydrodynamic injection) is applied. When the buffer reservoir has been replaced, the electric field is applied and the separation performed. Optical detection (UV-VIS and fluorescence absorption) can be done at the opposite end, directly through the capillary wall.

The versatility of CE is derived partially from its numerous modes of operation, that is, capillary zone electrophoresis (CZE), micellar electrokinetic chromatography (MEKC), capillary gel electrophoresis (CGE), capillary isoelectric focusing (CIEF), and capillary isotachophoresis (CITP). Of these, CZE is used most often for charged ions or molecules because of its great simplicity and versatility.

12.2.9 AUTOMATIC ANALYZERS AND MONITORING

One of the major advances in analytical chemistry in recent decades has been the emergence on the market of automated systems for analysis (automatic analyzers), which provide analytical data with minimal operator intervention. Automation implies the partial or complete replacement of human involvement in an operation or sequence of operations. The monographs by Valcárcel and Luque de Castro[49] offer detailed descriptions of automatic analyzers.

Initially, these systems were designed to address the needs of clinical laboratories, but at present they are used in such diverse areas as industrial process control (process analyzers) or routine determinations of various substances in the air, water, and soil (environmental monitoring). Automatic analyzers can be classified according to the way in which samples are transported and manipulated:

1. Discrete or batch analyzers: In these, each sample preserves its integrity in a vessel transported mechanically to various zones of the analyzer, where the different analytical stages are carried out in a sequential manner. Each sample finally arrives at the detector where the relevant signals are recorded.
2. Continuous analyzers: They are characterized by the use of a continuous stream of liquid. The samples are sequentially introduced at regular intervals into a channel carrying a liquid that does or does not merge with other channels carrying reagents, buffers, masking agents, and so on, upon reaching the detector. The final reaction mixture yields an analytical signal that is duly recorded. There are two types of continuous analyzers for segmented-flow analysis (SFA) and unsegmented-flow analysis. SFA are characterized by a segmented flow of air bubbles, the aim of which is to preserve the integrity of samples. Unsegmented-flow analysis, mainly flow injection analysis (FIA) and continuous flow analysis (CFA), provide a flow that is not segmented by air bubbles.

Batch analyzers can be classified according to the manner in which the sample is transferred to the final operation, that is, with or without final transfer of the sample. In analyzers with final transfer, the reaction mixture is transported to the detection system, where the analytical measurement is carried out in a fixed cuvette. In analyzers without final transfer of the reaction mixture, all the stages of the process take place in the same vessel.

In SFA, the bubbles prevent contamination between successive samples, reduce sample dispersion, and facilitate mixing of sample and reagents in a way that enables physical and chemical equilibrium to be attained before the sample reaches the detector. The main elements of these analyzers are as follows:

1. A sampling system: Sampling is carried out with the aid of a moving aspirating tip; air bubbles are also aspirated between samples.
2. A propelling system: This is a peristaltic pump used to propel fluids along continuous segmented systems—flexible plastic tubing is squeezed by a series of rollers, which starts the flow of the enclosed liquids as a result. These systems can work with several streams, and the flow rate is determined by the internal diameter of the tube and the pump rotation speed.

3. A reaction system: This consists of helically coiled tubes of a given diameter that are connected to the other elements of the system.
4. A separation unit: Devices that permit continuous separation techniques, for example, dialyzers, filters, adsorbent columns, and exchange resins.
5. A debubbler: A high-precision device that removes the bubbles. This device is not essential if certain software treatments that discriminate signal parasites caused by bubbles are used.
6. A detection system: This consists of a flow cell located in the optical or electroanalytical instrument. Single-channel (for the determination of one analyte) or multichannel configuration (for several analytes) can be used.

In FIA, the sample and reagents are incompletely mixed and there is a concentration gradient that varies along the system as a function of time. The continuous detectors provide a transient signal, which is recorded. Neither physical equilibrium (homogenization of a portion of the flow) nor chemical equilibrium (reaction completeness) is attained by the time the signal is detected. Hence, FIA can be considered a fixed-time analytical methodology. The main elements of these analyzers are

1. A propelling system: A peristaltic pump
2. An injection system: A six-way rotary valve (three inlets and three outlets) that can adopt two positions—in the filling position the sample fills the loop; in the injecting position, the carrier sweeps the sample toward the reactor)
3. A transport and reaction system: Small tubes of between 0.1 and 2 mm i.d
4. A detector system: Optical or electroanalytical instruments

CFA is similar to FIA. The main difference between CFA and FIA is the long analysis time required to stabilize the hydrodynamic conditions. In addition, other differences are the use of large i.d. tubes, faster flow rates, and special requirements for the rinsing stages.

Many of the automatic analyzers described have been adapted to automatic water analysis (off-line and on-line water analyzers) to facilitate the analysis and monitoring of the inorganic constituents of water (Chapter 18). On-line water analyzers can claim a number of advantages over off-line analyzers: sampling is more representative, the risk of contamination by containers or changes in sample composition during storage is minimal, and the evolution of the system under study can be continuously monitored. A great number of instruments are commercially available for monitoring parameters (multiparameter analyzers) such as pH, pCl, redox potential, conductivity, dissolved oxygen, temperature, cyanide, sulfide, ammonia, nitrites, nitrates, salinity, metals, and so on. In these automated analyzers, the sample is injected or merged into a carrier stream that subsequently mixes with one or more reagent streams. The mixture solution can be detected by a variety of flow-through detection techniques, such as spectrophotometry, fluorescence, atomic absorption, potentiometry, and voltammetry.

12.3   DETERMINATION OF INORGANIC CONSTITUENTS

A great number of different standard and nonstandard analytical methods are available for the determination of inorganic constituents in water. Since the concentrations of some inorganic constituents are relatively high in water, classical methods (gravimetry and titration) were mostly used in early experiments. These methods, however, have been largely replaced, chiefly by faster, more sensitive, and more sophisticated instrumental methods.

Tables 12.3 through 12.8 summarize different standard or official methods for the determination of inorganic constituents. Table 12.3 shows the main procedures for nutrient determinations; it can be seen that UV-VIS spectrophotometry[50–52,58–60] and ISE[50,53,54,56] are the most commonly used analytical techniques.

TABLE 12.3
Selected Standard and Official Methods for the Determination of Nutrients

Analyte	Analytical Technique	Details of Method	Analytical Characteristics	Reference
NO_3^-	UV	Direct measurement at 220 nm. Additional reagents not required	Applicable to unpolluted samples with a low content of organic matter. Samples should be filtered to avoid interference from suspended particles or colloidal matter and organic compounds. Several ions (Br^-, SCN^-, I^-, CO_3^{2-}, NO_2^-, Fe^{2+}, Fe^{3+}, and $Cr(VI)$) may be important interferents	50
	VIS	Nitrate is reduced to nitrite (by cadmium reduction reaction), which is then determined by diazotization (pH = 2.0–2.5) with sulfanilamide and coupling N-(1-naphthyl)-1,2-ethylendiamine hydrochloride to form an intensely pink colored azo dye (540 nm)	Metals at high concentrations (around mg L^{-1}) diminish reduction efficiency. Residual Cl_2 interferes through cadmium oxidation	50
	VIS	Brucine oxidation by nitrate ion in H_2SO_4 at 100°C to form a yellow compound (cacoteline), which is measured at 410 nm	Applicable to 0.1–2 mg L^{-1} in surface, saline, domestic, and waste water. Strong oxidizing and reducing agents interfere ($NaAsO_3$ in addition to residual Cl_2 removal). Effect of Fe^{2+}, Fe^{3+}, and Mn^{4+} is negligible at <1 mg L^{-1}	51
	ISE	Direct potentiometric determination by using a NO_3^-–ISE	Applicable to 10^{-5}–10^{-1} M. Interferents (Cl^-, Br^-, I^-, S^2, CN^-, NO_2^-, and organic acids) are minimized by using Ag_2SO_4 buffer, sulfamic acid, and $Al_2(SO_4)_3$	50
NO_2^-	VIS	Nitrite reacts with sulfanilamide to form a diazo compound; this couples with N-(1-naphthyl)-1,2-ethylendiamine hydrochloride in an acidic medium to form an intensely pink colored azo dye (540 nm)	Applicable up to 0.25 mg L^{-1} in surface, tap, and waste water. High alkalinity, Cl^-, Cl_2, $S_2O_3^{2-}$, PO_4^{3-}, and Fe^{3+} may be important interferents	50,52
NH_4^+	Acid–base titration	NH_4^+ is distilled after alkalinization. Titration with standardized 0.01 M H_2SO_4 and a mixed indicator (methylene blue and methylene red)	Applicable to 1.0–25 mg L^{-1} in surface, saline, domestic, and waste water. Volatile alkaline compounds interfere	50,51,53–55
	VIS	NH_4^+ is distilled from water after alkalinization. Ammonium reacts with Nessler's reagent (I_2Hg—2IK) to form a yellow-brown colored complex (410–425 nm)	Applicable to 0.05–1 mg L^{-1} in surface, saline, domestic, and waste water.	50,51

continued

TABLE 12.3 (continued)

Analyte	Analytical Technique	Details of Method	Analytical Characteristics	Reference
	ISE		Applicable to 0.03–1400 mg L^{-1} in surface, saline, domestic, and waste water	50,53,54
	ISE	Direct potentiometric determination by using an ammonia-sensing membrane probe	Applicable to ammonium nitrogen concentration of up to 50 mg L^{-1} in raw and waste water and sewage	56
Kjeldahl nitrogen	Acid–base titration and VIS	Kjeldahl nitrogen is the sum of nitrogenous organic compounds, free ammonia, and ammonium. In the presence of H_2SO_4, K_2SO_4, and selenium catalyzer; organic nitrogen is converted to ammonium. Free ammonia is also converted to ammonium. After the addition of a base, the ammonia is distilled from an alkaline medium and absorbed in boric acid. The ammonia may be determined colorimetrically at 655 nm or by titration with a standard mineral acid	Applicable up to 10 mg in tap, natural, and waste water. Nitrate and nitrite may interfere	51,57
Total dissolved nitrogen	VIS	Alkaline peroxodisulfate oxidation, all N (free ammonia, ammonium, nitrate, nitrite, and nitrogenous organic compounds) is converted into nitrate. Nitrate is reduced to nitrite (by cadmium reduction reaction). Nitrite reacts with sulfanilamide and N-(1-naphthyl)-1,2-ethylendiamine hydrochloride to form an azo dye, which is measured at 540 nm	Applicable to 0.02–5 mg L^{-1} in tap, surface, natural, sea, and waste water	50,58
PO_4^{3-}	VIS	Reaction of PO_4^{3-} with acidified molybdate reagent (in the presence of potassium antimonyl tartrate) to yield $H_3PO_4(MoO_3)_{12}$, which is reduced by ascorbic acid to the intensely blue-colored compound phosphomolybdenum blue and measured at 840 nm	Applicable to 0.01–0.5 mg L^{-1} in surface and saline waters. High concentrations of Cu, Fe, silicate, and arsenate do not interfere	50,51,59
Si	Gravimetry	This involves a combined process of evaporation, baking, and ignition. Organic matter is removed by ashing, followed by digestion with hydrochloric acid, leaving a residue of insoluble silica	This procedure gives the total silicon content in water samples; only approximate results are obtainable	51

continued

TABLE 12.3 **(continued)**

Analyte	Analytical Technique	Details of Method	Analytical Characteristics	Reference
	VIS	Reaction with molybdate to form yellow silicomolybdic heteropolyacid ($H_4SiMo_{12}O_{40}$), which is then reduced (by sodium metabisulfite and 1-amino-2-naphthol-4-sulfonic acid) to colored silicomolybdenum blue and measured at 810 nm	Applicable to >1.0 mg L^{-1} in natural and waste water. This procedure gives the soluble silicate content after filtration. Oxalic acid is added to minimize interference from phosphate	50,60

Ammonium and Kjeldahl nitrogen are determined by acid–base titration[50,51,53–55,57] and silica by gravimetry.[51] IC (Table 12.4) is also used for the simultaneous determination of nutrients and major ions, halides, cyanide, and sulfur compounds.[50,51,54,61–66]

Table 12.5 lists other standard methods for the determination of major ions: carbonate and bicarbonate (alkalinity)[50,51,67, 68] and calcium and magnesium (hardness).[50,51,70–72]

Table 12.6 sets out other standard methods for the determination of nonmetallic substances: chloride is determined by titration,[50,51,73] VIS spectrophotometry,[50,74] and ISE;[51] chlorine by titration and VIS spectrophotometry;[75–77] fluoride by VIS spectrophotometry[50] and ISE;[50,78,79] iodide by VIS spectrophotomtery;[50] cyanide by titration, VIS spectrophotometry, and ISE;[50,80,81] sulfate by gravimetry[50,51] and turbidimetry;[50] sulfite by titration and VIS spectrophotometry;[51] and sulfur by titration and VIS spectrophotometry.[50,51]

TABLE 12.4
Selected Standard and Official Methods Using IC for the Determination of Nutrients, Major Ions, Halides, Cyanide, and Sulfur Compounds

Analyte	Analytical Technique	Details of Method	Analytical Characteristics	Reference
Br^-, Cl^-, F^-, NO_2^-, NO_3^-, PO_4^{3-}, SO_4^{2-}	IC	Anions are separated using a guard column, separator column, and suppressor device; they are measured using a conductivity detector. UV and an amperometric detector are also used	Applicable to tap, rain, surface, ground, and waste water. Carboxylic acids (malonic, maleic, malic, formic, and acetic acid) at high concentrations and carbonate may be important interferents	51,54,61–63
Li^+, Na^+, NH_4^+, K^+, Mn^{2+}, Ca^{2+}, Mg^{2+}, Sr^{2+}, Ba^{2+}			Applicable to tap, surface, and waste water. Amino acid and aliphatic amines may be important interferents	64
CrO_4^{2-}, I^-, SO_3^-, SCN^-, $S_2O_3^{2-}$			Mono- and dicarboxylic acids and sulfate may be important interferents	65
ClO_3^-, Cl^-, $EClO_2^-$			Mono- and dicarboxylic acids, F^-, Br^- Cl^-, CO_3^{2-}, NO_2, and NO_3^- may be important interferents	66
Anions	CE	Anions are separated using CE and quantified by indirect UV detection		50

TABLE 12.5
Other Standard Methods for the Determination of Major Ions

Analyte	Analytical Technique	Details of Method	Analytical Characteristics	Reference
Alkalinity: CO_3^{2-}, HCO_3^-	Acid–base titration	Titrating with standardized 0.01 M H_2SO_4 using phenolphthalein as indicator until the pink color disappears at about pH 8.3; titration of the same sample is then continued until the end point is reached at pH 4.5 using a bromocresol green-methyl red mixed indicator (total alkalinity)	Applicable to natural and waste water. Concentrations of bicarbonate, carbonate, and hydroxide can be calculated. The titration end points can be measured potentiometrically	50,51,67–69
Ca^{2+}	EDTA titration	Titrating with standardized 10 mM EDTA at pH 12–13 using HSN as indicator	Applicable to 2–100 mg L^{-1} in tap, surface, ground, and waste water. Interferences from Al, Ba, Co, Cu, Fe, Mn, Pb, Sb, and Zn can be removed by the addition of cyanide and triethanolamine	50,70
Hardness: Ca^{2+}, Mg^{2+}	EDTA titration	Titrating with standardized 0.01 M EDTA at pH 10 using eriochrome black T or calmagite as indicators	Al, Ba, Cd, Co, Cu, Fe, Mn, Ni, Pb, Sr, and Zn interferents can be removed by the addition of NaCN or Na_2S	50,51,71
Ca^{2+}, Mg^{2+}	FAAS	Ca and Mg are measured by FAAS at 422.7 nm and 258.2 nm, respectively, using air/C_2H_2 or NO_2/C_2H_2 flames	Applicable to 3–50 mg Ca L^{-1} and 0.9–5 mg Mg L^{-1} (interferences due to ionization are removed mainly by the use of CsCl or $LaCl_3$)	72

TABLE 12.6
Other Selected Methods for the Determination of Non-Metallic Substances

Analyte	Analytical Technique	Details of Method	Analytical Characteristics	Reference
Chloride	Mercuric nitrate titration	Chloride titration with mercuric ions to form soluble and slightly dissociated $HgCl_2$ (pH 2.3–2.8 and diphenylcarbazone as indicator).	Chromate, Fe^{3+}, and SO_3^{2-} interfere when levels are >10 mg L^{-1}. Sulfite may be removed by the addition of H_2O_2.	50,52
	Argentometric titration	Titration with silver nitrate standardized (0.0141 M) against 0.014 M NaCl using potassium chromate as indicator. The end point is indicated by the formation of the yellowish pink precipitate of Ag_2CrO_4.	Applicable to 0.15–10 mg L^{-1} in clear waters. Br^-, I^-, CN^-, $H_2PO_4^-$, and Fe interfere. Sulfide, sulfite, or thiosulfate interferents can be removed by oxidation with a small amount of H_2O_2.	50,73

continued

TABLE 12.6   (continued)

Analyte	Analytical Technique	Details of Method	Analytical Characteristics	Reference
	VIS	Chloride ions react with mercury (II) thiocyanate to form a sparingly dissociating mercuric chloride complex and liberate a stoichiometrically equivalent amount of thiocyanate ions $(2Cl^- + Hg(SCN)_2 \rightarrow HgCl_2 + 2SCN^-)$; the thiocyanate reacts with iron (III) ions, yielding the intensely red ferric thiocyanate complex $(SCN^- + Fe^{3+} \rightarrow Fe(SCN)^{2+})$, which is determined at 460 nm.	The presence of Br^-, I^-, CN^-, $S_2O_3^{2-}$, S^{2-}, SCN^-, and NO_2^- interferes. The method can be applied in the 0.01—10 mg L^{-1} concentration range.	50,74
	Potentiometric titration	The sample is titrated with standardized silver nitrate (0.0141 M) in a continuously mixed beaker with an Ag electrode, an Ag/AgCl electrode, or a chloride--ISE. The reference can be a glass mercuric sulfate, calomel, or Ag/AgCl electrode.	Applicable to 0.15–10 mg L^{-1} in colored or turbid waters. Iodine and bromide interfere by forming a less soluble precipitate with the silver ions. $[Fe(CN)_6]^{3+}$, CrO_4^{2-}, $Cr_2O_7^{2-}$, and Fe^{3+} interfere.	50
Free chlorine and total chlorine	Titration	Free chlorine is defined as the concentration of residual chlorine in water present as the dissolved gas (Cl_2), hypochlorous acid (HOCl), and/or hypochlorite ion (OCl^-). Free chlorine reacts with N,N-(1-naphthyl)-1,2-ethylendiamine hydrochloride (at pH 6.2–6.5) to form a red compound, which is titrated with a standardized solution of ammonium and ferrous sulfate until the red color of the solution disappears. Total chlorine is the sum of free and combined chlorine (defined as the residual chlorine existing in water in chemical combination with ammonia or organic amines). Total chlorine reacts with N,N-(1-naphthyl)-1,2-ethylendiamine hydrochloride and potassium iodide; the compound formed is titrated as stated above.	Applicable to 0.03–5 mg L^{-1} (expressed as Cl_2) in surface and sea water.	75
	VIS	Free chlorine reacts with N,N-(1-naphthyl)-1,2-ethylendiamine hydrochloride (at pH 6.2–6.5) to form a red compound, which is determined at 510 nm. Total chlorine reacts with N,N-(1-naphthyl)-1,2-ethylendiamine hydrochloride and potassium iodide; the compound formed is determined at 460 nm.	Applicable to 0.03–5 mg L^{-1} (expressed as Cl_2) in surface and sea water.	76

continued

TABLE 12.6   (continued)

Analyte	Analytical Technique	Details of Method	Analytical Characteristics	Reference
Total chlorine	Iodometric titration	Total chlorine reacts with potassium iodide to produce iodine, which is titrated with an excess of a standardized sodium thiosulfate solution. The excess thiosulfate is back-titrated with standardized potassium iodate solution.	Applicable to 0.71–15 mg L^{-1} (expressed as Cl$_2$) in surface water.	50,77
Fluoride	ISE	Direct potentiometric measurement: europium-dosed lanthanum fluoride crystals electrode can be used for direct fluoride measurement. A high concentration of background electrolyte (TISAB) is added to the ionic-strength-adjusting buffer to maintain the pH at 5–7.0.	Applicable to 0.2 mg–20 g L^{-1} in tap, natural, surface, and waste waters. Cyclohexylenedi-aminetetraacetic acid (CDTA) is added to complex interfering cations such as Ca, Mg, Fe, and Al and release free fluoride ions.	50,78,79
	VIS	Fluoride reacts with colored zirconium-SPADNS complex: F$^-$ forms the colorless ZrF$_6^{2-}$ complex; absorbance of the zirconium-SPADNS complex measured at 570 nm decreases as the fluoride concentration increases.	Potential interferences in this method from high alkalinity, aluminum, iron, and phosphate.	50
Iodide	VIS	Iodide is selectively oxidized to iodine by the addition of KHSO$_5$. The liberated iodine selectively oxidizes leuco crystal violet to form crystal violet in neutral medium (pH = 6.5). The color of the dye is measured at 592 nm.	Applicable to 50–6000 µg L^{-1}. Chloride (at concentrations >200 mg L^{-1}) interferes.	50
Cyanide	Argentometric titration	Cyanide is titrated with silver nitrate standardized solution to form a soluble cyanide complex (Ag[Ag(CN)$_2$]). The end point is shown by a silver-sensitive indicator (p-dimethylaminobenzalrhodanine), which reacts with the small excess of Ag ion. The color of the solution turns from yellow to red.	Applicable to water containing <50 mg L^{-1} (<100 mg of total cyanide L^{-1}).	50,80,81
	VIS	Cyanide absorbed in sodium hydroxide (previous distillation) is treated with chloramines-T to produce cyanogen chloride (CNCl). Addition of pyridine-barbituric acid produces a red-blue product with an absorbance maximum at 575–582 nm. An alternative method involves the reaction of CNCl with chloramine-T, pyridine-pyrazolone, and a phosphate buffer to produce a blue product that is determined at 620 nm.	Applicable to water containing <50 mg L^{-1} (<100 mg of total cyanide L^{-1}). Thiocyanate and sulfide interfere. Sulfide is removed by precipitation (as cadmium sulfide) by the addition of cadmium carbonate.	50,80,81

continued

TABLE 12.6 (continued)

Analyte	Analytical Technique	Details of Method	Analytical Characteristics	Reference
	ISE	Direct potentiometric measurement: an AgI membrane electrode with a double junction reference electrode system must be used to quantify CN^-.	Transition metal cations interfere by forming stable complexes. Chloride and bromide also interfere. Interference from sulfide is removed by the addition of 10 M NaOH ISA .	50
Sulfate	Gravimetry	Analysis of precipitated barium sulfate after reaction with barium chloride.	Applicable to >10 mg L^{-1} in natural and waste water. This method is time-consuming and susceptible to interference from suspended particulate matter, silica, nitrate, sulfite, sulfide, and alkali metals.	50,51
	Turbidimetry	Sulfate is precipitated with $BaCl_2$ in dilute HCl with constant stirring (60 s). The turbidity of the suspension formed is related to the sulfate concentration.	Applicable to surface and industrial water and suitable for the detection of sulfate at 1 mg L^{-1} or more. Color, suspended matter, silica, and sulfite interfere. Some suspended matter is removed by filtration. Glycerine and NaCl solution are added to stabilize the slurry and minimize interferences.	50,51
Sulfite	Iodometric titration	Titration with standardized iodide–iodate (0.0125 N). Free iodine is produced when the iodide–iodate reagent reacts with sulfite, and in the presence of a starch indicator, the first excess produces a blue color that signals the end point.	Applicable to natural and waste water. There may be interference from S^{2-}, Fe^{2+}, and thiosulfate that leads to overestimation of sulfite, or from the presence of ions (Cu^{2+}) that catalyze the conversion of sulfite to sulfate.	50
	VIS	The method involves the purging of SO_2 from an acidified sample with nitrogen gas for about 1 h. The SO_2 is trapped in a solution of Fe^{3+} and 1,10-phenanthroline. The Fe^{2+} produced by the reaction with SO_2 is detected at 510 nm.		51
Sulfide	Iodometric titration	Back-titration of iodine with standard sodium thiosulfate. Iodine is added to oxidize sulfide under acidic conditions.	Applicable to >1.0 mg L^{-1} in natural, hot spring, and waste water. Reducing agents (thiosulfite, sulfite, and organic compounds) may interfere.	50,51
	VIS (methylene blue method)	Reaction of sulfide, ferric chloride, and dimethyl-p-phenylenediamine to produce methylene blue, which has λ_{max} at 660 nm.	Applicable to <20 mg L^{-1} in natural and waste water. Ammonium phosphate must be added to mask interferences of residual ferric chloride. The effect of other interferents (reducing agents and thiosulfate) may be important.	50

continued

TABLE 12.6   (continued)

Analyte	Analytical Technique	Details of Method	Analytical Characteristics	Reference
	ISE	Direct potentiometric measurement by using an Ag/AgS/ISE	Oxidation of sulfide by dissolved O_2 must be prevented and ionic strength must be adjusted. An alkaline antioxidant reagent (NaOH, ascorbic acid, and EDTA) is used for this purpose. Dissolved organic matter may interfere with Ag/AgS/ISE measurements.	50

Many of the UV-VIS spectrophotometric methods (shown in Tables 12.3 and 12.6) have been automated by using flow analyzers. Thus, nitrite and nitrate,[50,82] ammonium,[50,83] orthophosphate,[50,84,85] silicates,[50,86] chloride,[50,87] cyanide,[50,88] and sulfate[50,89] are measured by CFA and FIA. Oxygen is measured by iodometric titration[51,90] and electrochemical methods[91] (Table 12.7). Other dissolved gasses (Table 12.2) are measured by ISE-based gas sensors.

Finally, metals (Table 12.8) are determined by UV-VIS spectrophotometry[51,92–97] and atomic spectrometry techniques.[50,51,98–113]

Other nonstandard methods (including CE and ASV) commonly used for determining inorganic constituents in water are summarized in selected monographs.[114–117]

TABLE 12.7
Selected Standard Methods for the Determination of Dissolved Gases

Analyte	Analytical Technique	Details of Method	Analytical Characteristics	Reference
O_2	Iodometric titration	It is based on the addition of Mn^{2+} solution, followed by the addition of a strong alkali to the sample in a glass-stoppered bottle. Dissolved O_2 rapidly oxidizes an equivalent amount of the dispersed divalent manganous hydroxide precipitate to hydroxides of higher valence states. In the presence of iodide ions in an acidic solution, the oxidized manganese reverts to the divalent state, with the liberation of a quantity of iodine equivalent to the original dissolved O_2 content. The iodine is then titrated with a standard solution of thiosulfate. The titration end point can be detected visually with a starch indicator, or by potentiometric techniques. The liberated iodine can be determined colorimetrically.	Applicable to 0.2 mg–20 mg L^{-1}. Oxidizing and reducing species interfere. To minimize interference, several modifications of the iodometric method are given (azide modification, permanganate modification, and the copper sulfate-sulfamic acid flocculation modification).	51,90
	Dissolved oxygen meter	The O_2 diffused through the membrane is reduced at the cathode: $$O_2 + 4H_3O^+ + 4\,e^- \leftrightarrow 6H_2O;$$ the electrons originate from the anodic reaction: $$Ag + Cl^- \rightarrow AgCl(s) + e^-.$$	Suitable for polluted and highly colored waters	91

TABLE 12.8
Selected Standard Methods for the Determination of Metals

Analyte	Analytical Technique	Details of Method	Analytical Characteristics	Reference
Al	VIS	Aluminum reacts with pyrocatechol violet to form a colored complex, which can be determined colorimetrically	Applicable up to 100 μg L^{-1} (50 mm cells) and up to 500 μg L^{-1} (10 mm cells) in potable waters, ground waters, and lightly polluted surface and sea waters	92
As	VIS	Arsenic in the sample is converted to arsine, which is evolved and then complexed with silver diethyldithiocarbamate. The intensity of the color of the complex is measured at 510–525 nm.	Applicable to 0.001–0.1 mg L^{-1} in surface and waste waters. Interference from Sb may be important. Ag, Cr, Co, Mo, Ni, Hg, and Pt at concentrations <5 mg L^{-1} do not interfere.	93
Cr(VI) and total Cr	VIS	Cr(VI) reacts with 1,5-diphenylcarbazide to form a colored complex, which is measured at 540 nm. For total chromium determination, Cr(III) is first oxidized to Cr(VI) by $KMnO_4$.	Applicable to 0.005–0.2 mg L^{-1} in surface and waste waters. Hg, Fe, Mo, and V may interfere.	94
Fe	VIS	Iron (II) reacts with 1,10-phenanthroline to form a colored complex, which can be determined colorimetrically	Specific procedure for water and waste water. The procedures are applicable to concentrations between 0.01 and 5 mg L^{-1}.	95
Cr(VI)	VIS	Cr(VI) reacts with diphenylcarbazide to form a colored complex, which is determined at 540 nm	Applicable to 2.0–50 μg L^{-1} in slightly polluted tap, surface, and ground waters. S^{2-}, Cl_2, ClO_2, O_3, and H_2O_2 may interfere	96
Mn	VIS	Manganese reacts with formaldoxime to form a colored complex, which can be determined colorimetrically	Applicable to concentrations between 0.01 and 5 mg L^{-1} in surface and drinking water	97
Mn	VIS	Mn compounds react with ammonium persulfate/silver nitrate to form permanganate, which is determined at 525 nm	Applicable to 0.05–1.5 mg L^{-1} in natural and waste waters. Cl^{-}, Br^{-}, I^{-}, and organic matter may interfere.	51
Al	FAAS	Al is measured by aspirating the sample directly into the NO_2/C_2H_2 flame and measuring the absorbance at 309.3 nm	Applicable to 5–100 mg L^{-1}. SO_4^{2-}, Cl^{-}, PO_4^{3-}, Na, K, Mg, Ca, Fe, Ni, Co, Cd, Pb, Si, Ti, and F may interfere.	98
Cd	FAAS	Cd is measured by aspirating the sample directly into the air/C_2H_2 flame and measuring the absorbance at 228.8 nm	Applicable to 0.05–1.0 mg L^{-1}. SO_4^{2-}, Cl^{-}, PO_4^{3-}, Na, K, Mg, Ca, Fe, Ni, Co, Cd, Pb, Si, and Ti may interfere.	99

continued

TABLE 12.8 (continued)

Analyte	Analytical Technique	Details of Method	Analytical Characteristics	Reference
Cr	FAAS	Cr is measured by aspirating the sample directly into the NO_2/C_2H_2 flame and measuring the absorbance at 357.9 nm	Applicable to 0.5–20 mg L^{-1}. SO_4^{2-}, Cl$^-$, Na, K, Mg, Ca, Fe, Ni, Co, Al, and Zn may interfere. Interference due to ionization is generally removed by the addition of LaCl$_3$;	100
K	FAAS	K is measured by aspirating the sample directly into the air/C_2H_2 flame and measuring the absorbance at 766.5 nm	Applicable to 5–50 mg L^{-1}. Interference due to ionization is removed by the addition of CsCl	50,101
Na	FAAS	Na is measured by aspirating the sample directly into the air/C_2H_2 flame and measuring the absorbance at 589.0 nm	Applicable to 5–50 mg L^{-1}. Interference due to ionization is removed by the addition of CsCl	51,102
Ag, Au, Bi, Ca, Cd, Co, Cr, Cu, Fe, Ir, K, Li, Mg, Mn, Na, Ni, Pd, Pb, Pt, Rh, Ru, Sb, Sn, Sr, Tl, and Zn	FAAS	Metals are determined by FAAS: the sample is aspirated directly into the air/ C_2H_2 flame. Ag, Cd, Co, Cr, Cu, Fe, Mn, Ni, and Zn at low concentrations must be extracted prior to quantification	Applicable to natural waters	50
Al, Ba, Be, Ca, Mo, Os, Re, Si, Th Ti, and V	FAAS	Metals are determined by FAAS: the sample is aspirated directly into the NO_2/C_2H_2 flame. Al and Be at low concentrations must be extracted prior to quantification.	Applicable to natural waters	50
Cd, Cr, Cu, Fe, Pb, Mg, Mn, Ag, and Zn	FAAS	Metals are determined by FAAS: Pb and Cd at low concentrations are chelated (ammonium pyrrolidine dithiocarbamate at pH = 2.5), preconcentrated, and then extracted with methyl isobutyl ketone prior to quantification	Applicable to surface and saline waters Ionization interferences are generally removed by the addition of LaCl$_3$	51
Co, Ni, Cu, Zn, Cd, and Pb	FAAS	Metals are directly determined by FAAS; metals are determined after chelation with ammonium pyrrolidine dithiocarbamate and extraction in methyl isobutyl ketone; and metals are determined after chelation with hexamethyleneammonium–hexamethylenedithiocarbamate and extraction in di-isopropyl ketone-xylene	Applicable to natural waters	103

Analyte	Technique	Method	Applicability/Interferences	Reference
Na and K	FAES	Na and K are measured by aspirating the sample into a flame of sufficient thermal energy to cause the emission of characteristic radiation and measuring the intensity at 589.0 and 766.5 nm for Na and K, respectively	Applicable to water samples with Na and K concentrations up to 10 mg L^{-1}	104
Al	ETAAS	Al is measured by ETAAS at 309.3 nm using an injection volume of 20 μL	Applicable to 10–100 μg L^{-1}. Fe, Cu, Ni, Co, Cd, Pb, Si, Na, K, Ca, Cl$^-$, SO$_4^{2-}$, PO$_4^{3-}$, and acetate may interfere.	98
Cd	ETAAS	Cd is measured by ETAAS at 228.8 nm using an injection volume of 10 μL	Applicable to 0.3–3.0 μg L^{-1}. Fe, Cu, Ni, Co, Pb, Na, K, Ca, Mg, Cl$^-$, and SO$_4^{2-}$ may interfere.	99
Cr	ETAAS	Cr is measured by ETAAS at 357.9 nm using an injection volume of 20 μL	Applicable to 5.0–100 μg L^{-1}. SO$_4^{2-}$, Cl$^-$, Na, K, Mg, Ca, Fe, Ni, Co, Al, and Zn may interfere.	100
Ag, Al, As, Cd, Co, Cr, Cu, Fe, Mn, Mo, Ni, Pb, Sb, Se, Tl, V, and Zn	ETAAS	Metals are measured by ETAAS using the manufacturer's recommended conditions and an injection volume of 20 μL	Applicable to surface, ground, tap, and waste water. A high Cl$^-$ concentration interferes. To minimize the matrix effect, the chemical modification, standard addition method, and background correction systems may be used.	105
Ag, Al, As, B, Ba, Be, Bi, Ca, Cd, Co, Cr, Cu, Fe, K, Li, Mg, Mn, Mo, Na, Ni, P, Pb, S, Sb, Se, Si, Sn, Sr, Ti, V, W, Zn, and Zr	ICP-OES	Metals are measured by ICP-OES using the manufacturer's recommended conditions	Applicable to tap and waste water	50,106
Ag, Al, As, Au, B, Ba, Be, Bi, Ca, Cd, Ce, Co, Cr, Cs, Cu, Dy, Er, Eu, Fe, Ga, Gd, Ge, Hf, Ho, In, Ir, K, La, Li, Lu, Mg, Mn, Mo, Na, Nd, Ni, P, Pb, Pd, Pr, Pt, Rb, Re, Rh, Ru, S, Sb, Sc, Se, Si, Sm, Sn, Sr, Tb, Te, Th, Tl, Tm, Ti, U, V, W, Y, Yb, Zn, and Zr	ICP-MS	Metals are measured by ICP-MS using the manufacturer's recommended conditions	Applicable to 0.1–1.0 μg L^{-1} in tap, surface, ground, and waste water	50,107,108
Hg	CV-AAS	Organic Hg is oxidized to inorganic Hg by KMnO$_4$, K$_2$S$_2$O$_8$, and heating. The Hg is chemically reduced to the elemental state with stannous chloride or sodium tetrahydroborate; the vapor generated is transferred to the absorption cell and measured at 253.7 nm	Applicable to 0.1–10 μg L^{-1} in surface, ground, and waste waters	51,109

continued

TABLE 12.8   (continued)

Analyte	Analytical Technique	Details of Method	Analytical Characteristics	Reference
Hg	CV-AAS (amalgamation)	Hg is chemically reduced to the elemental state with stannous chloride or sodium tetrahydroborate; the vapor generated is collected on an amalgamation surface/Au or Pt. The concentrated mercury is revolatilized by rapid heating of the amalgamation surface and transferred to the absorption cell for measurement at 253.7 nm	Applicable to 0.1–1.0 μg L^{-1} in surface, ground, and waste waters	110
Hg	CV-AFS	Hg is chemically reduced to the elemental state with stannous chloride; the vapor generated is transferred and measured by AFS	Applicable to 0.001–10 μg L^{-1} in tap, rain, surface, ground, and waste waters	111
As	HG-AAS	As is chemically reduced to arsine with sodium tetrahydroborate; the vapor generated is transferred to the absorption cell and measured at 193.7 nm	Applicable to 1.0–10 μg L^{-1} in tap, surface, and ground waters	50,112
Se	HG-AAS	Se(IV) is chemically reduced to selenium hydride with sodium tetrahydroborate; the vapor generated is transferred to the absorption cell and measured at 196.0 nm	Applicable to selenium and organic selenium in drinking, ground, and surface waters, in a concentration range of 1–10 μg L^{-1}. To avoid errors in determination, other oxidation states need to be converted to Se(IV) prior to the determination	50,113

ACRONYMS AND ABBREVIATIONS

AAS	atomic absorption spectrometry
AES	atomic emission spectrometry
AFS	atomic fluorescence spectrometry
ASV	anodic stripping voltammetry
CCD	charge-coupled device
CE	capillary electrophoresis
CFA	continuous flow analysis
CGE	capillary gel electrophoresis
CID	charge injection device
CIEF	capillary isoelectric focusing
CITP	capillary isotachophoresis
CS	continuous source
CSV	cathodic stripping voltammetry
CV-AAS	cold vapor-atomic absorption spectrometry
CV-AFS	cold vapor-atomic fluorescence spectrometry
CVG	chemical vapor generation
CVG-AAS	chemical vapor generation-atomic absorption spectrometry
CVG-AFS	chemical vapor generation-atomic fluorescence spectrometry
CVG-ICP-MS	chemical vapor generation-inductively coupled plasma-mass spectrometry
CVG-ICP-OES	chemical vapor generation-inductively coupled plasma-optical emission spectrometry
CV-ICP-MS	cold vapor-inductively coupled plasma-mass spectrometry
CV-ICP-OES	cold vapor-inductively coupled plasma-optical emission spectrometry
CZE	capillary zone electrophoresis
DC	direct current
DME	dropping mercury electrode
EcHG	electrochemical hydride generation
EDL	electrodeless discharged lamp
EOF	electro-osmotic flow
ETAAS	electrothermal atomic absorption spectrometry
FAAS	flame atomic absorption spectrometry
FAES	flame atomic emission spectrometry
FIA	flow injection analysis
HCL	hollow cathode lamp
HG-AAS	hydride generation-atomic absorption spectrometry
HG-AFS	hydride generation-atomic fluorescence spectrometry
HG-ICP-MS	hydride generation-inductively coupled plasma-mass spectrometry
HG-ICP-OES	hydride generation-inductively coupled plasma-optical emission spectrometry
HI-HCL	high intensity-hollow cathode lamp
HPLC	high-performance liquid chromatography
HR-CS AAS	high resolution-continuous source atomic absorption spectrometry
IC	ion chromatography
ICP	inductively coupled plasma
ICP-MS	inductively coupled plasma-mass spectrometry
ICP-OES	inductively coupled plasma-optical emission spectrometry
ISE	ion-selective electrode
ISO	International Organization for Standardization
LEAFS	laser-excited atomic fluorescence spectrometry
LIF	laser-induced atomic fluorescence spectrometry

LS	line source
MEKC	micellar electrokinetic chromatography
OES	optical emission spectrometry
PTFE	polytetrafluoroethylene
RF	radio frequency
SFA	segmented-flow analysis
SPME	solid-phase microextraction
STAT	slotted tube atom trap
STPF	stabilized temperature platform furnace
THGA	transversal heated graphite atomizer
TOF	time of flight
UV	ultraviolet
VIS	visible
WCAT	water-cooled atom trap

REFERENCES

1. Millero, F.J. and M.L. Sohn. 1992. *Chemical Oceanography*. Boca Raton, FL: CRC Press.
2. Chapman, D. 1992. *Water Quality Assessments. A Guide to the use of Biota, Sediments and Water in Environmental Monitoring*. London: Chapman & Hall.
3. Skoog, D.A., D.M. West, F.J. Holler, and S.R. Crouch. 2005. *Fundamentals of Analytical Chemistry*, 8th edition. Belmont: Thomson Brooks/Cole.
4. Kellner, R., J.M. Mermet, M. Otto, M. Valcárcel, and H.M. Widmer. 2004. *Analytical Chemistry. A Modern Approach to Analytical Science*. Weinheim, Germany: Wiley-VCH Verlag.
5. Heinz-Helmut, P. 1992. *UV-VIS Spectroscopy and it Applications*. Berlin: Springer.
6. Harris, D.A. and C.L. Bashford. 1988. *Spectrophotometry & Spectrofluorimetry: A Practical Approach*. Oxford: IRL Press.
7. Welz, B. and M. Sperling. 1999. *Atomic Absorption Spectrometry*. Weinheim, Germany: Wiley-VCH Verlag.
8. Ebdon, L., E.H. Evans, A. Fisher, and S.J. Hill. 1998. *An Introduction to Analytical Atomic Spectrometry*. West Sussex, UK: Wiley.
9. Welz, B., H. Becker-Ross, S. Florek, and U. Heitmann. 2005. *High Resolution Continuum Source AAS*. Weinheim, Germany: Wiley-VCH Verlag.
10. Matusiewicz, H. 1997. Atom trapping and in situ preconcentration technique for flame atomic absorption spectrometry. *Spectrochim. Acta B* 52: 1711–1136.
11. Matusiewicz, H. and M. Kopras. 1997. Methods for improving the sensitivity in atom trapping flame atomic absorption spectrometry: Analytical scheme for the direct determination of trace elements in beer. *J. Anal. At. Spectrom.* 12: 1287–1291.
12. Yaman, M. 2005. The improvement of sensitivity in lead and cadmium determination using flame atomic absorption spectrometry. *Anal. Biochem.* 339: 1–8.
13. Smith, S.B. and G.M. Hieftje. 1983. A new background correction method for atomic absorption spectrometry. *Appl. Spectrosc.* 37: 419–424.
14. L'vov, B.V. 1978. Electrothermal atomization—the way toward absolute methods of atomic absorption analysis. *Spectrochim. Acta B* 33: 153–193.
15. Slavin, W., D.C. Manning, and G.R. Carnrick. 1981. The stabilized temperature platform furnace. *At. Spectrosc.* 2: 137–145
16. Ediger, R.D. 1975. Atomic absorption analysis with the graphite furnace using matrix modification. *At. Absorpt. Newsl.* 14: 127–130.
17. Schlemmer, G. and B. Welz. 1986. Palladium and magnesium nitrates, a more universal modifier for graphite furnace atomic absorption spectrometry. *Spectrochim. Acta B* 41: 1157–1165.
18. Voth-Beach, L.M. and D.E. Shrader. 1987. Investigation of a reduced palladium chemical modifier for graphite furnace atomic absorption. *J. Anal. At. Spectrom.* 2: 45–50.
19. Welz, B., D.L. Borges, F.G. Lepri, M.G.R. Vale, and U. Heitmann. 2007. High-resolution continuum source electrothermal atomic absorption spectrometry—an analytical and diagnostic tool for trace analysis. *Spectrochim. Acta B* 62: 873–883.

20. Montaser, A. and D.W. Golightly (eds). 1998. *Inductively Coupled Plasmas in Analytical Atomic Spectrometry*. New York, USA: VCH.
21. Nölte, J. 2003. *ICP Emission Spectrometry. A Practical Guide*. Weinheim, Germany: Wiley-VCH Verlag.
22. Dean, J.R. 2005. *Practical Inductively Coupled Plasma Spectroscopy*. Hoboken, NJ: Wiley.
23. Todolí, J.L. and J.M. Mermet. 2005. Elemental analysis of liquid microsamples through inductively coupled plasma spectrochemistry. *Trends Anal. Chem.* 24: 107–116.
24. Maestre, S.E., J.L. Todolí, and J.M. Mermet. 2004. Evaluation of several pneumatic micronebulizers with different designs for use in ICP-AES and ICP-MS. Future directions for further improvements. *Anal. Bioanal. Chem.* 379: 888–899.
25. Almagro, B., A.M. Ganan-Calvo, M. Hidalgo, and A. Canals. 2006. Flow focusing pneumatic nebulizers in comparison with several micronebulizers in inductively coupled plasma atomic emission spectrometry. *J. Anal. At. Spectrom.* 21: 770–777.
26. McLaughlin, R.L.J. and I.D. Brindle. 2005. *Multimode Sample Introduction System*. United States Patent US 6,891,605 B2, May 10, 2005.
27. Lowe, R.M. 1971. High-intensity hollow-cathode lamp for atomic fluorescence. *Spectrochim. Acta B* 26: 201–205.
28. Broekaert, J.A.C. 2005. *Analytical Atomic Spectrometry with Flames and Plasmas*, 2nd edition. Weinheim, Germany: Wiley-VCH Verlag.
29. Jarvis, K.E., A.L. Gray, and R.S. Houk. 1996. *Handbook of Inductively Coupled Plasma Mass Spectrometry*. London, UK: Blackie Academic & Professional.
30. Hill, S.J. (ed.). 2007. *Inductively Coupled Plasma Spectrometry and its Applications*, 2nd edition. Oxford, UK: Blackwell Publishing Ltd.
31. Sabe, R. and G. Rauret. 2004. Challenges for achieving traceability of analytical measurements of heavy metals in environmental samples by isotopic dilution mass spectrometry. *Trends Anal. Chem.* 23: 273–280.
32. Dědina J. and D.L. Tsalev. 1995. *Hydride Generation Atomic Absorption Spectrometry*. Surrey, UK: Wiley.
33. Sturgeon, R.E., X. Guo, and Z. Mester. 2005. Chemical vapor generation: Are further advances yet possible? *Anal. Bioanal. Chem.* 382: 881–883.
34. Pohl, P. 2004. Recent advances in chemical vapor generation via reaction with sodium tetrahydroborate. *Trends Anal. Chem.* 23: 21–27.
35. Pohl, P. 2004. Hydride generation—recent advances in atomic emission spectrometry. *Trends Anal. Chem.* 23: 87–101.
36. D'Ulivo, A. 2004. Chemical vapor generation by tetrahydroborate (III) and other borane complexes in aqueous media—A critical discussion of fundamental processes and mechanisms involved in reagent decomposition and hydride formation. *Spectrochim. Acta B* 59: 793–825.
37. Narsito, J. Agterdenbos, and S.J. Santosa. 1990. Study of processes in the hydride generation atomic absorption spectrometry of antimony, arsenic and selenium. *Anal. Chim. Acta* 237: 189–199.
38. Laborda, F., E. Bolea, and J.R. Castillo. 2007. Electrochemical hydride generation as a sample introduction technique in atomic spectrometry: Fundamentals, interferences and applications. *Anal. Bioanal. Chem.* 388: 743–775.
39. Tsalev, D.L. 2000. Vapor generation or electrothermal atomic absorption spectrometry? Both! *Spectrochim. Acta B* 55: 917–933.
40. Chaudhry, M.M., A.M. Ure, B.G. Cooksey, D. Littlejohn, and D.J. Halls. 1991. Investigation of in situ concentration of hydride forming elements in a graphite furnace atomizer. *Anal. Proc.* 28: 44–46.
41. Bard, A.J. and L.R. Faulkner. 2001. *Electrochemical Methods. Fundamentals and Applications*. New York: Wiley.
42. Wang J. 1985. *Stripping Analysis. Principles, Instrumentation and Applications*. Weinheim, Germany: VCH.
43. Brainina, Kh. and E. Neyman. 1993. *Electroanalytical Stripping Methods*. New York: Wiley.
44. Evans, A. 1987. *Potentiometry and Ion Selective Electrodes*. New York: Wiley.
45. Lough, W.J. 1995. *High Performance Liquid Chromatography. Principles and Practice*. London: Blackie Academic & Professional.
46. Meyer, V. 1994. *Practical High Performance Liquid Chromatography*. New York: Wiley.
47. Morteza, K. 1998. *High-Performance Capillary Electrophoresis. Theory, technique, and applications*. New York: Wiley.

48. Shintani, H. and J. Polonsky. 1997. *Handbook of Capillary Electrophoresis Applications*. London: Blackie Academic & Professional.

49. Valcárcel, M. and M.D. Luque de Castro. 1988. *Automatic Methods of Analysis*. Amsterdam: Elsevier.

50. APHA. 1998. Standard Methods for the Examination of Water and Wastewater, 20th edition. American Public Health Association (APHA), American Water Works Association (AWWA), Water Environment Federation publication (WPCF). APHA, Washington, DC.

51. Official methods of analysis of The Association of Official Analytical Chemists. 2005. Methods Manual, 18th Edition.

52. International Standard Organization. 1984. Water quality. Determination of nitrite. Molecular absorption spectrometric method. ISO 6777. International Organization for Standardization, Case Postale 56, CH-1211, Geneva 20 Switzerland.

53. Standing Committee of Analysis. 1981. *Methods for Examination of Waters and Associated Materials—Ammonia in Water*. London: Her Majesty's Stationery Office.

54. U.S. Environmental Protection Agency. 1991. *Methods for Chemical Analysis of Water and Wastes*. Washington, DC: United States Environmental Protection Agency, Office of Research and Development.

55. International Standard Organization. 1984. Water quality. Determination of ammonium nitrogen. Distillation and titrimetric method. ISO 5664. International Organization for Standardization, Case Postale 56, CH-1211, Geneva 20 Switzerland.

56. International Standard Organization. 1984. Water quality. Determination of ammonium. Potentiometric method. ISO 6778. International Organization for Standardization, Case Postale 56, CH-1211, Geneva 20 Switzerland.

57. International Standard Organization. 1984. Water quality. Determination of Kjeldahl nitrogen. Method after mineralization with selenium. ISO 5663. International Organization for Standardization, Case Postale 56, CH-1211, Geneva 20 Switzerland.

58. International Standard Organization. 1997. Water quality. Determination of nitrogen. Part 1: Method using oxidative digestion with peroxodisulfate. ISO 11905-1. International Organization for Standardization, Case Postale 56, CH-1211, Geneva 20 Switzerland.

59. International Standard Organization. 2004. Water quality. Determination of phosphorus. Ammonium molybdate spectrometric method. ISO 6878. International Organization for Standardization, Case Postale 56, CH-1211, Geneva 20 Switzerland.

60. USEPA. 1994. Methods for the Determination of Metals in Environmental Samples Supplement I. EPA 600/R 94/111.

61. ASTM Standard. 1997. D4327/97 Standard Test Method for Anions in Water by Chemically suppressed Ion Chromatography.

62. International Standard Organization. 2007. Water quality. Determination of dissolved anions by liquid chromatography of ions. Part 1: Determination of bromide, chloride, fluoride, nitrate, nitrite, phosphate and sulfate. ISO 10304-1. International Organization for Standardization, Case Postale 56, CH-1211, Geneva 20 Switzerland.

63. International Standard Organization. 1995. Water quality. Determination of dissolved anions by liquid chromatography of ions. Part 2: Determination of bromide, chloride, nitrate, nitrite, orthophosphate and sulfate in waste water. ISO 10304-2. International Organization for Standardization, Case Postale 56, CH-1211, Geneva 20 Switzerland.

64. International Standard Organization. 1998. Water quality. Determination of dissolved Li^+, Na^+, NH_4^+, K^+, Mn^{2+}, Ca^{2+}, Mg^{2+}, Sr^{2+} and Ba^{2+} using ion chromatography. Method for water and waste water. ISO 14911. International Organization for Standardization, Case Postale 56, CH-1211, Geneva 20 Switzerland.

65. International Standard Organization. 1997. Water quality. Determination of dissolved anions by liquid chromatography of ions. Part 3: Determination of chromate, iodide, sulfite, thiocyanate and thiosulfate. ISO 10304-3. International Organization for Standardization, Case Postale 56, CH-1211, Geneva 20 Switzerland.

66. International Standard Organization. 1997. Water quality. Determination of dissolved anions by liquid chromatography of ions. Part 4: Determination of chlorate, chloride and chlorite in water with low contamination. ISO 10304-4. International Organization for Standardization, Case Postale 56, CH-1211, Geneva 20 Switzerland.

67. International Standard Organization. 1994. Water quality. Determination of alkalinity. Part 1: Determination of total and composite alkalinity. ISO 9963-1. International Organization for Standardization, Case Postale 56, CH-1211, Geneva 20 Switzerland.

68. International Standard Organization. 1994. Water quality. Determination of alkalinity. Part 2: Determination of carbonate alkalinity. ISO 9963-2. International Organization for Standardization, Case Postale 56, CH-1211, Geneva 20 Switzerland.

69. International Standard Organization. 2008. Water quality. Determination of total alkalinity in sea water using high precision potentiometric titration. ISO 22719. International Organization for Standardization, Case Postale 56, CH-1211, Geneva 20 Switzerland.

70. International Standard Organization. 1984. Water quality. Determination of calcium content. EDTA titrimetric method. ISO 6058. International Organization for Standardization, Case Postale 56, CH-1211, Geneva 20 Switzerland.

71. International Standard Organization. 1984. Water quality. Determination of the sum of calcium and magnesium. EDTA titrimetric method. ISO 6059. International Organization for Standardization, Case Postale 56, CH-1211, Geneva 20 Switzerland.

72. International Standard Organization. 1986. Water quality. Determination of calcium and magnesium. Atomic absorption spectrometric method. ISO 7980. International Organization for Standardization, Case Postale 56, CH-1211, Geneva 20 Switzerland.

73. International Standard Organization. 1989. Water quality. Determination of chloride. Silver nitrate titration with chromate indicator (Mohr's method). ISO 9297. International Organization for Standardization, Case Postale 56, CH-1211, Geneva 20 Switzerland.

74. 705 ASTM. 1971. Annual book of ASTM Standards. Part 23. Philadelphia, American Society for Testing and Materials.

75. International Standard Organization. 1985. Water quality. Determination of free chlorine and total chlorine. Part 1: Titrimetric method using N,N-diethyl-1,4-phenylenediamine. ISO 7393-1. International Organization for Standardization, Case Postale 56, CH-1211, Geneva 20 Switzerland.

76. International Standard Organization. 1985. Water quality. Determination of free chlorine and total chlorine. Part 2: Colorimetric method using N,N-diethyl-1,4-phenylenediamine, for routine control purposes. ISO 7393-2. International Organization for Standardization, Case Postale 56, CH-1211, Geneva 20 Switzerland.

77. International Standard Organization. 1990. Water quality. Determination of free chlorine and total chlorine. Part 3: Iodometric titration method for the determination of total chlorine. ISO 7393-3. International Organization for Standardization, Case Postale 56, CH-1211, Geneva 20 Switzerland.

78. International Standard Organization. 1992. Water quality. Determination of fluoride. Part 1: Electrochemical probe method for potable and lightly polluted water. ISO 10359-1. International Organization for Standardization, Case Postale 56, CH-1211, Geneva 20 Switzerland.

79. International Standard Organization. 1994. Water quality. Determination of fluoride. Part 2: Determination of inorganically bound total fluoride after digestion and distillation. ISO 10359-2. International Organization for Standardization, Case Postale 56, CH-1211, Geneva 20 Switzerland.

80. International Standard Organization. 1984. Water quality. Determination of cyanide. Part 1: Determination of total cyanide. ISO 6703-1. International Organization for Standardization, Case Postale 56, CH-1211, Geneva 20 Switzerland.

81. International Standard Organization. 1984. Water quality. Determination of cyanide. Part 2: Determination of easily liberatable cyanide. ISO 6703-2. International Organization for Standardization, Case Postale 56, CH-1211, Geneva 20 Switzerland.

82. International Standard Organization. 1996. Water quality. Determination of nitrite nitrogen and nitrate nitrogen and the sum of both by flow analysis (CFA and FIA) and spectrometric detection. ISO 13395. International Organization for Standardization, Case Postale 56, CH-1211, Geneva 20 Switzerland.

83. International Standard Organization. 2005. Water quality. Determination of ammonium nitrogen. Method by flow analysis (CFA and FIA) and spectrometric detection. ISO 11732. International Organization for Standardization, Case Postale 56, CH-1211, Geneva 20 Switzerland.

84. International Standard Organization. 2003. Water quality. Determination of orthophosphate and total phosphorus contents by flow analysis (FIA and CFA). Part 1: Method by flow injection analysis (FIA). ISO 15681-1. International Organization for Standardization, Case Postale 56, CH-1211, Geneva 20 Switzerland.

85. International Standard Organization. 2003. Water quality. Determination of orthophosphate and total phosphorus contents by flow analysis (FIA and CFA). Part 2: Method by continuous flow analysis (CFA). ISO 15681-2. International Organization for Standardization, Case Postale 56, CH-1211, Geneva 20 Switzerland.

86. International Standard Organization. 2002. Water quality. Determination of soluble silicates by flow analysis (FIA and CFA) and photometric detection. ISO 16264. International Organization for Standardization, Case Postale 56, CH-1211, Geneva 20 Switzerland.

87. International Standard Organization. 2000. Water quality. Determination of chloride by flow analysis (CFA and FIA) and photometric or potentiometric detection. ISO 15682. International Organization for Standardization, Case Postale 56, CH-1211, Geneva 20 Switzerland.

88. International Standard Organization. 2002. Water quality. Determination of total cyanide and free cyanide by continuous flow analysis. ISO 14403. International Organization for Standardization, Case Postale 56, CH-1211, Geneva 20 Switzerland.

89. International Standard Organization. 2006. Water quality. Determination of sulfate. Method by continuous flow analysis (CFA). ISO 22743. International Organization for Standardization, Case Postale 56, CH-1211, Geneva 20 Switzerland.

90. International Standard Organization. 1983. Water quality. Determination of dissolved oxygen. Iodometric method. ISO 5813. International Organization for Standardization, Case Postale 56, CH-1211, Geneva 20 Switzerland.

91. International Standard Organization. 1990. Water quality. Determination of dissolved oxygen. Electrochemical probe method. ISO 5814. International Organization for Standardization, Case Postale 56, CH-1211, Geneva 20 Switzerland.

92. International Standard Organization. 1994. Water quality. Determination of aluminium. Spectrometric method using pyrocatechol violet. ISO 10566. International Organization for Standardization, Case Postale 56, CH-1211, Geneva 20 Switzerland.

93. International Standard Organization. 1982. Water quality. Determination of total arsenic. Silver diethyl-dithiocarbamate spectrophotometric method. ISO 6595. International Organization for Standardization, Case Postale 56, CH-1211, Geneva 20 Switzerland.

94. International Standard Organization. 1994. Water quality. Determination of chromium(VI). Spectrometric method using 1,5-diphenylcarbazide. ISO 11083. International Organization for Standardization, Case Postale 56, CH-1211, Geneva 20 Switzerland.

95. International Standard Organization. 1988. Water quality. Determination of iron. Spectrometric method using 1, 10-phenanthroline. ISO 6332. International Organization for Standardization, Case Postale 56, CH-1211, Geneva 20 Switzerland.

96. International Standard Organization. 2005. Water quality. Determination of chromium(VI). Photometric method for weakly contaminated water. ISO 18412. International Organization for Standardization, Case Postale 56, CH-1211, Geneva 20 Switzerland.

97. International Standard Organization. 1986. Water quality. Determination of manganese. Formaldoxime spectrometric method. ISO 6333. International Organization for Standardization, Case Postale 56, CH-1211, Geneva 20 Switzerland.

98. International Standard Organization. 1997. Water quality. Determination of aluminium. Atomic absorption spectrometric methods. ISO 12020. International Organization for Standardization, Case Postale 56, CH-1211, Geneva 20 Switzerland.

99. International Standard Organization. 1994. Water quality. Determination of cadmium by atomic absorption spectrometry. ISO 5961. International Organization for Standardization, Case Postale 56, CH-1211, Geneva 20 Switzerland.

100. European Standard. 1996. Water quality. Determination of chromium. Atomic absorption spectrometric methods. EN 1233. European Committee for Standardization, Brussels, Belgium.

101. International Standard Organization. 1993. Water quality. Determination of sodium and potassium. Part 2: Determination of potassium by atomic absorption spectrometry. ISO 9964-2. International Organization for Standardization, Case Postale 56, CH-1211, Geneva 20 Switzerland.

102. International Standard Organization. 1993. Water quality. Determination of sodium and potassium. Part 1: Determination of sodium by atomic absorption spectrometry. ISO 9964-1. International Organization for Standardization, Case Postale 56, CH-1211, Geneva 20 Switzerland.

103. International Standard Organization. 1986. Water quality. Determination of cobalt, nickel, copper, zinc, cadmium and lead. Flame atomic absorption spectrometric methods. ISO 8288. International Organization for Standardization, Case Postale 56, CH-1211, Geneva 20 Switzerland.

104. International Standard Organization. 1993. Water quality. Determination of sodium and potassium. Part 3: Determination of sodium and potassium by flame emission spectrometry. ISO 9964-3. International Organization for Standardization, Case Postale 56, CH-1211, Geneva 20 Switzerland.

105. International Standard Organization. 2003. Water quality. Determination of trace elements using atomic absorption spectrometry with graphite furnace. ISO 15586. International Organization for Standardization, Case Postale 56, CH-1211, Geneva 20 Switzerland.
106. International Standard Organization. 2007. Water quality. Determination of selected elements by inductively coupled plasma optical emission spectroscopy (ICP-OES). ISO 11885. International Organization for Standardization, Case Postale 56, CH-1211, Geneva 20 Switzerland.
107. International Standard Organization. 2004. Water quality. Application of inductively coupled plasma mass spectrometry (ICP-MS). Part 1: General guidelines. ISO 17294-1. International Organization for Standardization, Case Postale 56, CH-1211, Geneva 20 Switzerland.
108. International Standard Organization. 2003. Water quality. Application of inductively coupled plasma mass spectrometry (ICP-MS). Part 2: Determination of 62 elements. ISO 17294-2. International Organization for Standardization, Case Postale 56, CH-1211, Geneva 20 Switzerland.
109. European Standard. 2007. Water quality. Determination of mercury. Method using atomic absorption spectrometry. EN 1483. European Committee for Standardization, Brussels, Belgium.
110. European Standard. 1998. Water quality. Determination of mercury. Enrichment methods by amalgamation. EN 12338. European Committee for Standardization, Brussels, Belgium.
111. European Standard. 2001. Water quality. Determination of mercury by atomic fluorescence spectrometry. EN 13506. European Committee for Standardization, Brussels, Belgium.
112. International Standard Organization. 1996. Water quality. Determination of arsenic. Atomic absorption spectrometric method (hydride technique). ISO 11969. International Organization for Standardization, Case Postale 56, CH-1211, Geneva 20 Switzerland.
113. International Standard Organization. 1993. Water quality. Determination of selenium. Atomic absorption spectrometric method (hydride technique). ISO 9965. International Organization for Standardization, Case Postale 56, CH-1211, Geneva 20 Switzerland.
114. Burden F.R., I. McKelvie, U. Förstner, and A. Guenther. 2002. *Environmental Monitoring Handbook*. New York: McGraw-Hill.
115. Nollet, L.M.L. 2000. *Handbook of Water Analysis*. New York: Marcel Dekker.
116. Dean, J.R. 2003. *Methods for Environmental Trace Analysis*. New York: Wiley.
117. Quevauviller, P. and K.C. Thompson. 2006. *Analytical Methods for Drinking Water. Advances in Sampling and Analysis*. New York: Wiley.

13 Analytical Techniques for the Determination of Organic and Organometallic Analytes

Erwin Rosenberg

CONTENTS

13.1 Introduction 304
   13.1.1 Water as a Matrix for Organic and Organometallic Analytes 304
   13.1.2 The Human Impact on the Hydrosphere 305
   13.1.3 Classification of Organic and Organometallic Pollutants 305
13.2 Analytical Methods 306
   13.2.1 Classification 306
   13.2.2 Separation Techniques 309
      13.2.2.1 Gas Chromatography 309
      13.2.2.2 Liquid Chromatography 311
      13.2.2.3 Capillary Electrophoresis 313
      13.2.2.4 Other Separation Techniques 315
      13.2.2.5 Size Exclusion Chromatography 315
   13.2.3 Detection Techniques 315
      13.2.3.1 Quadrupole Mass Spectrometers 316
      13.2.3.2 Ion Trap Mass Spectrometers 316
      13.2.3.3 Triple Quadrupole Mass Spectrometers 316
      13.2.3.4 Time-of-Flight (TOF) Instruments 317
      13.2.3.5 Fourier-Transform Ion Cyclotron Resonance Instruments 317
      13.2.3.6 Ionization Techniques 317
      13.2.3.7 Inductively Coupled Plasma-Mass Spectrometry 318
   13.2.4 Sample Preparation for Chromatographic Analysis 318
   13.2.5 Analyte Derivatization 325
   13.2.6 Nonchromatographic Techniques 326
13.3 Applications to Different Classes of Pollutants 329
   13.3.1 Solvents and Volatile Compounds 329
   13.3.2 Pesticides 330
   13.3.3 Phenols and Other Industrial Contaminants 330
   13.3.4 Surfactants 331
   13.3.5 Sulfonates 333
   13.3.6 Estrogenic Substances 333
   13.3.7 Pharmaceuticals 334
   13.3.8 Personal Care and Cosmetic Products (PCCPs) 334

         13.3.9   Organometallic Species 335
                 13.3.9.1   Organotin (OT) Compounds 335
                 13.3.9.2   Organolead Compounds 339
                 13.3.9.3   Organogermanium Compounds 339
                 13.3.9.4   Organoselenium Compounds 340
13.4   Conclusions and Outlook 342
References 342

13.1   INTRODUCTION

13.1.1   WATER AS A MATRIX FOR ORGANIC AND ORGANOMETALLIC ANALYTES

Water covers almost three-quarters of the surface of our planet and is the most valuable resource for humans, fauna, and flora. However, only 0.65% of the water masses of our planet are fresh water and they are subject to severe environmental pressures. It is therefore to be expected that, without appropriate counteraction, about 2/3 of the world's population will be deprived of access to fresh water by 2025.[1] Water plays various roles in our lives and environment: rivers, lakes, and oceans provide important transportation routes; water is essential in industry (as a solvent) and in agriculture (for irrigation in intensive farming); water serves as a habitat for flora and fauna; and most importantly, water is an essential foodstuff that is consumed in large quantities. For all these reasons, the maintenance of water quality is of the utmost importance, and the monitoring of water quality is a task that requires powerful analytical methods. Only rarely, for example, in the case of accidental release, do concentration levels of organic and organometallic contaminants in water samples reach or exceed the ppm level at which some analytes may already exhibit acute toxicity. Nonetheless, chronic effects are observed already at concentrations that are typically three orders of magnitude lower. Moreover, since some analytes are known to be carcinogenic or to exhibit endocrine disrupting effects, organic and organometallic contaminants have to be monitored at the low and sub-ppb level in the hydrosphere. This calls for the development and application of sophisticated analytical methods. This chapter discusses the analytical strategies for determining organic and organometallic pollutants at these low concentration levels. It will cover the analysis of organic and organometallic pollutants exclusively in the aqueous (truly dissolved) phase; while analyses applicable to the colloidal phase, sediments, and biota are dealt with in dedicated reviews.[2–4] We shall therefore discuss the analysis of organic and organometallic pollutants at the different stages of the water cycle, including pristine (spring and river) water, ocean water, ground water, waste water, rain, and fog water (Figure 13.1). The justification for treating organic and (selected) organometallic contaminants

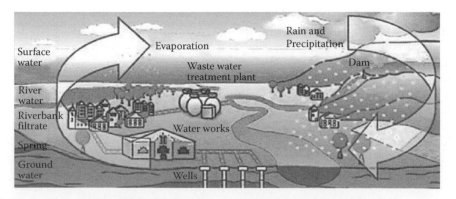

FIGURE 13.1   Biogeochemical cycle of water and possible threats to water quality.

in the same chapter is that the methods of analysis and the instrumentation used for this task are essentially the same for both groups of compounds.

13.1.2   The Human Impact on the Hydrosphere

The intensive use of water in its different functions exposes this resource to human impacts of various kinds. Industrial use leads to the discharge of industrial pollutants, mostly from point sources of high intensity: typical examples are solvents, intermediates, and other process chemicals. In contrast to this, the household use of water leads to widespread contamination at low concentrations. The probably most important group of compounds in this context is that of personal care and cosmetic products (PCCPs), which are practically in ubiquitous use, and enter water bodies despite having passed through communal waste water treatment plants (WWTPs). Agriculture has a further important impact on water quality: (inorganic) fertilizers and organic pesticides (mostly herbicides, insecticides, and fungicides) are used in large quantities in intensive farming, and antibiotics in cattle breeding. Transportation is the third major source releasing organic (various hydrocarbons) and organometallic compounds into the hydrosphere. When discussing these impacts and the analytical methods for the determination of analytes with such widely varying properties, it must be borne in mind that water bodies act not only as large reservoirs but also as reactors: The euphotic (well-illuminated) zone, typically representing the top few meters (depending on the content of particulate matter and sunlight intensity) of the water column, is the zone where photochemical reactions take place. Throughout the water column, organic and organometallic compounds may be transformed as a result of biological activity, photolysis, or hydrolysis, and in the anoxic zones of the sediments, reductive reactions may take place (Figure 13.2). All these processes lead to an increase in the complexity of water samples: these contain not only parent compounds, but also their metabolites, formed in the different types of reaction taking place in different zones of the water column.

13.1.3   Classification of Organic and Organometallic Pollutants

From the analytical point of view, it is appropriate to classify organic and organometallic compounds according to their physicochemical properties, as it is these that determine the analytical approach to be taken. The main criteria are volatility and polarity, although these factors are not completely independent. While polar compounds typically exhibit low to very low volatility,

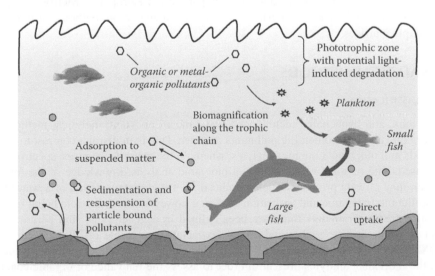

FIGURE 13.2   Interactions of organic and organometallic pollutants in different zones of an aquatic ecosystem.

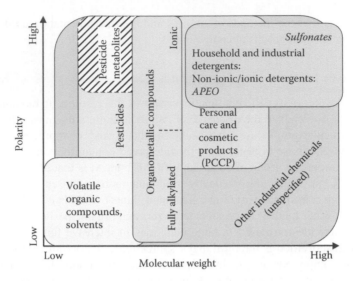

FIGURE 13.3 Common organic and organometallic pollutants in aquatic systems, classified according to the criteria of molecule size and polarity.

nonpolar organic and organometallic compounds are typically more volatile, as long as their molecular weight does not become too large (Figure 13.3). In the case of polar analytes, derivatization is a tool frequently used to increase volatility and thermal stability, and also to make them amenable to gas chromatographic separation.[5–7]

The so-called blacklists of compounds compiled by legislators for water quality monitoring in Europe or the United States, such as the one contained in the European Water Framework Directive,[8] do not follow such rational criteria of classification. Instead, such "blacklists" (derived from a combination of toxicological considerations and the scale on which these compounds are produced and discharged into the environment) appear to have been drawn up quite arbitrarily, specifying as they do both explicitly named individual compounds as well as classes of generic compounds (Table 13.1). This heterogeneous range of compounds requires a battery of different analytical techniques that will be described in the following pages. As a starting point, readers may also refer to the database on analytical methods established as part of the European "Metropolis" project (in 2002–2004), which provides a web interface for seeking out analytical methods for individual analytes and sample matrices.[9]

13.2 ANALYTICAL METHODS

13.2.1 Classification

Chromatographic and hyphenated techniques are the most important analytical methods used for analyzing organic and organometallic pollutants in water. This is in distinct contrast to the analysis of trace metals and inorganic (anionic) analytes, where, in the majority of cases, spectroscopic techniques are used. The reason for this is that in inorganic analysis, knowledge of the total element concentration may already provide sufficient indication of the contamination of a water body by a particular pollutant. In organic and speciation analytics, however, this is far from being sufficient, the only exception being parameters that have been defined by legislators as sum parameters for the assessment of water quality, such as adsorbable/extractable/purgeable organohalogen compounds (AOX/EOX/POX)[10–13] or the phenol index.[14] In all other cases, it is essential to correctly identify and quantitate the individual analytes present in order to assess the relevance of organic contaminants, and this inevitably requires efficient separation of the analytes from one another, and from the matrix

TABLE 13.1

Provisional Environmental Quality Standards (EQSs) as Proposed in the Common Position Adopted by the Council on December 20, 2007, with a View to the Adoption of a Directive of the European Parliament and of the Council on EQSs in the Field of Water Policy and Amending Directives 82/176/EEC, 83/513/EEC, 84/156/EEC, 84/491/EEC, 86/280/EEC, and 2000/60/EC

Number	Name of Substance	CAS-Number	EQS-AA[a] Inland Surface Waters[b]	EQS-AA[a] Other Surface Waters	EQS-MAC[c] Inland Surface Waters[b]	EQS-MAC[c] Other Surface Waters
1.	Alachlor	15972-60-8	0.3	0.3	0.7	0.7
2.	Anthracene	120-12-7	0.1	0.1	0.4	0.4
3.	Atrazine	1912-24-9	0.6	0.6	2.0	2.0
4.	Benzene	71-43-2	10	8	50	50
5.	Brominated diphenylether[d]	32534-81-9	0.0005	0.0002	Not applicable	Not applicable
6.	Cadmium and its compounds (depending on water hardness classes)[e]	7440-43-9	≤ 0.08 (Class 1) 0.08 (Class 2) 0.09 (Class 3) 0.15 (Class 4) 0.25 (Class 5)	0.2	≤ 0.45 (Class 1) 0.45 (Class 2) 0.6 (Class 3) 0.9 (Class 4) 1.5 (Class 5)	
6a.	Carbon tetrachloride[f]	56-23-5	12	12	Not applicable	Not applicable
7.	C10–C13 Chloroalkanes	85535-84-8	0.4	0.4	1.4	1.4
8.	Chlorfenvinphos	470-90-6	0.1	0.1	0.3	0.3
9.	Chlorpyrifos (Chlorpyrifos-ethyl)	2921-88-2	0.03	0.03	0.1	0.1
9a.	Aldrin[f] Dieldrin[f] Isodrin[f] Endrin[f]	309-00-2 60-57-1 465-73-6 72-20-8	Σ = 0.01	Σ = 0.005	Not applicable	Not applicable
9b.	DDT total[f,g] Para–para-DDT[f]	Not applicable 50-29-3	0.025 0.01	0.025 0.01	Not applicable Not applicable	Not applicable Not applicable
10.	1,2-Dichloroethane	107-06-2	10	10	Not applicable	Not applicable
11.	Dichloromethane	75-09-2	20	20	Not applicable	Not applicable
12.	Di(2-ethylhexyl)phthalate (DEHP)	117-81-7	1.3	1.3	Not applicable	Not applicable
13.	Diuron	330-54-1	0.2	0.2	1.8	1.8
14.	Endosulfan	115-29-7	0.005	0.0005	0.01	0.004
15.	Fluoranthene	206-44-0	0.1	0.1	1	1
16.	Hexachlorobenzene	118-74-1	0.01	0.01	0.05	0.05
17.	Hexachlorobutadiene	87-68-3	0.1[h]	0.1[h]	0.6	0.6
18.	Hexachlorocyclohexane	608-73-1	0.02[h]	0.002[h]	0.04	0.02
19.	Isoproturon	34123-59-6	0.3	0.3	1.0	1.0
20.	Lead and its compounds	7439-92-1	7.2	7.2	Not applicable	Not applicable
21.	Mercury and its compounds	7439-97-6	0.05[h]	0.05[h]	0.07	0.07
22.	Naphthalene	91-20-3	2.4	1.2	Not applicable	Not applicable
23.	Nickel and its compounds	7440-02-0	20	20	Not applicable	Not applicable
24.	Nonylphenol (4-nonylphenol)	104-40-5	0.3	0.3	2.0	2.0
25.	Octylphenol [(4-(1,1′,3,3′-tetramethylbutyl)-phenol)]	140-66-9	0.1	0.01	Not applicable	Not applicable

continued

TABLE 13.1   (continued)

Number	Name of Substance	CAS-Number	EQS-AA[a] Inland Surface Waters[b]	EQS-AA[a] Other Surface Waters	EQS-MAC[c] Inland Surface Waters[b]	EQS-MAC[c] Other Surface Waters
26.	Pentachlorobenzene	608-93-5	0.007	0.0007	Not applicable	Not applicable
27.	Pentachlorophenol	87-86-5	0.4	0.4	1	1
28.	Polyaromatic hydrocarbons (PAHs)[i]	Not applicable	Not applicable	Not applicable	Not applicable	Not applicable
	Benzo(a)pyrene	50-32-8	0.05	0.05	0.1	0.1
	Benzo(b)fluoranthene	205-99-2	$\Sigma = 0.03$	$\Sigma = 0.03$		
	Benzo(k)fluoranthene	207-08-9				
	Benzo(g,h,i)perylene	191-24-2	$\Sigma = 0.002$	$\Sigma = 0.002$	Not applicable	Not applicable
	Indeno(1,2,3-cd)pyrene	193-39-5				
29.	Simazine	122-34-9	1	1	4	4
29a.	Tetrachloroethylene[f]	127-18-4	10	10	Not applicable	Not applicable
29b.	Trichloroethylene[f]	79-01-6	10	10	Not applicable	Not applicable
30.	Tributyltin compounds (tributyltin cation)	36643-28-4	0.0002	0.0002	0.0015	0.0015
31.	Trichlorobenzenes	12002-48-1	0.4	0.4	Not applicable	Not applicable
32.	Trichloromethane	67-66-3	2.5	2.5	Not applicable	Not applicable
33.	Trifluralin	1582-09-8	0.03	0.03	Not applicable	Not applicable

Source: Adapted from Lepom, P. et al. 2009. *J. Chromatogr. A.* 1216: 302–315.

Note: Concentrations are given in µg/L. AA, annual average; CAS, chemical abstracts service; MAC, maximum allowable concentration.

[a]   This parameter is the EQS expressed as an annual average value (AA-EQS). Unless otherwise specified, it applies to the total concentration of all isomers.

[b]   Inland surface waters encompass rivers and lakes and related artificial or heavily modified water bodies.

[c]   This parameter is the EQS expressed as a maximum allowable concentration (MAC-EQS). Where the MAC-EQSs are marked as "not applicable," the AA-EQS values are considered protective against short-term pollution peaks in continuous discharges since they are significantly lower than the values derived on the basis of acute toxicity.

[d]   For the group of priority substances covered by brominated diphenyl ethers (No. 5) listed in Decision 2455/2001/EC, an EQS is established only for congener numbers 28, 47, 99, 100, 153, and 154.

[e]   For cadmium and its compounds (No. 6), the EQS values vary depending upon the hardness of the water as specified in five class categories (Class 1: <40 mg $CaCO_3$ L^{-1}, Class 2: 40 to <50 mg $CaCO_3$ L^{-1}, Class 3: 50 to <100 mg $CaCO_3$ L^{-1}, Class 4: 100 to <200 mg $CaCO_3$ L^{-1}, and Class 5: ≥200 mg $CaCO_3$ L^{-1}).

[f]   This substance is not a priority substance but one of the other pollutants for which the EQSs are identical to those laid down in the legislation that applied prior to the entry into force of this Directive.

[g]   DDT total comprises the sum of the isomers 1,1,1-trichloro-2,2 bis(*p*-chlorophenyl)ethane (CAS number 50-29-3; EU number 200-024-3); 1,1,1-trichloro-2 (*o*-chlorophenyl)-2-(*p*-chlorophenyl)ethane (CAS number 789-02-6; EU number 212-332-5); 1,1-dichloro-2,2 bis(*p*-chlorophenyl)ethylene (CAS number 72-55-9; EU number 200-784-6); and 1,1-dichloro-2,2 bis(*p*-chlorophenyl)ethane (CAS number 72-54-8; EU number 200-783-0).

[h]   If Member States do not apply EQSs for biota, they shall introduce stricter EQSs for water in order to achieve the same level of protection as the EQSs for biota set out in Article 3(2). They shall notify the Commission and other Member States, through the Committee referred to in Article 21 of Directive 2000/60/EC, of the reasons and basis for using this approach, the alternative EQSs for water established, including the data and the methodology by which they were derived and the categories of surface water to which they would apply.

[i]   For the group of priority substances of PAHs (No. 28), each individual EQS is applicable, that is, the EQS for benzo(a) pyrene, the EQS for the sum of benzo(b)fluoranthene and benzo(k)fluoranthene, and the EQS for the sum of benzo(g,h,i) perylene and indeno(1,2,3-cd)pyrene must be met.

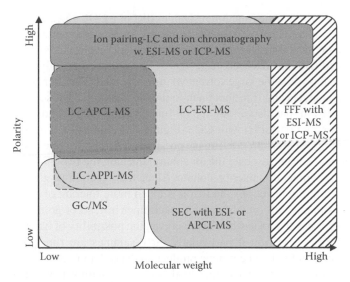

FIGURE 13.4 Important separation techniques for organic and organometallic contaminants, classified according to the criteria of molecular size and polarity and the possibility to be coupled with mass spectrometric detection.

constituents. Depending on the chemical nature and particularly the volatility of the analytes, either gas chromatography (GC) or liquid chromatography (LC) is used. In rare cases, ion chromatography (IC) is also used to separate organic analytes when these are ionic or ionizable compounds. However, the last technique has the drawback that its typical mobile phases are not very compatible with (molecular) mass spectrometric detection, which is nowadays the most sensitive and versatile technique. Capillary electrophoresis (CE) also covers a similar range of application as IC, but is used even less frequently in environmental analysis, for the following reason: it is capable of handling only very small sample volumes, so the concentration detection limits of this method are rather poor and often inadequate for the determination of organic and organometallic compounds at trace and ultratrace levels. Figure 13.4 gives an overview of the range of applicability of the various separation techniques in terms of molecular weight and polarity and their combination with mass spectrometric detection. In the following, important instrumental implications of the common separation and detection techniques will be discussed.

13.2.2 SEPARATION TECHNIQUES

13.2.2.1 Gas Chromatography

GC continues to play a key role in the separation of sufficiently volatile organic and organometallic compounds in environmental samples.[15] Current instruments provide various inlet options that can be selected according to the particular sample and matrix.[16] In order to reach maximum sensitivity, either splitless injection (in a split/splitless injection port) or on-column injection is used. The latter technique is particularly suitable for thermally labile compounds, for example, toxaphenes or organophosphorus pesticides, since the sample is injected directly to the head of the analytical column or a precolumn, where it is gently vaporized as the injector and column temperatures rise. Cold-on-column (COC) injection also minimizes sample discrimination, which occurs with split/splitless injectors when analyte mixtures spanning a broad range of boiling points, for example, polycyclic aromatic hydrocarbons (PAHs), are injected. Programmable temperature vaporizer (PTV) injectors are a third common injector option, particularly versatile in reducing the degradation of thermally labile compounds and for minimizing split discrimination.[17] They can be used for small volumes ≤5 μL (where their function resembles that of a COC injector) or with large injection volumes (10–400 μL). The latter is particularly useful for environmental trace analysis, since it allows sensitivity to be increased proportionally to the injection volume. For large injection volume, the PTV liner is filled with an adsorbent (e.g., Tenax) that retains the analytes at a low temperature during the injection step while the large solvent volume is vented. The increase in PTV temperature desorbs the trapped analytes and transfers them to the GC column. This setup is not only attractive for large volume injection, but also allows the on-line coupling of LC and GC[18] or the coupling of on-line solid-phase extraction (SPE) with GC.[19] Figure 13.5 illustrates typical instrumental implementations of this setup. Apart from offering the possibility of large volume injection, the PTV injector has several practical advantages over the on-column injector, the two most important being that it does not require a large inner diameter precolumn (since the injection needle does not have to enter the column as in on-column injection) and that it is much more tolerant of dirty samples. Any non-volatile sample constituents will remain in the packed liner instead of going to the head of the GC column, and the liner is easily replaceable. There is even the possibility of having the liner exchanged automatically after every injection if particularly dirty samples are run; probably, however, this approach is more useful in food and cosmetic product analysis than in environmental analysis.[20]

To meet the high demands of organic trace analysis,[21] GC columns have been subject to continuous refinement. This refers not only to the reduction in diameter of the nowadays almost exclusively used capillary columns (separation efficiency increases with decreasing capillary diameter), but also reflects the development in stationary phase technology: In order to reduce column bleed (which is essential for mass spectrometric detection), highly cross-linked stationary phases are used to

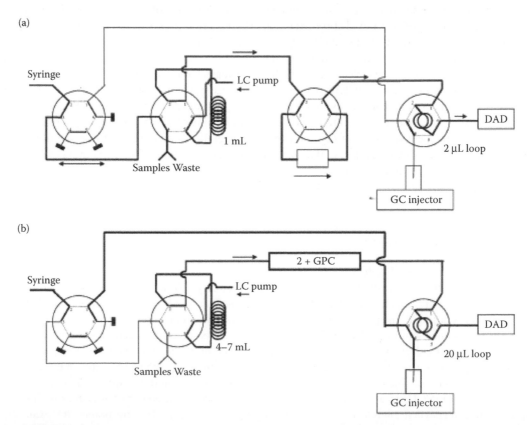

FIGURE 13.5 Schematic setup of system configurations for: (a) on-line SPE-GC and (b) on-line LC-GC with a large volume injection into GC. (Reprinted from Kerkdijk et al. 2007. *Anal. Chem.* 79: 7975–7983. With permission.)

provide clear blanks even at higher temperatures (columns of this type are often denoted by the suffix "MS," e.g., DB5MS). GC columns for polar analytes of the wax type [based on cross-linked polyethylene glycol (PEG) stationary phases] have always been limited to relatively low working temperatures to minimize thermal decomposition. Here, the introduction of stationary phases based on the sol-gel process with embedded PEG groups has substantially extended the application range, making their use at temperatures up to 350°C possible.[22] One of the most recent additions to the range of GC stationary phases are GC columns using ionic liquids as stationary phases.[23] Although the separation power in terms of the theoretical plate height of these columns is still inferior to siloxane-based columns, they offer a very interesting and unusual range of selectivity that may help to resolve mixtures that are inseparable on classical types of GC columns. Moreover, the very different separation behavior of GC columns with stationary ionic liquid phases as compared to that of typical apolar or even polar GC columns makes the former very interesting candidates for two-dimensional GC (GC × GC), providing a less correlated retention behavior in the two dimensions than with other types of GC stationary phases.[24]

In recent years, GC × GC has also increased significantly in importance and in the number of applications. The coupling of two dimensions of gas chromatographic separations with sufficiently different, and in the ideal case, orthogonal retention characteristics admits the possibility of resolving mixtures of a complexity, which cannot be resolved in one single chromatographic dimension. The key to this technique is the use of a suitable modulator that collects fractions of the eluate from the first separation dimension and transfers these quantitatively to the second dimension, where the mixture is resolved in a rapid separation (typically taking from one to a few seconds) on a relatively short column (1–2 m). The data of the many short chromatograms are combined to produce the typical "contour plots" (Figure 13.6).[25] Although GC × GC has been used extensively with "structured" samples, for example, mixtures of hydrocarbons of various homologous series in the petrochemical industry, it has rather rarely been used to resolve the complexity of unstructured environmental samples.[27] The enormous advantage of this technique for the analysis of complex environmental samples is that it can produce practically uninterfered chromatographic peaks, which enable straightforward identification by mass spectrometry (MS). However, only very fast MS detectors are capable of being used in GC × GC—typically only time-of-flight-MS (TOF-MS) instruments, fast scanning quadrupole MS instruments, or unspecific detectors such as flame ionization detector (FID), electron-capture detection (ECD), or flame photometric detection (FPD) if no structural information is required. Given a typical peak width in the second dimension chromatogram of 30–50 ms, plus the need to obtain about three data points along the peak for qualitative analysis and at least 10 data points for quantitative analysis, explains why data acquisition rates of 50–200 Hz are required.

13.2.2.2 Liquid Chromatography

It appears that over the last decade high-performance liquid chromatography (HPLC) with its different modes of operation is overtaking GC in its importance and the number of applications in environmental analysis. The reasons for this are various: first, many of the anthropogenic contaminants of the hydrosphere are not amenable to GC because of their polarity or molecular weight. Second, it must be borne in mind that the water column is not only a reservoir, but a reactor in which degradation and metabolization of organic compounds takes place. This means that even if the parent compounds initially released into the water body are apolar, their metabolites usually are not because metabolization or degradation render the compounds more hydrophilic. Third, it should be mentioned that for many years LC had a significant disadvantage in comparison to GC in that coupling with MS was difficult to implement. With the introduction of modern atmospheric pressure ionization (API) interfaces,[28,29] however, this is no longer the case, and simple (quadrupole) mass spectrometers at least can be used as detectors for LC in the same straightforward way as in gas chromatography-mass spectrometry (GC-MS).

Although LC separation does not normally provide the same high resolving power as GC with long capillary columns, it can be better tuned to achieve the necessary resolution. In addition to the

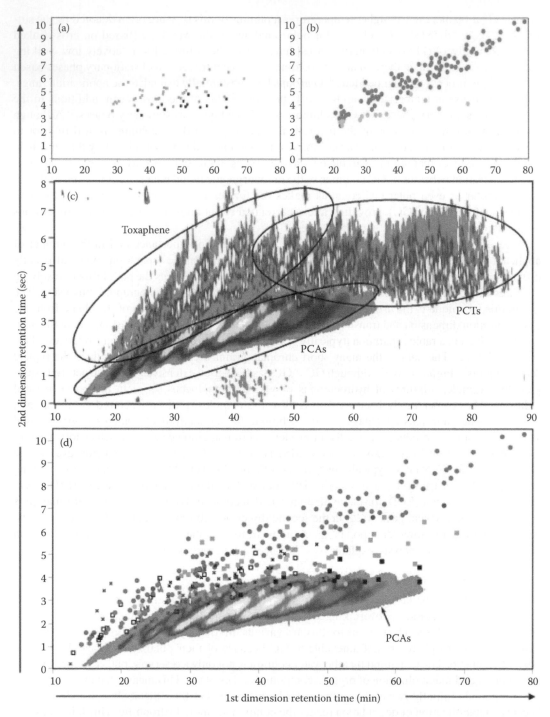

FIGURE 13.6 GC × GC–μECD overlay plot of various pollutant groups on a DB-1 × 007–65HT column combination. Pollutant groups: (a) (■) PCDTs and (■) PCDD/Fs; (b) (•) PCDEs and (•) PBDEs; (c) (•) PCBs, (•) PBBs, (•) PCDEs, (•) PBDEs, (•) PCDTs, (□) PCNs, (■) PCDD/Fs, (X) OCPs, (X) individual toxaphene standards, and PCAs (PCA-60); (d) PCAs (PCA-60), PCTs (Aroclors 5442 + 5460), and toxaphene technical mixture. (Reprinted from Korytar et al. 2005. *J. Chromatogr. A* 1086: 29–44. With permission.)

choice of stationary phase, mobile phase conditions can be varied widely in terms of composition, pH, temperature, and so on, thus allowing the separation to be optimized as required. Whereas the "classical" 25 cm × 4 mm ID columns with 5 µm particles used to be a widely used format, many laboratories have now moved to smaller formats, for example, 10 or 15 cm columns with 3 µm particles, and have also reduced the column diameter. In accordance with chromatographic scaling laws, the resolution should essentially remain unaltered if column length and particle diameter are reduced by the same factor. While the positive effect of reduced column length on the system pressure is normally more than offset by the reduced particle diameter, the reduction in separation time is evident. The decrease in column diameter (to 3 or 2 mm) brings about a significant reduction in solvent consumption (proportional to the square of the column diameter), which may be an important economic argument. Moreover, the slower flows through narrow-bore columns are often better compatible with the requirements of MS detection. Most recently, columns packed with particles of 1.7 or 1.8 µm diameter and 30–50 mm in length have come onto the market. These columns provide superior separation properties, but they create a backpressure that can be handled only with new generation HPLC systems, operating under trade acronyms such as UPLC ("ultrahigh pressure/ ultraperformance liquid chromatography") or RRLC ("rapid resolution liquid chromatography").

Stationary phase technology has also seen significant improvements over the past years. The silica base material is nowadays often a hybrid material, synthesized from tetraalkoxysilanes and functionalized trialkoxysilanes, for example, methyl-trimethoxysilane (MTMS). The introduction of alkyl-trialkoxysilanes into the silica backbone makes the material more resistant to hydrolytic attack and also improves their separation behavior for basic analytes.[30] C18 (= octadecylsilane) stationary phases are still the materials typically used in environmental analysis, and the enormous choice of materials with gradually different properties allows columns to be selected that are particularly well suited to a given separation task.[31] Reversed phase separations with materials of shorter alkylsilane chain length (C8, C4, and C1) are less frequently used.

Normal phase (NP) separations are comparatively rarely used in environmental analysis. Again, the reasons lie in the range of analytes amenable to this mode of separation, and in the limited compatibility of typical normal phase HPLC (NP-HPLC) mobile phases with mass spectrometric detection (this also applies to IC). Not only for this reason has interest recently grown in hydrophilic–lipophilic interaction chromatography (HILIC), which represents a viable alternative to the separation of very polar compounds with mobile phases that have a much better compatibility with MS detection, for example, acetonitrile/water with a low water content, typically below 10%.[32] Nonetheless, NP chromato-graphy retains its important role in sample preparation, particularly for the cleanup of complex environmental samples. In the off-line approach, fractions are collected and the relevant one is injected into the reversed phase HPLC (RP-HPLC) system, often after solvent exchange.

In parallel with recent developments in GC, multidimensional HPLC (LC × LC) is now also finding application in environmental analysis.[33] The combination of two sufficiently different separation dimensions (e.g., NP-HPLC × RP-HPLC or IC × RP-HPLC), however, remains difficult because of the solvent compatibility issues discussed above. Here, too, HILIC may bring about a significant improvement, since its mobile phase requirements are much closer to RP-HPLC than those of other liquid chromatographic techniques.[34] In contrast to GC × GC, LC × LC cannot be implemented with a (thermal) modulator that collects the analytes after the first separation dimension and reinjects them into the second column; it is most practically realized with a double-loop interface that alternately collects and transfers the analytes from the first to the second dimension (Figure 13.7). Even though the second dimension chromatogram is also very fast, detection is not normally a problem since the peak widths in the second dimension are usually still of the order of 1–2 s.

13.2.2.3 Capillary Electrophoresis

Compared to GC or HPLC, CE is used far less frequently for the environmental analysis of organic compounds.[35,36] In the case of organometallic speciation, there appears to be a gradual increase in the

(a)

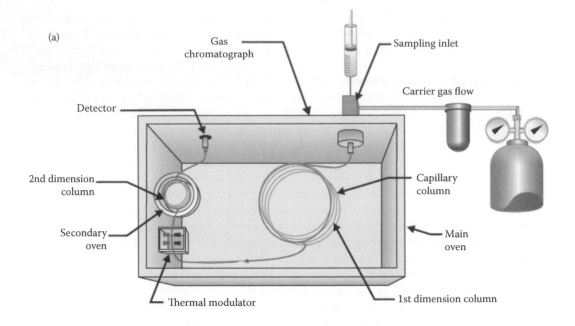

(b)

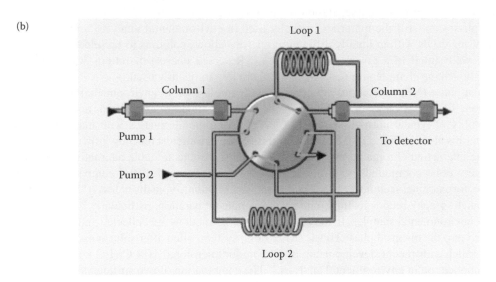

FIGURE 13.7 Instrumental setup of (a) a GC×GC system (Anon., GC×GC Comprehensive Two-Dimensional Gas Chromotography Form No. 209-184 R2.58-REV-1; LECO Corporation, St Joseph, MI 49085; P3, 2008. With permission.); (b) a LC × LC system based on a second-dimension column with two storage loops (T. Hyotylainen, LC x LC switching valves configuration; Chromedia Amsterdam; P "Two Dimensional LC (LC × LC)," 2008. With permission.).

use of this technique in comparison to the chromatographic methods.[37,38] This is easily explained, first, by the need for an analyte to be charged (at least under the conditions of separation) for it to be separable by CE, and, second, by the relatively low concentration sensitivity of CE. This calls for suitable preconcentration techniques which, as with isoelectric focusing, can even be performed on-line.[39] Only with (electrospray) mass spectrometric detection or with fluorescence detection is CE sufficiently sensitive to detect analytes at environmentally relevant (ppb- and sub-ppb-)concentration levels. Here, speciation analysis has an enormous advantage over organic analysis: it can be coupled to ultrasensitive element-specific detection by inductively coupled plasma-mass spectrometry (ICP-MS), rather than molecule-specific detection by electrospray ionization-mass spectrometry (ESI-MS). Moreover, because many environmentally relevant organometallic compounds are ionic, they would have to be derivatized to make them amenable to GC separation.[40]

13.2.2.4 Other Separation Techniques

Few other separation techniques have been used to a larger extent for the analysis of organic and organometallic compounds in water samples. Among these, micellar electrokinetic chromatography (MEKC)—a hybrid between electrophoresis and chromatography with a pseudostationary phase created by micelles dissolved in the electrophoresis buffer—is quite widely applied in water analysis.[41,42] Ion chromatography is less often employed for the analysis of organic pollutants, mostly because the mobile phase is not easily compatible with atmospheric pressure ionization-mass spectrometry (API-MS) detection unless a suppressor module is used to remove the additives of the mobile phase after it has passed through the ion exchange column.[43,44] In contrast, IC is a most versatile tool for speciation analysis, particularly in combination with ICP-MS detection. There are thus numerous reports on the speciation analysis of arsenic,[45,46] selenium,[47,48] chromium,[49,50] and antimony,[51] as well as on simultaneous multielemental speciation analytics.[52,53]

13.2.2.5 Size Exclusion Chromatography

This is a common separation technique in the analysis of natural and synthetic polymers. Since it separates compounds according to their molecular size (which is approximately proportional to the cubic root of their molecular weight) and its separation efficiency is relatively poor in terms of the number of peaks that can be resolved in one run, it tends to be used as a sample preparation and cleanup technique rather than for analytical chromatography. The few examples published on the use of size exclusion chromatography (SEC) as a single separation technique for the characterization of organic substances thus refer to the analysis of dissolved organic matter/humic substances in water,[54,55] rather than to the actual characterization of organic analytes in water samples.[56] In contrast to this, there are more reports on the use of SEC for speciation analysis, since this technique allows metals to be distinguished that are associated with the high- or low-molecular-weight fraction of a water sample. In the former case, this means that the metal is bound to larger humic or fulvic acid associates, whereas in the latter case, the metal is either present in free form or complexed by small inorganic or organic ligands.[57,58] Related to SEC separation in its separation ability is field flow fractionation (FFF). Being a nonchromatographic separation technique, FFF is also used to separate molecules and even colloids according to size in a long, very thin separation channel (typically, 20–50 cm in length and 100–250 μm in thickness) in which a laminar flow profile is established. Like SEC, FFF is also a low-resolution separation technique, which only allows the fractionation of samples or yields information about which fraction metals or certain metal species are bound to.[59,60]

13.2.3 DETECTION TECHNIQUES

The large majority of chromatographic separations in environmental analysis are nowadays performed by mass spectrometric detection. The reasons for this are the following:

- Its sensitivity.
- The structural information available from MS.
- Its general applicability.

All alternative detection schemes fail in at least one of the above criteria, and mass spectrometric detection has replaced other detectors even in applications in which they were well established, such as ECD for the gas chromatographic determination of organochlorine pesticides,[61] or atomic emission detection (AED) or FPD in the GC analysis of organotin (OT) compounds.[62]

Recent years have witnessed rapid advances in MS instrumentation; the most important features of the particular types of mass spectrometers will now be discussed briefly.

13.2.3.1 Quadrupole Mass Spectrometers

They are still the workhorses of coupled mass spectrometric applications, as they are relatively simple to run and service, relatively inexpensive (for a mass spectrometer), and provide unit mass resolution and scanning speeds up to approximately 10,000 amu/s. This even allows for simultaneous scan/ selected ion monitoring (SIM) operation, in which one part of the data acquisition time is used to scan an entire spectrum, whereas the other part is used to record the intensities of selected ions, thus providing both qualitative information and sensitive quantitation. They are thus suitable for many GC-MS and liquid chromatography-mass spectrometry (LC-MS) applications. In contrast to GC-MS with electron impact (EI) ionization, however, LC-MS provides only limited structural information as a consequence of the soft ionization techniques commonly used with LC-MS instruments [electrospray ionization (ESI) and atmospheric pressure chemical ionization (APCI)]. Because of this limitation, other types of mass spectrometers are increasingly gaining in importance for LC-MS.

13.2.3.2 Ion Trap Mass Spectrometers

These are compact ion storage devices that can be operated in such a way that they sequentially eject ions of a defined mass-to-charge ratio onto the detector. Control of the electrical field and voltages applied to the ion trap allows the entire spectrum to be scanned, typically at low (unit) resolution and scanning speeds comparable to those of a quadrupole mass analyzer. Ion traps can provide full-scan spectra for lower amounts of substance than quadrupole mass spectrometers, as they store a larger fraction of the ions formed. A further advantage of this type of mass spectrometer is that it can be used for higher-order MS (MS^n) experiments. In these, the so-called precursor ion is isolated in the trap and subjected to further fragmentation by the application of an RF pulse. The fragments thus formed are further analyzed. Since this sequence can be repeated several times, it is possible to obtain mass spectra of higher orders (typically, up to $MS^5 \ldots MS^6$) before the signal intensity becomes limiting.

13.2.3.3 Triple Quadrupole Mass Spectrometers

Triple quadrupole or MS/MS instruments have seen increased use in recent years for applications where qualitative and quantitative analyses are equally important. They consist of three quadrupoles coupled in series, the first and third of which (Q1 and Q3) can be scanned by variation of the DC and RF voltages applied. The second quadrupole (Q2) essentially functions as a collision chamber in which ions that have been passing through Q1 are fragmented by collision with the collision gas (argon) at reduced pressure. The fragments are then analyzed in Q3. The most common mode of operation for triple quadrupole instruments is the multiple reaction-monitoring (MRM) mode. In general, MRM is performed by scanning Q1 and Q3 while using Q2 as the collision cell. This mode of operation offers superior selectivity compared to single-quadrupole MS detection, since one characteristic precursor ion can be selected and thus isolated from the background. It is thus used particularly in multiresidue methods, even if chromatographic resolution cannot be achieved for all the relevant analytes.[63] Furthermore, fragmentation can be induced and characteristic mass spectra produced even with soft API techniques. This can in part also be realized by single-quadrupole instruments with in-source collision-induced dissociation (CID), but again with poor selectivity and also reduced sensitivity.[64,65] Further modes of operation of triple quadrupole instruments are the neutral loss scan, where Q1 and Q3 scan with a constant offset. This mode can be used for groupspecific detection in complex samples. It has, for example, been successfully used for the HPLC-ESI-MS/MS detection of aromatic carboxylic acids by monitoring the neutral loss of m/z 44 (corresponding to the loss of CO_2).[66] Finally, a precursor scan (with Q1) can be performed to detect, with Q3 held at a fixed m/z value, all precursors that produce a specific fragment. Provided the fragment ion is specific to a certain functional group, this type of analysis also provides screening for compounds containing a given functional group. An example of this mode of operation is the detection of aldehydes, dicarbonyl-, and hydroxycarbonyl compounds based on the monitoring of specific fragments of their 2,4-dinitrophenylhydrazine (DNPH) derivatives.[67] In the case of aldehydes, a fragment ion at m/z 163 is characteristic, whereas for dicarbonyl- or hydroxycarbonyl compounds, the DNPH derivatives produce fragment signals at m/z 182.

13.2.3.4   Time-of-Flight (TOF) Instruments

TOF-MSs have nowadays largely replaced the double-focusing sector field instruments earlier used for high-resolution measurements. Different from all other types of mass analyzers, TOF instruments utilize a field-free zone, the flight tube, in which, because of their different masses, ion packages travel with different velocities after they have taken up the same kinetic energy. The exact m/z ratio of the ion is calculated from the time needed to arrive at the detector. To achieve high mass resolution (typically, $R = m/\Delta m \geq 10{,}000$), reflectron-based TOF instruments are used. In these, the ion does not have a linear flight path, but is reflected at one end of the flight tube by a "magnetic mirror" and travels back to the detector. This minimizes the energy dispersion of the ions, which is necessary to achieve high resolution. TOF instruments always provide scan data at repetition rates of up to 20 kHz (although typically tens to hundreds of spectra are co-added on board the MS to reduce the amount of data and improve the signal-to-noise ratio, leading to 50–200 spectra per second). For the identification of unknowns, hybrid quadrupole-orthogonal acceleration time-of-flight mass spectrometers (Q-oa-TOF-MS or Q-TOF-MS) are used.[68–70] The determination of the exact fragment or precursor ion mass allows elemental formulas to be calculated and structures to be assigned for the unknowns, as shown for various metabolites.[70] Despite their evident advantages, the use of TOF and hybrid TOF instruments is still limited owing to their significantly higher investment costs than for quadrupole or ion trap instruments.

13.2.3.5   Fourier-Transform Ion Cyclotron Resonance Instruments

If ultrahigh resolution (R up to 100,000) is required, Fourier transform ion cyclotron resonance mass spectrometers (FT-ICR-MS) may provide the solution. These operate on the principle of detecting the minute currents induced in a detector by ion packages of a defined mass-to-charge ratio circulating in an orbit in a very strong magnetic field. The exact frequency of circulation depends on the m/z ratio and can be calculated from the Fourier-(back-)transformation of the detected signal. The extremely high mass resolution can be used to obtain exact mass data even for high-molecular-weight compounds. Although this technique has not yet been used for environmental contaminants in the hydrosphere, it has already been used to characterize dissolved organic matter (humic and fulvic acids) in aqueous samples.[71–73]

13.2.3.6   Ionization Techniques

For GC-MS, most applications use EI ionization. This produces information-rich spectra with an abundance of fragment ions and mostly readily detectable molecular ions that can be searched for in commercial or purpose-built libraries.[74] Chemical ionization (CI) is of limited use for screening analysis, since in the virtual absence of fragmentation it provides only molecular weight information, but hardly any structural information. For LC-MS, API techniques have almost completely replaced the earlier successful techniques of LC-MS interfacing and ionization due to their significantly greater ease of use and robustness. Their only drawback is that they are soft-ionization techniques that typically produce molecular ions (M^+ or M^-) or quasi-molecular ions ($[M + H]^+$ or $[M - H]^-$), and very few fragment ions, if any at all, under normal conditions. This disadvantage can be overcome by the above-discussed in-source CID, or by the use of hybrid instruments—the former at the cost of sensitivity, the latter for a significantly higher financial investment. Moreover, the two main API techniques, ESI and APCI have the disadvantage that they are essentially applicable only to polar–ionic analytes.[75] For less polar analytes, an atmospheric pressure photo-ionization (APPI) source has been developed,[76] which has successfully been used for the analysis of, for example, PAHs[77] or endocrine disruptors.[78] Modern LC/MS instruments also feature dual ionization sources, such as combined ESI/APCI[79] or APCI/APPI[80] sources.

13.2.3.7 Inductively Coupled Plasma-Mass Spectrometry

Inductively coupled plasma represents a further important ionization source for mass spectrometry (ICP-MS), but which, however, provides elemental rather than molecular information due to the fact that the sample is introduced in a highly energetic, high-temperature plasma (with temperatures reported to approach 10,000 K).[81] In this plasma, analytes are almost completely atomized and ionized, which enables their subsequent mass spectrometric detection with several of the above-mentioned mass analyzers, notably quadrupole- and TOF-MS. In many practical applications, however, it is found that this is not always the case and that molecular fragments survive in the plasma and contribute to the spectral background. For this reason, ICP-MS instruments have been developed, which have a collision or reaction cell between the ion source and the mass analyzer. Similar to a triple-quadrupole instrument, this cell can be used to collide the ions formed in the plasma with an inert gas (e.g., argon in the case of a collision cell) or a reactive gas (e.g., oxygen in the case of a reaction cell) to remove molecular interferences by either breaking up the molecular fragments or creating stable oxide ions that have a mass-to-charge ratio that is not interfered.[82] Most frequently, ICP-MS is used with quadrupole mass analyzers, although TOF instruments are also becoming popular for this range of applications. The molecular structural information for organometallic species has to be derived from chromatographic separation or the parallel use of molecular (ESI) and atomic mass spectrometry (ICP-MS). This, however, is not always an easy task owing to the significant difference in sensitivity of the two detection techniques (ICP-MS is typically two to three orders of magnitude more sensitive than ESI-MS in the full-scan mode), and the different tolerance of the ionization techniques to buffer concentrations and mobile phase compositions.[83,84]

13.2.4 Sample Preparation for Chromatographic Analysis

Despite the remarkable sensitivity of modern instrumental detection techniques, analysis of environmental water samples nearly always requires enrichment of the analytes. This, together with separation from the matrix, are the two main functions of sample preparation; appropriate sample preparation techniques address both issues at the same time, while striving to impose as few restrictions as possible on the subsequent instrumental determination (separation and detection). Sample preparation is strongly dependent on the nature of the analyte and the matrix, particularly with regard to its volatility and polarity. Figure 13.8 gives a general overview of common sample preparation (enrichment) techniques for aqueous and other matrices.

Volatile compounds (solvents, alkyl-element species, etc.) can be isolated and enriched from the headspace (HS) of an aqueous sample.[85] HS techniques can be divided into static and dynamic HS techniques. In the former, a fraction of the HS over an aqueous sample is withdrawn after its thermal equilibration and introduced into the gas chromatograph for analysis. The technique is very easy to perform and automate, but does not provide high sensitivity, since it relies on the equilibrium between the liquid and gas phases, which usually means that only a small fraction of the analyte can be transferred to the GC system.[86–88] It has nevertheless found widespread application in environmental analysis, for example, for the determination of halogenated and aromatic solvents in water samples.[89] To overcome the lack of sensitivity and the strong influence of the matrix in static HS analysis, which requires the use of appropriate internal or surrogate standards for quantitation,[90] dynamic HS is often preferred for sample preparation. In this technique, also called purge and trap (P&T) sample preconcentration, the analytes are exhaustively extracted by passing an inert gas stream (typically the carrier gas He) through or over the sample to liberate the purgeable compounds and to transport them to an adsorbent trap. On completion of the purge step, the analytes are thermally desorbed from this trap and injected into the GC.[91] The trap material has thus to be chosen with particular consideration of the analytes in order to ensure their efficient trapping during the purge step, and their complete recovery in the desorption step. Multiadsorbent traps are available to tailor the trapping properties as desired. Since extraction is normally performed to completion, P&T techniques are significantly more sensitive than static HS techniques. Even so, special measures

		Sample matrix		
	None	Direct aqueous injection (DAI)	Static headspace with gas syringe (SHS)	Dynamic headspace/ purge & trap (P&T)
Extractant phase	Cryogenic zone			• Purge-and-cryogenic trapping
	Solvent	• Liquid/liquid extraction (LLE) • Steam distillation/extraction (SDE) • Solvent microextraction (SME): – SDME – LPME – DLLME	• Headspace-solvent– microextraction (HS-SME) - Headspace single-drop microextraction (HS-SDME) - Headspace liquid phase microextraction (HS-LPME)	
	Sorbent/ solid phase	• Solid phase extraction (SPE) • Solid phase microextraction (SPME) • Stir bar sorptive extraction (SBSE) • INCAT/OTT/in-tube-SPME • SPDE	• Headspace-solid phase microextraction (HS-SPME) • Headspace stir-bar sorptive extraction (HS-SBSE)	• Purge-and-sorbent trapping • Spray-and-sorbent trapping
	Membrane	• SLM • MESI • MMLLE • MIMS • PME • HF-LPME • MASE • LGLME	• Headspace-solid phase dynamic extraction (HS-SPDE)	• Purge-and-membrane extraction • Headspace/membrane extraction w. sorbent interface (HS-MESI) • Headspace/membrane inlet mass spectrometry (HS-MIMS)

FIGURE 13.8 Graphical overview of sample preparation techniques for organic and organometallic analytes in aqueous samples. For explanation of the abbreviations, see Table 13.2. (After Demeestere et al. 2007. *J. Chromatogr. A* 1153: 130–144.)

have to be taken to handle the significant amounts of water vapor removed from the sample and trapped on the adsorbent. They include

- The use of hydrophobic adsorbents combined with a dry-purge step in which loosely bound water is removed from the adsorbent trap[92]
- The use of membrane driers to remove water vapor during the sampling step before arriving at the adsorbent trap[93]
- Cryogenic removal (freezing-out) of the water vapor[94]

LODs obtained with P&T techniques are typically in the ng/L to low µg/L range and thus often 100 to more than 1000 times lower than those achieved with static HS techniques. P&T does have shortcomings, however. The main ones are[95]

- The more complex instrumentation required than in static-HS-GC analysis, particularly if on-line and real-time monitoring is to be performed
- The necessity for a very efficient water management system
- The possibility of cross-contamination
- Foaming

The time required per analysis—in the 10–30 min range—is approximately the same as for static HS-GC but does not allow parallel sample preparation of multiple samples, since the trapped analytes have to be injected directly into the GC.[96]

In an attempt to overcome the significant difficulties that the presence of water vapor poses to the analysis of very volatile compounds, purge-and-membrane extraction techniques have been developed that largely prevent the introduction of water into the analytical system. Typical implementations of this form of sample introduction have been called by its developers "membrane extraction with a sorbent interface" (MESI),[97] or membrane introduction mass spectrometry (MIMS).[98,99] They are based on a silicone hollow-fiber membrane that is inserted into the sample to be monitored, and the passing of a certain volume of inert gas through the membrane. Volatile compounds permeate the membrane and are swept to the adsorbent trap from which they are desorbed into the GC. This method of sample introduction is particularly suited for field and process monitoring and for dirty samples, since it prevents any nonvolatile compounds from entering the analytical system.[100]

A revolutionary development in the field of sample preparation for volatile and semivolatile organic and organometallic compounds was the introduction of solid-phase microextraction (SPME) by Belardi and Pawliszyn[101] in 1989. This tool, which in the initial phase was a fused silica fiber with a polymer coating in a modified syringe-like holder, allowed organic analytes to be extracted from aqueous samples by either HS extraction (HS-SPME) or direct immersion (DI-SPME). After a defined time, or after partitioning, equilibrium between the sample and the fiber coating has been reached, the fiber is withdrawn from the sample and introduced into the injection port of the GC where it is thermally desorbed (Figure 13.9).

This procedure offers a number of advantages over conventional (static) HS and liquid extraction techniques for sample preparation. The major ones are the following:

- The elimination of organic solvent use
- The ability to tune the selectivity of enrichment by appropriate choice of the polymer coating of the fiber
- The relatively high degree of enrichment with appropriate choice of polymer material and dimensions (film thickness) despite being an equilibrium technique
- The introduction of relatively low amounts of water when using hydrophobic fiber materials
- The ability to extract from small to large sample volumes (from less than 1 mL to 1 L or even more)

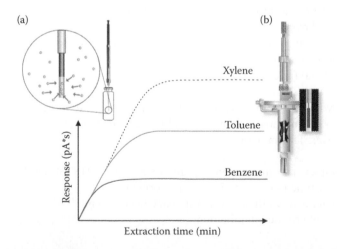

FIGURE 13.9 Principle of SPME. (a) Extraction in a closed vessel by DI or the use of an SPME device. (b) Desorption of analytes from the fiber in the GC injection port. The graph in the middle corresponds to the amount of substance introduced in the GC. The signal due to the analytes increases with increasing hydrophobicity and extraction time.

- The ability to sample from solid, liquid, or gaseous samples
- The possibility of automation
- No complicated or expensive instrumentation is needed (although the extraction fibers have to be regarded as relatively expensive consumables)

Because of these distinct advantages over classical extractive sample preparation techniques, it is easy to understand why SPME has become one of the most popular sample preparation techniques for water analysis. Numerous monographs and reviews document this great popularity and discuss in detail the theory and practice of SPME,[102,103] as well as applications in the field of environmental analysis[104,105] and speciation analytics.[106,107]

The general versatility of this technique is limited only by its unsuitability for efficiently extracting analytes with a low affinity for the polymer coating (corresponding to low K_{OW} partitioning coefficients), which reduces the sensitivity of this method. One way of overcoming this drawback is to increase the volume ratio of the extracting phase/water sample, which has been done in the form of a stir-bar coated with polydimethyl-siloxane (PDMS). This is essentially the same material as the apolar coating of SPME fibers but is used with a substantially larger volume (50–200 µL instead of approximately 0.5 µL of the SPME fiber coating). The extraction efficiency thus assumes ≥90% already for K_{OW} values of 10^3 (compared to ≤5% for SPME).[108] After reaching equilibrium (or after a predetermined time), the analytes are desorbed from the PDMS-coated stir bar (providing the technique with the acronym SBSE, "stir-bar sorptive extraction") either with a suitable solvent or by thermal desorption. A commercial unit is available for the thermal desorption of the PDMS-coated stir bars. In the case of solvent desorption, GC or LC analyses are possible. In comparison to thermal desorption, solvent desorption for SBSE has a significantly lower sensitivity, since the enriched analytes are desorbed in a large solvent volume of which only a small fraction can normally be introduced into the GC or LC. Several variations of the basic idea—a sorptive enrichment and sample introduction device—have been reported over the last decade that offer alternatives or sometimes improvements over the classical SPME technique. An overview of these techniques with their principle of operation is given in Table 13.2 and in two reviews.[108,109]

Although the miniaturization of extraction and enrichment steps is attractive for a variety of reasons, much work, particularly in routine laboratories, still relies on "classical", that is, conventional scale extraction techniques, which have the reputation of being more robust, particularly for the extraction of samples with difficult matrices. These are solvent extraction (liquid–liquid extraction, LLE) and SPE. Although LLE is probably the most versatile and flexible extraction technique in that it specifically allows the extraction medium to be selected according to the analytes of interest, it will most likely be phased out from the analytical laboratory to reduce the consumption of volatile and/or toxic solvents. There is a growing interest in the so-called green solvents (which are much less toxic, readily biodegradable, and have a negligible vapor pressure), but this topic needs to be developed still much further.[110]

In contrast to this, SPE in its different formats continues to be used widely for environmental analysis.[111–113] This allows the enrichment of large water volumes in off-line, at-line, and on-line modes, all of which can be automated by suitable instrumentation. Off-line SPE offers the greatest flexibility: different solvents can be used for conditioning, washing, and eluting the cartridges; the elution volume is not critical as it can be reduced; and the solvent can even be exchanged for subsequent analysis. At-line analysis (which in this context denotes the automated enrichment of samples on SPE cartridges, and their subsequent elution and injection into the chromatographic system, whereby only an aliquot of the eluate is injected) is similar in this respect, with only the minor reservation that the sample preparation step should not take much longer than the chromatographic run; otherwise, it will limit sample processing frequency. Both off-line and at-line SPE can be performed with the same format of SPE cartridges or SPE-disks, the latter being particularly advantageous with their larger surface area when large sample volumes containing suspended matter are to be enriched.[114]

TABLE 13.2

Overview of Extraction/Sample Preparation Techniques Derived from or Related to SPME

Acronym	Name of the Method	Principle/Characteristics	Application(s)
CME	Capillary microextraction	Fused silica capillary of up to 40 cm coated internally with the stationary phase; analyte elution by solvent	Suited for on-line SPE-HPLC, for example, with MTMS/PEO-coated capillary
ESD	Equilibrium sampling device	Use of PDMS membrane that equilibrates with the surrounding water sample	Extraction of halogenated compounds from sea and river water samples
HSPE	Headspace solvent microextraction	HS extraction of volatile compounds into a single drop of solvent, which is then injected into the GC	For volatile, GC-amenable compounds, particularly in water and flavor analysis
INCAT	Inside-needle capillary adsorption trap	Combination of SPME and P&T methods: a hollow needle with either a short length of GC capillary column placed inside it, or an internal coating of carbon, is used as the preconcentration device. Sampling by passing the gas or liquid through the device either actively with a syringe or passively via diffusion	Sampling of gaseous samples, solutions, or the solution HS
In-tube SPME (ITSPME) ITE	In-tube solid-phase microextraction In-tube extraction	Internal polymer-coated fused silica (GC) column through which the sample is drawn for enrichment while analytes are eluted on-line into the HPLC	Relatively small sample volumes ($\leq$1 mL) of medium- to nonpolar analytes are enriched and measured directly by HLC-UV or -MS
LPME	Liquid-phase microextraction	LLE with a minute amount of solvent, typically one microdrop, exposed at the tip of a microliter syringe, which is then withdrawn after extraction and injected into the GC	Mostly GC-amenable analytes
MEPS	Microextraction in a packed syringe	Sample enrichment on a small amount (1–2 mg) of solid material inserted into a syringe (100–250 μL) as a plug. Elution with a suitable solvent	Equally applicable to HPLC and GC analysis
Micro-LLE	Micro-liquid–liquid extraction	As LPME	
NT	Needle trap	Device with a hypodermic needle, whose tip is filled with a solid adsorbent onto which the sample is adsorbed. For desorption, which takes place inside the GC injection port, the carrier gas flow is forced through the needle, entering it through a hole at its side	Similar application profile as HS-SPME; however, the NT device is more robust and also has a higher capacity, potentially allowing exhaustive extraction. Also useful in P&T analysis
OTME	Open-tubular microextraction	As in-tube (micro)extraction	
OTT	Open tubular trap	As in-tube (micro)extraction	Uses not restricted to liquid-phase separation; the name suggests preferential use in conjunction with GC for volatile compounds

continued

TABLE 13.2   (continued)

Acronym	Name of the Method	Principle/Characteristics	Application(s)
SBSE	Stir-bar sorptive extraction	PDMS-coated stir bar used for direct or HS extraction and agitation of the sample. More efficient for analytes less hydrophobic than SPME owing to the larger phase ratio. Analytes are thermally desorbed in a dedicated desorption unit (for GC) or eluted with solvent (for GC or LC)	Organic compounds that are less hydrophobic to provide good recovery with SPME
SDE	Single-drop extraction	As LPME	
SDME	Single-drop microextraction	As LPME	
SME	Solvent microextraction	As LPME	
Sol-gel CME	Sol-gel capillary microextraction	Capillary microextraction with sol-gel coating	Typically for the enrichment of more polar analytes
Sol-gel OTME	Sol-gel open-tubular microextraction	Sol-gel pen tubular microextraction with sol-gel coating	Typically for the enrichment of more polar analytes
SPDE	Solid-phase dynamic extraction		
SPME	Solid-phase microextraction	Polymer-coated fused silica, polymer, or metal fiber that enriches analytes during exposure to the solid, liquid, or gaseous sample. Analytes are desorbed either thermally (in the GC injection port) or with solvent (for HPLC in a dedicated injector)	Wide range of applications owing to the great variety of commercially available polar to apolar fiber coatings and film thicknesses

On-line SPE is the most elegant, but at the same time the most difficult enrichment technique to implement. On-line SPE-GC requires a dedicated interface that allows the introduction of the large solvent volume needed for the elution of the SPE cartridge. For on-line SPE-HPLC, which can be performed with simple valve-switching systems as well as with dedicated instrumentation (Figure 13.10), a number of restrictions have to be taken into consideration.

Firstly, the sorbent trap must be small enough and designed in such a way that it minimizes peak broadening. This typically calls for enrichment columns of 2 mm ID or less and 1 cm in length, filled with less than 20 mg of sorbent material. As a consequence, breakthrough volumes must be carefully determined for the target analytes, since the amount of sorbent is significantly less than for off-line SPE cartridges (with typically 500–1000 mg for silica-based materials).[116] Secondly, elution must be accomplished in as narrow a zone as possible, which equates to the smallest possible eluent volume. This can be achieved by various means, such as elution with a high-organic-content eluent, an increase in temperature, elution in backflush mode, or a combination of these factors. In particular, elution with a high-organic-content eluent is critical because the third point of consideration regarding on-line SPE is that the analytes should be refocused at the head of the analytical column to yield sharp peaks. Chromatographic common sense requires that the analytical column should have a stronger retention than the SPE enrichment cartridge. In practice, however, the SPE cartridge is filled with a sorbent material that has the same or an even larger retention than the stationary phase of the analytical column. In order to avoid excessive peak broadening, the (strong) elution solvent must be mixed with a weaker eluent after leaving the SPE cartridge in order to reduce the

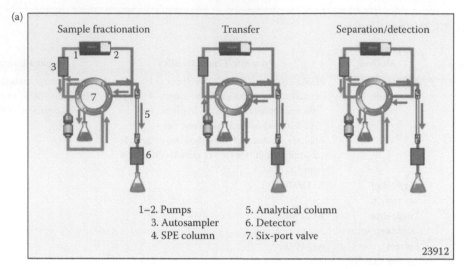

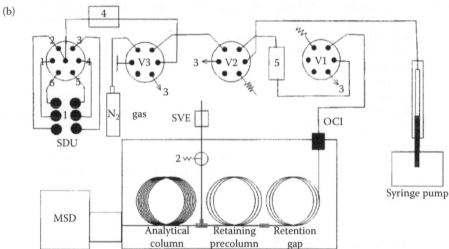

FIGURE 13.10 Generic setup for (a) on-line SPE-HPLC implemented with a simple valve-switching system (G. Maio, R. Morello, F. Arnold, and K.-S. Boos, Analysis of Antimycotic Drugs in Biofluids by On-Line SPE-LC; Application note LPN 1859-01 06/06; Diones Corporation, Sunnyvale, CA 94088-3603; Figure 2, p. 1, 2006. With permission.) and for (b) on-line SPE-GC implemented with a large volume injector with a solvent venting option.

elution strength of the mobile phase and to allow for a refocusing of the analytes at the analytical column head. Once these issues are resolved, on-line SPE does provide a highly sensitive method for the analysis of environmental contaminants such as estrogens,[117] pharmaceuticals,[118] and pesticides[119] at the trace level. The high sensitivity is a consequence of the fact that the complete eluate is transferred to the chromatographic system and analyzed, and not merely a small fraction of it (typically 0.1–1%), as is the case with off-line SPE techniques.

While the above-described sorbents are essentially nonspecific and designed to allow extraction of a wide range of analytes, there are also sorbent phases that are selective toward individual analytes, or at least classes of analytes. These are immunoaffinity (IA) sorbents and molecularly imprinted polymers (MIPs). In the first case, antibodies are immobilized on the solid support used for extraction, and the selective (in the ideal case: specific) biochemical interactions allow an antigen to bind selectively to the antibody, whereas the other sample constituents are not retained and can be washed out. Elution of the analyte(s) is achieved with an eluent that temporarily weakens the antigen–antibody interaction or that competes with the analyte for the binding sites.[120,121] MIPs, on the other hand, represent an interesting and economic alternative to IA sorbents in view of the tedious and sometimes difficult preparation of the antibodies required, their limited availability, and significant cost.[122,123] MIPs are polymer-based sorbents containing cavities introduced during the synthesis that have a very high affinity for the analyte with which they have been "imprinted." In comparison to IA sorbents they have a significantly lower affinity for the analyte, but can normally be produced faster, more easily, and at a lower cost than IA sorbents. Already commercially available, they hold great promise for environmental analysis.

13.2.5 ANALYTE DERIVATIZATION

Analyte derivatization is commonplace, particularly for chromatographic separation. Derivatization serves mainly three purposes in this context.[124]

- It leads to a better separation of the analyte.
- It improves the detectability of the analyte.
- In the case of GC, it also makes less volatile or thermally labile analytes amenable to GC separation.

In the ideal case, all the three objectives can be combined. Common examples are the derivatization of polar or ionic analytes for GC. In the case of organic analytes, molecules with active H atoms have to be derivatized in order to prevent the formation of hydrogen bonds between analyte molecules, which reduces their volatility or may even lead to thermal decomposition before volatilization. In the case of ionic organometallic analytes, derivatization may involve, for example, hydridization or alkylation, so that the ionic species can be transformed into neutral and thus volatile ones.

The improvement of detectability through derivatization may be due to improved analyte peak shape, but more importantly is due to the introductions of atoms (tags) that enhance the detectability of the derivatized analyte. Common examples are the introduction of a large number of halogen atoms when alcohols or amines are derivatized by means of perfluorinated or perchlorinated acyl chlorides or anhydrides. The resulting perfluoro- or perchloro-carboxylic acid derivatives can be detected by electron-capture detection with very high sensitivity and good selectivity. More exotic, but equally useful are elemental tags other than halogens. One example is the derivatization of alcohols, phenols, thiols, or amines by ferrocene carboxylic acid in order to introduce Fe into the molecule, which can then be detected by GC/AED with unique selectivity and sensitivity.[125] Although derivatization is considered primarily for gas chromatographic analysis, it may be equally important in LC (and CE) to ensure or enhance fluorescence detection.[124] Even for mass spectrometric detection, derivatization may improve detectability in that derivatives of higher molecular weight are formed, which can then be detected in a mass range with less chemical interference.

Derivatization may also be necessary as a means of preconcentrating very polar analytes, such as the herbicide glyphosate and its metabolite aminomethylphosphonic acid (AMPA), which are otherwise too polar to be retained on an SPE cartridge.[126]

A vast number of derivatization reactions and reagents can be classified into the following groups:

- Esterification reagents (e.g., MeOH/HCl) for acidic compounds
- Acylation reagents (e.g., trifluoroacetic acid anhydride) for amines and alcohols
- Silylation reagents (e.g., N,O-bis(trimethylsilyl)trifluoroacetamide, BSTFA) for compounds with acidic protons
- Hydride reagents (e.g., $NaBH_4$) for hydride-forming elements and compounds
- Alkylation reagents (e.g., Grignard reagents) for ionic organoelement species

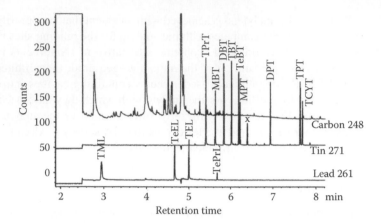

FIGURE 13.11 Element-specific GC-AED chromatograms of simultaneously detected propylated derivatives of OT and organolead compounds in a spiked buffer solution (conc. ~1 µg/L as tin or lead). Abbreviations: TML, trimethyllead, TeEL, tetraethyllead, TEL, triethyllead, TePrT, tetrapropyllead; TPrT, tripropyltin; MBT, monobutyltin; DBT, dibutyltin; TBT, tributyltin; TeBT, tetrabutyltin; MPT, monophenyltin; DPT, diphenyltin; TPT, triphenyltin; TCyT, tricyclohexyltin; X, unidentified OT substance as their propyl derivatives. (From Louter et al. 2005. *J. Chromatogr. A* 725: 67–83. With permission.)

Analyte derivatization has been the subject of a large number of books, book chapters, and reviews.[5,127–133]

The gas chromatographic analysis of organometallic species typically requires the transformation of the ionic species into alkylated, arylated, or hydride derivatives.[6,134] Grignard reagents have long been used for this task and provide high derivatization yields and robust methods. But they do require a nonaqueous medium, as solvent exchange is necessary prior to the derivatization reaction, and this adds at least one further step to the analytical procedure. For this reason, $NaBH_4$ (for hydridization) and $NaBEt_4$ or $NaBPr_4$ reagents have become increasingly popular as they can be used directly in the aqueous phase,[135] although they require a buffered reaction medium and are more susceptible to matrix interferences (Figure 13.11). In the case of OT compound analysis in water samples, the German standard DIN EN ISO 17353[136] (formerly DIN 38407-F13)[137] suggests the use of a surrogate standard for each substitution stage of the alkyl/aryltin chlorides (R_xSnCl_{4-x}) to compensate for fluctuating derivatization yields. The derivatized organometallic species are then extracted into a suitable solvent (or by SPME) and determined by GC/MS, GC/AED, or gas chromatography with flame photometric detection (GC/FPD). In an effort to bypass the crucial derivatization step, liquid chromatographic methods have also been developed for the determination of organometallic compounds.[138] In the RP mode of HPLC separation, however, chromatographic resolution is far inferior to GC; moreover, detection is problematic, as the relevant alkylelement species do not absorb in the accessible UV/VIS range and so still have to be derivatized, or be registered by mass spectrometric detectors.

13.2.6 Nonchromatographic Techniques

In typical cases the complexity of water samples requires chromatographic separation. In a few instances, however, techniques other than chromatographic, electrophoretic or hyphenated ones are used to determine organic pollutants in the hydrosphere. Many of these relate to the determination of groups of compounds rather than individual compounds, for which the selectivity of the determination method would not be sufficient. Important examples are the determination of organohalogen compounds after adsorption, extraction, or purging from water samples (AOX/EOX/POX).[139] After the initial preconcentration step, the organohalogen compounds are combusted, and the halogenide thus formed is collected in a buffer solution in which it is coulometrically titrated. Since this *sum parameter* does not distinguish the different organohalogen compounds, it can only give a rough estimate of the potential hazard posed by the sample. However, in the case of water samples with a relatively uniform composition (e.g., pulp or paper mill effluents), it may provide a good indication of the relative level of organohalogen compound pollution. Likewise, the "phenol index" (based on the color-forming reaction of phenols with 4-aminoantipyrine and photometric measurement at 460 nm) is a widely accepted estimate of the total phenol concentration in water samples, although it is hampered in its interpretability by the fact that different phenols have largely different absorption coefficients at the detection wavelength, and some substituted phenols do not react to the same extent.[140] Many of these techniques are implemented in the form of *flow injection analysis* (FIA) methods. This allows for high-throughput, automated analysis with the possibility of integrating sample preparation in the automated procedure.[141,142]

Sensors further extend the repertoire of measurement techniques for organic pollutants in the aquatic environment. Although there is still some debate about the exact definition of a sensor system, we can in this context consider a sensor to be a device capable of producing a continuous and reversible response to a single target analyte or a group of them. These characteristics make sensor systems particularly suitable for applications in situations with rapidly changing concentration profiles, such as the effluent line of a factory. A multitude of sensing principles is available (Table 13.3), of which electrochemical and optical ones are the most important.[143,144]

The underlying chemical reaction can be chemical or, taking advantage of the significantly higher selectivity, biochemical. In the latter case, immobilized enzymes catalyze a specific reaction whose products are detected, for example, by electrochemical or optical measurements; alternatively, antibodies are immobilized to induce a specific immunological reaction with the specific antigen, or DNA serves as a biorecognition element.[145,146] On the basis of the biorecognition principle, biosensors are classified as immunochemical, enzymatic, nonenzymatic receptor, whole-cell, and DNA. Biosensors can also be classified according to the type of signal transduction: A transducer converts

---

TABLE 13.3
Overview of Sensor Techniques Applicable to Organic and Related Analytes and Analyte Classes

Compound	Techniques
Metals and metal species	Voltammetry, color chemistry, and photometry
Oils/fuels/solvents	UV fluorescence, UV photometry, electromagnetic absorption, optical scattering and reflection, capacitive, vapor purging, and VOC gas sensor
BOD	Bacterial biosensor, biomass oxygen consumption
COD	Thermal/chemical oxidation and IR-based CO_2 detection, ozone oxidation/consumption, UV/visible spectrometry (inferential method)
TOC	Oxidation and IR-based CO_2 detection, UV/visible spectrometry (inferential)
Toxicity	Bacterial oxygen consumption, algal fluorescence, microbial respiration inhibition
DO	Clark electrode, fluorescence quenching
SS/turbidity	IR and visible light scattering, optical absorption
Conductivity	Current flow between two electrodes
Total ion concentration/TDS	Conductivity sensor
pH	Glass electrode
Algae/chlorophyll	UV fluorescence

Source: Modified from Bogue, R. 2008. *Sens. Rev.* 28: 275–282.

Notes: BOD, biological oxygen demand; COD, chemical oxygen demand; DO, dissolved oxygen; SS, suspended solids; TDS, total dissolved solids; TOC, total organic carbon; VOC, volatile organic compounds.

TABLE 13.4

Comparative Features of On-Line SPE-LC-MS Methods versus Biosensors for Environmental Analysis

On-Line SPE-LC-MS	Biosensors
Comparatively higher sample volumes of water are necessary	Small sample volumes are sufficient to obtain satisfactory sensitivity
Matrix effects; ionic suppression; or enhancement in MS spectrometry	Matrix effects variable, depending on biorecognition principle and transduction element
Preconcentration of the sample necessary (SPE)	Direct analysis of the sample. Minimal sample preparation
Multiresidue analysis	Limited multianalyte determination
Automatization and minimal sample handling	Possible automatization of the system
Direct and fast elution of the sample after preconcentration; minimal degradation	Direct analysis after sampling is possible. Minimal degradation
No biological stability restrictions	Low biological material stability
Determination of chemical composition	Determination of biological effect and of bioavailable pollutant content
Compound selectivity with the use of specific sorbents (MIPs and immunosorbents)	Compound selectivity with the use of a specific biological recognition element
Minimal consumption of organic solvents (elution with LC mobile phase)	Consumption of organic solvents avoided. Direct analysis of contaminants in water
Generation of organic solvent waste	Minimal, noncontaminating waste
Short analysis time and high throughput	Faster analysis; real-time detection and high throughput
Limited portability. Confined to the laboratory	Availability of portable biosensor systems
Applicable to early-warning and on-site monitoring	Applicable to early-warning and on-site monitoring
Qualified personnel required	Nonqualified personnel required; user-friendly
Equipment expensive	Equipment cost-effective

Source: Adapted from Rodriguez-Mozaz et al. 2007. *J. Chromatogr. A* 1152: 97–115.

the biochemical signal resulting from the interaction of an analyte with a biological component into an electronic signal. Physical transducers include electrochemical, spectroscopic, thermal, piezoelectric, and surface wave devices. Table 13.4 lists the advantages of biosensors over on-line SPE-LC with MS detection.

In contrast to biosensors, *biotests* for monitoring water quality make use of bacteria, microorganisms, or higher organisms and register their physiological response to exposure to a potentially hazardous sample. Typical parameters monitored are fluorescence/bioluminescence, respiratory rate, and vitality/mobility.[147–149] Biotests have the same kinds of advantages as biosensors for the analysis of water samples. The most significant difference between the two is that the latter typically respond to single compounds, or at least, to a defined class of compounds, whereas the former indicate the overall effect on an organism. Only under very well controlled conditions can the effect be related directly to the concentration of a single chemical in the matrix. Frequently, on the other hand, it is the potential hazard (bioeffect) of a water sample that is to be assessed, rather than the concentrations of all its constituents determined.

More recently, biotests have been coupled with chromatographic techniques in order to combine the separation/identification capabilities of hyphenated techniques with the assessment of biological effects. In such experimental setups, which typically involve the splitting of the HPLC effluent between a (mass) spectrometric detector and a flow that is directed to a modified biotest, it is possible to identify compounds and in near real time also to assess the potential hazard posed by this compound (Figure 13.12).[150,151]

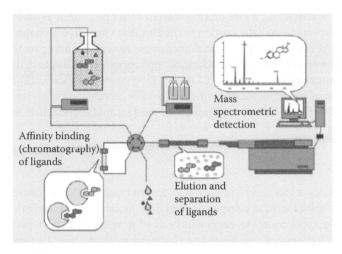

FIGURE 13.12   System for the bioeffect-related analysis of endocrine disrupting chemicals. (Modified from Seifert et al. 1999. *Fresenius J. Anal. Chem.* 363: 767–770. With permission.)

13.3   APPLICATIONS TO DIFFERENT CLASSES OF POLLUTANTS

In the first part of this chapter we discussed the general analytical approaches and techniques for the analysis of organic and organometallic pollutants. Now we will take a brief look at the individual classes of relevant compounds and illustrate the discussion with some selected examples; given the limited scope of this chapter, it is impossible to cover the entire field of organic and organometallic analysis in a comprehensive manner. As a justification of our selection of analytes, we will consider a recent study in which waters from 139 streams in the United States were analyzed for a wide range of contaminants.[152,153] It revealed that the highest concentrations were found for detergent metabolites, plasticizers, and steroids, although the concentrations were significantly lower than 1 µg/L. The compound classes detected most frequently were steroids, insect repellents, caffeine, triclosan, tri-(2-chloroethyl-)phosphate (a fire retardant), and the detergent metabolite 4-nonylphenol. This is important because both steroids and detergent metabolites are on the list of endocrine disrupting compounds (EDCs). Moreover, nonylphenols (NPs) and octylphenols (OPs), the degradation products of the widely used alkylphenol ethoxylate (APEO) surfactants, have been listed as priority hazardous substances in the European Water Framework Directive.[154] Also in European countries, awareness of the contamination of river and surface waters with these so-called emerging pollutants has increased in recent years.[155–157] The first results of a similar monitoring program, initiated by the German Ministry of the Environment, show the presence of pharmaceutical compounds at levels >10 ng/L in 39 out of 105 ground water samples.[158] Among pharmaceuticals, antibiotics are the substances that are probably giving the greatest cause for concern, in view of the possible resistance of environmental microorganisms to them.

13.3.1   SOLVENTS AND VOLATILE COMPOUNDS

Analytical methods for the analysis of volatile compounds in the environment have been extensively reviewed.[85,87,159,160] The volatility of this class of compounds—industrial solvents, emissions from the petrochemical industry and from combustion engines—suggests that GC should be used for their determination. Solvent-free sample preparation techniques, such as P&T (dynamic HS), static HS, and SPME or SBSE, in which the analytes are isolated from the aqueous matrix and simultaneously preconcentrated, are preferred. They also have the advantage that extraction solvents that could interfere with early-eluting, volatile analytes are avoided. If solvent extraction of volatile compounds from water samples is inevitable, a late-eluting solvent could be chosen as an alternative to prevent the solvent from "blackening out" the portion of the chromatogram in which the volatile compounds (e.g., methyl tert-butyl ether (MTBE) or trihalomethanes) elute. Depending on the detector used and the sample preparation/preconcentration technique applied, detection limits in the low ng/L level are easily achievable with sample volumes of ≤5 mL.[161] Despite its simplicity, direct aqueous injection for the determination of trihalomethanes and other volatile compounds is no longer used because of its inferior detection limits.[162] However, membrane-based sample introduction techniques have been successfully used for this task.[98,99]

13.3.2 PESTICIDES

The increased use of pesticides (herbicides, fungicides, and insecticides) in intensive farming has led to a growing need to monitor these compounds and their metabolites in the hydrosphere, into which they easily leach after application. An enormous body of work has thus been performed over the last three decades to develop, improve, and validate analytical methods, based mostly on GC or LC, and more rarely on other chromatographic or electrophoretic techniques.[163–167] The inhomogeneity of this class of chemicals makes it impossible to design one method to suit all of them. Clearly, GC- and especially GC-MS-based methods are very important for the more volatile pesticides, such as organochlorine and organophosphorus pesticides, pyrethroids, and triazines, to mention but a few. Some analytes that are not sufficiently volatile can be determined after derivatization (e.g., chlorophenoxyacid herbicides after esterification) or decomposition (e.g., dithiocarbamate fungicides that decompose to form CS_2, which can be detected with very good sensitivity). Since MS detection has essentially supplanted all other detectors for gas chromatographic trace analysis, this has also paved the way for multiresidue methods. With current instrumentation, it is easy to integrate the determinations of from tens to more than one hundred pesticides into one chromatographic run.[168–170]

Considering the likelihood of co-elution or matrix interference, additional identification or confirmation power is obtained by using either high mass resolution (with sector-field or TOF instruments), low-energy fragmentation of selected precursor ions (available with quadrupole MS/MS or ion trap instruments), or retention time locking (RTL). In this last technique, retention times are used as an additional parameter in the library search for pesticides. This requires an exceptional retention time stability, which is achieved by adjusting the actual retention times by the use of a lock compound.[171] The high data acquisition rate of modern TOF instruments allows, already after 1D-GC separation, the deconvolution of overlapping peaks by suitable mathematical procedures. This assumes that the overlapping peaks are not completely co-eluting, and that their spectra are sufficiently different (Figure 13.13). 2D-GC-MS provides significantly higher chromatographic resolution, which, in many cases, leads to peaks that are completely resolved from their neighbors. Nowadays, however, the majority of multiresidue methods are performed with LC/MS. This alleviates the restrictions resulting from the polarity or nonvolatility of the analytes. Because of the lower separation capability of LC in comparison to GC, such multiresidue methods are only successful with triple quadrupole or other hybrid mass spectrometers that offer enhanced selectivity.[67,172,173]

13.3.3 PHENOLS AND OTHER INDUSTRIAL CONTAMINANTS

Phenols are important industrial chemicals, for example, in the production of various plastics and resins, and may leach into surface and ground waters either during production or from the discharged products at landfill sites. Because of their ecological importance and their widespread use, methods for phenols and related compounds (e.g., anilines) were developed already at an early stage. Many of these methods rely on GC (or GC/MS), which normally requires derivatization prior to GC analysis. Standard methods for the derivatization of phenols are silylation, methylation, or acetylation.[5] The last mentioned has the advantage that it can be carried out in the aqueous sample directly. The derivatized phenols can thus be extracted more easily and with a higher yield from the aqueous sample by

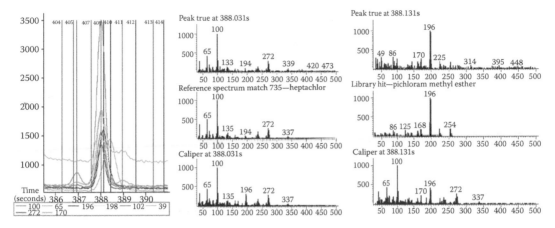

FIGURE 13.13   Left: Part of the GC/TOF-MS chromatogram from the analysis of pesticides (acquired with a Leco Pegasus III TOF MS). Right: Two mass spectra near the peak apex are reported (top), together with the library entries (middle) and the deconvoluted spectra (bottom).

solvent extraction, SPE, or SPME. Furthermore, derivatization may enhance sensitivity for the following reasons: firstly, the improved peak shape leads to a narrower, symmetrical peak resulting in better quantification. Secondly, by way of derivatization, several halogen atoms are introduced into the derivative of one analyte molecule if ECD is used, thereby overproportionally enhancing both sensitivity and selectivity. Many useful reagents exist for the introduction of halogens (e.g., perfluorinated carboxylic acid anhydrides or chlorides), but also other heteroatoms such as iron can be introduced as the "elemental tag" for less common derivatization reagents like ferrocenecarboxylic acid chloride. If this reagent is used for the derivatization of phenols in combination with AED, unmatched selectivity and sensitivity can be achieved.[174] In many instances, GC is the preferred separation method for phenols owing to its higher separation power. More recently, however, there has been a shift in favor of LC, and toward LC/MS in particular. This reflects the generally greater availability of this technique as well as its ability to determine underivatized phenols in water samples at ultratrace levels. This technique can be rendered even more sensitive by automated on-line SPE, which enables phenols to be quantitated in water samples even at ppt levels (Figure 13.14).[175] Similar considerations also apply to other phenols, such as the EDC bisphenol A[145] and anilines.[176]

13.3.4   SURFACTANTS

Nonionic surfactants represent the largest fraction of surfactants used nowadays. Among these, APEOs and their metabolites (alkylphenols and short ethoxy chain oligomers) are the compounds of greatest environmental concern due to their estrogenic potential. Even if their estrogenic activity appears to be low in comparison to natural estrogens, this is by far offset by the scale at which these compounds are used—some 10^6 tons/annum, with practically all of these being discharged into sewage and waste water. Despite the relatively high concentrations in waste water, a multistep sample preparation procedure is still required to reduce matrix interferences and to preconcentrate analytes. For aqueous samples, SPE is the technique of preference. With C18 SPE cartridges, extraction yields in excess of 70% have been reported.[177] To achieve a preliminary separation, sequential SPE schemes with different sorbent materials may be applied that allow the fractionation of neutral and acidic degradation products.[178]

The separation of APEOs on an RP column is only successful if there are differences in the hydrophobic part. Homologs with different numbers of ethoxy units co-elute in RP-HPLC. As a result of this co-elution, the chromatograms have a relatively simple structure and the signal intensity is increased. However, quantitation is severely compromised because of the different response

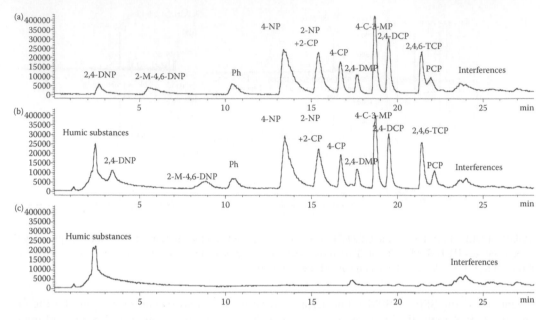

FIGURE 13.14 A typical LC-MS chromatogram obtained in the scan mode of a preconcentrated aqueous standard solution (a) and of a preconcentrated spiked river water sample (b), as well as of an unspiked preconcentrated river water sample (c). Peak labels denote: Ph, Phenol; 2,4-DMP, 2,4-dimethylphenol; 2-CP, 2-chlorophenol; 4-CP, 4-chlorophenol; 4-NP, 4-nitrophenol; 2-NP, 2-nitrophenol; 4-C-3-MP, 4-chloro-3-methylphenol; 2,4-DCP, 2,4-dichlorophenol; 2,4-DNP, 2,4-dinitrophenol; 2,4,6-TCP, 2,4,6-trichlorophenol; 2-M-4,6-DNP, 2-methyl-4,6-dinitrophenol; and PCP, pentachlorophenol. Spike level 1 μg/L; 10 mL preconcentrated.

factors of the oligomers, and also the isobaric interference of the doubly charged ions $[M + 2Na]^{2+}$ of highly ethoxylated APEOs and the singly charged ions of less ethoxylated APEOs.

Separation of nonylphenols takes place in accordance with the number of ethoxy groups. Homologs with the same number of ethoxy groups but different alkyl substituents co-elute. A prominent example is the separation of octylphenol ethoxylates (OPEO) and nonylphenol ethoxylates (NPEO). Since the mobile phase in NP-HPLC typically consists of one apolar organic solvent or a mixture of them, a polar solvent must be added to enhance ion formation for ESI-MS. Altogether, complete resolution of ethoxylate surfactants requires multidimensional separation, which can be performed on- or off-line by coupling two orthogonal separation mechanisms such as NP- and RP-HPLC, SEC and RP-HPLC, or RP-HPLC and HILIC.[179]

In LC-MS analysis of ethoxylates, ESI used to be the preferred method of ionization due to its higher sensitivity, particularly for APEOs. Under these conditions, APEOs display a marked tendency to form sodium adduct ions $[M + Na]^+$, predominantly with long-chain APEOs, but less frequently with shorter-ethoxy chain APEOs. This variable response of the homologs, depending on their ethoxy chain length, and the reduced fragmentation observed under CID conditions, is why the formation of ammonium adduct ions, being more suitable for MRM detection, is now preferred (Figure 13.15).

There are numerous reports on the use of ESI-MS after NP or RP separation for the determination of ethoxylates in environmental samples, providing both qualitative and quantitative information. For lack of appropriate reference substances, analysis of their metabolites is more difficult. As an example, mono- and dicarboxylated metabolites are formed in the biodegradation of APEOs, leading respectively to alkylphenol ethoxy carboxylic acids (APECs) and carboxy alkylphenol ethoxy carboxylic acids (CAPEC), which are detectable in negative ion-ESI-MS. Quantitation is tentatively based on MS/MS spectra, which exhibit intensive signals for APECs at m/z 219 (for nonylphenol ethoxylate carboxylic acids, NPEC) and m/z 205 (for octylphenol ethoxylate carboxylic

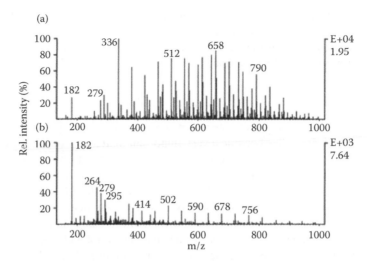

FIGURE 13.15 Flow injection–APCI-MS(+)spectra for waste water samples. (a) WWTP inflow and (b) conventional WWTP effluent after C18 SPE; eluent and methanol. Positive APCI. (Redrawn from Li et al. 2000. *J. Chromatogr. A* 889: 155–176. With permission.)

acids, OPECs).[180,181] GC and GC/MS, possibly following derivatization, are useful techniques for both the parent ethoxylates and their degradation compounds, and they offer even better opportunities for identification and quantitation.[182]

13.3.5 SULFONATES

Sulfonates are the most important group of anionic surfactants and probably the most widely used household detergents. Among these, the best known group is that of the linear alkylbenzene sulfonates (LAS), which are commonly analyzed by RP-HPLC-negative ion-ESI-MS methods.[183,184] This class of compounds can also be determined by GC techniques after suitable derivatization (e.g., by esterification with alkylation reagents).[185] Sulfophenyl carboxylates (SPCs) are the degradation products of LAS, but they are too polar to allow successful LC separation. Instead, their chromatographic separation requires IC or ion-pair RP with triethylamine or tetraethylammonium acetate.[186]

More recently, perfluorinated sulfonates, such as perfluorooctane sulfonate (PFOS), have attracted attention because of their resistance to biodegradation and the practically ubiquitous exposure of humans and wildlife to these compounds. From the analytical point of view, these compounds can also be determined by LC-MS with negative ion-ESI[187] or by GC/MS after derivatization.[188]

The importance and broad scope of this topic are reflected in numerous reviews and books covering analysis of ionic and nonionic surfactants[189–191]

13.3.6 ESTROGENIC SUBSTANCES

The increased use of contraceptive pills worldwide has made the release and the elevated levels of estrogens and steroid sex hormones consequently found in sewage and surface waters an important issue. Analytical methods for this group of compounds are well developed and have been extensively reviewed.[192–195]

They are based either on GC/MS after derivatization or on LC/MS. In order to achieve the sensitivity and selectivity required for the analysis of estrogens in waste water samples, triple quadrupole instruments are typically employed, and sample preparation—normally based on SPE enrichment and cleanup—must be well designed. Rigid quality control has to be applied to ensure the correctness of results in the complex sample matrix; it has to take into account the stability of chromatographic retention time and mass spectrometric peak intensity ratios, and make use of isotopically labeled standards where available. As is the case with other organic compounds in the hydrosphere, conjugated forms of the steroid hormones (e.g., sulfates or glucuronides) should also be taken into account in their determination.[196]

13.3.7 PHARMACEUTICALS

The occurrence of pharmaceuticals in waste water, sewage, surface, and ground water—and in rare cases also in drinking water—has become a very important matter as a result of the increased consumption of these compounds. About 3000 different substances with widely differing properties and structures are used today for human and veterinary health care.[197] Many pharmaceutical compounds and metabolites are polar, which makes LC and in particular LC/MS the method of choice for their analysis. Again, the selectivity and sensitivity of the analytical method depend principally on the sample preparation method, which usually involves SPE, using either "classical" RP materials or more modern polymeric materials that extract analytes over a wide range of polarities, based on the presence of different functional groups in the material ("mixed mode" SPE). Like the multiresidue methods for pesticide analysis, such methods have also been developed for pharmaceutical residues with the aim of determining one or several classes of pharmaceutical compounds in water samples.[198–200]

Particularly in the highly complex samples stemming from WWTP effluents, the structural elucidation capabilities of (single quadrupole-)LC/MS are limited. In these cases, the use of triple quadrupole-LC/MS (LC/MS/MS) or of LC with hybrid mass spectrometric detectors becomes indispensable. These are capable of providing second- or higher-order mass spectra, and/or high mass resolution, from which additional structural information can be generated and increased selectivity attained (Figure 13.16).[201,202]

Table 13.5 gives an overview of typical analytical procedures for important pharmaceutical compounds. Among the groups of compounds listed there, antibiotics are probably giving the greatest cause for alarm, as their occurrence in the environment can induce resistance in certain microorganisms to these compounds. X-ray contrast and magnetic resonance imaging (MRI) agents are a further class of compounds whose environmental effects cannot yet be clearly foreseen.

13.3.8 PERSONAL CARE AND COSMETIC PRODUCTS (PCCPs)

PCCPs constitute a broad class of compounds used for human and veterinary applications (e.g., food additives, sunscreens, insect repellents, body lotions, shampoos, and deodorants).[203,204] Some of these compounds are on the list of high production volume chemicals. The annual production of PCCPs in Germany, for example, exceeded 550,000 tons in the early 1990s.[205] These compounds are applied externally as skin, hair, and dental care products or soap additives, or they are directly or indirectly ingested, transformed in the body or not, and finally excreted as a combination of unaltered PCCPs and metabolites. Both parent compounds and their metabolites enter the aquatic environment, mainly through municipal WWTPs, in ppt to ppb concentrations. An alternative route to the hydrosphere is when PCCPs are released directly into surface waters from the skin during swimming or bathing; variable concentrations of these compounds are therefore detected in surface, ground, and coastal waters. Moreover, other degradation intermediates of these compounds may form as a result of biotic or abiotic processes in WWTPs or surface waters:[206] The analysis of PCCPs follows different strategies, depending on the chemical nature of the compounds investigated. After SPE or LLE for preconcentration and cleanup, the analytes are separated and determined by GC or LC with MS detection. Given the complexity of samples, two-dimensional chromatography appears to be an appropriate way of improving chromatographic resolution, while the use of MS/MS or TOF-MS brings additional spectrometric information that allows structural confirmation or elucidation.[202,206]

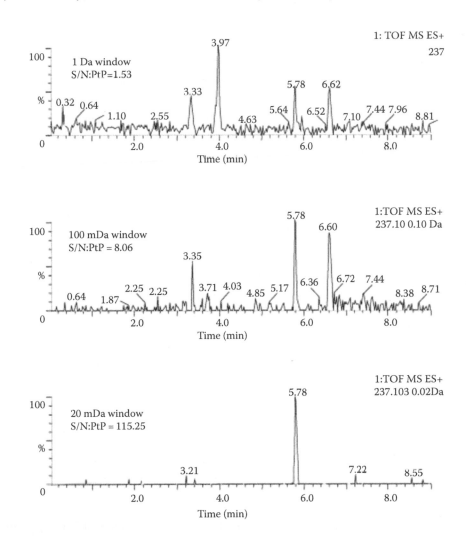

FIGURE 13.16 Enhanced selectivity of UPLC–TOF–MS analysis of the pharmaceutical compound carbamazepine (m/z 237.103) in an urban wastewater sample, as a result of varying mass windows in the reconstruction ion chromatogram. (Reprinted from Petrovic et al. 2006. *J. Chromatogr. A* 1124: 68–81. With permission.)

13.3.9 ORGANOMETALLIC SPECIES

13.3.9.1 Organotin (OT) Compounds

In the present context, we shall restrict our discussion of the determination of elemental species to OT compounds and other organic compounds of group IV elements (Pb and Ge) and Se, since the speciation of other elements is covered in Chapter 7.

In the aquatic environment, organometallic speciation is of particular relevance for OT compounds. These, and particularly tributyltin, have been intensively used for decades as the bioactive ingredient for antifouling paints applied to ships' hulls. Tributyltin compounds are highly toxic to molluscs, thereby preventing the growth of barnacles and shells on the hull. The surface roughness of the hull is an important factor affecting the cruising speed of sea-going vessels and their fuel consumption in a given route. It is thus of the greatest economic importance to prevent the growth of crustacea on hulls, and this is achieved by the controlled release of OT compounds from the

TABLE 13.5

Examples of Analytical Methods for Different Pharmaceutical Compound Classes

Class	Compounds	Extraction method	Separation	Detection
Lipid regulators	Clofibric acid, bezafibrate	SPE, LiChrolut RP-18	Genesis C18 (150 × 2.1 mm, 4 μm)	Negative ion-ESI-MS-MS (MRM)
Antiphlogistics	Diclofenac, ibuprofen, naproxen	SPE, LiChrolut EN, Oasis HLB	LiChrospher C18 (125 × 3 mm, 5 μm)	Negative ion-ESI-MS
Betablockers	Metoprolol, propanolol, betaxolol	SPE, PPL Bond Elut	Nucleosil C18 (250 × 2 mm, 3 μm)	Positive ion-ESI-MS-MS
β2-Sympathomimetics	Terbutalin, salbutamol	SPE	LiChrospher C18 (125 × 3 mm, 5 μm)	Positive ion-ESI-MS-MS (MRM)
Psychiatric drug	Diazepam	SPE, Isolute C18	LiChrospher RP-18 (125 × 3 mm, 5 μm)	Positive ion-ESI-MS-MS (MRM)
Antibiotics	Clarithromycin, sulfamethoxazol trimethoprim, ciprofloxacin	Tandem SPE Oasis HLB/MCX, mixed phase MPC	Luna C8 (100 × 4.6 mm, 3 μm) Luna RP-18 (150 × 2 mm, 3 μm) Nucleosil RP-18 (200 × 2 mm, 5 μm)	Positive ion-ESI-MS Positive ion-APCI-MS-MS, Positive ion-ESI-MS-MS
X-ray contrast media	Iopamidol, iopromide, iomeprol	SPE, Isolute ENV+	LiChrospher RP-18 (125 × 3 mm, 5 μm)	Positive ion-ESI-MS-MS (MRM)
Estrogens	Estradiol, 17β- estrone	SPE, C18	Purospher RP-18 (55 × 2 mm, 3 μm) LiChrospher RP-18 (250 × 4 mm, 5 μm)	Negative ion-ESI-ITMS (SIM), ESI-MS (SIM)

Source: Modified after Zwiener, C. and F.H. Frimmel. 2004. *Anal. Bioanal Chem.* 378: 851–861.

antifouling paint.[207,208] However, once released into the aquatic environment, OT compounds also affect nontarget organisms such as oysters or dogwhelks (a species of marine snail). It was actually the observation of malformations in oysters produced at oyster farms in southern France (in the vicinity of which there was a high intensity of yachting activities with OT-painted boats) that focused public attention on the undesirable and most alarming side effects of OT compounds. Later, the phenomenon of imposex (the induction of male sexual organs in female animals) was also observed, as a result of which the female gastropods become infertile. These findings indicated that OT compounds—mainly tributyltin and triphenyltin—represent a great environmental hazard and induce endocrine disrupting effects. As a consequence, numerous environmental studies were performed to assess the actual state of pollution[209] and methods developed for the analysis of OT compounds in aqueous, sediment, sludge, and biota samples.[6,210,211] It very quickly became evident that, owing to the hydrophobicity of trisubstituted OT compounds, measurements of their concentrations in the aqueous phase would only produce meaningful data in the case of recent OT input into the hydrosphere, and in the absence of suspended matter or biota to which the OT compounds could be adsorbed. In the case of biota (particularly fish and mussels) the high concentrations in relation to the aqueous concentrations of OT compounds are explained not only by partitioning between the two phases but also by bioamplification, as these animals are higher members of the trophic chain. The analysis of OT compounds from these matrices is clearly more demanding than that from aqueous samples because of the interferences in the sample preparation (extraction and derivatization) or determination steps. In this case, the sample preparation strategy has to be appropriately designed so as to reduce interferences from the matrix. This is, for example, achieved by silica-column cleanup or by SEC of the extracts. The typical analytical procedure for OT compound analysis involves a sample digestion and/or extraction step. In the case of biota (fish or mussel tissue), this is alternatively done by using acidic organic phases (e.g., methanolic HCl or AcOH/MeOH), which do not achieve complete digestion of the sample material, or by alkaline digestion with methanolic NaOH or Bu$_4$NOH that completely solubilizes the sample tissue and releases the physi- or chemisorbed OT compounds.[211] For inorganic (sediment and soil) samples, only organic/acidic extractant phases are normally used. For these, and also for the extraction of OT compounds from aqueous samples, complexing agents are added. Acetic acid or the chloride anion from the acid used during extraction may at the same time act as ligands for the OT compounds and enable its transfer to an organic phase (hexane or *iso*-octane). Transfer to the organic phase (followed by the drying of this phase) is required when derivatization is performed with a Grignard reagent. After partitioning of the reaction mixture with dilute sulfuric acid (to remove the excess of derivatization reagent), the organic phase can be dried again and injected directly into the GC for analysis.[6]

As an alternative to this route of sample preparation, derivatization with sodium tetraethylborate (NaBEt$_4$) has gained in popularity. In this case, no solvent exchange is required prior to derivatization; instead, this can be performed directly in the aqueous phase buffered to a pH of approximately 4.5.[135] The analytes are then extracted with a suitable solvent or by SPME prior to GC analysis. The practicality of this approach makes up for the somewhat lower stability of this method in comparison to Grignard derivatization performed under the optimum conditions of each method. In the early phase of OT speciation analysis, hydride formation of OT compounds was also a common route, and particularly suitable for the analysis of OT compounds in aqueous samples by P&T sample preconcentration.[212] Because of the high volatility of OT hydrides and the danger of losses during sample preparation, this procedure is nowadays no longer considered suitable. The derivatized OT compounds can be determined by GC with various detectors. GC with atomic emission detection (GC-AED) is one of the most versatile techniques, combining as it does very good sensitivity and selectivity for OT compound determination. Since this type of instrumentation is available only from one commercial supplier, GC-AED faces strong competition from GC-ICP-MS instruments. These are more expensive and also more demanding in their operation, but are often available in specialist laboratories and can be used for OT compound analysis with a sensitivity of about two orders of magnitude higher. Other instrumental solutions suitable for the routine monitoring of OT compounds at low trace (but perhaps not ultratrace) levels include GC with (pulsed) flame photometric detection [GC-(P)FPD] and GC-MS (Figure 13.17).[213]

Much in line with the general tendency to use LC-MS for the analysis of compounds, which otherwise would require derivatization for GC analysis, liquid chromatographic techniques are also used for OT compound analysis.[138] In the absence of chromophoric groups, HPLC analysis again requires either derivatization (e.g., fluorescence derivatization)[214] or the use of mass spectrometric detection. Molecule-specific mass spectrometric detection has been used successfully for this task,[215] but ICP-MS is several orders of magnitude more sensitive; it is thus the technique to be preferred, if available. Moreover, ICP-MS offers better compatibility than does ESI- (or APCI-)MS with separation modes or mobile phase compositions providing good chromatographic resolution (IC or ion-pairing chromatography on RP columns).[216] Table 13.6 presents an overview of common separation/detection options for OT compounds and some of their most important characteristics.

As a result of the 2003 ban on OT compounds in antifouling paints, the current input of OT compounds into the water column is slowly decreasing.[217] However, as there are also other important uses of OT compounds (as fungicides in agriculture, as stabilizers in plastics, as industrial catalysts, and as precursors in organic and inorganic syntheses), they will continue to occur in the aquatic environment, albeit at reduced levels. The lower the concentration of OT compounds in the aqueous phase becomes, the more re-emission from contaminated sludge or sediments by resuspension or re-equilibration has to be considered as a new and long-lasting source of OT compound occurrence in the aquatic environment.[218]

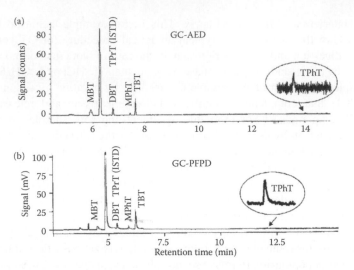

FIGURE 13.17 Typical chromatograms obtained from the analysis of a freeze-dried fish reference sample (NIES 11) with GC-MIP-AED (a) and PFPD (b); 2 mL of extract used in (a) and 0.5 mL of extract used in (b). Peak labels denote: MBT, monobutyltin; DBT, dibutyltin; TBT, tributyltin; TPrT (ISTD), tripropyltin (used as internal standard), MPhT, monophenyltin, TPhT, triphenyltin, all as ethylated derivatives. (Redrawn from Aguerre et al. 2001. *J. Anal. At. Spectrom.* 16: 263–269. With permission.)

TABLE 13.6
Overview of the Individual Steps of Procedures for OT Compound Speciation in Marine Samples (Sediments, Water, and Biota) as Reported in the Literature

Extraction	Derivatization	Preconcentration	Analysis

• Acid and polar solvent
  – HCl or HBr/MeOH
  – Acetic acid/MeOH
	• Tetraalkylborate reagents
	  – Ethylation
	  – Propylation
		• SPME
		• Solvent
		  extraction + solvent
		  volume reduction
			• GC-MS
			• GC-FPD
			• GC-PFPD

• Acid and other (less polar)
  solvent
  – HCl/toluene
  – HCl/diethyl ether
  – HCl/ethyl acetate
	• Grignard
	  – Pentylation
	  – Propylation
	  – Ethylation
		• SPE	• GC-ICP-MS
			• GC-MS/MS

• Extraction with apolar
  solvents + compleing agent
  – Toluene (HCl) + tropolone
  – Hexane + DDTC
			• HPLC-ICP-MS
			• HPLC-ESI/APCI-MS
			• HPLC-FD

• Basic digestion
  – TMAH
  – MeOH/NaOH			

• PLE (ASE):
  – Sodium acetate and acetic
    acid/MeOH			

• SPE (for aqueous samples)			

13.3.9.2 Organolead Compounds

The main source of organolead compounds in the environment is the tetraalkyllead compounds used as antiknocking agents in leaded gasoline.[219] Lead additives have been banned in the industrialized nations, but are still in common use in other parts of the world. These species enter the atmosphere as a result of incomplete combustion or evaporation during the production and distribution of antiknocking agents and leaded gasoline. Once released into the environment, the volatile organolead species can be transported over great distances in the gaseous state, or adsorbed to particles. The originally emitted tetraalkyllead compounds decompose via sequential dealkylation through the tri-, di-, and monosubstituted stages into inorganic lead.[220] The toxicity of organolead compounds depends on the actual chemical composition and increases with the number of alkyl groups, thus making speciation of organolead compounds necessary.

Commonly used techniques for this purpose are chromatography and hyphenated methods, mostly coupled with element-specific detection, for example, GC-AED,[221] GC coupled with atomic absorption detection (GC-AAS),[222] or ICP-MS. The last mentioned provides superior sensitivity as well as the possibility of using isotopically labeled standards for quantification and quality control.[223]

As organolead compounds are typically present in the environment in ionic form, their sample preparation is similar to that of OT compounds: alkylation is normally applied (occasionally also hydridization) to form GC-amenable derivatives of the formula $R_xPbR'_{4-x}$ or R_xPbH_{4-x}. Grignard derivatization (with alkyl reagents of various chain lengths) has long been the preferred derivatization technique. Derivatization with $NaBEt_4$ is not feasible because ethyllead compounds are used in addition to methyllead, so that their derivatization with $NaBEt_4$ will render speciation analysis meaningless. As soon as the analogous tetrapropylborate reagent ($NaBPr_4$) became commercially available, the simplified alkylation procedure was readily adopted for the direct derivatization of alkyllead compounds in aqueous samples.[135,222]

In contrast to OT compounds, organolead compounds were not directly released into the aquatic environment, at least not intentionally. As becomes clear from their profile of usage, however, they entered the water cycle through evaporation from their technical uses, became more water-soluble in the form of ionic compounds that were formed through dealkylation, and were transported over long distances. The analysis of water samples for organolead compound contamination can thus be considered an indicator of anthropogenic activity at the particular location of observation.[219] This explains why more often than river or surface water, samples of rainwater or snow from remote locations such as Antarctica are investigated for their organolead content. These concentration profiles clearly reflect the increased technical use of organolead compounds from the 1930s, and their eventual phasing out (at least in large parts of the world) from the early 1980s.[224] For the same reason, and because the use as an antiknocking additive to gasoline was the only relevant large-scale application of organolead compounds, interest in the speciation analysis of this element has virtually ceased.

13.3.9.3 Organogermanium Compounds

The biogeochemistry of germanium is not yet well understood.[225,226] Despite its position in the main group IV of the periodic table, which suggests that this element can form many organic compounds, there are actually only a few organogermanium compounds with technical or other relevance, and only two of environmental relevance.[227] These two—mono- and dimethylgermanium—have been detected at ultratrace concentrations (i.e., in the low picomolar range) in various waters, including surface, river, and sea water.[228] In contrast to OT and organolead compounds, the profiles of these two organogermanium compounds are very conservative, indicating that methylgermanium compounds may be converted on a timescale of 10^6 years. It is also not understood why, again in contrast to the organoelement compounds of the heavier members of this group, there is no naturally occurring trimethylgermanium. Apparently, the biosynthesis of methylgermanium compounds leads only to the mono- and disubstituted compounds; initial reports of trimethylgermanium detection can

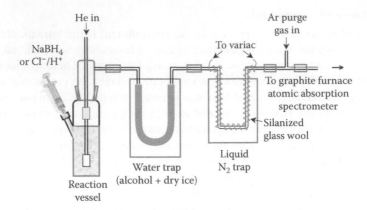

FIGURE 13.18 Typical experimental setup for the (P&T) generation collection and determination of volatile germanium hydrides from water samples after borohydride derivatization.

probably be attributed to analytical artifacts. Analytical methods for organogermanium compounds have not been particularly well investigated, but they follow the same strategy as for other organometallic compounds: the ionic methylgermanium compounds have to be derivatized for GC analysis by alkylation with Grignard or tetraalkylborate reagents, or by hydride formation. Given the fact that germanium compounds are more volatile than their tin or lead homologs, simultaneous matrix-separation and enrichment of the analytes is efficiently performed by P&T analysis.[228] In this early example of Ge speciation, the cold trap served simultaneously as a packed analytical column for separating individual methylgermanium species (Figure 13.18). Later, the analytical method was refined in that capillary columns were used for the GC separation. Alternatively, underivatized methylgermanium chlorides (which are sufficiently volatile and relatively stable) were determined directly,[229] or after Grignard derivatization.[230] Fully alkylated organogermanium compounds show a better recovery owing to their reduced interaction with active sites of the GC system, and also a better peak shape. Following the early work on organogermanium compounds in (sea) water in the 1980s[228,231] and some further methodological development in the mid-1990s,[229,230,232] which remained without application in the hydrosphere, there appears to have been no recent interest in the speciation analysis of organogermanium compounds in the aquatic environment, which may reflect the low level of hazard believed to be due to these species.

13.3.9.4 Organoselenium Compounds

Although not an element of the main group IV of the periodic table, selenium also forms stable organoelement species of some environmental relevance. Whereas selenium may occur in its inorganic compounds in the oxidation states $-II$, $+IV$, or $+VI$, the only stable organoselenium compounds are those of $Se(+II)$. Dimethyl selenide and dimethyl diselenide, and also the trimethylselenonium cation, have been reported in the natural environment.[233] The transformation of inorganic Se to the more volatile but less toxic methylated Se species by microbial action is an important link in the biogeochemical cycling of this element. In the presence of biota, the selenoaminoacid and selenomethionine can also occur.[234] Dimethyl selenide and dimethyl diselenide are volatile compounds that can be determined by GC with element- or molecule-specific detection. Due to their volatile nature, enrichment and matrix separation by P&T is particularly advantageous.[235] To determine the complete speciation of this element, the concentration of the inorganic Se species also has to be determined, which is at least indirectly possible with a modification of this technique: inorganic $Se(+IV)$ (selenite) can be derivatized directly with $NaBH_4$ or $NaBEt_4$ to form SeH_2 or Et_2Se. While the former species is well known from the determination of (total) selenium by hydride generation-atomic spectrometry (atomic absorption detection (AAS, OES, or ICP-MS), the latter derivative can be determined separately without interferences[236] or together with the dimethyl(di-)selenide species.[237] In a further report,[238] tetraethylborate derivatization was compared to the derivatization of Se(+IV) by 4,5-dichloro-1,2-phenylenediamine to form the corresponding piazselenol. The analytical figures of merit of the two derivatization schemes compare very well. NeBEt$_4$ derivatization is particularly advantageous since it can be performed directly in the aqueous phase, and is thus more expeditious, particularly when combined with SPME. As a general disadvantage of all these methods, it must be mentioned that only Se(+IV) can be derivatized directly. If Se(+VI) (selenate) is to be determined, it has to be prereduced, for example, with half-concentrated HCl.[239] The Se(+VI) concentration can thus be determined from the difference in the signal with and without the preceding reduction step; however, the analytical precision of the determination suffers significantly from this indirect determination and may be inadequate.

The determination of organic selenium compounds is done preferably by GC coupled to element- or molecule-specific detectors, such as GC-AED or molecular mass spectrometric detection (GC-MS).[240] In this case, ICP-MS detection does not yield the improvement in sensitivity otherwise seen, which is due to spectral interferences. Dietz et al.[241] have compared the analytical figures of merit of three detector systems for GC (AED, atomic fluorescence spectroscopy (AFS), and ICP-MS), arriving at the conclusion that GC-AED is the most sensitive and most practical

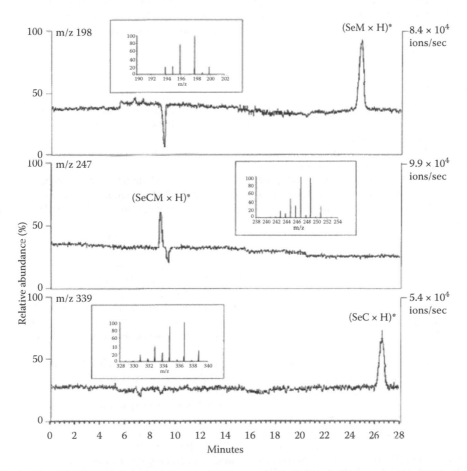

FIGURE 13.19 CE/ESI-MS electropherogram for the analysis of three selenium species (SeM, SeC, and SeCM) using 5% acetic acid as electrolyte buffer. The insets give information about the isotopic patterns of the species. (Reprinted from Michalke et al. 1999 *Fresenius' J. Anal. Chem.* 363: 456–459.)

hyphenated GC technique for volatile organoselenium species. When the inorganic species Se(+IV) and Se(+VI) are to be determined, the preferred method of analysis is IC.[242] With this technique, the Se amino acids selenomethionine and selenocysteine can also be determined in the same run. Alternative methods for the analysis of Se amino acids are gas chromatographic analysis with element- or molecule-specific detection after derivatization, for example, with chloroformate reagents[243] or in a two-step acylation/esterification reaction.[244] Furthermore, CE, coupled to ICP-MS or ESI-MS detection, can be used in the speciation of inorganic selenium species[245] and selenium amino acids (Figure 13.19).[246]

13.4 CONCLUSIONS AND OUTLOOK

Methods for the analysis of organic and organometallic compounds are discussed in this chapter. It has become evident that for the analysis of these two classes of compounds, the analyst can draw on a very similar repertoire of analytical techniques with respect to sample preparation, separation, and detection. Chromatographic and, in particular, hyphenated techniques are the workhorses of environmental water analysis. The various formats and technical realizations of mass spectrometers are the most versatile detectors. Their sensitivity and ability to provide structural information at the low and even sub-pg level are an asset and at the same time a prerequisite for (ultra)trace analysis in the aquatic environment. As further significant improvements in detector sensitivity are unlikely, the probable focus of attention in the future will again be on sample preparation. Here, the introduction of new approaches, techniques, and materials for sample preparation can be expected to make a significant impact in this field.

REFERENCES

1. UNEP (United Nations Environmental Programme). 2002. Vital water graphics—An overview of the state of the world's fresh and Marine waters. Available at: http://www.unep.org/dewa/assessments/ecosystems/water/vitalwater/. ISBN: 92-807-2236-0.
2. Adams, W.J., R.A. Kimerle, and J.W. Barnett, Jr. 1992. Sediment quality and aquatic life assessment. *Environ. Sci. Technol.* 26: 1864–1875.
3. Borja, A., V. Valencia, J. Franco, et al. 2004. The water framework directive: Water alone, or in association with sediment and biota, in determining quality standards? *Mar. Poll. Bull.* 49: 8–11.
4. Rifkin, E., P. Gwinn, and E. Bouwer. 2004. Chromium and sediment toxicity. *Environ. Sci. Technol.* 38: 267A–271A.
5. Wells, R.J. 1999. Recent advances in non-silylation derivatization techniques for gas chromatography. *J. Chromatogr. A* 843: 1–18.
6. Morabito, R., P. Massanisso, and P. Quevauviller. 2000. Derivatization methods for the determination of organotin compounds in environmental samples. *Trends Anal. Chem.* 19: 113–119.
7. Hušek, P. and P. Šimek. 2006. Alkyl chloroformates in sample derivatization strategies for GC analysis. Review on a decade use of the reagents as esterifying agents. *Curr. Pharmaceut. Anal.* 2: 23–43.
8. Lepom P., B. Brown, G. Hanke, et al. 2009. Needs for reliable analytical methods for monitoring chemical pollutants in surface water under the European Water Framework Directive. *J. Chromatogr. A.* 1216: 302–315.
9. Metropolis. 2004. Metrology in support of precautionary sciences and sustainable development policies, database of analytical methods created in the frame of the EU Project Metropolis (GTC2–2001–53008). Available at: http://metropolis.speciation.net/index.php.
10. DIN EN 1485. 1996. Water Quality—Determination of Adsorbable Organically Bound Halogens (AOX). German version EN 1485. Berlin: Beuth-Verlag.
11. EN ISO 9562. 2004. Water quality—Determination of adsorbable organically bound halogens (AOX) (ISO 9562:2004). German version EN ISO 9562. Berlin: Beuth-Verlag.
12. DIN 38409-H18. 1981. DIN 38409 H18. Determination of Hydrocarbons. Berlin: Beuth-Verlag.
13. DIN 38409-H22. 2001. DIN 38409 H22. Water Quality—Determination of Adsorbable Organically Bound Halogens in Highly Saline Waters After Solid Phase Extraction (SPE-AOX). Berlin: Beuth-Verlag.

14. BS EN 13370:2003. 2003. Characterization of waste. Analysis of eluates. Determination of Ammonium, AOX, conductivity, Hg, phenol index, TOC, easily liberatable CN⁻, F⁻. British Standards.
15. Santos, F.J. and M.T. Galceran. 2002. The application of gas chromatography to environmental analysis. *Trends Anal. Chem.* 21: 672–685.
16. Bailey, R. 2005. Injectors for capillary gas chromatography and their application to environmental analysis. *J. Environ. Monit.* 7: 1054–1058.
17. Engewald, W., J. Teske, and J. Efer. 1999. Programmed temperature vaporiser-based injection in capillary gas chromatography. *J. Chromatogr. A* 856: 259–278.
18. Hyötyläinen, T. and M.-L. Riekkola. 2003. On-line coupled liquid chromatography-gas chromatography. *J. Chromatogr. A* 1000: 357–384.
19. Kerkdijk, H., H.G.J. Mol, and B. Van Der Nagel. 2007. Volume overload cleanup: An approach for on-line SPE-GC, GPC-GC, and GPC-SPE-GC. *Anal. Chem.* 79: 7975–7983.
20. David, F., C. Devos, D. Joulain, et al. 2006. Determination of suspected allergens in non-volatile matrices using PTV injection with automated liner exchange and GC-MS. *J. Sep. Sci.* 29: 1587–1594.
21. Zeeuw, J.D. and J. Luong. 2002. Developments in stationary phase technology for gas chromatography. *Trends Anal. Chem.* 21: 594–607.
22. Cervini, R., G. Day, A. Hibberd, et al. 2001. Investigation of a novel, sol-gel derived stationary phase for gas chromatography. *LC-GC Europe* 14: 564–569.
23. Buszewski, B. and S. Studzinśka. 2008. A review of ionic liquids in chromatographic and electromigration techniques. *Chromatographia* 68: 1–10.
24. Seeley, J.V., S.K. Seeley, E.K. Libby, et al. 2008. Comprehensive two-dimensional gas chromatography using a high-temperature phosphonium ionic liquid column. *Anal. Bioanal. Chem.* 390: 323–332.
25. Semard, G., A. Bruchet, P. Cardinaël, et al. 2008. Use of comprehensive two-dimensional gas chromatography for the broad screening of hazardous contaminants in urban wastewaters. *Water Sci. Technol.* 57: 1983–1989.
26. Korytar, P., P.E.G. Leonards, J. De Boer, et al. 2005. Group separation of organohalogenated compounds by means of comprehensive two-dimensional gas chromatography. *J. Chromatogr. A* 1086: 29–44.
27. Skoczynśka, E., P. Korytár, and J. De Boer. 2008. Maximizing chromatographic information from environmental extracts by GC × GC-ToF-MS. *Environ. Sci. Technol.* 42: 6611–6618.
28. Reemtsma, T. 2001. The use of liquid chromatography-atmospheric pressure ionization-mass spectrometry in water analysis. Part I: Achievements. *Trends Anal. Chem.* 20: 500–517.
29. Zwiener, C. and F.H. Frimmel. 2004. LC-MS analysis in the aquatic environment and in water treatment—A critical review: Part I: Instrumentation and general aspects of analysis and detection. *Anal. Bioanal. Chem.* 378: 851–861.
30. Wyndham, K.D., J.E. O'Gara, T.H. Walter, et al. 2003. Characterization and evaluation of C18 HPLC stationary phases based on ethyl-bridged hybrid organic/inorganic particles. *Anal. Chem.* 75: 6781–6788.
31. Visky, D., Y. Vander Heyden, T. Iványi, et al. 2002. Characterisation of reversed-phase liquid chromatographic columns by chromatographic tests. Evaluation of 36 test parameters: Repeatability, reproducibility and correlation. *J. Chromatogr. A* 977: 39–58.
32. Nguyen, H.P. and K.A. Schug. 2008. The advantages of ESI-MS detection in conjunction with HILIC mode separations: Fundamentals and applications. *J. Sep. Sci.* 31: 1465–1480.
33. Dugo, P., F. Cacciola, T. Kumm, et al. 2008. Comprehensive multidimensional liquid chromatography: Theory and applications. *J. Chromatogr. A* 1184: 353–368.
34. Jandera, P. 2008. Stationary phases for hydrophilic interaction chromatography, their characterization and implementation into multidimensional chromatography concepts. *J. Sep. Sci.* 31. 1421–1437.
35. Dabek-Zlotorzynśka, E., V. Celo, and M.M. Yassine. 2007. Recent advances in CE and CEC of pollutants. *Electrophoresis* 29: 310–323.
36. Ravelo-Pérez, L.M., J. Hernández-Borges, and M.Á. Rodríguez-Delgado. 2006. Pesticides analysis by liquid chromatography and capillary electrophoresis. *J. Sep. Sci.* 29: 2557–2577.
37. Kubáň P., P. Houserová, P. Kubáň et al. 2007. Mercury speciation by CE: A review. *Electrophoresis* 28: 58–68.
38. Li, Y., X.-B. Yin, and X.-P. Yan. 2008. Recent advances in on-line coupling of capillary electrophoresis to atomic absorption and fluorescence spectrometry for speciation analysis and studies of metal-biomolecule interactions. *Anal. Chim. Acta* 615: 105–114.
39. Simpson, S.L. Jr., J.P. Quirino, and S. Terabe. 2008. On-line sample preconcentration in capillary electrophoresis. Fundamentals and applications. *J. Chromatogr. A* 1184: 504–541.

40. Álvarez-Llamas, G., M.D.R. Fernández De LaCampa, and A. Sanz-Medel. 2005. ICP-MS for specific detection in capillary electrophoresis. *Trends Anal. Chem.* 24: 28–36.

41. Ravelo-Pérez, L.M., J. Hernández-Borges, A. Cifuentes, et al. 2007. MEKC combined with SPE and sample stacking for multiple analysis of pesticides in water samples at the ng/L level. *Electrophoresis* 28: 1805–1814.

42. Liu, Y. and L. Jia. 2008. Analysis of estrogens in water by magnetic octadecylsilane particles extraction and sweeping micellar electrokinetic chromatography. *Microchem. J.* 89: 72–76.

43. Jin, M. and Y. Yang. 2006. Simultaneous determination of nine trace mono- and di-chlorophenols in water by ion chromatography atmospheric pressure chemical ionization mass spectrometry. *Anal. Chim. Acta* 566: 193–199.

44. Barron, L. and B. Paul. 2006. Simultaneous determination of trace oxyhalides and haloacetic acids using suppressed ion chromatography-electrospray mass spectrometry. *Talanta* 69: 621–630.

45. Karthikeyan, S. and S. Hirata. 2004. Ion chromatography-inductively coupled plasma mass spectrometry determination of arsenic species in marine samples. *Appl. Organomet. Chem.* 18: 323–330.

46. Chen, Z., N.I. Khan, G. Owens, et al. 2007. Elimination of chloride interference on arsenic speciation in ion chromatography inductively coupled mass spectrometry using an octopole collision/reaction system. *Microchem. J.* 87: 87–90.

47. Tirez, K., W. Brusten, S. Van Roy, et al. 2000. Characterization of inorganic selenium species by ion chromatography with ICP-MS detection in microbial-treated industrial waste water. *J. Anal. At. Spectrom.* 15: 1087–1092.

48. Wallschläger, D. and R. Roehl. 2001. Determination of inorganic selenium speciation in waters by ion chromatography-inductively coupled plasma-mass spectrometry using eluant elimination with a membrane suppressor. *J. Anal. At. Spectrom.* 16: 922–925.

49. Pantsar-Kallio, M. and P.K.G. Manninen. 1996. Speciation of chromium in waste waters by coupled column ion chromatography-inductively coupled plasma mass spectrometry. *J. Chromatogr. A* 750: 89–95.

50. Chen, Z., M. Megharaj, and R. Naidu. 2007. Speciation of chromium in waste water using ion chromatography inductively coupled plasma mass spectrometry. *Talanta* 72: 394–400.

51. Ulrich, N. 1998. Speciation of antimony(III), antimony(V) and trimethylstiboxide by ion chromatography with inductively coupled plasma atomic emission spectrometric and mass spectrometric detection. *Anal. Chim. Acta* 359: 245–253.

52. Pantsar-Kallio, M. and P.K.G. Manninen. 1999. Optimizing ion chromatography-inductively coupled plasma mass spectrometry for speciation analysis of arsenic, chromium and bromine in water samples. *Int. J. Environ. Anal. Chem.* 75: 43–55.

53. Seubert, A. 2001. On-line coupling of ion chromatography with ICP-AES and ICP-MS. *Trends Anal. Chem.* 20: 274–287.

54. Perminova, I.V., F.H. Frimmel, A.V. Kudryavtsev, et al. 2003. Molecular weight characteristics of humic substances from different environments as determined by size exclusion chromatography and their statistical evaluation. *Environ. Sci. Technol.* 37: 2477–2485.

55. Reemtsma, T. and A. These. 2005. Comparative investigation of low-molecular-weight fulvic acids of different origin by SEC-Q-TOF-MS: New insights into structure and formation. *Environ. Sci. Technol.* 39: 3507–3512.

56. Gremm, Th.J. and F.H. Frimmel. 2000. Characterization of AOX by fractionation analysis and size-exclusion chromatography. *Acta Hydrochim. Hydrobiol.* 28: 202–211.

57. Grant, T.D., R.G. Wuilloud, J.C. Wuilloud, et al. 2004. Investigation of the elemental composition and chemical association of several elements in fulvic acids dietary supplements by size-exclusion chromatography UV inductively coupled plasma mass spectrometric. *J. Chromatogr. A* 1054: 313–319.

58. Vachet, R.W. and M.B. Callaway. 2003. Characterization of Cu(II)-binding ligands from the Chesapeake Bay using high-performance size-exclusion chromatography and mass spectrometry. *Marter. Chem.* 82: 31–45.

59. Reszat, T.N. and M.J. Hendry. 2007. Complexation of aqueous elements by DOC in a clay aquitard. *Ground Water* 45: 542–553.

60. De Momi, A. and J.R. Lead. 2006. Size fractionation and characterisation of fresh water colloids and particles: Split-flow thin-cell and electron microscopy analyses. *Environ. Sci. Technol.* 40: 6738–6743.

61. van Leeuwen, S.P.J. and J. de Boer. 2008. Advances in the gas chromatographic determination of persistent organic pollutants in the aquatic environment. *J. Chromatogr. A* 1186: 161–182.

62. Morabito, R. and P. Quevauviller. 2002. Performances of spectroscopic methods for tributyltin (TBT) determination in the 10 years of the EU-SM&T organotin programme. *Spectrosc. Europe* 14: 18–23.

63. Asperger, A., J. Efer, T. Koal, et al. 2002. Trace determination of priority pesticides in water by means of high-speed on-line solid-phase extraction-liquid chromatography-tandem mass spectrometry using turbulent-flow chromatography columns for enrichment and a short monolithic column for fast liquid chromatographic separation. *J. Chromatogr. A* 960: 109–119.

64. Bure C. and C. Lange. 2003. Comparison of dissociation of ions in an electrospray source, or a collision cell in tandem mass spectrometry. *Curr. Org. Chem.* 7: 1613–1624.

65. Marquet, P., N. Venisse, E. Lacassie, et al. 2000. In-source CID mass spectral libraries for the "general unknown" screening of drugs and toxicants. *Analusis* 28: 925–934.

66. Ohlenbusch, G., C. Zwiener, R.U. Meckenstock, et al. 2002. Identification and quantification of polar naphthalene derivatives in contaminated groundwater of a former gas plant site by liquid chromatography-electrospray ionization tandem mass spectrometry. *J. Chromatogr. A* 967: 201–207.

67. Zwiener, C., T. Glauner, and F.H. Frimmel. 2002. Method optimization for the determination of carbonyl compounds in disinfected water by DNPH derivatization and LC-ESI-MS-MS. *Anal. Bioanal. Chem.* 372: 615–621.

68. Pozo, O.J., J.V. Sancho, M. Ibáñez, et al. 2006. Confirmation of organic micropollutants detected in environmental samples by liquid chromatography tandem mass spectrometry: Achievements and pitfalls. *Trends Anal. Chem.* 25: 1030–1042.

69. Ferrer, I. and E.M. Thurman. 2003. Liquid chromatography/time-of-flight/mass spectrometry (LC/TOF/MS) for the analysis of emerging contaminants. *Trends Anal. Chem.* 22: 750–756.

70. Bobeldijk, I., J.P.C. Vissers, G. Kearney, et al. 2001. Screening and identification of unknown contaminants in water with liquid chromatography and quadrupole-orthogonal acceleration-time-of-flight tandem mass spectrometry. *J. Chromatogr. A* 929: 63–74.

71. Sleighter, R.L. and P.G. Hatcher. 2007. The application of electrospray ionization coupled to ultrahigh resolution mass spectrometry for the molecular characterization of natural organic matter. *J. Mass Spectrom.* 42: 559–574.

72. Pinto, D.M., R.K. Boyd, and D.A. Volmer. 2002. Ultra-high resolution for mass spectrometric analysis of complex and low-abundance mixtures—The emergence of FTICR-MS as an essential analytical tool. *Anal. Bioanal. Chem.* 373: 378–389.

73. Stenson, A.C., A.G. Marshall, and W.T. Cooper. 2003. Exact masses and chemical formulas of individual Suwannee River fulvic acids from ultrahigh resolution electrospray ionization Fourier transform ion cyclotron resonance mass spectra. *Anal. Chem.* 75: 1275–1284.

74. Mondello, L., A. Salvatore, P.Q. Tranchida, et al. 2008. Reliable identification of pesticides using linear retention indices as an active tool in gas chromatographic-mass spectrometric analysis. *J. Chromatogr. A* 1186: 430–433.

75. Thurman, E.M., I. Ferrer, and D. Barceló. 2001. Choosing between atmospheric pressure chemical ionization and electrospray ionization interfaces for the HPLC/MS analysis of pesticides. *Anal. Chem.* 73: 5441–5449.

76. Robb, D.B. and M.W. Blades. 2008. State-of-the-art in atmospheric pressure photoionization for LC/MS. *Anal. Chim. Acta* 627: 34–49.

77. Moriwaki, H., M. Ishitake, S. Yoshikawa, et al. 2004. Determination of polycyclic aromatic hydrocarbons in sediment by liquid chromatography-atmospheric pressure photoionization-mass spectrometry. *Anal. Sci.* 20: 375–377.

78. Viglino, L., K. Aboulfadl, M. Prévost, et al. 2008. Analysis of natural and synthetic estrogenic endocrine disruptors in environmental waters using online preconcentration coupled with LC-APPI-MS/MS. *Talanta* 76: 1088–1096.

79. Yu, K., L. Di, E. Kerns, et al. 2007. Ultra performance liquid chromatography/tandem mass spectrometric quantification of structurally diverse drug mixtures using an ESI-APCI multimode ionization source. *Rapid Commun. Mass Spectrom.* 21: 893–902.

80. Hanold, K.A., J. Horner, R. Thakur, et al. 2002. Dual APPI/APCI source for LC/MS. *Proceedings of the 50th ASMS Conference on Mass Spectrometry and Allied Topics*, pp. 31–32.

81. Ammann, A.A. 2007. Inductively coupled plasma mass spectrometry (ICP MS): A versatile tool. *J. Mass Spectrom.* 42: 419–427.

82. Tanner, S.D., V.I. Baranov, and D.R. Bandura. 2002. Reaction cells and collision cells for ICP-MS: A tutorial review. *Spectrochim. Acta B* 57: 1361–1452.

83. Rosen, A.L. and G.M. Hieftje. 2004. Inductively coupled plasma mass spectrometry and electrospray mass spectrometry for speciation analysis: Applications and instrumentation. *Spectrochim. Acta B* 59: 135–146.

84. Szpunar, J., R. Lobinski, and A. Prange. 2003. Hyphenated techniques for elemental speciation in biological systems. *Appl. Spectrosc.* 57: 102A–112A.

85. Demeestere, K., J. Dewulf, B. De Witte, et al. 2007. Sample preparation for the analysis of volatile organic compounds in air and water matrices. *J. Chromatogr. A* 1153: 130–144.

86. Snow, N.H. and G.C. Slack. 2002. Head-space analysis in modern gas chromatography. *Trends Anal. Chem.* 21: 608–617.

87. Wardencki, W., J. Curyło, and J. Namieśnik. 2007. Trends in solventless sample preparation techniques for environmental analysis. *J. Biochem. Biophys. Methods* 70: 275–288.

88. Wang, Y., J. McCaffrey, and D.L. Norwood. 2008. Recent advances in headspace gas chromatography. *J. Liq. Chromatogr. Rel. Technol.* 31: 1823–1851.

89. Ketola, R.A., V.T. Virkki, M. Ojala, et al. 1997. Comparison of different methods for the determination of volatile organic compounds in water samples. *Talanta* 44: 373–382.

90. Jávorszky, E., G. Erdoedy, and K. Torkos. 2006. The choice of internal standards for measuring volatile pollutants in water. *Chromatographia* 63: S55–S60.

91. Huybrechts, T., J. Dewulf, and H. Van Langenhove. 2003. State-of-the-art of gas chromatography-based methods for analysis of anthropogenic volatile organic compounds in estuarine waters, illustrated with the river Scheldt as an example. *J. Chromatogr. A* 1000: 283–297.

92. Helmig, D. and L. Vierling. 1995. Water adsorption capacity of the solid adsorbents Tenax TA, Tenax GR, Carbotrap, Carbotrap C, Carbosieve SIII, and Carboxen 569 and water management techniques for the atmospheric sampling of volatile organic trace gases. *Anal. Chem.* 67: 4380–4386.

93. Silgoner, I., E. Rosenberg, and M. Grasserbauer. 1997. Determination of volatile organic compounds in water by purge-and-trap gas chromatography coupled to atomic emission detection. *J. Chromatogr. A* 768: 259–270.

94. Smith, G.A. and E.T. Heggs. 2001. Optimization of method parameters for the evaluation of USEPA method 524.2 using a Tekmar-Dohrmann 3100 purge and trap with GC/MS. *ACS National Meeting Book of Abstracts* 41: 25–30.

95. Dewulf, J. and H. Van Langenhove. 1999. Anthropogenic volatile organic compounds in ambient air and natural waters: A review on recent developments of analytical methodology, performance and interpretation of field measurements. *J. Chromatogr. A* 843: 163–177.

96. George, C. and R.E. Majors. 2001. High-speed analysis of volatile organic compounds in environmental samples using small-diameter capillary columns and purge-and-trap GC-MS systems. *LC-GC North America* 19: 578–588.

97. Luo, Y.Z. and J. Pawliszyn. 2000. Membrane extraction with a sorbent interface for headspace monitoring of aqueous samples using a cap sampling device. *Anal. Chem.* 72: 1058–1063.

98. Kostiainen, R., T. Kotiaho, I. Mattila, et al. 1998. Analysis of volatile organic compounds in water and soil samples by purge-and-membrane mass spectrometry. *Anal. Chem.* 70: 3028–3032.

99. Ojala, M., R. Ketola, T. Mansikka, et al. 1997. Detection of volatile organic sulfur compounds in water by headspace gas chromatography and membrane inlet mass spectrometry. *HRC—J. High Res. Chromatogr.* 20: 165–169.

100. Segal, A., T. Górecki, P. Mussche, et al. 2000. Development of membrane extraction with a sorbent interface-micro gas chromatography system for field analysis: *J. Chromatogr. A* 873: 13–27.

101. Belardi, R.P. and J.B. Pawliszyn. 1989. Application of chemically modified fused silica fibers in the extraction of organics from water matrix samples and their rapid transfer to capillary columns. *Water Poll. Res. J. Can.* 24: 179–191.

102. Pawliszyn, J. 1997. *Solid Phase Microextraction: Theory and Practice.* New York: Wiley-VCH.

103. Scheppers Wercinski, S.A. 1999. *Solid Phase Microextraction: A Practical Guide.* New York: Marcel Dekker.

104. Ouyang, G. and J. Pawliszyn. 2006. SPME in environmental analysis. *Anal. Bioanal. Chem.* 386: 1059–1073.

105. Aulakh, J.S., A.K. Mailk, V. Kaur, et al. 2005. A review on solid phase micro extraction—High performance liquid chromatography (SPME-HPLC) analysis of pesticides. *Crit. Rev. Anal. Chem.* 35: 71–85.

106. Mothes, S. and R. Wennrich. 2000. Coupling of SPME and GC-AED for the determination of organometallic compounds. *Mikrochim. Acta* 135: 91–95.

107. Mester, Z., R. Sturgeon, and J. Pawliszyn. 2001. Solid phase microextraction as a tool for trace element speciation. *Spectrochim. Acta B* 56: 233–260.

108. Baltussen, E., C.A. Cramers, and P.J.F. Sandra. 2002. Sorptive sample preparation—A review. *Anal. Bioanal. Chem.* 373: 3–22.

109. Hyötyläinen, T. and M.-L. Riekkola. 2008. Sorbent- and liquid-phase microextraction techniques and membrane-assisted extraction in combination with gas chromatographic analysis: A review. *Anal. Chim. Acta* 614: 27–37.

110. Armenta, S., S. Garrigues, and M. de la Guardia. 2008. Green analytical chemistry. *Trends Anal. Chem.* 27: 497–511.

111. Picó, Y., M. Fernández, M.J. Ruiz, et al. 2007. Current trends in solid-phase-based extraction techniques for the determination of pesticides in food and environment. *J. Biochem. Biophys. Methods* 70: 117–131.

112. Liška, I. 2000. Fifty years of solid-phase extraction in water analysis—Historical development and overview. *J. Chromatogr. A* 885: 3–16.

113. Thurman E.M. and M.S. Mills. 1998. *Solid-phase Extraction: Principles and Practice.* New York: Wiley-Interscience.

114. Thurman, E.M. and K. Snavely. 2000. Advances in solid-phase extraction disks for environmental chemistry. *Trends Anal. Chem.* 19: 18–26.

115. Louter A.J.H., C.A. van Beekvelt, P. Cid Montanes, et al. 2005. Analysis of microcontaminants in aqueous samples by fully automated on-line solid-phase extraction-gas chromatography-mass selective detection. *J. Chromatogr. A* 725: 67–83.

116. Wissiack, R., E. Rosenberg, and M. Grasserbauer. 2000. Comparison of different sorbent materials for on-line solid-phase extraction with liquid chromatography-atmospheric pressure chemical ionization mass spectrometry of phenols. *J. Chromatogr. A* 896: 159–170.

117. Wang, S., W. Huang, G. Fang, et al. 2008. On-line coupling of solid-phase extraction to high-performance liquid chromatography for determination of estrogens in environment. *Anal. Chim. Acta* 606: 194–201.

118. Kot-Wasik, A., J. Dębska, A. Wasik, et al. 2006. Determination of non-steroidal anti-inflammatory drugs in natural waters using off-line and on-line SPE followed by LC coupled with DAD-MS. *Chromatographia* 64: 13–21.

119. Koal, T., A. Asperger, J. Efer, et al. 2003. Simultaneous determination of a wide spectrum of pesticides in water by means of fast on-line SPE-HPLC-MS-MS—A novel approach. *Chromatographia* 57: S93–S101.

120. Hennion, M.-C. and V. Pichon. 2003. Immuno-based sample preparation for trace analysis. *J. Chromatogr. A* 1000: 29–52.

121. Cichna-Markl, M. 2006. Selective sample preparation with bioaffinity columns prepared by the solgel method. *J. Chromatogr. A* 1124: 167–180.

122. Pichon, V. 2007. Selective sample treatment using molecularly imprinted polymers. *J. Chromatogr. A* 1152: 41–53.

123. Pichon, V. and F. Chapuis-Hugon. 2008. Role of molecularly imprinted polymers for selective determination of environmental pollutants-A review. *Anal. Chim. Acta* 622: 48–61.

124. Moldoveanu, S.C. and V. David. 2002. Sample preparation in chromatography. *J. Chromatogr. Library,* Vol. 65, pp. 473–525. Amsterdam: Elsevier.

125. Seiwert, B. and U. Karst. 2008. Ferrocene-based derivatization in analytical chemistry. *Anal. Bioanal. Chem.* 390: 181–200.

126. Vreeken, R.J., P. Speksnijder, I. Bobeldijk-Pastorova, et al. 1998. Selective analysis of the herbicides glyphosate and aminomethylphosphonic acid in water by on-line solid-phase extraction-high-performance liquid chromatography-electrospray ionization mass spectrometry. *J. Chromatogr. A* 794: 187–199.

127. Knapp, D.R. 1979. *Handbook of Analytical Derivatization Reactions.* New York: Wiley-Interscience.

128. Rosenfeld, J.M. 1999. Solid-phase analytical derivatization: Enhancement of sensitivity and selectivity of analysis. *J. Chromatogr. A* 843: 19–27.

129. Waterval, J.C.M., H. Lingeman, A. Bult, et al. 2000. Derivatization trends in capillary electrophoresis. *Electrophoresis* 21: 4029–4045.

130. Halket, J.M. and V.G. Zaikin. 2003. Derivatization in mass spectrometry—1. Silylation. *Eur. J. Mass Spectrom.* 9: 1–21.

131. Halket, J.M. and V.G. Zaikin. 2004. Derivatization in mass spectrometry—3. Alkylation (arylation). *Eur. J. Mass Spectrom.* 10: 1–19.

132. Zaikin, V.G. and J.M. Halket. 2003. Derivatization in mass spectrometry—2. Acylation. *Eur. J. Mass Spectrom.* 9: 421–434.

133. Zaikin, V.G. and J.M. Halket. 2004. Derivatization in mass spectrometry—4. Formation of cyclic derivatives. *Eur. J. Mass Spectrom.* 10: 555–568.

134. de la Calle-Guntiñas, M.B., R. Scerbo, S. Chiavarini, et al. 1997. Comparison of derivatization methods for the determination of butyl- and phenyltin compounds in mussels by gas chromatography. *Appl. Organomet. Chem.* 11: 693–702.

135. Schubert, P., E. Rosenberg, and M. Grasserbauer. 2000. Comparison of sodium tetraethylborate and sodium tetra(*n*-propyl)borate as derivatization reagent for the speciation of organotin and organolead compounds in water samples. *Fresenius J. Anal. Chem.* 366: 356–360.

136. DIN EN ISO 17353. 2005. Water quality—Determination of selected organotin compounds—Method based on gas chromatography (ISO 17353:2004). German version EN ISO 17353:2005 (version: 2005-11). Berlin: Beuth-Verlag.

137. DIN 38407-13. 1999. Determination of selected organotin compounds by means of gas chromatography (F 13). Draft as of October 1999. Berlin: Beuth-Verlag.

138. González-Toledo, E., R. Compañó, M. Granados, et al. 2003. Detection techniques in speciation analysis of organotin compounds by liquid chromatography. *Trends Anal. Chem.* 22: 26–33.

139. Reichert, J.K. and J. Lochtman. 1983. Testing methods for halogenated organic compounds. Experiences with drinking and surface waters. *Environ. Technol. Lett.* 4: 15–26.

140. Licha, T., M. Herfort, and M. Sauter. 2001. Phenolindex—ein sinnvoller Parameter für die Altlastenbewertung? (in German). *Grundwasser* 6: 8–14.

141. Miró, M., J.M. Estela, and V. Cerdà. 2004. Application of flowing-stream techniques to water analysis: Part II. General quality parameters and anionic compounds: Halogenated, sulphur and metalloid species. *Talanta* 62: 1–15.

142. Miró, M. and W. Frenzel. 2004. What flow injection has to offer in the environmental analytical field. *Microchim. Acta* 148: 1–20.

143. Lynggaard-Jensen, A. 1999. Trends in monitoring of waste water systems. *Talanta* 50: 707–716.

144. Bogue, R. 2008. Environmental sensing: Strategies, technologies and applications. *Sens. Rev.* 28: 275–282.

145. Rodriguez-Mozaz, S., M.J. Lopez de Alda, and D. Barceló. 2007. Advantages and limitations of on-line solid phase extraction coupled to liquid chromatography-mass spectrometry technologies versus biosensors for monitoring of emerging contaminants in water. *J. Chromatogr. A* 1152: 97–115.

146. Kontana, A., C.A. Papadimitriou, P. Samaras, et al. 2008. Bioassays and biomarkers for ecotoxicological assessment of reclaimed municipal wastewater. *Water Sci. Technol.* 57: 947–953.

147. Girotti, S., E.N. Ferri, M.G. Fumo, et al. 2008. Monitoring of environmental pollutants by bioluminescent bacteria. *Anal. Chim. Acta* 608: 2–29.

148. Millán de Kuhn, R., C. Streb, R. Breiter, et al. 2006. Screening for unicellular algae as possible bioassay organisms for monitoring marine water samples. *Water Res.* 40: 2695–2703.

149. Tahedl, H. and D.-P. Häder. 1999. Fast examination of water quality using the automatic biotest ECOTOX based on the movement behavior of a freshwater flagellate. *Water Res.* 33: 426–432.

150. Seifert, M., G. Brenner-Weiß, S. Haindl, et al. 1999. A new concept for the bioeffects-related analysis of xenoestrogens: Hyphenation of receptor assays with LC-MS. *Fresenius J. Anal. Chem.* 363: 767–770.

151. Brenner-Weiß, G. and U. Obst. 2003. Approaches to bioresponse-linked instrumental analysis in water analysis. *Anal. Bioanal. Chem.* 377: 408–416.

152. Kolpin, D.W., E.T. Furlong, M.T. Meyer, et al. 2002. Pharmaceuticals, hormones, and other organic wastewater contaminants in U.S. streams, 1999–2000: A National Reconnaissance. *Environ. Sci. Technol.* 36: 1202–1211.

153. Erickson, B.E. 2002. Analyzing the ignored environmental contaminants. *Environ. Sci. Technol.* 36: 140A–145A.

154. European Commission. 2000. Water Framework Directive. *Off. J. Eur. Commun.* Directive 2000/60/EC, L327.

155. Rodil, R., J.B. Quintana, P. López-Mahía, et al. 2009. Multi-residue analytical method for the determination of emerging pollutants in water by solid-phase extraction and liquid chromatography-tandem mass spectrometry. *J. Chromatogr. A.* 1216: 2958–2969.

156. Wells, M.J.M., L.J. Fono, M.-L. Pellegrin, et al. 2007. Emerging pollutants. *Water Environ. Res.* 79: 2192–2209.

157. Barceló, D. and M. Petrovic. 2006. New concepts in chemical and biological monitoring of priority and emerging pollutants in water. *Anal. Bioanal. Chem.* 385: 983–984.

158. Sacher, F., F.T. Lange, H.-J. Brauch, et al. 2001. Pharmaceuticals in groundwaters. Analytical methods and results of a monitoring program in Baden-Württemberg, Germany. *J. Chromatogr. A* 938: 199–210.

159. Marczak, M., L. Wolska, W. Chrzanowski, et al. 2006. Microanalysis of volatile organic compounds (VOCs) in water samples—Methods and instruments. *Microchim. Acta* 155: 331–348.

160. Dewulf, J., H. Van Langenhove, and T. Huybrechts. 2006. Developments in the analysis of volatile halogenated compounds. *Trends Anal. Chem.* 25: 300–309.

161. Lara-Gonzalo, A., J.E. Sánchez-Uría, E. Segovia-García, et al. 2008. Critical comparison of automated purge and trap and solid-phase microextraction for routine determination of volatile organic compounds in drinking waters by GC-MS. *Talanta* 74: 1455–1462.

162. Wolska, L., C. Olszewska, M. Turska, et al. 1998. Volatile and semivolatile organo-halogen trace analysis in surface water by direct aqueous injection GC-ECD. *Chemosphere* 37: 2645–2651.

163. Cairns, T. 1992. *Emerging Strategies for Pesticide Analysis.* Boca Raton: CRC Press.
164. Barceló, D. (ed.). 1996. Applications of LC-MS in environmental chemistry. *J. Chromatogr. Library,* Vol. 59. Amsterdam: Elsevier.
165. Barceló, D. (ed.). 2000. Sample handling and trace analysis of pollutants. *Techniques and Instrumentation in Analytical Chemistry,* Vol. 21. Amsterdam: Elsevier.
166. Hennion, M.-C. and D. Barceló (eds). 1997. Trace determination of pesticides and their degradation products in water. *Techniques and Instrumentation in Analytical Chemistry,* Vol. 19. Amsterdam: Elsevier.
167. Fernández Alba, A.R. (ed.). 2004. Chromatographic-mass spectrometric food analysis for trace determination of pesticide residues. *Comprehensive Analytical Chemistry,* Vol. 43. Amsterdam: Elsevier.
168. Baugros, J. B., B. Giroud, G. Dessalces, et al. 2008. Multiresidue analytical methods for the ultra-trace quantification of 33 priority substances present in the list of REACH in real water samples. *Anal. Chim. Acta* 607: 191–203.
169. Forcada, M., J. Beltran, F.J. López, et al. 2000. Multiresidue procedures for determination of triazine and organophosphorus pesticides in water by use of large-volume PTV injection in gas chromatography. *Chromatographia* 51: 362–368.
170. Lacorte, S., I. Guiffard, D. Fraisse, et al. 2000. Broad spectrum analysis of 109 priority compounds listed in the 76/464/CEE Council Directive using solid-phase extraction and GC/EI/MS. *Anal. Chem.* 72: 1430–1440.
171. Cook, J., M. Engel, P. Wylie, et al. 1999. Multiresidue screening of pesticides in foods using retention time locking, GC-AED, database search, and GC/MS identification. *J. AOAC Int.* 82: 313–326.
172. García-Reyes, J.F., M.D. Hernando, C. Ferrer, et al. 2007. Large scale pesticide multiresidue methods in food combining liquid chromatography—time-of-flight mass spectrometry and tandem mass spectrometry. *Anal. Chem.* 79: 7308–7323.
173. Carvalho, J.J., P.C.A. Jerónimo, C. Gonçalves, et al. 2008. Evaluation of a multiresidue method for measuring fourteen chemical groups of pesticides in water by use of LC-MS-MS. *Anal. Bioanal. Chem.* 392: 955–968.
174. Rolfes, J. and J.T. Andersson. 2001. Determination of alkylphenols after derivatization to ferrocenecarboxylic acid esters with gas chromatography-atomic emission detection. *Anal. Chem.* 73: 3073–3082.
175. Wissiack, R. and E. Rosenberg. 2002. Universal screening method for the determination of US Environmental Protection Agency phenols at the lower ng l^{-1} level in water samples by on-line solid-phase extraction-high-performance liquid chromatography-atmospheric pressure chemical ionization mass spectrometry within a single run. *J. Chromatogr. A* 963: 149–157.
176. Patsias, J. and E. Papadopoulou-Mourkidou. 2000. Development of an automated on-line solid-phase extraction-high-performance liquid chromatographic method for the analysis of aniline, phenol, caffeine and various selected substituted aniline and phenol compounds in aqueous matrices. *J. Chromatogr. A* 904: 171–188.
177. Martinez, E., O. Gans, H. Weber, et al. 2004. Analysis of nonylphenol polyethoxylates and their metabolites in water samples by high-performance liquid chromatography with electrospray mass spectrometry detection. *Water Sci. Technol.* 50: 157–163.
178. Li H.Q., F. Jiku, and H.F. Schröder. 2000. Assessment of the pollutant elimination efficiency by gas chromatography/mass spectrometry, liquid chromatography–mass spectrometry and—tandem mass spectrometry comparison of conventional and membrane-assisted biological wastewater treatment processes. *J. Chromatogr. A* 889: 155–176.
179. Jandera, P., J. Fischer, H. Lahovská, et al. 2006. Two-dimensional liquid chromatography normal-phase and reversed-phase separation of (co)oligomers, *J. Chromatogr. A* 1119: 3–10.
180. Hao C., T.R. Croley, R.E. March, et al. 2000. Mass spectrometric study of persistent acid metabolites of nonylphenol ethoxylate surfactants. *J. Mass Spectrom.* 35: 818–830.
181. Jonkers N., R.W.P.M. Laane, and P. de Voogt. 2003. Fate of nonylphenol ethoxylates and their metabolites in two Dutch estuaries: Evidence of biodegradation in the field. *Environ. Sci Technol.* 37: 321–327.
182. Ventura, F. and P. de Voogt. 2003. Separation and detection. GC and GC-MS determination of surfactants. In: T.P. Knepper, D. Barceló, and P. de Voogt (eds), *Analysis and Fate of Surfactants in the Aquatic Environment,* Comprehensive Analytical Chemistry, Vol. 40, pp. 51–76. Amsterdam: Elsevier.
183. Di Corcia, A., L. Capuani, F. Casassa, et al. 1999. Fate of linear alkyl benzenesulfonates, coproducts, and their metabolites in sewage treatment plants and in receiving river waters. *Environ. Sci. Technol.* 33: 4119–4125.
184. Knepper, T.P. and M. Kruse. 2000. Investigations into the formation of sulfophenylcarboxylates (SPC) from linear alkylbenzenesulfonates (LAS) by liquid chromatography/mass spectrometry. *Tenside Surfact. Det.* 37: 41–47.

185. Reemtsma, T. 1996. Methods of analysis of polar aromatic sulfonates from aquatic environments. *J. Chromatogr. A* 733: 473–489.
186. Eichhorn, P. and T.P. Knepper. 2002. α,β-unsaturated sulfophenylcarboxylates as degradation intermediates of linear alkylbenzenesulfonates: Evidence for omega-oxygenation followed by beta-oxidations by liquid chromatography-mass spectrometry. *Environ. Toxicol. Chem.* 21: 1–8.
187. Weremiuk, A.M., S. Gerstmann, and H. Frank. 2006. Quantitative determination of perfluorinated surfactants in water by LC-ESI-MS/MS. *J. Sep. Sci.* 29: 2251–2255.
188. Scott, B.F., C.A. Moody, C. Spencer, et al. 2006. Analysis for perfluorocarboxylic acids/anions in surface waters and precipitation using GC-MS and analysis of PFOA from large-volume samples. *Environ. Sci. Technol.* 40: 6405–6410.
189. Murphy, R.E. and M.R. Schure. 2008. The analysis of surfactants by multidimensional liquid chromatography. In: S.A. Cohen and M.R. Schure (eds), *Multidimensional Liquid Chromatography*, pp. 425–446. Hoboken, New York: Wiley.
190. Gonzalez, S., D. Barceló, and M. Petrovic. 2007. Advanced liquid chromatography-mass spectrometry (LC-MS) methods applied to wastewater removal and the fate of surfactants in the environment. *Trends Anal. Chem.* 26: 116–124.
191. Knepper T.P., D. Barceló, and P. de Voogt (eds). 2003. Analysis and fate of surfactants in the aquatic environment. *Comprehensive Analytical Chemistry*, Vol. 40. Amsterdam: Elsevier.
192. Kuster, M., M.J. Lopez de Alda, and D. Barceló. 2005. Estrogens and progestogens in wastewater, sludge, sediments, and soil. In: D. Barceló (ed.), *Handbook of Environmental Chemistry*, Vol. 5, pp. 1–24. Berlin: Springer.
193. Kuster, M., M.J. Lopez de Alda, S. Rodriguez-Mozaz, et al. 2007. Analysis of steroid estrogens in the environment. In: G. Svehla and D. Barceló (eds), *Comprehensive Analytical Chemistry*, Vol. 50, pp. 219–264. Amsterdam: Elsevier.
194. Petrovic, M., E. Eljarrat, M.J. Lopez de Alda, et al. 2004. Endocrine disrupting compounds and other emerging contaminants in the environment: A survey on new monitoring strategies and occurrence data. *Anal. Bioanal. Chem.* 378: 549–562.
195. Lopez de Alda, M.J., S. Diaz-Cruz, M. Petrovic, et al. 2003. Liquid chromatography-(tandem) mass spectrometry of selected emerging pollutants (steroid sex hormones, drugs and alkylphenolic surfactants) in the aquatic environment. *J. Chromatogr. A* 1000: 503–526.
196. Gentili, A., D. Perret, S. Marcheseet, et al. 2002. Analysis of free estrogens and their conjugates in sewage and river waters by solid-phase extraction then liquid chromatography-electrospray-tandem mass spectrometry. *Chromatographia* 56: 25–32.
197. Petrovic M. and D. Barceló (eds). 2007. Analysis, fate and removal of pharmaceuticals in the water cycle. *Comprehensive Analytical Chemistry*, Vol. 50. Amsterdam: Elsevier.
198. Öllers, S., H.P. Singer, P. Fässler, et al. 2001. Simultaneous quantification of neutral and acidic pharmaceuticals and pesticides at the low-ng/l level in surface and waste water. *J. Chromatogr. A* 911: 225–234.
199. Castiglioni, S., R. Bagnati, R. Calamari, et al. 2005. A multiresidue analytical method using solid-phase extraction and high-pressure liquid chromatography tandem mass spectrometry to measure pharmaceuticals of different therapeutic classes in urban wastewaters. *J. Chromatogr. A* 1092: 206–215.
202. Petrovic, M., M. Gros, and D. Barceló. 2006. Multi-residue analysis of pharmaceuticals in wastewater by ultra-performance liquid chromatography–mass spectrometry. *J. Chromatogr. A* 1124: 68–81.
201. Dębska, J., A. Kot-Wasik, and J. Namieśnik. 2004. Fate and analysis of pharmaceutical residues in the aquatic environment. *Crit. Rev. Anal. Chem.* 34: 51–67.
202. Kot-Wasik, A., J. Dębska, and J. Namieśnik. 2007. Analytical techniques in studies of the environmental fate of pharmaceuticals and personal-care products. *Trends Anal. Chem.* 26: 557–568.
203. Daugthon, C.G., T. Jones-Lepp, and D. Washington. 2001. Pharmaceuticals and personal care products in the environment: Scientific and regulatory issues. *ACS Symposium Series* 791. Washington: American Chemical Society.
204. Peck, A. 2006. Analytical methods for the determination of persistent ingredients of personal care products in environmental matrices. *Anal. Bioanal. Chem.* 386: 907–939.
205. Ternes A.T., A. Joss, and H. Siegrist. 2004. Scrutinizing pharmaceuticals and personal care products in wastewater treatment. *Environ. Sci. Technol.* 38: 393A–398A. 206. Matamoros, V., E. Jover, and J.M. Bayona. 2008. Advances in the determination of degradation intermediates of personal care products in environmental matrixes: A review. *Anal. Bioanal. Chem.* in print (doi:10.1007/s00216-008-2371-7).
207. Champ, M.A. 2000. A review of organotin regulatory strategies, pending actions, related costs and benefits. *Sci. Total Environ.* 258: 21–71.

208. Rosenberg, E. 2005. Speciation of tin compounds. In: R. Cornelis, J.A. Caruso, H. Crews, and K.G. Heumann (eds), *Handbook of Elemental Speciation*, pp. 422–463. Weinheim: Wiley-VCH.

209. Antizar-Ladislao, B. 2008. Environmental levels, toxicity and human exposure to tributyltin (TBT)-contaminated marine environment. A review. *Environ. Int.* 34: 292–308.

210. Abalos, M., J.-M. Bayona, R. Compañó, et al. 1997. Analytical procedures for the determination of organotin compounds in sediment and biota: A critical review. *J. Chromatogr. A* 788: 1–49.

211. Pellegrino, C., P. Massanisso, and R. Morabito. 2000. Comparison of twelve selected extraction methods for the determination of butyl- and phenyltin compounds in mussel samples. *Trends Anal. Chem.* 19: 97–106.

212. Donard, O.F.X., S. Rapsomanikis, and J.H. Weber. 1986. Speciation of inorganic tin and alkyltin compounds by atomic absorption spectrometry using electrothermal quartz furnace after hydride generation. *Anal. Chem.* 58: 772–777.

213. Aguerre, S., G. Lespes, V. Desauziers, et al. 2001. Speciation of organotins in environmental samples by SPME-GC: Comparison of four specific detectors: FPD, PFPD, MIP-AES and ICP-MS, *J. Anal. At. Spectrom.* 16: 263–269.

214. Compañó, R., M. Granados, C. Leal, et al. 1995. Liquid chromatographic determination of triphenyltin and tributyltin using fluorimetric detection. *Anal. Chim. Acta* 314: 175–182.

215. Rosenberg, E., V. Kmetov, and M. Grasserbauer. 2000. Investigating the potential of high-performance liquid chromatography with atmospheric pressure chemical ionization-mass spectrometry as an alternative method for the speciation analysis of organotin compounds. *Fresenius J. Anal. Chem.* 366: 400–407.

216. White, S., T. Catterick, B. Fairman, et al. 1998. Speciation of organotin compounds using liquid chromatography-atmospheric pressure ionisation mass spectrometry and liquid chromatography-inductively coupled plasma mass spectrometry as complementary techniques. *J. Chromatogr. A* 794: 211–218.

217. Sonak, S., P. Pangam, A. Giriyan, et al. 2009. Implications of the ban on organotins for protection of global coastal and marine ecology. *J. Environ. Manage.* 90, Suppl. 1: S96–S108.

218. Fent, K. 2006. Worldwide occurrence of organotins from antifouling paints and effects in the aquatic environment. In: I.K. Konstantinou (ed.), *Handbook of Environmental Chemistry*, Vol. 5, pp. 71–100. Berlin: Springer.

219. Van Cleuvenbergen, R.J.A., D. Chakraborti, and F.C. Adams. 1986. Occurrence of tri- and dialkyllead species in environmental water. *Environ. Sci. Technol.* 20: 589–593.

220. Radojevic, M. and R.M. Harrison. 1987. Concentrations and pathways of organolead compounds in the environment: A review. *Sci. Total Environ.* 59: 157–180.

221. Paneli, M., E. Rosenberg, M. Grasserbauer, et al. 1997. Assessment of organolead species in the Austrian Danube-basin using GC-MIP-AED. *Fresenius J. Anal. Chem.* 357: 756–762.

222. Bergmann, K. and B. Neidhart. 2001. In situ propylation of ionic organotin and organolead species in water samples—extraction and determination of the resulting tetraorganometallic compounds by gas chromatography-atomic absorption spectrometry. *J. Sep. Sci.* 24: 221–225.

223. Baena, J.R., M. Gallego, M. Valcárcel, et al. 2001. Comparison of three coupled gas chromatographic detectors (MS, MIP-AES, ICP-TOFMS) for organolead speciation analysis. *Anal. Chem.* 73: 3927–3934.

224. Łobiński, R. 1995. Organolead compounds in archives of environmental pollution. *Analyst* 120: 615–621.

225. Froelich Jr., P.N. and M.O. Andreae. 1981. The marine geochemistry of germanium: Ekasilicon. *Science* 213: 205–207.

226. Lewis, B.L., M.O. Andreae, P.N. Froelich, et al. 1988. A review of the biogeochemistry of germanium in natural waters. *Sci. Total Environ.* 73: 107–120.

227. Rosenberg, E. 2008. Germanium: Environmental occurrence, importance and speciation. *Rev. Environ. Sci. Bio/Technol.* in print (doi:10.1007/s11157-008-9143-x).

228. Hambrick III, G.A., P.N. Froelich Jr., M.O. Andreae, et al. 1984. Determination of methylgermanium species in natural waters by graphite furnace atomic absorption spectrometry with hydride generation. *Anal. Chem.* 56: 421–424.

229. Jiang, G.B. and F.C. Adams. 1997. Direct determination of trimethylgermanium in water by on-column capillary gas chromatography with flame photometric detection using quartz surface-induced germanium emission. *J. Chromatogr. A* 759: 119–125.

230. Jiang, G.B. and F.C. Adams. 1997. Evaluation of gas chromatography with a flame photometric detector based on quartz surface-induced emission for determining the speciation of inorganic and methylgermanium compounds. *Anal. Chim. Acta* 337: 83–91.

231. Lewis, B.L., M.O. Andreae, and P.N. Froelich. 1989. Sources and sinks of methylgermanium in natural waters. *Marter. Chem.* 27: 179–200.

232. Jiang, G. B., M. Ceulemans, and F.C. Adams. 1996. Optimization study for the speciation analysis of organotin and organogermanium compounds by on-column capillary gas chromatography with flame photometric detection using quartz surface-induced luminescence. *J. Chromatogr. A* 727: 119–129.

233. Tanzer, D. and K.G. Heumann. 1990. GC determination of dimethyl selenide and trimethyl selenonium ions in aquatic systems using element specific detection. *Atmos. Environ.* 24: 3099–3102.

234. Pyrzynska, K. 2002. Determination of selenium species in environmental samples. *Microchim. Acta* 140: 55–62.

235. de la Calle Guntiñas, M.B., M. Ceulemans, C. Witte, et al. 1995. Evaluation of a purge-and-trap injection system for capillary gas chromatography-microwave induced plasma-atomic emission spectrometry for the determination of volatile selenium compounds in water. *Mikrochim. Acta* 120: 73–82.

236. de la Calle Guntiñas, M.B., R. Łobiński, and F.C. Adams. 1995. Interference-free determination of selenium(IV) by capillary gas chromatography-microwave-induced plasma atomic emission spectrometry after volatilization with sodium tetraethylborate. *J. Anal. At. Spectr.* 10: 111–115.

237. Gómez-Ariza, J.L., J.A. Pozas, I. Giráldez et al. 1999. Use of solid phase extraction for speciation of selenium compounds in aqueous environmental samples. *Analyst* 124: 75–78.

238. Pérez-Sirvent, C. and M.-J. Martínez-Sánchez. 2007. Comparison of two derivatizing agents for the simultaneous determination of selenite and organoselenium species by gas chromatography and atomic emission detection after preconcentration using solid-phase microextraction. *J. Chromatogr. A* 1165: 191–199.

239. Brunori, C., M.B. de la Calle-Guntiñas, R. Morabito. 1998. Optimization of the reduction of Se(VI) to Se(IV) in a microwave oven. *Fresenius J. Anal. Chem.* 360: 26–30.

240. Uden, P.C. 2002. Modern trends in the speciation of selenium by hyphenated techniques. *Anal. Bioanal. Chem.* 373: 422–431.

241. Dietz, C., J. S. Landaluze, P. Ximenez-Embun. 2004. SPME-multicapillary GC coupled to different detection systems and applied to volatile organo-selenium speciation in yeast. *J. Anal. At. Spectrom.* 19: 260–266.

242. Afton, S., K. Kubachka, B. Catron, et al. 2008. Simultaneous characterization of selenium and arsenic analytes via ion-pairing reversed phase chromatography with inductively coupled plasma and electrospray ionization ion trap mass spectrometry for detection. Applications to river water, plant extract and urine matrices. *J. Chromatogr. A* 1208: 156–163.

243. Haberhauer-Troyer, C., G. Álvarez-Llamas, E. Zitting, et al. 2003. Comparison of different chloroformates for the derivatisation of seleno amino acids for gas chromatographic analysis. *J. Chromatogr. A* 1015: 1–10.

244. Peláez, M.V., M.M. Bayón, J.I.G. Alonso, et al. 2000. Comparison of different derivatization approaches for the determination of selenomethionine by GC-ICP-MS. *J. Anal. At. Spectrom.* 15: 1217–1222.

245. Michalke, B. and P. Schramel. 1998. Selenium speciation by interfacing capillary electrophoresis with inductively coupled plasma-mass spectrometry. *Electrophoresis* 19: 270–275.

246. Michalke, B., O. Schramel, and A. Kettrup. 1999. Capillary electrophoresis coupled to inductively coupled plasma mass spectrometry (CE/ICP-MS) and to electrospray ionization mass spectrometry (CE/ESI-MS): An approach for maximum species information in speciation of selenium. *Fresenius J. Anal. Chem.* 363: 456–459.

14 Introducing the Concept of Sustainable Development into Analytical Practice: Green Analytical Chemistry

Waldemar Wardencki and Jacek Namieśnik

CONTENTS

14.1 Introduction 353
14.2 History of Green Analytical Chemistry 354
14.3 Implementation of Greener Analytical Chemistry Approaches in Laboratory Practice 355
14.4 General Characteristics of the Analytical Process 356
14.5 Greening Analytical Chemistry in Sample Pretreatment: General Characteristics 356
    14.5.1   Accelerated Solvent Extraction 356
    14.5.2   Ultrasonic and Microwave Extraction 357
    14.5.3   Solid-Phase Microextraction 357
    14.5.4   Stir Bar Sorptive Extraction 358
    14.5.5   Thin-Film Microextraction 358
    14.5.6   Single-Drop Microextraction 359
    14.5.7   Liquid-Phase Microextraction 359
    14.5.8   Pressurized Hot Water Extraction 360
    14.5.9   Supercritical Fluid Extraction 360
    14.5.10 Alternative Solvents: The Application of ILs 361
14.6 Greener Separation Techniques 361
14.7 Advances in Electrochemistry Sensing Technology Relevant to Green Analytical Chemistry 362
14.8 Miniaturization in Analytical Chemistry Methods 362
14.9 Conclusions 363
References 364

14.1 INTRODUCTION

Since the Rio de Janeiro conference in 1992 the issue of sustainable development has become a matter of increasing global concern. The main goals of sustainable development are focused on sound economic development in accordance with an intact environment and a fair global social balance. The trend of sustainable development requires chemistry to be cleaner and greener. The beginning of green chemistry is frequently considered to be a response to the need to reduce the damage to the environment caused by man-made materials and the processes for producing them.

The term "green chemistry" was first used in 1991 by P.T. Anastas in a special program launched by the US Environmental Protection Agency (EPA) to encourage industry, academia, and governments to introduce sustainable development to chemistry and chemical technology. In 1995, the annual US Presidential Green Chemistry Challenge was announced. Shortly afterwards, similar awards were established in European countries. In 1996, the Working Party on Green Chemistry came into existence within the framework of the International Union of Applied and Pure Chemistry (IUPAC). One year later, the Green Chemistry Institute (GCI) was formed with chapters in 20 countries to facilitate contacts between governmental agencies and industrial corporations on the one hand, and universities and research institutes on the other with regard to the design and implementation of new technologies.

Green chemistry—otherwise referred to as environmentally benign chemistry, clean chemistry, the atom economy, and benign-by-design chemistry—brings a fresh approach to the synthesis, processing, and application of chemical substances, the idea being to reduce the threats to health and the environment. It is usually presented as a set of 12 principles put forward by Anastas and Warner[1] embodying instructions on how to implement new chemical compounds, syntheses, and technological processes; the rules for introducing green chemical synthesis on a technological scale were set out by Winterton.[2] The significance of green chemistry was highlighted in a cover story in *Chemical Engineering News*.[3]

The first books and journals on the subject of green chemistry were published in the 1990s, including *Journal of Clean Processes and Products* (Springer-Verlag) and *Green Chemistry*, sponsored by the Royal Society of Chemistry. Some journals, such as *Environmental Science and Technology* and *Journal of Chemical Education*, have regular dedicated sections on green chemistry, whereas others devote complete issues to this new approach.[4,5] All these publications reflect the importance attached to the different aspects of green chemistry in current research activities.

The first conference highlighting green chemistry was held in Washington in 1997, since when other, similar conferences have been held on a regular basis in order to raise the awareness of and participation in this greener vision.[6]

14.2   HISTORY OF GREEN ANALYTICAL CHEMISTRY

The 12 principles of green chemistry proposed by Anastas and Warner[1] address mainly aspects of synthetic chemistry; the idea of green analytical chemistry was conceived only later.[7,8] At first, it did not catch on, but as the "green chemistry movement" has gained in momentum, the green approach to analytical chemistry has become a key part of green chemistry.

Analytical chemists have long been environmentally sensitive, but the word "green" has rarely been used in descriptions of their activities in the last 15 years. Since the year 2000, however, there has been a dramatic increase in the number of papers dealing with green or clean analytical chemistry. A few interesting reviews have been published recently, some addressing general issues of green analytical chemistry,[9–11] and others particular methodologies.[12–14]

The irony is that the methods used in laboratories to analyze the state of environmental pollution, as well as the analytical chemists applying them, through the uncontrolled disposal of reagents, solvents, and other chemical wastes, may themselves be the source of large amounts of pollutants entering the environment. Traditional analytical procedures require considerable quantities of chemical compounds; sampling, and especially the preparation of samples for their final determination, frequently involves the formation of large amounts of pollutants (vapors, liquid effluents— waste reagents and solvents, and solid waste). If environmental pollution by analytical reagents and so on is to be avoided, the rules of green chemistry must be introduced into chemical laboratories on a large scale. Analytical chemists strive for the traditional goals of accuracy, precision, sensitivity, and low detection limits; but by implementing green chemistry rules in laboratory practice, they are demonstrating their awareness of the impact of their work on the environment.

Based on a survey of the recent analytical literature, this chapter reviews a number of innovatory methodologies for sample preparation, as well as the development and application of new approaches to traditional analytical methods in order to make this field of chemistry greener.

14.3 IMPLEMENTATION OF GREENER ANALYTICAL CHEMISTRY APPROACHES IN LABORATORY PRACTICE

Analytical chemistry is still a relatively poorly examined area of green chemistry. The main objective of green analytical chemistry is to apply analytical procedures and devices that generate less hazardous waste, are safer to use, and kinder to the environment. This objective can be achieved through the development of entirely new analytical methodologies or the modification of old ones to incorporate procedures using either fewer hazardous chemicals or at least smaller quantities of them. The general strategy toward making analytical methodologies greener involves not only changing or modifying reagents and solvents, or reducing the amounts of chemicals used, but also the miniaturization and even the elimination of sampling by measuring the analytes of interest *in situ* in real time and on-line.

In accordance with the 12 principles of green chemistry, the actions to be taken to achieve a greener analytical chemistry are easily stated. The following are priority issues with regard to analytical procedures:

- Eliminating or minimizing the use of chemical reagents, particularly organic solvents.
- Eliminating chemicals with a high toxicity and ecotoxicity.
- Reducing labor- and energy-intensive steps (per analyte).
- Reducing the impact of chemicals on human health.

X-ray fluorescence,[15,16] surface acoustic waves (SAW) for determining volatile organic compounds (VOCs),[17,18] and immunoassays[19–21] are examples of direct analytical techniques (in which a sample preparation step is unnecessary) that are environmentally friendly. In addition, there are environmentally benign procedures from which reagents and solvents have been eliminated or their quantities minimized (calculated per analytical cycle):

- Solid-phase extraction (SPE)[22,23]
- Accelerated solvent extraction (ASE)[24]
- Solid-phase microextraction (SPME)[25]
- Stir bar sorptive extraction (SBSE)[26]
- Thin-film microextraction[27]
- Single-drop microextraction (SDME)[28]
- Liquid-phase microextraction (LPME)[29]
- Supercritical fluid extraction (SFE)[30,31]
- Extraction in automated Soxhlet apparatus[32]
- Vacuum distillation of VOCs[33]
- Mass spectrometry with membrane interface (MIMS).[34]

The next important challenge of green analytical chemistry is in-process monitoring. Developing and using in-line or on-line analyzers enable analytes to be determined in real time, and disturbances to be detected already in the initial steps of a process. This means of analysis provides rapid information and the opportunity for preventive measures to be taken—the process can be stopped or its operational parameters altered—with an overall improvement in efficiency.

The following sections discuss recent advances in green analytical chemistry in the context of the whole analytical process, that is, from sample collection, sample preparation, to sample analysis, as well as the characteristics of some traditional methodologies that have always been environmentally benign but were never described as "green."

14.4 GENERAL CHARACTERISTICS OF THE ANALYTICAL PROCESS

Analytical chemistry is important in that it supports the development of chemical engineering and technology for the production of chemicals and confirms (or not) the environmental friendliness of new methods, processes, and products. Analytical procedures deliver information about chemical substances, their occurrence in the environment and in organisms. Analytical chemistry also affects sociopolitical decisions: The results of chemical analysis provide strong arguments for enacting new laws and administrative regulations with regard to food quality, the freshness of raw materials, and the nutritive values of processed food. Accurate assessments of these parameters, as well as the determination of food additives and contaminants, are especially important to reassure consumers that the products they purchase are safe. In order to provide the relevant data, analysts have to develop better, less labor-intensive, faster, and more accurate procedures.

The analytical process consists of a series of steps: sampling, sample handling, laboratory sample preparation, separation and quantitation, and statistical evaluation. Each one of these steps is important if accurate results are to be obtained, but the key component of the analytical process is sample preparation. It is important to bear in mind that these analytical steps are consecutive: the next step cannot begin until the preceding one has been completed. If any one of these steps is not carried out properly, the overall performance of the procedure will be poor, errors will be introduced, and inconsistency in the results can be expected.

The application of green chemistry rules in the design of new analytical methods is the key to diminishing the adverse effects of analytical chemistry on the environment. The same ingenuity and innovation that were applied earlier to obtain excellent sensitivity, precision, and accuracy are now invoked to reduce or eliminate the application of hazardous substances in analysis.

Analytical chemistry is generally considered a small-scale activity; but in view of the very large numbers of runs performed in controlling and monitoring laboratories, the scale becomes comparable with that of the fine chemicals or pharmaceutical industries.

14.5 GREENING ANALYTICAL CHEMISTRY IN SAMPLE PRETREATMENT: GENERAL CHARACTERISTICS

The choice of sample preparation method is crucial in chemical analysis because it is often the most critical and time-consuming step of an analytical process.[35] There is a wide choice of methods for sample pretreatment and preparation for further analysis. Unfortunately, however, there are no universal methods of sample treatment because analytical samples come in a huge variety of forms. Ideally, the sample preparation methodology should be solvent-free, simple, inexpensive, efficient, selective, and compatible with final analytical methods.

There is an urgent need to evaluate the analytical procedures in current use, not only with respect to the reagents, instrumental costs, and analytical parameters, but also in the context of their adverse effect on the environment. The continual development of new solventless sample preparation methods is a good example of activities in this field. Indeed, recent years have witnessed particularly rapid progress in the development of these techniques,[36,37] which provide higher yields, better sample cleanup, cost effectiveness and chemist safety, and are also less harmful to the environment.

Some examples of modern analytical techniques used for sample pretreatment are now presented.

14.5.1 ACCELERATED SOLVENT EXTRACTION

ASE—also referred to as pressurized fluid extraction (PFE)—offers an order of magnitude of additional reductions in solvent use with faster sample processing time, and with the potential of automated, unattended extraction of multiple samples. Briefly, with ASE a solid sample is enclosed in a cell containing an extraction solvent; after the cell has been sealed, the sample is permeated by the extracting solvent under elevated temperatures and pressure for short periods (5–10 min). Typically, the samples are extracted under static conditions, where the fluid is held in the cell for controlled time periods to allow sufficient contact between the solvent and the solid for efficient extraction. Alternatively, dynamic or flow-through techniques can be used. Compressed gas is used to purge a sample extract from the cell into a collection vessel. ASE achieves rapid extraction with small volumes of conventional organic solvents by using high temperatures (up to 200°C) and high pressures (up to 20 MPa) to maintain the solvent in a liquid state. The use of liquid solvents at elevated temperatures and pressures improves efficiency compared with extractions at or near room temperature and atmospheric pressure because of enhanced solubility and mass-transfer effects and the disruption of surface equilibrium. A number of review papers with very detailed descriptions and evaluations of this sample pretreatment technique have been published.[38,39] ASE has been used to extract hydrophobic organic compounds from different environmental samples.[40–42] Some studies have compared ASE with conventional techniques such as SFE and Soxhlet extraction: the performance of ASE was consistently equivalent to or better than conventional techniques such as Soxhlet and sonication extraction.

14.5.2 Ultrasonic and Microwave Extraction

Ultrasonic and microwave extractions are relatively simple and inexpensive techniques for greening extractions.

Ultrasonic extraction uses high-frequency acoustic waves to create microscopic bubbles in liquids. When these bubbles burst, small shock waves and cavitations occur that are particularly well suited for breaking up solids and promoting their dissolution. This technique has been used to extract a variety of organic compounds, for example, nicotine from pharmaceutical samples into heptane for gas chromatography (GC) analysis (reducing the amount of solvent required by 5/6 compared to the conventional method),[43] and phthalates from cosmetics into ethanol/water for high-performance liquid chromatography (HPLC),[44] and also inorganic analytes, such as mercury from milk samples.[45]

Microwave-assisted extractions (MAE) can be performed in open (focused MAE) or closed (pressurized MAE) flasks. This technique is commonly used for extractions from complex and difficult sample matrices, replacing time- and solvent-intensive Soxhlet extractions or hydrodistillations.[46] MAE is also widely applied to environmental samples, for example, for extracting polycyclic aromatic hydrocarbons (PAH) from soil, methylmercury from sediments, and trace metals and pesticide residues from plant material.[47,48] The use of microwave treatment instead of hydrodistillation offers a solvent-free separation technique: essential oils are heated and dry-distilled.[46]

14.5.3 Solid-Phase Microextraction

SPME is a fast, universal, sensitive, solventless, and economical method of preparing samples for GC or HPLC analysis, enabling detection limits at a level of 5–50 ppt for volatile, semivolatile, and nonvolatile compounds to be achieved. The approximate sample preparation time is usually 2–15 min.[49,50]

The effectiveness of analyte preconcentration using SPME depends on a number of parameters, for example, the type of fiber, sample stirring, extraction time, and ionic strength.

The sensitivity of the technique depends mainly on the value of the partition coefficient of the analytes partitioned between the sample and the fiber stationary phase. The efficiency of preconcentration depends on both the type of fiber used and its thickness (amount). This type of fiber affects the amount and character of the sorbed species.[51] The general rule "like dissolves like" applies here, that is, polar compounds are sorbed on polar fibers, and nonpolar compounds on nonpolar fibers. A broad range of standard fibers is commercially available.

A great number of compounds can be determined using this technique: it enables the isolation of pesticides from different matrices[52–54] and of solvent residues,[55] and also the analysis of complex mixtures such as aroma compounds.[56–58]

An SPME fiber can be exposed in two modes—by immersing it directly in the liquid sample to be analyzed (direct immersion SPME—DI-SPME), or by exposing it to the headspace (HS-SPME). In the latter case, the fiber is inserted into the headspace, above a liquid or solid sample.

Sampling volatile analytes from samples having complex matrices usually takes place in the HS-SPME mode. This variant yields decidedly better results in the determination of aroma compounds[59] and other volatile components.[60] Moreover, HS-SPME prolongs the life of the fiber because it is not in direct contact with the sample. On the other hand, the direct extraction of less volatile compounds from solution is possible using DI-SPME. But in this case, the fiber deteriorates more quickly, increasing the cost of analysis. Headspace sampling is therefore employed whenever possible.

100 μm polydimethylsiloxane (PDMS)[61] and divinylbenzene-PDMS (DVB-PDMS)[62] are undoubtedly the most frequently and most universally used fiber materials. The preferred final method of analyzing enriched compounds is GC coupled with mass spectrometry (MS) (GC-MS),[63] although HPLC is sometimes used as an alternative.[64]

Another important mode of operation in SPME is in-tube SPME.[65] In this system, usually coupled on-line to HPLC, a finite portion of sample is drawn through an internally coated capillary tube and then ejected into the sample vial. This technique requires more complex instrumentation than that used for standard SPME, but a greater sensitivity is obtainable with a longer tube (and consequently more sorbent). Two solvent desorption modes—are usually applied for introducing species into HPLC: off-line desorption and on-line desorption. In the latter, the HPLC mobile phase is used for desorbing the analytes.

14.5.4 Stir Bar Sorptive Extraction

In 1999, a new technique of sorptive extraction called SBSE was introduced into analytical practice.[66] Developed to extract organic analytes from liquid samples, it is based on the sorption of analytes onto a thick film of PDMS coated on an iron stir bar.[67,68] Originally, the stir bars were prepared by removing the Teflon® coating from existing stir bars, reducing the outer diameter of the magnet, and enclosing the magnet in a glass tube to give a 1.2 mm outer diameter. Silicone tubing with an internal diameter of 1.5 mm and an outer diameter of 3 mm was then slid over the magnetic glass tube. However, as the stir plate is itself magnetic, the use of a magnetic stir bar is not required. Nonmagnetic stir bars were prepared from stainless steel rods with an outer diameter of 0.8 mm and a length of 40 mm. The total amounts of PDMS material present on the 10 and 40 mm stir bars were 75.7 and 300.9 mg, respectively, which convert with a density of 0.825 g mL^{-1} to volumes of 92 and 365 μL; as the PDMS tubing contains approximately 40% (v/v) of fumed silica as filling material, as determined with solid-state nuclear magnetic resonance (NMR) and thermogravimetric analysis (TGA), the effective volumes of PDMS are 55 and 219 μL, respectively. The stir bar is placed in an aqueous sample and extraction takes place during stirring. Because of the low phase ratio (the volume of the water phase divided by the volume of the PDMS phase), very high recoveries were obtained, especially for volatile compounds. The efficiency of SBSE has been compared with other sorptive techniques.[69] This technique has been applied to the extraction of different types of organic compounds in aqueous solutions,[70,71] wine,[72] and in fruits and vegetables.[73,74] In combination with thermodesorption-GC-MS,[75] SBSE enables low detection limits to be achieved. As an alternative, the analytes from the stir bar can be desorbed by liquid extraction and the extract injected into a liquid chromatography (LC) system.[76]

14.5.5 Thin-Film Microextraction

To obtain a greater volume of the extraction phase, the surface area of the polymer is extended, which has been done by using membranes instead of fiber coatings. The use of a thin membrane has the advantage that enhanced extraction efficiency and hence high sensitivity can be achieved without the disadvantage of longer equilibrium times, as is the case when thick phase coated stir bars are used. A cross-linked commercial PDMS membrane[27] was successfully evaluated for the extraction of PAH in headspace mode. An in-house prepared membrane of PDMS[77] was applied to both nonpolar PAH extraction and to polar phenolic compounds. A commercial porous polysulfone hollow fiber membrane, coated with a variety of hydroxylated polymethacrylate compounds,[78] tended to swell when used in water. This is an advantage over classical SPME for the extraction of alkyl-substituted phenols in seawater samples. One of the drawbacks of the system, however, is the necessity for a thermal desorption system or a high-volume GC injector.

14.5.6 SINGLE-DROP MICROEXTRACTION

In 1997, Jeannot and Cantwell[79] as well as He and Lee[80] independently introduced a simple kind of microextraction in which an organic drop hangs from the tip of a GC syringe needle. SDME is a simple method of reducing solvent consumption; indeed, the small amounts of solvent used in SDME are an advantage of this technique. Pure solvents or mixtures can be used for the selective extraction of different organic species.[81] This technique therefore represents an inexpensive and attractive alternative to SPME requiring a standard GC syringe only. SPME does not give a solvent peak in GC, but analyte desorption from the polymer in a hot injector is significantly slower than solvent evaporation, resulting in peaks with a tendency to tail. Alternatively, stirring the sample increases extraction efficiency by SPME, but stirring or sonification of samples in SDME experiments can cause the organic drop to fall off the needle. Consequently, these two methods cannot be applied together with SDME. Nevertheless, the adequate precision, linearity, and repeatability of SDME indicate that this virtually solventless extraction method is reliable for routine analysis.

A review of SDME including 27 references[28] summarizes investigations in this rapidly growing field up till 2002. In a more recent work, a benzyl alcohol microdrop was found to be the optimum solvent for extracting solvent residuals from vegetable oils.[82] Octanol provided optimum extraction efficiency for a variety of short-chain alcohols[83] from water samples. The field of applications has even been extended to the determination of organometallic compounds such as tributyltin,[84] which was extracted into a decane microdrop. Using hexane as a solvent, volatile halohydrocarbons could be extracted in this way from aqueous samples,[85] with limit of detection (LOD) as low as $0.001\ \mu g\ L^{-1}$ for CCl_4 using GC with electron-capture detection. An ionic liquid (IL) (1-octyl-3-methylimidazolium hexafluorophosphate) as extraction solvent was found to be suitable for the extraction of substituted phenols[86] and formaldehyde from mushrooms[87] following derivatization with 2,4-dinitrophenylhydrazine. Even though this is a virtually solventless, inexpensive, fast, and simple method for analyte extraction and/or preconcentration, frequent problems with drop stability and lack of sensitivity have been reported. In an attempt to overcome these limitations, a microliter-size liquid membrane was placed between the sample (octane) and the microdrop (water); simultaneous extraction/back extraction was found to be taking place.[88]

14.5.7 LIQUID-PHASE MICROEXTRACTION

This technique can be considered as a further development of SDME. To give one example: the organic phase was contained within the lumen of the fiber and the sample solution was filled into a vial with a screw top/silicone septum. Two conventional medical syringe needles (guiding needles) were inserted through the silicon septum in the screw top and the two ends connected to each other by a piece of Q3/2 Accurel KM polypropylene hollow fiber. The latter served to contain the microliter volume of extracting solution. For extraction in combination with GC the hollow fiber was filled with n-octanol. For extraction in combination with capillary electrophoresis (CE) or HPLC the hollow fiber mounted on the guiding needles was first dipped into n-octanol for 5 s to immobilize the solvent in pores. Then, the fiber was placed in the sample and the extraction was performed. After extraction, the acceptor solution was collected in microvials by the application of a small head pressure on one of the guiding steel needles for automated analysis by GC or CE.

The disposable nature of the hollow fiber eliminates the possibility of carry-over effects and cross-contamination, thus providing enhanced reproducibility. Further, the small pore size prevents large molecules and particles present in the donor solution from entering the acceptor phase, providing effective matrix/analyte separation. Psillakis and Kalogerakis[89] reviewed the hollow fiber configurations, LPME sampling modes, and different parameters to be taken into account during method optimization, while another review focused on the use of LPME for drug analysis.[90] Rasmussen and Pedersen-Bjergaard[91] examined the basic extraction principles, technical setup, recovery, enrichment, extraction speed, selectivity, and applications. A disadvantage of LPME is the lack of precision, which may be because the whole operation, from fiber preparation and conditioning to the handling of the extract, is done manually.

14.5.8 PRESSURIZED HOT WATER EXTRACTION

Techniques that reduce solvent consumption during sample preparation have also become more commonly applied, for example, pressurized hot water extraction (PHWE), which has replaced conventional organic solvents in a variety of extraction processes.[92,93] Because PHWE employs water, it can also be classified as a "solventless" technique. Selective extraction can be achieved by temperature tuning. Temperatures below the critical value of water but usually above 100°C are usually employed. When working with a liquid phase, the pressure must be sufficient to prevent the water from vaporizing. In the vapor phase some pressure is generally needed for the effective transportation of water. A high temperature increases the initial desorption of the compounds from the sample particles. In addition, rapid diffusion, low viscosity, and low surface tension are achieved at higher temperatures. On the other hand, thermally labile compounds may decompose, and the quantities of coextracted compounds may be greater than at lower temperatures.

14.5.9 SUPERCRITICAL FLUID EXTRACTION

The superior solvation qualities of supercritical fluids over conventional liquids have been known for more than a century, since 1879 in fact, when Hannay and Hogarth investigated the solubility of different inorganic salts in supercritical ethanol. But it was not until the late 1960s that the extraction potential of supercritical fluids was recognized.

Several liquids and gases can be brought into the supercritical phase. Different solvents can be selected as extraction media for use in analytical-scale SFE. Carbon dioxide is most commonly used as an SFE medium because of its desirable properties and easy handling; it is relatively inexpensive and commercially available at a purity grade acceptable for most analytical applications. Another advantage of carbon dioxide is that the polarity can easily be adjusted by adding modifiers such as methanol to the supercritical fluid or the extraction vessel.

SFE is superior to traditional extraction and cleanup for organic compounds in samples in every respect: Solvent use is reduced to a minimum, analysis time is reduced to 2–3 h, a large sample throughput is possible by using automated systems, repeatability is better than with traditional techniques, and optimization for different compound classes is possible, not to mention simultaneous analyses of many different organic compounds in one sample. The considerable reduction in analysis time and cost opens up the possibility of performing large monitoring studies covering many different compounds.

There are review articles[94,95] examining the different aspects of the introduction of SFE into analytical practice.

Studies of new approaches to SFE and of fresh applications of this efficient extraction technique are on-going. The following aspects have been accorded special attention:

- Restrictor plugging in off-line SFE.[96]
- A new analyte collection method for off-line SFE based on mixing an expanding supercritical effluent with overheated organic solvent vapor.[97]

- The collection capacity of a solid-phase trap in SFE.[98]
- The design of an SFE-GC system with quantitative transfer of extraction effluent to a megabore capillary column.[99]
- The application of SFE to physicochemical studies (e.g., the determination of partition coefficients).[100]

14.5.10 ALTERNATIVE SOLVENTS: THE APPLICATION OF ILs

The properties of ILs (ILs are composed of ions with a melting point close to or below room temperature) are very promising for green chemistry applications; they are nonvolatile and good solvents for many organic and inorganic materials. They behave very differently from conventional molecular liquids, and one of their advantages is their thermal robustness, which offers a broad thermal operation range (typically from $-40°C$ to $200°C$).

ILs can be applied not only to existing methods whose sensitivity and selectivity of analysis should be improved, but their behavior and properties offer new opportunities in chemical analysis.

Their main advantage over organic solvents for applications in analytical chemistry is their low volatility coupled with high thermal stability up to $260°C$, which make them useful as solvents for working at high temperatures, and also as stationary phases in GC.[101,102] They provide symmetrical peak shapes, and because their ranges of solvation-type interaction are different for anions and cations, they exhibit a dual selectivity behavior.

The low volatility of ILs makes them useful as solvents working under high vacuum, and together with their more amorphous solid analogs they are convenient in matrix-assisted laser-desorption/ionization-mass spectroscopy (MALDI-MS) analysis.[103,104]

ILs have good solvating properties which, together with their excellent spectral transparency, make them suitable solvents for spectroscopic measurements of a wide range of organic and inorganic species.[105] Reliable solution spectra have been reported for highly charged complex ions with high- and low-oxidation states in the UV, visible, and IR regions.

They are also superior to the standard solvents used as separation media in liquid–liquid extraction in that high separation efficiencies and selectivities are achievable, for example, for the extraction of sulfur and nitrogen compounds from gasoline and diesel oil.[106]

With the use of ILs as an electrolyte medium, it is possible to achieve a wider range of operational temperatures and conditions relative to the more conventional electrolytic media. They are, moreover, promising materials in a variety of electrochemical devices such as batteries, fuel cells, sensors, and electrolytic windows.[107]

14.6 GREENER SEPARATION TECHNIQUES

The main requirements for analysts using chromatographic methods wishing to implement the principles of green analytical chemistry are as follows[108]:

- Utilizing direct chromatographic analysis whenever possible, as it permits analytes in a sample to be determined without the need for pretreatment or sample preparation
- Reducing labor and energy demands, for example, reducing sample preparation time when direct chromatographic analysis is not possible
- Eliminating or reducing the amount of solvent from sample preparation steps applied before final chromatographic analysis
- Conducting all operations with solvents in hermetic systems
- Reducing matrix interferences
- Shortening chromatographic run times
- Integrating the steps of analytical procedures, for example, by using hyphenated techniques

The principles of green chemistry can be implemented in GC in many ways. First and foremost, eliminating or minimizing the amount of solvent in sample preparation techniques prior to final chromatographic analysis is strongly recommended. Therefore, techniques using gas and supercritical fluids for the extraction of numerous pollutants are very common. Rapid chromatography, especially with field-portable instrumentation, has gained in importance,[109] as has the coupling of GC to techniques with a high identification potential, for example, to MS (GC-MS), because confirmation can be achieved in the same step as analysis with a second dimension of information. This provides increased confidence in the result in conjunction with increased effectiveness.

The principles of green analytical chemistry have also been implemented in electrically driven separation methods in that solvent and sample consumption have been reduced, selectivity increased, analysis times shortened, and mechanically simpler instruments designed.

14.7 ADVANCES IN ELECTROCHEMISTRY SENSING TECHNOLOGY RELEVANT TO GREEN ANALYTICAL CHEMISTRY

The advantages of electrochemical systems include high sensitivity and selectivity, a wide linear range, as well as minimal space, power requirements, and instrumentation.

Stricter environmental control and effective process monitoring have created considerable demands for innovative analytical methodologies. For meeting the requirements of green analytical chemistry, new devices and procedures with negligible waste generation or no hazardous substances, and *in situ* real-time monitoring capability, are needed.

The combination of modern electrochemical techniques with the breakthrough in microelectronics and miniaturization has enabled the introduction of powerful analytical devices for effective process or pollution control.

The performance of electrochemical measurements depends intimately on the material of the working electrode. For many years, the mercury working electrode was frequently used owing to its attractive behavior and its highly reproducible, renewable, and smooth surface; at the same time, it posed a toxicity hazard. New developments in the electrode field have dramatically improved the greenness of electrochemistry. Various nonmercury electrodes have been suggested, such as bismuth-film electrodes and new ion-selective electrodes.[110,111] These provide better-quality trace-metal measurements than those achievable with mercury electrodes. In addition, bismuth is a green element—it has a low toxicity and for this reason it is widely used in the pharmaceutical industry. Also, carbon (and other solid)-based electrodes have shown themselves to be readily adaptable as electrochemical biosensors, for example, for the determination of glucose, nitrate, and polyphenol.

Another way of implementing the rules of green chemistry is the miniaturization of solid electrodes; this offers many practical advantages, including a reduction in sample consumption. Furthermore, the significant decrease in resistance makes for easier voltammetric measurements in low-ionic-strength water samples.[112]

14.8 MINIATURIZATION IN ANALYTICAL CHEMISTRY METHODS

Miniaturization, one of basic trends in analytical chemistry, is playing a very important part in "greening" analytical methodologies.[113–116] The main objectives of miniaturization are to

- Improve existing methodologies
- Address new analytical problems
- Provide support for automation and simplification
- Reduce solvent and reagent consumption (hazard minimization)
- Reduce energy consumption
- Reduce the dimensions of analytical devices

Initially, the main goal of miniaturization was to enhance the analytical performance of devices rather than to reduce their size. However, it was found that such tools had the advantage of reducing the consumption of sample and reagents, for example, the smaller consumption of carrier and mobile phases in separation systems (the first analytical system to be miniaturized was the gas chromatograph). Research in this area focused on the development of components such as micropumps, microvalves, and chemical sensors.

In the last 10 years considerable interest has been shown in the production of microfluidic analytical devices that integrate multiple sample-handling processes with the actual measurement step on a microchip platform. Such devices are referred to as "micro total analysis systems" (μTAS), also called "labs-on-a-chip."[114] The whole analytical procedure, including sample pretreatment, derivatization, and separation, is carried out on a single platform. The high degree of integration of the analytical procedure implies that the principles of green chemistry can be applied to all steps of the analytical process.

Microminiaturization is an especially important approach for minimizing the waste generated and is essential for analysis when the amount of sample available is extremely small, usually at a scale of less than microliters. The amount of waste generated can be reduced by 4–5 orders of magnitude in comparison to conventional liquid chromatographic analysis (e.g., 10 μL per day as against 1 L per day). Such a considerable reduction in waste generation and material usage has enormous implications for green chemistry.

Additionally, the downscaling and integration of the analytical process make such microsystems particularly attractive as green chemistry analytical tools, especially for on-site environmental or industrial applications.

14.9 CONCLUSIONS

Growing public concern over protecting our environment is obliging all chemists to modify their chemical activities in such a way that they will be conducted in an environmentally friendly manner. This can be realized within the framework of the principles of green chemistry.

Green chemistry is not a new branch of science, and emphasizing its importance is not a political slogan. It is a new philosophical approach which, through the dissemination and application of green chemistry principles, can contribute to sustainable development. New analytical methodologies are being developed for implementation according to green chemistry standards, and they will be useful in evaluating the effects of chemical processes on the environment.

The application of green chemistry rules to the design of analytical methods is a key factor in efforts to diminish the negative effect of analytical chemistry on the environment. The same ingenuity and innovation applied earlier to obtain excellent sensitivity, precision, and accuracy is now being used to reduce or eliminate the application of hazardous substances in analysis.

Great interest is being shown in solventless techniques of sample preparation. This is for both ecotoxicological and economic reasons: the emission of toxic solvents into the environment is avoided, as can the high costs of recycling expensive, high-purity solvents, for example, by distillation. Furthermore, most of these techniques can be automated and quite easily coupled to "green" final methods of analysis, for example, GC. Efforts are still being undertaken to shorten sample preparation time so that a given preparation method can be linked up with high-speed chromatography.

An important aspect of green chemistry research is the development of new analytical methodologies, for example, the design of new analytical tools for real-time industrial process monitoring and for preventing the formation of toxic materials. Similarly, a real-time field measurement capability is being developed to replace the traditional method of sample collection and transport to a laboratory. This can be done by developing and using in-line or on-line analyzers, which allows analytes to be determined in real time and disturbances to be detected already in the initial steps of a process. Such a mean of analysis supplies rapid information and enables effective preventive measures to be taken, for example, the technological process can be halted or the operational parameters changed, thereby improving overall efficiency.

REFERENCES

1. Anastas, P.T. and J.C. Warner. 1998. *Green Chemistry: Theory and Practice*. New York: Oxford University Press.
2. Winterton, N. 2001. Twelve more green chemistry principles. *Green Chem.* 3: G73–G75.
3. Ritter, S.K. 2001. Green chemistry. *Chem. Eng. News* 79: 24–37.
4. *Pure Appl. Chem.* 2000. Special topic issue on Green Chemistry 72.
5. *Chem. Rev.* 2007. Special issue on Green Chemistry 1007.
6. *Pure Appl. Chem.* 2006. Papers based on presentations at the 2nd International Symposium on Green/ Sustainable Chemistry, January 10–13, 2006, Delhi, India, 78.
7. Anastas, T. 1999. Green chemistry and the role of analytical methodology development. *Crit. Rev. Anal. Chem.* 29: 167–175.
8. Namieśnik, J. 2001. Green analytical chemistry—some remarks. *J. Sep. Sci.* 24: 151–153.
9. Koel, M. and M. Kaljurand. 2006. Application of the principles of green chemistry in analytical chemistry. *Pure Appl. Chem.* 78: 1993–2002.
10. Keith, H.L., L.U. Gron, and J.L. Young. 2007. Green analytical methodologies. *Chem. Rev.* 107: 2695–2708.
11. Wardencki, W., J. Curyło, and J. Namieśnik. 2005. Green chemistry—current and future issues. *Pol. J. Environ. Stud.* 14: 389–395.
12. Wang, J. 2002. Real-time electrochemical monitoring: Toward green analytical chemistry. *Acc. Chem. Res.* 35: 811–816.
13. He, Y., L. Tang, X. Wu, et al. 2007. Spectroscopy: The best way toward green analytical chemistry. *Appl. Spectr. Rev.* 42: 119–138.
14. Curyło, J., W. Wardencki, and J. Namieśnik. 2007. Green aspects of sample preparation—a need for solvent reduction. *Pol. J. Environ. Stud.* 16: 5–16.
15. Bamford, S.A., D. Wegrzynek, E. Chinea-Cano, et al. 2004. Application of X-ray fluorescence techniques for the determination of hazardous and essential trace elements in environmental and biological materials. *Nukleonika*, 49: 87–95.
16. Alvarez, J., M.L. Marco, J. Arroyo, et al. 2003. Determination of calcium, potassium, manganese, iron, copper and zinc levels in representative samples of two onion cultivars using total reflection X-ray fluorescence and ultrasound extraction procedure. *Spectrochim. Acta B*, 58B: 2183–2189.
17. Biswas, S., K. Heindselmen, H. Wohltjen, et al. 2004. Differentiation of vegetable oils and determination of sunflower oil oxidation using a surface acoustic wave sensing device. *Food Control* 15: 19–26.
18. Martin, S.P., D.J. Lamb, J.M. Lynch, et al. 2004. Enzyme-based determination of cholesterol using the quartz crystal acoustic wave sensor. *Anal. Chim. Acta* 487: 91–100.
19. Nichkova, M., P. Eun-Kee, M.E. Koivunen, et al. 2004. Immunochemical determination of dioxins in sediment and serum samples. *Talanta* 63: 1213–1223.
20. Kumar, K., A. Thompson, A.K. Singh, et al. 2004. Enzyme-linked immunosorbent assay for ultra trace determination of antibiotics in aqueous samples. *J. Environ. Qual.* 33: 250–256.
21. Suter, M.J.-F. 2008. Effect oriented environmental analysis. *Anal. Bioanal. Chem.* 390: 1957–1958.
22. Alumbaugh, R.E., L.M. Gieg, and J.A. Field. 2004. Determination of alkylbenzene metabolites in groundwater by solid-phase extraction and liquid chromatography-tandem mass spectrometry. *J. Chromatogr. A* 1042: 89–97.
23. Dąbrowska, H., Ł. Dąbrowski, M. Biziuk, et al. 2003. Solid-phase extraction clean-up of soil and sediment extracts for the determination of various types of pollutants in a single run. *J. Chromatogr. A* 1003: 29–42.
24. Giergielewicz-Możajska, H., Ł. Dąbrowski, and J. Namieśnik. 2001. Accelerated solvent extraction (ASE) in the analysis of environmental solid samples—some aspects of theory and practice. *Crit. Rev. Anal. Chem.* 31: 149–165.
25. Wardencki, W., M. Michulec, and J. Curyło. 2004. A review of theoretical and practical aspects of solid-phase microextraction in food analysis. *J. Food Sci. Technol.* 39: 703–717.
26. Benanou, D., F. Acobas, and M.R. De Roubin. 2004. Optimization of stir bar sorptive extraction applied to the determination of odorous compounds in drinking water. *Water Sci. Technol.* 49: 161–170.
27. Bruheim, I., X.C. Liu, and J. Pawliszyn. 2003. Thin-film microextraction. *Anal. Chem.* 75: 1002–1010.

28. Psillakis, E. and N. Kalogerakis. 2001. Solid-phase microextraction versus single-drop microextraction for the analysis of nitroaromatic explosives in water samples. *J. Chromatogr. A* 938: 113–120.

29. Zhao, L. and H.K. Lee. 2002. Liquid-phase microextraction combined with hollow fiber as a sample preparation technique prior to gas chromatography/mass spectrometry. *Anal. Chem.* 74: 2486–2492.

30. Shimmo, M., A. Piia, K. Hartonem, et al. 2004. Identification of organic compounds in atmospheric aerosol particles by on-line supercritical fluid extraction-liquid chromatography-gas chromatography-mass spectrometry. *J. Chromatogr. A* 1022: 151–159.

31. Wang, S., Y. Lin, and C.M. Wai. 2003. Supercritical fluid extraction of toxic heavy metals from solid and aqueous matrices. *Sep. Sci. Technol.* 38: 2279–2282.

32. U.S. EPA. 1994. Method 3541 Automated Soxhlet Extraction 1.0. Scope and application.

33. Akhtar, J., S.K. Durrani, N.A. Chughtai, et al. 2002. Determination of oxygen and nitrogen impurities in magnesium metal by vacuum distillation and ICP-OES techniques. *Turk. J. Chem.* 26: 681–690.

34. Ferreira, B.S., F. van Keulen, and M.M.M. da Fonseca. 2002. A microporous membrane interface for the monitoring of dissolved gaseous and volatile compounds by on-line mass spectrometry. *J. Membr. Sci.* 208: 49–56.

35. Theodoridis, G. and I.N. Papadoyannis. 2001. Modern sample preparation methods in chemical analysis. *Mikrochim. Acta* 136: 199–204.

36. Namieśnik, J. and W. Wardencki. 2000. Solventless sample preparation techniques in environmental analysis. *J. High Resolut. Chromatogr.* 23: 297–303.

37. Wardencki, W., J. Curyło, and J. Namieśnik. 2007. Trends in sample preparation techniques for environmental analysis. *J. Biophys. Methods* 70: 275–288.

38. Brumley, W.C. 1995. Techniques for handling environmental samples with potential for capillary electrophoresis. *J. Chromatogr. Sci.* 33: 670–685.

39. Dąbrowski, Ł., H. Giergielewicz-Możajska, M. Biziuk, et al. 2002. Some aspects of the analysis of environmental pollutants in sediments using pressurized liquid extraction and gas chromatography—mass spectrometry. *J. Chromatogr. A* 957: 59–67.

40. Gallagher, P.A., S. Murray, X. Wei, et al. 2002. An evaluation of sample dispersion media used accelerated solvent extraction for the extraction and recovery of arsenicals from LFB and DORM-2. *J. Anal. At. Spectrom.* 17: 581–586.

41. Lee, H.B. and T.E. Peart. 2002. Determination of bisphenol A in sewage effluent and sludge by solid-phase and supercritical fluid extraction and gas chromatography/mass spectrometry. *J. AOAC Int.* 83: 290–297.

42. Ferrer, I. and E.T. Furlong. 2002. Accelerated solvent extraction followed by on-line solid-phase extraction coupled to ion trap LC/MS/MS for analysis of benzalkonium chlorides in sediment samples. *Anal. Chem.* 74: 1275–1280.

43. Zuo, Y.G., L.L Zhang, J.P. Wu, et al. 2004. Ultrasonic extraction and capillary gas chromatographic determination of nicotine in pharmaceutical formulations. *Anal. Chim. Acta* 526: 35–39.

44. De Orsi, D., L. Gagliardi, R. Porra, et al. 2006. An environmentally friendly reversed-phase liquid chromatography method for phthalates determination in nail cosmetics. *Anal. Chim. Acta* 555: 238–241.

45. Cava-Montesinos, P., E. Rodenas-Torralba, A. Morales-Rubio, et al. 2004. Sampling using multicommutation. *Anal. Chim. Acta* 506: 145–153.

46. Lucchesi, M.E., F. Chemat, and J. Smadja. 2004. Solvent-free microwave extraction of essentials oil from aromatic herbs: Comparison with conventional hydro-distillation. *J. Chromatogr. A* 1043: 323–327.

47. Barriada-Pereira, M., M.J. Gonzalez-Castro, S. Muniategui-Lorenzo, et al. 2007. Comparison of pressurized liquid extraction and microwave assisted extraction for the determination of organochlorine pesticides in vegetables. *Talanta* 71: 1345–1351

48. Barriada-Pereira, M., I. Iglesias-Garcia, and M.J. Gonzalez-Castro. 2008. Pressurized liquid extraction and microwave-assisted extraction in the determination of organochlorine pesticides in fish muscle samples. *J. AOAC Int.* 91: 174–180.

49. Pawliszyn, J. 1997. *Solid Phase Microextraction. Theory and Practice.* New York: Wiley-VCH.

50. Zhang, Z. and J. Pawliszyn. 1999. Headspace solid-phase microextraction. *Anal. Chem.* 65: 1843–1952.

51. Górecki, T., X. Yu, and J. Pawliszyn. 1999. Theory of analyte extraction by selected porous polymer SPME fibres. *Analyst* 124: 643–649.

52. Boy-Boland, A.A. and J.B. Pawliszyn. 1995. Solid-phase microextraction of nitrogen containing herbicides. *J. Chromatogr. A* 704: 163–172.

53. Jimenez, J.J., J.L. Bernal, M.J. del Nozal, et al. 1998. Solid-phase microextraction applied to the analysis of pesticide residues in honey using gas chromatography with electron-capture detection. *J. Chromatogr. A* 829: 269–277.

54. Hwang, B.H. and M.R. Lee. 2000. Solid-phase microextraction for organochlorine pesticide residues analysis in Chinese herbal formulations. *J. Chromatogr. A* 898: 245–256.

55. Evans, T.J., C.E. Butzke, and S.E. Ebeler. 1999. Analysis of 2,4,6-trichloroanisole in wines using solid-phase microextraction coupled to gas chromatography mass spectrometry. *J. Chromatogr. A* 786: 293–298.

56. Mestres, M., M.P. Marti, O. Busto, et al., 2000. Headspace solid phase microextraction of higher fatty acid ethyl esters in white rum aroma. *J. Chromatogr. A* 881: 583–590.

57. Mestres, M., O. Busto, and J. Gausch. 2000. Analysis of organic sulfur compounds in urine aroma. *J. Chromatogr. A* 881: 569–581.

58. Fitzgerald, G., K.J. James, K. MacNamara, and M.A. Stack. 2000. Characterization of whiskeys using solid-phase microextraction with gas chromatography-mass spectrometry. *J. Chromatogr. A* 896: 351–359.

59. Augusto, F., A.S. Pires Valente, T.E. dos Santos, et al. 2000. Screening of Brazilian fruit aromas using solid-phase microextraction-gas chromatography-mass spectrometry. *J. Chromatogr. A* 873: 117–127.

60. Ligor, M. and B. Buszewski. 1999. Determination of menthol and menthone and pharmaceutical products by solid-phase microextraction-gas-chromatography. *J. Chromatogr. A* 847: 161–169.

61. Fernandez, M., C. Padron, and L. Marconi. 2001. Determination of organophosphorus pesticides in honeybees after solid-phase microextraction. *J. Chromatogr. A* 922: 257–265.

62. Hwang, B.H. and M.R. Lee. 2000. Solid-phase microextraction for organochlorine pesticide residues analysis in Chinese herbal formulations. *J. Chromatogr. A* 898: 245–256.

63. Eisert, R., S. Jackson, and A. Krotzky. 2001. Application of on-site solid-phase microextraction in aquatic dissipation studies of profoxydim in rice. *J. Chromatogr. A* 909: 29–36.

64. Zambonin, C.G. 2003. Coupling solid-phase microextraction to liquid chromatography. A review. *Anal. Bioanal. Chem.* 375: 73–80.

65. Bagheri, H. and A. Salemi. 2004. Coupling of a concentric in-two-tube solid phase microextraction technique with HPLC-fluorescence detection for the ultratrace determination of polycyclic hydrocarbons in water samples. *Chromatographia* 59: 501–505.

66. Baltussen, E., P. Sandra, F. David, et al. 1999. Stir bar sorptive extraction (SBSE), a novel extraction technique for aqueous samples: Theory and principles. *J. Microcolumn Sep.* 11: 737–747.

67. Kolahgar, B., A. Hoffmann, and A.C. Heiden. 2002. Application of stir bar sorptive extraction to the determination of polycyclic aromatic hydrocarbons in aqueous samples. *J. Chromatogr. A.* 963: 225–230.

68. Ochiai, N., K. Sasamoto, H. Kanda, et al. 2008. Sequential stir bar extraction for uniform enrichment of trace amounts of organic compounds in water sample. *J. Chromatogr. A* 1200: 72–79.

69. Blasco, C., G. Font, and Y. Pico. 2002. Comparison of microextraction procedures to determine pesticides in oranges by liquid chromatography-mass spectrometry. *J. Chromatogr. A* 970: 201–212.

70. David, F., B. Tienpont, and P. Sandra. 2003. Stir-bar sorptive extraction of trace organic compounds from aqueous matrices. *LC-GC Europe* 16: 14–16.

71. Lipinski, J. 2000. Automated multiple solid phase micro extraction. An approach to enhance the limit of detection for the determination of pesticides in water. *Fresenius J. Anal. Chem.* 367: 445–449.

72. Sandra, P., B. Tienpont, and J. Vercammen. 2001. Stir bar sorptive extraction applied to the determination of dicarboximide fungicides in wine. *J. Chromatogr. A* 928: 117–126.

73. Kende, A., Z. Csizmazia, T. Rikker, et al. 2006. Combination of stir bar sorptive extraction—retention time locked gas chromatography-mass spectrometry and automated mass spectral deconvolution for pesticide identification in fruits and vegetables. *Microchem. J.* 84: 63–69.

74. Liu, W., Y. Hu, J. Zhao, et al. 2005. Determination of organophosphorus pesticides in cucumber and potato by stir sorptive extraction. *J. Chromatogr. A* 1095: 1–7.

75. Roy, G., R. Vuillemin, and J. Guyomarch. 2005. On-site determination of polynuclear aromatic hydrocarbons in sea water by stir bar extraction (SBSE) and thermal desorption (GC-MS). *Talanta* 66: 540–546.

76. Popp, P., C. Bauer, and L. Wennrich. 2001. Application of stir bar sorptive extraction in combination with column liquid chromatography for the determination of polycyclic aromatic hydrocarbons in water samples. *Anal. Chim. Acta* 436: 1–9.

77. Hu, Y., Y. Yang, and J. Huang. 2005. Preparation and application of poly(dimethylsiloxane)/β-cyclodextrin solid-phase microextraction membrane. *Anal. Chim. Acta* 543: 17–24.

78. Basher, C., A. Parthiban, A. Jayaraman, et al. 2005. Determination of alkylphenols and bisphenol-A. A comparative investigation of functional polymer-coated membrane microextraction and solid-phase microextraction techniques. *J. Chromatogr. A* 1087: 274–282.

79. Jeannot, M.A. and A.A. Cantwell. 1997. Mass transfer characteristics of solvent extraction into a single drop at the tip of a syringe needle. *Anal. Chem.* 69: 235–239.

80. He, Y. and H.K. Lee. 1997. Application of a 32-microband electrode array detection system for liquid chromatography analysis. *Anal. Chem.* 69: 4634–4640.

81. Yamini, Y., M. Hojjati, and M. Haji-Hosseini. 2004. Headspace solvent microextraction. *Talanta* 62: 265–270.

82. Michulec, M. and W. Wardencki. 2005. Selected aspects of chromatographic solvents residues determination using HS, SPME an SDE techniques for isolation and preconcentration of analytes. Book of abstracts of 11th International Symposium on Separation Sciences, September 12–14, 2005, Pardubice, Czech Republic.

83. Saraji, M. 2005. Dynamic headspace liquid-phase microextraction of alcohols. *J. Chromatogr. A* 1062: 15–21.

84. Arthur, C.L. and J. Pawliszyn. 1990. Solid phase microextraction with thermal desorption using fused silica optical fibers. *Anal. Chem.* 62: 2145–2148.

85. Li, Y., T. Zhang, and P. Liang. 2005. Application of continuous-flow liquid-phase microextraction to the analysis of volatile halohydrocarbons in water. *Anal. Chim. Acta* 536: 245–247.

86. Liu, J.F., Y.G. Chi, and G.B. Jiang. 2004. Ionic liquid-based liquid-phase microextraction, a new sample enrichment procedure for liquid chromatography. *J. Chromatogr. A* 1026: 143–147.

87. Liu, J.F., J.F. Peng, Y.G. Chi, and G.B. Jiang. 2005. Determination of formaldehyde in shiitake mushroom by ionic liquid-based phase microextraction coupled with liquid chromatography. *Talanta* 65: 705–709.

88. Ma, M. and F.L. Cantwell. 1999. Solvent microextraction with simultaneous back-extraction for sample cleanup and preconcentration: Preconcentration into a single microdrop. *Anal. Chem.* 71: 388–393.

89. Psillakis, E. and N. Kalogerakis. 2003. Developments in liquid-phase microextraction. *Trends Anal. Chem.* 22: 565–574.

90. Pedersen-Bjergaard, S. and K.E. Rasmussen. 2005. Bioanalysis of drugs by liquid-phase microextraction. *J. Chromatogr. B* 817: 3–12.

91. Rasmussen, K.E. and S. Pedersen-Bjergaard. 2004. Developments in hollow fibre-based, liquid-phase microextraction. *Trends Anal. Chem.* 23: 1–10.

92. Eskilsson, C.S., K. Hartonen, L. Mathiasson, et al. 2003. Pressurized hot water extraction of insecticides from process dust—comparison with supercritical fluid extraction. *J. Sep. Sci.* 27: 59–64.

93. Deng, C., N. Yao, A. Wang, et al. 2005. Determination of essential oil in a traditional Chinese medicine, *Fructus amoni*, by pressurized hot water extraction followed by liquid–liquid extraction and gas chromatography—mass spectrometry. *Anal. Chim. Acta* 536: 237–242.

94. Hedrick, J.L. and L.J. Mulcahey. 1992. Supercritical fluid extraction. *Microchim. Acta* 108: 115–132.

95. Kirschner, C.H. and L.T. Taylor. 1993. Recent advances in sample introduction for supercritical fluid chromatography. *J. High Resolut. Chromatogr.* 16: 73–84.

96. Page, S.H., B.A. Benner, J.A. Small, et al. 1999. Application of stir bar sorptive extraction in combination with column liquid chromatography for the determination of polycylic aromatic hydrocarbons in water samples. *J. Supercrit. Fluids* 14: 257–270.

97. Vejrosta, J., P. Karasek, and J. Planeta. 1999. Analyte collection in off-line supercritical fluid extraction. *Anal. Chem.* 71: 905–909.

98. Björklund, E., L. Mathiasson, P. Persson, et al. 2001. Collection capacity of a solid phase trap in supercritical fluid extraction for the extraction of lipids from a model fat sample. *J. Liquid Chromatogr. Rel. Technol.* 24: 2133–2143.

99. Stone, M.A. and L.T. Taylor. 2000. SPE-GC with quantitative transfer of the extraction effluent to a megabore capillary column. *Anal. Chem.* 72: 3085–3092.

100. Karasek, P., J. Pol, J. Planeta, et al. 2002. Partition coefficients of environmentally important phenols in a supercritical carbon dioxide-water system from concurrent extraction without analysis of the compressible phase. *Anal. Chem.* 74: 4294–4297.

101. Anderson, J.L. and D.W. Armstrong. 2003. A new class of stationary phases for gas chromatography. *Anal. Chem.* 75: 4851–4858.

102. Ding, J., T. Welton, and D.W. Armstrong. 2004. Chiral ionic liquids as stationary phases in gas chromatography. *Anal. Chem.* 76: 6819–6822.

103. Armstrong, D.W., L.-K. Zhang, L. He, et al. 2001. Ionic liquids as matrixes for matrix-assisted laser desorption/ionization mass spectrometry. *Anal. Chem.* 73: 3679–3686.

104. Zabet-Moghaddam, M., E. Heizle, and A. Tholey. 2004. Qualitative and quantitative analysis of low molecular weight compounds by ultraviolet matrix-assisted laser desorption/ionization mass spectrometry using ionic liquid matrices. *Rapid Commun. Mass Spectrom.* 18: 141–148.

105. Appleby, D., C.L. Hussey, K.R. Seddon, et al. 1986. Room-temperature ionic liquids as solvents for electronic absorption spectroscopy of halide complexes. *Nature* 323: 614–616.
106. Bösman, A., L. Datsevich, A. Jess, et al. 2001. Deep desulfurization of diesel fuel by extraction with ionic liquids. *Chem. Commun.* 23: 2494–2495.
107. Seddon, K.R. 2003. Ionic liquids: A taste of the future. *Nat. Mater.* 2: 363–365.
108. Wardencki, W. and J. Namieśnik. 2002. Some remarks on gas chromatographic challenges in the context of green analytical chemistry. *Pol. J. Environ. Stud.* 11: 185–187.
109. Cramers, C.A., H.-G. Jannssen, M.M. Van Deurse, et al. 1999. High-speed gas chromatography: An overview of various concepts. *J. Chromatogr. A* 856: 315–327.
110. Wang, J., J. Lu, S. Hocevar, et al. 2000. Bismuth-coated carbon electrodes for anodic stripping voltammetry. *Anal. Chem.* 72: 3218–3822.
111. Vamvakai, V., K. Tsagaraki, and N. Chaniotakis. 2006. Carbon nanofiber-based glucose biosensor. *Anal. Chem.* 78: 5538–5542.
112. Wang, J. 2002. Electrochemical detection for microscale analytical systems: A review. *Talanta* 56: 223–231.
113. *Fresenius J. Anal. Chem.* 2001. Special issue on Miniaturization and Chip Technology in Analytical Chemistry 371.
114. Greenwood, P.A. and G.M. Greenway. 2002. Sample manipulation in micro total analytical systems. *Trends Anal. Chem.* 21: 726–740.
115. *Trends Anal. Chem.* 2002. Special issue on Fast Response Analytical Systems 21.
116. *Anal. Bioanal. Chem.* 2002. Special issue on Biosensors and Biochips 372.

15 Chemometrics as a Tool for Treatment Processing of Multiparametric Analytical Data Sets

Stefan Tsakovski and Vasil Simeonov

CONTENTS

15.1 Introduction 369
15.2 Basic Chemometric Methods 370
    15.2.1 Cluster Analysis 370
        15.2.1.1 Theoretical Principles 370
        15.2.1.2 Case Study (Struma River) 373
    15.2.2 Self-Organizing Maps 376
        15.2.2.1 Theoretical Principles 376
        15.2.2.2 Case Study (Struma River) 377
    15.2.3 Principal Component Analysis 380
        15.2.3.1 Theoretical Principles 380
        15.2.3.2 Case Study (Struma River) 382
    15.2.4 Receptor Modeling (PCA with Multiple Linear Regression Analysis) 383
        15.2.4.1 Theoretical Considerations 383
        15.2.4.2 Case Study (Struma River) 385
15.3 Conclusions 385
Acknowledgment 386
References 386

15.1 INTRODUCTION

In the last three decades chemometrics has undergone enormous development as a result of the ever greater attention accorded by scientists to environmental data treatment, intelligent instrument signal interpretation, the design and optimization of analytical procedures, and, last but not least, the new metrological aspects of the methods and procedures of analysis. However, it is our deep conviction that the pathways and the effects of numerous polluting species in all environmental compartments have turned out to be the most specific reason for the predominant application of chemometrics in the environmental research field. The opportunities offered by chemometric approaches to classify, model, and reliably interpret large, multiparametric data sets (delivered mainly by the procedures for the environmental monitoring of water, air, and soil) are indeed remarkable.

Modern chemometrics exploits the capabilities of multivariate statistics, already regarded as the classical approach, but also newer, less common methods. It is generally assumed that well-developed and very widely applied basic methods such as cluster analysis (CA), principal components analysis, discriminant analysis, and neuron net approaches are entirely sufficient to cover the main objectives of chemometrics as novel multiparametric information tools helping to acquire a better understanding of the information hidden in large data sets. With their aid one can perform effective data mining, exploratory data analysis, and intelligent data analysis in order to achieve the satisfactory classification, dimension reduction, projection, and appropriate modeling of principally complex objects described by many chemical, physical, and biological variables (or parameters). The undoubted advantage of data interpretation and modeling by the use of chemometric methods has been thoroughly demonstrated by a large number of case studies relating to different locations, time periods, and environmental compartments. It has become accepted that environmental quality assessment achieved by chemometrics is the only reliable basis for decision-making in critical situations and the most helpful instrument for solving problems at different levels of responsibility.

According to its already classical definition, chemometrics is a branch of modern analytical chemistry that involves the application of mathematical, statistical, and other methods employing formal logic

- To evaluate and to interpret chemical and analytical data.
- To optimize and to model chemical and analytical processes and experiments.
- To extract a maximum of chemical and analytical information from complex data sets.

Chemometric methods are traditionally divided into

- Unsupervised multivariate statistical methods [CA, principal components analysis, Kohonen's self-organizing maps (SOMs), nonlinear mapping, etc.], which perform spontaneous data analysis without the need for special training (learning), levels of knowledge, or preliminary conditions.
- Supervised (learning) methods where a priori information is needed, for example, demarcation of a certain number of classes in the classification process.

The main goal of this chapter is to present the theoretical background of some basic chemometric methods as a tool for the assessment of surface water quality described by numerous chemical and physicochemical parameters. As a case study, long-term monitoring results from the watershed of the Struma River, Bulgaria, are used to illustrate the options offered by multivariate statistical methods such as CA, principal components analysis, principal components regression (models of source apportionment), and Kohonen's SOMs.

15.2 BASIC CHEMOMETRIC METHODS

15.2.1 Cluster Analysis

15.2.1.1 Theoretical Principles

The notion of CA or clustering incorporates a broad class of methods used to classify variables (usually chemical components) or objects (usually sampling locations) into groups. This exploratory approach is very useful for revealing and displaying the structure of the data investigated. In this respect, CA is an exploratory, unsupervised pattern cognition technique that does not need *a priori* knowledge about the investigated objects.[1,2]

In environmental studies, responses to the following questions are frequently required:

- What are the factors controlling the data structure obtained?
- Which is the proper set of variables for reliable environmental modeling?
- What is the optimal monitoring scheme?

For an appropriate modeling approach and source assessment, we need correct answers to these questions: this makes exploratory data techniques such as CA a necessary preliminary step prior to undertaking modeling studies.

Two big families of clustering algorithms can be distinguished: hierarchical and nonhierarchical ones. Hierarchical clustering can be performed in an agglomerative and in a divisive way. At the beginning of the agglomerative procedure each object is located in a separate cluster. In the agglomerative algorithms the aims are to combine similar objects into a cluster, add objects to an already formed "closest" cluster, or combine similar clusters. The divisive algorithms, on the other hand, start with one single cluster containing all the objects and then, step by step, the most "inhomogenous" ones are stripped away to form smaller, "more homogenous" clusters. The hierarchical clustering output is a tree-like diagram called a dendrogram.

The aim of classification by nonhierarchical clustering is to classify the objects under consideration into a certain number of preliminary intended clusters. The clusters are formed simultaneously by partitioning methods, which allow the objects to be rearranged between the clusters. The main disadvantage of nonhierarchical clustering is the absence of a graphical output.

The most commonly used procedures in environmental studies are clustering performed using hierarchical agglomerative procedures because of the comprehensible graphical output and clear "hierarchical" relations between clusters.[2,3]

In CA the *input data matrix* $\mathbf{X}$ (measurement data) is presented as

$$\mathbf{X} = \begin{pmatrix} x_{11} & x_{12} & \cdots & x_{1m} \\ x_{22} & x_{22} & \cdots & x_{2m} \\ \cdots & \cdots & x_{ij} & \cdots \\ x_{n1} & x_{n2} & \cdots & x_{nm} \end{pmatrix}, \tag{15.1}$$

where n is the number of objects and m their variables. Usually the data are autoscaled by the following z-transformation formula to avoid the influence of dimension on classification:

$$z_{ij} = \frac{x_{ij} - \bar{x}_j}{s_j}, \tag{15.2}$$

where

$$\bar{x}_j = \frac{1}{2}\sum_{i=1}^{n} x_{ij} \quad \text{and} \quad s_j = \frac{1}{n-1}\sum_{i=1}^{n}(x_{ij} - \bar{x}_j)^2.$$

To find the structures of the objects in the data set, we need a measure of similarity. Although many types of measures can be applied, the Euclidean distance is the most frequently used similarity measure. According to the law of Pythagoras, the distance between two points O_1 and O_2 characterized by variables x and y can be presented as follows (Figure 15.1):

$$d(O_1, O_2) = \sqrt{(y_1 - y_2)^2 + (x_1 - x_2)^2}. \tag{15.3}$$

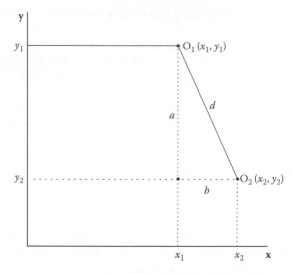

FIGURE 15.1   Distance between two objects in two-dimensional (2D) space.

The extension to more than two dimensions is given by the Euclidean distance and the distance between two objects, i and k, and can be written as

$$d(i,k) = \sqrt{\sum_{j=1}^{m} (x_{ij} - x_{kj})^2},$$

(15.4)

where m is the number of variables (dimensions). Then a *similarity matrix* is calculated, which includes all the distances between the objects to be classified. The matrix is symmetrical with zero values on the main diagonal:

$$\mathbf{D} = \begin{pmatrix} 0 & d_{12} & d_{13} & \cdots & d_{1n} \\ d_{21} & 0 & d_{23} & \cdots & d_{2n} \\ \cdots & \cdots & \cdots & \cdots & \cdots \\ d_{n1} & d_{n2} & d_{n3} & \cdots & 0 \end{pmatrix}.$$

(15.5)

There is a wide variety of hierarchical algorithms, but the typical ones include the *single linkage*, *complete linkage*, and *average linkage* methods. In the similarity matrix one seeks out the two most similar, linked objects p and q (with the smallest distance D_{qp} in order to start constructing the dendrogram. The process is repeated until all the objects are linked in the hierarchical classification scheme. In principle, the most similar objects are considered to form a new object p^* out of these two. In this way, the similarity matrix is reduced by one column and one row. In *average linkage*, the similarities between the new object and the rest are obtained by averaging the similarities of the two most similar objects with the others (e.g., $D_{ip^*} = (D_{iq} + D_{ip})/2$). In *single linkage*, D_{ip^*} is the distance between some object i and the nearest of the linked objects, that is, D_{ip^*} is equal to the smaller of the two distances D_{iq} and D_{ip}. In *complete linkage*, the opposite rule is obtained: D_{ip^*} is the distance between object i and the most remote object q or p. Thus, the only difference in the different hierarchical clustering algorithms is the way in which the linkage sequence is determined. Generally, average linkage is the preferred procedure for larger data sets. However, the above-mentioned algorithms are not to be recommended, because they often form inversions, mainly because of space reduction. Special attention should be paid to *Ward's method* of clustering: This is based on a heterogeneity criterion defined as the sum of the squared distances of each member of a cluster to the centroid of that cluster. The objects and clusters are joined on the basis of the criterion that the sum of the heterogeneities of all clusters (formed in the next step) should increase as little as possible. In general, Ward's method is space conserving and seems to give better results. If squared Euclidean distances are used as a similarity measure, Ward's method also tends to optimize the data set variance.[2–4] As classification results depend on the clustering procedure, it is highly recommended that CA be combined with display methods like principal component analysis (PCA).[3,4]

15.2.1.2   Case Study (Struma River)

The Struma River is located in the southern part of Bulgaria. It flows from north to south and has a length of 290 km as far as the Greek border. From that point to the Aegean Sea the river is about 110 km long. Its total watershed in Bulgaria is nearly 10,250 km² and covers the Vitosha Mountains and the Rila, Pirin, and surrounding mountains (Figure 15.2). Being a crossborder river, the Struma basin is of substantial importance to both Bulgaria and Greece. That is why the careful monitoring of water quality in the long or short term at different sampling sites is not only an ecological but also a political issue. Figure 15.2 shows a map of the Struma in Bulgaria along with the locations of the sampling sites from the Struma River monitoring network controlled by the Ministry of Environment and Waters.

The monitoring system covers a large number of sites where water quality is tested regularly on a daily, weekly, or monthly basis. The industrial and agricultural activity in the Struma River basin is relatively great. The population of this basin is 532,000 (6.47% of the population of Bulgaria) with nearly 300,000 (71%) in urban areas. The main towns (populations >20,000) are Pernik, Blagoevgrad, Kyustendil, Dupnitsa, Petrich, and Sandanski. As far as land use is concerned, some 29,700 ha of land is under irrigation and the natural conditions in this region favor the cultivation of vegetables, fruits, tobacco, cotton, and almonds. The water in the basin is used for irrigation and the

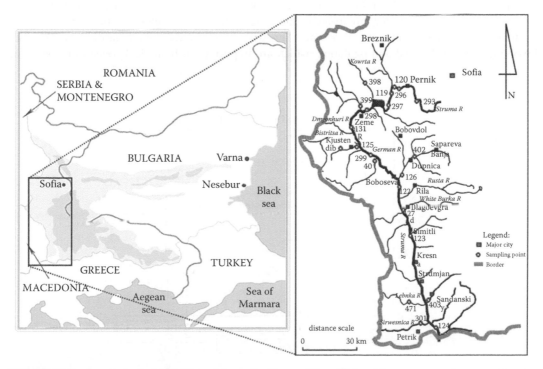

FIGURE 15.2   Location of monitoring points on the Struma River (Bulgaria).

production of electricity. Electricity is generated at power stations (Kalin, Kamenitsa, Pastra, Rila, and others) on some of the Struma's tributaries. There are just a few sites where there is industrial activity: sites 120, 296—steel works, food industry, and coal mining; site 125—agricultural activity and local food industry; site 122—coal mining; site 127—food industry, coal mining, and metal manufacturing; site 123—coal mining; and site 403—food industry. There are also several uncompleted irrigation systems: the Dolna Dikanya–Kovachevzi–Radomir schemes, which are intended to use water from the Pchelin reservoir; the Dyakovo–Dupnitsa scheme should use water from the Dyakovo reservoir. At present, the Pirinska Bistritsa irrigation system is in the process of reconstruction and modernization.

The data set used for the exploration of multivariate methods consists of more than 15,000 measurements on the Struma River. The sites chosen almost completely cover the length of the river from its source to the Greek border. Water samples were collected between 1989 and 1998. For three locations the data set included seven complete years of measurements (provisionally denoted as 3–7: number of sampling locations–number of years of measurements). For the other locations this relationship was as follows: 3–6, 6–5, 1–4, 9–2, and 1–1. The chemical indicators involved were pH, dissolved oxygen (O_2) (mg O_2 L^{-1}), oxidation ability (OXIS) (mg O_2 L^{-1}), biological oxygen demand (BOD) (mg O_2 L^{-1}), chemical oxygen demand (COD) (mg O_2 L^{-1}), dissolved matter (DISS) (mg L^{-1}), nondissolved matter (N-DISS) (mg L^{-1}), chloride (Cl^-) (mg Cl L^{-1}), sulfate (SO_4^{2-}) (mg S-SO_4 L^{-1}), ammonium (NH_4^+) (mg N-NH_4 L^{-1}), nitrate (NO_3^-) (mg N-NO_3 L^{-1}), nitrite (NO_2^-) (mg S-SO_4 L^{-1}), iron (Fe^{2+}) (mg Fe L^{-1}), and calcium (Ca^{2+}) (mg Ca L^{-1}). The chemical analyses were performed according to standard analytical methods as routinely applied in the monitoring network's laboratories. Potentiometry, titrimetry, gravimetry, and spectrophotometry are the standard methods used in surface water quality analysis, especially for major indicators like those mentioned above. Sample preparation and sample measurements are described in detail elsewhere.[5]

CA indicates three clusters of variables according to the less restrictive criterion of Sneath's index of cluster significance (2/3 of D_{max}, where D_{max} is the maximum distance): NO_2 and O_2; Fe^{2+}, SO_4^{2-}, NH_4^+, N-DISS, COD, OXIS, and BOD; and Cl^-, NO_3^-, DISS, Ca^{2+}, and pH (Figure 15.3).

Interpretation of the hierarchical dendrogram for the clustering of chemical variables shows clearly that three major clusters are formed. One of them shows up the close relation between nitrite content and oxygen content in the river water. This is an important indicator regarding the processes involving oxygen demand and oxidation, which are important in biological transformations. The

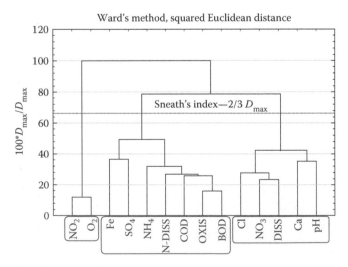

FIGURE 15.3   Classification of variables.

second cluster involves iron, sulfate, ammonium, nondissolved solids, COD, oxidation ability, and BOD, that is, again parameters representing mainly oxidation processes in the water body that are related directly to anthropogenic activity. The last cluster contains chloride, nitrate, dissolved oxygen, calcium, and pH: these are the parameters that determine the hardness and acidity of water.

Using the CA dendrogram (Figure 15.4) to assess the similarity patterns in the space of the sampling locations is a complex task.

The dendrogram shows that all the monitoring locations can be generally grouped into two main clusters (I′ and II′) according to the less restrictive significance criterion of Sneath's index (2/3 of D_{max}) or four clusters (I–IV) according to the more restrictive one (1/3 of D_{max}).

As already mentioned, the interpretation of the dendrogram for linking the sampling locations (including their time parameters) is quite a complicated task, especially in the case of large numbers of data. In this case study, four major clusters should be considered. Careful inspection of the content of each cluster, however, suggests the following explanation:

Cluster I: This contains mostly sampling sites located in urban environments (with provisional numbers 122, 123, and 126), which are characterized by elevated values of BOD and COD, chlorides, sulfates, calcium, and ammonium. Such site discrimination allows this cluster to be ascribed to the provisional "urban site location" pattern. In principle, these are sites with elevated levels of anthropogenic pollution.

Cluster II: This includes mostly sites located on the main tributaries (398, 399, and 403) of the Struma River and has therefore been provisionally named the "tributary" pattern. Here, the concentrations of chlorides, nitrates, calcium, and ammonium are low, and site

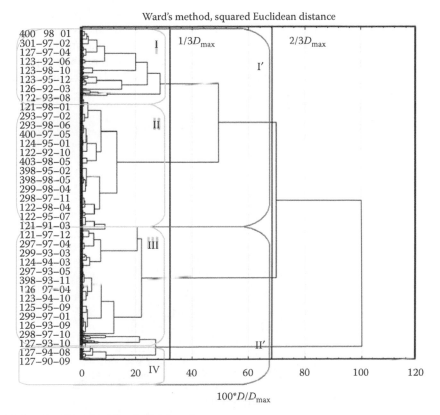

FIGURE 15.4   Classification of sampling locations.

discrimination is based on elevated concentrations of nitrite and total oxygen. These sampling sites indicate a lower anthropogenic impact and an enhanced biological (natural) impact.

Cluster III: This group of sites forms the "rural" pattern of sampling locations (predominantly sites 297 and 299), where the concentrations of the chemical variables are on an intermediate level with respect to both anthropogenic and natural impacts.

Cluster IV: This is a relatively small cluster, which reflects the behavior of a background site (127) with the lowest levels of all the chemical and physicochemical parameters measured.

In summary, CA enables a large data set (over 1100 monitoring results) to be classified into four major site patterns. They are related to the geographical positions of the sampling sites and facilitate the optimization of the whole monitoring process; for instance, instead of sampling each site in the monitoring network, a rapid monitoring procedure selects a representative from each separate pattern in order to make decisions or solve problems along in the watershed.[6–8]

15.2.2   SELF-ORGANIZING MAPS

15.2.2.1   Theoretical Principles

The SOM is an algorithm used to visualize and interpret large high-dimensional data sets[9]; it is an unsupervised pattern cognition method similar to CA. The main advantage of SOM is the simultaneous classification of variables and objects (sampling locations). Typical applications are visualizations of process states or financial results by representing the central dependencies within the data on the map. The map consists of a regular grid of processing units called neurons (Figure 15.5).

A model of some multidimensional observations, possibly a vector consisting of features (variables), is associated with each unit. The map attempts to represent all available observations with optimal accuracy using a restricted set of models. At the same time the models become ordered on the grid so that similar models are close to each other and dissimilar models far from each other. Fitting of the model vectors is usually carried out by a sequential regression process, where $t = 1, 2, \ldots$ is the step index. For each sample $x(t)$, the winner index c (best matching unit—BMU) is first identified by the condition

$$\forall i, \left\| x(t) - m_c(t) \right\| \leq \left\| x(t) - m_i(t) \right\|. \tag{15.6}$$

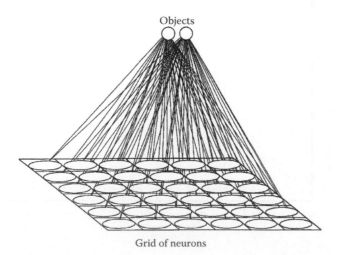

FIGURE 15.5   SOM architecture.

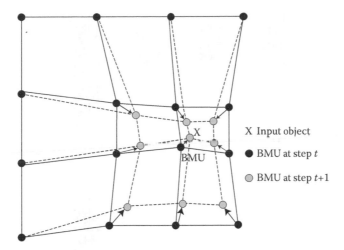

FIGURE 15.6 Updating the BMU and its neighbors toward the input object.

When the BMU has been found, the weight vectors of the SOM are updated so that the BMU is moved closer to the input vector in the input space (Figure 15.6).

Then, all the model vectors or a subset of them belonging to the nodes centered around node $c = c(\mathbf{x})$ are updated as

$$m_i(t + 1) = m_i(t) + h_{c(x),i}(x(t) - m_i(t)). \tag{15.7}$$

Here, $h_{c(x),i}$ is the "neighborhood function," a decreasing function of the distance between the ith and cth nodes on the map grid. This regression is usually reiterated over the available objects.

The trained map can be graphically presented by 2D planes for each variable, with the variable distribution values being indicated by different colors on the different regions of the map. Additionally, the node "coordinates" (vectors) can be clustered by the nonhierarchical K-means classification algorithm.

15.2.2.2   Case Study (Struma River)

The data set from the previous case study, in which the data were interpreted by clustering, is now treated using the SOM approach.

In the first step, the classification of variables was tried. The ordering of 2D SOM planes (Figure 15.7) indicates a high similarity of OXIS, BOD, COD, N-DISS, and NH_4^+. The O_2 and NO_2^- group is located near the aforementioned variables, which indicates a high level of similarity. Visible differences in the gray-scale filled pattern suggest the existence of inversely proportional correlations between O_2, and NO_2^- on the one hand, and OXIS, BOD, COD, N-DISS, and NH_4^+ on the other. These two groups include all the "oxygen-connected" variables and reveal links between the parameters responsible for the oxygen content of the water body (O_2, COD, BOD, and OXIS) and the various biological processes in the water. Dissolved oxygen levels are considered to be the most important indicator of a water body's ability to support aquatic life. COD is commonly used as an indirect measure of the oxygen required to oxidize all compounds, both organic and inorganic, in water. BOD reflects the amount of oxygen consumed by the biological processes that break down organic matter in water. A third, clearly distinguished, group of similar variables consisting of Cl^-, DISS, and NO_3^- represents human activity in the Struma River basin. This last group, not marked in Figure 15.7, can be recognized as a link between pH, and Fe^{2+} and Ca; it is not interpretable in such a simple way, as it includes acidity and hardness components.

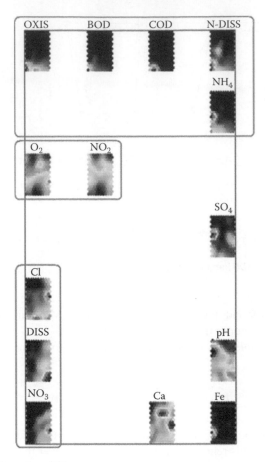

FIGURE 15.7   Classification of variables.

The SOM visualization is in good agreement with the CA results. Minor differences are related mainly to the linkage or the lack of linkage between pH and Fe^{2+} in the CA and SOM algorithms. CA attributes Fe^{2+} and pH to different clusters because of their negative correlation. Similar considerations can be ascribed to the position of SO_4^{2-}, which in SOM has the position of an "outlier." The visualization abilities of CA and SOM are clearly comparable, but SOM has three additional advantages:

- The projection of variables' similarity also contains semiquantitative information about the distribution of a given chemical parameter in the space of the sampling locations.
- SOM visualization is able to present similarity between both positively and negatively correlated variables.
- SOM visualization can indicate "outliers," that is, those chemical variables or sampling locations that do not belong to a well-organized group.

One advantage of hierarchical CA over SOM is the clear "hierarchical" relation between clusters and the possibility of interpreting more than one classification scheme.

In the next step, the similarity between the sampling sites was investigated using SOM. Different values of k (a predefined number of clusters) were tried and the sum of squares for each run was calculated. Finally, the best classification with the lowest Davies–Bouldwin index (also shown graphically in Figure 15.8) was chosen. It is seen that the five-cluster configuration has the lowest

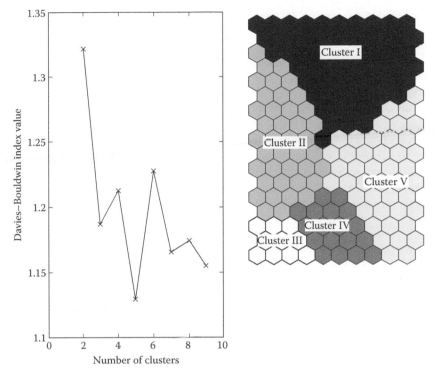

FIGURE 15.8   Classification of sampling locations.

index. For monthly averages, the dimensionality of the Kohonen map was 10×17, and it is clear that more than one case from the initial data set (1104) was related to a particular unit (hexagon). Cases included in each hexagon were grouped in agreement with cluster borders. The SOM classification obtained was then compared with the real data from the initial data set (all sites, all measurements, and monthly averages). Setting up the real values (mean value and SD) of the chemical indicators along with the classification results enabled a fairly reasonable interpretation to be obtained of the clustering pattern (Equation 15.8) connected with the water quality of the Struma River. Clusters I–V contain different numbers of cases out of the total of 1104: I–412, II–261, III–49, IV–101, and V–281. In general, clusters III and IV are linking cases characterized by increasing values of BOD and OXIS as well as increasing concentrations of Cl^-, SO_4^{2-}, Ca^{2+}, and NH_4^+, in parallel with decreasing levels of O_2, NO_2^-, and pH. In contrast, clusters I, II, and V include sites characterized by elevated levels of O_2 and NO_2^-. Clusters IV and V contain sites having the highest content of DISS as well as the highest NO_3^- concentration. SOM clustering yields a number of clusters (five) similar to the number of clusters indicated by CA (tour) according to the more restrictive criterion of Sneath's index (1/3 of D_{max}). This small difference may be due to the fact that the squared Euclidean distance was used to measure similarity among clusters in CA, whereas the Euclidean distance was applied in SOM. Detailed analysis of the variation of chemical indicators across clusters gives the impression that five clusters contain two distinctive patterns of sampling locations, which can be designated as "nonpolluted" (clusters I, II, and V) and "polluted" (clusters III and IV). This is in agreement with clustering using the less restrictive criterion of Sneath's index (clusters I′ and II′). It is easy to show the statistical significance of the differences between the monthly mean values of the variables between particular clusters using the nonparametric Kruskal–Wallis and Mann–Whitney U tests. This comparison is omitted here, however, because of the univariate nature of these procedures.

15.2.3 Principal Component Analysis

15.2.3.1 Theoretical Principles

There are different variants of PCA, but basically one feature they all have in common is that they produce linear combinations of the original columns in a data matrix (data set) responsible for describing the variables characterizing the objects of observation. These linear combinations represent a type of abstract measurement [factors and principal components (PCs)] that is a better descriptor of the data structure (data patterns) than the original (chemical or physical) measurements. Usually, the new abstract variables are called *latent variables*, and they differ from the original ones (called *manifest variables*). It is a common finding that just a few of the latent variables account for a large part of the data set variation. Thus, the data structure in a reduced space can be observed and studied. Generally, when analyzing a data set (matrix) $\mathbf{X}$ consisting of n objects for which m variables have been measured, PCA can extract s PCs (factors or latent variables) where $s < m$:

$$\begin{pmatrix} x_{11} & x_{12} & \cdots & x_{1n} \\ x_{21} & x_{22} & \cdots & x_{2n} \\ \cdots & \cdots & \cdots & \cdots \\ x_{m1} & x_{m2} & \cdots & x_{mn} \end{pmatrix} = \begin{pmatrix} a_{11} & \cdots & a_{1s} \\ a_{21} & \cdots & a_{2s} \\ \cdots & \cdots & \cdots \\ a_{m1} & \cdots & a_{ms} \end{pmatrix} \times \begin{pmatrix} f_{11} & f_{12} & \cdots & f_{1n} \\ f_{21} & f_{22} & \cdots & f_{2n} \\ \cdots & \cdots & \cdots & \cdots \\ f_{s1} & f_{s2} & \cdots & f_{sn} \end{pmatrix} + \begin{pmatrix} e_{11} & e_{12} & \cdots & e_{1n} \\ e_{21} & e_{22} & \cdots & e_{2n} \\ \cdots & \cdots & \cdots & \cdots \\ e_{m1} & e_{m2} & \cdots & e_{mn} \end{pmatrix}, \quad (15.8)$$

$$\mathbf{X} = \mathbf{A} \times \mathbf{F} + \mathbf{E},$$

where $\mathbf{X}$ is the data matrix, $\mathbf{A}$ is the factor loadings matrix, $\mathbf{F}$ is the factor scores matrix, and $\mathbf{E}$ is the residual matrix produced by the loss of information because of dimension reduction.

PC 1 represents the direction in the data containing the largest variation. PC 2 is orthogonal to PC 1 and gives the direction of the largest residual variation around PC 1. The next PCs follow this rule. It is important to note that all PCs are uncorrelated. Figure 15.9 shows a graphical presentation of the first two PCs.

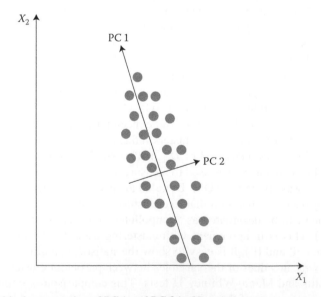

FIGURE 15.9 Graphical presentation of PC 1 and PC 2 in 2D space.

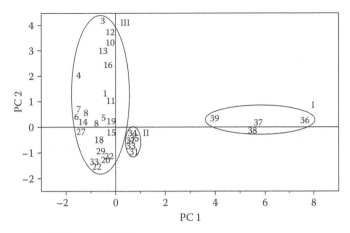

FIGURE 15.10   Score plot PC 1 versus PC 2.

The projections of the data on the plane of PC 1 and PC 2 can be computed and shown on a biplot (Figure 15.10). Known as a *score plot*, such a plot enables the similarity of groups of objects to be determined and classification trends to be detected.[1–4]

PCA makes it possible not only to analyze relationships among objects (e.g., sampling sites or periods in environmental monitoring studies) but also to reveal relationships between variables. According to PCA theory, the scores on the PCs (the new coordinates of the data space) are the weighted sum of the original variables (e.g., chemical and physical measurements): Score (value of object k along the PC p) $= a_{1p}v1 + a_{2p}v2 + \cdots + a_{mp}V_m$, where V denotes the variable value (e.g., concentration) and a represents the weights containing information about the variables. This score can be written for each of the PCs considered in a study. These weights are called *loadings*. Thus, the scores are linear combinations of the manifest variables. The information present in the loadings can also be displayed in respective *loadings plots* (Figure 15.11).

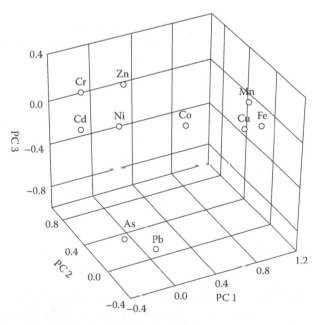

FIGURE 15.11   Loadings plot (PC 1 versus PC 2 versus PC 3).

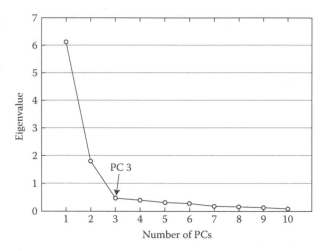

FIGURE 15.12   Scree plot.

These plots reveal important information about correlations and relationships among variables. The factor loadings also represent the relations between "old" variables and new latent variables. Factor loadings with absolute values equal to or higher than 0.7 are indicative of a strong association. Loadings between 0.4 and 0.7 are indicative of a moderate "participation" of initial variables in a PC. Factor loadings lower than 0.4 indicate that the variables are not associated with the PCs. The factor loadings of manifest variables are an indication of the origin of extracted PCs and provide important information on "hidden" factors that determine data set structure.

It is also worth mentioning that the correct application of PCA very often requires *scaling* of the input data in order to eliminate dependence of the analytical results on the scale of the original variables.

One of the key aspects of PCA analysis is the question of how many PCs should be retained. There are certain recommendations that are sometimes contradictory:

- Retained PCs should explain at least 70% of data set variation.
- Retained PCs should have eigenvalues higher than 1 (i.e., PCs accounting for a greater amount of variance than one original variable).
- The number of retained PCs should be assessed by the scree test.

The last approach is widely used for practical data exploration and yields a biplot in which PC eigenvalues are plotted against PC numbers (Figure 15.12). Usually the PCs retained are those on the slope of the graph before the decrease in eigenvalues levels off to the right of the plot. In the example presented, three PCs can be retained according to the scree test.

It is very important in environmental studies to identify the "nature" of a PC. A retained PC that does not represent "real" anthropogenic or natural factors governing an environmental system is useless and does not provide any useful information about the data set under scrutiny.

15.2.3.2   Case Study (Struma River)

The data set treated in this study is a part of the above-mentioned measurements in the Struma River basin. The period of observation in this region was 10 years (1989–98), and the chemical indicators were pH, dissolved oxygen, BOD5, COD, conductivity, acidity, DISS, N-DISS, total hardness, chloride, sulfate, ammonium, nitrate, nitrite, iron, magnesium, and calcium.

Four latent parameters (factors or PCs) explain a substantial part of the total variation of the set (the variance explained by all four PCs is about 79%; Table 15.1). The first factor could be provisionally

TABLE 15.1
Factor Loadings (Only Statistically Significant Figures are Given)

Variables	PC 1	PC 2	PC 3	PC 4
pH				0.65
O_2	0.76			
BOD	0.86			
COD			0.86	
DISS		0.94		
N-DISS	0.76			
Cl^-	0.93			
SO_4^{2-}	0.88			
NH_4^+	0.74			
NO_2^-			0.62	
NO_3^-			0.87	
Fe^{3+}			0.76	
Ca^{2+}		0.92		
Mg^{2+}		0.87		
Hardness		0.65		
Explained variance (%)	32.52	21.74	14.87	9.72

named "anthropogenic." It consists of chemical components that are related to the products of human activity such as dissolved oxygen, BOD5, free acidity, suspended matter, chloride, ammonium, and sulfate. The significance of this latent parameter for water quality is quite high as it explains over 30% of the total variance of the system. The second latent factor is probably of natural origin and is provisionally termed "water hardness." It contains the chemical parameters responsible for water hardness, that is, calcium, magnesium, and the total hardness parameter itself and, in addition, DISS. The third latent factor is related to the content of nitrite, nitrate, and iron, but also to the COD. This factor is termed "biological" since these parameters are predominantly linked to biological activity. Related as it is to the pH value of the water, this last factor has been provisionally named "acidity."

This chemometric procedure is of considerable importance in identifying the data set structure, as it reveals hidden factors (sources of pollution or natural sources) that are important for the data set structure.

15.2.4 RECEPTOR MODELING (PCA WITH MULTIPLE LINEAR REGRESSION ANALYSIS)

15.2.4.1 Theoretical Considerations

In many chemical studies, the measured properties of the system can be regarded as the linear sum of the fundamental effects or factors in that system. The most common example is multivariate calibration. In environmental studies, this approach, frequently called receptor modeling, was first applied in air quality studies. The aim of PCA with multiple linear regression analysis (PCA-MLRA), as of all bilinear models, is to solve the factor analysis problem stated below:

$$x_{ij} = \sum_{k=1}^{p} a_{ik} f_{kj} + e_{ij}, \tag{15.9}$$

where x_{ij} is the ith elemental concentration in the jth sample, a_{ik} is the contribution to the concentration of the ith element from the kth factor (source), f_{kj} is the elemental concentration from the kth source, and e_{ij} denotes residuals unexplained by the model. This model was first described by

Thurston and Spengler.[10] Originally used to model mainly aerosol data, it is now the most widely used receptor model in environmental studies and is applied to water studies as well.

The main advantage of PCA-MLRA is that PCA provides unique solutions to the system, and the interpretation of the variance results is straightforward since factor scores and loadings are forced to be orthogonal (in order to explain the maximum variance). Therefore, pollution sources can be interpreted directly from the factor scores and loadings. The main drawback, however, is that solutions may not always have a direct physical interpretation, as negative factor scores may be obtained. The fact that PCA searches for a linear combination of factors (sources) to fulfill the orthogonality constraints implies that the solutions have good mathematical properties but may not always have a physical meaning. A number of solutions are generally applied in order to correct for negative scores, such as rotation of PCA factor matrices to simplify interpretation as in the Varimax orthogonal rotation, scores uncentering (to make them positive by introducing a "zero day"), and regression to total sample mass.[10] The methodology of PCA-MLRA is described step by step as follows:

Step 1. PCA is usually performed as the Varimax orthogonal rotation of PCs. This rotation gives a more straightforward interpretation of extracted PCs by increasing higher factor loadings and decreasing lower ones.

Step 2. Introducing a "zero day" for each variable:

$$(Z_0)_i = \frac{(0 = \bar{C}_i)}{S_i}. \tag{15.10}$$

Step 3. Calculating the absolute zero PC score for each PC:

$$F_{0p} = \sum_{i=1}^{n} B_{pj}(Z_0)_i, \tag{15.11}$$

where B_{pj} is the factor score coefficient matrix produced by performing PCA. It is related to the factor loadings matrix $\mathbf{A}$: $\mathbf{B} = \mathbf{A}/\lambda_i$, where λ_i are the eigenvalues of the extracted PCs.

Step 4. Calculating the absolute PC score for each PC:

$$[APCS]_{p\times j} = [F]_{p\times j} - [F_0]_{p\times j}. \tag{15.12}$$

Step 5. Performing MLRA where the absolute PC scores (APCS) are independent variables (descriptors) and the sum of measured concentrations is a dependent variable:

$$M_j = \zeta_0 + \sum_{k=1}^{p} \zeta_k APCS_{kj}, \tag{15.13}$$

where M_j is the total concentration in observation j, $\zeta_k APCS_{kj}$ is the concentration identified with source k, and ζ_0 is the concentration due to sources unaccounted for in the PCA.

Step 6. Performing MLRA where the concentrations identified with sources are independent variables and each elemental concentration is a dependent variable:

$$C_j = a_0 + \sum_{k=1}^{p} a_k S_{kj}, \tag{15.14}$$

where C_j is the concentration of variable I during observation j, $S_{kj} = \zeta_k APCS_{kj}$ is the concentration of source k in sample j, and a_k is the mean concentration of source k represented by the element.

TABLE 15.2
Source-Apportioning Results (in % Contribution)

Variables	Anthropogenic Factor	Hardness	Biological Factor	Acidic Factor	R^2
pH	15.8	—	—	84.2	0.39
O_2	73.4	—	17.7	8.9	0.68
BOD	81.7	—	18.3	—	0.64
COD	13.7	—	71.7	14.6	0.66
DISS	15.8	69.6	14.6	—	0.76
N-DISS	73.7	12.3	14.0	—	0.84
Cl^-	88.1	—	11.9	—	0.58
SO_4^{2-}	74.2	12.8	—	10.0	0.64
NH_4^+	86.7	—	13.3	—	0.67
NO_2^-	10.1	—	79.9	10.0	0.41
NO_3^-	11.0	—	82.4	6.6	0.52
Fe^{3+}	8.8	—	91.2	—	0.44
Ca^{2+}	—	92.4	7.6	—	0.81
Mg^{2+}	—	94.2	5.8	—	0.82
Hardness	—	100	—	—	0.81

15.2.4.2 Case Study (Struma River)

After evaluating the factors responsible for the data structure (the same one used in the PCA case study), an apportioning procedure is carried out to assess the contribution of each possible source to the total mass of the surface water. The modeling was performed according to the apportioning approach of Thurston and Spengler,[10] where the total mass of the sample is distributed between the sources identified by PCA (four in this case) after multiple linear regression of the total mass (see Equation 15.13). The apportioning more or less reflects the weight of each latent factor on the sample mass. Thus, it is possible to determine the impact of different factors, both anthropogenic and natural, on surface water quality. Table 15.2 presents the results of source-apportioning for each water quality parameter according to Equation 15.14.

It is evident that the contribution of each latent factor to the portion of mass for each chemical parameter varies according to the different impact of the source on the concentration. For instance, the chloride concentration is distributed between the anthropogenic factor (88.1%) and the biological factor (11.9%), but the anthropogenic impact is much higher. Similar conclusions can be drawn for any chemical parameter involved in water quality. The last column of the table shows the multiple correlation coefficient R^2. This gives an idea of the suitability of the respective models for each of the chemical parameters. The nonsignificant coefficients are underlined. As a whole, most of the models are statistically appropriate and can be used for predictive purposes.[11,12]

15.3 CONCLUSIONS

The case study presented convincingly illustrates the application of chemometrics as a tool for exploratory data analysis, the aim of which is to extract specific information about river water assessment in a large watershed of national and international significance. A transboundary watercourse, the Struma River flows through Bulgaria and Greece, and the monitoring data should contribute to the mutual understanding of water quality on both sides of the border. The chemometric methods involved (CA, principal components analysis, principal components regression, and

Kohonen's SOMs) have made it possible to classify, model, and interpret long-term monitoring data sets and to offer the following significant and practically important conclusions:

- Data classification demonstrated the formation of several significant clusters linking different sampling locations with respect to their specific position within the Struma River monitoring network, that is, background sites, rural sites, urban sites, and sites located on the main tributaries; this classification scheme makes it possible to optimize the sampling procedures in the monitoring network; the monitoring can be organized in such a way that a certain number of sites can be neglected, especially when rapid monitoring is needed; each pattern of sites may offer a more limited number of sites for sampling.
- The SOM classification takes an additional step toward a better understanding of the links between the sampling sites and introduces greater specificity into the classification scheme; sites with similar levels of pollution can be located on a map in such a way that the distance between the patterns of similar sites becomes obvious and ready for interpretation; additional discriminating parameters can be determined since they are responsible for the formation of the different patterns on the map.
- Data projection by principal components analysis have helped to identify hidden factors responsible for the data structure and to interpret these factors accordingly; the source-apportioning models constructed by a variation of the principal components regression (using the approach of the absolute principal component scores of Thurston and Spengler) reveal a qualitative relationship between identified sources (anthropogenic or natural pollution sources) and the total concentration of each chemical parameter involved in the monitoring procedure.
- The classification and the chemometric modeling options have enabled certain information to be obtained on the seasonal patterns of the chemical and physicochemical parameters.
- All classification patterns (for both sampling sites or monitoring parameters) as well as the regression models of source apportionment can be further used for problem-solving and correct decision-making on a local or international scale.

ACKNOWLEDGMENT

The authors would like to express their sincere gratitude to the National Science Fund, Ministry of Education and Science, Bulgaria, for the financial support to carry out this study [Projects VUH 02/05 (2437) and VUH—203/06 (2472)].

REFERENCES

1. Einax, J.W., H.W. Zwanziger, and S. Geiss. 1997. *Chemometrics in Environmental Analysis*. Weinheim: VCH.
2. Massart, D.L. and L. Kaufman. 1983. *Interpretation of Analytical Chemical Data by the Use of Cluster Analysis*. New York: Wiley.
3. Vandeginste, B.G.M., D.L. Massart, L.M.C. Buydens, S. De Jong, P.J. Lewi, and J. Smeyers-Verbeke. 1998. *Handbook of Chemometrics and Qualimetrics: Part B*. Amsterdam: Elsevier.
4. Simeonov, V. 2001. *Classification. Encyclopedia of Environmetrics*. New York: Wiley.
5. Bulgarian State Standards. 1985. Water Analysis: Bulgarian Ministry of Environment and Waters.
6. Simeonova, P., V. Simeonov, and G. Andreev. 2003. Water quality study of the Struma River basin, Bulgaria. *Centr. Europ. J. Chem.* 2: 121–136.
7. Simeonova, P. 2007. Chemometric treatment of missing elements. *Ann. Univ. Sof. Fac. Chem.* 100: 138–145.
8. Astel, A., S. Tsakovski, P. Barbieri, and V. Simeonov. 2007. Comparison of self-organizing maps classification approach with cluster and principal components analysis for large environmental data sets. *Water Res.* 41: 4566–4578.

9. Kohonen, T. 2001. *Self-Organizing Maps*, 3rd edition. Berlin: Springer.
10. Thurston, G. and J. Spengler. 1985. A quantitative assessment of source contributions to inhalable particulate matter pollution in metropolitan Boston. *Atmos. Environ.* 19: 9–25.
11. Simeonova, P. 2007. Multivariate statistical assessment of the pollution sources along the stream of Kamchia River, Bulgaria. *Ecol. Chem. Eng.* 14: 867–874.
12. Simeonova, P., V. Lovchinov, D. Dimitrov, and I. Radulov. 2007. Quality assessment of the Yantra River water monitoring data. *Ecol. Chem. Eng.* 14: 693–705.

16 Quality Assurance and Quality Control of Analytical Results

Ewa Bulska

CONTENTS

16.1 Introduction 389
16.2 General Aspects of QA and QC 389
16.3 Quality in Chemical Measurements 390
16.4 Assuring the Quality of Analytical Results 391
    16.4.1 Validation of Analytical Procedure 393
    16.4.2 Traceability of Analytical Results 394
16.5 Monitoring the Quality of Analytical Results 394
    16.5.1 Internal QC 395
    16.5.2 External QC 395
16.6 Conclusions 396
Glossary 397
References 397

> If the result of a chemical measurement cannot be trusted, then it has little value and the analysis might as well not have been carried out.

16.1 INTRODUCTION

The quality of chemical measurements, and thus of analytical results, is an important issue in modern society, influencing as it does to a great extent the quality of life as well as global trade. The quality of analytical results is also important in a whole range of scientific disciplines in which chemical measurements are made, for example, biology, geology, medicine, microbiology, mineralogy, ecology, pharmacy, and toxicology. Although it is difficult to evaluate accurately the real impact of chemical measurements on all aspects of economic and social activities, it is clear that they are playing an increasingly important role in decision-making at the official, legal, or private level. It has therefore been recognized by those who need analytical data, and in particular, by those interested in environmentally related investigations, that the quality of the analytical data should be guaranteed. Clearly, it is important to deliver accurate results and to be able to show that they are correct. The importance of quality assurance (QA) and quality control (QC) is therefore well established and accepted in analytical chemistry.

16.2 GENERAL ASPECTS OF QA AND QC

QA is defined as all planned and systematic actions, implemented within a management system and demonstrated as required, that are deemed necessary to engender confidence that a product, process, or service will fulfill given requirements for quality. In this respect, "Quality" is regarded as the totality of features and characteristics of a product or service (in analytical chemistry this means the results delivered by a laboratory) that bear on its ability to satisfy the stated or implied needs of customers. In the particular case of chemical measurements, this could be expressed as follows: QA covers all the actions undertaken for planning the proper execution of the analytical task in order to obtain accurate and precise measurements.[1]

The original set of well-defined, strict rules for conducting chemical measurements was developed as the Good Laboratory Practice (GLP) concept by the US Food and Drug Administration (FDA) in the late 1970s and implemented in order to control the quality of analytical results. Later, the US Environmental Protection Agency (EPA) issued similar regulations for the production of agricultural and industrial toxic chemicals. In general, GLP is regarded as the set of conditions under which analytical laboratories should plan, perform, monitor, record, and report their work in such a way that for all samples the history of the results can be traced back.

The concepts of quality management (QM) and QA in analytical laboratories were developed primarily to harmonize the world market and in connection with the globalization of the world's major trading zones; they have now been formally established in relevant directives and standards—formerly ISO 25 and EN 45001, more recently ISO/IEC 17025. The requirements of these standards have become widely accepted as market-regulating tools by both the chemical industry and independent laboratories for routine chemical analysis. At present they are extensively implemented in the form of accreditation, a universally accepted process by which an authoritative body at national or international level gives formal recognition that a laboratory or a person is competent to carry out specific tasks.

It is important to point out that the quality of analytical results is not immediate; it can only be achieved if an extensive set of measures are adopted and complied with. Therefore, in parallel to the development of the QA concept, QC systems were introduced as an important tool supporting the QA of chemical measurements. The QC process of examination of laboratory performance in time should always follow QA. QC thus comprises a set of operational techniques and activities used to check whether the requirements for quality are fulfilled. In practice, QC in an analytical chemistry laboratory implies operations carried out daily during the collection, preparation, and analysis of samples, which are designed to ensure that the laboratory can provide accurate and precise results. QC procedures are intended to ensure the quality of results for specific samples or batches of samples and include the analysis of reference materials (RMs), blind samples, blanks, spiked samples, duplicate, and other control samples.[2]

Although several QA and QC activities are closely related, it should be stressed that QA and QC are not synonymous. QA covers in a broad sense all activities and procedures (managerial and technical) established in the laboratory to assure the overall quality of the delivered results, whereas QC describes the measures used to ensure the quality of individual results or a batch of results. QC is a means of evaluating the current performance of the method being used in the laboratory: it can not only be performed internally in a laboratory (internal QC), but also by external assessment of the results obtained by participation in interlaboratory comparisons (ILCs).

The need to carry out both QA and QC in order to achieve the expected quality of analytical results immediately generates the requirement for clearly defined performance criteria. These criteria enable comparability to be achieved via the traceability of analytical results to national or international standards along an unbroken chain of comparisons. Validation is the central task in the development of any analytical method whose capabilities in specific applications can be assessed with the aid of measurement uncertainty. Finally, proficiency testing (PT) serves to demonstrate comparability in terms of the scatter of the results.[3]

16.3   QUALITY IN CHEMICAL MEASUREMENTS

As mentioned before, chemical measurements are essential in different fields, for example, environmental protection, geology, medicine, and biology. Important decisions are often based on these measurements, for example, whether environmental compartments are polluted or not, food can be eaten, goods can be sold, a patient should be treated, and also in support of legislation (related to health care and trade), production processes, and social problems. This underscores not only the importance of the chemical measurements themselves, but also the need to guarantee their validity to their users by assuring the quality of the results. The term "users of results" has to be understood in a very broad sense, since it may mean anybody who needs information based on analytical measurements, for example, a client of a testing or research laboratory; a production department; national or local authorities; the judiciary; and customs and excise.

To achieve the required quality, the chemist should be involved from the beginning of the process, when the needs of the users of results are defined, until the final report is delivered. In practice, therefore, the analytical chemist has to be consulted at every stage of the process: sample selection, sample storage, and transport procedures, the parameters to be analyzed, and the level of accuracy and precision necessary for an adequate response to be given. This will enable the analyst to set up a scientifically and economically adapted and accepted measurement procedure for the intended purpose as required by the QA system. Moreover, implementation of this system must guarantee that all necessary QC measures can be anticipated, so that the entire quality cycle is under control.[4]

Nowadays, analytical chemists should be able to demonstrate a sufficient level of expertise to support the end users of the analytical results. This is an important requirement for the proper tailoring of the QA system and QC measures to the intended use of the results. Hence, the laboratory and its staff should be in a position to justify the validity of the delivered results by providing the right answer to the analytical part of the problem; in other words, results have to have demonstrable quality and be fit for a given purpose. Implicit in this is that the measurements carried out are appropriately designed for the given problem.[5] Method validation and properly established traceability of results enable analytical chemists to demonstrate this. Then, the QC process designed for particular analytical tasks should concentrate on those parameters that matter most, that is, the ones that have been identified as critical for the given method. Control charts should always be applied, for example, to monitor the stability of the instrument's calibration and to compare the stability of the values obtained with certified reference materials (CRMs), in both the short term and the long term.

16.4   ASSURING THE QUALITY OF ANALYTICAL RESULTS

It is now internationally recognized that for any laboratory to produce reliable data, an appropriate scheme of QA must be implemented. As a minimum, this must ensure that the laboratory is using methods that have been validated as fit for the purpose before their application to a specific task. These methods should be fully documented, staff should be trained, the laboratory infrastructure should be appropriate to the measurements to be made, and mechanisms ensuring that the procedure is under statistical control should be present. Implementation of appropriate QC measures ensures that the data produced and reported are of known quality and uncertainty. Last but not least, the laboratory should participate in PT schemes in order to demonstrate its competence.[6]

Present-day analytical laboratories are increasingly under pressure to supply objective evidence of their technical competence, of the reliability of their results and performance, and to seek formal certification or accreditation. This pressure may come from the laboratory's customers (e.g., industry and national bodies) but may also be due to scientific considerations. A QM system in place, validation of methods, uncertainty evaluation, the use of primary standards and CRMs, participation in ILCs, and PT, all serve to assure and demonstrate the quality of measurements. Compared to, say, 30 years ago, the stability of the equipment now available is much improved, and a greater range of RMs for method validation and calibration is accessible. Nevertheless, to achieve mutual (international) acceptance of various bodies of evidence for QA activities, a number of protocols have been developed. The most widely recognized protocols used in chemical measurements and testing are the ISO Guide 9000:2000, ISO/IEC 17025:2005, and OECD Guidelines for GLP, as well as its national and sector equivalents.

To comply with such requirements, a laboratory has itself to develop a QM system describing its QA activities and the management thereof. Since a quality system and its management often imply a dramatic change in everyday attitudes to laboratory work, sufficient time is needed for awareness building. Time and the relevant training are required to guide a process in which a laboratory is able to build up adequate mechanisms of general QM.[7] Implementation of QA rules, together with acceptance of the proper application of metrological principles, are the key issues underpinning the quality of results and is of great importance and benefit to the laboratory. This means acceptance of full traceability and the harmonization of all operations performed. The development of a quality system is and should always be considered a key process toward assuring quality of results. The final step, for example, formal recognition of the quality system by certification or accreditation should be seen as the end-point of this process.

A QA system describes the overall measures that a laboratory uses to ensure the quality of its operation. Typical items include suitable equipment, trained and skilled staff, documented and validated methods, calibration requirements, standards and RMs, traceability, internal QC, PT, nonconformance management, internal audits, and statistical analysis.

There are a number of QA issues related to the general management systems used within the whole organization: proper supervision of documents and records; in-depth, relevant reviews of contracts with customers; control of nonconforming work; appropriate procedures for carrying out corrective and preventive actions when needed; confidentiality; and competent data handling. In general, QA is that part of the management system within a laboratory responsible for the demanded quality requirements being satisfied. A laboratory should be able to show that its overall organization fulfills the requirements of appropriate standards, which could be ISO 9001:2001 (a general standard that applies to all types of organization) or more specifically, ISO/IEC 17025:2005, which applies to all laboratories carrying out tests and/or calibrations.

The QM system should be designed in such a way as to ensure customer satisfaction by meeting customer requirements. This means that the laboratory management should keep a full history of every single sample, from the contract review (agreement between the laboratory and customer on analytical tasks, accuracy, and precision of results, as well as the time and cost of the analysis), through the receipt of the samples to the final report. All samples and related information should be uniquely identified. Moreover, the laboratory should design the process, making sure that any data related to measurements of given item, for example, validation, calibration, QC, and raw data, are identified and retained for a stated period of time. It is also important to demonstrate the competence of the staff performing measurements, which means keeping records of their training and authorization.

QA requires a management system to be in place (as described by ISO 9001 and Chapter 4 of ISO/IEC 17025); but the laboratory should also assure quality by fulfilling specific technical requirements. This is well described in Chapter 5 of ISO/IEC 17025, which covers various issues related to accommodation and ambient conditions in the laboratory, validation of the methods used, the need to estimate the uncertainty of measurements, and the need to demonstrate not only the traceability of results by using the proper standards and RMs but also the integrity of the sample. Another important issue as regards QA of results is the laboratory environment. This applies to both the storage of samples, reagents, standards, and RMs and the performance of measurements on particular instruments. In the case of samples, it is important to maintain the integrity of the delivered item: its identity must be safeguarded at all costs; samples must be protected against contamination, destruction, or loss of the compound of interest. Ambient conditions in the laboratory, for example, temperature, humidity, and a particle-free atmosphere, must be controlled and monitored, as they may affect analytical results. Some measurement instruments are sensitive to variations in ambient conditions, so the relevant restrictions should be imposed. In general, the laboratory should have a certificate for all the volumetric glassware and standards. Moreover, all measurement instruments should be appropriate to their intended uses; they should be calibrated and maintained in good order to ensure accuracy of measurements. All these procedures should be documented.

With an appropriate QA system in place, a laboratory is in a position to assure customers that it has adequate facilities and equipment for carrying out particular measurements. Moreover, the laboratory should be able to demonstrate that analyses are performed by competent and authorized staff according to well-described procedures and validated methods. Well-designed and properly implemented QA supports a laboratory to ensure that the delivered results are valid and fit for a stated purpose.

It is therefore clear that a laboratory must take appropriate QA measures to ensure that it is capable of providing data of the required quality and that it actually does so. Method validation is therefore an absolutely essential component of the measures that a laboratory should introduce in order for it to produce reliable analytical data.

16.4.1 VALIDATION OF ANALYTICAL PROCEDURE

Validation of the analytical procedure is regarded as one of the most important issues of QA. Before selecting the measurement procedure (analytical method) for a particular purpose, the laboratory should consider its experience, the technical infrastructure at its disposal, and the expected time frame and financial outlay. Validation of the analytical procedure provides necessary information on its performance characteristics and raises the confidence of users in the results.[8]

According to ISO/IEC 17025, validation is confirmation by the examination and provision of objective evidence that the particular requirements for a specific intended use are fulfilled. Performance parameters can be divided into two groups. The first group refers to the properties of the measurement procedure: detection limit and determination limit, working range, linearity, and sensitivity. The second group covers the properties of the results obtained with this particular measurement procedure, that is, traceability and uncertainty (including recovery, robustness, precision, and accuracy).[9]

Full validation of an analytical method usually comprises an examination of its characteristics in interlaboratory method performance studies. However, before a method is subjected to validation by collaborative studies, the method must be validated by a single laboratory, usually by the laboratory that developed or modified this particular measurement procedure. Method validation can be described as the set of tests used to establish and document the performance characteristics of a method and against which it may be judged, thereby demonstrating that the method is fit for a particular analytical purpose.

There are two approaches to single-laboratory method validation: The traditional one that identifies and then evaluates the set of analytical parameters, and a more recent one that is based on the evaluation of uncertainty.

The newer approach places strong emphasis on measurement uncertainty being evaluated using a "component-by-component" approach: the variance or uncertainties inherent in an analytical method are identified and quantified as input quantities. These input quantities are then combined to give an estimate of the overall uncertainty of the analytical procedure. This approach can be regarded as a development of the traditional approach but with several components of overall uncertainties being identified and quantified together.

Validation should cover the whole analytical procedure—from the preparation of the laboratory sample to the evaluation of the result, that is, the whole range of intended matrices, and should be performed within the expected range of concentrations. The intended use of the analytical results should also be considered. This means that the result can be used to evaluate compliance with regulations, to maintain quality and process control, to make regulatory decisions, to support national and international trade, and, last but not least, to support research. It should be clearly understood that validation is carried out in order to evaluate the performance of the applied analytical procedure, not the performance of the analyst or the laboratory.

Several techniques can be used for validation, the most highly recommended ones being (i) evaluation of uncertainty (i.e., a systematic assessment of the quantities influencing the result); (ii) performing CRM analysis; (iii) participation in ILCs/PTs; and (iv) comparison of results with other analytical methods. When a fully validated method is available, the analyst can envisage starting a statistical control system (QC), including the follow-up of performance with the aid of control charts.

If the laboratory develops the validation method in-house, there always needs to be some sample to be used for this purpose; a sample that best mimics routine samples is the most suitable. The usual practice is that a routine sample is used for this purpose as knowledge of the true value is not a critical issue at this stage. Next, the trueness of a method is usually determined by analyzing an appropriate CRM and/or participating in an ILC, one with an externally defined reference value.[10]

16.4.2 TRACEABILITY OF ANALYTICAL RESULTS

Worldwide acceptance of analytical results requires reliable, traceable, and comparable measurements. A key property of a reliable result is its traceability to a stated reference. Traceability basically means that a laboratory knows what is being measured and how accurately it is measured. It is also an important parameter where comparability of results is concerned and is usually achieved by linking the individual result of chemical measurements to a commonly accepted reference or standard. The result can therefore be compared through its relation to that reference or standard. Every link in the traceability chain must be based on the comparison of an unknown value with a known value. The stated reference might be an International System of unit (SI) or a conventional reference scale such as the pH scale, the delta scale for isotopic measurements, or the octane number scale for petroleum fuel. In order to be able to state the uncertainty of the measurement result, the uncertainty of the value assigned to that standard must be known. Therefore a traceability chain should be designed and then demonstrated using the value of the respective standard with its uncertainty.[11]

As already mentioned, the analytical parameters required for method validation and for the estimation of measurement uncertainty can be evaluated without assigned values. But to assess the accuracy of delivered results, as stated in ISO/IEC 17025, there is a requirement for assigned values with a stated uncertainty, which are traceable to the same reference as the analytical results of the method used. In physics, measurements are made in accordance with the SI units, which were introduced under the convention of the meter. In chemical measurements, traceability of results to SI units is not always possible. Therefore, the role of chemical standards is decisive in establishing the comparability of results between laboratories. During the validation of the analytical procedure, traceability of the result can be demonstrated by comparison against the certified value of a CRM, which provides exactly this traceable assigned value with a stated uncertainty.

16.5 MONITORING THE QUALITY OF ANALYTICAL RESULTS

In analytical laboratories that are expected to deliver results with a defined level of accuracy and uncertainty, QC basically involves examining at regular intervals whether the QA system was well designed and executed in such a way as to fulfill the requirements over time. In practice, in accordance with ISO/IEC 17025, the laboratory undertakes QC procedures for monitoring the validity of a test. The resulting data are recorded in such a way that trends are detectable and, where practicable, statistical techniques are applied to the reviewing of the results.[12] This monitoring is planned with respect to the frequency of performing QC measurements and reviewed in order to assure quality over time. Monitoring of the QC process should include the regular use of CRMs or RMs and replicate tests for internal QC, as well as participation in ILC schemes for external quality assessment.

Clearly, QC activities are an essential element of a QA system. Moreover, all these activities should be planned and well documented, and QC data should be analyzed so that corrective action can be taken whenever needed. QC activities mean comparisons of results and their uncertainties with quality criteria and/or reference data, and typically are done by

- The use of control charts of the results obtained for RMs, CRMs, in-house control samples, blanks, and so on

- Regular checks of instrument performance by calibration or adjustments
- Checks on the purity and stability of reagents and solutions used
- Monitoring the ambient conditions in the whole laboratory or in part of it, whenever this is relevant to the measurements performed
- Examination of the repeat results obtained by the same procedure on the same sample in order to examine the influence of any factors on the results

16.5.1 INTERNAL QC

Internal QC involves a continuous evaluation of the selected measures on the basis of various kinds of data. The usual way of performing QC is to use control samples, which should be applied throughout the analytical process, starting with the sample entering the laboratory and ending with the measurement report. The best practice is to analyze control samples in parallel to the routine samples in the same way. The results obtained for the control samples should be plotted on a control chart to show that the results lie within given limits. If the result falls outside the limits, no analytical results are reported and corrective action has to be taken. It is commonly accepted that for QC purposes various kinds of RMs can be used as control samples. It should be stressed that the term "reference material" is of a generic nature and describes any material, sufficiently homogenous and stable with respect to one or more specific properties, which has been established as being fit for its intended use in a measurement process. RMs can be used for calibrating a measurement system, assessing a measurement procedure, assigning values to other materials, and for QC. Even so, a single RM cannot be used in the same measurements for two different purposes. It can only be used for a single purpose in a given measurement at any one time, for example, either for calibration of instruments or for QC. With the use of in-house RMs for method validation, calibration and QC are common practices in environmental laboratories. CRMs are a special kind of RMs—they are materials possessing special characteristics—a certificate, and traceable, assigned values with an uncertainty statement.

The typical way of examining QC data is to use various kinds of control charts, on which results are plotted versus time. Control charts have been developed for monitoring production as a means of statistical process control (SPC). Control charts have also been adopted for analysis as statistical quality control (SQC), where they serve as a warning sign for the laboratory. Quality is no longer guaranteed whenever a measurement exceeds the alarm limit. At regular intervals, the analyst determines the substance to be monitored in the control material and reports the result graphically. When starting a chart, the analyst has to determine from several replicate measurements the mean value, as well as the standard deviation, that represents the reproducibility of the method. This reproducibility value will allow acceptance limits to be predefined, for example, "warning" and "alarm" levels, expressed as 2 or 3 times the standard deviation, respectively.

Apart from the standard Shewart charts, the analyst can also apply X-charts, on which the mean of several replicate measurements is plotted, or R-charts, where the difference between two replicate measurements is plotted. X- and R-charts give an indication of the reproducibility of the method. Drift in analytical procedure, for example, slows changes in the system caused by the aging of parts of instruments, decalibration in wavelength, or the aging of calibration stock solutions, can be detected early when a Cusum chart (cumulative sum) is applied. In Cusum charts, the analyst reports the cumulative sum of the differences between delivered and reference values. If this reference value is certified (CRM), the Cusum chart allows the accuracy of the determination to be monitored.

16.5.2 EXTERNAL QC

All the above discussion was focused on various activities performed within the laboratory, referred to as internal QC; but it is also extremely important for a laboratory to obtain an independent assessment of its performance. This can be achieved by participation in ILC schemes such as PT or collaborative studies. Laboratories participating in an ILC carry out chemical measurements using their routine analytical procedures on the ILC test sample with an undisclosed assigned value. Participation in external quality assessment schemes provides an external measure of the laboratory's performance by comparison with other laboratories carrying out similar types of analysis. The main objective of this external assessment is to obtain an indication of laboratory performance on the basis of a number of commonly accepted scores. In general, all scores involve the difference between a laboratory's result and the assigned value, which is determined by the target range, for example, the standard deviation or uncertainty of the assigned value.

One important task of the organizers of any ILC is to carry out a statistical evaluation of participants' results, as laid down by the ISO 13528:2005 standard. At first, a clear statement should be given on how the assigned value was delivered: by formulation, as a consensus value, or by CRMs being used for the particular ILC. Although CRMs are seldom used as ILC samples, their advantage is that the assigned value can be used for ensuring the traceability of the results in the laboratory.

The range (denominator) used for scoring can be defined as the target range or calculated standard deviation of all data. The drawback of the latter is that it may not reflect the reproducibility of the analytical method being assessed. The evaluation of laboratory results usually applied is the z-score, where the difference is related to the target range. When the z-score is below 2, the result is considered satisfactory; when the z-score lies between 2 and 3, the result is considered questionable; and a z-score above 3 indicates an unsatisfactory performance by the laboratory. Another type of scoring, known as the zeta-score, takes into account the standard uncertainty of the assigned value and the uncertainty of the laboratory result. This approach requires that the laboratory should provide a valid estimate of its uncertainty. The zeta-score can be interpreted using the same criterion as the z-score. A third type of scoring, which takes into account the expanded uncertainties of the assigned value and the laboratory's result, is the En number. When the coverage factor of expanded uncertainty is equal to 2, the En score below 1 indicates that the result is satisfactory.

An important aspect of external QC performed via ILC participation is that the laboratory is assured of the confidentiality of its identity and that of the other participating laboratories. The laboratory is identified only by a code number, which is shown in the report. It is recommended that laboratories take part in a number of ILCs; in this way, they will be able to monitor their performance over time, evaluate trends, and take corrective action when needed.

16.6   CONCLUSIONS

Modern society depends on the skills of analytical chemists to reliably measure the concentrations of compounds of interest. Apart from posing a risk to our health and the environment, an incorrect measurement is a waste of time and money. The processes of QA and QC have therefore been established as tools for ensuring the reliability of results delivered by laboratories to their customers. QA describes the overall measures that a laboratory uses to ensure the quality of its operations. In practice, QA covers a set of managerial and technical procedures implemented in a laboratory, supported by interacting working systems that include QC. QC covers all the operational techniques and activities for monitoring the overall performance of a given laboratory against stated requirements for quality. QC procedures are intended to ensure the quality of results delivered for specific samples or batches of samples. They include the analysis of RMs and/or measurement standards; analysis of blind samples, blanks, spiked, or duplicate samples; the use of QC samples and control charts; and participation in ILCs.

To conclude, the use of RMs and the regular participation in ILC schemes have become fundamental pillars of the assurance and control of analytical data quality in terms of precision and accuracy, thus proving the competence of analytical laboratories.

GLOSSARY

Accuracy: The closeness of agreement between a test result and the accepted reference value.

Interlaboratory comparison (ILC): The organization, performance, and evaluation of tests on the same sample by two or more laboratories in accordance with predetermined conditions to determine testing performance. According to purpose, they can be classified as collaborative studies or proficiency studies.

Performance characteristic: A functional quality that can be attributed to an analytical method, for example, specificity, accuracy, trueness, precision, repeatability, reproducibility, recovery, detection capability, and ruggedness.

Performance criteria: Requirements for a performance characteristic according to which it can be judged that the analytical method is fit for the intended use and generates reliable results.

REFERENCES

1. Neidhart, B. and W. Wegscheider (eds). 2001. *Quality in Chemical Measurements.* Berlin: Springer.
2. Kellner, R., J.M. Mermet, M. Otto, and H.M. Widmer (eds). 1998. *Analytical Chemistry.* Weinheim: Wiley-VCH.
3. Otto, M. 1999. *Chemometrics: Statistical and Computer Application in Analytical Chemistry.* Weinheim: Wiley-VCH.
4. Robouch, P., E. Bulska, S. Duta, M. Lauwaars, I. Leito, N. Majcen, J. Norgaard, M. Suchanek, E. Vassileva, and P. Taylor. 2003. *TrainMiC—Training in Metrology in Chemistry.* Luxemburg: European Communities. EUR Report 20841 EN.
5. ISO/IEC 17025. 2005. General requirements for the competence of testing and calibration laboratories, ISO, Geneva.
6. ISO 5725 Part 1–6. 1998. Accuracy (trueness and precision) of measurement methods and results, ISO, Geneva.
7. ISO/IEC Guide 99. 2007. International vocabulary of metrology—basic and general concepts and associated terms (VIM), ISO, Geneva.
8. Eurachem/CITAC Guide. 1998. Quality assurance for research and development and non-routine analysis.
9. Eurachem/CITAC Guide. 1998. A fitness for purpose of analytical measurements: A laboratory guide to method validation and related topics.
10. Bulska, E. and P. Taylor. 2003. Do we need education in metrology in chemistry? *Anal. Bioanal. Chem.* 377: 588–589.
11. King, B. 2000. The practical realization of the traceability of chemical measurements standards. *Accred. Qual. Assur.* 5: 429–436.
12. Prichard, E. and V. Barwick. 2007. *Quality Assurance in Analytical Chemistry.* Chichester, Hoboken, NJ: Wiley.

17 Analytical Procedures for Measuring Precipitation Quality Used within the EMEP Monitoring Program

Wenche Aas

CONTENTS

17.1 Scope of the Rural Monitoring Network in Europe 399
17.2 Analytical Methods Used for Precipitation Samples
    in Rural Areas in Europe 401
    17.2.1 Collection of Precipitation 401
    17.2.2 Measurements of the Main Ions, pH, and Conductivity in Precipitation 402
        17.2.2.1 Introduction 402
        17.2.2.2 Determination of pH in Precipitation 403
        17.2.2.3 Determination of Conductivity 403
        17.2.2.4 Ion Chromatography 404
        17.2.2.5 Spectrophotometry 404
        17.2.2.6 Determination by Flame Atomic Spectroscopy 405
    17.2.3 Measurement of Heavy Metals and Metalloids in Precipitation 405
        17.2.3.1 Introduction 405
        17.2.3.2 Sample Preparation 405
        17.2.3.3 Inductively Coupled Plasma Mass Spectrometry 406
        17.2.3.4 Graphite Furnace Atomic Absorption Spectroscopy 407
        17.2.3.5 Flame-Atomic Absorption Spectroscopy 408
        17.2.3.6 Cold Vapor Atomic Fluorescence Spectroscopy 408
    17.2.4 Measurements of POPs in Precipitation Using GC-MS 408
17.3 Data Quality Control 409
17.4 Future Perspectives 410
Acknowledgments 411
References 411

17.1 SCOPE OF THE RURAL MONITORING NETWORK IN EUROPE

The "Cooperative program for monitoring and evaluation of long-range transmission of air pollutants in Europe" (EMEP) was launched in 1977 as a response to the growing concern over the environmental effects of acid deposition. EMEP was organized under the auspices of the United

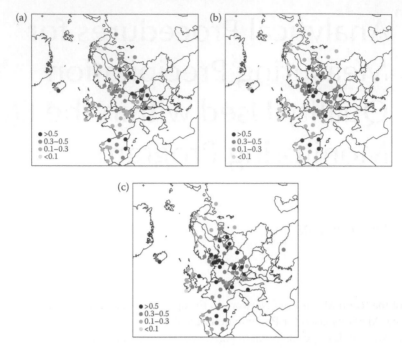

FIGURE 17.1 Annual volume-weighted concentration of (a) ssc SO_4^{2-}, (b) NO_3^-, and (c) NH_4^+ in precipitation in EMEP 2006, units mg S or N/L.[1]

Nations Economic Commission for Europe (ECE). Today, EMEP is an integral component of the cooperation under the Convention on Long-range Transboundary Air Pollution (CLTRAP) launched in 1979. Including EMEP, there are eight protocols under the Convention that identify specific measures to be taken by Parties to cut their emissions of air pollutants.

The main objective of EMEP is to provide governments with information on the deposition and concentrations of air pollutants, as well as on the quantity and significance of the long-range transmission of pollutants and transboundary fluxes. Monitoring of atmospheric concentrations and deposition is one of the basic ways of achieving the objectives of EMEP. In addition to measurements, the program includes official reporting of national emissions, the development of atmospheric dispersion models, and integrated assessment. The EMEP measurements are important for model validation and compliance monitoring. In other words, it is necessary to have measurements that are robust and useful for trend analysis, which is needed to see whether the reductions in emissions defined under the different protocols are also observed in the deposition. The monitoring requirements provide important data for the assessment of environmental issues also considered by other conventions, including local air quality, climate change, water quality, and biodiversity.

Of the different regional monitoring networks in the world, EMEP is the one with the longest time series; it is also one of the largest. The different monitoring networks have harmonized their methods as far as possible to avoid duplication and noncomparable measurements. The methods and procedures described in this chapter are generally derived from the development and experience gained within EMEP, as well as information provided by similar programs in North America, by the World Meteorological Organization/Global Atmospheric Watch (WMO/GAW), and various other research programs.

The EMEP monitoring network of precipitation chemistry consists of about a hundred stations distributed in almost 30 countries across Europe.[1] All of these measure inorganic ions as well as pH and conductivity. Figure 17.1 illustrates the concentration levels of sulfate (corrected for sea salt), nitrate, and ammonium in 2006. The monitoring sites of heavy metals and persistent organic pollutants (POPs) are less densely distributed;[2] in 2006, there were around 50 for heavy metals such as lead

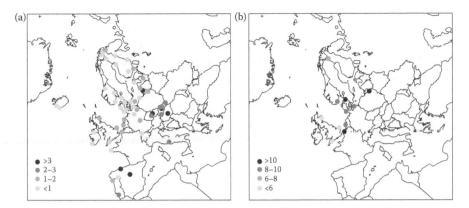

FIGURE 17.2 Annual volume-weighted concentration of (a) Pb (in μg/L) and (b) Hg (in ng/L) in precipitation in EMEP 2006.[2]

and cadmium, 13 for measuring mercury, and 10 for POPs (Figure 17.2). In addition to precipitation chemistry, the EMEP program covers air measurements of major ions, photo-oxidants, particulate matter, POPs, and heavy metals.[3]

17.2 ANALYTICAL METHODS USED FOR PRECIPITATION SAMPLES IN RURAL AREAS IN EUROPE

For the majority of the methods, the necessary quality assurance is provided by a combination of simple and robust sampling techniques with well-described sampling equipment, and the use of synthetic control samples for the chemical analyses. The methods defined in EMEP[4] are harmonized whenever possible with recommendations in other networks such as WMO/GAW[5] and standardization organizations such as the European Committee for Standardization (CEN).

17.2.1 COLLECTION OF PRECIPITATION

The purpose of the sampling and chemical analysis of precipitation in the EMEP network is generally to obtain an accurate indication of the chemical composition of precipitation, which can be used to derive wet deposition estimates on both a short-term (day–month) and a long-term basis. Precipitation is collected in a vessel with a horizontal opening of defined dimensions. The sampling equipment consists of in principle a funnel and a receiving vessel. In order for the sample not to be contaminated from the ground during heavy rain, the rim of the funnel should be positioned 1.5–2 m above the ground level. It is recommended that the sampler be further protected from the settlement of dust and the adsorption of gases during dry periods by an automatic lid, which opens after activation of a precipitation sensor—a wet-only collector. Bulk samplers are recommended only if it can be shown that contamination by the dry deposition of dust and gases, for example, ammonia, is negligible.

The collecting vessel must be constructed from a material that does not alter the chemical composition of the sample. Polyethylene, tetrafluoroethylene, and tetrafluoroethylene-fluorinated ethylpropylene copolymer are generally recommended. Glass and metal containers are not good for measuring the major ions and must be avoided, as these materials are liable to produce both positive and negative artifacts for cations. The sample should give a reliable measure of the amount of precipitation on a daily basis. It is recommended to equip sites with a rain gauge in addition to the wet-only collector.

Collecting a representative sample of snow for precipitation chemistry measurements poses special problems. Most electronic sensors on precipitation chemistry samplers do not detect snow, particularly light, dry snow, as efficiently as rain. Light, dry snow may also fall into and then be blown out of an open container or funnel. Snow may stick to sampler parts and later be blown into the sample container. Ice may coat sampler parts and prevent proper operation. Heavy snow may even fill the container to overflowing and block sampler operation. Some samplers are specially adapted to improve snow collection. Heating the collector lid and other moving parts to about 4–5°C may help prevent snow and ice buildup from interfering with sample collection or sampler operation. For samplers with funnels, applying enough heat to melt snow and ice may be necessary, if the funnel depth is too shallow to accommodate the entire volume. Care should be taken when applying heat to avoid increased sample loss due to evaporation or sublimation. One way to avoid having to heat the sample is to use an open container instead of a funnel.[5]

For the inorganic ions in precipitation, the precipitation volume of the daily (or weekly) samples is measured either by weight or by volume, after which an aliquot is transferred to a sample storage and transport bottle. To a large extent, the sampling of heavy metals follows the same procedures as for the main components in the precipitation, but because of the sensitivity of heavy metal samples to contamination, extra precautions need to be taken. The precipitation volume in mm is calculated from the weight of the water and its density, and the sampler is shipped to the laboratory without transferring any sample to smaller transport bottles.

Mercury is collected in special precipitation samplers. Two alternative materials may be used for funnels and collection bottles: borosilicate glass and a halocarbon such as Teflon or PFA. Borosilicate glass is often preferred because of its lower cost and general availability. Quartz glass can also be used but is generally avoided owing to its high cost. For extended sampling periods, diffusion of Hg^0 into the precipitation sample collected has to be prevented, since it could contribute to the mercury content of the sample via oxidation to water-soluble forms. This is easily done by inserting a capillary tube between the funnel and the bottle. The sample bottles also have to be shielded from light to avoid photo-induced reduction of the mercury in the precipitation sample.[6,7]

Regardless of the duration of the sampling period, there is always the possibility of chemical degradation of the sample in the field during the course of sample collection, during shipment from the field to the laboratory, and prior to analysis at the laboratory. The sample preservation practices followed by most networks often do not completely prevent chemical degradation. One recommended practice is to store samples at <4°C in the laboratory before analysis. Alternatively, chemical biocides can be added to the samples to prevent degradation. But this requires strict quality control procedures to ensure that these additives contain nothing that will contaminate the samples. To date, biocides have been used primarily for research purposes and only on a limited basis.[5]

17.2.2 MEASUREMENTS OF THE MAIN IONS, pH, AND CONDUCTIVITY IN PRECIPITATION

17.2.2.1 Introduction

In connection with the determination of transboundary fluxes and deposition of air pollutants, the concentrations of sulfate, ammonium, and nitrate in precipitation are particularly important. However, determination of one or more sea-salt constituents (Na, Cl, and Mg) is also necessary in order to determine the fraction of the sulfate concentration due to marine sea-spray aerosols. Moreover, determination of the base cations Ca, K, and Mg is desirable in order to obtain an indication of the large-scale deposition of bases; this is needed in connection with the determination of critical loads. Finally, pH and conductivity should be determined in order to obtain some idea of the overall composition of the samples, and to check the consistency of the chemical analyses.

Most of the major ions in precipitation samples can be determined by ion chromatography, which is the generally recommended method for anions such as chloride, nitrate, and sulfate, although other methods may be used for some compounds. Table 17.1 gives a list of alternative recommended methods.

These last three anions are not part of the standard EMEP measurement program. They are included here, however, because they are found in precipitation samples in concentrations comparable to those of some of the other ions, and may be necessary to explain the ion balance and measured conductivities, particularly for samples with pH > 5. In other parts of the world, for example, the tropics, organic acids may be of great importance and should be included.[5]

TABLE 17.1
Recommended Methods for Chemical Analysis of Main Ions in Precipitation

Component or Parameter	Recommended Methods	Alternative
Conductivity	Conductivity cell and resistance bridge	
Hydrogen ion (H$^+$)	Potentiometry (glass electrode) pH < 5.0	Titration
Ammonium ion (NH$_4^+$)	Ion chromatography	Spectrophotometry (indophenol blue color reaction)
Sodium ion (Na$^+$)	Atomic absorption spectroscopy (AAS)	Ion chromatography
Potassium ion (K$^+$)	AAS	Ion chromatography
Magnesium ion (Mg^{2+})	AAS	Ion chromatography
Calcium (Ca^{2+})	AAS	Ion chromatography
Sulfate ion (SO$_4^{2-}$)	Ion chromatography	
Nitrate ion (NO$_3^-$)	Ion chromatography	Spectrophotometry, Griess method (reduction to nitrite and diazotation)
Chloride ion (Cl$^-$)	Ion chromatography	Spectrophotometry [displacement of SCN$^-$ in Hg(SCN)$_4^{2-}$, determination of colored Fe(SCN) complex]
Bicarbonate ion (HCO$_3^-$)	Titration	
Formate ion (HCOO$^-$)	Ion exclusion chromatography	Ion chromatography
Acetate ion (CH$_3$COO$^-$)	Ion exclusion chromatography	Ion chromatography

Note that most of the components can be determined by ion chromatography, which is strongly recommended for sulfate, nitrate, and chloride anions. However, ion chromatography holds no advantages over conventional methods when it comes to the determination of ammonia and base cations.

17.2.2.2 Determination of pH in Precipitation

The method is based on determining the potential difference between an electrode pair, consisting of a glass electrode sensitive to the difference in the hydrogen ion activity in the sample solution and the internal filling solution, and a reference electrode, which is supposed to have a constant potential independent of the immersing solution.

The pH of precipitation varies between about 3.0 and 7.5. Past experience from regional networks and laboratory intercomparisons has shown that measuring pH in precipitation is difficult, mainly because of the low ionic strength of the samples.[5] Samples may also degrade as a result of biological activity and should therefore be kept refrigerated until the time of analysis, when they are brought to room temperature. The pH measurements should be carried out within 2 days of a sample's arrival at the laboratory.

It is strongly recommended that the electrode system be checked at regular intervals by comparison of the "apparent pH" of a low-ionic-strength solution with a known pH or hydrogen ion concentration. The pH readings should be within 0.02 or 0.05 pH units of the "theoretical" result. The pH meter should have both an intercept and a slope adjustment and should be capable of measuring to within ±0.01 pH unit.

17.2.2.3 Determination of Conductivity

Conductance is the inverse of resistance in a solution and conductivity is the inverse of specific resistance. Conductivity is measured with a bridge and a measuring cell; it is dependent on the distance between the electrodes and their area in the measurement cell.

The conductivity of precipitation samples depends on the concentrations of various ion species and their different abilities to transport electric charges in solution, that is, the equivalent conductivity of the ion species. By comparison with estimated conductivity and in combination with ion balance calculations, conductivity measurements can help identify ion concentrations that are wrong or inaccurate.

17.2.2.4 Ion Chromatography

Basic information about ion chromatography will be found in Weiss.[8] The ISO norm 10304-1[9] sets out the details of the ion chromatographic determination of anions in solution in lightly contaminated waters.

A small volume of the sample, typically <0.5 mL, is introduced into the injection system of an ion chromatograph. The sample is mixed with an eluent and pumped through a guard column, a separation column, a suppressor device, and a detector, normally a conductivity cell. The separation column is an ion exchange column that has the ability to separate the ions of interest. The separation column is often preceded by a shorter guard column containing the same substrate as the separation column in order to prevent the separation column from becoming overloaded and/or blocked by particles. Different types of separation columns, eluents, and suppression devices have to be used for anions and cations. Each ion is identified by its retention time within the separation column.

Any species with a retention time similar to that of the main ions can interfere. Large amounts of one of the ions may interfere by reducing the peak resolution of the next ion in the elution sequence. Sample dilution may then be necessary. In some systems the so-called negative water dip at the start of the chromatogram may interfere with the Cl^- determination. This can be avoided by adding a small amount of concentrated eluent to all samples and calibration standards to match the eluent concentration.

The ion chromatograph is calibrated with standard solutions containing known concentrations of the target ions. Calibration curves are constructed from which the concentration of each ion in the unknown sample is determined. It is strongly recommended to match the calibration solutions with the sample matrix. Five calibration solutions and one zero standard (blank, normally water) are needed to generate a suitable calibration curve. The range to be used will depend on the concentration range for the different samples.

All reagents must be of recognized analytical grade. The water used for dilution should be deionized and filtered. The water should have a resistance of >10 MΩ/cm and not contain particles >0.20 μm. The bottles that are to contain sample, calibration standards, and reagent solutions should be made of polyethylene or polypropylene. For the anions, borosilicate glass may also be used. Stock standard solutions may be purchased as certified solutions from different manufacturers or NIST (National Institute for Standards and Technology, USA), or else prepared from salts or oxide that are dried, dissolved, and diluted.

17.2.2.5 Spectrophotometry

Nitrate, ammonium, and chloride may be determined spectrophotometrically if an ion chromatograph is not available.

17.2.2.5.1 Griess Method for Nitrate

This method can be used to determine the nitrate content in precipitation within the range from 0.02 to 0.23 mg NO_3–N/L (0.1–1.0 mg NO_3/L). Nitrate is reduced to nitrite using cadmium treated with copper sulfate as reducing agent in the presence of ammonium chloride. With this method the sum of nitrate and nitrite is determined. Nitrite and sulfanilamide form a diazo compound that couples with N-(1-naphthyl)-ethylenediamine dihydrochloride to form a red azo dye. The concentration in the solution is determined spectrophotometrically at 520 nm. Note that nitrite will interfere with the determination of nitrate.

17.2.2.5.2 Indophenol Blue Method for Ammonium

This method is applicable to the determination of the ammonium content in precipitation within the 0.04–2.0 mg NH_4/L range. In alkaline solution (pH 10.4–11.5), ammonium ions react with hypochlorite to form monochloramine. In the presence of phenol and excess hypochlorite, the monochloramine will form a blue-colored compound, indophenol, when nitroprusside is used as catalyst. The concentration of ammonium is determined spectrophotometrically at 630 nm.

17.2.2.5.3 Mercury Thiocyanate-Iron Method for Chloride

The method can be used for the direct determination of the chloride ion content in precipitation samples within the 0.05–5 mg/L range. Chloride ions will replace the thiocyanate ions in undissociated mercury thiocyanate. The thiocyanate ions thus released react with ferric ions to form a dark red iron–thiocyanate complex.

17.2.2.6 Determination by Flame Atomic Spectroscopy

Sodium, potassium, magnesium, and calcium in precipitation can be analyzed by atomic spectroscopic methods or by ion chromatography. Both flame- (AAS and AES) and plasma (ICP-AES and inductively coupled plasma mass spectrometry, ICP-MS)-based methods can be used. For these ions, ion chromatography has no special advantage concerning sensitivity, precision, and accuracy over the spectroscopic methods, but analysis of all the ions in one run is not possible with flame AAS or AES. The method can normally be used to determine sodium, magnesium, potassium, and calcium in precipitation within the 0.01–2 mg/L range, but this will depend to a certain degree on the commercial instruments used.

The ions in the sample solution are converted to neutral atoms in an air–acetylene flame. Light from a hollow cathode or an electrodeless discharge lamp (EDL) is passed through the flame. The light absorption of the atoms in the flame, which is proportional to the ion concentration in the sample, is measured by a detector following a monochromator set at the appropriate wavelength. This principle holds for measurements performed in the AAS mode. In the AES mode, the light emitted from the atoms excited in the flame is measured. Most commercial instruments can be run in both modes. Sodium may be measured more favorably in the emission mode.

In atomic absorption spectroscopy (AAS) both ionization and chemical interferences may occur. These interferences are caused by other ions in the sample and result in a reduction of the number of neutral atoms in the flame. Ionization interference is avoided by adding a relatively high amount of an easily ionized element to the samples and calibration solutions. For the determination of sodium and potassium, cesium is added. To eliminate chemical interferences from, for example, aluminum and phosphate, lanthanum can be added to the samples and calibration solutions.

EDL or hollow cathode lamps are used to determine Na, K, Mg, and Ca. Single-element lamps are preferred, but multielement lamps may be used. EDLs are more intense than hollow cathode lamps, and are preferred for K and Na. When performing analyses in emission mode, no lamps are needed.

17.2.3 MEASUREMENT OF HEAVY METALS AND METALLOIDS IN PRECIPITATION

17.2.3.1 Introduction

In EMEP, ICP-MS is defined as the reference technique. The exception is mercury, where cold vapor atomic fluorescence spectroscopy (CV-AFS) is chosen. Other techniques may be used, if they are shown to yield results of a quality equivalent to that obtainable with the recommended method. These other methods include graphite furnace atomic absorption spectroscopy (GF-AAS), flame-atomic absorption spectroscopy (F-AAS), and CV-AFS. The choice of technique depends on the detection limits desired. ICP-MS has the lowest detection limit for most elements and is therefore suitable for remote areas. The techniques described in this manual are presented with minimum detection limits. Table 17.2 lists the detection limits for the different methods.

17.2.3.2 Sample Preparation

After measuring the sampling volume by weighing the storage bottles, nitric acid should be added (this can also be added before sampling)—1 mL of supra-pure conc. HNO_3 per 100 mL of precipitation. This will dissolve the metals that could be adsorbed on the walls of the container and will also

TABLE 17.2
Minimum Detection Limit for Different Analytical Methods

Element	ICP-MS[10] (ng/mL)	GF-AAS[11] (ng/mL)	F-AAS[12] (ng/mL)
As	<0.01	0.056	0.02
Cd	<0.01	0.0014	0.5
Cr	<0.01	0.0038	2
Cu	<0.01	0.015	1
Ni	<0.03	0.072	2
Pb	<0.001	0.007	10
Zn	<0.02	0.006	0.8
Hg		0.2	0.001

prevent the growth of microorganisms. Samples for the analysis of total mercury should be preserved with low blank HCl (5 mL 30% acid/L).

The precipitation sample may contain a large fraction of undisclosed material. Such nonhomogenous samples should be filtered before analysis to avoid problems with the analytical instrument. Filtration also ensures that the precipitation sample is homogenous, which makes the analysis more reproducible. Either a disposable syringe filter or vacuum filtration equipment should be used to filter the acidified samples.

As it is easy to contaminate samples for heavy metal analysis, they must be handled with care; gloves must always be worn.

17.2.3.3 Inductively Coupled Plasma Mass Spectrometry

ICP-MS is a multielement technique that is suitable for trace analysis; it offers a long linear range and low background for most elements. ICP-MS is a technique where the ions produced in inductively coupled plasma are separated in a mass analyzer and detected. The sample solution is fed into a nebulizer by a peristaltic pump. The nebulizer converts the liquid sample into a fine aerosol that is transported into the plasma by an Ar gas flow. In the plasma the sample is evaporated, dissociated, atomized, and ionized to varying extents. The positive ions and molecular ions produced are extracted into the mass analyzer. Detailed descriptions of the ICP-MS technique can be found in a number of textbooks.[13,14]

In ICP-MS analysis it is necessary to consider interferences such as isobar overlap and physical interference (Table 17.3). Isobar overlap exists when two elements have isotopes of essentially the same mass. Isobar overlap may also occur as a result of the formation of polyatomic species consisting of two or more atomic species, for example, ArO^+. They are formed by rapid ion–molecule reactions between the components of the solvent or sample matrix and the constituents of the plasma. The dominant species in the plasma and its surroundings are Ar, O, N, and H. These elements can combine with each other to give a variety of polyatomic ions. The extent to which polyatomic ions form depends on several parameters, including sampling geometry, plasma and nebulizer conditions, choice of acids and solvents, and the nature of the sample matrix. By careful optimization of the ICP-MS instrument, it is possible to keep the formation of polyatomic species to a minimum and the elemental sensitivity close to the maximum. Isobar overlap may also be caused by doubly charged ions; such ions are detected at half mass ($m/2$). The elements that could produce doubly charged ions are typically the alkaline metals, alkaline earth metals, and some transition metals. Physical interferences are associated with nebulization and transport processes as well as with ion-transition efficiencies. The efficiency of the nebulization and transport processes depends on the viscosity and surface tension of the aspirated solution. Therefore, physical interference (matrix effect) may occur when

TABLE 17.3
Isotopes of the Main Heavy Metals and Some Possible Interferences

Element	Isotope Mass	Relative Abundance	Isobar Overlap (% Abundance)	Polyatomic Species
Cr	**52**	83.76		ArC^+, $^{35}ClOH^+$
	53	9.55		$^{37}ClOH^+$
Ni	58	67.88	^{58}Fe	^{42}CaO
	60	26.23		^{44}CaO
	61	1.19		
	62	3.66		^{46}CaO
	64	1.08		^{48}CaO
Cu	**63**	69.09		TiO^+, $ArNa^+$, PO_2^+
	65	30.91		$ArMg^+$
Zn	64	48.89	^{64}Ni (1.8)	SO_2^+, SS^+, $ArMg^+$
	66	27.81		$ArMg^+$
As	**75**	100		$Ar^{35}Cl^+$
Cd	**111**	12.75		$^{93}MoO^+$
	114	28.86	^{114}Sn (0.66)	
Pb	204	1.48	^{204}Hg (6.85)	
	206	23.6		
	207	22.6		
	208	52.3		

the samples and calibration standards have different matrices. In addition to matrix matching of samples and calibration standards, the use of an internal standard may reduce these problems.

If there is a considerable difference in concentration between samples or standards that are analyzed in sequence, a memory effect may occur. This effect is caused by sample deposition on the cones and in the spray chamber; it also depends on which type of nebulizer is being used. The washout time between samples must be long enough to bring the system down to a blank value.

Three calibration blank standards should be analyzed to establish a representative blank level, after which the calibration standards are analyzed. After calibration, the quality control standard should be analyzed to verify the calibration. The sample introduction system is flushed with rinse blank, and the blank solution is analyzed to check for carry-over and the blank level. If the blank level is acceptable, the samples can be analyzed. If the blank values are too high, the flushing of the sample introduction system and analysis of the blank solution should be repeated until an acceptable blank level is reached. The calibration blank value, which is the same as the absolute value of the instrument response, must be lower than the method's detection limit.

17.2.3.4 Graphite Furnace Atomic Absorption Spectroscopy

GF-AAS is a powerful technique suitable for trace analysis. The technique is highly sensitive (analyte amounts 10^{-8}–10^{-11} g absolute), is capable of handling micro samples (5–100 µL), and has a low noise level from the furnace. Matrix effects from components in the sample other than the analyte are more serious in this technique than in F-AAS. The precision of GF-AAS is typically 5–10%. A graphite tube is located in the sample compartment of an AA spectrometer with the light from an external source passing through it. A small volume of sample is placed inside the tube, which is then heated by applying a voltage across its ends. The analyte dissociates, and the fraction of analyte atoms in the ground state absorbs portions of light. The attenuation of the light beam is measured. As the analyte atoms are formed and diffuse out of the tube, the absorption rises and falls in a peak-shaped signal.

The Beer–Lambert law describes the relation between the measured attenuation and the analyte concentration. Detailed descriptions of the GF-AAS technique can be found in various textbooks.[14]

With this technique, problems may arise with interference, such as background absorption—the nonspecific attenuation of radiation at the analyte wavelength caused by matrix components. To compensate for background absorption, correction techniques such as a continuous light source (D_2-lamp) or the Zeeman or Smith–Hieftje method should be used. Enhanced matrix removal due to matrix modification may reduce background absorption. Nonspectral interference occurs when components of the sample matrix alter the vaporization behavior of the particles that contain the analyte. To compensate for this kind of interference, the method of standard addition can be used. Enhanced matrix removal by matrix modification or the use of a L'vov platform can also reduce nonspectral interferences. Hollow cathode lamps are used for As, Cu, Cr, Ni, Pb, and Zn; single-element lamps are preferred, but multielement lamps may be used if no spectral interference occurs.

Calibration standards are prepared by single or multiple dilutions of the stock metal solutions. A reagent blank and at least three calibration standards should be prepared in graduated amounts in the appropriate range of the linear part of the curve. The calibration standards must contain the same acid concentration as in the samples after processing.

17.2.3.5 Flame-Atomic Absorption Spectroscopy

F-AAS is a very specific technique, subject to few interference effects. F-AAS is a single-element technique, and analyte determinations in the mg/L region are routine for most elements. A liquid sample is nebulized to form a fine aerosol, which is mixed with fuel and oxidant gases, then carried into a flame. In the flame the sample is dissociated into free ground-state atoms. A light beam from an external source emitting specific wavelengths passes through the flame. The wavelength is chosen to correspond with the absorption energy of the ground-state atoms of the target element. The parameter measured in F-AAS is the attenuation of light. Unfortunately, the detection limits (Table 17.3) are too high to make this technique very useful for precipitation samples in rural areas, except in the case of Zn, which usually has a relatively high concentration level.

17.2.3.6 Cold Vapor Atomic Fluorescence Spectroscopy

The most common procedure for analyzing mercury in precipitation is oxidation with BrCl, prereduction with $NH_2OH \cdot HCl$, followed by reduction of the aqueous Hg to Hg^0 using $SnCl_2$. Hg is purged onto gold traps, thermally desorbed, and analyzed using CV-AFS.[6,15,16]

The traps should be dried at about 40°C in a mercury-free N_2 flow for 5 min prior to analysis, after which they should be connected to the AFS detector on-line with the helium gas flow. The mercury is then thermally desorbed either directly into the detector or onto an analytical trap. If an analytical trap is used, a second heating step should be performed before the detection. The advantages of dual amalgamation are that the influence of any interfering substances adsorbed on the first trap may be reduced and that the mercury adsorbed on the second analytical trap will be more easily desorbed, thus yielding a sharper peak.

The calibration step is critical. In general, the basic principle is always to use two independent calibration solutions. One of these can be made from pure chemicals, for example, Hg^0 dissolved in concentrated HNO_3 and diluted to the appropriate volume. For mercury, commercially available standard solutions can be used, but regular checks against a reference standard must be made. Certified reference materials (CRFs) should be used if available, but reference standards can also be prepared from pure mercury compounds. In the absence of aqueous-phase reference standards, solid materials may be used.

17.2.4 Measurements of POPs in Precipitation Using GC-MS

Measurement of POPs in precipitation is a very difficult task because of problems with contamination and the very low concentration levels. Monitoring networks have usually focused on air samples, which to some extent are easier both from an analytical point of view and with regard to applying the data to study transport and sources. Nevertheless, precipitation measurements are of great importance for a better understanding and quantification of the deposition of these compounds.

GC-MS can be used to analyze organochlorine pesticides, for example, α-, β-, γ-HCH, HCB, and polychlorinated biphenyls (PCB). The components are quantified by using an internal standard. Furthermore, a calibration is performed with a standard mixture containing known concentrations of the components to be measured and one or more components not contained in the sample (internal standards). The calibration is followed by injection of the sample containing known amounts of internal standards. Quantification is relative to the internal standard. In this way, the sample extract volume will not be included in the calculations, and it is not necessary to accurately determine the final sample volume after evaporation of the injection volume. The GC-MS instrument should be calibrated every day. The sensitivity of the mass spectrometer can, for instance, be controlled daily by determining the signal-to-noise ratio for a given amount of a chosen component (PCB-101 could be one such component). For further details of the method, the reader is referred to different manuals and papers on the subject.[4,17]

17.3 DATA QUALITY CONTROL

Measurements should be standardized as far as possible so that the data obtained are comparable and of sufficient quality. Traceability is an important concept for documenting the quality of the measurements; every standard solution must be regularly checked against a reference material.

Documents describing the equipment and procedures should be available to the operators and technicians responsible for the sampling and chemical analysis, and these documented procedures should be followed to the letter. All the personnel involved should be adequately trained and instructed.

Frequent use should be made of field and laboratory blanks; these are essential for discovering the weak links in the sampling, handling, and analytical procedures. The blank results should also be used to correct measurements when necessary. The detection limits for the methods need to be quantified as 3 times the standard deviations of the blanks. The chemicals used may themselves be a source of contamination for some elements and have to be checked.

A clean laboratory and equipment are undoubtedly crucial to all analytical methods. For trace element and POP measurements, however, additional precautions need to be taken. Glassware and other materials used for storing samples may act as both a source and a sink for some transition and heavy metal ions. Consequently, it is important to clean glassware and polyethylene equipment several times with dilute solutions of nitric acid followed by deionized water. Gloves must be worn whenever working with samples and sampling equipment.

Interlaboratory exercises have to be a part of the measurement program in order to ensure, as far as possible, a consistent data set. CRMs of artificial precipitation samples and solid samples are available from various organizations, for example, BCR, NIST, and IAEA. In addition, laboratory intercomparisons are arranged annually by, for example, WMO/GAW and EMEP. Artificial precipitation samples are distributed to different laboratories. EMEP laboratory comparisons of the main components in precipitation have been conducted for 25 years, and they have provided important documentary evidence for the evolution of data quality in EMEP during this period. The results show that laboratory performance has improved during this period, so that at present most laboratories manage to be within the 10% relative standard deviation (RSD) for all the major ions (Table 17.4).

For heavy metals, laboratory intercomparisons have been conducted for about 10 years; there has been an improvement in respect of these elements, too. The relative uncertainty here, however, is greater for the major ions (Table 17.5). For some laboratories, low concentrations are particularly difficult to measure; the general problem is that detection limits are too high.

Field intercomparisons are another important quality assurance step for quantifying the uncertainty of methods; they also include the sampling uncertainty and not just the analytical uncertainty

TABLE 17.4
RSD of Major Ions in the Laboratory Intercomparison in 2005[18]

	SO_4^{2-}	NO_3^-	NH_4^+	pH	Mg^{2+}	Na^+	Cl^-	Ca^{2+}	K^+	Cond
AT	0.6	**0.2**	1.4	0.3	3.1	1.3	4.2	2.8	1.6	0.9
CH	0.7	**0.6**	1.0	0.0	1.2	0.4	1.1	0.7	2.9	0.3
CZ	0.9	**0.3**	10.5	0.5	1.2	2.3	1.3	3.2	1.3	1.2
DE	0.6	**0.5**	1.1	0.1	0.8	1.2	0.9	0.7	1.0	3.0
DE Leipz	0.1	**0.3**	0.7	0.0	0.8	0.4	0.8	0.5	0.3	1.4
DK	0.3	**0.3**	1.2	0.0	1.2	5.1	3.2	3.0	2.1	1.8
EE	1.2	**1.4**	32.7	1.3	2.0	2.7	4.1	6.7	0.8	3.1
ES	6.3	**7.0**	4.1	0.3	0.4	1.8	12.0	1.0	0.7	0.9
FI	0.9	**1.7**	2.6	0.3	2.8	11.2	9.6	2.3	2.8	0.8
FR	0.4	**0.9**	1.0	0.2	3.5	1.6	2.1	3.5	1.3	1.9
GB	0.9	**0.9**	1.9	0.3	18.5	15.2	1.8	6.2	10.3	4.0
HU	2.7	**2.9**	1.2	0.3	0.8	2.1	18.2	22	8.5	2.1
HR	1.2	**2.0**	0.8	0.2	9.8	3.7	1.3	8.8	8.3	1.1
IE	0.5	**1.1**	2.6	0.2	2.0	1.3	1.8	2.0	2.1	0.4
IS	2.0	**6.0**	11.4	0.3	2.4	0.7	12.5	1.5	5.2	1.6
IT	1.0	**0.7**	3.8	0.5	2.4	3.2	3.3	1.8	2.8	2.1
IT	0.5	**3.4**	11.4	1.0	1.2	0.3	3.6	1.7	9.5	1.2
LT	3.2	**0.6**	3.0	0.1		2.1	3.1	45.1	1.6	1.0
LV	2.3	**2.4**	1.2	0.2	1.6	0.2	6.8	1.8	0.7	0.6
MK		**9.9**	89.2	1.3	31.3	1.1		183.1	7.1	16.6
NL	0.5	**3.5**	0.5	0.3	3.9	2.0	5.6	1.8	7.0	1.2
NO	0.5	**0.7**	1.2	0.2	3.5	1.4	1.1	1.5	0.5	1.4
PL	0.9	**0.7**	3.4	0.2	2.0	2.7	1.6	3.7	4.1	1.0
PL05	1.5	**2.4**	0.8	0.4	0.4	0.5	2.5	0.8	0.8	1.3
PT	11.2	**2.5**	4.9	0.9	5.1	4.1	22.3	3.5	5.4	1.9
RU	3.9	**6.7**	1.8	0.2	9.1	10.5	31.0	24.1	7.8	0.9
SE	0.1	**0.2**	2.9	0.2	2.4	0.5	1.4	2.8	0.7	1.4
SI	0.6	**2.1**	2.2	0.2	2.0	1.3	7.0	1.3	0.3	1.3
SK	4.5	**1.3**	38.4	0.2	10.2	5.4	3.9	13.5	17.3	0.5
TR	0.7	**2.6**	11.8	0.3	4.7	2.0	2.9	4.5	3.6	6.6
YU	0.4	**0.7**	1.0	0.1	2.8	1.4	2.1	2.2	1.3	0.5

5–10%   10–25%   >25%

that is measured in the laboratory intercomparison. This is especially important in the initial phase of the measurement program so as to prevent erroneous data being produced over a long period of time. For example, a comparison between bulk- and wet-only collectors should be done if the bulk collector is the preferred one, in order to evaluate the influence of dry deposition.

17.4 FUTURE PERSPECTIVES

The analytical methods for the main ions and heavy metals are well established and there is no special need for further improvements. The greatest uncertainties in these methods lie in the sampling procedures and site representativeness. For POPs, reference methods are at present available only for air measurements; such methods need to be established for precipitation samples as well. Even though the analytical method may be the same, sample preparation and the sampling itself

TABLE 17.5
RSD of Heavy Metals in 2006[2]

Lab	As Low	As High	Cd Low	Cd High	Cr Low	Cr High	Cu Low	Cu High	Pb Low	Pb High	Ni Low	Ni High	Zn Low	Zn High
1	4	2	<DL	4	3	6	0	4	0	3	<DL	5	<DL	11
2	12	4	8	3	26	2	22	10	1	5	16	14	16	12
3	15	1	13	7	7	12	25	14	5	0	20	4	0	1
4	25	11	146	11	<DL	26	100	34	15	21	<DL	31	<DL	<DL
5	21	20	13	12	28	11	7	8	9	11	6	8	10	11
6	<DL	2	<DL	<DL	<DL	<DL	<DL	<DL	<DL	11	<DL	<DL	<DL	10
7	11	8	12	3	1	0	5	0	3	0	<DL	0	6	0
8	1	0	2	2	3	2	1	3	3	6	1	5	1	2
10	<DL	<DL	17	2	<DL	<DL	<DL	<DL	25	8	<DL	<DL	<DL	<DL
13	<DL	<DL	16	4	<DL	<DL	10	4	51	10	<DL	<DL	11	865
14	5	2	7	2	49	3	3	3	1	2	6	1	4	3
15	1	3	2	1	2	0	1	1	6	2	6	2	11	3
16	<DL	<DL	0	0	0	0	4	0	9	3	12	0	0	4
23	8	6	18	2	20	6	5	4	2	3	2	1	4	7
32	4	3	23	5	10	13	21	3	19	4	11	5	6	5
33	41	8	15	1	9	10	3	3	4	6	9	18	9	9
36	1	1	5	4	7	2	1	1	10	10	11	2	1	7
38	<DL	5	<DL	12	<DL	3	<DL	7	21	3	<DL	9	<DL	90
39	<DL	0	<DL	0	0	1	6	3	6	5	<DL	0	<DL	0

☐ 10–25% RSD ☐ 25–50% RSD ☐ >50% RSD

Note: DL means lower than the detection limit, and low and high indicate concentration levels of the sample.

need meticulously elaborated operating procedures. Furthermore, there might be a need to include other species, such as phosphate, organic acids, and black carbon. The last two are important for studying carbon fluxes, and phosphates are significant in the nutrient balance. Standardized protocols for these compounds may be developed in the future.

ACKNOWLEDGMENTS

The methods described in this chapter have been developed continuously since 1977, the beginning of the EMEP Programme. There are many scientists who have contributed to defining reference methods for EMEP: I would especially like to mention Jan Erik Hanssen, Jan Schaug, Arne Semb, and Hilde Thelle Uggerud.

REFERENCES

1. Hjellbrekke, A.G. 2008. Data Report 2006. Acidifying and eutrophying compounds and particulate matter. EMEP/CCC-Report 1/2008.
2. Aas, W. and K. Breivik. 2008. Heavy metals and POP measurements 2006. EMEP/CCC-Report 4/2008.
3. UN-ECE. 2004. EMEP Monitoring Strategy and Measurement Programme 2004–2009, EB.AIR/GE.1/2004/5.
4. EMEP. 2001. *Manual for Sampling and Chemical Analysis*. Revised November 2001. Kjeller, Norway: Norwegian Institute for Air Research. EMEP/CCC Report 1/95. URL: http://www.nilu.no/projects/ccc/manual/index.html.

5. WMO/GAW. 2004. Manual for the precipitation chemistry programme. WMO Report No. 160.
6. Munthe, J. 1996. Guidelines for the sampling and analysis of mercury in air and precipitation. Gothenburg. IVL-Report L 96/204.
7. OSPAR. 1997. JAMP guidelines for the sampling and analysis of mercury in air and precipitation. London.
8. Weiss, J. 1994. *Ion Chromatography*, 2nd edition. Weinheim: Wiley-VCH.
9. ISO. 1992. ISO norm 10304-1, publication date 1992-11. Method for determination of soluted anions fluoride, chloride, nitrite, *o*-phosphate, bromide, nitrate, sulphate with liquid ion chromatography: Low polluted water.
10. Fisons Scientific equipment, VG Instrument Group, Bulletin No. 5M/AMSG/390, England.
11. Perkin Elmer, "new Analyst™ 800 detection limits," Technical note, Norwalk, USA, 1998.
12. Parsons, M.L. and A.L. Forster. 1983. Trace element determination by atomic spectroscopic methods—state of the art. *Appl. Spectros.* 37: 411–418.
13. Jarvis, K.E., A.L. Gray, and R.S. Houk. 1992. *Handbook of Inductively Coupled Plasma Mass Spectrometry*. Glasgow: Blackie.
14. Montaser, A. 1998. *Inductively Coupled Plasma Mass Spectrometry*. New York: Wiley.
15. Bloom, N.S. and E.A. Crecelius. 1983. Determination of mercury in seawater at subnanogram per litre levels. *Mar. Chem.* 14: 49–59.
16. Bloom, N.S. and W.F. Fitzgerald. 1988. Determination of volatile mercury species at the picogram level by low temperature gas chromatography with cold-vapour atomic fluorescence detection. *Anal. Chim. Acta* 208: 151–161.
17. Vogelsang, J. 1991. The quality control chart principle: Application to the routine analysis of pesticide residues in air. *Fresenius J. Anal. Chem.* 340: 384–388.
18. Aas, W. 2007. Data quality 2005, quality assurance, and field comparisons. EMEP/CCC-Report 3/2007.

18 Life Cycle Assessment of Analytical Protocols

Helena Janik and Justyna Kucińska-Lipka

CONTENTS

18.1 General Idea of LCA 413
18.2 Methodology of LCA 415
    18.2.1 Goal and Scope 415
    18.2.2 LCI Analysis 416
    18.2.3 Life Cycle Impact Assessment 418
    18.2.4 Life Cycle Interpretation and Application 421
18.3 LCA for Solvent Use in Analytical Protocols 424
18.4 Conclusions 427
Acknowledgments 428
References 428

18.1 GENERAL IDEA OF LCA

Life cycle assessment (LCA) is a relatively new tool for environmental management, which is becoming more and more important owing to the globalization of the world economy, where there is a need to develop standards in protecting the environment.

LCA is one of the most valuable analytical tools that governments, businesses, and environmentalists can use to assess the environmental load and impact caused by any kind of human undertaking on the Earth.

LCA originated in the early 1970s. At the beginning, studies in this field were carried out in only a few countries—Sweden, the United Kingdom, Switzerland, and the United States. The products that dominated LCA discussion for a long time were beverage containers.[1]

Later, in the 1970s and the 1980s, more studies were carried out using different methods but without a common theoretical framework. The consequences of this approach were negative, as the results differed greatly, even though the objects of the study were the same. This is what prevented LCA from becoming a more generally accepted analytical tool. Moreover, these results were taken up by firms in order to substantiate marketing claims.

Since about 1990, under the coordination of the Society of Environmental Toxicology and Chemistry (SETAC), the discussion and exchange of ideas between LCA experts has increased, and efforts have been undertaken to harmonize the methodology and establish a "Code of Practice."[2]

Complementary to the efforts of SETAC, the International Organization for Standardization (ISO) (Technical Committee 207, Subcommittee 5) has played a role in LCA improvement since 1994. While SETAC has focused on the development of methodology, the ISO has begun work on its standardization (ISO 14040, 1997 E, ISO 14041, 1998E, ISO 14042, 2000E, and ISO 14043,

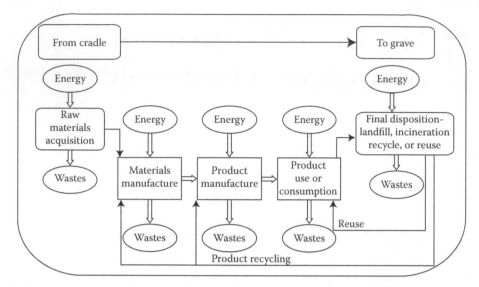

FIGURE 18.1   The general material and product flow showing the idea of LCA—"From the Cradle to the Grave." (Reworked on the basis of ISO 14 040.[3])

2000E). LCA[3,4] is a method used to identify the authentic environmental impact of a product (material, service), taking into account its effect on the environment at every stage of its life cycle (the "From the Cradle to the Grave" viewpoint—Figure 18.1). It attempts to analyze each step in the life cycle of products, services, or activities by identifying the energy, materials, and other components used in order to assess their impacts on the environment.

This applies to the impact caused by the extraction of raw materials (e.g., energy and water usage, farming practices, renewable resources, and greenhouse gas emissions), through their processing into the final product, transporting the product, using (and reusing) it, and its eventual disposal (including residues and greenhouse gas potentials) or recycling. The intention of LCA is to present a methodology and framework within which quantitative criteria to support policy decisions can be generated on a systematic basis. These criteria encompass the set of materials and energy inputs and outputs of the life of a product, process, or activity. LCA results are the background to be considered in the actions for the conversion of current production and consumption (in a broad sense) patterns to environmentally less burdensome patterns. LCA is currently being implemented by industry to help governments establish certification criteria for different fields.

There is already quite a substantial literature on the LCA of materials or processes, for example:

- LCA of the anaerobic digestion of waste products[5] (comparison of anaerobic and aerobic digestion processes)
- LCA of forestry products[6] (watchdogs for the sustainable harvesting of forestry products)
- LCA of extractive industries[7] (minimization of environmental threats)
- LCA of communal waste[8]
- LCA of product impact on the environment[9]
- LCA of ship lifetime[10]

In this chapter we shall discuss LCA in the context of protocols for the analysis of polychlorinated biphenyl (PCB) and polycyclic aromatic hydrocarbon (PAH) in surface water using two different extraction techniques,[11] and the LCA of the utilization of different solvents.[12–26] Generally speaking, this is the beginning of the use of LCA to assess the environmental impact of analytical protocols; at the moment, there are not many papers on this subject, but this situation is sure to improve in the near future.

18.2 METHODOLOGY OF LCA

According to ISO 14040-14049, LCA should consist of the following four fundamental and separate phases (Figure 18.2):

1. Definition of goal and scope
2. Life cycle inventory (LCI) analysis
3. Life cycle impact assessment (LCIA)
4. Life cycle interpretation.

18.2.1 GOAL AND SCOPE

In this first phase of LCA, the following aspects must be defined/described: the purpose and extent of the assessment; the descriptive functional unit (represented by a product or service) that is formed and its limits; the basis of comparison; the components of the product's life cycle; and assumptions and possible limitations.

CASE STUDY

Assessment goal: Identification of the main streams of environmental loading in the phases of materials production and extraction processes with two different preparation techniques applied to the analysis of PCB and PAH in surface water.[11]

Assessment scope: Comparison of liquid–liquid extraction (LLE) and solid-phase extraction (SPE).

Functional group: The preparation, over a period of 10 years, of 30,000 final extractions (27,500 samples—water analysis, 2500—blank analysis) to establish the amount of PCB and PAH in surface water using gas chromatography with an attached mass spectrometer (GC-MS).

Limitation: This particular analysis is demonstrated for scientific and teaching reasons; marketing purposes are thus excluded.

System studied: Figure 18.3 shows a diagram of the elements of the product cycle (here: a service) for both systems studied.

Assumptions: Elements that are the same for both techniques (e.g., syringes and pipettes) are excluded from the LCA.

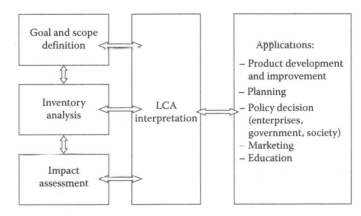

FIGURE 18.2   Life cycle assessment phases and the use of results. (Reworked according to ISO 14040.[3])

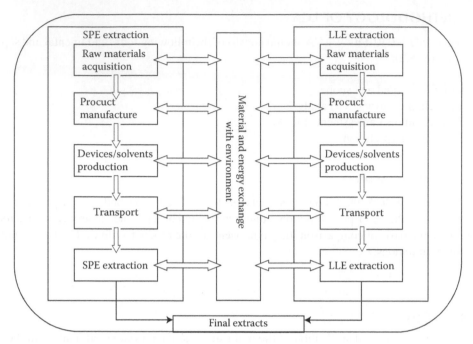

FIGURE 18.3   A general view of the systems (case studies) with their life cycles.[11]

18.2.2 LCI Analysis

The second phase of LCA considers the collection and quantification of system inputs and outputs that are of extreme environmental importance (land use, emissions, waste generation, and use of resources).[27–30]

Inputs include the materials and energy entering the system studied, whereas outputs take the form of energy, materials, emissions, and waste products that cross over from the system to the environment.[3,4] There are many programs (Umberto, GaBi, TEAM, Eco Manager, Eco Pro, Chalmers, EarthShift, ATROiD, SimaPro, and others not mentioned here) for doing LCA, including the construction of inventory tables and impact assessment with the aid of a computer.

CASE STUDY: CONTINUATION OF THE ABOVE ANALYSIS WITH THE USE OF SimaPro 6.04

The first step in this phase, with the use of the computer procedure, is to create all the input and output data using the database library and one's own information (Tables 18.1 through 18.4).

From the database library, one can select the relevant data for material, energy, transport, processing, and so on. In this way the model assembly can be constructed, taking into account all the elements of the system studied.

An assembly contains a list of materials and production processes, as well as transport processes. Assemblies do not contain environmental data; instead, they link to processes that contain such data. Some parts, such as the mains cable, can be defined in subassemblies. A convenient way to visualize the structure and contents of an assembly is to use the "process tree" function (Figure 18.4).

TABLE 18.1
Input Elements for SPE Method[11]

Element	Amount	Mass (g)	Volume (L) (for Solvents)
Glass vacuum chamber	1	1800	—
Polyamide cover	1	841	—
Polyethylene sealings	10	7.2	—
Polytetrafluoroethylene ("teflone") taps	120	614	—
Polytetrafluoroethylene ("teflone") drains	60	574	—
Polytetrafluoroethylene ("teflone") stand	1	283	—
Polytetrafluoroethylene ("teflone") adapters	60	315	—
Polypropylene extraction tiny columns	30,000	66,480	—
Metal pump	1	5000	—
Dichloromethane	—	518,700	390
Methanol	—	142,200	180
Deionized water	—	90,000	90
Washing deionized water	—	300,000	300
Washing acetone	—	237,000	300
Washing dichloromethane	—	399,000	300
Running water	—	1,000,000	1000
Nitrogen	—	554,000	—
Energy	1000 kWh	—	—
Transport 1 (solvents)	1008 km	1,296,000	—
Transport 2 (pump)	1259 km	5000	—
Transport X (etc.)	...	...	—

Once a product assembly has been defined, SimaPro can immediately calculate the so-called LCI results or inventory table (Table 18.5). This is a list of all raw material extractions and emissions that occur in the production of the assembly and the materials and processes that link to it. SimaPro enables the LCI results to be specified as one table or per compartment, such as airborne or waterborne emissions. The LCI results provide the most detailed level of specification. At this step of LCA, it is not easy to interpret these long lists of data, as it is unclear what the environmental relevance of each raw material extraction or emission is. ISO 14042 on impact assessment specifies a number of procedures that can be used to achieve a better understanding of LCI results.

TABLE 18.2
Output of Elements for SPE Method (Wastes to the Environment)[11]

Type	Mass (kg)
Dichloromethane	532
Methanol	142
Acetone	237
Nitrogen	554

TABLE 18.3
Input Elements for LLE Method[11]

Element	Amount	Mass (g)	Volume (L) (for Solvents)
Glass separator	15	5,190,000	—
Polytetrafluoroethylene ("teflone") parts	2	566	—
Polyethylene parts	2	148	—
Dichloromethane	—	119,700	900
Washing deionized water	—	15,000,000	
Running water	—	45,000,000	
Washing dichloromethane	—	300,000	
Washing acetone	—	300,000	
Nitrogen	—	1,756,000	—
Transport 1 (glass separator)	10,008 km	5,804,000	
Transport (solvents)	10,008 km	119,700	
Transport X (etc.)	…	…	

TABLE 18.4
Output of Elements for LLE Method[11]

Type	Mass (kg)
Dichloromethane	1497
Acetone	300
Nitrogen	1756

18.2.3 LIFE CYCLE IMPACT ASSESSMENT

This element plays a crucial role in the whole LCA as it assesses the influences of environmental impacts (interventions) in regard to the results of LCI analysis. Its goal is to examine the product from the environmental point of view using information collected through LCI analysis. Many aspects during the whole life cycle of the system under consideration can be analyzed. This can be done for the production step, the processing step, recycling of wastes, solvent use, and solvent emission, and in the end one can analyze the entire life cycle. There are many different impact assessment methods to choose from; they are denoted in the literature by characteristic abbreviations, for example, CML92, CML2-2000, Eco-indicator 95, Ecopoints-Swiss 97, Eco-indicator 99, EDIP/UMIP 96, EDIP 99, EPS 2000, and IMPACT 2002+. The results of LCIA are shown for all impact categories in different ways: characterization, normalization, weighting, and single score (accumulated indicator). Table 18.6 shows the environmental impact categories.[27–30]

The categories may differ (Table 18.7) with respect to the LCA of different products (or the object of LCA) or assessment methods.[12–14]

For every impact category the proper category indicator is calculated.

CASE STUDY: CONTINUATION

With the inventory tables (Table 18.5) drawn up in the previous phase of LCA in mind, it is time to move on to the next step, which is called characterization. The results of this step are presented in a graph showing a number of impact category indicator results calculated from the LCI results. It is an obligatory step in impact assessment. In this example, we use the

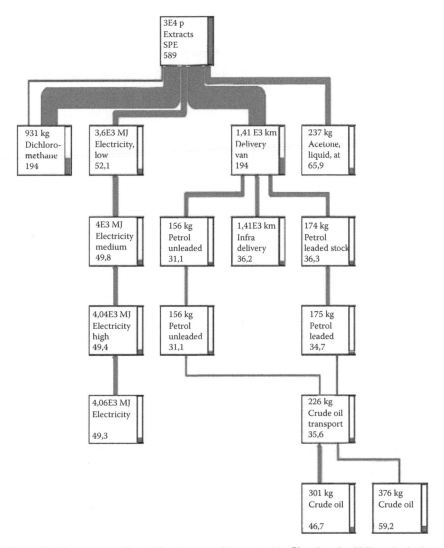

FIGURE 18.4 The "process tree" used for the assembly created in SimaPro for SPE analytical protocol.[11]

Eco-indicator 99 method, but it is also possible to choose one of the other methods that are available in SimaPro. All the results of this step of LCIA are scaled to 100% (Figure 18.5).

Each bar (column) represents the impacts arising from different subassemblies of the analytical protocol studied. With all impact category indicator results scaled to 100%, it is not easy to see which parts of the assembly have the highest overall environmental impact. Each bar on the histogram could represent 100% of a very large impact, or, equally, 100% of a small one. To obtain a better picture, the normalization procedure is used (Figure 18.6).

Normalization is an optional step in impact assessment. The impacts are now compared on a scale of inhabitant equivalents rather than that of 100%. Normalization only reveals which effects are large and which effects are small in relative terms. A weighting procedure (Figure 18.7) can be applied to the normalization results. This scales the results to a certain level of seriousness. Weighting is a subjective step. According to ISO 14042, weighting may not be used in the case of public comparisons between products. In the final step, SimaPro can add up all the evaluation scores to give a total impact score for each subassembly element.

Table 18.5
Inventory Table for SPE Technique Created in SimaPro[11]

Material	1	2	3	4	5	6	Amount
Teflon	2195 g	0		500 g			2695 g
Glass	1800	2600					4400 g
Polyamide (PA)	841						841 g
Polyethylene (PE)	7.2 g	2.6 kg		400 g			3021 g
Polypropylene (PP)		66.5 kg					66.5 kg
Gel		15 kg					15 kg
Dichloromethane			518 kg			399 kg	919 kg
Methanol			142 kg				142 kg
Steel				5 kg			5 kg
Azote					554 kg		554 kg
Water millipore			90		500	300 kg	890 kg
Dichloromethane washing					13.3 kg		13.3 kg
Acetone						237 kg	
Running water					1000 kg		1000 kg
Energy					1000 kWh		1000 kWh
Transport	1.6 tkm	84.8 tkm	666 tkm	7.4 tkm	13.4 tkm	641 tkm	1414 tkm

TABLE 18.6
Environmental Impact Categories Considered in LCA

Impact Category	Unit
Carcinogens	DALY[a]
Respirability organics	DALY
Respirability inorganics	DALY
Climate change	DALY
Global warming	DALY
Radiation	DALY
Ozone layer	DALY
Human toxicity	PAF*m^2 yr
Ecotoxicity	PAF*m^2 yr
Photochemical oxidant	PAF*m^2 yr
Acidification	PDF*m^2 yr[b]
Eutrophication	PDF*m^2 yr
Degradation of ecosystems	PDF*m^2 yr
Degradation of landscapes	PDF*m^2 yr
Land use	PDF*m^2 yr
Resource depletion—abiotic and biotic	MJ surplus[c]
Minerals	MJ surplus
Fossil fuels	MJ surplus

[a] DALY, disability adjusted life years.
[b] PDF, potentially disappeared fraction of plant species.
[c] MJ surplus, additional energy requirement to compensate for lower future ore grade.

TABLE 18.7
Impact Categories Comparison between Eco-Indicator 99[12] and Impact 2002+[14]

Eco-Indicator 99	Impact 2002+
Carcinogens (C)	Carcinogens
	Noncarcinogens
Respiratory inorganics (RI)	Respiratory inorganics
Respiratory organics (RO)	Respiratory organics
	Global warming potential (GWP) (DF)
Ozone depletion potential (OL)	Ozone depletion potential
Ecotoxicity (E)	Aquatic and terrestrial toxicity
Acidification and eutrophication (A/E)	Terrestrial acidification and nutrification
	GWP (CF)
	Aquatic acidification
	Terrestrial eutrophication
Resource (energy) (R)	Resource (energy)
Resource (mineral) (R)	Resource (mineral)
Radiation (R)	
Land use (LU)	
Resource (fossil fuels) (FF)	
Climate change (CC)	

18.2.4 LIFE CYCLE INTERPRETATION AND APPLICATION

In the final phase of LCA, inferences are drawn especially from LCI analysis and LCIA. From an analysis of the results, conclusions can be drawn and limitations defined, and recommendations for producers and policy-makers can be made. In general, the purpose of an LCA is to make inferences that can support a decision or provide a basis for a viewpoint. This means that the process of drawing conclusions is perhaps the most important step in any LCA. The relevant issue of interpretation is dealt with in ISO 14043. This phase will be exemplified as a case study comparing the results of LCA for two analytical protocols.

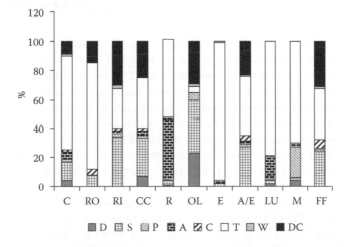

FIGURE 18.5 Characterization graphs in the LCIA phase of LCA with all impact categories for SPE[11] (D, device; S, solvent; P, pump; A, nitrogen; C, columns; T, transport; W, washing; and DC, drain cleaning; for the remaining abbreviations, see Table 18.7).

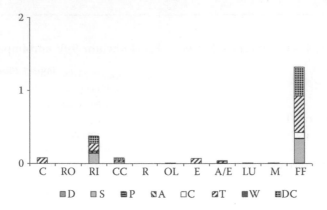

FIGURE 18.6   Normalization step in the LCIA phase of LCA for SPE[11] (D, device; S, solvent; P, pump; A, nitrogen; C, columns; T, transport; W, washing; and DC, drain cleaning; for the remaining abbreviations, see Table 18.7).

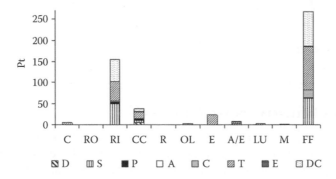

FIGURE 18.7   Weighting step in the LCIA phase of LCA for SPE[11] (D, device; S, solvent; P, pump; A, nitrogen; C, columns; T, transport; W, washing; and DC, drain cleaning; for the remaining abbreviations, see Table 18.7).

CASE STUDY: COMPARISON OF TWO DIFFERENT ANALYTICAL TECHNIQUES (SERVICES—SPE AND LLE—USING THE ECO-INDICATOR 99 METHOD IN SimaPro

From Figure 18.8 it is clear that all kinds of impact categories are involved in both extraction techniques. At this stage of the assessment, it looks as if SPE has the lower environmental load in almost all categories with the exception of minerals. This emerges directly from the types of devices used in SPE, as opposed to LLE, where only a simple glass separator is used.

After normalization of the results (Figure 18.9), three impact categories turned out to be very important for both techniques: fossil fuels, respiratory inorganic, and climate change. According to the normalization graph, the environmental load of SPE is slightly lower than that of LLE.

It is also possible to display the results for one product with all the categories in one graph (Figure 18.10). This way of presenting results is very useful for comparing two different products (here: extraction techniques).

From Figure 18.10 it is again clear that the environmental load of SPE is lower than that of LLE. Similar steps in the LCIA of LCA can be done to assess the damage category (human health, ecosystem quality, and resources). Figure 18.11 shows the details of the LCIA phase for these assessment categories.

Table 18.8 compares the quantitative impact and damage categories for the two different extraction techniques.

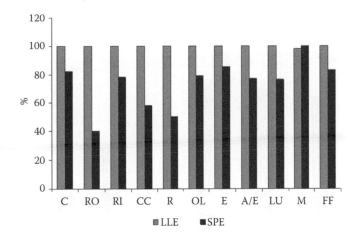

FIGURE 18.8 Characterization graphs of the LCIA phase in LCA for two analytical techniques (SPE and LLE) used to estimate PCB and PAH in surface water.[11]

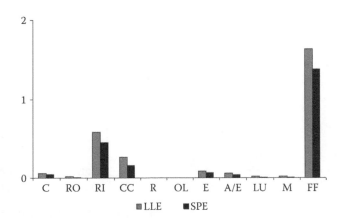

FIGURE 18.9 Normalization graphs of the LCIA phase in LCA for two analytical techniques (SPE and LLE) used to estimate PCB and PAH in surface water.[11]

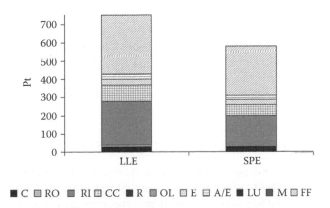

FIGURE 18.10 Environmental loading for all impact categories: for two different extraction techniques with the use of LCA in SimaPro and Eco-indicator 99 (single score).

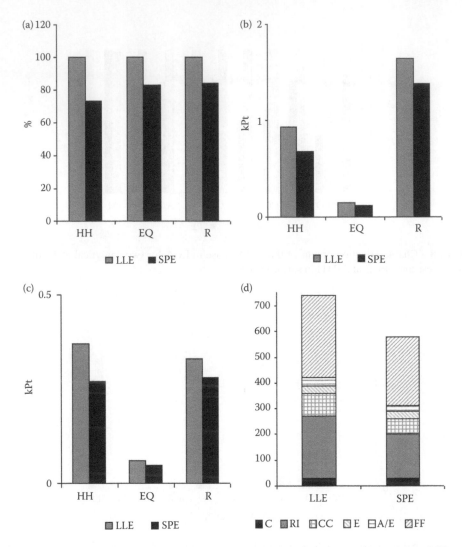

FIGURE 18.11 Comparison of LCIA results for SPE and LLE techniques used to estimate PCB and PAH levels in surface water within damage categories: (a) characterization graphs, (b) normalization graphs, (c) results of LCIA after weighting, and (d) results of LCIA-cumulated indicator (single score).

The comparison of two extraction techniques with the use of LCA allows one to assess which technique is the more environmentally friendly and which impact or damage categories carry the heaviest environmental load. The results can be used in many ways, for example, to modify elements of the devices used, to estimate the amount and the environmental loading of the solvent to be used in both techniques, and to consider modifications of the techniques used in a particular laboratory.

18.3 LCA FOR SOLVENT USE IN ANALYTICAL PROTOCOLS

Nowadays, solvents are used in large quantities in the chemical industry and chemical laboratories. The selection of solvents and subsequent waste-solvent management are based mostly on considerations of economy, safety, and logistics. Environmental concerns are often of minor importance for decision-makers.

TABLE 18.8

Quantitative Data for Impact Categories and Damage Categories for Two Different Extraction Techniques[11]

Category	Environmental Loading (Pt)		Category	Environmental Loading (Pt)	
Damage category	LLE	SPE	Impact category	LLE	SPE
Human health	371	266	Respiratory inorganic	230	180
			Climate change	109	64
Resources	329	275	Fossil fuels	327	273
Ecosystem quality	58.3	47	Ecotoxicity	32	27
			Acidification/eutrophication	21	16

Note: Results of LCA with the use of Eco-Indicator 99 in SimaPro program—teaching use.

The use of a solvent in any application is associated with a variety of indirect environmental impacts, such as the depletion of nonrenewable resources as a consequence of petrochemical production, atmospheric emissions as a result of solvent incineration, or the high energy demand for solvent recycling by distillation.[18–21]

The assessment of a solvent's environmental impact should consider not only impacts arising from its industrial production, recycling, and disposal, but also its ESH (environmental, safety, and health) characteristics. This kind of thinking has led to the idea of a "green solvent" or ecosolvent.[22,31–33]

Four approaches toward green solvents have been developed recently:

- The substitution of hazardous solvents by those with better EHS properties, such as enhanced biodegradability or low ozone depletion potential[23–25]
- The use of "biosolvents," that is, solvents produced from renewable resources (starch and cellulose[26]), so that the use of fossil resources can be avoided
- The substitution of organic solvents by supercritical fluids with a consequent reduction in ozone depletion[34]
- The substitution of organic solvents by ionic liquids that have a low vapor pressure and are thus less likely to be emitted into the atmosphere[34,35]

Numerous solvents are also used in the analysis of aquatic media (Table 18.9), so the "green chemistry" approach in this field is highly appropriate.

Capello et al.[16] applied LCA to 26 organic solvents (acetic acid, acetone, acetonitrile, butanol, butyl acetate, cyclohexane, cyclohexanone, diethyl ether, dioxane, dimethylformamide, ethanol, ethyl acetate, ethyl benzene, formaldehyde, formic acid, heptane, hexane, methyl ethyl ketone, methanol, methyl acetate, pentane, *n*- and isopropanol, tetrahydrofuran, toluene, and xylene). They applied the EHS Excel Tool[36] to identify potential hazards resulting from the application of these substances. It was used to assess these compounds with respect to nine effect categories: release potential, fire/explosion, reaction/decomposition, acute toxicity, irritation, chronic toxicity, persistency, air hazard, and water hazard. For each effect category, an index between zero and one was calculated, resulting in an overall score between zero and nine for each chemical. Figure 18.12 shows the life cycle model used by Capello et al.[16]

One functional unit was defined as the use of 1 kg of solvent as a reaction medium in a chemical production process. To calculate the environmental impact of specific solvents, the Ecosolvent-Tool[31] was used, which combines the LCIs of 45 petrochemical products.

Overall high scores for formaldehyde, dioxane, formic acid, acetonitrile, and acetic acid were obtained. Formaldehyde scored relatively low for fire/explosion hazards but high with regard to

TABLE 18.9

Examples of Solvent Use in Chemical Analysis of Aquatic Environments

Chemical Analyzed	Method	Solvent Use
Azote (nitrate)	Colorimetry	Chloroform, ethanol
Cyanides	1. Colorimetry	1. n-butyl alcohol or isoamyl alcohol
	2. Argentometric titration	2. Chloroform, hexane or iso-octane
Zinc	Colorimetry	Carbon tetrachloride
Chlorine dioxide	Iodometry	Chloroform
Phosphates	Colorimetry	Benzene isobutanol
Iodides	1. Colorimetry	Chloroform
	2. Titration	Ethanol
Molybdenum	1. Colorimetry	Isoamyl acetate or another
	2. Atomic absorption spectrometry (ASA)	Organic solvent
Naphtha products	IR	Carbon tetrachloride
Mercury	Ditizone method	Chloroform
Selenium	Colorimetry	Toluene
Silver	Colorimetry	Acetone, carbon tetrachloride, or chloroform

acute and chronic toxicity, irritation, and air hazard. Dioxane had a high persistency, while both acetic acid and formic acid scored high on irritation. Low overall scores were obtained for methyl acetate, ethanol, and methanol: the environmental hazards they represent are particularly low and the health hazards relatively low. The use of tetrahydrofuran, butyl acetate, cyclohexanone, and 1-propanol is not recommended for a life cycle perspective: these solvents have a high environmental impact, especially during the production of petrochemicals. At the other end of the scale, hexane, heptane, and diethyl ether are the most environmentally favorable solvents.

Capello et al.[16] also assessed the environmental impacts of the life cycles of four solvent mixtures (methanol–water, ethanol–water, methanol–ethanol, and n-propyl alcohol–water of different compositions w/w) that can be used for the solvolysis of p-methoxybenzoyl chloride. Different waste treatment scenarios for these binary mixtures (incineration and distillation) were analyzed. It appears that a solvent mixture with a high water content has a low environmental impact because the cumulative energy demand (CED) for the production of water is about three orders of magnitude lower than that for organic solvents.[37]

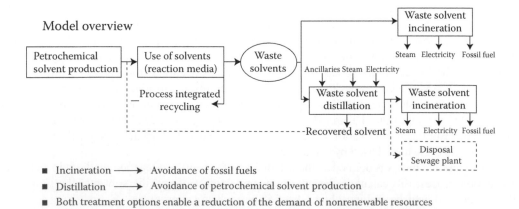

- Incineration ⟶ Avoidance of fossil fuels
- Distillation ⟶ Avoidance of petrochemical solvent production
- Both treatment options enable a reduction of the demand of nonrenewable resources

FIGURE 18.12   System model for solvent assessment using the life cycle assessment method.[14]

Therefore, the solvent mixtures with water as a component are superior to the methanol–ethanol mixture. Generally, incineration as a treatment option is superior to distillation for these solvent mixtures, particularly for those containing a high proportion of water. Only with regard to the *n*-propyl alcohol–water mixture is distillation the better treatment option, especially for high concentrations of this alcohol, because of its high environmental impact in petrochemical production.

18.4 CONCLUSIONS

In spite of the on-going development of LCA as an environmental tool, there is still very little available information on LCA for analytical protocols. There are drawbacks to the existing LCA programs, which mean that the information they furnish is incomplete. Nonetheless, a number of research projects are currently focused on LCA, which is sure to become an essential tool in environmental protection. Even though LCA is an excellent environmental tool in concept, in practice, it is still hard to tell whether it can be applied quantitatively. This is because there are no universal standards of comparison; in any case, the precise environmental costs, not to mention the practical ones, are hard to estimate.

The aim of LCA is to reduce the overall environmental impact by providing directional environmental indicators. These should be examined carefully using other analytical techniques.[38,39] However, minimizing the impacts of subsystems does not ensure that the impacts of an entire system are minimized or even reduced.[40] In many cases, a reduction or change in one part of the system merely shifts the burden.

Thus, the results obtained by LCA for a particular product must be analyzed very carefully before final conclusions can be drawn about the environmental impact of the product. For example, methyl tertiary butyl ether (MTBE) is a fuel additive, initially introduced in the late 1970s as an octane booster.[41] By reducing the vapor pressure of gasoline, MTBE reduces emissions of carbon monoxide (CO), other products of incomplete combustion, and evaporative emissions. Reformulated gasoline containing MTBE is especially effective in reducing emissions in older vehicles when engines are cold and subjected to heavy loads. However, one physical property of MTBE not possessed by most other petroleum hydrocarbon (HC) additives is its excellent solubility in water; its transport is therefore not limited to groundwater as a result of soil adsorption. In the event of a gasoline spill, other HCs would tend to stay put, whereas MTBE would travel relatively quickly, potentially polluting nearby lakes, streams, and drinking water sources. It has also been found that MTBE is an animal carcinogen with the potential to cause cancer in humans.[40,42] Despite the potential carcinogenicity and the physical properties of MTBE, a recent LCA study concluded that reformulated gasoline containing MTBE had a significant advantage over conventional gasoline as far as the reduction of hazardous air pollutants was concerned. Moreover, the study's conclusion resulted in the recommendation that "MTBE blended gasoline be considered for use in areas where population density is relatively high and concerns regarding hazardous air pollutants exist."[43] Now, the findings and recommendations of this study are not false within the context of reducing hazardous air pollutants. But the study presented in the chapter did not consider the scenario of increased potential harm to humans and the impact on the environment via the groundwater pathway. The study focused solely on the problem to be solved, namely, the reduction of hazardous air emissions. Hence, it is very important to examine the scope and aim of the LCA results. Sometimes there is lack of information on the influence of data pertaining to other media, that is, soil and water.[44] Although the study's recommendations were correct within the context of reducing hazardous air pollutants, they may have serious environmental repercussions should a potential carcinogen enter drinking water supplies.[42] The boundaries of the LCI and the LCIA were clearly stated by the study. What remained implicit was the boundaries of the recommendations, and the shortcoming of the study was that the recommendations did not explicitly describe the context that they targeted.

LCA shows precisely that the price one pays for a product rarely reflects the environmental cost of producing it.

It is worth noting that in Europe more and more governmental organizations are drawing attention to the significance of developing LCA. The Council of the European Union, for instance, is promoting just such an approach.

ACKNOWLEDGMENTS

We are indebted to Lidia Wolska and Jacek Namieśnik for the consultations they kindly held with us.

REFERENCES

1. Udo de Haes, H. 1993. Application of life cycle assessment: Expectations, drawbacks and perspectives. *J. Cleaner Prod.* 1: 131–137.
2. Consoli, F., D. Allen, I. Boustead, et al. 1993. *Guidelines for life-cycle assessment: A "Code of Practice"*, J. Séguin and B. Vigon (eds). Report of SETAC Workshop—Sesimbra, Portugal, March 31 to April 3. Society of Environmental Toxicology and Chemistry (SETAC), Sesimbra, Portugal.
3. ISO 14 044 (DIN EN ISO 14040). 2006. Environmental management—life cycle assessment—principles and framework.
4. Boguski, T., R. Hunt, J. Cholakis, and W. Franklin. 1996. LCA Methodology. In: M. Curran (ed.), *Environmental Life-cycle Assessment*, pp. 21–37. New York: McGraw-Hill.
5. Raysoni, A. 2002. Life cycle assessment for anaerobic digestion of waste products. *Int. J. LCA* 7: 187.
6. Verma, M., S. Dubey, and R. Bharadwaj. 2002. Application of life cycle assessment to forestry products. *Int. J. LCA* 7: 187–188.
7. Durucan, S. and A. Korre. 2000. Life cycle assessment of mining projects for waste minimization and long term control of rehabilitated sites. *Proceedings of the 3rd Annual Workshop EUROTHEN*, p. 257.
8. De Boer, J., J. Jager, E. Szpadt, et al. 2003. Life cycle assessment based tools for the modeling of development of integrated waste management strategies for cities and regions with rapidly growing economies. Conference Proc. Technical, Economical and organizing aspects of waste management. Poznan-Gniezno, Poland, May 18–21, 2003.
9. Dudek, M., L. Wolska, and J. Namiesnik. 2005. Assessment of the life cycle—multiphase analysis of the product impact on environment. *Ecol. Tech.* 13: 66–75.
10. Dudek, M., L. Wolska, H. Walk, A. Stachowiak-Wencek, and J. Namiesnik. 2004. Studies on evaluation of the ship lifetime cycle. *Ecol. Tech.* 15: 141–153.
11. Stoklosa, M. 2006. The use of life cycle assessment (LCA) for ecological comparison of isolation protocols of organic compounds from aquatic samples (in Polish). MSc thesis, Gdansk University of Technology.
12. Goedekop, M., and R. Spriensma. 2000. The Eco-Indicator 99: A damage oriented method for life-cycle impact assessment. Methodology Report 2000a. Available at www.pre.nl.
13. Ekvall, T., and B.P. Weidema. 2004. LCA methodology, system boundaries and input data in consequential life cycle inventory analysis. *Int. J. LCA* 9: 161–171.
14. Jolliet, O., M. Margni, R. Charles, S. Humbert, J. Payet, and G. Rebitzer. 2003. IMPACT 2002+: A new life cycle impact assessment methodology. *Int. J. LCA* 8: 324–330.
15. Cooper, J., and J. Fava. 2006. The life cycle assessment practitioners survey: Assessment methods for evolutionary and revolutionary electronic products. *Proceedings of the 2006 IEEE International Symposium on Electronics and the Environment*, pp. 1–5.
16. Capello, C., U. Fischer, and K. Hungerbühler. 2007. What is a green solvent? A comprehensive framework for the environment assessment of solvents. *Green Chem.* 9: 927–934.
17. Capello, C., S. Hellweg, C. Seyler, and K. Hungerbühler. 2005. Ecosolvent: A tool for waste-solvent management in the chemical industry. 27th LCA Discussion Forum (17th November). Swiss Federal Institute of Technology.
18. Seyler, C. 2005. Waste-solvent incineration plant. *J. Clean. Prod.* 13: 1211–1224.
19. Capello, C., S. Helleweg, B. Badertscher, and K. Hungerbühler. 2005. Life cycle inventory of waste solvent distillation: Statistical analysis of empirical data. *Environ. Sci. Technol.* 39: 5885–5892.
20. Seyler, C., T. Hofstetter, and K. Hungerbühler. 2005. Life cycle inventory for thermal treatment of waste solvent from chemical industry: A multi—input allocation model. *J. Clean. Prod.* 13: 1211–1224.

21. Seyler, C., S. Helleweg, S. Monteil, and K. Hungerbühler. 2004. Life cycle inventory for waste solvent as fuel substitute in the cement industry: a multi-input allocation model. *Int. J. LCA.* 10: 120–130.
22. Seyler, C., C. Capello, S. Hellweg, et al. 2006. Waste-solvent management as an element of Green Chemistry. *Ind. Eng. Chem. Rev.* 45: 7700–7709.
23. Curzons, A., C. Constable, and V. Cunningham. 1999. A guide to the integration of environmental, health and safety criteria into the selection of solvents. *Clean Prod. Process.* 1: 82–90.
24. Curran, P., J. Maul, P. Ostrowski, G. Ublacker, and B. Linclau. 1999. Benzotrifluoride and derivatives: Useful solvents for organic synthesis and fluorous synthesis. *Topics in Current Chemistry* 206: 79–105.
25. Gani, R., C. Jimenez-Gonzales, A. Kate, et al. 2006. A modern approach to solvent selection. *Chem. Ing.* 1: 30–43.
26. Sawaiko, B. 2004. A promising future for ethanol. *World Ethanol Biofuels Rep.* 2: 20–28.
27. Thomas, P.G. 2000. An approach to dynamic environmental life-cycle assessment by evaluating structural economic sequences. Dissertation, Tufts University.
28. ISO International Standard 14041, 1999E. Environmental management—life cycle assessment—goal and scope, definition and inventory analysis. International Organization for Standardization (ISO), Geneva.
29. Bennet, R.M., R.H. Phipps, and A.M. Strange. 2006. The use of life cycle assessment to compare the environmental impact of production and feeding of conventional and genetically modified maize for broiler production in Argentina. *J. Anim. Feed Sci.* 15: 71–82.
30. Steen, B. 1999. A systematic approach priority strategies in product development (EPS). Version 2000. Center for Environmental Assessment of Products and Materials System. Chalmers University of Technology, Technical Environmental Planning, Göteborg.
31. Capello, C., S. Hellweg, and K. Hungerbühler. 2007. Environmental assessment of waste-solvent treatment options: Part I: The ecosolvent tool. *J. Ind. Ecol.* 11: 26–38. (The Ecosolvent Tool, ETH Zurich, Safety & Environmental Technology Group, Zurich, http://www.sust-chem.ethz/tools/ecosolvent.)
32. Slater, C. and J. Savelski. 2007. A method to characterize the greenness of solvents used in pharmaceutical manufacture. *J. Environ. Sci. Health* 42: 1595–1605.
33. Jimenez-Gonzalez, C., A. Curzons, and V. Cunningham. 2004. Expanding GSK's solvent selection guide-application of life cycle assessment to enhance solvent selections. *J. Clean. Tech. Environ. Policy* 7: 42–50.
34. Noyori, R. 1999. Supercritical fluids: Introduction. *Chem. Rev.* 99: 353–354.
35. Leveque, J. and G. Cravatto. 2006. Microwaves power ultrasound, and ionic liquids. A new synergy in green organic synthesis. *Chimia* 60: 313–320.
36. Sugiyama, H., U. Fischer, and K. Hungerbühler. What is a green solvent? The EHS Tool, ETH Zurich, Safety & Environmental Technology Group, Zurich. Available at http://www.sust-chem.ethz/tools/EHS.
37. Dones, R., T. Heck, and M. Faist Emmenegger. 2004. Final Reports Ecoinvent 2000, No. 1-15, CD-ROM, Swiss Centre for Life Cycle Inventories, Dubendorf, CH. Available at www.ecoinvent.ch.
38. Owens, J.W. 1996. LCA impact assessment—case study using a consumer product. *Int. J. LCA* 1: 209–217.
39. Owens, J.W. 1997. Life cycle assessment: Constraints on moving from inventory to impact assessment. *J. Ind. Ecol.* 1: 37–49.
40. Belpoggi, F., M. Soffritti, and C. Maltoni. 1998. Pathological characterization of testicular tumours and lymphomas-leukemias, and of their precursors observed in Sprague-Dawley rats exposed to methyl-tertiary-butyl-ether (MTBE). *Eur. J. Oncol.* 3: 201–206.
41. Thomas, P.G. 2000. An approach to environmental life-cycle assessment by evaluating structural economic sequences. PhD dissertation, Tufts University.
42. Bird, M., H. Burleigh-Flayer, J. Chun, and J. Douglas. 1997. Oncogenicity studies of inhaled methyl tertiary butyl ether (MTBE) in CD-1 mice and F-344 rats. *J. Appl. Toxicol.* 17: S45–S55.
43. Raynolds, M., D. Checkel, and R. Fraser. 1998. Life cycle value assessment (LCVA). Comparison of conventional gasoline and reformulated gasoline. Design and manufacture for environment. SAE International Papers. Doc. No. 980468, SP-1342: 111–130.
44. Peereboom, E., R. Kleijn, S. Lemkowitz, et al. 1998. Influence of inventory data sets on life-cycle assessment results: A case study on PVC. *J. Ind. Ecol.* 2: 109–130.

19 Preparation of Samples for Analysis: The Key to Analytical Success

Jacek Namieśnik and Piotr Szefer

CONTENTS

19.1 Introduction 431
19.2 Types of Analytical Data 434
19.3 Speciation Analytics: An Important Task for Analytical Chemists 435
    19.3.1 Physical Speciation 437
    19.3.2 Chemical Speciation 437
19.4 Problems Associated with Trace Element Analysis 439
19.5 Stages of the Analytical Procedure 440
19.6 New Methodological Developments in Preparing Samples for Analysis 442
19.7 Application of Membrane Techniques 442
    19.7.1 Membrane Extraction 445
19.8 Matrix Solid-Phase Dispersion 448
19.9 Supercritical Fluid Extraction 449
19.10 Unique Phase Separation Behavior of Surfactant Micelles 451
19.11 Ionic Liquids: A New Type of Solvent and Extractant 452
19.12 Microwave-Enhanced Chemistry (MEC) 454
19.13 Application of Ultrasound (US) in the Sample Preparation Process 455
19.14 Green Chemistry: Introduction of the Concept of Sustainable Development
    to Chemical Laboratories 458
    19.14.1 History 458
    19.14.2 Green Analytical Chemistry 459
19.15 Summary and Conclusion 463
References 463

19.1 INTRODUCTION

Recent decades have witnessed a sharply growing demand for information. This also pertains to information obtainable from the analytical examination of samples of material objects.

The desire to satisfy the need for analytical data stimulates actions toward

- Developing new analytical methodologies
- Designing and implementing new technical solutions for the measuring instruments used in analytical practice

Analytical methodologies and measuring instruments are the tools for obtaining reliable data on the composition of the material objects being studied. The science of the construction and operating rules of measuring instruments is often referred to as instrumentation. The successive stages in the development of this science are easily discernible.

Access to a variety of information sources facilitates decision-making not only in politics, but also in economy and in technology (related to control over the processes of manufacturing consumer goods). A new type of market has emerged in which information is bought and sold.[1]

Analytical data on the material objects in question are a specific kind of information, which is based on the analysis not so much of the whole objects as of representative samples of such objects. Samples therefore have to be collected in such a way that the most important criterion—representativeness—is met.

To satisfy the growing demand for analytical data, more and more intensive research is taking place with the aim of developing new methodological and instrumental solutions so that analytical results can be a copious source of information; in other words, they can possess the greatest possible information capacity.

Measurement results must be reliable (credible), that is to say, they must accurately (both precisely and truly) reflect the real content (amount) of analytes in a sample that is representative of the material object under study. This leads to the conclusion that all developments in analytical chemistry are derived from the desire to obtain in-depth analytical data. Analytical chemistry uses a very broad spectrum of measurement methods and techniques: Table 19.1 presents a basic classification.

Depending on the objective of the measurement, one of two fundamental procedures has to be selected. The first is the classical procedure recommended and even required by the official

TABLE 19.1
The Basic Classification of Modern Chemical Analytical Methods

Basis for Categorization	Types of Analytical Methods	Comments
1	2	3
Relation to the current international system of units (SI) (location in the comparison chain ensuring traceability)	Primary methods Relationship methods Secondary methods	Used for direct measurement of units in the SI system Isotope dilution mass spectrometry (IDMS)
Measurement principle	Absolute methods	Based on such units as mass, volume, time, and electric current intensity, which do not require calibration
	Relative methods	By comparing signals from analytes in the model sample and in the examined sample; the calibration stage is necessary
	Direct methods	An appropriate measurement device (sensor) is inserted into the examined object in order to obtain analytical data (measurement of pH and electrical conductivity)
Means of examining the sample	Indirect methods	In most cases used because of • Very low analyte concentration levels • The complicated matrix composition and the presence of INTERFERENTS; the sample must be prepared properly, and the analyte concentration is measured in an appropriate extract

continued

TABLE 19.1   (continued)

Basis for Categorization	Types of Analytical Methods	Comments
1	2	3
Type of analytical data	Methods for determining the instantaneous concentration of analytes in the examined material object	Methods used in examining the quality of the environment and determining individual exposure
	Methods for determining time-weighted concentrations while taking the sample	
Location of the analytical process	Methods of making *in situ* measurements	Appropriate mobile laboratories, movable or portable measurement devices are used
	Laboratory methods	
Means of obtaining analytical data	Methods using devices that can read the amount/concentration of analyte directly	Usually used in field research in order to obtain analytical data (often semiquantitative) quickly
	Methods with previously prepared samples and the amount/concentration of analyte calculated from laboratory measurements	
Means of taking a representative sample	Sedimentation methods	Analyte sample is collected by free migration of the analyte onto the collecting surface
	Isolation methods	The sample is put into a container (probe) of specified volume
	Aspiration methods	Analyte samples are collected by running a stream of medium through a trap (e.g., a sorption tube)
Level of automation	Manual methods	Most of the operations and actions (both in the field and in the laboratory) connected with sample preparation are performed manually
	Automatic methods	All or part of the operations are performed without the participation (intervention) of an operator (analyst)
	Monitoring methods	Specific type of automatic methods; the devices used must have the following features:
		• They must be able to obtain data in real time or with only a slight time delay
		• They must be capable of performing continuous measurements
		• They must be able to operate autonomously for extended periods of time

regulations applying to environmental programs. Based on sampling and laboratory analysis, it involves a number of steps between sampling and analysis, such as conditioning, storage, transportation, and pretreatment. The other procedure, carried out on-site, makes use of on-line measurement systems, field-portable devices, or test kits. In actual fact, the two approaches are often used in tandem, combining as they do the scientific relevance of certain practices (e.g., on-site measurement of dissolved gases and temperature) and the availability of systems for on-line monitoring.[2]

The problem of preparing samples for analysis has been presented in a large number of both original and review papers.[3–29] These discuss universal problems in chemical analytics, the problems and challenges concerning the most appropriate ways of preparing material samples for analysis, and the concrete requirements regarding the preparation of samples to be analyzed using a specific technique.[30,31]

It must be borne in mind, however, that

- Despite further development in instrumentation and the availability on the market of many complex hyphenated devices, the basic principle that a device is merely a necessary and useful tool in the process of obtaining analytical data is often forgotten. Possession of the tool itself will not solve any analytical problems. Without an understanding of the chemistry of the analytical process, no reliable and credible results can be obtained. People who treat an analytical device as a typical black box deserve to be called "operators of analytical devices" rather than "analytical chemists." In such a situation, erroneous analytical information is all too easily produced, notwithstanding the amount of time and work expended on the analytical process.
- There is a need to educate specialists who will be able to make use of these innovative methodologies and devices.

Analytical research draws on various procedures and analytical techniques. Some of the measurement devices used are referred to as "monitors" and should have the following operating parameters:

- High measurement sensitivity.
- Immediate, or at worst only slightly delayed, delivery of analytical information on the investigated object.
- High resolution of results characterized by a short response time.
- Long period of unsupervised operation.

Monitoring also imposes several requirements regarding

- Instrument zeroing and calibration.
- Protecting the instrument against power surges.
- Providing the instrument with an independent power supply.
- Automatic replenishment with solution and reagents (electrochemical monitors).
- Installation of devices preventing the flame from going out (in certain detectors, e.g., flame ionization detector (FID) and flame photometric detector (FPD).
- Exchange and regeneration of spent filters.

19.2   TYPES OF ANALYTICAL DATA

Data obtained through analysis of samples may prove useful in different fields of science, technology, and human life. They play a particularly vital role when it is necessary to

- Describe the condition of the examined material object and discover the changes it is subject to.
- Confirm a new theory or scientific hypothesis.
- Take a decision concerning the law and the economy.
- Plan and implement educational campaigns in order to raise social awareness.

Various types of data may be obtained as a result of sample analysis. There is no doubt, however, that in the majority of cases quantitative data (the amount or concentration of analyte in a sample) are most important. It is therefore worth learning the basic terminology of chemical metrology with reference to the quantitative determination of analytes. The diagram in Figure 19.1 will help put these terms in the correct order on the analyte concentration axis [expressed in the same units as the standard deviation of analytical noise (δ)].

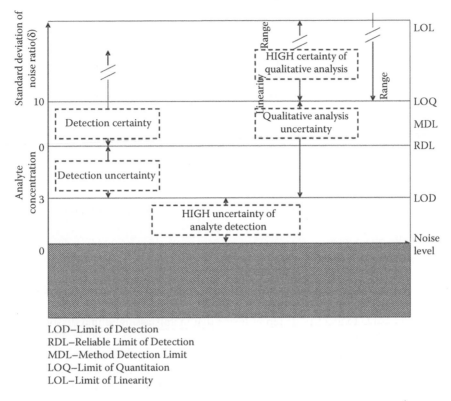

LOD–Limit of Detection
RDL–Reliable Limit of Detection
MDL–Method Detection Limit
LOQ–Limit of Quantitaion
LOL–Limit of Linearity

FIGURE 19.1   Basic metrological terminology relating to the quantitative analytics of trace elements.

Many elements and compounds occur in a variety of matrices at concentrations that could not be detected by the analytical methods first developed in the nineteenth century. As analytical technology improved, and it became known that elements were present at these very low concentrations, the term "trace" was coined to describe them. Although modern analytical methods permit the accurate, repeatable determination of elements at such low levels, the generic terms "trace" and "trace element" are still in use.

The boundaries of trace analysis are described by the definition of "trace element" in the IUPAC Compendium of Chemical Terminology, 2nd edition: "Any element having an average concentration of less than about 100 parts per million atoms and less than $100\,\mu g\,g^{-1}$." As analytical techniques have become more sophisticated and detection capabilities have improved, this upper boundary of the definition of "trace" is now so far away from the capabilities of analysis in a number of fields that new terms such as "ultratrace analysis" have entered common parlance. There is no agreement, however, on the range of ultratrace analysis, and this term has no rigorous definition. In the literature, the term is used to define the presence of elements at mass fractions less than 10^{-6} and 10^{-8} ($1\,\mu g\,g^{-1}$ and $0.01\,\mu g\,g^{-1}$).

19.3   SPECIATION ANALYTICS: AN IMPORTANT TASK FOR ANALYTICAL CHEMISTS

Attempts at environmental or health protection can yield only dubious results, if any, if they are based on suspect data. Therefore, a rigid quality control program is required for speciation analysis. Species alterations have to be avoided or minimized; information on the degree of possible species changes must therefore be elucidated.[32–39]

In general, there are at least three approaches to the use of the term "speciation analytics" in the analytical context:

- Local concentration differences of a particular element or compound in a given structure of a material
- The physical distribution of an element or compound in different phases that are in contact with each other
- The presence of different chemical conditions or binding states of a particular element within a single phase

Initially, speciation analytics was associated only with the biogeochemical cycles of metals in aquatic environments.

Even in the 1950s, geochemistry distinguished between two forms that metals could assume:

- Metals in dissolved form
- Metals bound to suspended matter

At that time, passing samples of water through a filter with 45 μm diameter pores was sufficient to properly separate the two phases. Later, following the development of electrochemical analytical methods, it was possible to identify different forms that metals assumed in a dissolved state—free metal ions and complex ion forms.

Simultaneously conducted simulation studies on the possible equilibria between ions and organic or nonorganic ligands have led to the conclusion that a wide variety of chemical compounds and metals can exist in aquatic environments. Nowadays, speciation analytics deals not only with metals, but also with other elements and different types of tests.

It is well known that the toxicity of many elements depends on the physicochemical forms they assume. So, for instance, determining the total content of a certain element in a sample is definitely not sufficient to measure its toxicity. Selenium is a case in point: in small amounts this element is essential to human health. But the transition from the necessary amount (about 70 μg of selenium per day for an adult) to a toxic dose (about 800 μg of selenium per day) is relatively easy. In rats, moreover, the fatal dose of Se(IV) compounds is 3.2 mg kg^{-1} of body mass, whereas for dimethyl selenide it is 1600 mg kg^{-1} of body mass. Nonorganic selenium compounds [Se(IV) and Se(VI)] are believed to be the most toxic ones, whereas in the environment selenium occurs most commonly bound to amino acids (selenomethionine and selenocysteine). The least toxic forms seem to be the volatile methyl compounds of selenium, which are metabolites of a detoxication process.

The question concerning what "speciation" actually means is very often asked: the answer can be found in IUPAC recommendations. Table 19.2 presents the most frequently used terms. Generally speaking, speciation analytics plays a very important role in

- Studies of the geochemical cycles of elements and chemical compounds
- Determining the toxicity and ecotoxicity of given compounds
- The quality control of food products
- Research into the environmental impact of technological installations
- The examination of occupational exposure
- The control of medicines and pharmaceutical products
- Clinical analysis

Different chemical species and their physical forms behave differently in geochemical, ecological, and metabolic cycles. This applies in particular to

- Deposition
- Accumulation

TABLE 19.2
Terms Connected with Speciation Analytics

Term	Definition
Chemical (species)	Specific form of an element defined according to isotopic composition, electronic or oxidation state, and/or complex or molecular structure
Speciation	Distribution of an element among its chemical species in a system
Speciation analysis	Analytical activities of identifying and/or measuring the quantities of one or more individual species in a sample
Fractionation	Classification process of an analyte or a group of analytes from a certain sample according to physical size/solubility or chemical properties (bonding and reactivity)

- Mobility/transportation
- Phase transfer
- (Re)mobilization
- (Bio)availability
- Resorption/excretion
- Essentiality/toxicity.

The physicochemical properties of particular species strongly influence their behavior in complex multiphase systems such as specific ecosystems. Special attention should be paid to

- Solubility → mobility, remobilization, resorption, deposition
- Volatility → phase transfer, transportation
- Oxidation state → bioavailability, essentiality, toxicity
- Reactivity → remobilization, bioavailability
- Polarity/charge → accumulation, bioavailability
- Molecular weight → mobility, phase transfer, deposition

A search of the literature reveals that several types of speciation analytics can be distinguished (see below).

19.3.1 PHYSICAL SPECIATION

Physical speciation takes place when different forms of the same chemical species have to be determined in a sample. Examples include adsorbed forms, dissolved forms, complex forms, and so on.

19.3.2 CHEMICAL SPECIATION

Chemical speciation occurs when different chemical species should be determined in the sample under investigation.

It is possible to distinguish five types of chemical speciation:

- Screening speciation—the detection and determination of one particular analyte, for example, one known for its especially high environmental toxicity
- Group speciation—determination of the concentration level of a specific group of compounds or of elements existing in different compounds in a specific oxidation state, and their physical forms

- Distribution speciation—this takes place when the same chemical species needs to be determined in different compartments of the material object under investigation
- Chiral speciation—determination of the enantiomers of the given compound
- Individual speciation—the most difficult type encountered in speciation analytics, involving the broadest range of analytical work. Its purpose is to separate, detect, determine, and identify all species of an element in a sample.

Specialists in speciation analytics are interested in various chemical species and the physical forms that certain elements assume. Here are some examples of the physicochemical forms of trace element species in water bodies:

Dissolved	• Simple hydrated ions
	• Inorganic complexes
	• Organic complexes
	• Molecules and polymeric compounds
	• Ion pairs
Colloidal	• Mineral substances
	• Products of hydrolysis and precipitation
	• Biopolymers
Suspended particles	• Mineral substances
	• Precipitates and agglomerated colloids
	• Plankton
	• Bacteria and microorganisms

A set of important factors affects the formation, stability, and transformation of dissolved elemental species in the samples under investigation, namely,

- Shift of pH
- Change of redox potential
- Presence of reactants (e.g., inorganic and organic ligands)
- Catalytic effects
- Presence of particulate matter and microorganisms (adsorption and biotransformation).

Speciation analysis comes into its own mainly in environmental, nutritional, and biomedical research. The sample matrices are generally highly complex and the requirements for reliable (trace) element determinations are stringent (even for total amounts). The most important challenges in this context involve

- The often very low concentration of an individual species
- Large concentration differences between the elemental species
- Small structural differences in the elemental species
- The low thermodynamic and kinetic stability of a species
- Preserving the integrity of the sought-after species throughout the analytical procedure
- The existence of as yet unidentified species
- The nonavailability of suitable reference materials

Certain specific analytical methodologies are available for speciation analytics. The most common of them are

- Direct *in situ* detection of species (e.g., ion selective electrodes and electron spectrometry)
- Chemical derivatization of individual species (optical molecular spectrometry)

- Separation of individual species and element-specific detection (extraction, sorption, ion exchange, gas permeation, and electrolysis)
- Separation of all species and their determination (chromatographic and electrophoretic methods)

Chromatographic techniques are the techniques of choice for speciation analytics. In specific cases, however, nonchromatographic techniques may need to be applied in view of the characteristic properties of the analytes. They are as follows:

- Differentiation of oxidation states of inorganic metal compounds
  - Electrochemical methods
  - Selective derivatization and molecular spectrometry
  - Ion-exchange separation of anionic and cationic species
- Separation of inorganic ions from organic species
  - Solvent extraction
  - Solid-phase extraction (SPE) using reversed-phase materials
- Separation of volatiles from nonvolatiles
  - (Isothermal) distillation
  - Dynamic headspace analysis (purge-and-trap)
- Separation of low- and high-molecular-weight compounds
  - Membrane techniques (dialysis, ultrafiltration)

Speciation is one of the forces driving development in the field of chemical analytics and instrumentation, and the following novel approaches in this field have come into use:

- High-resolution separation techniques in hyphenation with high-sensitivity detectors (two-dimensional separations)
- Separation techniques in hyphenation with elucidation of the structure of organometallic compounds [electrospray ionization-mass spectrometry (ESI-MS) and matrix assisted laser desorption ionization-mass spectrometry (MALDI-MS)]
- New *in situ* techniques with enhancements in sensitivity and selectivity (sensors based on molecular imprinted polymers)
- Selective sampling for species, making use of "biological receptors"
- (New) reference materials and round robin tests for quality control

19.4   PROBLEMS ASSOCIATED WITH TRACE ELEMENT ANALYSIS

Many analysts are faced with the problem of determining the content of trace and microtrace components in samples with complex and often varying matrix compositions. There is no doubt that this kind of analytical work poses a special challenge. The end result of analysis is influenced by a number of additional factors, which are not taken into account when the presence of higher content components is determined (these issues have been discussed in a great many publications).[40–52] But the lack of awareness of these specific requirements when performing analytical research on various types of samples for trace elements may lead to situations where the obtained result, instead of being a reliable source of analytical data, will supply erroneous information.

Contamination of the sample with the analyte and/or losses of the analyte from the sample are the most important systematic errors that can occur during preparatory steps such as[53]

- Sampling
- Storage
- Sample pretreatment

- Separation of constituents
- Final determination.

The first steps of an analytical procedure have been a largely neglected area in trace analysis for a long time, research having been focused mainly on improving the sensitivity and selectivity of the procedure. But over the last 20 years, analysts have come to recognize that the majority of systematic errors may be introduced at the very beginning rather than toward the end of a combined analytical procedure. A good analytical strategy will also include a sampling procedure free of contamination and losses, as well as proper stabilization and storage of the sample. As analytical chemistry is a discipline that helps other disciplines and solves their problems, close cooperation is clearly necessary.

In practice, the analytical chemist is often not involved in the actual sampling procedure; indeed, he/she is mostly not even informed of the sample's origin. This state of affairs must inevitably give rise to serious systematic errors already in the first steps of an analytical procedure.

The influence of contamination and losses on analytical results becomes increasingly important with diminishing analyte concentrations. These effects depend not only on the concentration range, but also on the nature of the analyte. One should bear in mind that while contamination and/or losses can never be completely eliminated, it is imperative that they be reduced to an acceptably low level.

19.5   STAGES OF THE ANALYTICAL PROCEDURE

Every analytical procedure is a series of stages that take place in a specific sequence. It can therefore be compared with a chain consisting of a great number of links, where it is obvious that the entire chain is as strong as its weakest link. This is illustrated diagrammatically in Figure 19.2.

The final step (interpretation and evaluation of analytical results) should provide the definitive answer to the initial problem, generally stated by a client of the laboratory. If the answer is not satisfactory, the analytical cycle can be repeated, after a change to or adaptation of one or more steps. Sometimes this leads to the development of a new method or the modification of part of the procedure in order, for example, to achieve better separation of certain components or to attain a lower detection limit for specific compounds.

Generally speaking, the weakest link in a chain of chemical analysis is not the one usually regarded as a part of such a process, for example, chromatographic separation or spectrometric detection. It is more likely to be one of the preceding steps, often taking place outside the analytical laboratory, such as the selection of object(s) to be sampled, the design of the sampling plan, and the selection and use of techniques and facilities for obtaining, transporting, and storing samples.[54]

When the analytical laboratory is not responsible for sampling, the quality management system often does not even take these weak links in the analytical process into account. Furthermore, if sample preparation (extraction, cleanup, etc.) has not been carried out carefully, even the most advanced, quality-controlled analytical instruments and sophisticated computer techniques cannot prevent the results of the analysis from being called into question. Finally, unless the interpretation and evaluation of results are underpinned by solid statistical data, the significance of these results is unclear, which in turn greatly undermines their merit. We therefore believe that quality control and quality assurance should involve all the steps of chemical analysis as an integral process, of which the validation of the analytical methods is merely one step, albeit an important one. In laboratory practice, quality criteria should address the rationality of the sampling plan, validation of methods, instruments and laboratory procedures, the reliability of identifications, the accuracy and precision of measured concentrations, and the comparability of laboratory results with relevant information produced earlier or elsewhere.

On the basis of a wide range of information, it can be stated that extracting and preparing samples for analysis are the weakest links in this chain. This leads to one very obvious conclusion,

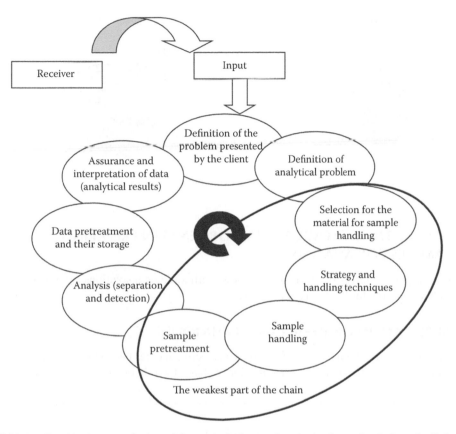

FIGURE 19.2 Graphical representation of the analytical procedure in the form of a chain—the links are the particular stages and operations.

namely, that it is necessary to pay particular attention to these two stages so that the outlay of time, labor, and money produces the desired effect, that is, reliable analytical data, for which there is a great demand. The extracted samples must be appropriately prepared for the final stage of analysis. The various operations performed *in situ* and/or in the laboratory yield a sample for analysis that is characterized by appropriate values of the following parameters:

- Size (mass, volume)
- State of matter
- Analyte concentration range
- Presence of interferants.

Figure 19.3 shows the contribution of different parts of analytical procedures to the whole uncertainty budget and the duration of analysis. The information used for preparing these diagrams was collected in a questionnaire sent to over 250 respondents (analytical laboratories in Central European countries).

Keeping documentation up-to-date is also a significant aspect of the sample preparation stage.[55]

Chromatographic and related techniques play a vital role in chemical analytics. They should be regarded as a tool with a very high decomposition potential. Appropriately prepared samples for analysis may still extend the practical range of applications for chromatographic techniques.

In analytical practice, analytes in organic samples, the matrix compositions of which are often very complex and variable, are isolated and enriched using a wide spectrum of techniques based on the mass transport phenomenon.[56,57]

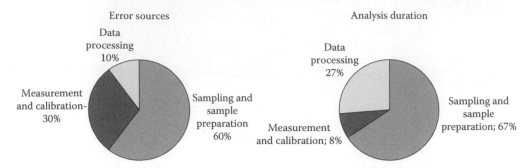

FIGURE 19.3 Contribution of different parts of analytical procedures to the whole uncertainty budget and the duration of analysis.

19.6 NEW METHODOLOGICAL DEVELOPMENTS IN PREPARING SAMPLES FOR ANALYSIS

Table 19.3 compares published information on new methodological solutions for preparing samples with complex matrices for final determinations.

19.7 APPLICATION OF MEMBRANE TECHNIQUES

Expressed simply, a membrane can be treated as a selective barrier between two phases. The phase in which mass transfer takes place is called the donor phase, the other phase being called the acceptor

TABLE 19.3
General Information on New Methodological Developments in Preparing Samples for Analysis

No.	Innovatory Examples (New Solutions)	Operations Related to the Sample Preparation Stage	Reference
1	2	3	4
1.	Cloud point phenomenon (cloud point extraction—CPE)	Extracting analytes (both organic and inorganic) from water samples	58–65
2.	Pressure-assisted chelating extraction (PACE)	Novel technique for the digestion of metals in solid matrices	66
3.	Sequential solid-phase extraction (SSPE)	Extraction of nonsulfonic acids from coastal water samples	67–69
4.	Development of matrix solid-phase dispersion (MSPD) concerning: • New sorbents • Temperature and pressure of extraction • Cleanup of extracts • Miniaturization	Extraction of organic xenobiotics from a variety of solid, semisolid, and viscous environmental and biological matrices	70–75
5.	Application of pressurized hot water (subcritical water) as the extraction medium	• Extraction of moderately and nonvolatile, thermally stable organic pollutants from a variety of solid and semisolid environmental matrices • Extraction of metals such as copper and lead from spent industrial oils with acidified pressurized hot water extraction (PHWE)	76–78,79

continued

TABLE 19.3　(continued)

No.	Innovatory Examples (New Solutions)	Operations Related to the Sample Preparation Stage	Reference
1	2	3	4
6.	New achievements and development of new techniques: pressurized liquid extraction, PLE; accelerated solvent extraction, ASE; pressurized fluid extraction, PFE)	Extraction of different microcontaminants from a variety of semisolid samples	77,80–85
7.	Microemulsion-mediated *in situ* derivatization—extraction (INDEX)	Derivatization—extraction of acidic compounds in a water matrix with alkylbromides in a homogenous reaction mixture produced by mixing water, hydrophilic alkyl bromide, and cosolvent	86
8.	Enzyme catalyzed esterification of phenolic acids in a surfactantless microemulsion system (SLME)	Derivatization of phenolic acid in water prior to its chromatographic determination	87
9.	Application of ultraviolet (UV) radiation at different stages of sample preparation	Postcolumn UV irradiation to destroy the structure of organic compounds leaving the chromatographic column	88
		Oxidizing organic matter contained in the sample	77,89
		Hybrid photocatalysis/membrane treatment of water	90
		UV digestion of the sample	91
10.	New achievements in wet digestion techniques	Sample matrix digestion with the use of chemical reagents	91–95
11.	New applications of supercritical fluids	*In situ* derivatization reactions prior to SFE with CO_2	96
		Sequential supercritical fluid extraction (SSFE) for estimating the availability of PAHs in a solid	97
		Supercritical water oxidation technology (SCWO) applied to the treatment of industrial wastes and sludges	98
12.	Application of ultrasound (US) (sonochemistry)	Ultrasonic treatment of wastes and waste-activated sludges	99,100
		New US-assisted extraction techniques for both inorganic and organic sample constituents under investigation	31,34,101–109
		US-assisted cold vapor generation	110
		Focused sonic probe for speciation analytics	108
		Mineralization of organic compounds by a heterogeneous US/catalyst process	111,112
		Ultrasonic atomization applied to the removal of endocrine disrupting compounds (EDCs) from an aquatic environment	113
		Sono-sorption as a new technique for removing lead ions from an aqueous solution	114
		UV disinfection of water	115
13.	Miniaturization of extraction with a solvent	Single-drop extraction of different types of analytes from liquid and gaseous matrices	116–123
		Liquid–liquid microextraction of organic micropollutants from water	124,125

continued

TABLE 19.3   (continued)

No.	Innovatory Examples (New Solutions)	Operations Related to the Sample Preparation Stage	Reference
1	2	3	4
14.	Application of new types of membrane-based devices as suitable techniques for the extraction of a broad spectrum of analytes from various matrices	Application of semipermeable membrane devices for evaluating persistent organic pollutant (POP) bioavailability in water	126
		Permeable environmental leaching capsules (PELCAPs) for *in situ* evaluation of contaminant immobilization in soil	127
		Application of cellulose membrane and chelator to differentiate labile and inert metal species in aquatic systems	128
		Supported liquid hollow fiber membrane microextraction of analytes from water samples	129–134
		New type of heated membrane introduction mass spectrometry interface	135
		New achievements in the application of permeation liquid membranes (PLMs)	136,137
		Application of different types of filtration to evaluate the distribution of size-fractionated particulate matter	138–142
		Chromatomembrane cells as a unit for advanced sample pretreatment in the monitoring of different types of organic compounds in water	143
15.	Derivatization of analytes with new agents	Optimization of different derivatization approaches for determining pentachlorophenol (PCP) in wastewater irrigated soil	144
		Application of ion-pair extraction and derivatization of analytes from groups of aliphatic and aromatic amines in various environmental matrices prior to gas chromatography-mass spectrometer (GC-MS) determination	145
16.	New applications of the SPE technique	Solid-phase microextraction (SPME) for sampling analytes from different matrices and introducing them to the analytical device	102,146–155
		Application of stir bar sorptive extraction (SBSE) in environmental analytics	156–162
		Miniaturization and automation of SPE devices	163–167
		Validation of the fluidized-bed extraction (FBE) technique for determining POPs in solid samples	63,168
		Molecularly imprinted polymers for extracting organic compounds from environmental and biological samples	169–176
		Development and characterization of an immunoaffinity SPE sorbent for trace analysis	177
		Extraction syringe—a device connecting sample preparation and GC	178,179
		Evaluation of multiwalled carbon nanotubes as an adsorbent for trapping volatile organic compounds (VOCs) from environmental samples	180–182
		Studies of extraction techniques based on the application of polydimethylsiloxane (PDMS) as a trapping medium	183–185
		Hemicelle- and admicelle-based SPE of linear alkylbenzene sulfonates (LASs) and phthalate esters from water	186,187
		Colorimetric solid-phase extraction (CSPE) in speciation analytics	188
		New extraction materials used for isolating analytes from complex samples and cleaning up extracts	
		• Polymeric materials	189–200
		• Inorganic sorbents	201,202–205
		• Natural sorbents	17,201,202, 205,206

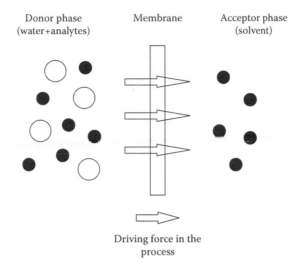

Donor phase   Membrane   Acceptor phase
(water+analytes)   (solvent)

Driving force in the
process

FIGURE 19.4   Schematic representation of transport across membranes.

phase.[56] The general principle for separating liquid mixture components using membranes is shown schematically in Figure 19.4. The main factors affecting mass transfer across a membrane are

- The type of membrane
- The force driving the extraction process

A number of criteria are used for classifying membranes. The ones most often taken into account are[56]

- The state of the membrane
- The morphology of the membrane (closely related to porosity and internal structure)
- The shape of the membrane

Figure 19.5 shows a diagram illustrating membranes classified according to the above criteria, and Table 19.4 summarizes information on the morphology of various types of membranes that could find application in environmental analytical chemistry.

The separation of components in the membrane process is due to differences in the transfer rate of chemical compounds across the barrier. It is a nonequilibrium process, in which the flow of a component depends on the driving force. Table 19.5 provides some basic information on the forces driving membrane processes.

Various analytical techniques make use of both porous and nonporous (semipermeable) membranes. For porous membranes, components are separated as a result of a sieving effect (size exclusion), that is, the membrane is permeable to molecules with diameters smaller than the membrane pore diameter. The selectivity of such a membrane is thus dependent on its pore diameter. The operation of nonporous membranes is based on differences in solubility and the diffusion coefficients of individual analytes in the membrane material. A porous membrane impregnated with a liquid or a membrane made of a monolithic material, such as silicone rubber, can be used as nonporous membranes.

19.7.1   MEMBRANE EXTRACTION

The membrane extraction process mostly makes use of nonporous membranes. Such a membrane can be a liquid or a solid phase (a polymer impregnated with a liquid), which is placed between two

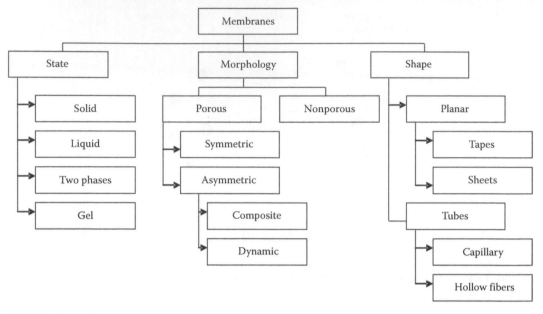

FIGURE 19.5 Classification of membranes with respect to their state, morphology, and shape.

TABLE 19.4
Information on Membrane Morphology

Porous Membranes

Symmetric	Asymmetric
• Capillary or irregular pores	• Increase in porosity perpendicular to the surface
• Identical porosity perpendicular to external surfaces	• Smallest porosity in the surface layer
• Preparation methods:	• Separation layer—surface layer
• Sintering	• Support layer (reinforcing)
• Radiation with etching	• The rest of the membrane
• Phase inversion	• Methods of formation:
	• Thermal gelation
	• Vapor adsorption
	• Loeb–Sourirajan phase inversion
	• Composite asymmetric membranes:
	• Two- or multilayer
	• Different composition of individual layers
	• Formed by coating a layer of selective properties onto a porous protective layer
	• Dynamic asymmetric membranes:
	• Formed dynamically
	• Formed by coating colloids or macromolecular compounds onto a porous bed under pressure
	• Support—filtration foil made from an organic material; plate, tube, or molder made from a ceramic, carbon, or metal sinter

continued

TABLE 19.4 (continued)

Nonporous Membranes

Symmetric   **Asymmetric**

- Lack of conventional pores (pores of molecular dimensions)
- Continual variations in the number, size, and location of pores as a result of thermal molecular motions in the membrane material

Solid   **Liquid**

Inorganic membranes:
- Material: metals, metal alloys, sintered ceramics, glass
- Organic membranes:
- Material: natural and synthetic polymers, for example, cellulose acetate, silicone rubber, polyethylene

- Thin liquid layer with a dissolved mediator
- Separates a donor solution from an acceptor solution
- Kinds:
  - Thick layer
  - Emulsion
  - Reinforced

Ion-Exchange Membranes

- Nonporous, microporous, and porous membranes of symmetric or asymmetric structure
- Kinds:
  - Cationic (cation exchange)—cations pass toward the cathode through a constant electrical field and exclude anions
  - Anionic (anion exchange)—anions pass toward the anode through a constant electrical field and exclude cations

---

other phases, usually liquid, but sometimes also gaseous. A review of the available literature indicates that the term "membrane extraction" includes the following types of apparatus and procedures[56,207–210]:

- Supported liquid membrane extraction (SLM)
- Microporous membrane liquid–liquid extraction (MMLLE)
- Polymeric membrane extraction (PME)
- Membrane extraction with a sorbent interface (MESI).

Table 19.6 gives basic information on these techniques.

TABLE 19.5
Basic Information on the Driving Forces in Membrane Processes

No.	Driving Force of the Extraction Process	Name and Mathematical Form of the Equation Used to Describe Mass Transfer	Membrane Techniques in which the Mass Transfer Equations are Used
1	Concentration gradient	Fick's law of diffusion: $J_m = -D^a A (dC/dx)$	Dialysis, membrane extraction
2	Pressure difference	Hagen–Poisseuille equation: $J_v = -K^b A (dP/dx)$	Filtration
3	Potential difference	Ohm's law: $J_c = -R^c A (dE/dx)$	Electrodialysis

a Diffusion coefficient.
b Hydrodynamic permeability.
c Resistance.
d A, diffusion surface (membrane surface).

TABLE 19.6
Basic Information on the Membrane Extraction Techniques Applied in the Analysis of Liquid Samples

Acronym	Name of Technique	Type of Membrane	Combination of Phases Used: Donor/Membrane/Acceptor
SLM	Supported liquid membrane extraction	Nonporous	Aqueous/organic/aqueous
MMLLE	Microporous membrane liquid–liquid extraction	Nonporous (microporous)	Aqueous/organic/organic
			Organic/organic/aqueous
PME	Polymeric membrane extraction,	Nonporous	Aqueous/polymer/aqueous
MASE	Membrane-assisted sorbent extraction		Organic/polymer/aqueous
			Aqueous/polymer/organic
MESI	Membrane extraction with a sorbent interface	Nonporous	Gaseous/polymer/gaseous
			Liquid/polymer/gaseous

19.8 MATRIX SOLID-PHASE DISPERSION

In 1989, solid sorbents were used for the first time to extract analytes from solid samples. Samples were placed in a glass mortar and blended with modified silica.[14] Sorbent was added in order to[211]

- Grind the sample evenly (the sorbent acts as an abrasive material)
- Bind solvents that cause lysis of cellular membranes for biological material
- Improve the mechanical properties of the sample blended with sorbent, which makes the fractionated extraction of analytes possible (the mixture of sample and sorbent is a new type of filling)

The liquid extraction technique from ground solid samples may be used to isolate analytes from a solid. This extraction method is particularly useful when extracting analytes from dense materials, such as food, plant and animal tissues, or fats.

The extraction column with a paper filter at the bottom was filled with the sample prepared as described above. The deposit was secured with a paper filter at the top as well, and the whole was compressed with a special piston. Next, the column was filled with a known volume of solvent for extraction. Solvent flow was forced with a rubber pipette syringe, and the solvent was collected in special receivers.

This method of extraction is different from SPE in that for the latter the samples put in the column must be in the form of a liquid solution. The interactions between the various components of the dispersed sample are stronger and to some extent different from those occurring in SPE. Specific interactions among all elements of the system, that is, analytes, interferants, the sample matrix, the solid sorbent added to the sample, and the solvent used for extraction, have been observed. The obtained extracts were purified using SPE or were subjected to final analysis using chromatographic techniques without purification.

This extraction technique is similar to classic sample homogenization techniques, which usually involve grinding, pounding, or crushing samples. The effects of mechanical grinding are often enhanced by adding solvents, acids, alkalis, detergents, or chelating agents, which usually leads to the partial extraction of analytes—an unintended effect. The extracted compounds may adsorb on the walls of the vessels and instruments used. Emulsion formation may be another negative effect. In such a case, it is necessary to centrifuge and re-extract analytes from the sample, which is an additional obstacle to carrying out an analytical procedure. *Medium pressure liquid extraction* (MPLE) is an extraction technique intermediate between MSPD and ASE. In this case, the ground sample mixed with solid sorbent fills the chromatographic column through which the solvent is pumped by means of a special, low-pressure pump. The column discharge (extract) may be subjected to final analysis without further purification.

19.9 SUPERCRITICAL FLUID EXTRACTION

A supercritical fluid is a substance that comes into existence after the so-called critical point has been exceeded, that is, when it simultaneously exhibits the properties of a gas and a liquid, but is actually neither the one nor the other. In 1962, Klesper, Corwin, and Turner were the first researchers to use supercritical fluids for analytical purposes. A supercritical fluid was used in high-pressure fluid chromatography, where it was part of the mobile phase. Extraction with a supercritical fluid was first achieved in 1978, since when the supercritical fluid extraction (SFE) technique has been undergoing active development, finding many applications in laboratory analysis and industry.[212]

On exceeding the critical point, the substance shows certain characteristics of both a gas and a liquid at the same time, but also a number of properties characteristic only of this form of matter, namely,

- It does not condense.
- It does not boil.
- It does not form a meniscus (a characteristic property of liquids), but it does have the capability to dissolve, which is characteristic of liquids.
- High "diffusibility": Dissolved substances spread in a supercritical fluid at speeds between those of liquids and real gases.
- No surface tension: A supercritical fluid can thus penetrate even the smallest pores of the sample matrix.
- The low viscosity of supercritical fluids ensures effective penetration of the entire sample.

This combination of the aforementioned properties of supercritical fluids accounts for the fact that they penetrate the sample matrix like a gas and at the same time dissolve analytes like liquids.

Every substance has its own individual critical pressure and temperature that is often difficult to obtain under laboratory conditions. Because of this, and despite attempts to use various substances as extraction media during the development of SFE, most of these substances have proved useless.

A substance with favorable critical parameter values and that best matches the other aforementioned criteria is carbon dioxide (CO_2). The critical temperature of CO_2 is +31.3°C, which is especially important for thermally unstable analytes, and its critical pressure of 72.9 bar (1 bar = 10^5 Pa) is easy to obtain under laboratory conditions. Moreover, CO_2 is nonflammable, nontoxic, does not pose any additional, serious threat to the environment, and is relatively inexpensive. For on-line solutions, it is important that CO_2 be compatible with most chromatographic detectors.

Because CO_2 has weak dissolving capabilities, it is suitable as an extraction medium in SFE only for compounds of small and medium molecular mass and of low polarity. As a result, suitable modifiers must be added in order to extract polar substances. Modifiers are polar organic solvents, that is, with a nonzero dipole moment (methanol, acetonitrile, tetrahydrofuran, or water are the most commonly used) that enhance the diffusibility of polar analytes in nonpolar extraction media such as CO_2.

SFE is carried out above the solvent critical point, and the properties of a supercritical fluid depend on pressure and change along with its density. These criteria determine the selectivity of the extraction medium. One fluid can therefore be used to extract a whole series of compound groups (depending on the pressure in the system, the temperature, extraction medium volume flow, and extraction time) and to separate the obtained extract into appropriate fractions. Selective fractionation is used, for example, to separate olfactory and gustatory substances in the extraction of hops for beer production.

Fractions of three groups of substances used in beer production are extracted from hops using supercritical CO_2. The first fraction, the so-called oil essence, was obtained via extraction with CO_2 at a density of $0.30\,g\,mL^{-1}$ and a temperature of 50°C. Bitter substances were collected as the second fraction at a CO_2 density of $0.70\,g\,mL^{-1}$ (50°C); that fraction overlapped only slightly the third and last fraction of neutral fats, extracted at a CO_2 density of $0.90\,g\,mL^{-1}$ (50°C).

The extraction medium in SFE is supplied from a cylinder to a pump where it is compressed to the desired pressure in the critical range. Next, the fluid in this form reaches the vessel containing a sample situated in a chamber heated to the critical temperature. Here, the substance, already in the supercritical fluid state, extracts the analytes, and the extract is collected in a special receiver. Figure 19.6 shows a diagram of the instruments used for SFE.

The sample for extraction is situated in a special container that is then introduced into the chamber. These are the two preparatory stages before sample extraction. For solid samples, an additional homogenization stage is necessary, which facilitates the diffusion of analytes in the whole sample volume. Desiccants, such as Na_2SO_4 or $MgCl_2$, are often added to a sample in order to remove moisture. Soils and sediments usually contain certain amounts of organosulfur compounds, which decompose under the influence of temperature, and the products of this decomposition may cause fluctuations in flow volume and even choke the outflow from the extraction chamber. To avoid such complications, acid-cleaned copper granules, which react with organosulfur compounds to produce copper sulfide, are placed in the sample-containing chamber.

SFE may be carried out in both *off-line* and *on-line* systems. In the off-line case, the receiver may be an empty container, a trap, an analytical column with which further analysis will be carried out, or a container with the solvent. There are several variants of SFE in the off-line system: extraction in a dynamic or static system, or in a supercritical fluid recirculating system.

Extraction under static conditions consists of flooding the sample with a supercritical fluid, where it is "drenched" for some time, and then the solvent, together with the enriched analytes, is taken to a receiver. The "drenching" stage is useful when the analytes are difficult to isolate from the matrix owing to the low dissolution rate or the compact structure of the sample. In a supercritical fluid

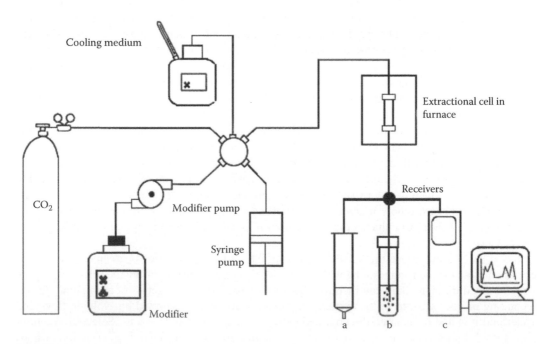

FIGURE 19.6   Diagram of apparatus used for SFE.

recirculating system, one dose of solvent is pumped many times through the sample container. After some time, the solvent with the isolated analytes is entirely or partly collected in the receiver. For *on-line* SFE, the extract in the container is not collected in the container, but is supplied directly to the analytical apparatus.

In a dynamic extraction system, the supercritical fluid is pumped only once through the container with the sample to the receiver. In the receiver, the liquid is vaporized, leaving concentrated analytes that are then dissolved in a small volume of the solvent. Such extracts are analyzed to determine selected analytes. This manner of extraction is effective if the analytes are well soluble in the solvent and the sample matrix is penetrable. Apart from the aforementioned possibility of fractionated extraction, SFE has many other advantages accruing from the special properties of supercritical fluids:

- Amounts of solvents (usually harmful to the environment and human health) are substantially reduced compared with "classical" extraction methods, such as extraction by shaking or Soxhlet extraction.
- Extraction samples are small.
- Extraction times are shorter.
- The costs of the process are lower compared with "classical" techniques.
- Low temperatures are used, which favors the extraction of thermally unstable compounds.

19.10 UNIQUE PHASE SEPARATION BEHAVIOR OF SURFACTANT MICELLES

Aqueous solutions of neutral (i.e., nonionic or zwitterionic) surfactants can form micellar assemblies in which a certain number of surfactant molecules aggregate to form an assembly possessing a central core region comprised of long alkyl (or alkylaryl) hydrocarbon chains with their more polar polyethyleneoxide (or zwitterionic) headgroups extending outward and interacting with the bulk water.[213]

Aqueous solutions of neutral surfactants have two particularly important properties—their solubilizability and phase separation behavior—that can be exploited in order to develop a new viable extraction–preconcentration technique.

Firstly, it is well known that micellar aggregates in water can solubilize and bind hydrophobic solute molecules that are typically insoluble or only sparingly soluble in bulk water. For example, although the solubility of pyrene and anthracene in water is in the 0.1–0.6 micromolar range, their solubility can easily be increased to the 10 millimolar range in the presence of micelles. The amount of solute solubilized and bound to the micellar aggregate in an aqueous solution is typically proportional to the surfactant concentration up to the limiting value.

In view of its superior solubilizing power, the addition of a known volume of solution containing micellar surfactant to either a given volume of an aqueous solution of a sample or a given mass of a solid sample provides micelles capable of binding and concentrating (in the former) or desorbing and then binding (in the latter) in the micellar entity the organic species that were originally present in the aqueous or solid sample. For the extraction technique for solids, an aqueous concentrated neutral surfactant micellar solution is merely placed in contact with or passed through the solid sample containing the organic component(s). The organic solute(s) present is/are desorbed and solubilized into the micelles in the bulk solution, which are then further enriched (as will be described shortly) by the phase separation behavior of the surfactant solution. The desorption process is thought to be similar to the molecular mechanism reported for the solubilization of water-insoluble solids by micellar solutions, which involves direct micelle diffusion to and from the surfactant-modified solid surface, in series with interfacial steps including adsorption and desorption of the micellar organic species.[214–216]

A significant advantage of this bulk micellar extraction technique is that once the initial "extraction" from the solid matrix has been performed, the organic component(s) now present in the extractant micellar solution can be further enriched and preconcentrated prior to final quantitation of the workup (as can be any organic species originally present in a waterborne sample to which a small amount of a concentrated surfactant micellar solution has been added). This preconcentration is made possible by the phase separation ability of micellar solutions. Aqueous solutions of neutral surfactant micellar compositions can exhibit the so-called critical phenomena and clouding following a change in temperature: Upon increasing the temperature of such isotropic aqueous micellar solutions, a critical temperature is eventually reached at which the aqueous solution suddenly becomes turbid (clouding point) owing to the reduced solubility of the surfactant micelles present in the bulk water. After some time interval (which can be speeded up by centrifugation), separation into two transparent liquid phases occurs (i.e., the formation of a wet surfactant-rich micellar phase in equilibrium with almost pure water containing the same surfactant molecules). Any organic species present that can bind and partition to the micellar entity will be extracted into and thus concentrated in the small volume element of the surfactant-rich micellar phase.

A plot of the temperatures required for clouding versus surfactant concentration typically exhibits a minimum in the case of nonionic surfactants (or a maximum in the case of zwitterionics) in its coexistence curve, with the temperature and surfactant concentration at which the minimum (or maximum) occurs being referred to as the critical temperature and concentration, respectively. This type of behavior is also exhibited by other nonionic surfactants, that is, nonionic polymers, n-alkylsulfinylalcohols, hydroxymethyl or ethyl celluloses, dimethylalkylphosphine oxides, or, most commonly, alkyl (or aryl) polyoxyethylene ethers. Likewise, certain zwitterionic surfactant solutions can also exhibit critical behavior in which an upper rather than a lower consolute boundary is present. Previously, metal ions (in the form of metal chelate complexes) were extracted and enriched from aqueous media using such a cloud point extraction approach with nonionic surfactants. Extraction efficiencies in excess of 98% for such metal ion extraction techniques were achieved with enrichment factors in the range of 45–200. In addition to metal ion enrichments, this type of micellar cloud point extraction approach has been reported to be useful for the separation of hydrophobic from hydrophilic proteins, both originally present in an aqueous solution, and also for the preconcentration of the former type of proteins.

19.11 IONIC LIQUIDS: A NEW TYPE OF SOLVENT AND EXTRACTANT

Ionic liquids, a new type of solvent, are salts that contain ions (an organic cation and an anion, usually inorganic) and occur as liquids at room temperature.

There are three main types of ionic liquids[218]:

- Quaternary ammonium salts $[R_xNH_{4-x}]^+Y^-$
- Iminium salts

Imidazolium salt   Pyridinium salt

- Phosphonium salts $[R_xPH_{4-x}]^+Y^-$,

where $x = 1, 2, 3, 4$ and $Y = BF_4, PF_6, NO_3, SbF_6, AlCl_4, CuCl_2$.

The physical properties of ionic liquids depend on the kind of cation and anion. Quaternary ammonium salts, which have been known about for more than a century, exhibit interesting and unique characteristics. In the 1940s, their powerful antibacterial properties were discovered; later, these compounds started to be used as phase transfer catalysts.

At present, ionic liquids, also known as room-temperature ionic liquids, nonaqueous ionic liquids, molten salts, liquid organic salts, and fused salts, are considered to be the new generation of solvents. In chemical abstracts, they can be found under the headings "ionic liquid" or "liquids ionic." Publications on ionic liquids are increasing in number.

The first ammonium salt to be recognized as an ionic liquid was obtained in 1914. It was a nitrate with the structure $[C_2H_5NH_3]^+NO_3^-$. Many ionic liquids at present being examined were tested in the USA in the 1970s as electrolytes for batteries (the research was financed by the *Air Force Office of Scientific Research*). It was found that ionic liquids

- Dissolve both organic compounds (from simple solvents to polymers) and inorganic ones (including some rocks and carbon)
- Are thermally stable: their boiling point is high, often in excess of 350°C
- Do not usually mix with water
- Are nonvolatile (the vapor pressure at 25°C is very low)
- Dissolve catalysts, especially transition metal complexes, without damaging the walls of a glass or steel reactor

The temperature range within which ionic liquids occur in the liquid state is very characteristic; it is assumed that it is never greater than 300°C. No other type of commonly used solvent occurs as a liquid over such a great range. Table 19.7 lists the physicochemical properties of frequently used solvents.

With the use of ionic liquids, it is possible to investigate the kinetics of reactions over a much greater temperature range than before (it has been noted that some reactions occur faster at lower temperatures, a topic open to debate).

These liquids are not normally miscible with water and are heavier than it, which is why there is no distinct phase border. Salts containing a tetrafluoroborate or hexafluorophosphate (V) anion are stable in air and in contact with water, but salts containing a tetrachloroaluminum ion are sensitive to water, reacting violently with it to produce toxic hydrogen chloride.

The extraction of organic compounds from water using ionic liquids takes place in the same way as with traditional organic solvents.

TABLE 19.7
Physicochemical Properties of Popular Solvents

Solvent	T_T (°C)	T_W (°C)	$T_W - T_T$ (°C)
Ammonium	− 78	− 34	44
Benzene	5	80	75
Water	0	100	100
Chloroform	− 63	61	124
Acetone	− 94	56	150
Ethyl acetate	− 84	77	161
Methanol	− 98	65	163
Hexane	− 95	69	164
Nitrobenzene	6	211	205
N,N-dimethylformamide	− 61	153	215
Ionic liquid	~ −96	>200	>300

Notes: T_T, melting point; T_W, boiling point; and $(T_W - T_T)$, temperature range in which solvent is a liquid.

19.12 MICROWAVE-ENHANCED CHEMISTRY (MEC)

The literature underscores the increasing importance of microwave radiation in the chemical laboratory. 2450 MHz radiation was first applied in microwave ovens, but some time later it was observed that this radiation can be used to heat different liquids and solids. It may be said that microwave ovens were the precursors of this new direction in analytics. In an analytical laboratory, at different stages of preparation of samples for analysis, microwave radiation[28,218] supports the process of analyte extraction. Microwave-assisted extraction (MAE) was introduced to the scientific community in 1986.[43] Initially used in the food and agricultural industries for conditioning food products, microwaves have been used for sample digestion since the mid-1980s. More recently, they have been used in the solvent extraction of organic analytes from a solid sample. Enhancement is based on the absorption of microwave energy by molecules of chemical compounds. The most frequently used solvents include dichloromethane and acetone–hexane mixtures.

MAE can be carried out in two ways:

1. Pressurized MAE in closed vessels: This technique employs a microwave-transparent vessel for the extraction and a solvent with a high dielectric constant (electrical permittivity). Such solvents absorb microwave radiation and can thus be heated to a temperature exceeding solvent boiling points under standard conditions. Boiling does not occur, however, because the vessel is pressurized. This mode of operation is very similar to ASE—the elevated pressure and temperature facilitate extraction of the analyte from the sample.
2. Atmospheric MAE system: This second technique employs solvents with low dielectric constants. Such solvents are essentially microwave-transparent; they thus absorb very little energy, and extraction can therefore be performed in open vessels. The temperature of the sample increases during extraction because it usually contains water and other components with high dielectric constants: the process is thereby enhanced. Because extraction conditions are milder, this mode of operation can be used to extract thermolabile analytes.

This technique is known as focused microwaves (FMW), and it also yields satisfactory results for polycyclic aromatic hydrocarbons, polychlorinated biphenyls (PCBs), organochlorine pesticides, and alkanes with the same advantages of security and ease of manipulation.

Microwave heating is very efficient and can basically be explained by the interactions of an electric field with charged particles and polar molecules in a solution involving two mechanisms of energy absorption—ionic conductance and dipole rotation. However, problems arise in MAE when using apolar solvents, because microwave energy can only be effectively absorbed by molecules with dipole moments. For the extraction of organic contaminants this is a drawback, but the problem can be solved by increasing the polarity.

The most important instruments for microwave extraction are a microwave radiation source, a waveguide, a resonant cavity, and an energy source. As waveguides are made of materials that strongly reflect electromagnetic waves (e.g., metal foils or metal sheets), microwave radiation passes from the magnetron to the receiver (resonator). The resonant cavity (resonator) is the part of the microwave apparatus from which microwave radiation is repeatedly reflected from its walls. The container with the sample to be extracted is placed in the resonator. MAE can use not only solvents that absorb microwave energy and have a high dielectric constant, but also those with a low dielectric constant that do not absorb microwave energy.[220]

In analyte extraction using solvents capable of absorbing a high level of microwave radiation (with a high dielectric constant), the extraction takes place at high temperatures. Because of the high pressure in the extraction vessel, the temperature of the solvent usually exceeds its boiling point, and may be as high as 300°C. In view of this, the vessel containing the sample and solvent must have special properties. Attention should be paid to the following points: (1) the chemical and thermal resistance of the material the vessel is made from, (2) its permeability to microwave radiation, and (3) its resistance to solvent action. The aforementioned requirements are fulfilled by containers made of polytetrafluoroethylene (PTFE), quartz, and certain composite materials. MAE has several advantages:

- A shorter heating and extraction time
- Compact devices
- Easy control of the sample heating process
- Reduced amount of solvent used for extraction
- Efficient use of energy (it is used solely to heat the sample and solvent)

It should be pointed out that several additional operations must usually be performed prior to the final determination:

- Separation of the extract from the matrix (by filtration or decantation)
- Concentration of the extract (removal of excess solvent)
- Purification and drying of the extract

Using the dynamic approach for extraction is generally advantageous, especially with respect to partitioning the solvent into the extraction media. This can be highly efficient when fresh solvent is continuously introduced into the extraction cell, that is, the rate constant for desorption need not be large compared with the rate constant for adsorption in order for the target solute to be removed efficiently.

Another new approach combines MAE with the use of an aqueous surfactant solution as the extracting phase. This new technique is called microwave-assisted micellar extraction (MAME). This procedure is based on the well-known solubilization capacity of aqueous micellar solutions toward water-insoluble or sparingly soluble organic compounds. As a general rule, nonionic surfactants are usually the most effective, showing greater solubilization capacities that rapidly increase with the solubilization kinetics as the cloud-point temperature of the solution is raised.

Table 19.8 presents information on the application of microwave radiation as an enhancing factor for other operations associated with sample preparation for analysis.

The main characteristics of commercially available FMW technology are

- Safety due to operation at atmospheric pressure.
- The ability to handle large samples (mainly of organic materials) that may generate very large quantities of gas.
- The use of different types of materials to construct reaction vessels, such as borosilicate glass, quartz, and PTFE.
- The programmable addition of reagents at any time during the digestion, which makes allowance for sequential acid attack.
- A low-power FMW field can be employed either to accelerate leaching of organometallic species without affecting carbon–metal bonds or to extract organic compounds. The focused nature of microwave energy is highly efficient and avoids the need for high power levels.
- Multiple methods for different samples can be simultaneously applied, which allows for the independent operation of each reaction vessel.

19.13   APPLICATION OF ULTRASOUND (US) IN THE SAMPLE PREPARATION PROCESS

Sound waves are mechanical vibrations in a solid, liquid, or gas and are intrinsically different from electromagnetic waves. While the latter (radio waves; infrared, visible, or ultraviolet light; X-rays; and gamma rays) can pass through a vacuum, sound waves must travel through matter, as they involve expansion and compression cycles traveling through a medium. Expansion pulls molecules apart, compression pushes them together.

TABLE 19.8
Applications of Microwave Radiation in Sample Preparation for Analysis

Sample Preparation Operation	Reference
Desiccating	The process may be carried out both under normal pressure and in a vacuum. Polar water particles are selectively heated[204,220] and water evaporates[221,222]
Water (moisture) content determination	Special instruments are used—microwave scale dryer
Quick sample heating	
Microscopic sample preparation	Microwave radiation greatly enhances the fixing of biological material
Sample incineration and melting	Microwave ovens use ceramic heaters, which are remotely heated by microwave radiation[221]
Plasma incineration	Plasma induced by microwave radiation (pressure of 10 mbar (1 kPa)) significantly decreases the incineration temperature
Activation of sample components in plasma	The microwave-induced plasma (MIP) is used
Carrying out chemical reactions, including derivatization	Microwave radiation can be used to accelerate the rate of chemical reactions[62,223,224]
Steam distillation	99,225
Pyrolysis	226,227,228
Sample digestion	Oxidation takes place in the medium of active reagents (nitric acid, hydrofluoric acid, hydrogen peroxide)[229–233]
Evaporization of aqueous solutions	221
Microwave thermal inertization of wastes	Applied to asbestos-containing waste[234]
Heating of GC column	Negative temperature programming can be employed to enhance separation of compounds during the separation process[235]
MAE of samples	The high efficiency of this type of extraction process enables it to be applied to the extraction of a wide spectrum of analytes from various matrices[236,237–242]

US is simply sound with a frequency higher than the range audible to humans (>16 kHz). The lowest US frequency is normally taken to be 20 kHz. The top end of the frequency range is limited only by the ability to generate the signals, so frequencies in the gigahertz (GHz) range have been used in some applications.

The use of US in science has expanded in recent years, including such fields as medicine and industry where US has had the most impact; it continues to do so, with new uses appearing at regular intervals. Analytical chemistry has also availed itself of US energy—two of its aspects have been exploited [59]:

a. It facilitates the development of different steps in the analytical process, related mainly to the preliminary steps involving solid samples.
b. It improves detection (US has even been used as the actual means of detection).

There are two common US devices for sample preparation applications: bath and probe units.
Although US baths are more widely used, they have two main disadvantages that substantially reduce experimental repeatability and reproducibility[243]:

a. A lack of uniformity in the distribution of US energy (only a small fraction of the total liquid volume in the immediate vicinity of the US source experiences cavitations)
b. A decline in power with time

US probes are superior to US baths, however, in that they focus their energy on a localized sample zone, thereby providing more efficient cavitations in the liquid. It has been demonstrated that, because of the formation of standing waves, the local intensity in a flask fixed in a US cleaner is highly susceptible to changes in experimental conditions, so precision is considerably affected as a result. Most US-assisted sample preparation applications are developed in discrete systems; nevertheless, continuous approaches mean that a given step can be automated, as it can be interfaced with others that are also automated.

The performance of baths and probes in US-assisted digestion under soft or strong conditions depends on a number of factors that are rarely optimized in the development of US-assisted digestion methods. The variables affecting US-assisted digestion common to baths and probes are[244]

- The shape of the vessel containing the target chemical system, a factor usually ignored. Flat-bottomed vessels (e.g., conical flasks) are to be preferred because the energy transfer is more efficient.
- The stirring of the target suspension; this can be advantageous because it ensures effective contact between the solid and the liquid during sonication.
- The temperature of the medium can have a strong influence on the rate of digestion. In the case of thermolabile analytes, operation over very short periods of time or circulation of thermostatted cold water in the tank may be alternative means of controlling the temperature.
- The influence of pressure on US-assisted digestion has hardly been studied at all. There are only a few cases of chemical reaction acceleration where high pressure has been applied in closed ultrasonic reactors. These devices can also be used as ultrasonic digestors.
- Solvent properties affect US-assisted digestion, as they impose a cavitation threshold above which sonochemical effects are "perceived," as it were, by the medium. Therefore, any phenomenon altering the same solvent property can modify such a threshold.
- The radiation amplitude is directly related to the amount of energy applied to the system. Exhaustive treatments require high irradiation amplitudes, for which probes are more suitable than baths.
- Particle size is a key variable, so digestion mechanisms are influenced by particle diameter. In fact, depending on particle size, simultaneous microstreaming and microjetting or some other effect can determine the efficiency of US-assisted digestion to a variable extent.

There are other specific physical variables that influence digestion assisted by an ultrasonic probe, namely,[244]

- The depth of immersion into the sample vessel or bath containing the transmitting liquid has a decisive influence on the effect of ultrasonic probes. This is because virtually no sonication exists alongside the tip or above it. Therefore, if the probe is immersed only slightly, it will cause foaming at the liquid surface, resulting in a loss of US energy. On the other hand, if the probe is immersed too deeply, the energy supplied will be inadequately transmitted through the liquid and digestion efficiency will suffer as a result.
- The probe-tip/sample-cell distance should be considered when the probe is inserted into the sample vessel. The shorter the distance, the less the attenuation, and the higher the energy applied to the sample as a result.
- When ultrasonic energy is applied in a pulsed mode, pulse duration can be an important variable.

US can be used as an enhancing agent during different steps of sample preparation for analysis:

- Analyte extraction[119,156,175,245,246]: US-assisted extraction is an effective way of removing a number of analytes from different types of samples, as it combines several effects, namely,

a. Extremely high effective temperatures, which result in increased solubility and diffusivity.
b. High pressures, which favor penetration and transport at the interface between an aqueous or organic solution subject to US energy, and an organic or aqueous phase, or a solid matrix (which is more common).
c. The oxidative energy of radicals created during sonolysis of the solvent (hydroxyl and hydrogen peroxide for water).

- Digestion[235]
- Dissolution[247]
- Homogenization and emulsification[84,158]
- Filtration[248]
- Analytical reactions (e.g., derivatization)[137]
- Reagent generation[36]
- Slurry formation[59]
- Cleaning[46]
- Degassing.[249]

It is worth emphasizing two main constraints in the use of US in the field:

a. US energy can lead to undesirable deterioration of sample components, so special care must be taken to ensure that only the desired effect is produced in a US-treated system.
b. The use of the appropriate US device is key to obtaining a given effect that cannot be produced by devices designed for other, different tasks. For example, US baths may be designed for cleaning purposes, for which power stability or uniformity in the distribution of US energy is not mandatory.

Even considering that microwave technology has improved some traditional operations in chemistry, there is still a long road ahead, since only some 10% of laboratories throughout the world are equipped with laboratory-designed microwave ovens.[232]

19.14 GREEN CHEMISTRY: INTRODUCTION OF THE CONCEPT OF SUSTAINABLE DEVELOPMENT TO CHEMICAL LABORATORIES

19.14.1 HISTORY

The term "green chemistry" was first used in 1991 by P.T. Anastas in a special program launched by the US Environmental Protection Agency (EPA) to implement sustainable development in chemistry and chemical technology by industry, academia, and government. In 1995, the annual US Presidential Green Chemistry Challenge was announced. Similar awards were soon established in European countries. In 1996, the Working Party on Green Chemistry was created, acting within the framework of the International Union of Applied and Pure Chemistry. One year later, the Green Chemistry Institute (GCI) was formed with chapters in 20 countries. Its aim was to facilitate contact between governmental agencies/industrial corporations and universities/research institutes in the design and implementation of new technologies. The first conference highlighting green chemistry was held in Washington in 1997. Since that time, other similar scientific conferences have been held on a regular basis. The first books and journals on the subject of green chemistry were introduced in the 1990s, including the *Journal of Clean Processes and Products* (Springer-Verlag) and *Green Chemistry*, sponsored by the Royal Society of Chemistry. Other journals, such as *Environmental Science and Technology* and the *Journal of Chemical Education*, have sections devoted to green chemistry. The latest information can also be found on the Internet.

The concept of green chemistry arose in the United States as a joint research program resulting from the interdisciplinary cooperation of university teams, independent research groups, industry, scientific societies, and governmental agencies, each with their own programs aiming at reducing pollution.

Green chemistry incorporates a new approach to the synthesis, processing, and application of chemical substances in such a manner as to reduce threats to health and the environment. This new approach is also known as

- Environmentally benign chemistry
- Clean chemistry
- Atom economy
- Benign-by-design chemistry.

Green chemistry is commonly presented as a set of Twelve Principles proposed by Anastas and Warner—they are a set of instructions for professional chemists to be adhered to when synthesizing new chemical compounds and implementing new technological processes.[250]

Green chemistry is not a new branch of science. It is a new philosophical approach which, through its application and extension of principles, can contribute to sustainable development. Numerous interesting examples of the use of green chemistry rules lie to hand. In addition, new analytical methodologies that can be implemented in accordance with green chemistry standards are being developed; they will be useful in carrying out chemical processes and in the evaluation of their effects on the environment.

19.14.2 Green Analytical Chemistry

The irony is that the analytical methods used by analytical chemists in laboratories to assess the state of environmental pollution, through uncontrolled disposal of reagents and solvents or chemical waste, may in fact be a source emitting large amounts of pollutants that adversely affect the environment. This is because considerable quantities of chemical compounds need to be used in the successive steps of analytical procedures. Sampling, and especially the preparation of samples for their final determination, frequently involves the formation of large amounts of pollutants (vapors, reagent and solvent wastes, and solid waste). The rules of green chemistry therefore need to be introduced into chemical laboratories right across the board.

If we look at the Twelve Principles of Green Chemistry, it is easy to indicate the issues that should determine the "green" character of analytical chemistry. The following should be treated as priorities[251,252]:

- Eliminating or minimizing the use of chemical reagents, particularly organic solvents, from analytical methods.
- Eliminating highly toxic and ecotoxic chemicals from analytical procedures.
- Reducing labor- and energy-intensive steps in particular analytical methods (per single analyte).
- Reducing the impact of chemicals on human health.

The development of analytics and environmental monitoring leads to better knowledge of the state of the environment and the processes that take place in it. As a result of the introduction of new methodologies and new measuring techniques for identifying and determining trace and microtrace components in samples with complex compositions into analytical practice, the following important circumstances have been established:

- The acidification of certain components of the environment
- The existence of stratospheric ozone depletion

- The existence of long-term trends in the changes of trace components in atmospheric air
- The presence of elevated concentrations of POPs, for example, dioxins [polychlorinated dibenzodioxin (PCDD) and polychlorinated dibenzofuran (PCDF)] and PCBs
- The bioaccumulation of pollutants in the tissues of organisms at different levels of the trophic pyramid

This branch of analytical chemistry presents many challenges. The most important are

- The low and very low concentration levels of analytes
- The existence of temporal and spatial fluctuations of analytes in the investigated media
- A broad range of concentrations of analytes belonging to the same group of compounds
- The possibility of the presence of interfering compounds, frequently with similar chemical structures and properties

The techniques of sample preparation, extraction (isolation), and/or preconcentration of analytes are usually applied in the analysis of trace components of gaseous, liquid, and solid samples. During this operation, transport of analytes from primary matrices (donors) to the secondary matrix (the acceptor) takes place. It should be remembered, however, that the extraction and preconcentration steps could be a source of environmental pollution. The techniques of sample preparation introduced in this chapter have the following advantages[253]:

- They are solvent-free or virtually so—solvent usage per one analysis is reduced to a minimum.
- The transport of analytes to the matrix is characterized by simplicity of composition compared with primary matrices, and is more suitable and compatible with the analytical technique used in the final determination step.
- The removal, or at least reduction, of interfering substances as a result of the selective transfer of sample components to the acceptor matrices.
- Increased concentration of analytes in the acceptor matrix to levels over the limit of quantitation for the chosen analytical technique.

There is an urgent need to evaluate the applied analytical methods, with respect not only to the reagent, instrumental costs, and analytical parameters, but also to their negative influences on the environment. A good tool for such evaluation is life cycle assessment (LCA).

The introduction of LCA as a tool for examining the environmental burden of various analytical techniques may revolutionize the field of analytics. In rating analytical techniques according to their influence on the environment, one should apply a holistic approach that will include

- The output of raw materials
- The production of reagents and energy
- The transport of raw materials and final products
- The influence of a given reagent on the environment in the process of its usage
- The storage and recycling of chemical waste and out-of-date reagents

Green analytical chemistry is an essential component of green chemistry. The ongoing development of new solvent-free techniques is a good example of activities in this field. The following direct analytical techniques (in which a preparation step is not necessary) may be treated as typical examples of environmentally friendly procedures:

- X-ray fluorescence
- Surface acoustic waves (SAW) for the determination of VOCs
- Immunoassay.

In addition, other techniques, in which quantities of reagents and solvents per one analytical cycle are limited, are part of environmentally benign procedures. These include

- SPE
- ASE
- SPME
- Micro liquid-liquid extraction (MLLE) and other microextraction techniques
- Ultrasonic extraction
- SFE
- Extraction in an automated Soxhlet apparatus
- Vacuum distillation of VOCs
- Mass spectrometry with membrane interface (MIMS)

The extraction of pesticides from soil samples using accelerated solvent extraction is a good example of an analytical procedure fulfilling the rules of green chemistry. This procedure has many advantages over the classical techniques used for extracting analytes from complex matrices.

The main advantages regarding green chemistry are that

- Solvent use is reduced (by up to 95%).
- Analysis time is shortened (from 16 h to 10 min).
- Energy is saved (the extraction cell of an ASE instrument heats up to 100°C in 10 min in comparison to the 16 h required to heat a plate in a Soxhlet apparatus), thereby reducing exposure to solvents because of the shorter extraction time and the smaller amounts of solvents.
- Similar analytical characteristics for smaller samples (ASE).

This procedure can be treated as an alternative to the commonly used Soxhlet extraction.

Table 19.9 compares various solvent extraction techniques with regard to the duration and the amount of the solvent used (per 1 sample), and Figure 19.7 presents basic information on the main groups of solvent-free techniques in the context of sample preparation for analysis.

Thermal desorption enables the exchange of solvent into a more environmentally friendly stream of gas at the stage where analytes are being released into a suitable trap (sorption tube, denuder, and passive dosimeter). Figure 19.8 illustrates the basic mechanisms of the thermal desorber.

The ability to rapidly assess or monitor the disposition of environmental contaminants at purported or existing hazardous waste sites is an essential component of green chemistry. Soil samples, which represent approximately half the total number, are extracted with solvents, and then further separated using additional solvent to produce chemical-specific fractions. Each fraction is then analyzed using an appropriate method. The new technology proposed at the Tufts

TABLE 19.9
Comparison of Solvent Extraction Techniques

Extraction Technique	Amount of Solvent Used (mL)	Mean Duration of the Extraction Process
Soxhlet extraction	200–500	4–48 h
Automated Soxhlet extraction	50–100	1–4 h
US-enhanced solvent extraction	100–300	30 min–1 h
Microwave-enhanced solvent extraction	25–50	30 min–1 h
Accelerated solvent extraction	15–40	12–18 min
SFE	8–50	30 min–2 h

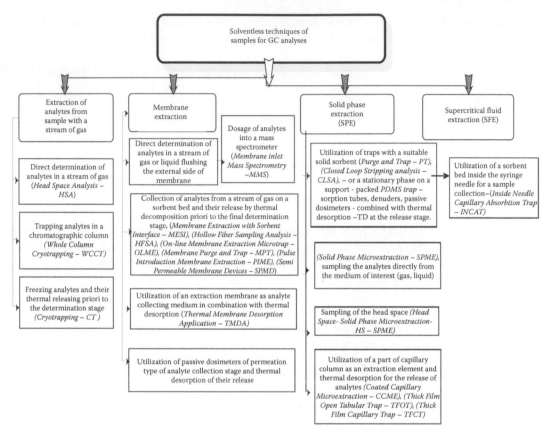

FIGURE 19.7 Classification of solvent-free techniques in the context of sample preparation for analysis.

University (MA, USA) aims at reducing or eliminating solvent usage during sample collection and analysis by collecting and detecting organic pollutants *in situ* without bringing the actual soil sample to the surface. A thermal extraction cone penetrometry probe coupled to an ultrafast gas chromatography-mass spectrometer (TECP-TDGC-MS) has been developed to collect and analyze subsurface organic contaminants *in situ*. The TECP is capable of heating the soil to 300°C, which is sufficient to collect volatile and semivolatile organics bound to the soil, in the presence of a soil–water content as high as 30%. Rather than using solvents to extract organics from soil, the TECP

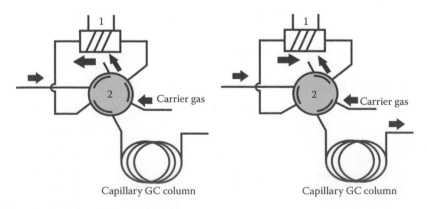

FIGURE 19.8 Diagram of a thermal desorber.

uses heat, then traps the hot vapor in a Peltier-cooled thermal desorption GC sample inlet for on-line analysis. In addition, this technology reduces solvent usage during the decontamination of sample collection probes and the apparatus used to homogenize samples. No other technology exists that is capable of thermally extracting organics as diverse as PCBs, explosives, or polyaromatic hydrocarbons (PAHs) under these conditions. When combined with the Ion Fingerprint Detection TM software, ultrafast TDGC-MS is capable of analyzing complex environmental samples in less than 5 min.

The next important challenge of green analytical chemistry is in-process monitoring. Developing and using in-line or on-line analyzers allow analytes to be determined in real time, in turn enabling disturbances to be detected already in the initial steps of a process. Such a means of analysis provides rapid information and the chance of an appropriate response—stopping the process or changing its operational parameters—and improves overall efficiency.

The application of green chemistry rules in designing greener analytical methods is key to diminishing the negative effects of analytical chemistry on the environment. The same ingenuity and innovatory skills, applied earlier to obtain excellent sensitivity, precision, and accuracy, are now being used to reduce or eliminate the application of hazardous substances in environmental analytics.

19.15   SUMMARY AND CONCLUSION

The chapter reviews the literature on the individual stages of environmental sample preparation up to the stage of final determinations with regard to analytes occurring in low concentrations. Special attention is paid to

- Challenges related to speciation analytics
- State-of-the-art techniques of extraction and analyte enrichment
- The use of US and microwave radiation at each stage of analytical procedures
- The implementation of principles associated with the concept of sustainable development in the procedures used in analytical laboratories

Analytical research is usually labor- and time-consuming as well as expensive, and thus its end result should provide authoritative and reliable information. Because of this, the following conditions must be fulfilled:

- A sample must be representative with regard to the examined object material.
- A sample must be properly prepared for analysis.
- A sample analysis should be carried out with suitable control and measurement instruments that cannot be treated as a blank.

REFERENCES

1. Konieczka, P. and J. Namieśnik. 2007. *Ocena i kontrola jakośći wyników pomiarów analitycznych.* Warszawa: WNT.
2. Thomas, O. and M.-F. Pouet. 2005. Wastewater quality monitoring: On-line/on-site measurement. *Handbook Environ. Chem.* 5: 245–272.
3. Saito, Y. 2003. Miniaturization of separation systems and its applications. *Chromatography* 24: 7–17.
4. Wang, J. and E.H. Hansen. 2003. On-line sample-pre-treatment schemes for trace-level determinations of metals by coupling floe injection or sequential injection with ICP-MS. *Trends Anal. Chem.* 22: 836–845.
5. Theodoridis, G. and I.M. Papadoyannis. 2001. Modern sample preparation methods in chemical analysis. *Microchim. Acta* 136: 199–200.
6. de Oliviera, E. 2003. Sample preparation for atomic spectroscopy: Evolution and future trends. *J. Braz. Chem. Soc.* 14: 174–182.

7. Ryba, S.A. and R.M. Burgess. 2002. Effects of sample preparation on the measurement of organic carbon, hydrogen, nitrogen, sulfur, and oxygen concentrations in marine sediments. *Chemosphere* 48: 139–147.

8. Eljarrat, E. and D. Barcelo. 2003. Priority lists for persistent organic pollutants and emerging contaminants based on their relative toxic potency in environmental samples. *Trends Anal. Chem.* 22: 655–665.

9. Nobrega, J.A., M.C. Santos, R.A. de Sousa, S. Cadore, R.M. Barnes, and M. Tatro. 2006. Sample preparation in alkaline media. *Spectrochim. Acta B* 61: 465–495.

10. Santos, F.J., J. Parera, and M.T. Galceran. 2006. Analysis of polychlorinated n-alkanes in environmental samples. *Anal. Bioanal. Chem.* 386: 837–857.

11. Munoz-Olivas, R. 2004. Screening analysis: An overview of methods applied to environmental, clinical and food analyses. *Trends Anal. Chem.* 23: 203–216.

12. Al-Gailani, B.R.M., G.M. Greenway, and T. McCreedy. 2007. Miniaturized flow-injection-analysis (μFIA) system with on-line chemiluminescence detection based on the luminol-hypochlorite reaction for the determination of ammonium in river water. *Int. J. Environ. Anal. Chem.* 87: 425–436.

13. Kemmei, T., S. Kodama, A. Yamamoto, Y. Inoue, and K. Hagakawa. 2007. Determination of low-level ethylenediaminetetraacetic acid in water samples by ion chromatography with ultraviolet detection. *Chromatographia* 65: 229–232.

14. Cheng, Ch.Y., Ch.Y. Wu, Ch.H. Wang, and W.H. Ding. 2006. Determination and distribution characteristics of degradation products of nonylphenol polyethoxylates in the rivers of Taiwan. *Chemosphere* 65: 2275–2281.

15. Al-Gailani, B.R.M., G.M. Greenway, and T. McCreedy. 2007. A miniaturized flow-injection analysis (μFIA) system with on-line chemiluminescence detection for the determination of iron in estuarine water. *Int. J. Environ. Anal. Chem.* 87: 637–646.

16. Thomaidis, N.S., A.S. Stasinakis, and T.D. Lekkas. 2007. A screening method for the determination of toluene extractable organotins in water samples by electrothermal atomic absorption spectrometry and rhenium as chemical modifier. *Appl. Organometal. Chem.* 21: 425–433.

17. Mohan, D. and Ch. U. Pittman. 2006. Activated carbons and low cost adsorbents for remediation of tri- and hexavalent chromium from water. *J. Hazard. Mater. B* 137: 762–811.

18. Katarina, R.K., N. Lenghor, and S. Motomizu. 2007. On-line preconcentration method for the determination of trace metals in water samples using a fully automated pretreatment system coupled with ICP-AES. *Anal. Sci.* 23: 343–350.

19. Dietz, Ch., J. Sanz, E. Sanz, R. Munos-Olivas, and C. Camara. 2007. Current perspectives in analyte extraction strategies for tin and arsenic speciation. *J. Chromatogr. A* 1153: 114–129.

20. Cram, S., C.A. Ponce de Leon, P. Fernandez, J. Sommer, H. Rivas, and L.M. Morales. 2006. Assessment of trace elements and organic pollutants from a marine oil complex into the coral reef system of Cayo Arcas, Mexico. *Environ. Monit. Assess.* 121: 127–149

21. Flores, G., M. Herraiz, G.P. Blanch, and M.L.R. del Castillo. 2007. Polydimethylsiloxane as a packing material in a programmed temperature vaporizer to introduce large-volume samples in capillary gas chromatography. *J. Chromatogr. Sci.* 45: 33–37.

22. Allabashi, R., M. Arkas, G. Hormann, and D. Tsiourvas. 2007. Removal of some organic pollutants in water employing ceramic membranes impregnated with cross-linked silylated dendritic and cyclodextrin polymers. *Water Res.* 41: 476–486.

23. Dimou, A.D., T.M. Sakellarides, F.V. Vosniakos, N. Giannoulis, E. Leneti, and T. Albanis. 2006. Determination of phenolic compounds in the marine environment of Thermaikos Gulf, Northern Greece. *Int. J. Environ. Anal. Chem.* 86: 119–130.

24. Filby, M.H. and J.W. Steed. 2006. A modular approach to organic, coordination complex and polymer based podand hosts for anions. *Coor. Chem. Rev.* 250: 3200–3218.

25. Vanerkova, D., P. Jandera, and J. Hrabica. 2007. Behaviour of sulphonated azodyes in ion-pairing reversed-phase high-performance liquid chromatography. *J. Chromatogr. A* 1143: 112–120.

26. Aparicio, I., J.L. Santos, and E. Alonso. 2007. Simultaneous sonication-assisted extraction, and determination by gas chromatography–mass spectrometry, of di-(2-ethylhexyl)phthalate, nonylphenol, nonylphenol ethoxylates and polychlorinated biphenyls in sludge from wastewater treatment plants. *Anal. Chim. Acta* 584: 455–461.

27. Leong, K.H., L.L. Benjamin Tan, and A.M. Mustafa. 2007. Contamination levels of selected organo-chlorine and organophosphate pesticides in the Selangor River, Malaysia between 2002 and 2003. *Chemosphere* 66: 1153–1159.

28. Smith, F.E. and E.A. Arsenault. 1996. Microwave-assisted sample preparation in analytical chemistry. *Talanta* 43: 1207–1268.

29. Stefaniak, A.B., C.A. Brink, R.M. Dickerson, G.A. Day, M.J. Brisson, M.D. Hoover, and R.C. Scripsick. 2007. A theoretical framework for evaluating analytical digestion methods for poorly soluble particulate beryllium. *Anal. Bioanal. Chem.* 387: 2411–2417.

30. Fontanals, N., R.M. Marcé, and F. Borrall. 2007. New materials in sorptive extraction techniques for polar compounds. *J. Chromatogr. A* 1152: 14–31.

31. Hernansez-Borges J., T.M. Borges-Miquel, M.A. Rodriguez-Delgado, and A. Cifuentes. 2007. Sample treatments prior to capillary electrophoresis–mass spectrometry. *J. Chromatogr. A* 1153: 214–226.

32. Namieśnik, J. and T. Górecki. 2001. Quality of analytical results. *Rev. Roum. Chim.* 46: 953–962.

33. Batley, E.G., S.C. Apte, and J. L. Stauber. 2004. Speciation and bioavailability of trace metals in water: Progress since 1982. *Aust. J. Chem.* 57: 903–919.

34. Dash, K., S. Thangavel, N.V. Krishnamurthy, S.V. Rao, D. Karunasagar, and J. Arunachalam. 2005. Ultrasound-assisted analyte extraction for the determination of sulfate and elemental sulfur in zinc sulfide by different liquid chromatography techniques. *Analyst* 130: 498–501.

35. Rasmussen, K.E. and S. Pedersen-Bjergaard. 2004. Development in hollow fibre-based, liquid-phase microextraction. *Trends Anal. Chem.* 23: 1–10.

36. Kalembkiewicz, J. and E. Soco. 2005. Ekstrakcja sekwencyjna metali z próbek środowiskowych. *Wiad. Chem.* 59: 698–710.

37. Vaisainen, A. and R. Suontomo. 2002. Comparison of ultrasound-assisted extraction, microwave-assisted acid leaching and reflux for the determination of arsenic, cadmium and copper in contaminated soil samples by electrothermal atomic absorption spectrometry. *J. Anal. At. Spectrom.* 17: 739–742.

38. Doig, L.E. and K. Liber. 2007. Nickel speciation in the presence of different sources and fractions of dissolved organic matter. *Ecotoxicol. Environ. Saf.* 66: 169–177.

39. Vermillon, B.R. and R.J.M. Hudson. 2007. Thiourea catalysis of MeHg ligand exchange between natural dissolved organic matter and a thiol-functionalized resin: A novel method of matrix removal and MeHg preconcentration for ultratrace Hg speciation analysis in freshwaters. *Anal. Bioanal. Chem.* 388: 341–352.

40. Simster, C., F. Caron, and R. Gedye. 2004. Determination of the thermal degradation rate of polystyrene-divinyl benzene ion exchange resins in ultra-pure water at ambient and service temperature. *J. Radioanal. Nucl. Chem.* 261: 523–531.

41. Toulhoat, P. 2005. Defis actuels et a venir en matiere d'analyse de traces et d'ultra-traces. *Oil Gas Sci. Technol.* 60: 967–977.

42. Gwozdz, R. and F. Grass. 2004. Contamination by human fingers: The Midas touch. *J. Radioanal. Nucl. Chem.* 259: 173–176.

43. Gallagher, P.A., C.A. Schegel, A. Parks, B.M. Gamble, L. Wymer, and J.T. Creed. 2004. Preservation of As(III) and As(V) in drinking water supply samples from across the United States using EDTA and acetic acid as a means of minimizing iron-arsenic coprecipitation. *Environ. Sci. Technol.* 38: 2919–2927.

44. Ganzler, K., A. Salgo, and K. Valko. 1986. Microwave extraction. A novel sample preparation method for chromatography. *J. Chromatogr.* 371: 299–306.

45. Feldmann, J. 2003. Sample preparation for the analysis of volatile metal species. *Compr. Anal. Chem.* XLI: 1211–1232.

46. Feldmann, J. 2005. What can the different current-detection methods offer for element speciation? *Trends Anal. Chem.* 24: 228–242.

47. Moretto, L.M., N.S. Bloom, P. Scopece, and P. Ugo. 2003. Application of ultra clean sampling and analysis methods for the speciation of mercury in the Venice lagoon (Italy). *J. Phys. IV France* 107: 887–890.

48. Moore, M.R., W. Vetter, C. Gaus, G.R. Shaw, and J.F. Muller. 2002. Trace organic compounds in the marine environment. *Mar. Pollut. Bull.* 45: 62–68.

49. Lorrain, A., N. Savoye, L. Chauvaud, Y-M. Paulet, and N. Naulet. 2003. Decarbonation and preservation method for the analysis of organic C and N contents and stable isotope ratios of low-carbonated suspended particulate material. *Anal. Chim. Acta* 491: 125–133.

50. van Look, G. and V.R. Meyer. 2002. The purity of laboratory chemical with regard to measurement uncertainty. *Analyst* 127: 825–829.

51. Amouroux, D., Ch. Pecheyran, and O.F.X. Donard. 2000. Formation of volatile selenium species in synthetic seawater under light and dark experimental conditions. *Appl. Organomet. Chem.* 14: 236–244.

52. Skjevrak, I., A. Due, K.O. Gjerstad, and H. Herikstad. 2003. Volatile organic components migrating from plastic pipes (HDPE, PEX and PVC) into drinking water. *Water Res.* 37: 1912–1920.

53. Niemela, M., H. Kola, P. Peramaki, J. Piispanen, and J. Poikolainen. 2005. Comparison of microwave-assisted digestion methods and selection of internal standards for the determination of Rh, Pd and Pt in dust samples by ICP-MS. *Microchim. Acta* 150: 211–217.

54. Jakubowska, N., Ż. Polkowska, J. Namiesńik, and A. Przyjazny. 2005. Analytical application and environmental liquid sample preparation. *Crit. Rev. Anal. Chem.* 35: 217–235.

55. Prasad, S.S. 1994. Trends in quality assurance in the characterization of chemical contaminants in the environment. *Trends Anal. Chem.* 13: 157–168.

56. Basheer, Ch., M. Vetrichelvan, S. Valiyaveettil, and H.K. Lee. 2007. On-site polymer-coated hollow fiber membrane microextraction and gas chromatography-mass spectrometry of polychlorinated biphenyls and polybrominated diphenyl ethers. *J. Chromatogr. A* 1139: 157–164.

57. Delgado, B., V. Pino, J.H. Ayala, V. Gonzalez, and A.M. Alfonso. 2004. Nonionic surfactant mixtures: A new cloud-point extraction approach for the determination of PAHs in seawater using HPLC with fluorimetric detection. *Anal. Chim. Acta* 518: 165–172.

58. Padron Sanz, C., Z. Sosa Ferrera, and J.J. Santana Rodriguez. 2002. Extraction and preconcentration of polychlorinated dibenzo-p-dioxins using the cloud-point methodology. Application to their determination in water samples by high-performance liquid chromatography. *Anal. Chim. Acta* 470: 205–214.

59. Giokas, D.L., V.A. Sakkas, T.A. Albanis, and D.A. Lampropoulou. 2005. Determination of UV-filter residues in bathing waters by liquid chromatography UV-diode array and gas chromatography-mass spectrometry after micelle mediated extraction-solvent back extraction. *J. Chromatogr. A* 1077: 19–27.

60. Yuan, Ch. G., G.B. Jiang, B. He, and J.F. Liu. 2005. Preconcentration and determination of tin in water samples by using cloud point extraction and graphite furnace atomic absorption spectrometry. *Microchim. Acta* 150: 329–334.

61. Carabias-Martinez, R., E. Rodriguez-Gonzalo, J. Dominiquez-Alvaro, C. Garcia Pinto, and J. Hernandez-Mendez. 2003. Prediction of the behaviour of organic pollutants using cloud point extraction. *J. Chromatogr. A* 1005: 23–34.

62. Giokas, D.L., J. Antelo, E.K. Paleologos, F. Arce, and M.I. Karayannis. 2002. Copper fractionation with dissolved organic matter in natural waters and wastewater-a mixed micelle mediated methodology (cloud point extraction) employing flame atomic absorption spectrometry. *J. Environ. Monit.* 4: 505–510.

63. Farajzadeh, M.A. and M.R. Fallahi. 2006. Simultaneous cloud-point extraction of nine cations from water samples and their determination by flame atomic absorption spectrometry. *Anal. Sci.* 22: 635–640.

64. Halko, R., C. Padron Sanz, Z. Sosa Ferrera, and J.J. Santana Rodriquez. 2004. Determination of benzimidazole fungicides by HPLC with fluorescence detection after micellar extraction. *Chromatographia* 60: 151–156.

65. Wanekaya, A.K., S. Myung, and O.A. Sadik. 2002. Pressure assisted chelating extraction: A novel technique for digesting metals in solid matrices. *Analyst* 127: 1272–1276.

66. Castillo, M., M.C. Alonso, J. Riu, and D. Barcelo. 1999. Identification of polar, ionic, and highly water soluble organic pollutants in untreated industrial wastewaters. *Environ. Sci. Technol.* 33: 1300–1306.

67. Alonso, M.C., E. Pocurull, R.M. Marcé, F. Borrull, and D. Barcelo. 2002. Monitoring of aromatic monosulfonic acids in coastal waters by ion-pair liquid chromatography followed by electrospray-mass spectrometric detection. *Environ. Toxicol. Chem.* 21: 2059–2066.

68. Loos, R., G. Hanke, and S. Eisenreich. 2003. Multi-component analysis of polar water pollutants using sequential solid-phase extraction followed by LC-ESI-MS. *J. Environ. Monit.* 5: 384–394.

69. Kristenson, E.M., L. Ramos, and U.A.Th. Brinkman. 2006. Recent advances in matrix solid-phase dispersion. *Trends Anal. Chem.* 25: 96–111.

70. Michel, M. and B. Buszewski. 2004. Optimization of a matrix solid-phase dispersion method for the determination analysis of carbendazin residue in plant material. *J. Chromatogr. B* 800: 309–314.

71. Li, Z.Y., Z. Ch. Zhang, Q.L. Zhou, R.Y. Gao, and Q.S. Wang. 2002. Fast and precise determination of phenthoate and its enantiomeric ratio in soil by the matrix solid-phase dispersion method and liquid chromatography. *J. Chromatogr. A* 977: 17–45.

72. Curren, M.S.S. and J.W. King. 2001. Ethanol-modified subcritical water extraction combined with solid-phase microextraction for determining atrazine in beef kidney. *J. Agric. Food. Chem.* 49: 2175–2180.

73. Bogialli, S., R. Curini, A. Di Corcia, M. Nazzari, and D. Tamburo. 2004. A simple and rapid assay for analyzing residues of carbamate insecticides in vegetables and fruits: Hot water extraction followed by liquid chromatography-mass spectrometry. *Agric. Food Chem.* S2: 665–671.

74. Kristenson, E.M., E.G.J. Haverkate, C.J. Slooten, L. Ramos, R.J.J. Vreuls, and U.A.Th. Brinkman. 2001. Miniaturized automated matrix solid-phase dispersion extraction of pesticides in fruit followed by gas chromatographic–mass spectrometric analysis. *J. Chromatogr. A* 917: 277–286.

75. Andersson, T., T. Pihtsalmi, K. Hartonen, T. Hyotylainem, and M.L. Riekkola. 2003. Effect of extraction vessel geometry and flow homogeneity on recoveries of polycyclic aromatic hydrocarbons in pressurised hot water extraction. *Anal. Bioanal. Chem.* 376: 1081–1088.

76. Ramos, L., E.M. Kristenson, and U.A.Th. Brinkman. 2002. Current use of pressurised liquid extraction and subcritical water extraction in environmental analysis. *J. Chromatogr. A* 975: 3–29.

77. Boch, K., M. Schuster, G. Risse, and M. Schwarzer. 2002. Microwave-assisted digestion procedure for the determination of palladium in road dust. *Anal. Chim. Acta* 459: 257–265.

78. Fernandez-Perez, V., M.M. Jimenez-Carmona, and M.D. Luque de Castro. 2001. Continuous liquid–liquid extraction using modified subcritical water for the demetalisation of used industrial oils. *Anal. Chim. Acta* 433: 47–52.

79. Perraudin, E., H. Budzinski, and E. Villenave. 2005. Analysis of polycyclic aromatic hydrocarbons adsorbed on particles of atmospheric interest using pressurised fluid extraction. *Anal. Bioanal. Chem.* 383: 122–131.

80. Rudel, H., W. Bohmer, and C. Schroter-Kermani. 2006. Retrospective monitoring of synthetic musk compounds in aquatic biota from German rivers and coastal areas. *J. Environ. Monit.* 8: 812–823.

81. Wang, X., X. Piao, J. Chen, J. Hu, F. Xu, and S. Tao. 2006. Organochlorine pesticides in soil profiles from Tianjin, China. *Chemosphere* 64: 1514–1520.

82. Lundstedt, S., P. Haglung, and L. Oberg. 2006. Simultaneous extraction and fractionation of polycyclic aromatic hydrocarbons and their oxygenated derivatives in soil using selective pressurized liquid extraction. *Anal. Chem.* 78: 2993–3000.

83. Lara-Martin, P.A., A. Gomez-Parra, and E. Gonzalez-Mazo. 2006. Simultaneous extraction and determination of anionic surfactants in waters and sediments. *J. Chromatogr. A* 1114: 205–210.

84. Prycek, J., M. Ciganek, and Z. Simek. 2004. Development of an analytical method for polycyclic aromatic hydrocarbons and their derivatives. *J. Chromatogr. A* 1030: 103–107.

83. Pardasani, D., M. Palit, A.K. Gupta, P.K. Kanaujia, K. Sikhar, and D.K. Dubey. 2006. Microemulsion mediated *in situ* derivatization-extraction and gas chromatography-mass spectrometric analysis of alkylphosphonic acids. *J. Chromatogr. A* 1108: 166–175.

86. Topakas, E., H. Stamatis, P. Biely, D. Kekos, B.J. Macris, and P. Christakopoulos. 2003. Purification and characterization of a feruloyl esterase from *Fusarium oxysporum* catalyzing esterification of phenolic acids in ternary water–organic solvent mixtures. *J. Biotechnol.* 102: 33–44.

87. Liang L., S. Mo, P. Zhang, et al. 2006. Selenium speciation by high-performance anion-exchange chromatography-post-column UV irradiation coupled with atomic fluorescence spectrometry. *J. Chromatogr. A* 1118: 139–143.

88. Komjarova, I. and R. Biust. 2006. Comparison of liquid–liquid extraction, solid-phase extraction and co-precipitation preconcentration methods for the determination of cadmium, copper, nickel, lead and zinc in seawater. *Anal. Chim. Acta* 576: 221–228.

89. Le-Clech, P., E.-K. Lee, and V. Chen. 2006. Hybrid photocatalysis/membrane treatment for surface waters containing low concentrations of natural organic matters. *Water Res.* 40: 323–330.

90. Matusiewicz, H. 2003. Wet digestion methods. In: Z. Mester and R. Sturgeon (eds), *Sample Preparation for Trace Element Analysis*, .pp. 193–232. Amsterdam: Elsevier Science Publishers.

91. Pinho, J., J. Canario, R. Cesario, and C. Vale. 2005. A rapid acid digestion method with ICP-MS detection for the determination of selenium in dry sediments. *Anal. Chim. Acta* 551: 207–212.

92. Loska, K. and D. Wiechuła. 2006. Comparison of sample digestion procedures for the determination of arsenic in bottom sediment using hydride generation AAS. *Microchim. Acta* 154: 235–240.

93. Sun, Y.C., P.II. Chi, and M.Y. Shiue. 2001. Comparison of different digestion methods for total decomposition of siliceous and organic environmental samples. *Anal. Sci.* 17: 1395–1399.

94. de Godoi Pereira, M. and M.A.Z. Arruda. 2003. Trends in preconcentration procedures for metal determination using atomic spectrometry techniques. *Microchim. Acta* 141: 115–131.

95. Bayona, J.M. 2000. Supercritical fluid extraction in speciation studies. *Trends Anal. Chem.* 19: 107–112.

96. Szolar, O.H., H. Rost, D. Hermann, M. Hasinger, R. Braun, and A.P. Loibner. 2004. Sequential supercritical fluid extraction (SSFE) for estimating the availability of high molecular weight polycyclic aromatic hydrocarbons in historically polluted soils. *J. Environ. Qual.* 33: 80–88.

97. Bermejo, M.D. and M.J. Cocero. 2006. Supercritical water oxidation: A technical review. *AIChE J.* 11: 3933–3951.

98. Dewil, R., J. Baeyens, and R. Goutvrind. 2006. Ultrasonic treatment of waste activated sludge. *Environ. Progress* 25: 121–128.

99. Casadonte, D.J., Jr., M. Flores, and C. Petrier. 2005. The use of pulsed ultrasound technology to improve environmental remediation: A comparative study. *Environ. Technol.* 26: 1411–1417.

100. Ryno, M., L. Rantanen, E. Papaioannou, A.G. Konstandopoulos, T. Koskentalo, and K. Savela. 2006. Comparison of pressurized fluid extraction, Soxhlet extraction and sonication for the determination of polycyclic aromatic hydrocarbons in urban air and diesel exhaust particulate matter. *J. Environ. Monit.* 8: 488–493.

101. Goncalves, C., A. Dimou, V. Sakkas, M.F. Alpendurada, and T.A. Albanis. 2006. Photolytic degradation of quinalphos in natural waters and soil matrices under simulated solar irradiation. *Chemosphere* 64: 1375–1382.

102. Ferguson, P. L., Ch.R. Iden, and B.J. Brownawell. 2001. Analysis of nonylphenol and nonylphenol ethoxylates in environmental samples by mixed-mode high-performance liquid chromatography-electrospray mass spectrometry. *J. Chromatogr. A* 938: 79–92.

103. Mzoughi, N., F. Hellal, M. Dacharaui, et al. 2002. Methodologie de l'extraction des hydrocarbures aromatiques polycycliques. Application a des sediments de la lagune de Bizerte (Tunise). *C. R. Geosci.* 334: 893–901.

104. Krasnodębska-Ostręga, B., M. Kaczorowska, and J. Golimowski. 2006. Use of ultrasound-assisted extraction for the evaluation of element mobility in bottom sediment collected at mining and smelting Pb–Zn ores area in Poland. *Microchim. Acta* 05–907: 1–5.

105. Lavilla, I., P. Vilas, J. Millos, and C. Bendicho. 2006. Development of an ultrasound-assisted extraction method for biomonitoring of vanadium and nickel in the coastal environment under the influence of the *Prestige* fuel spill (North east Atlantic Ocean). *Anal. Chim. Acta* 577: 119–125.

106. Yebra, M.C., S. Cancela, and A. Moreno-Cid. 2005. Continuous ultrasound-assisted extraction of cadmium from vegetable samples with on-line preconcentration coupled to a flow injection-flame atomic spectrometric system. *Int. J. Environ. Anal. Chem.* 85: 305–313.

107. Sanz, E., R. Munos-Olivas, and C. Camara. 2005. Evaluation of a focused sonication probe for arsenic speciation in environmental and biological samples. *J. Chromatogr. A* 1097: 1–8.

108. Richter, P., M. Jimenez, R. Salazar, and A. Marican. 2006. Ultrasound-assisted pressurized solvent extraction for aliphatic and polycyclic aromatic hydrocarbons from soils. *J. Chromatogr. A* 1132: 15–20.

109. Gil, S., I. Lavilla, and C. Bendicho. 2006. Ultrasound-promoted cold vapor generation in the presence of formic acid for determination of mercury by atomic absorption spectrometry. *Anal. Chem.* 78: 6260–6264.

110. Molina, R., F. Martinem, J.A. Melero, D.H. Bremner, and A.G. Chakinala. 2006. Mineralization of phenol by a heterogeneous ultrasound/Fe-SBA-15/H_2O_2 process: Multivariate study by factorial design of experiments. *Appl. Catal. B: Environ.* 66: 198–207.

111. Väisänen, A. and A. Ilander. 2006. Optimization of operating conditions of axially and radially viewed plasmas for the determination of trace element concentrations form ultrasound-assisted digests of soil samples contaminated by lead pellets. *Anal. Chim. Acta* 570: 93–100.

112. Maruyama, H., H. Seki, Y. Matsukawa, A. Suzuki, and N. Inoue. 2006. Removal of bisphenol A and diethyl phthalate from aqueous phases by ultrasonic atomization. *Ind. Eng. Chem. Res.* 45: 6383–6386.

113. Entezari, M.H. and T.R. Bastami. 2006. Sono-sorption as a new method for the removal of lead ion from aqueous solution. *J. Hazard. Mater. B* 137: 959–964.

114. Hinze, W.L. 1992. Report UNC-WRRI-92-269. Water Research Institute of University of North Carolina, Winston-Salem, NC.

115. Bagheri, H. and A. Salemi. 2006. Headspace solvent microextraction as a simple and highly sensitive sample pretreatment technique for ultra trace determination of geosmin in aquatic media. *J. Sep. Sci.* 29: 57–65.

116. Bagheri, H., A. Saber, and S.R. Mousavi. 2004. Immersed solvent microextraction of phenol and chlorophenols from water samples followed by gas chromatography-mass spectrometry. *J. Chromatogr. A* 1046: 27–33.

117. Lopez-Blanco, M.C., S. Blanco-Cid, B. Cancho-Grande, and J. Simal-Gandara. 2003. Application of single-drop microextraction and comparison with solid-phase microextraction and solid-phase extraction for the determination of α- and β-endosulfan in water samples by gas chromatography-electron-capture detection. *J. Chromatogr. A* 984: 245–252.

118. Yazdi, A.S. and A. Assai. 2004. Determination of trace of methyl *tert*-butyl ether in water using liquid drop headspace sampling and GC. *Chromatographia* 60: 699–702.

119. Vidal, L., A. Canals, N. Kalogerakis, and E. Psillakis. 2005. Headspace single-drop microextraction for the analysis of chlorobenzenes in water samples. *J. Chromatogr. A* 1089: 25–30.

120. Tor, A. and M.E. Aydin. 2006. Application of liquid-phase microextraction to the analysis of trihalomethanes in mater. *Anal. Chim. Acta* 575: 138–143.

121. Tor, A. 2006. Determination of chlorobenzenes in water by drop-based liquid-phase microextraction and gas chromatography-electron capture detection. *J. Chromatogr. A* 1125: 129–132.

122. Psillakis, E. and N. Kalogerakis. 2003. Developments in liquid-phase microextraction. *Trends Anal. Chem.* 22: 565–574.

123. Yazdi, A.S. and Z. Es'haghi. 2005. Liquid–liquid–liquid phase microextraction of aromatic amines in water using crown ethers by high-performance liquid chromatography with monolithic column. *Talanta* 66: 664–669.

124. Shen, S., Z. Chang, and H. Liu. 2006. Three-liquid-phase extraction systems for separation of phenol and p-nitrophenol from wastewater. *Sep. Purif. Technol.* 49: 217–222.
125. Bergqvist, P.-A., L. Augulyte, and V. Jurjoniene. 2006. PAH and PCB removal efficiencies in Umea (Sweden) and Siauliai (Lithuania) municipal wastewater treatment plants. *Water Air Soil Pollut.* 175: 291–303.
126. Spalding, B.P. and S.C. Brooks. 2005. Permeable environmental leaching capsules (PELCAPs) for *in situ* evaluation of contaminant immobilization in soil. *Environ. Sci. Technol.* 39: 8912–8918.
127. Rosa, A.H., I.C. Bellin, D. Goveia, et al. 2006. Development of a new analytical approach based on cellulose membrane and chelator for differentiation of labile and inert metal species in aquatic systems. *Anal. Chim. Acta* 567: 152–159.
128. Vora-Adisak, N. and P. Varanusupakul. 2006. A simple supported liquid hollow fiber membrane microextraction for sample preparation of trihalomethanes in water samples. *J. Chromatogr. A* 1121: 236–241.
129. Rios, A., A. Escarpa, M.C. Gonzalez, and A.G. Crevillen. 2006. Challenges of analytical microsystems. *Trends Anal. Chem.* 25: 467–479.
130. Yazdi, A.S. and Z. Es'haghi. 2005. Two-step hollow fiber-based, liquid-phase microextraction combined with high-performance liquid chromatography: A new approach to determination of aromatic amines in water. *J. Chromatogr. A* 1082: 136–142.
131. Kuosmanen, K., T. Hyotylainen, K. Hartonen, J.A. Jonsson, and M.L. Riekkola. 2003. Analysis of PAH compounds in soil with on-line coupled pressurised hot water extraction-microporous membrane liquid-liquid extraction-gas chromatography. *Anal. Bioanal. Chem.* 375: 389–399.
132. Einsle, T., H. Paschke, K. Bruns, S. Schrader, P. Popp, and M. Moeder. 2006. Membrane-assisted liquid–liquid extraction coupled with gas chromatography-mass spectrometry for determination of selected polycyclic musk compounds and drugs in water samples. *J. Chromatogr. A* 1124: 196–204.
133. Zhou, Q., G. Jiang, J. Liu, and Y. Cai. 2004. Combination of microporous membrane liquid–liquid extraction and capillary electrophoresis for the analysis of aromatic amines in water samples. *Anal. Chim. Acta* 509: 55–62.
134. Thompson, A.J., A.S. Creba, R.M. Ferguson, E.T. Krogh, and C.G. Gill. 2006. A coaxially heated membrane introduction mass spectrometry interface for the rapid and sensitive on-line measurement of volatile and semi-volatile organic contaminants in air and water at parts-per-trillion levels. *Rapid Commun. Mass Spectrom.* 20: 2000–2008.
135. Tomaszewski, L., J. Buffle, and J. Galceran. 2003. Theoretical and analytical characterization of a flow-through permeation liquid membrane with controlled flux for metal speciation measurements. *Anal. Chem.* 75: 893–900.
136. Parthasarathy, N., M. Pelletier, and J. Buffle. 2004. Permeation liquid membrane for trace metal speciation in natural waters. Transport of liposoluble Cu (II) complexes. *J. Chromatogr. A* 1025: 33–40.
137. Weinstein, S.E. and S.B. Moran. 2004. Distribution of size-fractionated particulate trace metals collected by bottles and in-situ pumps in the Gulf of Maine-Scotian Shelf and Labrador Sea. *Mar. Chem.* 87: 121–135.
138. Van der Bruggen, B. and L. Braeken. 2006. The challenge of zero discharge: From water balance to regeneration. *Desalination* 188: 177–183.
139. Kloepfer, A., J.B. Quintana, and T. Reemtsma. 2005. Operational options to reduce matrix effects in liquid chromatography-electrospray ionization-mass spectrometry analysis of aqueous environmental samples. *J. Chromatogr. A* 1067: 153–160.
140. Kumar, M., S. S. Adham, and W. R. Pearce. 2006. Investigation of seawater reverse osmosis fouling and its relationship to pretreatment type. *Environ. Sci. Technol.* 40: 2037–2044.
141. Kiene, R.P. and D. Slezak. 2006. Low dissolved DMSP concentrations in seawater revealed by small-volume gravity filtration and dialysis sampling. *Limnol. Oceanogr. Meth.* 4: 80–95.
142. Simon, J., A. Kirchoff, and O. Gultzow. 2002. Advanced sample pretreatment for the monitoring of polycyclic aromatic hydrocarbons and extractable organic halogens in waste water. Flow based procedures with chromatomembrane cells. *Talanta* 58: 1335–1341.
143. Liu, Y., B. Wen, and X.-Q. Shan. 2006. Determination of pentachlorophenol in wastewater irrigated soils and incubated earthworms. *Talanta* 69: 1254–1259.
144. Akyuz, M. and S. Ata. 2006. Simultaneous determination of aliphatic and aromatic amines in water and sediment samples by ion-pair extraction and gas chromatography–mass spectrometry. *J. Chromatogr. A* 1129: 88–94.
145. Diaz, A., F. Ventura, and M.T. Galceran. 2006. Analysis of odorous trichlorobromophenols in water by in-sample derivatization/solid-phase microextraction GC/MS. *Anal. Bioanal. Chem.* 384: 1447–1461.

146. Popp, P., C. Bauer, B. Hauser, P. Keil, and L. Wennrich. 2003. Extraction of polycyclic aromatic hydro-carbons and organochlorine compounds from water: A comparison between solid-phase microextraction and stir bar sorptive extraction. *J. Sep. Sci.* 26: 961–967.

147. Heringa, M.B. and J.L.M. Hermens. 2003. Measurement of free concentrations using negligible depletion-solid phase microextraction (nd-SPME). *Trends Anal. Chem.* 22: 575–587.

148. Ouyang, G., Y. Chen, and J. Pawliszyn. 2005. Time-weighted average water sampling with a solid-phase microextraction device. *Anal. Chem.* 77: 7319–7325.

149. Zhao, Y.-Y., S. Hrudey, and X.-F. Li. 2006. Determination of microcystins in water using integrated solid-phase microextraction with microbore high-performance liquid chromatography-electrospray quadruple time-of-flight mass spectrometry. *J. Chromatogr. Sci.* 44: 359–365.

150. Salado-Petinal, C., M. Garcia-Chao, M. Llompart, C. Garcia-Jares, and R. Cela. 2006. Headspace solid-phase microextraction gas chromatography tandem mass spectrometry for the determination of brominated flame retardants in environmental solid samples. *Anal. Bioanal. Chem.* 385: 637–644.

151. Sakamoto, A., T. Niki, and Y.W. Watanabe. 2006. Establishment of long-term preservation for dimethyl sulfide by the solid-phase microextraction method. *Anal. Chem.* 78: 4593–4597.

152. Lam, K.-H., H.-Y. Wai, K.M.Y. Leung, et al. 2006. A study of the portioning behavior of Irgarol-1051 and its transformation products. *Chemosphere* 64: 1177–1184.

153. Kayali, N., F.G. Tamayo, and L.M. Polo-Diez. 2006. Determination of diethylhexyl phthalate in water by solid phase microextraction coupled to high performance liquid chromatography. *Talanta* 69: 1095–1099.

154. Tölgyessy, P. and J. Hrivnak. 2006. Analysis of volatiles in water using headspace solid-phase micro-column extraction. *J. Chromatogr. A* 1127: 295–297.

155. Ochiai, N., K. Sasamoto, H. Kanda, and S. Nakamura. 2006. Fast screening of pesticide multiresidues in aqueous samples by dual stir bar sorptive extraction-thermal desorption-low thermal mass gas chroma-tography-mass spectrometry. *J. Chromatogr. A* 1130: 83–90.

156. Llorca-Porcel, J., G. Martinez-Sanchez, B. Alvarez, M.A. Cobollo, and I. Valor. 2006. Analysis of nine polybrominated diphenyl ethers in water samples by means of stir bar sorptive extraction-thermal desorption-gas chromatography-mass spectrometry. *Anal. Chim. Acta* 569: 113–118.

157. Serodio, P. and J.M.F. Nogueira. 2005. Development of a stir-bar-sorptive extraction-liquid desorption-large-volume injection capillary gas chromatographic-mass spectrometric method for pyrethroid pesti-cides in water samples. *Anal. Bioanal. Chem.* 382: 1141–1151.

158. Garcia-Falcon, M.S., B. Rancho-Granda, and J. Simal-Gandara. 2004. Stirring bar sorptive extraction in the determination of PAHs in drinking waters. *Water Res.* 38: 1679–1684.

159. Jayaraman, S., R.J. Pruell, and R. Mc Kinney. 2001. Extraction of organic contaminants from marine sediments and tissues using microwave energy. *Chemosphere* 44: 181–191.

160. Kawaguchi, M., R. Ito, N. Endo, et al. 2006. Stir bar sorptive extraction and thermal desorption-gas chromatography-mass spectrometry for trace analysis of benzophenone and its derivatives in water sample. *Anal. Chim. Acta* 557: 272–277.

161. Leon, V.M., J. Llorca-Porcel, B. Alvarez, M.A. Cobollo, S. Munoz, and J. Valor. 2006. Analysis of 35 priority semivolatile compounds in water by stir bar sorptive extraction-thermal desorption-gas chroma-tography-mass spectrometry Part II: Method validation. *Anal. Chim. Acta* 558: 261–266.

162. Saito, Y., M. Nojiri, M. Imaizumi, et al. 2002. Polymer-coated synthetic fibers designed for miniaturized sample preparation process. *J. Chromatogr. A* 975: 105–112.

163. Alonso, M.C. and D. Barcelo. 2002. Stability study and determination of benzene- and naphthalenesul-fonates following an on-line solid-phase extraction method using the new programmable field extraction system. *Analyst* 127: 472–279.

164. van Zoonen, P., H.A. van't Klooster, R. Hoogerbrugge, S.M. Gort, and H.J. van de Weil. 1998. Validation of analytical methods and laboratory procedures for chemical measurements. *Arch. Hig. Rada. Toxicol.* 19: 355–370.

165. Long, X., M. Miro, R. Jensen, and E.H. Hansen. 2006. Highly selective micro-sequential injection lab-on-valve (μSI-LOV) method for the determination of ultra-trace concentrations of nickel in saline matrices using detection by electrothermal atomic absorption spectrometry. *Anal. Bioanal. Chem.* 386: 739–748.

166. Marin, J.M., J.V. Sancho, O.J. Pozo, F.J. Lopez, and F. Hernandez. 2006. Quantification and confirmation of anionic, cationic and neutral pesticides and transformation products in water by on-line solid phase extraction-liquid chromatography-tandem mass spectrometry. *J. Chromatogr. A* 1133: 204–214.

167. Gfrerer, M., B.M. Gawlik, and E. Lankmayr. 2004. Validation of a fluidised-bed extraction method for solid materials for the determination of PAHs and PCBs using certified reference materials. *Anal. Chim. Acta* 527: 53–60.

168. Rubio, S. and D. Perez-Bendito. 2003. Supramolecular assemblies for extracting organic compounds. *Trends Anal. Chem.* 22: 470–485.

169. Sanchez-Barragan, I., J.M. Costa-Fernandez, R. Pereiro, et al. 2005. Molecularly imprinted polymers based on iodinated monomers for selective room-temperature phosphorescence optosensing of fluoranthene in water. *Anal. Chem.* 77: 7005–7011.

170. Zhu, X., J. Yang, Q. Su, J. Cai, and Y. Gao. 2005. Molecularly imprinted polymer for monocrotophos and its binding characteristics for organophosphorus pesticides. *Ann. Chim.* 95: 877–885.

171. Zhu, Q. Z., P. Degelmann, R. Niessner, and D. Knopp. 2002. Selective trace analysis of sulfonylurea herbicides in water and soil samples based on solid-phase extraction using a molecularly imprinted polymer. *Environ. Sci. Technol.* 36: 5411–5420.

172. Merino, F., S. Rubio, and D. Perez-Bendito. 2005. Supramolecular systems-based extraction-separation techniques coupled to mass spectrometry. *J. Sep. Sci.* 28: 1613–1627.

173. Caro, E., R.M. Marcé, F. Borrull, P.S.G. Cormack, and D.C. Sherrington. 2006. Application of molecularly imprinted polymers to solid-phase extraction of compounds from environmental and biological samples. *Trends Anal. Chem.* 25: 143–154.

174. Carabias-Martinez, R., E. Rodriquez-Gonzalo, and E. Herrero-Hermandez. 2005. Determination of triazines and dealkylated and hydroxylated metabolites in river water using a propazine-imprinted polymer. *J. Chromatogr. A* 1085: 199–206.

175. Caro, E., R.M. Marcé, P.S.G. Cormack, D.C. Sherrington, and F. Borrull. 2003. On-line solid-phase extraction with molecularly imprinted polymers to selectively extract substituted 4-chlorophenols and 4-nitrophenol from water. *J. Chromatogr. A* 995: 233–238.

176. Pichon, V., A.I. Krasnova, and M.C. Hennion. 2004. Development and characterization of an immunoaffinity solid-phase-extraction sorbent for trace analysis of propanil and related phenylurea herbicides in environmental waters and in beverages. *Chromatographia* 60: S221-S226.

177. Norberg, J. and E. Thordarson. 2000. Extracting syringe-connecting sample preparation and gas chromatography. *Analyst* 125: 673–676.

178. Barri, T., S. Bergström, A. Hussen, J. Norberg, and J.-Å. Jönsson. 2006. Extracting syringe for determination of organochlorine pesticides in leachate water and soil-water slurry: A novel technology for environmental analysis. *J. Chromatogr. A* 1111: 11–20.

179. Li, Q.-L., D.-X. Yuan, and Q.-M. Lin. 2004. Evaluation of multi-walled carbon nanotubes as an adsorbent for trapping volatile organic compounds from environmental samples. *J. Chromatogr. A* 1026: 283–288.

180. Zhou, Q., J. Xiao, and W. Wang. 2006. Using multi-walled carbon nanotubes as solid phase extraction adsorbents to determine dichlorodiphenyltrichloroethane and its metabolites at trace level in water samples by high performance liquid chromatography with UV detection. *J. Chromatogr. A* 1125: 152–158.

181. Zhou, Q., Y. Ding, and J. Xiao. 2006. Sensitive determination of thiamethoxam, imidacloprid and acetamiprid in environmental water samples with solid-phase extraction packed with multiwalled carbon nanotubes prior to high-performance liquid chromatography. *Anal. Bioanal. Chem.* 386: 1520–1525.

182. Bragg, L., Z. Qin, M. Alaee, and J. Pawliszyn. 2006. Field sampling with a polydimethylsiloxane thin-film. *J. Chromatogr. Sci.* 44: 317–323.

183. Popp, P., C. Bauer, A. Paschke, and L. Montero. 2004. Application of a polysiloxane-based extraction method combined with column liquid chromatography to determine polycyclic aromatic hydrocarbons in environmental samples. *Anal. Chim. Acta* 504: 307–312.

184. Zhao, W., G. Ouyang, M. Alaee, and J. Pawliszyn. 2006. On-rod standardization technique for time-weighted average water sampling with a polydimethylsiloxane rod. *J. Chromatogr. A* 1124: 112–120.

185. Lopez-Jimenez, F.J., S. Rubio, and D. Perez-Bendito. 2005. Determination of phthalate esters in sewage by hemimicelles-based solid-phase extraction and liquid chromatography-mass spectrometry. *Anal. Chim. Acta* 551: 142–149.

186. Lunar, L., S. Rubio, and D. Perez-Bendito. 2006. Analysis of linear alkylbenzene sulfonate homologues in environmental water samples by mixed admicelle-based extraction and liquid chromatography/mass spectrometry. *Analyst* 131: 835–841.

187. Steiner, S.A., M.D. Porter, and J.S. Fritz. 2006. Ultrafast concentration and speciation of chromium (III) and (VI). *J. Chromatogr. A* 1118: 62–67.

188. Fontanals, N., R.M. Marcé, and F. Borrull. 2006. Improved polymeric materials for more efficient extraction of polar compounds from aqueous samples. *Curr. Anal. Chem.* 2: 171–179.

189. Fontanals, N., R.M. Marcé, and F. Borrull. 2005. New hydrophilic materials for solid-phase extraction. *Trends Anal. Chem.* 24: 394–406.

190. Pinheiro, J.P. and W. Bosker. 2004. Polystyrene film-coated glassware: A new means of reducing metal losses in trace metal speciation. *Anal. Bioanal. Chem.* 380: 964–968.

191. Anthemidis, A.N. and K.-I.G. Ioannou. 2006. Evaluation of polychlorotrifluoroethylene as sorbent material for on-line solid phase extraction systems: Determination of copper and lead by flame atomic absorption spectrometry in water samples. *Anal. Chim. Acta* 575: 126–132.

192. Mitrovic, B., R. Milacic, B. Pihlar, and P. Simoncic. 1998. Speciation of trace amounts of aluminium in environmental samples by cation-exchange FPLC-ETAAS. *Analysis* 26: 381–388.

193. Ehmann, T., C. Mantler, D. Jensen, and R. Neufang. 2006. Monitoring the quality of ultra-pure water in the semiconductor industry by online ion chromatography. *Microchim. Acta* 154: 15–20.

194. Zhou, R., L. Zhu, K. Yang, and Y. Chen. 2006. Distribution of organochlorine pesticides in surface water and sediments from Qiantang River, East China. *J. Hazard. Mater. A* 137: 68–75.

195. Bielicka-Daszkiewicz, K., A. Voelkel, M. Szejner, and J. Osypisk. 2006. Extraction properties of new polymeric sorbents in SPE/GC analysis of phenol and hydroquinone from water samples. *Chemosphere* 62: 890–898.

196. Carabias-Martinez, R., E. Rodriguez-Gonzalo, E. Miranda-Cruz, J. Dominguez-Alvarez, and J. Hernandez-Mendez. 2006. Comparison of a non-aqueous capillary electrophoresis method with high performance liquid chromatography for the determination of herbicides and metabolites in water samples. *J. Chromatogr. A* 1122: 194–201.

197. Silva, E., S. Batista, P. Viana, et al. 2006. Pesticides and nitrates in groundwater from oriziculture areas of the "Baixo Sado" region (Portugal). *Int. J. Environ. Anal. Chem.* 13: 955–972.

198. Tedetti, M., K. Kawamura, B. Charriere, N. Chevalier, and R. Sempere. 2006. Determination of low molecular weight dicarboxylic and ketocarboxylic acids in seawater samples. *Anal. Chem.* 78: 6012–6018.

199. Allaire, S.E., S.R. Yates, F. Ernst, and S.K. Papiernik. 2003. Gas-phase sorption–desorption of propargyl bromide and 1,3-dichloropropene on plastic materials. *J. Environ. Qual.* 32: 1915–1921.

200. Spivakov, B.Y., G.I. Malofeeva, and O.M. Petrukhin. 2006. Solid-phase extraction on alkyl-bonded silica gels in inorganic analysis. *Anal. Sci.* 22: 503–519.

201. Katsumata, H., S. Kaneco, T. Suzuki, and K. Ohta. 2006. Determination of atrazine and simazine in water samples by high-performance liquid chromatography after preconcentration with heat-treated diatomaceous earth. *Anal. Chim. Acta* 577: 214–219.

202. Terada, K. 1992. Preconcentration by sorption. In: Z.B. Alfassi and Ch.M. Wai (eds), *Preconcentration Techniques for Trace Elements*, pp. 211–241. Boca Raton-Ann Arbor-London: CRC Press.

203. Lemic, J., D. Kovacevic, M. Tomasevic-Canovic, D. Kovacevic, T. Stanic, and R. Pfend. 2006. Removal of atrazine, lindane and diazinone from water by organo-zeolites. *Water Res.* 40: 1079–1085.

204. Zarpon, L., G. Abate, L.B.O. Dos Santos, and J.C. Masini. 2006. Montmorillonite as an adsorbent for extraction and concentration of atrazine, propazine, deethylatrazine, deisopropylatrazine and hydroxyatrazine. *Anal. Chim. Acta* 579: 81–87.

205. Soylak, M., L. Elci, and M. Dogan. 2003. Uses of activated carbon columns for solid phase extraction studies prior to determinations of traces heavy metal ions by flame atomic absorption spectrometry. *Asian J. Chem.* 15: 1735–1738.

206. Khan, E. and S. Subramania-Pillai. 2007. Interferences contributed by leaching from filters on measurements of collective organic constituents. *Water Res.* 41: 1841–1850.

207. Daud, W.R.W. 2004. Rate-based design of non-fouled cross-flow hollow fiber membrane modules for ultratitration. *Sep. Sci. Technol.* 39: 1221–1238.

208. Lorain, O., B. Hersant, F. Persin, A. Grasmick, N. Brunard, and J. M. Espenan. 2007. Ultrafiltration membrane pre-treatment benefits for reverse osmosis process in seawater desalting. Quantification in terms of capital investment cost and operating cost reduction. *Desalination* 203: 277–285.

209. Daud, W.R.W. 2006. Shortcut design method for reverse osmosis tubular module: The effect of varying transmembrane pressure and concentration polarization. *Desalination* 201: 297–305.

210. Giergielewicz-Możdajska, H., Ł. Dąbrowski, and J. Namieśnik. 2000. Przegląd technik ekstrakcyjnych wykorzystywanych na etapie wyodrębnienia analitów z próbek stałych. Część I: Klasyczne techniki ekstrakcji i ekstrakcja w stanie nadkrytycznym. *Ekologia i Technika* 8: 159–167.

211. Zorita, S., R. Westbom, L. Thorneby, E. Bjorklund, and L. Mathiasson. 2006. Development of a combined solid-phase extraction-supercritical fluid extraction procedure for the determination of polychlorinated biphenyls in wastewater. *Anal. Sci.* 22: 1455–1459.

212. Bahram, M., T. Madrakian, E. Bozorgzadeh, and A. Afkhami. 2007. Micelle-mediated extraction for simultaneous spectrophotometric determination of aluminum and beryllium using mean centering of ratio spectra. *Talanta* 72: 408–414.

213. Shemirani, F., R.R. Kozani, and Y. Assai. 2007. Development of a cloud point extraction and preconcentration method for silver prior to flame atomic absorption spectrometry. *Microchim. Acta* 157: 81–85.
214. Madrakian, T., A. Afkhami, R. Moeina, and M. Bahram. 2007. Simultaneous spectrophotometric determination of Sn(II) and Sn(IV) by mean centering of ratio kinetic profiles and partial least squares methods. *Talanta* 72: 1847–1852.
215. Takayanagi, T. and S. Motomizu. 2007. Pseudo-homogeneous micelle extraction of ion-associates formed between tetrabutylammonium ion and some aromatic sulfonate ions into nonionic surfactant micelle studied through the mobility measurements in capillary zone electrophoresis. *J. Chromatogr. A* 1141: 295–301.
216. Hernansez-Borges, J., M.A. Rodriquez-Delgado, and F.J. Garcia-Montelongo. 2006. Optimization of the microwave-assisted saponification and extraction of organic pollutants from marine biota using experimental design and artificial neural networks. *Chromatographia* 63: 155–160.
217. Pernak J., 2000. Ciecze jonowe—rozpuszczalniki XXI wieku. *Przem.Chem.* 79: 150–153.
218. Giergielewicz-Możdajska, H., Ł. Dąbrowski, and J. Namieśnik. 2001. Przegląd technik ekstrakcyjnych wykorzystywanych na etapie wyodrębnienia analitów z próbek stałych. Część´ II. Ekstrakcja wspomagana mikrofalami oraz przyspieszona ekstrakcja za pomocą rozpuszczalnika. *Ekologia i Technika* 9: 3–11.
219. Link, D.D., H.M. Kingston, G.J. Havrilla, and L.P. Coletti. 2002. Development of microwave-assisted drying methods for sample preparation for dried spot micro-X-ray fluorescence analysis. *Anal. Chem.* 74: 1165–1170.
220. Meunier, L., S. Canonica, and U. Von Gunten. 2006. Implications of sequential use of UV and ozone for drinking water quality. *Water Res.* 40: 1864–1876.
221. Mills, M.A., T.J. Mc Donald, J.S. Bonner, M.A. Simon, and R.L. Autenrieth. 1999. Method for quantifying the fate of petroleum in the environment. *Chemosphere* 39: 2563–2582.
222. Criado, M.R., S. Pombo da Torre, J.R. Pereiro, and R.C. Torrijos. 2004. Optimization of a microwave-assisted derivatizationextraction procedure for the determination of chlorophenols in ash samples. *J. Chromatogr. A* 1024: 155–163.
223. Foster, B.L. and M.F. Cournoyer. 2005. The use of microwave ovens with flammable liquids. *Chem. Health Saf.* 12: 27–32.
224. During, R.A., X. Zhang, H.E. Hummel, J. Czynski, and S. Gath. 2003. Microwave-assisted steam distillation with simultaneous liquid/liquid extraction of pentachlorophenol from organic wastes and soils. *Anal. Bioanal. Chem.* 375: 584–588.
225 Dominguez, A., J.A. Menendez, M. Inguanzo, P.L. Bernad, and J.J. Pis. 2003. Gas chromatographic-mass spectrometric study of the oil fractions produced by microwave-assisted pyrolysis of different sewage sludges. *J. Chromatogr. A* 1012: 193–206.
226. Daus, B., H. Wiess, J. Mattusch, and R. Wennrich. 2006. Preservation of arsenic species in water samples using phosphoric acid-limitations and long-term stability. *Talanta* 69: 430–434.
227. Menendez, J.A., A. Dominquez, M. Inguanzo, and J.J. Pis. 2005. Microwave-induced drying, pyrolysis and gasification (MWDPG) of sewage sludge: Vitrification of the solid residue. *J. Anal. Appl. Pyrolysis* 74: 406–412.
228. De Boer, J., J. Klungsoyr, G. Nesje, et al. 1999. MATT: Monitoring, analysis and toxicity of Toxaphene—improvement of analytical methods. *Organohalogen Compd.* 41: 569–574.
229. Van Emon, J. 2001. Immunochemical applications in environmental science. *J. AOAC Int.* 84: 125–133.
230. Eilola, K. and P. Peramaki. 2003. Development of a modified medium pressure microwave vapor-phase digestion method for difficult to digest organic samples. *Analyst* 128: 194–197.
231. Navarro, P., J.C. Raposo, G. Arana, and N. Etxebarria. 2006. Optimisation of microwave assisted digestion of sediments and determination of Sn and Hg. *Anal. Chim. Acta* 566: 37–44.
232. Sastre, J., A. Sahuquillo, M. Vidal, and G. Rauret. 2002. Determination of Cd, Cu, Pd and Zn in environmental samples: Microwave-assisted total digestion versus aqua regia and nitric acid extraction. *Anal. Chim. Acta* 462: 59–72.
233. Leonelii, C., P. Veronesi, D.N. Boccaccini, et al. 2006. Microwave thermal inertisation of asbestos containing waste and its recycling in traditional ceramics. *J. Hazard. Mater. B* 135: 149–155.
234. Bao, J., N. Nazem, L.T. Taylor, J. Cynko, and K. Kyle. 2006. Negative temperature programming using microwave open tubular gas chromatography. *J. Chromatogr. Sci.* 44: 108–112.
235. Flotron, V., J. Houesson, A. Bosio, C. Delteil, A. Bermond, and V. Camel. 2003. Rapid determination of polycyclic aromatic hydrocarbons in sewage sludges using microwave-assisted solvent extraction. Comparison with other extraction methods. *J. Chromatogr. A* 999: 175–184.

236. Hildebrandt, A., S. Lacorte, and D. Barcelo. 2006. Sampling of water, soil and sediment to trace organic pollutants at a river-basin scale. *Anal. Bioanal. Chem.* 386: 1075–1088.

237. Sun, Y., M. Takaoka, N. Takeda, T. Matsumoto, and K. Oshita. 2006. Application of microwave-assisted extraction to the analysis of PCBs and CBzs in fly ash from municipal solid waste incinerators. *J. Hazard. Mater. A* 137: 106–112.

238. Pino, V., J.H. Ayala, A.M. Afonso, and V. Gonzalez. 2003. Micellar microwave-assisted extraction combined with solid-phase microextraction for the determination of polycyclic aromatic hydrocarbons in a certified marine sediment. *Anal. Chim. Acta* 477: 1–91.

239. Serrano, A. and M. Gallego. 2006. Continuous microwave-assisted extraction coupled on-line with liquid–liquid extraction: Determination of aliphatic hydrocarbons in soil and sediments. *J. Chromatogr. A* 1104: 323–330.

240. Morales-Munoz, S., J.L. Luque-Garcia, and M.D. Luque de Castro. 2004. Screening method for linear alkylbenzene sulfonates in sediments based on water Soxhlet extraction assisted by focused microwaves with on-line preconcentration/derivatization/detection. *J. Chromatogr. A* 1026: 41–46.

241. During, R.A. and S. Gath. 2000. Microwave assisted methodology for the determination of organic pollutants in organic municipal wastes and soils: Extraction of polychlorinated biphenyls using heat transformer disks. *Fresenius J. Anal. Chem.* 368: 684–688.

242. Capadoglio, C. 1997. Sampling techniques for sea water and sediments. In: A. Gianguzza, E. Pelizzetti, and S. Sammartano (eds), *Marine Chemistry,*. pp. 115–130. Dordrecht: Academic Kluwer Publisher.

243. Priego-Capote, F. and M.D. Luque de Castro. 2007. Ultrasound-assisted digestion: A useful alternative in sample preparation. *J. Biochem. Biophys. Meth.* 70: 299–310.

244. Capelo, J.L., P. Ximenez-Embun, Y. Madrid-Albarran, and C. Camara. 2004. Enzymatic probe sonication: Enhancement of protease-catalyzed hydrolysis of selenium bound to proteins in yeast. *Anal. Chem.* 76: 233–237.

245. Xiao, H.B., M. Krucker, K. Albert, and X.M. Liang. 2004. Determination and identification of isoflavonoids in *Radix astragali* by matrix solid-phase dispersion extraction and high-performance liquid chromatography with photodiode array and mass spectrometric detection. *J. Chromatogr. A* 1032: 117–124.

246. Yang, Z., S. Matsumoto, and R. Maeda. 2002. *Comparison of dynamic transient- and steady state measuring methods in a batch type BOD sensing system. Sens. Actuat. A* 95: 274–280.

247. Wang, R.Y., J.A. Jarratt, P.J. Keay, J.J. Hawkes, and W.T. Coakley. 2000. Development of an automated on-line analysis system using flow injection, ultrasound filtration and CCD detection. *Talanta* 52:129–139.

248. Yang, L. and J.W. Lam. 2001. Microwave-assisted extraction of butyltin compounds from PACS-2. Sediment for quantitation by high-performance liquid chromatography inductively coupled plasma mass spectrometry. *J. Anal. At. Spectrom.* 16: 724–731.

249. Namiesńik, J. and W. Wardencki. 2000. Solventless sample preparation techniques in environmental analysis. *J. High Resol. Chromatogr.* 23: 297–303.

250. Jermak, S., B. Pranaityte, and A. Padarauskas. 2007. Ligand displacement, headspace single-drop microextraction, and capillary electrophoresis for the determination of weak acid dissociable cyanide. *J. Chromatogr. A* 1148: 123–127.

251. Afzali, D., A. Mostafavi, M.A. Taher, and A. Moradian. 2007. Flame atomic absorption spectrometry determination of trace amounts of copper after separation and preconcentration onto TDMBAC-treated analcime pyrocatechol-immobilized. *Talanta* 71: 971–975.

252. Curyło, J., W. Wardencki, and J. Namiesńik. 2007. Green aspects of sample preparation—a need for solvent reduction. *Polish J. Environ. Stud.* 16: 5–16.

253. Harmel, R.D., R.J. Cooper, R.M. Slade, R.L. Haney, and J.G. Arnold. 2006. Cumulative uncertainty in measured streamflow and water quality data for small watersheds. *Amer. Soc. Agri. Biol. Eng.* 49: 689–701.

Index

A

AAS. *See* Atomic absorption spectroscopy
Absorbents, 45
Absorption, 45, 74
Absorption spectrophotometry, 264
Absorptivity, 264
Accelerated solvent extraction (ASE), 356–357
Acidic herbicides, 28
AdCSV. *See* Adsorptive cathodic stripping voltammetry
Addition of poisons (biocides), 21
Adsorption, 45, 74
Adsorptive cathodic stripping voltammetry (AdCSV), 126
Advanced oxidation processes (AOPs)
    for water sample preparation, 96
        ozone oxidation, 98–99
        UV photo-oxidation, 97–98
AEs. *See* Alkyl ethoxylates
AES. *See* Alkyl ether sulfates
AFS. *See* Atomic fluorescence spectrometry
Agriculture, 305
Air–acetylene flame, 267
Algal toxins, passive sampling, 55–56
Aliphatic hydrocarbons, 30
Alkyl ether sulfates (AES), 145
Alkyl ethoxylates (AEs), 145, 146
Alkyl sulfonates (AS), 145
Alkyl-trialkoxysilanes, 313
Alkylphenol ethoxylate, 145, 329, 332
Aluminum, 13, 285
Ammonium determination, indophenol blue method for, 404–405
Amphenicols detection, 162–163
Analyte derivatization, 325–326
Analyte extraction, modern techniques of
    liquid membrane based techniques
        microporous membrane liquid–liquid extraction, 84–87
        supported liquid membrane (SLM) extraction, 79–84
        two-phase HF LPME, 87–88
    miniaturized nonmembrane-based extraction techniques
        miniaturized liquid-phase extraction techniques, 70–72
        miniaturized SPE techniques, 72–73
        solid-phase microextraction, 73–75
        stir bar sorptive extraction, 75
    nonporous polymeric membranes based techniques, 75–76
        membrane-assisted solvent extraction, 78–79
        membrane extraction with sorbent interface, 76–77

membrane inlet (introduction) mass spectrometry (MIMS), 76
Analytical data, types of, 434–435
Analytical procedure
    for organic compounds determination, 27–28
    stages of, 440–442
Analytical protocols, 7–8
Analytical results, QA/QC
    monitoring of, 394–395
        external QC, 395–396
        internal QC, 395
    traceability, 394
    validation of analytical procedures, 393–394
Analytical techniques, of inorganic constituents, 261
    atomic spectrometry, 265–272
    automatic analyzers and monitoring, 281–282
    chemical vapor generation-atomic spectrometry (CVG-AS), 273–275
    electrochemical techniques, 275–278
    gravimetric measurements, 262
    mass spectrometric techniques, 272–273
    separation techniques, 278–281
    spectrophotometric technique
        instrumentation, 264–265
        on molecular absorption radiation, 263
    trimetric measurement, 262
Androgens, immunochemical determination of, 168–169
Animals, differential tissue analysis, 110–111
Anodic stripping voltammetry (ASV), 126, 275–276
Antibiotics, detection of, 158–160, 165–166
    amphenicols, 162–163
    β-lactams, 164
    fluoroquinolones, 161–162
    macrolides, 165
    sulfonamides, 160–161
    tetracyclines, 163–164
Aquatic ecosystem
    inorganic constituents in, 260
    speciation analytics in. *See* Speciation analytics
AOC. *See* Assimilable organic carbon
AOPs. *See* Advanced oxidation processes
API. *See* Atmospheric pressure ionization
APPI. *See* Atmospheric pressure photo-ionization
ARGE-Elbe project, 209
Arsenic speciation, 122
    electrochemical speciation, 126–127
    hyphenated methods
        chemiluminescence and colorimetric reactions, 128
        chromatography, 127–128
        exchange columns, 127
    in situ speciation, in aquatic system, 123–125
    sample preservation and storage, 126
    sampling, 125

AS. *See* Alkyl sulfonates
As speciation. *See* Arsenic speciation
ASE. *See* Accelerated solvent extraction
Assimilable organic carbon (AOC), 232
ASV. *See* Anodic stripping voltammetry
At-line solid phase extraction, 321
Atmospheric microwave-assisted extractions, 454
Atmospheric pressure ionization (API), 311, 317
Atmospheric pressure photo-ionization (APPI), 317
Atomic absorption spectroscopy (AAS), 224, 243, 405
Atomic emission spectrometry (AES), 270
Atomic fluorescence spectrometry (AFS), 228, 230,
    271–272
    atomizers, 272
    radiation, source of, 271
Atomic spectrometry, 265
    atomic emission spectrometry, 270
    atomic fluorescence spectrometry, 271–272
    electrothermal atomic absorption spectrometry,
        268–269
    flame atomic absorption spectrometry, 266–268
    high-resolution continuous source atomic absorption
        spectrometry, 269–270
    inductively coupled plasma-optical emission
        spectrometry (ICP-OES), 270–271
Atomizer, 268, 272, 275
Automated flowing MMLLE
    off-line systems, 84, 86–87
        nonautomated, nonflowing design, 86–87
    on-line systems, 84–85
    syringe device extraction, 85–86
Automated water analyzer computer supported system
        (AWACSS), 160
Automatic analyzers and monitoring, 281–282
Automatic sampling system, 42
Automatic water sampling systems, 14
    basic components, 15
Average linkage method, 372–373
AWACSS. *See* Automated water analyzer computer
        supported system
AZUR Environmental, 196

B

Background correction, 267
Badge-type samplers, 46
Batch analyzer. *See* Discrete analyzer
Beer–Lambert law, 408
Benzalkonium surfactants, 32
Benzio(a)pyrene, 143
β-lactams (BL), detection of, 164
BEWS. *See* Biological Early Warning Systems
Biacore Q, 163
Biacore SPR sensor, 164
Bias, 47
Bioassays, 56
Biochemical oxygen demand (BOD), 14, 224, 232
    determination of, 225
    limitations of, 225
Biocides, 21
Bioconcentratable hydrophobic estrogen receptor
        agonists, 56
Biofouling, 48–50
Bioindicator organisms, 198

Bioindicator techniques, 104, 196
    standard and guidelines for performing toxicity tests
        using, 194–195
Biological Early Warning Systems (BEWS), 192–193
Biological oxygen demand (BOD), 192, 374, 377
Bioluminescent bacteria, 196
Biomonitoring, 42, 104
    passive sampling, 57
Biosensors, 144
Biota, 57
    mineralization of, 251
Biota analysis as information source, in state of aquatic
        environments, 103
    assessment strategies, 109–110
        animals, differential tissue analysis, 110–111
        plants, contaminants accumulation and
            partitioning, 110
    species choice
        pelagic species, 108–109
        primary producers, 105–106
        sediment dwellers, 107–108
        suspension feeders, 106–107
    supplementary methodologies
        metallothioneins, 113–114
        oxidative stress, 112–113
Biotests, 189, 328
    environmental monitoring, bioassay application in, 210
        hot spots, identification of, 211–212
        monitoring parameters, revision of, 214–216
        polluted areas managed by specific areas, ranking
            of problems in, 212–214
        toxic compounds, identification of, 210–211
    environmental pollution extent assessment, chemical
        monitoring in, 190–192
    environmental samples, ecotoxicological classification
        of, 201, 207–210
    legal regulations applicable to toxicity measurement,
        201, 205–207
    toxicity tests, importance of, 192–195
        Toxkit tests, 195–199
    water pollution assessment, integrated system of, 200
Biotoxins adsorption, 55
Bisphenol A (BPA), 150–151
BL. *See* β-Lactams
BOD. *See* Biochemical oxygen demand; Biological
        oxygen demand
Borosilicate glass, 13
Botanical pesticides, 28
Bouguer–Lambert–Beer law, 263
BPA. *See* Bisphenol A
Brass, 13
British Standards Institute Publicly Available
        Specification (BSI PAS-61), 57
BSI PAS-61. *See* British Standards Institute Publicly
        Available Specification

C

C18 stationary phases, 313
CA. *See* Cluster analysis
CAP detection. *See* Chloramphenicols detection
Capillary electrophoresis (CE), 127, 280–281, 309,
        313–315
Capillary gel electrophoresis (CGE), 281

Capillary isoelectric focusing (CIEF), 281
Capillary isotachophoresis (CITP), 281
Capillary microextraction (CME), 127
Capillary zone electrophoresis (CZE), 281
Carbamates, 28
Carbon steel, 13
Carcinus maenas, 111, 114
Cathodic stripping voltammetry (CSV), 126, 231
Cationic surfactants, 145
CCDs. *See* Charge-coupled devices; Chlorinated
          cyclodienes
CE. *See* Capillary electrophoresis
Ceramic dosimeters, 15
Ceramics, 13
Ceriodaphnia dubia, 198
Certified reference materials (CRMs), 391, 395, 396
Cesium determination, 246–248
    activity calculation, 247
    separation by adsorption on AMP, 246–247
CFA. *See* Continuous flow analysis
CFC. *See* Chlorofluorocarbons
CFME. *See* Continuous-flow microextraction
CGE. *See* Capillary gel electrophoresis
Charge-coupled devices (CCDs), 266–270
Charm II RIA, 160, 164
Chelex-based sorption phase, 55
Chemcatcher, 46, 48, 54
Chemical ionization (CI), 317
Chemical measurements, quality in, 390–391
Chemical modification. *See* Matrix modification
Chemical oxygen demand (COD), 192, 225, 231–232, 377
    limitations of, 225
Chemical speciation, 224
Chemical vapor generation (CVG), 266, 274
Chemical vapor generation-atomic spectrometry
          (CVG-AS), 273–275
Chemiluminescence and colorimetric reactions, 128
Chemiluminescence enzyme immunoassay (CLEIA), 167
Chemometrics, 369
    definition, 370
    methods, 370
        cluster analysis (CA), 370–376
        principal component analysis (PCA), 380–383
        receptor modeling, 383–385
        self-organizing maps (SOM), 376–379
Cherenkov counting technique, 242, 249
Chiral speciation, 438
Chloramphenicols (CAPs) detection, 158, 162–163, 164
Chloride
    mercury thiocyanate-iron method for, 405
Chlorinated cyclodienes (CCDs), 152
Chlorinated hydrocarbons, 28
Chlorofluorocarbons (CFC), 13
Chlorophenoxy acid herbicides, 156
Chlorsulfuron, detection of, 157–158
Chromatography, 127–128, 318–325, 439
Chromium speciation, 26, 122
    electrochemical speciation, 126–127
    hyphenated methods
        chemiluminescence and colorimetric reactions,
          128
        chromatography, 127–128
        exchange columns, 127
    in situ speciation, in aquatic system, 123–125

    sample preservation and storage, 126
    sampling, 125
CI. *See* Chemical ionization
CIEF. *See* Capillary isoelectric focusing
CITP. *See* Capillary isotachophoresis
Clarias gariepinus, 113
Clark electrode, 276
Class weight score, calculation of, 208
Classical analytical methods, 191
CLEIA. *See* Chemiluminescence enzyme immunoassay
CLTRAP. *See* Convention on Long-range Transboundary
          Air Pollution
Cluster analysis (CA)
    case study (Struma River), 373–376
    theoretical principles, 370–373
Clustering algorithms, 371
CME. *See* Capillary microextraction
Coacervates, 32
COC. *See* Cold-on-column
COD. *See* Chemical oxygen demand
Cold vapor atomic fluorescence spectrometry, 130, 408
Cold-on-column (COC) injection, 309
Cold-vapor atomic absorption spectrometry, 231
Colonization of organisms, 48
Complete linkage method, 372
Composite water sample, 3–4
Conductivity, of precipitation samples, 403–404
Containers, 20
Continuous analyzer, 281
Continuous flow analysis (CFA), 281
Continuous-flow microextraction (CFME), 71–72
Control charts, 391, 395
Convention on Long-range Transboundary Air Pollution
          (CLTRAP), 400
Conversion reactions, 27
Copper, 13
Corbicula fluminea, 111
Corticosteroids, immunochemical determination of, 169
Cosolvent method, 47
Council Directive 76/464/EEC, 141, 191–192
Council Regulation 793/93, 193
Cr speciation. *See* Chromium speciation
Crassostrea gigas, 114
CRMs. *See* Certified reference materials
^{137}Cs determination. *See* Cesium determination
CSV. *See* Cathodic stripping voltammetry
Cusum charts, 395
CV/HG methods. *See* CVG methods
CVG. *See* Chemical vapor generation
CVG-AS. *See* Chemical vapor generation-atomic
          spectrometry
CZE. *See* Capillary zone electrophoresis

D

Daphni pulex, 198
Daphnia magna, 198, 207
Data quality control, 409–410
DDT. *See* Dichloro-diphenyl-trichloroethane
Deca-BDE, 148
Deep-water samplers, 14
Derivatization. *See* Analyte derivatization
DET. *See* Diffusive Equilibrium in Thin Films
Detection limit, for different analytical methods, 406

Detection techniques, for organic and organometallic
    pollutants, 315
    Fourier-transform ion cyclotron resonance
        instruments, 317
    inductively coupled plasma-mass spectrometry, 318
    ion trap mass spectrometers, 316
    ionization techniques, 317
    quadrupole mass spectrometers, 316
    time-of-flight (TOF) instruments, 317
    triple quadrupole mass spectrometers, 316–317
DGT sampler. *See* Diffusive gradients in thin films
DIC. *See* Dissolved inorganic carbon
Dichloro-diphenyl-trichloroethane (DDT), 14
Diclofenac, detection of, 165
Diffusion barrier, 45
Diffusion flames, 272
Diffusion-based samplers, 46
Diffusive Equilibrium in Thin Films (DET) sampler, 123
Diffusive gradients in thin films (DGT) sampler, 46, 55,
    123, 125
DIHN. *See* Direct injection high efficiency nebulizers
Dimethyl mercury (DMHg), 26, 129
    analytical methods for determination, 130
DIN 38412, 197
Dioxin, 30, 142
Direct immersion SPME (DI-SPME), 358
Direct injection high efficiency nebulizers (DIHN), 271
Direct injection nebulizers, 271
Direct toxicity assessment (DTA), 200
Discharge-proportional sampling, 4
Discontinuous sampling techniques, 4
Discrete analyzer, 281
Discrete water sample, 3
Dispersive liquid–liquid microextraction (DLLME), 72
Dispersive solvent, 72
DI-SPME. *See* Direct immersion SPME
Dissolved gases determination methods, 290
Dissolved inorganic carbon (DIC), 232
Dissolved organic carbon (DOC), 30, 47, 225, 233
Dissolved organic matter (DOM), 96
Dissolved organic sulfur, 233
Distribution speciation, 438
Divinylbenzene polymeric resins, 55
Divisive algorithms, 371
DLLME. *See* Dispersive liquid–liquid microextraction
DLPME. *See* Dynamic liquid-phase microextraction
DME. *See* Dropping mercury electrode
DMHg. *See* Dimethyl mercury
DOC. *See* Dissolved organic carbon
DOM. *See* Dissolved organic matter
Double check-valve bailer with a bottom-emptying
    device, 14
Double-focusing magnetic sector mass analyzer, 273
Dredges, as sediment samplers, 12, 15
Dropping mercury electrode (DME), 275
DTA. *See* Direct toxicity assessment
Dynamic liquid-phase microextraction (DLPME), 71

E

ECD. *See* Electron capture detection
EcHG. *See* Electrochemical hydride generation
Eco-indicator, 99 method, 419
Ecological/environmental risk assessment (ERA), 112

Ecosolvents. *See* Green solvents, 425
Ecotoxicity assessment, 209
    passive sampling, 56
EDCs. *See* Endocrine disrupting chemicals
EDL. *See* Electrodeless discharge lamp
EDTA. *See* Ethylenediaminetetraacetic acid (EDTA)
EINECS. *See* European Inventory of Existing
    Commercial Chemical Substances
Electrochemical hydride generation (EcHG), 274
Electrochemical magneto immunosensing strategy, 156
Electrochemical speciation, 126–127
Electrochemical techniques, 275–278
    anodic stripping voltammetry, 275–276
    potentiometric sensors, 276–278
Electrochemistry sensing technology, advances in, 362
Electrodeless discharge lamp (EDL), 266
Electron capture detection (ECD), 130
Electro-osmotic flow (EOF), 280
Electrospray ionization (ESI), 332
Electrothermal Atomic Absorption Spectrometry
    (ETAAS), 268–269
    atomizer, 268
    matrix modification, 268–269
ELISAs. *See* Enzyme-linked immunosorbent assays
EMEP, 399
    measurements, 400
    monitoring network of precipitation chemistry,
        400–401
    objective of, 400
Emerging contaminants, 140
Empore disk, 54
En score, in evaluation of laboratory results, 396
Endocrine disrupting chemicals (EDCs), 166
Endocrine disruptors, 200–201
Enteromorpha intestinalis, 105
Environmental analysis, total parameters in, 227–233
Environmental analytics, 223, 224
Environmental monitoring
    bioassay application in, 210
        hot spots, identification of, 211–212
        monitoring parameters, revision of, 214–216
        polluted areas managed by specific areas, ranking
            of problems in, 212–214
        toxic compounds, identification of, 210–211
    legal regulations applicable to toxicity measurement,
        201, 205–207
Environmental pollution extent assessment, chemical
    monitoring in, 190–192
Environmental Protection Agency (EPA), 142, 200, 227,
    354, 390
Environmental quality standards (EQS), 59
Environmental samples, ecotoxicological classification
    of, 201, 207–210
Environmental water pollution, main sources of, 200
Enzyme-linked immunosorbent assays (ELISAs), 143,
    144
    for corticosteroids detection, 169
    for heavy metals analysis, 150
EOF. *See* Electro-osmotic flow
EOX. *See* Extractable organic halides
EPA. *See* Environmental Protection Agency
EQS. *See* Environmental quality standards
Equilibrium sampling, 43–44, 47
Equilibrium sampling devices (ESDs), 84

Equilibrium sampling through membranes (ESTM), 83
ERA. *See* Ecological/environmental risk assessment;
    Estrogen receptor agonists
ESDs. *See* Equilibrium sampling devices
ESI. *See* Electrospray ionization
ESTM. *See* Equilibrium sampling through membranes
Estrogen receptor agonists (ERA), 56
Estrogenic substances, 333–334
Estrogens, immunochemical determination of, 166–168
ETAAS. *See* Electrothermal Atomic Absorption
    Spectrometry
Ethylenediaminetetraacetic acid (EDTA), 25
Euphotic zone, 305
EUROCAT project, 211
Europe
    precipitation samples in, 401–409
        collection of, 401–402
        heavy metals and metalloids in precipitation,
            measurement of, 405–408
        ions, pH, and conductivity in precipitation,
            measurements of, 402–405
        POPs in precipitation using GC-MS,
            measurements of, 408–409
    rural monitoring network in, 399–401
European Inventory of Existing Commercial Chemical
    Substances (EINECS), 141
European "Metropolis" project, 306
Event-controlled sampling, 4, 5
Exchange columns (Trap and Elute), 127
Excited state, 263
External QC, 395–396
Extractable organic halides (EOX), 228
Extractable organohalogens. *See* Extractable organic
    halides (EOX)
Extraction solvent, 72

F

F-AAS. *See* Flame-atomic absorption spectroscopy
FAAS. *See* Flame atomic absorption spectrometry
Fabrication controls, 57
FAS. *See* Fatty alcohol sulfates
Fatty alcohol sulfates (FAS), 145
FDA. *See* Food and Drug Administration
Fe. *See* Iron
[55]Fe. *See* Iron
Fenamiphos, 28
Fenitrothion, 28, 31
FFF. *See* Field flow fractionation
FIA. *See* Flow injection analysis
Fiber-in-tube SPE (FIT-SPE), 72–73
Field blanks, 7–8
Field controls, 57
Field flow fractionation (FFF), 315
Field validation, 58–59
FI-ICP-ES. *See* Flow injection inductively coupled
    plasma-emission spectrometry
Fission products (^{90}Sr and ^{137}Cs) determination, 246–248
FIT-SPE. *See* Fiber-in-tube SPE
Flame atomic absorption spectrometry (FAAS), 266–268
    background correction, 267–268
    flame atomizer, 267
    nebulizers, 267
    radiation, source of, 266–267

Flame-atomic absorption spectroscopy (F-AAS), 405, 408
Flame atomic spectroscopy, 405
Flame atomizer, 267
Flat sheet Microporous membrane liquid–liquid
    extraction, 84
Flow injection analysis (FIA) methods, 281, 282, 327
Flow injection inductively coupled plasma-emission
    spectrometry (FI-ICP-ES) system, 127
Flow-weighted sampling. *See* Quantity-proportional
    sampling
Fluorescence polarization immunoassay (FPIA), 146, 150
Fluorocarbon polymers, 13
Fluoroquinolones (FQs), detection of, 161–162
FMW. *See* Focused microwaves
Focused microwaves (FMW), 454
Food and Drug Administration (FDA), 390
Fourier-transform ion cyclotron resonance instruments, 317
FPIA. *See* Fluorescence polarization immunoassay
FQs. *See* Fluoroquinolones
Freezing, water preservation method, 21
FS-MMLLE. *See* Flat sheet Microporous membrane
    liquid–liquid extraction
Fucus, 112
Fucus vesiculosus, 105
Furanes, 30

G

G immunoglobulins (IgG), 142
Galvanized steel, 13
Gas chromatograph coupled to an isotope ratio mass
    spectrometer (GCIRMS), 232
Gas chromatographic separation methods, 130
Gas chromatography (GC), 71, 309–311
    with atomic emission detection (GC-AED), 224
    -electron capture detection (GC-ECD), 86
    improvements, 131
    -mass spectrometry (GC-MS), 228
    -mass spectroscopy (LC-MS), 316, 317
Gas-permeable membrane sensors, 278
GC. *See* Gas chromatography
GC × GC, 311, 312, 314
GCI. *See* Green Chemistry Institute
GCIRMS. *See* Gas chromatograph coupled to an isotope
    ratio mass spectrometer
Gestagens, immunochemical determination of, 169
GF-AAS. *See* Graphite furnace atomic absorption
    spectrometry
Glass containers, 13, 26
Gloves, 13
GLP. *See* Good Laboratory Practice
Good Laboratory Practice (GLP), 193, 390, 391
Gracilaria verrucosa, 105
Gradient elution, 279
Graphite furnace atomic absorption spectrometry
    (GF-AAS), 125, 407–408
Gravimetric methods, 262
Gravity corers, 16
Green analytical chemistry, 70
    characteristics of, 356
    electrochemistry sensing technology, advances in, 362
    history of, 354–355
    implementation of, 355
    miniaturization in, 362–363

Green analytical chemistry (*Continued*)
  objective of, 355
  principles of, 362
  requirements for, 361
  in sample pretreatment, 356
    accelerated solvent extraction (ASE), 356–357
    ILs, application of, 361
    liquid-phase microextraction (LPME), 359–360
    pressurized hot water extraction (PHWE), 360
    single-drop microextraction (SDME), 359
    solid-phase microextraction (SPME), 357–358
    stir bar sorptive extraction (SBSE), 258
    supercritical fluid extraction (SFE), 360–361
    thin-film microextraction, 358–359
    ultrasonic and microwave extraction, 357
Green Chemistry Institute (GCI), 354
Green chemistry, 354, 458
  green analytical chemistry, 459–463
  history, 458–459
Green solvents, 425
Griess method, 404
Ground water samplers, 14
Ground water sampling, 11
Groundwater monitoring, 191
Group speciation, 437

H

Hand-held open-mouth bottle sampler, 9
HCL. *See* Hollow cathode lamp
Headspace (HS) techniques, 318
Headspace (HS-SPME), 358
Heavy metals and metalloids
  determination, 149–150
  in precipitation, 405
    cold vapor atomic fluorescence spectroscopy (CV-AFS), 408
    flame-atomic absorption spectroscopy (F-AAS), 408
    graphite furnace atomic absorption spectroscopy (GF-AAS), 407–408
    inductively coupled plasma mass spectrometry (ICP-MS), 406–407
    sample preparation, 405–406
Hediste diversicolor, 107
HELCOM. *See* Helsinki Commission
Helsinki Commission (HELCOM), 201
Herbicides, detection of, 154–158
Hexavalent chromium Cr(VI), 26
Hg speciation, 129
  DMHg, analytical methods for, 130
  handling and storage, of samples, 129–130
  Hg(0), analytical methods for, 130
  Hg-R, analytical methods for, 130
  Hg-T, analytical methods for, 130
  MMHg, analytical methods for, 130–132
  modeling of, 132
Hg(0), analytical methods for, 130
Hg-R, analytical methods for, 130
Hg-T, analytical methods for, 130
Hierarchical clustering, 371
HIFU. *See* High-intensity focused ultrasound
High-performance liquid chromatography (HPLC), 56, 73, 127, 279, 311

High-performance liquid chromatography tandem mass spectrometry (HPLC-MS/MS), 168
High-intensity focused ultrasound (HIFU), 230
High-intensity hollow cathode lamps (HI-HCL), 271
High-performance immunochromatographic (HPIAC) procedure, 147
High-resolution continuous source atomic absorption spectrometry (HR-CS AAS), 269–270
High-temperature precipitate digestion, 262
HI-HCL. *See* High-intensity hollow cathode lamps
Hollow cathode lamp (HCL), 266
Hot spots, identification of, 211–212
HPIAC. *See* High-performance immunochromatographic
HPLC. *See* High-performance liquid chromatography
HPLC-MS/MS. *See* High performance liquid chromatography tandem mass spectrometry
HR-CS AAS. *See* High-resolution continuous source atomic absorption spectrometry
HS. *See* Headspace; Humic substances
HS-SPME. *See* Headspace
Humic substances (HS), 231
Hydrocarbons storage, 31
Hydrogels, 55
Hydrophobic organic compounds, passive sampling, 50
Hydrosphere, human impact on, 305
Hyphenation, of MMLLE, 84

I

IA. *See* Immunoaffinity
IC. *See* Ion chromatography
ICE. *See* Integrated Control of Effluents
ICES. *See* International Council for the Exploration of the Sea
ICP-AES. *See* Inductively coupled plasma atomic emission spectrometry
ICP-MS. *See* Inductively coupled plasma mass spectrometry
ICP-OES. *See* Inductively coupled plasma optical emission spectroscopy
IgG. *See* G immunoglobulins
ILCs. *See* Interlaboratory comparisons
ILs. *See* Ionic liquids
Immunoaffinity (IA) sorbents, 324
Immunochemical analytical methods, for monitoring aquatic environment, 140–142
  industrial contaminants, determination of, 142
    bisphenol A, 150–151
    heavy metals and metalloids, 149–150
    organohalogenated compounds, 147–149
    polycyclic aromatic hydrocarbons, 142–145
    surfactants, 145–147
  for pesticides, 151–152
    herbicides and plant growth regulators, 154–158
    insecticides, 152–154
  pharmaceutical and personal care products, 158
    antibiotics, 158–166
    steroid hormones, 166–169
Immunosensors, 142
In situ speciation, in aquatic system, 123–125
In-tube extraction (ITEX), 73
In-tube SPME, 358
Indicator electrode, 277
Individual speciation, 438

Indomethacin, detection of, 165, 166
Indophenol blue method, 404–405
Inductively coupled plasma atomic emission spectrometry
    (ICP-AES), 233
Inductively coupled plasma mass spectrometry (ICP-MS),
    125, 132, 224, 406–407
Inductively coupled plasma optical emission spectroscopy
    (ICP-OES), 224, 270–271
    instrumentation, 270
    nebulizers and spray chambers, 271
    optics and detectors, 271
    plasma torch, 270
    RF generators, 270–271
Inductively coupled plasma-mass spectrometry, 272–273,
    318
    instrumentation, 272–273
    ion detectors, 273
    mass analyzers, 273
Industrial contaminants, determination of, 142
    bisphenol A, 150–151
    heavy metals and metalloids, 149–150
    organohalogenated compounds, 147–149
    polycyclic aromatic hydrocarbons, 142–145
    surfactants, 145–147
Injector, 270
Inorganic compounds determination, water samples
    preservation and storage for, 19–22
Inorganic constituents
    analytical techniques, 261
        gravimetric measurements, 262
        spectrophotometric technique, 263
        titrimetric measurement, 262–263
    characteristics of, 260
    classification, of aquatic ecosystem, 260–261
    determination of, 282
    dissolved gases determination methods, 290
    ion determination methods, 286
    metals determination methods, 291–294
    non-metallic substances determination methods,
        286–290
    nutrient determination methods, 283–285
    using IC, 285
Inorganic pesticides, 28
Inorganic sampling, 13
Insecticides, detection of, 152–154
Inside-needle SPE, 73
Integrated Control of Effluents (ICE), 201
Interlaboratory comparisons (ILCs), 390, 391, 396
Internal QC, 395
International Council for the Exploration of the Sea
    (ICES), 57
International Odra Project (IOP), 210
International Organization for Standardization (ISO), 143
International Union of Applied and Pure Chemistry
    (IUPAC), 354
    speciation analysis definition, 224
Ion chromatography (IC), 127, 278–280,
    309, 315, 404
Ion detectors, 273
Ion determination methods, 286
Ion trap mass spectrometers, 316
Ionic liquids (ILs), 361, 452–453
Ionization techniques, 317
Ions, in precipitation samples, 402–403

Ion-selective electrodes (ISEs), 263
    glass membranes, 277
    potentiometric dissolved gas sensors, types of, 278
IOP. *See* International Odra Project
Iron, 122–123
    coprecipitation of, in natural water with iron
        hydroxide, 243
    determination, 242, 243, 244
    mineralization, separation, and electrolysis in aquatic
        sediments and biota samples, 243–244
    separation and determination, 243
    speciation, 13, 122
        electrochemical speciation, 126–127
        hyphenated methods
            chemiluminescence and colorimetric
                reactions, 128
            chromatography, 127–128
            exchange columns, 127
        in situ speciation, in aquatic system, 123–125
        sample preservation and storage, 126
        sampling, 125
ISEs. *See* Ion-selective electrodes
ISO. *See* International Organization for Standardization
ISO Guide 9000:2000, 391
ISO/IEC 17025:2005, 391
Isocratic elution, 279
ITEX. *See* In-tube extraction
IUPAC. *See* International Union of Applied and Pure
    Chemistry

J

Judgemental sampling pattern, 5

K

^{40}K determination. *See* Potassium determination
Kin ExA™ 3000, 150
Kinetic sampling, 44–45, 47
Kjeldahl nitrogen, 285
Kremmer sampler, 14

L

Labs-on-a-chip, 363
LA-ICP-MS. *See* Laser ablation inductively coupled
    plasma mass spectrometry
LAS. *See* Linear alkylbenzene sulfonates
Laser ablation inductively coupled plasma mass
    spectrometry (LA-ICP-MS), 125
Laser-excited AFS (LEAFS), 272
Laser-induced atomic fluorescence spectrometry (LIF), 272
Latent variables, 380
LC. *See* Liquid chromatography
LC × LC, 313, 314
LCA. *See* Life cycle assessment
LC-MS. *See* Liquid chromatography-mass spectroscopy
LDPE. *See* Low-density polyethylene
Lead determination
    coprecipitation of, in natural water with manganese
        dioxide, 249–250
    radiochemical methods, 249
    radionuclide activity determination, 253–255
    separation and determination, 251

LEAFS. *See* Laser-excited AFS
Legal regulations, to measure toxicity, 201, 205–207
LIF. *See* Laser-induced atomic fluorescence spectrometry
Life cycle assessment (LCA), 427–428, 460
    general idea, 413–414
    methodology, 414–424
        goal and scope, 415–416
        life cycle impact assessment, 418–421
        life cycle interpretation and application, 421–424
        life cycle inventory analysis, 416–418
    for solvent use in analytical protocols, 424–427
Limits of detection (LODs)
    for trace metal species, in aquatic samples, 124
Linear alkylbenzene sulfonates (LAS), 145
Liquid chromatography (LC), 72, 131–132, 311–313
Liquid chromatography-mass spectroscopy (LC-MS), 316, 337
Liquid extraction techniques, 448
Liquid membrane based techniques
    microporous membrane liquid–liquid extraction, 84–87
    supported liquid membrane (SLM) extraction, 79–84
    two-phase HF-LPME, 87–88
Liquid–liquid extraction (LLE), 31, 32, 321
Liquid-phase microextraction (LPME), 71, 359–360
Littrow prism, 269
Liza aurata, 113
LLE. *See* Liquid–liquid extraction
LMWOM. *See* Low-molecular-weight organic molecules
Loadings plots, 381
LODs. *See* Limits of detection
LOQ. *See* Quantification limit
Low-density polyethylene (LDPE), 48, 75
Low-molecular-weight organic molecules (LMWOM), 30, 31
LPME. *See* Liquid-phase microextraction
L'vov platform, 268
Lyophilization, 31

M

Macoma balthica, 107
Macrolides, detection of, 165
MAE. *See* Microwave-assisted extractions
MALDI-MS. *See* Matrix-assisted laser-desorption/ionization-mass spectroscopy
MAME. *See* Microwave-assisted micellar extraction
Manganese speciation, 122–123
    electrochemical speciation, 126–127
    hyphenated methods
        chemiluminescence and colorimetric reactions, 128
        chromatography, 127–128
        exchange columns, 127
    in situ speciation, in aquatic system, 123–125
    sample preservation and storage, 126
    sampling, 125
Manifest variables, 380
Manual corers, 12, 16
Manual surface water samplers, 13–14
Marine strategy directive, 141
Marsh reaction, 274
MASE. *See* Membrane-assisted solvent extraction
Mass analyzers, 273
Mass spectrometric techniques, 272–273

inductively coupled plasma-mass spectrometry, 272–273
Mass spectroscopy (MS) detectors, 311
Matrix modification, 268–269
Matrix solid-phase dispersion, 448–448
Matrix-assisted laser-desorption/ionization-mass spectroscopy (MALDI-MS) analysis, 361
MCGC. *See* Multicapillary GC
MEC. *See* Microwave-enhanced chemistry
Medium pressure liquid extraction, 448
MEKC. *See* Micellar electrokinetic chromatography
Membrane electrode, 277
Membrane enclosed sorptive coating (MESCO) sampler, 46, 48
Membrane extraction with sorbent interface (MESI), 76–77, 320
Membrane inlet (introduction) mass spectrometry (MIMS), 76, 320
Membrane morphology, information on, 446–447
Membrane permeability, 46
Membrane techniques, application of, 442
    membrane extraction, 445–448
Membrane/water partition coefficient, 46
Membrane-assisted solvent extraction (MASE), 78–79
MEPS. *See* Microextraction in packed syringe
Mercury thiocyanate-iron method, 405
Mercury, speciation analysis, 26
MESCO. *See* Membrane enclosed sorptive coating
MESI. *See* Membrane extraction with sorbent interface
Metallothioneins (MTs), 113–114, 150
Metals, passive sampling, 55
Metals determination methods, 291–294
Methyl tertiary butyl ether (MTBE), 427
Methyl-trimethoxysilane (MTMS), 313
Micellar electrokinetic chromatography (MEKC), 281, 315
Micro total analysis systems (μTAS), 363
Microbics Corporation. *See* AZUR Environmental
Microbioassays, 198
Microextraction in packed syringe (MEPS), 73
Micro-immune-supported liquid membrane assay (μ-ISLMA), 156
Microporous membrane liquid–liquid extraction (MMLLE), 84–87
    automated flowing MMLLE
        off-line systems, 84, 86–87
        on-line systems, 84–85
        syringe device extraction, 85–86
    hyphenation, 84
    miniaturization, 84
    principle, 84
Microporous membranes, 46
Microtox, 196
Microwave heating, 454
Microwave-assisted extractions (MAE), 357, 454
Microwave-assisted micellar extraction (MAME), 455
Microwave-assisted mineralization
    of natural waters, 99–100
Microwave-enhanced chemistry (MEC), 454–455
MIMS. *See* Membrane inlet (introduction) mass spectrometry
Mineralization
    of biota, 251
    of sediment, 251
    of suspended matter, 250–251

Miniaturization
    in analytical chemistry methods, 362–363
        goal, 363
        objectives, 362
    of MMLLE, 84
Miniaturized liquid-phase extraction techniques, 70–72
    background, 71
    continuous-flow microextraction (CFME), 71–72
    dispersive liquid–liquid microextraction
        (DLLME), 72
    dynamic liquid-phase microextraction (DLPME), 71
Miniaturized nonmembrane-based extraction techniques
    miniaturized liquid-phase extraction techniques, 70–72
    miniaturized SPE techniques, 72–73
    solid-phase microextraction, 73–75
    stir bar sorptive extraction, 75
Miniaturized SPE techniques, 72–73
    fiber-in-tube SPE (FIT-SPE), 72–73
    inside-needle SPE, 73
    microextraction in packed syringe (MEPS), 73
MIPs. *See* Molecularly imprinted polymers
MMHg. *See* Monomethylmercury
MMLLE. *See* Microporous membrane liquid–liquid
        extraction
Mn speciation. *See* Manganese speciation
Molar absorptivity, 264
Molecular spectrophotometry, 264
Molecularly imprinted polymers (MIPs), 324, 325
Monitoring parameters, revision of, 214–216
Monitors, 434
Monomethylmercury (MMHg), 129
    analytical methods for, 130–132
        derivatization and validation, 130–131
        detection methods, 132
        extraction procedures, 130
        gas chromatographic separation methods, 130
        GC improvements, 131
        liquid chromatographic separation methods,
            131–132
MRM. *See* Multiple reaction-monitoring
MS. *See* Mass spectroscopy
MTBE. *See* Methyl tertiary butyl ether
Mtethyl mercury, 26
MTMS. *See* Methyl-trimethoxysilane
MTs. *See* Metallothioneins
μ-ISLMA. *See* Micro-immune-supported liquid
        membrane assay
μTAS. *See* Micro total analysis systems
Multicapillary GC (MCGC), 131
Multidimentional high-performance liquid
        chromatography, 313
Multi-IAC. *See* Multi-immunoaffinity chromatography
Multi-immunoaffinity chromatography (multi-IAC), 168
Multiple reaction-monitoring (MRM) mode, 316
Mussels, 57
Mytilus edulis, 57
Mytilus galloprovincialis, 112, 114
Mytilus, 106

N

Nafion, 50
Natural and transuranic alpha radionuclides
        determination, 249–256

Natural water, 95
    classification system for, 208
    microwave-assisted mineralization, 99–100
Near-infrared spectroscopy, 233
Nebulizers, 267
    and Spray chambers, 271
Nephelometry, 265
Nernst equation, 277
Neutron activation products (^{55}Fe and ^{63}Ni) determination,
        242–246
^{63}Ni determination. *See* Nickel determination
Nickel determination
    coprecipitation of, in natural water with hydroxide,
        245–246
    determination, 242–246
    mineralization, separation, and electrolysis of, in
        aquatic sediments and biota samples, 246
    separation and determination, 244–245
Nitrate
    griess method for, 404
Nitric acid, 25
Nitrofurans, detection of, 166
Nitrofurantoin, detection of, 165, 166
Nitrogen, 260. *See also* Nutrients
Nitrous oxide–acetylene flame, 267
Nonchromatographic techniques, for organic and
        organometallic pollutants, 326–328
Nonhierarchical clustering, 371
Noninorganic surfactants (NS), 31
Nonionic surfactants, 30, 145
Non-metallic substances determination methods, 286–290
Nonporous membranes, 46
Nonporous polymeric membranes based techniques, 75–76
    membrane-assisted solvent extraction, 78–79
    membrane extraction with sorbent interface, 76–77
    membrane inlet (introduction) mass spectrometry
        (MIMS), 76
Nonylphenol (NP), 145
Normal phase (NP) separations, 313
Normal phase HPLC (NP-HPLC) mobile phases, 313
Normalization, 419
NP. *See* Nonylphenol; Normal phase
NP-HPLC. *See* Normal phase HPLC
NS. *See* Noninorganic surfactants
Nuclear weapons, 242
Nutrient determination methods, 283–285
    using IC, 285
Nutrients, 260
Nylon, 13

O

OC. *See* Organic carbon; Organochlorine
Octa-BDE, 148
Octylphenol (OP), 145
Off-line SPE, 321
Off-line systems, of HF-MMLLE, 84, 86–87
    nonautomated, nonflowing design, 86–87
On-column injection, 309
On-line ESy-GC instrument, 85
On-line SFE, 451
On-line SPE, 323, 324, 331
On-line SPE-GC, 323
On-line SPE-GC-HPLC, 323

On-line SPE-LC-MS methods versus biosensors, 328
On-line systems, of MMLLE, 84–85
OP. *See* Octylphenol; Organophosphorus
Optical SPR immunosensors, 151
Optical waveguide lightmode spectroscopy (OWLS), 158
Orchestia gammarellus, 114
Organic and organometallic pollutants, 306
    classification, 305–306
    interactions, 305
Organic and organometallic pollutants, analytical
            methods for analyzing, 306
    analyte derivatization, 325–326
    chromatographic analysis, sample preparation for,
            318–325
    classification, 306–307
    detection techniques, 315
        Fourier-transform ion cyclotron resonance
                instruments, 317
        inductively coupled plasma-mass spectrometry, 318
        ion trap mass spectrometers, 316
        ionization techniques, 317
        quadrupole mass spectrometers, 316
        time-of-flight (TOF) instruments, 317
        triple quadrupole mass spectrometers, 316–317
    nonchromatographic techniques, 326–328
    separation techniques, 309
        capillary electrophoresis, 313–315
        gas chromatography, 309–311
        liquid chromatography, 311–313
        micellar electrokinetic chromatography
                (MEKC), 315
        size exclusion chromatography, 315
Organic carbon (OC)
    oxidation of, 225
Organic compounds determination, water samples
            preservations and storage for, 27–32
Organic polymers, 12
Organic sampling, 13
Organochlorine (OC), 151
Organogermanium compounds, 339–340
Organohalogenated compounds, 147–149
Organolead compounds, 339
Organometallic compounds, passive sampling, 54–55
Organometallic species
    organogermanium compounds, 339–340
    organolead compounds, 339
    organoselenium compounds, 340–342
    organotin (OT) compounds, 335–338
Organophosphates, 28
Organophosphorus (OP) insecticides, 151
Organoselenium compounds, 340–342
Organotin (OT) compounds, 335–338
OT. *See* Organotin
OWLS. *See* Optical waveguide lightmode spectroscopy
Oxidative stress, 112–113
Ozonation. *See* Ozone oxidation
Ozone oxidation, 98–99

P

Pachygrapsus marmoratus, 114
Paclitaxel, detection of, 166
PAHs. *See* Polycyclic aromatic hydrocarbons; Polynuclear
            aromatic hydrocarbons
Particulate matter (PM), 8

Passive sampling, 41–42
    affecting factors, 47–48
    applications, 50
        in algal toxins, 55–56
        in biomonitoring, 57
        in ecotoxicity assessment, 56
        in hydrophobic organic compounds, 50
        in metals, 55
        in organometallic compounds, 54–55
        in polar organic compounds, 50, 54
        in volatile organic compounds, 54
    biofouling, 48–50
    concept of, 42–43
    data validation, 57–59
    equilibrium sampling, 43–44
    future trends, 60–61
    kinetic sampling, 44–45
    quality assurance, 57
    quality control, 57
    in regulatory monitoring, 59–60
    samplers, 14–15, 51–53
        diffusion barrier, 45–46
        modeling and calibration, 46–47
        sorption phase, 45
Pasteurization, water preservation method, 21
[210]Pb determination. *See* Lead determination
PBDEs. *See* Polybrominated diphenylethers
PCA. *See* Principal component analysis
PCA with multiple linear regression analysis
            (PCAMLRA), 383
    advantage of, 384
    methodology of, 384
PCAMLRA. *See* PCA with multiple linear regression
            analysis
PCB RaPID Assay®, 147
PCBs. *See* Polychlorinated biphenyls
PCCPs. *See* Personal care and cosmetic products
PCDDs. *See* Polychlorinated dibenzo-para-dioxins
PCDFs. *See* Polychlorinated dibenzofurans
PDMS. *See* Polydimethyl-siloxane
PE containers. *See* Polyethylene containers
PEEK. *See* Polyetheretherketone
Pelagic species, 108–109
Penta-BDE, 148
Pentavalent arsenate As(V), 25
Perfluorinated sulfonates, 333
Perfluorooctane sulfonate (PFOS), 333
Performance parameters, 393
Performance reference compounds (PRCs), 47
Permeable polymeric membrane, 46
Permeation-based samplers, 46
Perna, 106
Persistent organic pollutants (POPs)
    measurement in precipitation, 408–409
Personal care and cosmetic products (PCCPs), 305, 334
Pesticides, 330
    immunochemical methods for, 151–152
        herbicides and plant growth regulators, 154–158
        insecticides, 152–154
    sample preservation and storage, 28, 29
PFE. *See* Pressurized fluid extraction
PFOS. *See* Perfluorooctane sulfonate
pH in precipitation, determination of, 403
Pharmaceutical and personal care products (PPCPs),
            immunochemical determination of, 158

antibiotics, 158–160, 165–166
    amphenicols, 162–163
    β-lactams, 164
    fluoroquinolones, 161–162
    macrolides, 165
    sulfonamides, 160–161
    tetracyclines, 163–164
    steroid hormones, 166
        androgens, 168–169
        corticosteroids, 169
        estrogens, 166–168
        gestagens, 169
Pharmaceuticals, 334
Phenanthrene, 58, 143, 144
Phenol index, 227
Phenols, 330–331
    sample preservation and storage, 28, 29
Phosphorus, 260. *See also* Nutrients
Photometer, 264
Photons, 263
Phragmites, 105
PHWE. *See* Pressurized hot water extraction
Physical speciation, 224
Piezoelectric quartz crystal (PQC) immunosensor, 147
PIME. *See* Pulse introduction (flow injection-type)
        membrane extraction
Plants, contaminants accumulation and partitioning, 110
Plasma torch, 270
Plastics, 13
Plutonium determination
    activity determination, 256
    coprecipitation of, in natural water with manganese
        dioxide, 249–250
    radiochemical methods, 249
    radionuclide activity determination, 253–255
    radionuclide activity measurement, 253
    separation, purification, and electrolysis of, 252
PM. *See* Particulate matter
Pneumatic nebulizers, 267
^{210}Po determination. *See* Polonium determination
POCIS. *See* Polar Organic Chemical Integrative Sampler
Polar Organic Chemical Integrative Sampler (POCIS),
        54, 56
Polar organic compounds, passive sampling, 50, 54
Pollutants, applications to different classes of, 329
    estrogenic substances, 333–334
    organometallic species
        organogermanium compounds, 339–340
        organolead compounds, 339
        organoselenium compounds, 340–342
        organotin (OT) compounds, 335–338
    personal care and cosmetic products (PCCPs), 334
    pesticides, 330
    pharmaceuticals, 334
    phenols, 330–331
    solvents and volatile compounds, 329–330
    sulfonates, 333
    surfactants, 331–333
Polluted areas managed by specific areas, ranking of
        problems in, 212–214
Polonium determination
    coprecipitation of, in natural water with manganese
        dioxide, 249–250
    radiochemical methods, 249
    radionuclide activity determination, 253–255

radionuclide activity measurement, 253
    separation and determination, 251
Polyacrylamide gel, 125
Polybrominated diphenylethers (PBDEs), 147, 148
Polychlorinated biphenyls (PCBs), 86, 147
    sample preservation and storage, 28, 29
Polychlorinated dibenzofurans (PCDFs), 147, 148
Polychlorinated dibenzo-para-dioxins (PCDDs), 147, 148
Polycyclic aromatic hydrocarbons (PAHs), 32, 71,
        142–145, 309
Polydimethyl-siloxane (PDMS), 321
Polyetheretherketone (PEEK), 71
Polyethersulfone, 49
Polyethylene containers, 13, 26
Polymers, 45
Polynuclear aromatic hydrocarbons (PAHs), 28, 30–31, 45
Polypropylene, 13
Polytetrafluoroethylene (PTFE), 26, 79, 276
Polyvinyl chloride, 13
Polyvinylidine difluoride (PVDF), 79
POPs. *See* Persistent organic pollutants
Posidonia, 105
Posidonia oceanica, 105
Potassium determination, 242
Potentiometric sensors, 277
POX. *See* Purgeable organic halides
PPCPs. *See* Pharmaceutical and personal care products
PQC. *See* Piezoelectric quartz crystal
PRCs. *See* Performance reference compounds
Precipitation quality measurement, analytical procedures for
        used within EMEP monitoring program
            data quality control, 409–410
            future perspectives, 410–411
            precipitation samples, in Europe, 401–409
            rural monitoring network in Europe, scope of,
                399–401
Precipitation samples, in Europe, 401
    collection of, 401–402
    conductivity, of precipitation samples, 403–404
    flame atomic spectroscopy, determination by, 405
    heavy metals and metalloids in precipitation,
            measurement of, 405
        cold vapor atomic fluorescence spectroscopy
                (CV-AFS), 408
        flame-atomic absorption spectroscopy (F-AAS), 408
        graphite furnace atomic absorption spectroscopy
                (GF-AAS), 407–408
        inductively coupled plasma mass spectrometry
                (ICP-MS), 406–407
        sample preparation, 405–406
    ion chromatography, 404
    ions, in precipitation samples, 402–403
    pH in precipitation, determination of, 403
    POPs in precipitation using GC-MS, measurements
            of, 408–409
    spectrophotometry, 404–405
Preservation and storage of water samples
    for inorganic compounds determination, 19–22
    for organic compounds determination, 27–32
    for speciation analysis of metals, 22–25
        arsenic, 25
        chromium, 26
        mercury, 26
        selenium, 26–27
        tin, 27

Pressurized fluid extraction (PFE). *See* Accelerated solvent extraction
Pressurized hot water extraction (PHWE), 360
Pressurized MAE, 454
Primordial radionuclide (^{40}K) determination, 242
Principal component analysis (PCA), 373
    case study (Struma River), 382–383
    theoretical principles, 380–382
Proficiency testing (PT) scheme, 390, 391
Programmable temperature vaporizer (PTV) injectors, 309
Propanil, detection of, 156
Prorocentrum lima, 56
PT. *See* Proficiency testing
PTFE. *See* Polytetrafluoroethylene
PTV. *See* Programmable temperature vaporizer
^{238}Pu determination. *See* Plutonium determination
$^{239+240}$Pu determination. *See* Plutonium determination
^{241}Pu determination. *See* Plutonium determination
Pulse introduction (flow injection-type) membrane extraction (PIME), 77
Pumps, 14
Purgeable organic halides (POX), 228
Purified water, 95
PVDF. *See* Polyvinylidine difluoride
Pyrethroids, 28

Q

QA/QC. *See* Quality assurance/quality control
QM. *See* Quality management
Quadrupole mass spectrometers, 316
Quadrupoles, 273
Quality assurance/quality control, 7, 8, 57, 389–390
    of analytical results, 391–393
        monitoring of, 394–396
        traceability, 394
        validation of analytical procedures, 393–394
    quality in chemical measurements, 390–391
Quality management (QM), 390, 392
Quantification limit (LOQ), 8
Quantity-proportional sampling, 4
Quantum theory, 263

R

^{222}Ra determination. *See* Radium determination
^{226}Ra determination. *See* Radium determination
Radiation source, 266–267
Radio frequency generator (RF generator), 270–271
Radiolead determination, radiochemical methods, 249
    coprecipitation of, in natural water with manganese dioxide, 249–250
    radionuclide activity measurement, 253–255
Radionuclides determination, 242
    ^{222}Ra in water, 249
    ^{226}Ra in water, 248–249
    fission products (^{90}Sr and ^{137}Cs), 246–248
    natural and transuranic alpha radionuclides (^{210}Po, ^{234}U, ^{235}U, ^{238}U, ^{238}Pu, $^{239+240}$Pu, and ^{241}Pu), 249–256
    neutron activation products (^{55}Fe and ^{63}Ni), 242–246
    primordial radionuclide (^{40}K), 242
Radium determination, 248–249
Random sampling pattern, 5

RBMP. *See* River Basin Management Plan
R-Biopharm GmbH, 163
R-charts, 395
Reagent blanks, 57
Reagent-free ion chromatography (RF-IC), 232, 233
Receptor modeling
    case study (Struma River), 385
    theoretical principles, 383–384
Reference materials (RMs), 395
Regulatory monitoring, passive sampling in, 59–60
Relative standard deviation (RSD)
    of heavy metals, in 2006, 411
    of major ions in laboratory intercomparison, in 2005, 410
Replicate samples, 8
Retention time locking (RTL), 330
Reversed phase HPLC (RP-HPLC), 313
RF generator. *See* Radio frequency generator
RF-IC. *See* Reagent-free ion chromatography
Rhizoclonium tortuosum, 112
RIANA. *See* River ANAlyzer
RIDASCREEN®, 163
River ANAlyzer (RIANA), 151, 157, 160
River Basin Management Plan (RBMP), 141
RMs. *See* Reference materials
Royal Society of Chemistry, 354
RP-HPLC. *See* Reversed phase HPLC
RSD. *See* Relative standard deviation
RTL. *See* Retention time locking
Rural monitoring network, in Europe, 399–401
Ruttner sampler, 10, 14

S

Sample collection strategies, 1–2
    general considerations, 2–5
        samples types
            composite sample, 3–4
            discrete sample, 3
        sampling patterns, 5
    sampling equipment
        automatic water sampling systems, 14
        compatibility of sampler material, 12–13
        ground water samplers, 14
        manual surface water samplers, 13–14
        passive samplers, 14–15
        sediment samplers, 15–16
        traditional techniques, 13
    sampling-related uncertainty, 5–8
    sediments samples, 11–12
    water samples, 3–4, 8–11
Sample size, 8
Sampler material, compatibility with water samples, 12–13
Sampler recovery spikes, 57
Samples preparation, for analysis, 431
    analytical data, types of, 434–435
    analytical procedure, stages of, 440–442
    green chemistry, 458
        green analytical chemistry, 459–463
        history, 458–459
    ionic liquids, 452–453
    matrix solid-phase dispersion, 448–448
    membrane techniques, application of, 442
        membrane extraction, 445–448

microwave-enhanced chemistry (MEC), 454–455
new methodological developments in, 442–444
speciation analytics, 435
    chemical speciation, 437–439
    physical speciation, 437
supercritical fluid extraction, 449–451
surfactant micelles, unique phase separation behavior
    of, 451–452
trace element analysis, problems associated with,
    439–440
ultrasound (US) application in, 455–458
Sampling patterns, 5, 6
Sampling rate, 45, 46
Sampling uncertainty, 5–8
SAs. *See* Sulfonamides
SBME. *See* Solvent bar microextraction
SBSE. *See* Stir bar sorptive extraction
SC. *See* Stripping chronopotentiometry
Score plot, 381
Scott chamber design, 271
Scree plot, 382
Screening speciation, 437
Scrobicularia plana, 107
SDME. *See* Single-drop microextraction
SDS. *See* Sodium dodecyl sulfate
Sediment sample collection, 11–12
Sediment samplers, 15–16
Sediment, mineralization of, 251
Segmented flow analysis (SFA), 281
Selected ion monitoring (SIM), 86
Selenium, 436
    speciation analysis, 26–27
Self-organizing maps (SOM)
    advantage of, 376
    architecture, 376
    case study (Struma River), 377–379
    theoretical principles, 376–377
Semipermeable membrane device (SPMD), 46, 49, 54,
    56, 58
Sensors, 327
Separation techniques, 278–281
    capillary electrophoresis, 280–281
    ion chromatography, 278–280
    for organic and organometallic pollutants, 309
        capillary electrophoresis, 313–315
        gas chromatography, 309–311
        liquid chromatography, 311–313
        micellar electrokinetic chromatography (MEKC),
        315
        size exclusion chromatography, 315
Sequential injection analysis (SIA), 145–146
Sequential injection/flow injection analysis (SIA/FIA)
    system, 230–231
SETAC. *See* Society of Environmental Toxicology and
    Chemistry
SFA. *See* Segmented flow analysis
SFE. *See* Supercritical fluid extraction
SIA. *See* Sequential injection analysis
SIA/FIA. *See* Sequential injection/flow injection analysis
SID-MS. *See* Speciated isotope dilution mass
    spectrometry
Silica, 285
Silica base material, 313
Silicon, 261. *See also* Nutrients

Silicone, 13
Silicone rubber, 45
Silicone strip sampler, 48
SIM. *See* Selected ion monitoring
SimaPro, 416–417, 419
Single linkage method, 372
Single-drop microextraction (SDME), 71, 359
Size exclusion chromatography, 315
SLM. *See* Supported liquid membrane
Slotted tube atom traps (STATs), 267
Small-depths samplers, 13–14
SME. *See* Solvent microextraction
Smith–Hieftje background correction, 268
Society of Environmental Toxicology and Chemistry
    (SETAC), 413
Sodium dodecyl sulfate (SDS), 145
Solid-phase adsorption toxin tracking (SPATT) bags, 55
Solid-phase dynamic extraction (SPDE), 73
Solid-phase extraction (SPE), 31, 32, 321, 331
Solid-phase microextraction (SPME) sampler, 31, 32, 46,
    55, 58, 73–75, 274, 320, 357–358
    applications, recent trends in, 74–75
    calibration in, 74
Solid-phase microextraction capillary gas
    chromatography (SPME-GC), 131
Solvent bar microextraction (SBME), 87
Solvent microextraction (SME), 71
Solvents and volatile compounds, 329–330
SOM. *See* Self-organizing maps
Sorption, 27
Sorption phase, 45
Sorption phase-water partition coefficient, 43, 44
Spartina, 105
SPATT bags. *See* Solid-phase adsorption toxin tracking
    bags
SPC. *See* Statistical process control
SPCs. *See* Sulfophenyl carboxylates
SPDE. *See* Solid-phase dynamic extraction
SPE. *See* Solid phase extraction
Speciated isotope dilution mass spectrometry (SID-MS),
    132
Speciation analysis, 439
    analytics, 121, 435
        chemical speciation, 437–439
        physical speciation, 437
        Cr, Fe, Mn, and As speciation, 122–127
        Hg speciation, 129–132
    definition of, 224
    of metals, 22–25
        arsenic, 25
        chromium, 26
        mercury, 26
        selenium, 26–27
        tin, 27
Spectinomycin, detection of, 166
Spectrophotometry, 263, 264, 404–405
    griess method, for nitrate, 404
    indophenol blue method, for ammonium, 404–405
    mercury thiocyanate-iron method, for chloride, 405
    on molecular absorption radiation
        ultraviolet-visible spectroscopy, 263
Split/splitless injection, 309
SPM. *See* Suspended particulate matter
SPMD sampler. *See* Semipermeable membrane device

SPME. *See* Solid-phase microextraction
SPME-GC. *See* Solid-phase microextraction capillary gas chromatography
Spoons/scoops, as sediment samplers, 11, 15
Spot water samples, 42, 59
SPR. *See* Surface plasmon resonance
Spray chambers
    nebulizers and, 271
SQC. *See* Statistical quality control
^{90}Sr determination. *See* Strontium determination
Stabilization procedures, 20, 21, 23–25
Stabilized temperature platform furnace (STPF), 268
Stainless steel, 242
STATs. *See* Slotted tube atom traps
Stationary phase technology, 313
Statistical process control (SPC), 395
Statistical quality control (SQC), 395
Steroid hormones, immunochemical determination of, 166
    androgens, 168–169
    corticosteroids, 169
    estrogens, 166–168
    gestagens, 169
Stir bar sorptive extraction (SBSE), 75, 258
    development, 75
    modes of operation, 75
STPF. *See* Stabilized temperature platform furnace
Stripping chronopotentiometry (SC), 126
Strontium determination, 246, 247–248
    activity calculation, 248
    separation by adsorption on AMP, 246–247
Styrenedivinylbenzene copolymer, 278
Sulfonamides (SAs), detection of, 160–161
Sulfonates, 333
Sulfophenyl carboxylates (SPCs), 145, 333
Supercritical fluid extraction (SFE), 360–361, 449–451
Supervised (learning) methods, 370
Supported liquid membrane (SLM) extraction, 79–84
    equilibrium sampling through, 83–84
    principle, 79
    selectivity, 83
    transport mechanism, 81
        diffusive transport, 81–82
        facilitated transport, 82
    unit and system configuration, 79–81
Surface plasmon resonance (SPR), 144
    SPR Biacore sensor, 146
Surface water monitoring, 190–191
Surface water sampling, 9–11, 13
Surfactant micelles, unique phase separation behavior of, 451–452
Surfactants, 145–147, 331–333
Suspended matter, mineralization of, 250–251
Suspended particulate matter (SPM), 107
Sustainable development, into analytical practice
    goals of, 353
    green analytical chemistry
        characteristics of, 356
        electrochemistry sensing technology, advances in, 362
        greener separation techniques, 361–362
        history of, 354–355
        implementation of, 355
        miniaturization in, 362–363
        in sample pretreatment, 356–361

SWIFT-WFD project, 59
Swiss Center for Electronics and Microtechnology (CSEM), 161
Synthetic organic pesticides, 28
Syringe device extraction, of MMLLE, 85–86
Systematic sampling pattern, 5

T

Tank exposure studies, 47
Target compounds, methods to minimize loss from water samples, 20
TBTs. *See* Trisubstituted organotins
TC. *See* Total carbon
TCDD. *See* 2,3,7,8-Tetrachlorodibenzo-p-dioxin
TCs. *See* Tetracyclines
TDC. *See* Total dissolved carbon
TDN. *See* Total dissolved nitrogen
TECP. *See* Thermal extraction cone penetrometry probe
2,3,7,8-Tetrachlorodibenzo-p-dioxin (TCDD), 148
Tetractenos glaber, 111
Tetracyclines (TCs), detection of, 163–164
TH. *See* Total hydrocarbons
Thermal desorber, 462
Thermal extraction cone penetrometry probe (TECP), 462
Thermal ionization mass spectrometry (TIMS), 242
THGAs. *See* Transversal heated graphite atomizers
Thief samplers, 14
TIMS. *See* Thermal ionization mass spectrometry
Thin-film microextraction, 358–359
304-grade stainless steel, 13
316-grade stainless steel, 13
Tigriopus brevicornis, 114
Time of flight (TOF), 273, 317
Time-proportional sampling, 4
Time-resolved fluoro-immunoassay (TR-FIA), 157
Time-weighted average (TWA) sampling, 42, 44, 45
Tin, speciation analysis, 27
Titration, 263
Titration error, 263
Titrimetric measurement, 262–263
TOC. *See* Total organic carbon
TOF. *See* Time of flight
Total carbon (TC), 192, 233
    fractions in liquid and solid samples, 226
Total dissolved carbon (TDC), 233
Total dissolved nitrogen (TDN), 233
Total hydrocarbons (TH)
    parameters and techniques, used for determination of, 227
Total inorganic mercury, 231
Total organic carbon (TOC), 192, 225–227
    definition of, 225
Total organic halides (TOX), 228, 232
Total parameters
    for evaluation of xenobiotics, in environment, 223
        biochemical oxygen demand (BOD), 224–225
        chemical oxygen demand (COD), 225
        in environmental analysis, 227–233
        total organic carbon (TOC), 225–227
Total sulfur (TS), 233
Total volatile organic carbon (TVOC), 228
Total volatile organic halogen (TVOX), 232
TOX. *See* Total organic halides

Toxic compounds, identification of, 210–211
Toxicity tests, importance of, 192–195
Toxkit tests, 195–199
Trace element analysis, 261, 439–440
Trace metal species in aquatic systems, concentration
    levels of, 122
Traditional techniques, 13–14
Transversal heated graphite atomizers (THGAs), 268
TR-FIA. *See* Time-resolved fluoro-immunoassay
Tributyltin compounds, 335
Trifluralin, 158
Triple quadrupole mass spectrometers, 316–317
Trisubstituted organotins (TBTs), 27
Triton X, 145
Trivalent arsenite As(III), 25
Trivalent chromium Cr(III), 26
TS. *See* Total sulfur
Tube-type samplers, 46
Turbidimetry, 265
TVOC. *See* Total volatile organic carbon
TVOX. *See* Total volatile organic halogen
TWA. *See* Time-weighted average
Twisters, 75
2D-GC-MS, 330
Two-phase HF-LPME, 87–88
    automation, 88
    development, 87

U

[234]U determination. *See* Uranium determination
[235]U determination. *See* Uranium determination
[238]U determination. *See* Uranium determination
Ultra performance liquid chromatography (UPLC), 141,
    313, 335
Ultrasonic extraction, 357
Ultrasound (US) application, in sample preparation
    process, 455–458
Ultraviolet photo-oxidation, 97–98
Ultraviolet-visible (UV-VIS), 263
Ulva, 105
Uncertainty, of analytical results, 5–8
Unsupervised multivariate statistical methods, 370
UPLC. *See* Ultra performance liquid chromatography
Uranium determination
    coprecipitation of, in natural water with manganese
        dioxide, 249–250
    radiochemical methods, 249
    radionuclide activity determination, 253–255
    radionuclide activity measurement, 253
    separation, purification, and electrolysis, 252–253
US. *See* Ultrasound
US Presidential Green Chemistry Challenge, 354
USA, 106, 200, 329, 459
UV photo-oxidation. *See* Ultraviolet photo-oxidation
UV-VIS. *See* Ultraviolet-visible

V

Validation of analytical procedures, 393–394
Van Dorn sampler, 14
Vibrio fischeri, 196, 207
Vibro-corers, 16

VOCs. *See* Volatile organic compounds
Volatile organic compounds (VOCs), 30, 31
    passive sampling, 54
    sampler, 14
Volatilization, 27
Voltammetry, 275
Voltammetry oxygen electrode, voltammetry sensor. *See*
    Clark electrode

W

Ward's method, of clustering, 373
Water, 1–2
    biogeochemical cycle of, 304
    as matrix for organic and organometallic analytes,
        304–305
Water Framework Directive (WFD), 59, 140–141, 190,
    193
Water pollution assessment, integrated system of,
    200–201, 202–204
Water sample preparation step, used mineralization
    techniques, 95
    advanced oxidation processes for, 96
        ozone oxidation, 98–99
        UV photo-oxidation, 97–98
Water samples, 3, 8–11
    collection of, 9–11
    composite sample, 3–4
    discontinuous sampling, 4
    discrete sample, 3
    location and sites, 9
    preservation and storage. *See* Preservation and
        storage of water samples
    sample size, 8
    sampling frequency, 11
Water-cooled atom traps (WCAT), 267
Water-soluble organic carbon (WSOC), 232
Waveguide interrogated optical system
    (WIOS), 161
WCAT. *See* Water-cooled atom traps
WEA. *See* Whole Effluent Assessment
WEER. *See* Whole Effluent Environmental Risk
Weighting, 419
WET. *See* Whole Effluent Toxicity
WFD. *See* Water Framework Directive
Whole Effluent Assessment (WEA), 200
Whole Effluent Environmental Risk (WEER), 201
Whole Effluent Toxicity (WET), 200
WIOS. *See* Waveguide interrogated optical system
Working Party on Green Chemistry, 354
WSOC. *See* Water-soluble organic carbon

X

X-charts, 395
XAD-2, 31

Z

z-score, in evaluation of laboratory results, 396
Zeeman effect, 268
Zeta-score, in evaluation of laboratory results, 396
Zostera, 105